FRESH-WATER INVERTEBRATES OF THE UNITED STATES

FRESH-WATER INVERTEBRATES OF THE UNITED STATES

PROTOZOA TO MOLLUSCA

THIRD EDITION

ROBERT W. PENNAK

Emeritus Professor of Biology
University of Colorado
Boulder, Colorado

WILEY

A Wiley-Interscience Publication

JOHN WILEY & SONS, INC.

New York · Chichester · Brisbane · Toronto · Singapore

Library of Congress Cataloging in Publication Data:

Pennak, Robert W. (Robert William)
    Fresh-water invertebrates of the United States : Protozoa to
Mollusca / Robert W. Pennak. — 3rd ed.
        p. cm.

    "A Wiley-Interscience publication."
    Includes bibliographies and index.
    ISBN 0-471-63118-3
    1. Freshwater invertebrates—United States.   I. Title.
QL141.P45   1991   88-18570
592.0973—dc19   CIP

Printed in the United States of America

10 9 8 7 6

To the memory of
C. Juday

"What's the use of their having names," the Gnat said, "if they won't answer to them?"
"No use to *them*," said Alice; "but it's useful to the people that name them, I suppose."

LEWIS CARROLL

"All cold-blooded animals . . . spend an unexpectedly large proportion of their time doing nothing at all, or at any rate, nothing in particular."

CHARLES ELTON

PREFACE

Although only about eleven years have elapsed since the appearance of the second edition of this book, it seems to me that it is now time for this third edition, especially in view of the rapid journal publication rate of new material. During this interval I have examined more than 5000 new references pertaining to freshwater invertebrates of the United States, and have accordingly added, corrected, and revised both text and keys, wherever necessary. Major revisions have been necessary for such chapters as the Introductory Essay, Porifera, Annelida, Gastropoda, and Copepoda, but only minor revisions have been required for such taxa as the Protozoa, Turbellaria, Nematoda, Nematomorpha, Ostracoda, Pelecypoda, and Hydracarina. From the standpoint of nomenclature and taxonomy, some of these are "unpopular" taxa of genuine interest to few researchers.

With the advent of biochemical taxonomy and scanning electron microscopy, vast new fields of invertebrate systematics are being brought to attention. More specifically, ultra-details of structure are being discovered that appear to have taxonomic significance, resulting in increasing "splitting" of taxa into new subcategories. For routine identifications, however, it is mandatory that we rely on structural details that can be seen with an ordinary light microscope. The horrendous expense involved in using the SEM rules it out for such purposes. Ordinarily, biochemical techniques are also not practical.

Two editorial changes have been incorporated into this edition. First, I have numbered all figures separately for each chapter. Thus the first figure in each chapter is "Fig. 1." We think this device will simplify, avoid confusion, and speed the use of keys. Second,

we are using only standard abbreviations of state names in the keys; this should likewise save space and time.

The decision to omit the aquatic insects from this edition was made only after long soul-searching, but it has the approval of John Wiley & Sons and also has the approval of many aquatic entomologists with whom I have had discussions.

I have always felt that the first two editions of this book have given unjustifiably short shrift to the insects. We should have included, for example, much more material on anatomy, ecology, behavior, and, above all, keys that go to the species level. Using today's level of insect information, however, we would end up with a 1500-page volume if insects were to be included in this third edition! The new information, "splitting," and taxonomic revisions for the Trichoptera, Ephemeroptera, Diptera, Plecoptera, and Coleoptera, for example, are reaching fantastic proportions.

And if I were to keep the insect chapters and expand them properly, this volume would not appear before 1994—an unconscionable delay which I cannot justify in my present functional retirement.

Fortunately, our aquatic entomologists are now producing excellent monographs, manuals, and text-references that will soon fill our needs.

To add a final bit of realism, I have done my stint, and it is now time for me to spend more of my days trout fishing.

ROBERT W. PENNAK

Boulder, Colorado
January 1989

PREFACE TO THE SECOND EDITION

About 25 years have elapsed since the publication of the first edition of this book. During this long interval I have done my best to gather the pertinent references and literature that would be necessary for the production of this second edition. This period has been characterized by a greatly increased interest in the biology of our fresh-water invertebrates. As a result, the germane literature since 1952 amounts to more than 5000 titles in technical journals. Colleagues and professional acquaintances everywhere have been generous in sending me reprints of their works, and I am greatly in their debt for such thoughtfulness. I am grateful to Dr. John Bushnell for his comments and patience in helping me with anatomical and nomenclatural problems in the Ectoprocta. Dr. Perry Holt generously supplied the key to the genera of Branchiobdellida.

Several of the shortest chapters have required only few changes, but most chapters have had a major overhaul. Nearly all keys have been rewritten and made longer in accordance with maturing systematics. A few of the old figures have been deleted, especially where I could find or make more suitable illustrations. Many new figures have been added, especially to clarify difficult key couplets. Lists of references at the end of each chapter have been revised and enlarged; about half of the citations are new.

When we refer to fresh-water invertebrates of the United States, we are including complete data only for the 48 contiguous states. In general, however, this volume covers essentially all the Alaskan fauna. The Hawaiian fresh-water fauna is impoverished and poorly known.

The Ronald Press Co. and I have been surprised and continuously gratified by the acceptance of the first edition by such a very broad spectrum of users. We hope this second edition will receive similar acceptance.

My greatest personal reward derived from producing the two editions of this book is the host of professional acquaintances I have made, not only in the United States but also in many foreign countries.

ROBERT W. PENNAK

Boulder, Colorado
March 1978

PREFACE TO THE FIRST EDITION

No comprehensive work on American fresh-water invertebrates has appeared since 1918, and in the meantime much new information has steadily accumulated in a wide variety of biological periodicals. To me, therefore, the need for the present volume is obvious. I hope it will prove useful to biologists and zoologists generally, and to aquatic zoologists, limnologists, fish biologists, and entomologists particularly. It will also serve as a text and reference pitched toward the level of the college senior and beginning graduate student.

Original material resulting from my own field and laboratory research forms only a small fraction of this book. Rather, my role as an author has been one of arrangement, organization, and selection from a very large mass of published information. Indeed, my most perplexing problem has been how *much* material on each taxonomic group to include within the covers of a single volume. General policies, however, have been clear from the outset; I have emphasized natural history, ecology, and taxonomy; I have minimized details in the sections on anatomy and physiology; I have tried to be complete, accurate, concise, and consistent from one chapter to another. During the preparation of this volume all of the chapters have been continually revised and kept up to date in accordance with new material appearing in the literature.

For the most part, only free-living, fresh-water invertebrates that occur in the United States are included. Cestodes, trematodes, parasitic nematodes, etc. are all omitted; their inclusion would have necessitated a much more extensive work and additional years of preparation. Furthermore, most aquatic biologists seldom need to refer to parasitic groups as contrasted with free-living groups.

It will be noted that some chapters contain keys to species, while others contain keys only

as far as genera. In general, the chapters in the former category deal with small or comparatively stable taxonomic groups that are fairly well known and in which few new species are being described from time to time in the United States. The chapters in the latter category, however, are all concerned with larger taxonomic groups, as well as those that are less well known in the United States and in which new forms are being described so abundantly that any key to species is quickly out of date. It is assumed that users of this manual are familiar with the fundamentals of taxonomy and the construction of keys. Incidentally, each half of a key couplet consists of but a single sentence; changes in subject within a sentence are indicated by semicolons.

The list of references at the end of each chapter consists only of especially significant and comprehensive works; these include older classical papers as well as recent contributions. Except where sections are especially valuable, I have not listed the larger standard reference works such as Hyman's *The Invertebrates,* Schulze's *Biologie der Tiere Deutschlands,* Bronn's *Tierreich,* Kükenthal and Krumbach's *Handbuch der Zoologie,* Dahl's *Die Tierwelt Deutschlands,* Grassé's *Traité de Zoologie,* and *Faune de France.*

Almost all of the chapters (with or without figures) were sent to at least one specialist for criticisms, corrections, and suggestions. I am deeply grateful to all these specialists, who were most encouraging, generous, and cooperative. Without their help and assurance I would have had much less confidence in my efforts. Following is a list of these specialists, with the sections examined by each.

John L. Brooks (Cladocera); Royal B. Brunson (Gastrotricha); C. F. Byers (Odonata); Fenner A. Chace, Jr. (Crustacea Introduction, Mysidacea); B. G. Chitwood (Nematoda); W. R. Coe (Nemertea); R. E. Coker (Copepoda); Ralph W. Dexter (Eubranchiopoda);

W. T. Edmondson (Rotatoria); T. H. Frison [deceased] (Plecoptera); R. E. Gregg (Insecta Introduction); H. H. Hobbs, Jr. (Decapoda); C. C. Hoff (Ostracoda); Leslie Hubricht (Amphipoda, Isopoda); H. B. Hungerford (Hemiptera); Libbie H. Hyman (Coelenterata, Turbellaria); F. P. Ide (Ephemeroptera); O. A. Johannsen (Diptera); M. W. de Laubenfels (Porifera); H. B. Leech (Coleoptera); James E. Lynch (Eubranchiopoda); J. G. Mackin (Isopoda); Ruth Marshall (Hydracarina); H. B. Mills (Collembola); J. Percy Moore (Hirudinea); L. E. Noland (Protozoa); Mary D. Rogick (Bryozoa); H. H. Ross (Plecoptera, Trichoptera); Henry van der Schalie (Gastropoda, Pelecypoda); Waldo L. Schmitt (Crustacea Introduction, Mysidacea); L. H. Townsend (Megaloptera, Neuroptera); H. C. Yeatman (Copepoda).

The introductory chapter was criticized by A. S. Pearse and by several of my graduate students. Other people have helped me in one way or another. A group of eight protozoologists, for example, selected the 300 most appropriate and common genera of Protozoa to be included in the Protozoa key. These men are: William Balamuth, Gordon H. Ball, A. M. Elliot, Harold Kirby [deceased]; J. B. Lackey, L. E. Noland, K. L. Osterud, and D. H. Wenrich.

Obviously, a book of this sort is bound to have errors and ambiguities, and certainly such faults are clearly my own responsibility. I hope readers will bring them to my attention.

In a sense, this book constitutes a plea and encouragement for more work on our rich but poorly known fresh-water invertebrate fauna. I hope these chapters will kindle sparks and raise questions in the minds of our students and beginning investigators in zoology, for it is among these young people that we shall find our authorities and specialists of tomorrow.

ROBERT W. PENNAK

Boulder, Colorado
1952

CONTENTS

FRESH-WATER
INVERTEBRATES
OF THE UNITED STATES

1

INTRODUCTORY ESSAY*

Excluding Protozoa and all parasitic classes but including aquatic insects, it is estimated that the fresh-water invertebrate fauna of the United States consists of about 11,000 described species. Ten years ago the figure was 10,000 species. Probably no aquatic biologist is of the opinion that this fauna is well known; certainly the total is bound to increase markedly in coming years, especially when the aquatic fauna of the western half of the country is more thoroughly studied. Although a few groups are fairly well known and reasonably stabilized, such as the Rotifera, Bryozoa, and Pelecypoda, the majority of American fresh-water invertebrate groups are in the process of taxonomic refinement. In expanding taxa composed mostly of non-cosmopolitan species we are lagging behind European fresh-water taxonomy by about 10 to 20 years. All taxonomic questions aside, we are seriously lacking in our understanding of geographic distribution patterns, physiology, natural history, and ecology of fresh-water invertebrates in the United States. In a few groups, however, some remarkable progress has been made since the publication of the first edition of this book.

It is a shortcoming of the fresh-water invertebrate literature that information bearing on the fundamental mechanisms for colonization and ecesis in the rigorous fresh-water environment is generally fragmentary, scattered, ignored, or mentioned incidentally (Macan, 1961). I should like, therefore, to integrate such information under one title. Unfortunately, space limitations forbid a detailed discussion, and I must be satisfied with briefly presenting a series of topics, and only minimal documentation and analysis. I have no *new* generalizations, but I hope to present some of them in a different light. I am also taking the liberty of making certain germane speculations. Bear in mind that each generalization may have notable exceptions. Thus, this contribution is written in the form of an essay.

I shall not recite long lists of taxa inhabiting fresh waters, comparing them with those found in salt waters, but remember that only aquatic spiders, mites, insects, pulmonate gastropods, and perhaps the rotifers, cladocerans, and eubranchiopods, are not derived directly from the sea. Rather I am interested in considering the morphological and physiological adaptations that have appeared on a grand scale within the fresh-water kingdom. In other words, what are the major evolutionary problems that have been solved as prerequisites to an existence within the rigorous fresh-water environment, and how do they operate? Some of these items are obvious; others are more subtle and obscure.

Furthermore, I am restricting my remarks to the "true" fresh-water invertebrate fauna on the one hand and the "true" marine invertebrate fauna on the other hand, even though the brackish environment is the site where fresh-water adaptations begin and where salt-water adaptations may be lost. A discussion of these comparative and complicated marine–brackish-fresh-water faunal relationships and ecological transitions must await another effort. Some such material has already been brought together in the classical work by Remane and Schlieper (1971). Inci-

*Many paragraphs in this chapter are reproduced in either a verbatim or modified form from Pennak (1985), The Fresh-water Invertebrate Fauna: Problems and Solutions for Evolutionary Success. *Am. Zool.* 25:671–687. This material is used with the approval of the American Society of Zoologists.

1

dentally, these authors emphasize the fact that the number of "true" brackish species, that is, those that complete their entire life cycles in brackish water, is insignificantly small compared with the numbers of true marine and true fresh-water species.

Similarly, I am avoiding a comprehensive discussion of the faunas of hypersaline ponds and playas of closed basins. This also requires separate treatment.

For the most part, the topics considered in this essay are *not* generally discussed in other texts dealing with fresh-water biology. I hope users of this volume will give these ideas a careful reading.

Most of my remarks are centered around north temperate and mid-latitude conditions as found in the United States. The Literature Cited list is selected and intentionally modest.

MAGNITUDE OF THE FRESH-WATER ENVIRONMENT

Compared with the massive marine environment, the fresh-water environment of the world is almost trivial. According to Todd (1970), the waters of the world are distributed as shown in Table I.

The oceans, in relation to lotic and lentic habitats together, constitute an unusual ratio 97.3:0.0091. This means that the mass of salt water is more than 10,000 times the mass of inhabitable fresh water. In terms of surface area, inland waters cover less than 2% of Earth's surface (Wetzel, 1975), compared with the salt-water environment, which covers about 71%. Nevertheless, this small fraction of Earth's water has been thoroughly colonized by fresh-water invertebrates.

This brings us to another point. Excluding the atmosphere, salt waters constitute our finest example of a physically continuous environment. That is, any (small) part of the seas is theoretically accessible to any other part. There are no formidable dry land barriers—only other ecological barriers such as

temperature, distance, food conditions, substrate, and competition, which, I submit, can often be at least temporarily tolerated.

The fresh waters of the world, on the other hand, are hopelessly fragmented into an enormous array of isolated habitats. To be sure, a large river system or a large lake is a considerable habitat-mass, though not without internal ecological barriers, but consider the millions of reservoirs, small lakes, and ponds of the world, ranging down to such restricted habitats as a stock tank completely isolated on our short grass plains, 10 km from its nearest neighboring stock tank or water hole, and dependent on a windmill and well for its source of water. This, of course, is an extreme example of the environmental situation to which a segment of the fresh-water fauna must be adapted. The small pond as an isolated microcosm is a neglected concept that was first emphasized a century ago in a classical paper by Forbes (1987).

THE CHEMICAL DIVERGENCE

The most obvious marine–fresh-water distinction lies, of course, in the salt content of these two regimes. The great bulk of marine habitat mass ranges between only 3.30 and 3.70% salt, of which 80% consists of sodium chloride. The vast majority of "true" marine species therefore are restricted to a chemically monotonous environment (Fig. 1), and they may be said to constitute the "chloride fauna."

Most fresh-water species, on the other hand, tolerate and occur naturally over a surprising range of total dissolved solids

(TDS). From a chemical standpoint, and unlike the oceans, there is no such thing as an "average" fresh-water habitat (Table II). Each lotic or lentic situation has its own chemistry, but customarily the higher the altitude and the nearer a body of water is to the headwaters of its flowage system, the lower the TDS. Nevertheless, the dominant and most distinctive ions are the carbonate–bicarbonate complex, which usually accounts for more than 55% of the TDS. We may thus refer to fresh waters generally and collectively as "car-

TABLE I. Waters of the World

Water item	Volume (km^3)	Percentage of the world's total water
Lakes, ponds, and reservoirs	125,000	0.009
Average instantaneous volume of rivers	1,250	0.0001
Saline inland waters	104,000	0.008
Soil moisture and vadose water	67,000	0.005
Ground water	8,350,000	0.61
Icecaps and glaciers	29,200,000	2.14
Atmosphere	13,000	0.001
World oceans	1,320,000,000	97.3

bonate" waters, and to the fauna as the "carbonate fauna." It must be acknowledged, however, that a significant minority of fresh waters are dominated by sulfates, rather than by the carbonate complex. Sodium chloride is present only in small quantities in most fresh waters, seldom exceeding 5% of the TDS.

Aside from the carbonate and sodium chloride situation, I have long felt that another major ionic and biological distinction between the two types of aquatic environments is centered around potassium as a key element. The average potassium content of sea water is about 380 mg/l, while fresh waters usually contain less than 10 mg/l, and often only traces. Large quantities of potassium salts in fresh waters are known to be toxic to many invertebrates, and conversely marine groups making the evolutionary passage into fresh waters must have made an extreme physiological potassium adjustment.

The fresh-water fauna is clearly distinguished by its occurrence in waters that, while dilute, have a wide range of dissolved salts, and thus it is, relatively speaking, a euryhaline fauna (Fig. 1). The true marine fauna, on the other hand, cannot reproduce and complete life cycles over wide ranges of salinity (Dorgelo, 1976).

In general, we may arbitrarily set 1000 mg/l as the usual upper limit of TDS for the fresh-water fauna and 10 mg/l as the lower limit. Within this wide range are permanent populations of many common species that tolerate, for example, a 10-fold range of 50 to 500 mg/l.

Running waters are especially notable for their annual variations in dissolved materials, depending on the local geochemistry and seasonal changes in discharge. Generally, an annual variation of ±50% in dissolved load is typical (Pennak, 1977), and in exceptional cases ±80%. Most stream invertebrates endure such variations without difficulty.

Sea water seldom exceeds a range of pH 7.8 to 8.3 in surface samples, but the "normal" biological range for most fresh waters is about pH 4.4 to 8.6. A single habitat, during the course of a year, may vary as much as two full pH units, depending on the vagaries of photosynthesis, light, current, respiratory processes, biota, circulation, and so on. The very large literature on this subject shows that many common species adjust to such variations with no difficulty.

TABLE II. A Comparison of the Total Dissolved Salt Content of Fresh Waters, Brackish Waters, and Sea Water. Maximum and minimum values are merely approximations. The most abundant salts in brackish waters are chlorides, carbonates, or bicarbonates, depending on the degree of brackishness. In fresh waters the most abundant salts are usually carbonates and bicarbonates. In alkali and saline lakes they are usually carbonates, bicarbonates, chlorides, and sulfates in varying proportions.

	Percentage concentration		Concentration in mg/l	
	Minimum	Maximum	Minimum	Maximum
Sea water	3.5		35,000	
Brackish waters	0.05	3.2	500	32,000
Fresh waters	0.001	0.05	10	500
Alkali and saline lakes	0.05	25.0	500	250,000

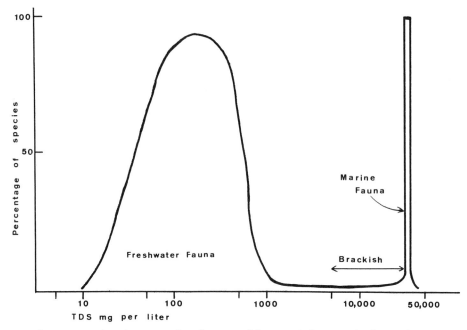

FIG. 1.–Salt content and environmental preferences of the aquatic faunas. This figure shows only *natural* occurrence of the faunas; it does not show limits of tolerance.

ORIGINS OF THE FRESH-WATER FAUNA

The generalization that most major fresh-water invertebrate groups originated from marine ancestors is firmly established. Only a few groups, such as insects, mites, and pulmonate snails, are presumed to have clearly originated from terrestrial habitats. The fresh-water fauna is therefore appropriately termed an immigrant fauna.

The fundamental problem that must be solved before any marine animal can make its way into fresh water involves major physiological readjustments. Body fluids of most marine invertebrates are roughly isotonic with sea water; that is, the internal dissolved salt concentrations are similar to or slightly higher than the 3.5% average salt concentration of sea water. It is further true that the majority of marine invertebrates cannot endure much dilution of sea water. Fresh waters commonly contain about 0.01 as much salts as the ocean, and the internal fluids of fresh-water invertebrates usually contain 0.03 to 0.40 as much salts as the ocean. From an osmotic standpoint, water therefore tends to

pass into the hypertonic tissues of fresh-water animals. Consequently any successful fresh-water animal must have developed physiological mechanisms for maintaining a proper salt and water balance against this strong gradient.

It is difficult to imagine the appearance of such mechanisms de novo, and it is assumed that the transitions from marine to fresh-water environments were not sudden and rapid processes, but rather series of slow evolutionary processes occurring by way of psammolittoral and phreatic waters, littoral zones, marshes, swamps, and river estuaries where there are transition zones between salt and fresh water.

It should be borne in mind, however, that a river estuary is not a constant environment. Salinities, currents, tides, food, temperatures, and other ecological factors vary widely from time to time, and these conditions are by no means favorable to the evolution and gradual development of forms suited to fresh water. Pearse (1950) has stated the problem well:

An estuary has been called the doorway by which marine forms have populated fresh water. This statement is perhaps in part true, but an estuarine doorway is not wide open and easily passed. There are many difficulties to be surmounted. Many animals struggle long ages to get through and fail. Only a few attain fresh water by this route.

Ideal conditions for the invasion of fresh water are afforded by such places as the Baltic Sea, where there is a large area involved and where there is a permanent and very gradual transition from the sea to fresh water. Yet the Baltic has not a single endemic brackish water species of metazoan that has evolved since the most recent glaciation, although there are a very few subspecies or physiological varieties that may be endemic.

The generalization that marine invertebrates are isotonic with sea water is actually a slight exaggeration. Even pelagic, deep-sea, and the most primitive marine species maintain a dynamic steady state by which a difference in concentration of several ions commonly obtains across external membranes and which must be maintained through physiological regulatory processes. The internal concentrations of magnesium and sulfate ions are often much lower than those in sea water. Potassium concentrations may be considerably higher or lower than in sea water, and calcium and chloride are also variable. Many marine invertebrates have the remarkable ability to concentrate astonishing amounts of elements that are present in the sea water only as traces. Examples are vanadium, iodine, strontium, and bromine.

A wide range in type and degree of osmoregulatory control is found among littoral and estuarine invertebrates. Many species have limited ability to regulate the relative amounts of internal salt and water; their membranes are easily permeable so that the body fluids become more or less isotonic in diluted sea water, and death occurs rather promptly. A few stenohaline species can endure some dilution of sea-water by regulating body volume and taking up more water into the tissues. A few other species can regulate their internal osmotic concentration only to a limited degree and venture into slightly brackish water; these species have a slight ability to remain hypertonic in diluted

sea water. "Typical" brackish water species are euryhaline and can persist in habitats containing 30 to 85% fresh water. Most of these invertebrates maintain a more or less hypertonic internal salt concentration, regardless of the degree of brackishness of their surroundings. They include representatives of many groups, especially arthropods, mollusks, and various kinds of worms. The active absorption of salts is an important mechanism, and perhaps in some species there is actually physiological control of the amounts of water absorbed. Marine and brackish species transferred directly to fresh water usually live a few hours at most, but when the transition is made very slowly, over days or weeks, they may live for weeks or months in fresh water.

Figure 2 is a diagram showing the relative composition of the aquatic fauna in relation to salinity. It is one of the most striking paradoxes of fresh-water, brackish, and marine faunas that animals thrive well in either an environment that is very low in salts or one that is high in salts, but environments of intermediate salinities (brackish) have a poor fauna. From the standpoint of total number of all types of species present, it should be noted that the minimum appears at a salinity of $7^0/_{00}$ (7 parts per thousand), which is well toward the fresh-water end of the diagram in Fig. 2. This situation results from the fact that the number of fresh-water species drops very rapidly with a slight rise in salinity, whereas the number of marine species drops less rapidly with a decrease in salinity.

This diagram also poses a second paradox. It shows that specific brackish-water species, which occur chiefly or exclusively in such environments, are most abundant at salinities of 7 to $10^0/_{00}$ or about the place where the fresh-water forms decrease abruptly and where the total number of all species is smallest. This condition is not to be expected since most brackish-water species have obviously been derived from marine relatives and not from fresh-water relatives. It would be more logical to expect the maximum number of brackish species much farther toward the right in the diagram.

The invasion of fresh waters is a continuing process, and some American species are undoubtedly new arrivals or are in the process of becoming adapted to fresh waters. The colonial coelenterate *Cordylophora* and a few

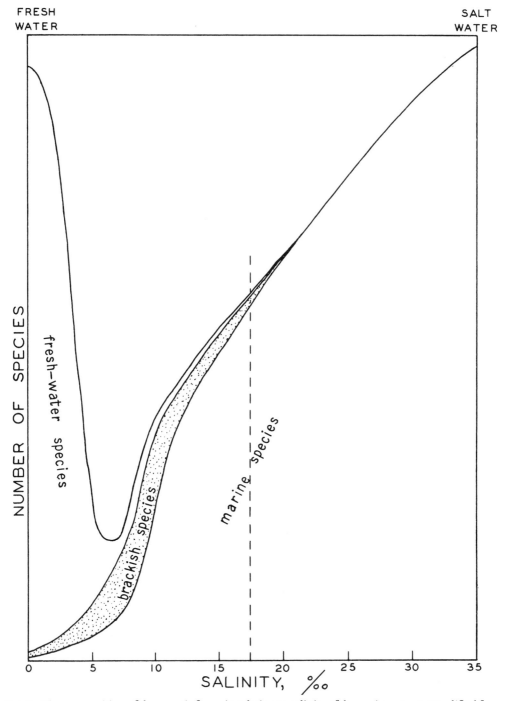

FIG. 2.–The composition of the aquatic fauna in relation to salinity of the environment. (Modified from Remane and Schlieper.)

species in each of the following groups are all typical examples: polychaetes, grapsoid crabs, shrimps, isopods, clams, and snails. A comparable list may be cited for Europe, where a few species are known to have invaded fresh waters within historical times. The mitten crab of the Old World has taken up permanent residence in rivers and returns to the sea only to breed. With few exceptions, the new fresh-water forms are thought to be physiological varieties of their close marine relatives.

Several species of European crustaceans are marine in northern Europe and both brackish and fresh alone, in central and southern Europe. Certain tropical areas, especially around the Bay of Bengal, Indonesia, the Malay Archipelago, Madagascar, and tropical America are relatively rich in species that have only recently made their way into fresh waters from the sea.

In a few areas clusters of species have been able to adapt to changing salinity conditions. At Lake Merced, CA, for example, the channel to the Pacific Ocean was closed between 1869 and 1894. Subsequently it slowly converted to fresh water, and a few species of marine ancestry were able to adapt. As of 1958, Miller reported the hydroid *Cordylophora,* the polychaete *Neanthes,* the mysid *Neomysis,* the isopod

Exosphaeroma, and the amphipod *Corophium. Cordylophora* and *Neomysis* are typical brackish species; the others are definitely marine.

It is presumed that constant temperatures and heavy rainfalls facilitate fresh-water invasions via estuaries. An alternate, newer theory emphasizes temperature and subarctic regions as more likely areas for entry into fresh water, chiefly because lower temperatures minimize the physiological effects of the various ions.

One other item should be discussed briefly in this section. The present writer (Pennak, 1963) has postulated a means by which micrometazoans could colonize fresh waters from the sea via the psammolittoral and phreatic habitats. As shown in Fig. 3, micrometazoans could possibly move from intertidal and subtidal zones into the marine psammolittoral and (1) hence into the phreatic ground water and thereafter inland to distant ponds or lakes via their psammolittoral, or (2) into the psammolittoral of an estuary, and hence progressively "upstream" in the river psammolittoral or deeper phreatic zone to inland fresh-water localities. Porous substrates below typical stream beds have long been known to have well-developed phreatic currents.

The phreatic brackish interstitial zone of

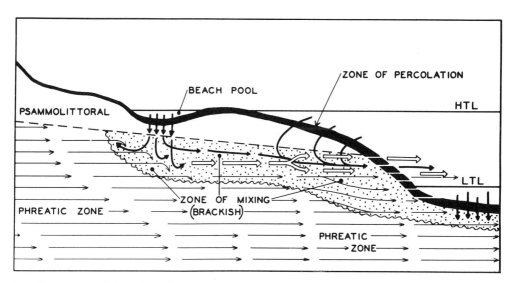

FIG. 3.–Diagrammatic section of a marine shoreline in an area where the fresh phreatic waters are seeping into the salt water. HTL, high tide level; LTL, low tide level; brackish zone of mixing shown by stippling; thick, open, and thin arrows show relative amounts of salt, brackish, and fresh waters, respectively. (From Pennak, 1963, after Delamare Deboutteville, 1960.)

mixing at the edge of the sea is relatively thin, but it is also relatively constant in position, and species in this habitat are in a much less hazardous situation than those species that must swim or crawl about on the substrate of an estuarine area. Pennak (1963) also emphasizes the relative rarity of the invasion of fresh waters from the standpoint of mutations and genetics. Ax and Ax (1970) discuss other aspects of the marine–fresh-water transition.

FRESH-WATER EMIGRANTS TO THE SEA

Once established in fresh water, few invertebrates have returned to brackish water and only rarely to undiluted sea water. Little is known about the major barriers to such migrations, but presumably they are physiological, and perhaps chiefly osmoregulatory. Potassium salts, which are abundant in the sea, are toxic to fresh-water invertebrates, but this effect may be partly neutralized by magnesium compounds. An impervious exoskeleton is advantageous for migrations into salt waters.

Only a few rotifers, hemipterans, beetle larvae and adults, dipteran larvae and pupae, and several caddis larvae have successfully invaded estuaries, brackish waters, and the intertidal zone. Diptera larvae have occasionally been dredged from 10 to 15 fathoms off the coast of England. In Samoa there is a peculiar dipteran that is submarine for its entire life cycle. A few marine water striders occur regularly on the surface film of the seas far away from land.

Biologically, insects are considered a most successful group. They exhibit endless adaptations, occur in enormous numbers, and are found in almost every type of habitat. Yet they have not been successful in colonizing the seas. With the exception of some of the groups just noted, a salt content in excess of 2.5% appears to be toxic. However, brackish ponds with salt contents ranging up to 1.0% are quickly colonized by "fresh-water" insects.

Occasionally typical fresh-water mollusks are reported from brackish estuaries. A few species of *Physella* and *Goniobasis* can endure up to 50% sea water, *Amnicola* and *Planorbis* up to 30%, and *Lymnaea* up to 25%.

MAJOR DISTINCTIONS BETWEEN MARINE AND FRESH-WATER INVERTEBRATES

Osmoregulation. Continuing the discussion of the foregoing sections, one of the basic physiological distinctions centers around the occurrence of more specialized osmoregulatory mechanisms in fresh-water organisms. Since their internal fluids are hypertonic to the greatly dilute surrounding water, there is a continuous and rapid inflow of water through all permeable surfaces of the body. Such surfaces may be epithelia, cuticle, chitin, gills, and other structures. The relative amounts of water absorbed through these various types of surfaces differ greatly from one taxon to another. In general, chitin and thick cuticle are relatively impermeable.

Contractile vacuoles, flame bulb systems, nephridia, and different types of glandular structures have the function of forming a highly dilute urine, which is excreted to the outside in considerable quantities. This urine is greatly hypotonic to the body fluids. By such means excessive water is prevented from accumulating in the body, and the relative concentrations of internal salts and water remain essentially constant. In a general way, fresh-water invertebrates are specialized for retaining their low ionic concentrations, even during long periods of inactivity. Figure 4 shows these relationships in a diagrammatic fashion.

Marine invertebrates usually have excretory organs that are basically more or less similar to those of their fresh-water relatives, but the urine they excrete is approximately isotonic with sea water and their body fluids. The fundamental problem that must be solved when a euryhaline animal invades fresh water is the physiological conversion of the osmoregulatory process to one that secretes dilute urine from hypertonic body fluids. Other

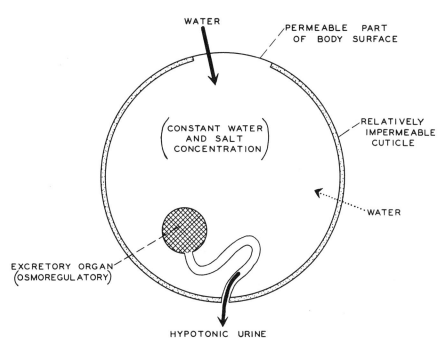

WATER

PERMEABLE PART
OF BODY SURFACE

CONSTANT WATER
AND SALT
CONCENTRATION

RELATIVELY
IMPERMEABLE
CUTICLE

WATER

EXCRETORY ORGAN
(OSMOREGULATORY)

HYPOTONIC URINE

FIG. 4.–Diagram showing the osmoregulatory water balance mechanism in a typical fresh-water invertebrate. (Greatly modified from Baldwin, 1948.)

things being equal, fresh-water invertebrates use more oxygen than their close marine relatives, probably because of the greater energy needed to maintain the proper internal osmotic pressure.

It is significant that contractile vacuoles are absent from most marine protozoans. Fresh-water isopods, amphipods, and decapods have larger excretory (osmoregulatory) glands than their marine relatives, and presumably this indicates greater activity and greater physiological and histological complexity.

Although the urine excreted by fresh-water invertebrates is highly dilute, it does contain traces of body salts. In most animals this loss is negligible, and is apparently made up from food and by the slow selective absorption of ions through body surfaces. In a few groups of fresh-water invertebrates, however, it has been shown that there is active absorption of ions by special parts of the body, particularly in snails, mussels, leeches, dragonfly nymphs, crayfish, and some Diptera larvae. At the same time, the ability to retain body salts is sometimes quite striking. Many insects and crustaceans may be placed in

distilled water and kept there for weeks at a time without any apparent harm.

Although fresh-water invertebrates are generally considered stenohaline organisms, this generalization is true only in the gross sense. "Fresh water" is highly variable in total salt content and in the relative quantities of different ions present, but many common species in almost all orders of fresh-water invertebrates may be found in waters whose total salt content ranges from 0.002 to 0.04%. This is actually a 20-fold difference in dissolved salts! With the exception of a few species adapted to highly saline inland waters, however, the osmoregulatory mechanisms of most fresh-water invertebrates quickly break down when they are placed in solutions containing in excess of 25% sea water (0.9% salt).

One of the most striking osmotic conditions may be found among invertebrates in our high mountain tarns, where there are usually populations of cladocerans, copepods, protozoans, rotifers, nematodes, oligochaetes, many immature insects, and even seed clams (Sphaeriidae). Here the total dissolved solids may be only 10 mg/l, distributed as follows:

Dissolved organics	4 mg/l
Carbonate–bicarbonate complex	3 mg/l
Silicon complex	2 mg/l
All other salts	1 mg/l

Such water is more dilute than what passes for the usual "distilled water" in our laboratories. Within 1 mg of "all other salts" are the chlorides, sulfates, nitrates, phosphates, magnesium, and iron, plus a host of other essential ions—all present in trace quantities in such dilute mountain water. Evolutionary perfection of osmotic systems and integuments gives these animals the ability (1) to absorb from such water, in various ways, all (essential) ions on a basis that is seemingly atom by atom, and (2) to retain these absorbed internal ions as well as those obtained through ingested food. Indeed, this is illustrated by the fact that many fresh-water arthropods will live in distilled water for weeks at a time.

A few marine species are specialized for absorbing and concentrating such exotic elements as zinc, vanadium, copper, and iodine (Galtsoff, 1934; Vinogradov, 1953), but I know of no comparable examples of American fresh-water invertebrates that selectively concentrate any unusual ions. Instead, inhabitants of exceptionally soft waters must take up essentially *all* ions against an unfavorable osmotic gradient.

Macan (1963) points out that the fauna of the most dilute fresh waters tends to be ultimately of terrestrial origin. This has not been my experience, as shown for extremely soft high altitude waters.

Eggs. No one seems to have collected extensive data to show conclusively whether there is a broad difference between fecundity in marine invertebrates and their fresh-water counterparts, but most of us are familiar with a few examples. Thus, in a few fresh-water hydroids such as *Hydra,* reproduction is restricted to one or a few fertilized eggs at a time. Its relative *Cordylophora* produces twice or more as many embryos in brackish as in fresh waters. Making allowances for the size difference, lobsters in berry carry proportionately more eggs than fresh-water crayfish. Fresh-water flatworms produce few eggs, and the same is often true, comparatively, for fresh-water Mysidacea, Amphipoda, and

Gastropoda. Needham (1930) and Barrington (1967) point out that the common oyster releases 1,800,000 eggs at once and that oviparous oysters may produce 100,000,000 eggs per spawning. The fresh-water *Anodonta cygnea,* however, produces only 15,000 to 2,000,000 (rarely) eggs. As a whole, fresh-water mussels produce several thousand to 3,000,000 embryos, depending on the size of the species. The marine gastropod *Buccinum* may deposit 12,000 eggs and *Nucella* about 245 capsules containing 400 to 600 eggs each, while fresh-water snails lay only a trivial 10 to hundreds at a time.

There has been some speculation about this fundamental difference in reproductive potential between the chloride fauna and the carbonate fauna. One such speculation argues that trace element nutrients are more difficult to accumulate in necessary quantities in fresh-water invertebrates, and as compensation their eggs must be larger (to store such materials for early growth), and hence the total number of eggs is much smaller. I consider this a weak explanation, especially in view of the ability of fresh-water forms to live and reproduce in extremely dilute waters. Furthermore, most fresh-water microcrustaceans are notable exceptions in having small eggs. And Thorson (1946) presents abundant evidence to show that eggs of marine invertebrates range widely in their yolk content, and in their small to large size. Careful experimental work on euryokous species might be revealing.

Egg-carrying habits may be related to this problem. In fresh-water species the eggs are often carried by the parent until hatching and release of the young. This situation occurs notably in certain rotifers and most cladocerans, mussels, copepods, and malacostracans, as well as in other taxa. The egg-carrying habit is also true for some marine species, but it is by no means as common as it is in fresh waters. If eggs must be carried until the hatchling is sufficiently grown to assume an independent existence, we may reason that there is less room on the female body for the clutch of eggs. Hence the small number of (larger) eggs.

Another explanation for fewer (larger) eggs is that the fresh-water organism must eclose with fully developed osmoregulatory capabilities, that is, must be at a more advanced stage to cope with the highly dilute surroundings.

In marine forms the production of (lightweight) planktonic eggs is common, while fresh-water eggs are heavy and remain or sink to the bottom.

Distinctive larval stages. Unlike most of their marine relatives, fresh-water invertebrates produce special larval stages in only a few major taxa, as shown in Table III. To this table could be added the Eubranchiopoda exception; this taxon is restricted to fresh waters and inland saline waters and does have nauplius stages.

Barrington (1967) states that "the disadvantage of fresh-water larval stages are so clear," but does not actually discuss the problem. I believe that there is probably a variety of evolutionary reasons, operating through a very long geological time frame, to explain the paucity of special larval stages in fresh-water invertebrates. It is, however, truly unusual that the fresh-water Copepoda, an eminently successful group, have retained the basic nauplius stages in the life history. It is, furthermore, a remarkably euryokous group, very closely tied to its salt-water relatives.

Undoubtedly the greater abundance of yolky stored food in fresh-water eggs is correlated with the general lack of special larval stages. In marine groups where there is little yolk the newly hatched larvae must begin foraging promptly.

One theory advanced to explain the infrequency of fresh-water invasion by marine species is based on the vulnerability of marine larvae. Although free swimming, they are unable to make their way upstream against river currents. Even if the adults were to breed in an estuary or river, the unattached larvae would be swept downstream. It is also thought that the larval metabolism and hypertonic tissues are unable to withstand the severe and variable ecological conditions in fresh waters.

Nine major taxa are well represented in both fresh and salt waters, but special larval stages are present in neither environment, the young being hatched in an advanced stage of development; these are Gastrotricha, Rotifera, Nematoda, Tardigrada, Oligochaeta, Hirudinea, Mysidacea, Isopoda, and Amphipoda.

Phylogenetic differences. It is quite true that the seas are phylogenetically much richer than fresh waters, and there are many major taxonomic categories that are confined to salt water or are predominantly marine (lists A and B of Table IV). Only four small groups, on the other hand, are exclusively fresh water. These are the Eubranchiopoda, the Syncarida, the several species of Thermosbaenacea, and one species of Spelaeogriphacea (known only from southern Africa). The Eubranchiopoda are unusual because some species are found in salt-poor waters (e.g.,

TABLE III. COMPARISON OF SPECIAL LARVAL STAGES IN THOSE MAJOR TAXA OCCURRING IN BOTH MARINE AND FRESH-WATER ENVIRONMENTS

Taxon	Special free-swimming marine larval stages	Special free-swimming fresh-water larval stages
Demospongiae	Coeloblastula, stereogastrula	"Flagellated embryo" in some species
Hydrozoa	Medusa, planula, frustule (creeping), actinula	Frustule (creeping) in a few species
Turbellaria	Müller's larva, Götte's larva	
Nemertea	Pilidium	
Entoprocta	"Ciliated larva"	
Ectoprocta	Cystid, cyphonautes	"Ciliated larva"
Copepoda	Nauplius stages, copepodid stages	Nauplius stages, copepodid stages
Ostracoda	Atypical nauplius	Atypical nauplius
Decapoda	Nauplius, protozoea, zoea, mysis, phyllosoma, etc.	
Gastropoda	Trochophore, veliger	
Pelecypoda	Trochophore, veliger	Glochidium (a secondary adaptation for parasitism)

TABLE IV. A CLASSIFICATION OF THE INVERTEBRATE KINGDOM ACCORDING TO ENVIRONMENTS. Various taxa are listed, including phyla, classes, and orders.

A. Exclusively marine and brackish
  Amphineura
  Amphionidacea
  Anthozoa
  Archiannelida (?)
  Brachiopoda
  Caridea
  Cephalocarida
  Cephalopoda
  Chaetognatha
  Cirripedia
  Ctenophora
  Cumacea
  Dendrobranchiata
  Echinodermata
  Echiuroidea
  Enteropneusta
  Euphausiacea
  Gnathostomulida
  Kinorhyncha
  Leptostraca
  Monoplacophora
  Mystacocarida
  Nebaliacea
  Nectiopoda
  Phoronidea
  Pogonophora
  Priapuloidea
  Pterobranchia
  Pycnogonida
  Scaphopoda
  Scyphozoa
  Sipunculoidea
  Stomatopoda
  Tanaidacea
  Xiphosura
B. Predominantly marine, with
  few to rare fresh-water species
  Bryozoa
  Hydrozoa
  Mysidacea
  Polychaeta
  Porifera
	C. Chiefly marine but well
	  represented in fresh waters
	  Amphipoda
	  Decapoda
	  Pelecypoda
	D. Abundant representatives in
	  both marine and fresh waters
	  Copepoda
	  Gastrotricha
	  Ostracoda
	  Sarcodina
	  Suctoria
	E. Abundant representatives in
	  both marine and fresh waters;
	  few terrestrial species
	  Tardigrada
	  Turbellaria
	F. With more fresh-water species
	  than marine
	  Ciliata
	  Mastigophora
	G. Predominantly fresh water;
	  marine species few to rare
	  Cladocera
	  Hydracarina
	  Nematomorpha
	  Rotifera
	H. Exclusively fresh-water species
	  Eubranchiopoda
	  Spelaeogriphacea
	  Syncarida
	  Thermosbaenacea
		J. Chiefly marine but with a few
		  fresh-water and terrestrial species
		  Nemertea
		K. Numerous marine, fresh-water,
		  and terrestrial species
		  Gastropoda
		  Hirudinea
		  Isopoda
		  Nematoda
		  Oligochaeta
		L. Chiefly terrestrial, but with
		  many fresh-water representatives;
		  few marine species
		  Insecta
		M. Exclusively terrestrial
		  Acarina (excluding Hydracarina
		  in the broad sense)
		  Other arachnid orders
		  Chilopoda
		  Onychophora
		  Progoneata (a few rare marine
		  millipedes)
		N. Exclusively parasitic
		  Acanthocephala
		  Cestoidea
		  Linguatulida
		  Mesozoa
		  Myzostoma
		  Sporozoa
		  Trematoda

tundra ponds), whereas others are restricted to highly saline inland waters. The Rotifera, Nematomorpha, Cladocera, and Hydracarina have so few marine representatives, however, that these might just as well be considered fresh-water groups. Because colonization of fresh waters probably began earlier than Devonian times, it is significant that many more fresh-water groups have not made their appearance.

With few exceptions, those taxonomic categories that are well represented in both environments are much richer in species in marine waters than in fresh waters. This is especially true of the Porifera, Hydrozoa, Bryozoa, Polychaeta, Mysidacea, Amphipoda, Decapoda, Pelecypoda, Copepoda, Isopoda, and Gastropoda.

For the sake of completeness, Table IV also includes strictly terrestrial and parasitic taxa, and it is notable that these lists are quite short.

Body size. In my own mind, I cannot rationalize the fundamental difference in size (body weight) of marine as compared with fresh-water invertebrates. In general, the temperate marine macrofauna consists of taxa whose average size is larger than that of their close fresh-water relatives. It would seem, however, to be reasonable to expect this situation. Fresh-water forms would have a metabolic and osmotic disadvantage in being larger since area (body surface) increases as the square, while body volume increases as the cube, thereby conferring a theoretical osmoregulatory disadvantage upon (large) fresh-water forms in the dilute milieu. We suspect, however, that the answer might lie in the fact that food in fresh waters is less available and less abundant than in salt waters, and that this limiting factor has taken precedence over osmoregulatory and other physiological problems. At any rate, the evolutionary processes have appeared to screen out those large species having the potential of making the transition into fresh waters. In this respect, Brandt (1897) showed that the size of Baltic Sea mollusks is correlated directly with salinity. In fresh waters there are only two taxa having bodies of any size, mussels and crayfish.

Anabiotic strategies. It is in the production of anabiotic devices that fresh-water invertebrates excel, and we are here using the term "anabiosis" (Greek *recovery of life*) in the broad sense to include all special resistant eggs, cysts, and other small resting stages that are produced to tide the animal over periods of desiccation, extreme cold, heat, anaerobic situations, lack of food, and other adverse conditions (Figs. 5 and 6). It seems logical to suppose that such fresh-water mechanisms have arisen in response to variable and unpredictable environmental situations usually not found in the more stable and predictable marine surroundings. The fact that such devices are not present in marine species but are present in fresh-water species in the same ancient taxon is indicative of the probable early appearance of these devices in the transitory primordial fresh-water environment. Table V is an attempt to summarize the anabiotic devices for taxa that occur in both the chloride and carbonate environments.

It is difficult to find accurate information as to the supposed existence of resistant devices among marine invertebrates, and the data in the middle column of Table V therefore

are subject to minor modifications. Recently, I resorted to asking a group of five experienced protozoologists about the relative frequency of resistant cyst stages in fresh waters as compared with salt water. They were surprisingly in agreement that: (1) cysts are known to occur (or probably occur) in more than 90% of fresh-water protozoans, but (2) are essentially unknown or rare in "true" stenohaline marine protozoans (as opposed to brackish forms).

In general, the third column in Table V shows the striking tendency for evolutionary development of thick-walled anabiotic stages, even though the associated instigating environmental conditions may be quite different from one time or one taxon to another. Indeed, this is a classical display of parallel adaptive evolution on a massive scale. Note the sharp break between ostracods and isopods in Table V. It clearly separates the microscopic anabiotic groups from the macroscopic nonanabiotic groups. In addition to the material in this table, there are a few taxa essentially restricted to fresh waters (and originating in fresh waters?) that have also developed special anabiotic devices, including the highly resistant resting eggs of the Eubranchiopoda and the ephippial eggs of Cladocera (ignoring the very few marine Cladocera). Many Eubranchiopoda are macroscopic, and in this respect they are an exception to the size versus anabiotic stage principle just mentioned. Most, however, are restricted to astatic waters. And many aquatic insects, being largely restricted to fresh waters, have a diapause or diapauselike period in the larval, pupal, or nymphal stages that are resistant to unfavorable environmental conditions (see Mansingh, 1971). Anabiotic stages of many fresh-water taxa actually *require* a period of cold, freezing, drying, or anaerobiosis before they will hatch.

Many years ago I acquired a quantity of dried mud that had just been collected from the bottom of a small pond in the Galapagos Islands. Annually thereafter for 10 years I put some of the mud in filtered pond water and periodically examined this microcosm for active invertebrates. Within a few days to weeks I always found active excysted protozoans, rotifers, tardigrades, nematodes, and microcrustaceans. Any student may duplicate these observations with dried pond mud.

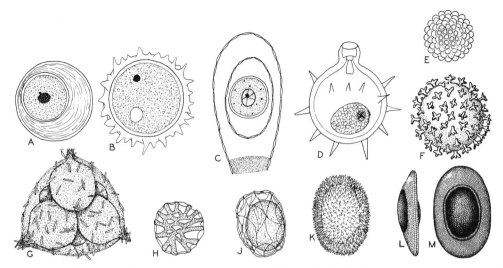

FIG. 5.–Examples of fresh-water invertebrate anabiotic devices. Various magnifications. A–D, protozoan cysts; E and F, tardigrade resting eggs; G, cluster of sponge gemmules, *Eunapius fragilis*; H, resting egg of clam shrimp, *Eulimnadia*; J and K, rotifer resting eggs; L and M, bryozoan statoblasts, *Hyalinella punctata*. (E and F from Ramazotti, 1972.)

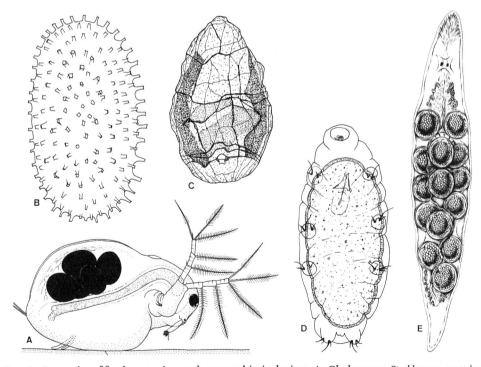

FIG. 6.–Examples of fresh-water invertebrate anabiotic devices. A, Cladoceran, *Streblocerus*, carrying five ephippial eggs, ×25; B, opsiblastic egg of gastrotrich, *Chaetonotus*, ×120; C, desiccated rotifer, *Philodina*, ×490; D, cyst of tardigrade, *Hypsibius*, ×80; E, turbellarian, *Mesostoma*, carrying thick-shelled anabiotic eggs, ×50. (A from Fryer, 1974; E from Riser and Morse, 1974, *Biology of the Turbellaria*, with permission of McGraw-Hill Book. Co., New York.)

TABLE V. RELATIVE OCCURRENCE OF ANABIOTIC STAGES IN MARINE AND FRESH-WATER INVERTEBRATES. Only those taxa occurring commonly in both environments are listed.

Fresh-water and marine taxon	Anabiotic devices in marine species	Anabiotic devices in at least some fresh-water species
Protozoa	Resistant stages rare	Many kinds of resistant cyst stages
Porifera	"Gemmulelike" structures rare	Gemmules, "reduction bodies"
Coelenterata		"Thecated embryos"
Turbellaria		Cocoons, fragmentation cysts, "winter" eggs
Gastrotricha		Opsiblastic eggs
Rotifera		Resting eggs, winter eggs, anabiotic stage of adult
Nematoda		Resting eggs, desiccated individuals
Tardigrada		Various stages described as: tun, cyst stage, anabiotic stage, cryptobiotic stage
Nemertea		Cysts
Ectoprocta		Statoblasts (at least six types), hibernacula
Entoprocta		Hibernacula
Oligochaeta		Cysts
Copepoda		"Resting eggs" in a few spp., diapause cocoon stage
Ostracoda		Resting eggs, advanced instars able to withstand cold and drying in diapause
Isopoda		
Amphipoda		
Decapoda		
Mysidacea		
Pelecypoda		
Gastropoda		

Dried intertidal marine mud, however, produces no such comparable populations when given similar treatment.

Also, I have many unpublished observations on the anabiotic effectiveness of pond invertebrates at high altitudes where the entire water mass and as much as 0.5 m of the subtrate are annually frozen solid for 6 months or more. Each summer, beginning with ice melt, the normal population of pond invertebrates is restored.

In the laboratory, the fantastic tolerances of cyst stages for drying, anaerobiosis, and temperature conditions far exceeding normal ranges are legendary (Von Brand, 1946; Ashwood-Smith and Farrant, 1980). Rotifer, protozoan, crustacean, and tardigrade anabiotic stages are especially resistant, and a surprising fraction of anabiotic stages recover and become active after such drastic treatment.

Low temperature tolerances. Aside from the resistant inactive anabiotic stages, the *active* stages of many fresh-water invertebrates also withstand and recover from freezing and thawing of the substrates of lake shores and streamsides, as well as in the substrate of ponds that freeze solidly for several weeks to as much as 9 months (Neldner and Pennak, 1955; Olsson, 1981; Kenk, 1949; Dougherty and Harris, 1963; Scholander et al., 1953; Sernov, 1927; Danell, 1981). In reality, ice and frozen substrates typically contain anabiotic stages in addition to immature and adult stages, depending on the particular species.

In marine habitats, on the other hand, subtidal species are unable to recover from freezing, although some intertidal forms, especially mollusks, can withstand being frozen in ice for days or weeks (Kanwisher, 1955).

Aestivation. Aestivation is a phenomenon closely associated with the production of anabiotic stages, but usually involves an inactive state of the whole mature animal, rather than an inactive state of an anabiotic device produced by that animal. We also prefer to think of aestivation as applying only to the desiccation conditions of heat and dryness. In marine habitats temporary aestivation is restricted to a relatively few species inhabiting rocky shores near the high tide line, such as barnacles and mollusks.

Fresh-water aestivation is most clearly shown for populations of temporary pools and ponds; see especially the extensive study by Wiggins et al. (1980). These authors designate aestivation forms as "overwintering residents," including juveniles and adults of such taxa as oligochaetes and leeches that secrete a protective coat of mucus, snails that secrete a mucoid epiphragm, a few copepods in the diapause stage, and a few immature and adult insects. These are species that actually survive true dry conditions, brought on before cold weather begins. In addition, there is always a large population of species that overwinter in crevices, in the water of crayfish burrows, in damp vegetable debris, and in damp pondside soil. These are not true examples of aestivation, however, in the sense that we are using the term here.

Anaerobiosis. There seems to be no pervasive and consistent distinction between fresh-water invertebrates and their marine relatives with respect to an ability to withstand anaerobic conditions. In both environments there are many related and unrelated taxa that have developed the ability to live for a few hours to a few months in the absence of oxygen, or where oxygen is present in concentrations below 10% saturation. It is probable, nevertheless, that the fresh-water fauna has more frequently developed the physiological mechanisms for withstanding anaerobic conditions for long periods than is the case for marine invertebrates. Examples include the fauna of sewage treatment waters and the benthic and plankton faunas of eutrophic lakes and ponds whose bottom waters are anaerobic or near-anaerobic for months in both summer and winter. Both marine and fresh-water environments, however, are commonly and continuously anaerobic or near-anaerobic below the surface of organic bottom muds. Perhaps our best marine example involves the subtidal thiobiotic microfauna.

Neuston. The surface film of salt water is virtually uninhabited by small metazoans, but the numerous fresh-water organisms associated with the surface film either constantly or intermittently constitute a special community, the neuston. Many hemipterans, collembolans, and even a few spiders run about on the upper face of the surface film, and some cladocerans and Protozoa are regularly attached to the underface of the film. Visitors to the film for varying lengths of time include dipteran larvae and pupae, hemipterans, beetles, planarians, hydras, a few ostracods, and some snails. The surface film is visited chiefly for food or to make brief contact with air in order to renew the supply of gaseous oxygen held on the surface of the body or in the tracheal system.

The silt and current problem. Most running waters (except for spring brooks) exhibit periodic silting. Heavy loads of suspended silt may be present for only a few hours after a heavy shower, or they may be present continuously for 2 or 3 months during spring and summer runoff. Suspended loads commonly reach levels of 1.0 g/l or more, and the water transparency approaches zero. Fine organic and inorganic particles have both abrasive and suffocative action, and while spates always have some decimating effects on lotic faunas, the stream bottom metazoans, especially insects, have remarkable powers of survival and resistance, so that "normal" population levels may soon be restored (Pennak, 1977; Hynes, 1970). Indeed, fresh-water pelecypods are dependent on the continuous (but moderate) existence of a load of silt as a food source. Continuously clear streams seldom have an appreciable population of clams.

Stream species exhibit another cluster of adaptations that are minimally developed in marine invertebrates found in intertidal habitats. Bottom insects, for example, are often greatly flattened and can thus hide in narrow rubble crevices. They also remain sufficiently active to dislodge silt that accumulates on the body. Special devices to combat silting and swift currents include streamlining, suckers, friction pads, hooks, sticky secretions, and protective ballast cases (Hynes, 1970). To be sure, the intertidal zone may be temporarily silted by waves, tides, and storms, but, except for major storms and hurricanes, silt periods are brief and have not generally brought about major evolutionary adaptations. Morphological adaptation to turbulence in marine habitats seems to be mostly in the nature of crustose or flattened growth forms, for example, a few sponges, tunicates, bryozoans, limpets, and sea anemones. Of additional importance, however, is the relatively small size of stream invertebrates as compared with marine invertebrates of the intertidal zone, and this may be a protective adaptation.

Another important distinction is the fact that the current in a swift stream microhabitat is constant and relatively unidirectional in the force exerted upon a metazoan, while intertidal turbulence is intermittent and multidirectional (Denny, 1983). Possibly this distinction is responsible for the wide divergence in adaptations in streams as compared with tidal zone turbulence.

EVOLUTIONARY SHORTCOMINGS IN FRESH WATERS

Thus far we have been concerned chiefly with adaptations that have come about as the result of migrations of marine taxa ancestors into fresh waters, where the new adaptations have permitted the incipient fauna to become established. In so doing, however, it appears that some features of the marine ancestors have become lost, and, for the most part, are not to be found in fresh waters. Three of these features are briefly discussed in the following paragraphs.

Bioluminescence. We need give only passing mention of the sharp separation of fresh- and salt-water faunas with respect to bioluminescence. It occurs commonly in many marine taxa, and a wide variety of functions has been ascribed to the phenomenon (Harvey, 1952). In striking contrast, it is rare in freshwater invertebrates. The proper combinations of anatomical adaptations, physiological mechanisms, bacteria, and essential enzyme systems have been lost in the environmental transition, or have not yet generally evolved in the shallower carbonate waters.

In some forms bioluminescence is produced by bacteria on the body surface; in other forms the phenomenon is produced by luciferin–luciferase mechanisms in the animal's tissues. Several genera of Coleoptera have bioluminescent aquatic larvae; most of these records come from Japan, India, Thailand, and New Zealand. A bioluminescent shrimp is known from Japan, and in New Zealand streams there is a small limpet that secretes light-producing mucus. It is *Latia neritoides,* first discovered in 1890. A single rare light-producing enchytraeid annelid is known from the nearshore sandy bottoms of Russian streams.

Morphological embellishments. Many marine species are notable for an abundance of external growths on the body surface, such as excessive setation, spines, palps, fleshy protuberances, and respiratory devices (Fig. 7). Such embellishments occur generally but are most common among Crustacea, Polychaeta, and Mollusca. Some devices are presumed to retard sinking in plankton species; others are sensory; a few are feeding devices; and perhaps most frequently they are assumed to be respiratory. It seems to me, however, that excessive respiratory surfaces, such as a rich array of palps and antennae in polychaetes, are an unnecessary waste of energy, and that most marine forms should be able to get along quite well with more modest respiratory surface areas, even where the species is a burrower. There seems to be no correlation between the total expanded body surface and the usual oxygen content of the habitat of marine invertebrates.

In making the evolutionary adjustments to become established in fresh waters, the great majority of fresh-water invertebrates have either lost or never had an imposing array of external protuberances. Exceptions (Figs. 8 and 9) include: a few species of fresh-water gastrotrichs that have excessive spination; an oligochaete *Branchiura sowerbyi* with excessive "gills"; and several species of polychaetes that breed in fresh waters, that are probably recent invaders, and that have retained their excess parapodia, palps, and cirri. In the United States these polychaetes include one species each of *Lycastoides, Namanereis, Nereis,* and *Manayunkia.* Entoprocts and ectoprocts in both environments have equally developed tentacular crowns but these are chiefly food-gathering devices.

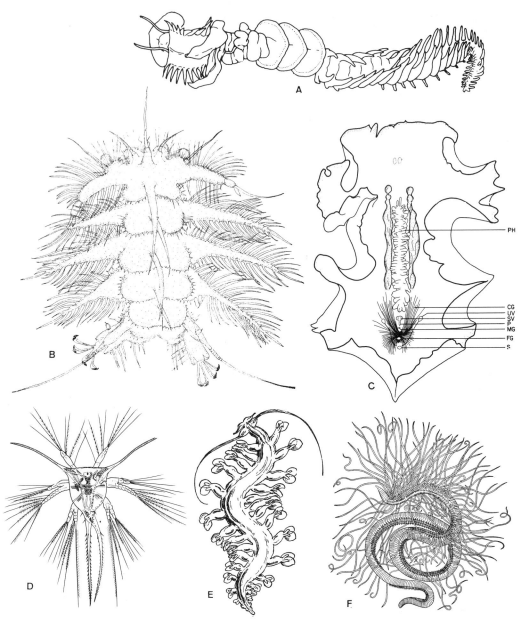

FIG. 7.–Examples of marine species showing excessive morphological embellishments. A, polychaete, *Chaetopterus,* ×0.5; B, tardigrade, *Neostigarcus acanthophorus,* ×200; C, turbellarian *Pericelis,* ×3; D, nauplius larva, ×40; E, plankton polychaete *Tomopteris,* ×11; F, polychaete, *Cirratulus grandis,* ×0.7. (B from Grimaldi de Zio, 1982; C from Riser and Morse, copyright 1974, *Biology of the Turbellaria,* with permission of McGraw-Hill, Book Co., New York; E from Hardy, 1956, *The Open Sea,* with permission of William Collins Sons, London; F from Miner, copyright 1950, *Field Book of Seashore Life,* with permission of G. P. Putnam's Sons, New York.)

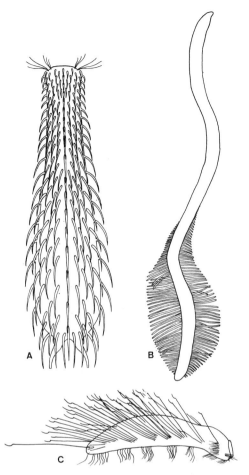

of fresh-water invertebrates, none of which appear to have wasted significant growth energetics on overgrowths or external structures that have no obvious or critical functions. Indeed, it is remarkable that fresh-water invertebrates have not developed a whole series of special respiratory embellishments, especially since so many species are exposed to anaerobic or near-anaerobic conditions at some time during the life cycle. A modest complement of gills and other respiratory structures seems to be adequate.

Let me cite an anomalous example occurring in swift streams. Often the bottom fauna contains several species of stonefly nymphs occupying similar spatial niches. Here it is common to find, in the same sample, nymphs with no external gills (e.g., *Isoperla*), as well as nymphs with a great abundance of fingerlike gills (e.g., *Pteronarcys*). Such streams are always at or close to 100% saturation with oxygen, yet there is a pronounced adaptive difference. Similar parallel examples may be found among the Trichoptera and Ephemeroptera, and, for that matter, similar parallel examples are abundant in marine taxa.

The whole problem of minimal, adequate, and excessive respiratory surfaces for both marine and fresh-water faunas needs investigation.

FIG. 8.–Examples of some of the few fresh-water invertebrates having morphological embellishments. Various magnifications. A, gastrotrich, *Chaetonotus;* B, oligochaete, *Branchiura sowerbyi;* C, gastrotrich, *Dasydytes.*

The only other fresh-water taxa having significant morphological embellishments are (1) some insect larvae and nymphs with spines or abundant gills, and (2) the Eubranchiopoda (fairy shrimps, tadpole shrimps, and clam shrimps), all of which possess 10 to 71 pairs of complicated flat respiratory appendages on the segments. Insects are of terrestrial origin, and the Eubranchiopoda are presumed to have originated in fresh waters, so neither group has immediate marine relatives, and the modifications are fresh-water acquisitions.

We are therefore left with the great majority

Coloration and color patterns. One of the most pervasive differences between marine and fresh-water invertebrates is the general lack of bright colors and contrasting patterns among the latter. Only a few exceptions can be cited: Fresh-water sponges are often bright green owing to their intracellular algal populations. Some high-altitude copepods are bright scarlet or bluish. A few gammarids have an orange coloration. Some water mites are bright red, blue, yellow, or green. And some dipteran larvae are red or greenish. These colors serve a variety of functions. Red copepods are said to be protected against short-wave radiation (Hairston, 1976); red water mites are said to be aposematic (first noted by Elton, 1922); and the red in some chironomid larvae is produced by hemoglobin. In addition, most small and translucent metazoans may have transient colors

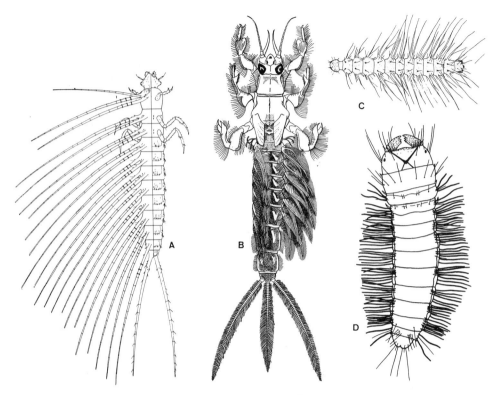

FIG. 9.–Examples of some of the very few immature fresh-water insects having morphological embellishments. Various magnifications. A, spinous *Peltodytes* larva, a beetle; B, gilled *Hexagenia* nymph, a burrowing mayfly; C, spinous *Atrichopogon* larva, a dipteran; D, gilled *Parargyractis* larva, a lepidopteran.

produced by food in the gut (e.g., eubranchio-pods). Some crayfish are purple, reddish, or orange, but hues are drab and seldom can be called "bright" colors.

Distinct and contrasting patterns in coloration are similarly lacking in nearly all fresh-water forms. Some leeches and planarians have a finely dotted coloration pattern, and many water mites have elaborate patterns produced by pigments in the body wall and by the varied contents of the visceral organs. Otherwise most distinctive patterns are to be found among immature aquatic insects, and especially in Ephemeroptera and Plecoptera, where there are yellow, brown, and tan pigment deposits in the integument. Personally, I think it is regrettable that in view of the enormous diversity of fresh-water insects, the evolutionary processes have not been more generous in distributing brightly patterned coloration among the aquatic stages in the life history.

Marine invertebrates, on the other hand, are often brilliantly colored and patterned. One need look no further than the abundance of color plates in identification manuals, natural history magazines, and decorative "coffee table" volumes.

Variable, or chromatophore, coloration is almost restricted to marine forms where it sometimes reaches astounding levels, especially in certain mollusks and crustaceans. I know of very few American fresh-water species that can make (rapid) changes in coloration. One example is *Palaemonetes* where chromatophores can color-adapt to the background in less than 24 hours, but this is a far cry from what we see in so many marine species.

No one has suggested the fundamental evolutionary reasons for the notable coloration disparity between marine and fresh-water species. It must stem from the differing results of an evolutionary complex of taxon age, genetics, behavior, food, excretory mechanisms, and so on. Indeed, the whole matter of colors and color patterns among aquatic forms is a fundamental problem that has scarcely been touched.

The cellulase problem. The sporadic occurrence of cellulose digestion in the fresh-water invertebrate kingdom is a continuing problem. Usually the criterion is the presence of cellulase in the digestive tract, either elaborated by the cells of the gut wall or by the normal bacterial flora of the gut. The "strongest" cellulase has been demonstrated (Monk, 1976) for a few *Gammarus* and some snails and limpets. "Weak" cellulase has been found in a few aquatic insects, especially beetles and caddis larvae. Unfortunately, however, there is no correlation between the occurrence of cellulase and the abundance of cellulose in the diet. In view of the fact that there are so very many taxa that ingest large quantities of cellulose (e.g., leaves and other organic detritus), we must conclude that nutrients are derived chiefly from the smaller bulk of noncellular organic matter associated and ingested along with the true cellulose organic matter.

Euryoky of fresh-water invertebrates. If estuarine habitats are omitted, even the most casual comparison of marine and fresh-water environments emphasizes the much greater variations in ecological factors in the latter. This generalization holds true from one comparable fresh-water habitat to another as well as from time to time within the same habitat. Seasonal and vertical temperature gradients in lakes and the greater variations in quiet shallows of ponds are much more pronounced than in salt water. Currents may vary seasonally from imperceptible to torrential in the same stream with accompanying variations in suspended silt. Stream invertebrates in general are capable of withstanding the severity of violent spring runoffs, intermittent cloudburst waters, and even short-term industrial pollution incidents produced by ions that are normally considered highly toxic. Other chemical conditions, including dissolved salts, dissolved organic matter, dissolved oxygen, and pH are highly variable and cannot be compared with the more monotonous conditions in typical marine habitats. Drying up of the habitat is a frequent and important hazard.

In general, therefore, our fresh-water invertebrates consist of a group of euryokous species that are physiologically, morphologically, and behavioristically adapted for maintaining themselves under trying circumstances that would be disastrous or impossible for the great majority of marine forms.

ATYPICAL FRESH-WATER HABITATS

As a whole, fresh-water habitats are greater in variety (and often in complexity) than those of the marine environment, especially when the former include the complete series of running water habitats and many small, peculiar types of bodies of water.

If one is interested in collecting a wide variety of fresh-water invertebrates, the most favorable locations are small, weedy lakes and ponds. Such habitats afford large numbers of niches occupied by a correspondingly large variety of animals, especially mollusks, aquatic insects, crustaceans, and many representatives of the worms and wormlike groups. Food and population relationships are intricate, and presumably this complexity is the chief reason why so little comprehensive ecological work has been done on ponds and small lakes.

At the other extreme are highly specialized bodies of water in which one or more ecological conditions are so severe that the fauna is restricted to a small number of species. Frequently, however, these few species attain great population densities, perhaps because of the lack of interspecific competition.

There is little advantage to be gained by outlining the many divisions and subdivisions of fresh-water communities. Several such outlines may be found in ecology texts. It may be worth while, nevertheless, to characterize some of the more peculiar habitats that represent extremes in ecological conditions. As compared with the usual sort of fresh-water habitat, these peculiar habitats are featured by less variable ecological conditions.

Alkali and salt lakes. Alkali and salt lakes and ponds are common in many areas of the western states, especially where there are closed drainage basins, semiarid conditions, and soils containing large quantities of soluble salts. Depending on surrounding soil chemistry, these waters may contain large quantities of chlorides, sulfates, or carbonates, especially those of sodium, potassium, and calcium. The expressions "salt" and "alkali" lakes are by no means sharply defined and are often interchangeable, but in general where the dissolved salts are of such composition and concentration as to produce persistent pH readings in excess of 8.8, a body of water is referred to, in common usage, as an "alkali lake."

At low concentrations, up to about 1 to 3% of dissolved salts, the fauna is still varied in species but much more restricted than in nonalkaline waters. Above this concentration species diminish rapidly in numbers, and the metazoan plankton becomes characterized by a few copepods and highly alkaline rotifers, particularly some species of *Brachionus* and *Hexarthra,* which may be extremely abundant. The insect fauna becomes highly restricted and consists mostly of Chironomidae, occasionally one species of *Aedes,* and a few beetles.

Great Salt Lake, which contains about 25% dissolved salt (chiefly sodium chloride), has been studied relatively carefully and has an extremely limited fauna. Protozoa include several ciliates, flagellates, and amoeboid forms; metazoans consist of only two species of *Ephydra* (the brine fly) and *Artemia salina* (the brine shrimp). *Artemia* populations, sometimes attain an average density of 50 individuals per liter. In contrast to the situation in most normal lake beaches, no interstitial micrometazoan fauna has been found in the sandy beaches of Great Salt Lake.

Polluted waters. Waters grossly polluted with organic matter, especially domestic or "city" sewage, likewise have a highly restricted fauna, one capable of thriving in very low concentrations of oxygen and high concentrations of dissolved and particulate organic matter. The chief metazoans are a few species of tubificid oligochaetes, red chironomid dipteran larvae, psychodid larvae, and the rat-tailed maggot larva, *Eristalis.* Sometimes the bottom of polluted rivers is literally covered with a waving, writhing mass of tubificids. This habitat also supports large populations of ciliates and colorless flagellates.

At a certain distance below the entry of domestic pollutants a river enters a "zone of recovery" in which the dissolved oxygen progressively increases and the polluting materials are oxidized or otherwise sufficiently removed from the water, especially as the result of bacterial action, so that the pollution fauna is slowly replaced by a "normal" river fauna. Depending on the intensity of pollution, size of the river, and so on, the zone of recovery may begin anywhere between one-quarter mile and 100 miles downstream from the source of pollution.

Biologists are prone to overlook the fact that the zone of recovery is frequently the most productive part of a flowage. It is here that the periphyton is most dense. Dense metazoan populations include such forms as *Physella,* isopods, planarians, leeches, amphipods, dipterans, caddis larvae, and mayfly nymphs.

In addition to organic pollution, the situation has been further severely aggravated by industrial toxicants, continuous silting, and more and more irrigation system water withdrawals. The history lesson of Lake Erie demonstrates the inherent recovery potential of our fresh-water fauna, but many of our waters have been continuously fouled for such long periods, especially by industry, that they are beyond recovery and are simply biological deserts.

For this reason those species indicated as having highly restricted ranges in this volume should not be taken at face value, especially in areas where the tradition of lake and stream despoliation has had a long and continuing history.

Aquatic biologists have been searching many years for the ideal common species or group of "indicator" species that would be (1) present unequivocally in the absence of pollution and (2) absent unequivocally in the presence of pollution. Unfortunately, this search appears to be a lost cause chiefly because of the euryokous nature of most species and because of individual differences within the same species. We are therefore forced to rely on facies or clusters of clean-water species and polluted-water species, differing in composition from one area to another, but nevertheless serving as rough indicators of relative purity or pollution.

Subterranean waters. Subterranean streams, cave ponds, and artesian wells some-times contain endemic, primitive, or highly specialized amphipods, isopods, decapods, copepods, and turbellarians. The Ozarks and similar limestone areas are particularly rich in underground aquatic habitats. Most subter-ranean species are colorless, translucent, or whitish; eyes are nonfunctional or absent; tactile structures are often well developed and include long antennae and abundant long setae. The subterranean environment is completely monotonous, with constant eco-logical conditions. Basic food materials con-sist of fungi and small bits of organic debris washed underground. Populations are gener-ally small, owing to the greatly limited food supply.

Endemic subterranean species are thought to have been derived from ancestral surface species that underwent evolutionary change and specialization concomitantly with the gradual submergence of their surface drainage habitats. Compared with Europe, little is known of the subterranean microcrustacean fauna of the United States, but we do have a rich fauna. It should be added that accessible subterranean streams and ponds usually also contain common species that are regular members of aquatic communities on the surface of the ground.

Phreatic and psammolittoral waters. These waters are ecologically closely related to subterranean waters, but are characterized by species that are generally smaller, including rotifers, tardigrades, gastrotrichs, mites, cope-pods, syncarids, and microturbellarians. These groups form the "interstitial community" living in the capillary water between sand and gravel particles 10 to 30 cm deep on the sandy beaches of lakes and streams (Fig. 10), and

also in the waters of the hyporheic zone below the surface of stream beds. Unfortunately, little work is being done on these habitats in the United States.

Cold springs, spring brooks. Typical cold spring communities are restricted to a small area at and just below the point where the water issues from the ground. As soon as it flows any distance as a spring brook it becomes warmer, loses it characteristic growth of water-cress, and contains a different animal com-munity. Depending on local conditions, the typical spring and spring brook fauna zone usually extends from 10 to 200 m below the source. Cold springs have an average tem-perature of about 8°C in temperate areas, and the associated animals are mostly cold steno-therms. The list of characteristic animals is relatively small, including planarians, amphi-pods, and some of the following: beetles, caddis larvae, leeches, isopods, sphaeriid clams, small snails, and blackfly larvae. Fre-quently two or more of these species are numerically dominant. If the spring emerges from appropriate underground water courses in limestone areas it occasionally brings up typical subterranean species. The ecology of springs and spring brooks has been sadly neglected in the United States.

Hot springs. In addition to their high temperatures (about 35°C), hot springs are often characterized by large quantities of dissolved salts. Sometimes the water is also high in dissolved carbon dioxide, hydrogen sulfide, or sulfur dioxide. Thus it is often difficult to distinguish between the limiting effects of temperature and the limiting effects of other ecological factors. In general, how-ever, 25 to 30°C seems to be the dividing line between a rich fauna at lower temperatures and a poor fauna at higher temperatures. Unlike cold springs, there do not seem to be any typical genera or species in hot springs, nor are hot spring animals ever abundant. Beetles are among the most characteristic metazoan inhabitants, many species having been reported from a wide variety of hot springs in the western states. Diptera larvae, especially Chironomidae, are also common inhabitants, along with an occasional culicid

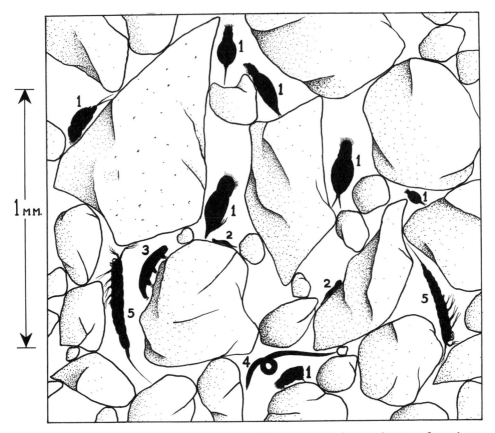

FIG. 10.–Optical section of a small portion of a sandy beach, 15 cm deep and 150 cm from the water's edge, showing relative sizes of sand grains and some of the common micrometazoans. *1*, rotifers; *2*, gastrotrichs; *3*, tardigrade; *4*, nematode; *5*, harpacticoid copepod. (From Pennak, 1939.)

or stratiomyid. Sometimes a few Hemiptera and Odonata may be collected. Most such records of insects are from springs whose temperatures range from 37 to 50°C. Other arthropods include amphipods, ostracods, and Hydracarina reported from springs and with temperatures between 32 and 51°C.

Micrometazoans are often present in exceptionally hot water, a few nematodes and rotifers having been taken as high as 60°C. Since the thermal death point of most fresh-water invertebratess is between 30 and 40°C, many hot spring animals must have developed considerable thermal acclimation.

DISPERSAL AND BARRIERS

The successful establishment of the ancient marine ancestors of fresh-water animals in swamps and near estuaries was actually only the beginning of the story. From such footholds they must have developed a higher degree of euryoky as they expanded to occupy new inland habitats.

Migration up and downstream throughout a single drainage system is comparatively easy

for many fresh-water invertebrates over a long period of time. Superficially, it would appear that they are "landlocked" in separate drainage systems, with little migration from one to another. In the long-time viewpoint, however, land barriers between different lakes and rivers are not effective for most groups within the confines of a single continental land mass. The ability of almost all microscopic species to form thick-shelled eggs, cysts, or resting stages that withstand partial drying or desiccation is a primary factor in ease of dispersal. Such small resting bodies are easily picked up and wind blown overland for long distances. If they drop into a favorable aquatic habitat with the appropriate combination of ecological conditions, they again become active and may produce a new population nucleus.

The literature contains scattered but authentic records of live fish being transported overland by storms, tornadoes, and other strong winds, and there is no reason to doubt that plankton and bottom invertebrates are transported in the same manner. Much of the "dust" of dust storms is organic debris, and it is not improbable that a small fraction consists of cysts and eggs. Ducks, shore birds, and aquatic insects are now known to be important for increasing the ranges of many groups, and the literature contains many accounts of such transport. Both resting and active stages are transported in the detritus and mud that frequently sticks to the appendages. Snails are frequently found on migratory birds, and there is one authentic record of a 2½-oz. live mussel being transported overland by a duck! Many microzoan cysts and resting eggs are carried to new habitats via the intestinal tract of birds.

Adult aquatic insects capable of flight easily migrate overland, often in what appears to be a completely random fashion, and come to occupy suitable new areas. Sometimes such flights are greatly aided by winds. Entomologists report collecting adult aquatic beetles and hemipterans 5 miles or more from the nearest open ponds. Occasionally such reports are for desert or semiarid regions. The surface of a concrete walk near the writer's home has a slight depression about 80 cm in diameter, which fills with water with every shower. On several occasions aquatic beetles have been found in this puddle, and the nearest spot from which they could have come is a pond more than a kilometer distant.

Anyone who follows the biological development of a new, man-made reservoir is bound to be impressed with the promptness with which the reservoir becomes colonized with aquatic animals. Stock tanks and stock ponds, newly constructed in grazing lands where the nearest water is 10 km away, quickly acquire a varied invertebrate population.

The literature of anabiotic transport by wind and vagile animals consists chiefly of fragmentary and widely scattered papers; a few of the more significant are Niethammer, 1953; Lansbury, 1955; Löffler, 1963; Maguire, 1963; Malone, 1965; Proctor and Malone, 1965; Dundee et al., 1967; Stewart et al.; 1970; Carroll and Viglierchio, 1981; Fernando, 1958.

It is not easy to explain the faunal composition of small spring brooks on the basis of anabiotic structures and overland transport. Such habitats are abundant and widely scattered over the United States, but are effectively isolated from each other by a matter of a few to many kilometers. The community makeup is readily predictable. A dense growth of watercress (*Nasturtium officinale*) is almost invariably present, the substrate is usually gravel, temperature is constant and cool, and the whole community, extending from the emergence point of water from the ground to the mouth of the brook, or to the point where watercress is no longer present, usually ranges from 10 to 200 m in length. The macrofauna is composed of planarians, isopods, amphipods, small snails (*Physella*), leeches, and a few insects, mostly dipteran and beetle larvae, and mayfly nymphs. Few of these species have anabiotic stages that would facilitate geographic distribution. Adult spring brook insects can, of course, easily migrate from one site to another, but most of the other spring brook species are presumably transported on the bodies of flying insects, water birds, and shore birds. Since spring brooks are first-order streams, it would be difficult for the various species to migrate actively and extensively downstream and upstream throughout large flowages, and through unfavorable habitats in so doing. We have seen many small spring brooks on steep, rocky canyon walls; active migrations into such sites would be impossible, except for adult insects.

It is tempting to theorize that spring brooks are stable habitats, and that their corresponding faunal community therefore is composed largely of species that do not produce anabiotic or diapause stages.

While it is now convenient to explain the

colonization of aquatic areas by disseminules distributed especially by insects and birds, I wonder about the zoogeographic spread of fresh-water invertebrates in the long-time sense. How, for example, was it brought about before the evolutionary appearance of insects, birds, and other terrestrial vertebrates in Earth's history? Surely a fresh-water fauna existed before birds and insects. Were winds, flooding, and changing drainage patterns then the chief means of distributing disseminules?

Some highly euryokous and vagile species, mostly microscopic, have spread so effectively that they are essentially cosmopolitan and occur the world over in suitable habitats. The list includes a great many Protozoa and Rotatoria, several sponges, *Paracyclops fimbriatus*, perhaps *Plumatella repens*, a few oligochaetes, some Tardigrada, and several Cladocera, such as *Daphnia longispina* and *Chydorus sphaericus*. Many more species could be added to this list.

The fresh-water invertebrate faunas of North America and Europe, and to some extent northern Asia, have many of the species in the following groups in common: Protozoa, Rotatoria, Tardigrada, Bryozoa, Oligochaeta, and Cladocera. American gastrotrichs and nematodes are so poorly known that their European affinities are not established. In all the other major taxa with low vagility, however, the great majority of species that occur in the United States are endemic and are not found in Europe.

The zoogeographic distribution of fresh-water invertebrates has been rather well investigated in Europe, and zoogeographic regions have been suggested for several taxonomic groups, especially some of the orders of crustaceans and insects, and the mollusks. Numerous distribution patterns have been correlated with the effects of the advances and retreats of glaciers on the European continent. In the United States, however, comprehensive studies of fresh-water invertebrate zoogeography are largely lacking, with the notable exceptions of the mollusks, decapods, and perhaps several other groups.

Europe, Asia, and Africa have certain very ancient lakes where primitive species have been preserved and where geographic isolation has been effective for such a long time that independent speciation has resulted in many unique and endemic species, especially among Porifera, Turbellaria, Oligochaeta, Gastropoda, and crustaceans. Some of the better known lakes of this type are Baikal in Asia, Tanganyika and Nyasa in Africa, Ochrid in Jugoslavia, and the lakes of the Malili River system in the central Celebes. No such ancient lakes occur in the United States.

FOOD WEBS

During the past two generations the literature of aquatic biology has been replete with "productivity" and "trophic level" studies. It is unfortunate that most of these studies are misleading because they (1) are seasonal rather than year round, (2) have been conducted on complex (rather than simple) ecosystems, and (3) are based on only a fraction of the species making up a particular ecosystem. As a consequence, the generalizations postulated in many of these investigations are open to doubt and modification. Actually a much more fruitful approach of population dynamics and interrelationships involves habitats characterized by few species and few niches. Examples are alkali and saline ponds, relatively barren alpine lakes, spring brooks, very small ponds, warm springs, and small mountain streams. Using such bodies of water, it should be relatively easy to determine food habits of individual species and to make quantitative estimates of populations. Productivity indexes, food cycles, energy transfers from one trophic level to another, species interactions, and seasonal population fluctuations are all important ecological concepts that should be derived from such studies on a *year-round basis*. It is important to note that the measurement of associated physical and chemical factors in small streams, lakes, and ponds is simpler than in more complex types of aquatic habitats and in terrestrial habitats.

Nevertheless, even in the simplest aquatic habitats the determination of pyramids of numbers and food chains (or better "food webs") is an involved problem, and only a very few thorough and inclusive studies have been attempted. The great majority of trophic investigations are concerned with only a few energy or biomass transpositions, the remainder being only roughly estimated. In any such study the basic role played by detritus and bacterial action should not be minimized. In a schematic fashion Fig. 11 shows the fundamental food interrelationships for any aquatic habitat, regardless of size and complexity. Only recently have aquatic ecologists come to realize the overwhelming role of detritus in the food web. The significance of the various segments of this diagram vary enormously from one habitat to another. Some habitats, for example, have few or no fishes; other habitats have no rooted aquatics or a poor substrate fauna.

The solid lines in Fig. 11 indicate consumption or utilization as food. Broken lines indicate death and disintegration, or defecation by animals. The dotted line indicates disintegration of detritus as the result of certain bacterial activities. The circle arrows originating and ending at "zooplankton," "fish," and "substrate fauna" are meant to show that certain species in each of these categories are carnivorous and feed on other individuals within the same category. Actually each such category may represent two, three, or more trophic levels. "Substrate fauna" includes all bottom invertebrates as well as the protozoans, micrometazoans, and macrometazoans associated with the surface of rooted aquatic plants and other objects. By inference, "bacteria" also includes molds and their activities.

For simplicity, certain minor aspects of this food web have been omitted. For example, the contribution of excretory materials to the

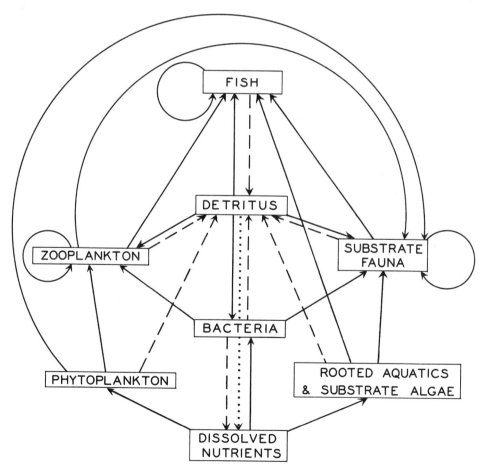

FIG. 11.–Basic features of aquatic food webs. See text for explanation.

main reservoir of dissolved nutrients is not shown. By inference, "fish" should include other vertebrates such as frogs, salamanders, and turtles.

Not shown, but nevertheless intimately concerned with the continuing balance within the web, are the nitrogen, sulfur, and phosphorus cycles. These are fundamentally similar to the same cycles operating in terrestrial habitats.

Following is a selected list of general references. It includes (1) items referring specifically to the content of this introductory chapter, (2) texts covering some of the same fields as the present manual, (3) important accessory texts covering fields relating to the study of fresh-water invertebrates, (4) periodical references and texts covering major phases of fresh-water invertebrate biology, and (5) a few references giving laboratory and field methods. Some references are recent; others are old classics.

COMMENT

By way of a brief summary, we must accept the proposition that incipient colonizers of fresh waters must have first acquired, in varying degrees, whole clusters of special devices and mechanisms in order to establish successful and permanent populations in the harsh and variable fresh-water environment. At the same time, certain specializations for the marine existence must have been lost. Our understanding of evolutionary and genetic processes does not permit the de novo establishment of fresh-water taxa. Rather we think of these groups as only very rarely and precariously being successful in the new environment. We visualize the age of fresh-water taxa as being quite different from each other. Some are probably "new"; others are "ancient." But, in my own estimation, and regardless of age, the presiding requisite for life in fresh waters in that of euryoky.

At the beginning of this essay, I mentioned the fact that there are exceptions to many of my generalizations. A few of the most striking of these are as follows: Encystment by a marine copepod (Coull and Grant, 1981) and by a marine tardigrade (Schulz, 1935); polychaetes living in fresh waters (Foster, 1972); bioluminescence of fresh-water oligochaetes (Bogatov et al., 1980), limpets, shrimp, and flies (Harvey, 1952); a veliger stage known for one (European) fresh-water mussel, *Dreissensia polymorpha*, a recent invader of marine ancestry.

A number of fresh-water invertebrate rarities and ecological curiosities in the United States are associated in various ways with the topics we have enumerated. Some involve "escapes," that is, certain uncommon fresh-water invertebrates with close marine ancestry that have made the marine (brackish?) to fresh-water transition accidentally, and contrary to the situation among their closest relatives. Space does not permit discussion or extended reference lists, but some of the more intriguing items may be listed as follows: hot spring faunas, hyporheic faunas, Acoela, ice worms, cave species, Nemertea, Archiannelida, glacial marine relicts, Thermosbaenacea, Bathynellacea, *Cordylophora, Urnatella*, intertidal and shoreline insects. Investigations of these topics and taxa may be instrumental in answering some of the questions posed in the foregoing paragraphs.

Undoubtedly many who read this essay will disagree with some of my generalizations and suggestions. If so, I shall be pleased in knowing that I have been responsible for such attention being accorded to the fresh-water invertebrate biota. I hope such controversy will generate significant field and laboratory research.

REFERENCES

Adolph, E. F. 1925. Some physiological distinctions between fresh-water and marine organisms. *Biol. Bull.* 48:327–335.

Ashwood-Smith, M. J., and J. Farrant. 1980. *Low temperature preservation in biology and medicine.* 323 pp. Pittman Medical, Kent, England.

Ax, P., and R. Ax. 1970. Das Verteilungsprinzip des subterranen Psammon am Übergang Meer—Susswässer. *Mikrofauna des Meeresbodens* 1:1–51.

Baldwin, E. 1948. *An introduction to comparative biochemistry.* 164 pp., Cambridge Univ. Press, Cambridge, England.

Bardach, J. E., J. O. Ryther, and W. O. McLarney. 1972. *Aquaculture.* 868 pp. Wiley–Interscience, New York.

Barnes, T. C., and T. L. Jahn. 1934. Properties of water of biological interest. *Q. Rev. Biol.* 9:292–341.

Barrington, E. J. W. 1967. *Invertebrate Structure and Function.* 549 pp. Houghton Mifflin, New York.

Beadle, L. C. 1943. Osmotic regulation and the faunas of inland waters. *Biol. Rev.* 18:172–183.

_____. 1957. Osmotic and ionic regulation in aquatic animals. *Ann. Rev. Physiol.* 19:329–358.

_____. 1969. Osmotic regulation and the adaptation of freshwater animals to inland saline waters. *Proc. Int. Assoc. Theor. Appl. Limnol.* 17:421–429.

Bick, H., et al. 1972. Das Zooplankton der Binnengewässer. *Die Binnengewässer* 26(1):1–294.

Bogatov, V. V. et al. 1980. The luminescence of freshwater Oligochaeta. *Hydrobiol. J.* 13:42–47.

Bond, R. M. 1933. A contribution to the study of the natural food cycle in aquatic environments. *Bull. Bingham Oceanogr. Coll.* 4:1–89.

Brandt, K. 1897. Die Fauna der Ostsee, in besondere die der Kieler Bucht. *Verh. Deutsch Zool. Gesellschaft* (1897):10–34.

Brooks, J. L. 1950. Speciation in ancient lakes. *Q. Rev. Biol.* 25:30–60, 131–176.

Brues, C. T. 1932. Further studies on the fauna of North American hot springs. *Proc. Am. Acad. Arts Sci.* 67:185–303.

Cairns, J., and K. L. Dickson (eds.). 1973. *Biological methods for the assessment of water quality.* ASTM Special Tech. Publ. 528:1–256.

Carpenter, K. E. 1928. *Life in inland waters.* 267 pp. Sidgwick and Jackson, London.

Carroll, J. J. and D. R. Viglierchio. 1981. On the transport of nematodes by the wind. *J. Nematol.* 13:476–482.

Coker, R. E. 1954. *Streams lakes ponds.* 327 pp. Univ. North Carolina Press, Chapel Hill.

Coull, B. B. and J. Grant. 1981. Encystment discovered in a marine copepod. *Science* 212:342–343.

Danell, K. 1981. Overwintering of invertebrates in a shallow northern Swedish lake. *Int. Rev. gesamten Hydrobiol.* 66:837–845.

Delamare Debouteville, C. 1960. *Biologie des Eaux Souterraines Littorales et Continentales.* 740 pp. Hermann, Paris.

Denny, M. W. 1983. A simple device for recording the maximum force exerted on intertidal organisms. *Limnol. Oceanogr.* 28:1269–1274.

Dorgelo, J. 1976. Salt tolerance in Crustacea and the influence of temperature upon it. *Biol. Rev.* 51:255–290.

Dougherty, E. C., and L. G. Harris. 1963. Antarctic micrometazoa: fresh-water species in the McMurdo Sound area. *Science* 140:497–498.

Dundee, D. S., et al. 1967. Snails on migratory birds. *Nautilus* 80:89–91.

Edmondson, W. T., and G. G. Winberg. 1971. *Secondary productivity in fresh waters.* I.B.P. Handbook 17:1–358.

Elton, C. 1922. On the colours of water-mites. *Proc. Zool. Soc. London* 1922:1231–1239.

Fernando, C. H. 1958. The colonization of small freshwater habitats by aquatic insects. 1. General discussion, methods and colonization in the aquatic Coleoptera. *Ceylon J. Sci. Biol. Sci.* 1:117–154.

Forbes, S. A. 1887. The lake as a microcosm. *Bull. Peoria Sci. Assoc.* 1887:77–87.

Foster, N. 1972. Freshwater polychaetes (Annelida) of North America. Biota of Freshwater Ecosystems Identification Manual, No. 4:1–15.

Fox, H. M. and B. G. Simmons. 1933. Metabolic rates of aquatic arthropods from different habitats. *J. Exp. Biol.* 10:67–74.

Frey, D. G. (ed.). 1962. *Limnology in North America.* 734 pp. Univ. Wisconsin Press, Madison.

Fryer, G. 1974. Evolution and adaptive radiation in the Macrothricidae (Crustacea: Cladocera): a study in comparative functional morphology and ecology. *Philos. Trans. R. Soc. London, B, Biol. Sci.* 269:142–272.

Galtsoff, P. S. 1934. The biochemistry of the invertebrates of the sea. *Ecol. Monogr.* 4:481–490.

Golterman, H. L. 1969. *Methods for chemical analysis of fresh waters.* I.B.P. Handbook 8:1–166.

Grimaldi de Zio, S. et al. 1982. *Neostygarctos acanthophorus* n. gen., n. sp., nuovo Tardigrado marino del Mediterraneo. *Cah. Biol. Mar.* 23:319–323.

Hairston, N. G. 1976. Photoprotection by carotenoid pigments in the copepod Diaptomus nevadensis. *Proc. Natl. Acad. Sci. USA* 73:971–974.

Harnisch, O. 1951. Hydrophysiologie der Tiere. *Die Binnengewässer* 19:1–299.

Hardy, D. 1956. *The open sea.* William Collins, London.

Hart, C. W., and S. L. H. Fuller (eds). 1974. *Pollution ecology of freshwater invertebrates.* 389 pp. Academic Press, New York.

Harvey, E. N. 1952. *Bio-luminescence.* 649 pp. Academic Press, New York.

Hora, G. L. 1930. Ecology, economics, and evolution of the torrential fauna. *Philos. Trans. R. Soc. London* 218B:171–282.

Hubalt, E. 1927. Contribution a l'étude des invertebres torrenticoles. *Bull. Biol. Fr. Belg., Suppl.* 9:1–390.

Husmann, S. 1967. Klassifizierung mariner, brackiger und limnischer Grundwasserbiotope. *Helgol. Wiss. Meeresunters.* 16:271–278.

_____. 1975. *First international symposium on groundwater ecology.* 232 pp. Swets Publ. Service, Amsterdam.

Hutchinson, G. E. 1957. *A treatise on limnology. Vol. I. Geography, physics, and chemistry.* 1015 pp. Wiley, New York.

_____. 1967. *Vol. II. Introduction to lake biology and the limnoplankton.* 1115 pp. Wiley, New York.

_____. 1975. *Vol. III. Limnological botany.* 660 pp. Wiley, New York.

Hynes, H. B. N. 1960. *The biology of polluted waters.* 199 pp. Liverpool Univ. Press, Liverpool.

_____. 1970. *The ecology of running waters.* 555 pp. Univ. Toronto Press, Toronto.

Illies, J. (ed.). 1978. *Limnofauna Europaea.* 532 pp. Gustav Fischer Verlag, Stuttgart.

Jacobs, W. 1935. Das Schweben der Wasserorganismen. *Erg. Biol.* 11:131–218.

Johannes, R. E., and M. Satomi. 1967. Measuring organic matter retained by aquatic invertebrates. *J. Fish. Res. Board Can.* 24:2467–2471.

Kajak, Z., and A. Hillbricht-Ilkowska (eds.). 1972. *Productivity Problems of Freshwater.* 918 pp. P.W.N. Polish Scientific Publishers, Warsaw, Poland.

Kanwisher, J. W. 1955. Freezing in intertidal animals. *Biol. Bull.* 109:56–63.

Kenk, R. 1949. The animal life of temporary and permanent ponds in southern Michigan. *Misc. Publ. Mus. Zool. Univ. Mich.* 71:1–66.

Krogh, A. 1939. *Osmotic regulation in aquatic animals.* 242 pp. Cambridge University Press, England.

———. 1941. *The comparative physiology of respiratory mechanisms.* 172 pp. Univ. Pennsylvania Press, Philadelphia.

Lansbury, I. 1955. Some notes on invertebrates other than Insecta found attached to water-bugs (Hemiptera-Heteroptera). *Entomologist* 88:139–140.

Löffler, H. 1963. Vogelzug und Crustaceen Verbreitung. *Verh. Deut. Zool. Gesell. München* 1963:311–316.

Lucas, C. E. 1947. The ecological effects of external metabolites. *Biol. Rev.* 22:270–295.

Macan, T. T. 1961. Factors that limit the range of freshwater animals. *Biol. Rev.* 36:151–198.

———. 1963. *Freshwater ecology.* 338 pp. Longmans, London.

———. 1975. *Life in lakes and rivers.* 3rd ed. 320 pp. Collins New Naturalist, London.

Mackenthun, K. M. 1973. *Toward a cleaner aquatic environment.* 273 pp. U.S.E.P.A. Office of Air and Water Programs.

Maguire, B. 1963. The passive dispersal of small aquatic organisms and their colonization of isolated bodies of water. *Ecol. Monogr.* 33:161–175.

Malone, C. R. 1965. Killdeer (*Charadrius vociferus* Linnaeus) as a means of dispersal for aquatic gastropods. *Ecology* 46:551–552.

Mansingh, A. 1971. Physiological classification of dormancy in insects. *Can. Ent.* 103:983–1009.

Melchiorri-Santolini, and J. W. Hopton (eds.). 1972. Detritus and its role in aquatic ecosystems. *Mem. Ist. Ital. Idrobiol.* 29 (Suppl.):1–540.

Miller, R. C. 1958. The relict fauna of Lake Merced, San Francisco. *J. Mar. Res.* 17:375–382.

Miner, D. 1950. *Field book of seashore life.* Putnam, New York.

Monk, D. C. 1976. The distribution of cellulase in freshwater invertebrates of different feeding habits. *Freshwater Biol.* 6:471–475.

Naumann, E. 1931. Limnologische. Terminologie, *Handbuch der biol. Arbeitsmeth.,* Sect. 9, Part 8, Nos. 1–5:1–776.

Needham, J. 1930. On the penetration of marine organisms into freshwater. *Biol. Zentralbl.* 50:504–509.

Neldner, K. H. and R. W. Pennak. 1955. Seasonal faunal variations in a Colorado alpine pond. *Am. Midl. Nat.* 53:419–430.

Nichol, E. A. T. 1935. The ecology of a salt marsh. *J. Mar. Biol. Assoc. U.K.* 20:203–261.

Nielsen, A. 1950. The torrential invertebrate fauna. *Oikos* 2:176–196.

Niethammer, G. 1953. Zum Transport von Süsswassertierchen durch Vogel. *Zool. Anz.* 151:41–42.

Olsson, T. I. 1981. Overwintering of benthic macroinvertebrates in ice and frozen sediment in a north Swedish river. *Holarctic Ecol.* 4:161–166.

Palmer, A. R. 1960. Miocene copepods from the Mojave Desert, California. *J. Paleontol.* 34:447–452.

Park, T., et al. 1946. Dynamics of production in aquatic populations. *Ecol. Monogr.* 16:311–391.

Pax, F. 1951. Die Grenzen tierischen Lebens in mitteleuropäischen Thermen. *Zool. Anz.* 147:275–284.

Pearse, A. S. 1928. On the ability of certain marine invertebrates to live in diluted sea water. *Biol. Bull.* 54:405–409.

———. 1932. Animals in brackish water ponds and pools at Dry Tortugas. *Publ. Carnegie Inst. Wash.* 435:125–142.

———. 1950. *The emigrations of animals from the sea.* 210 pp. Dryden, New York.

Pennak, R. W. 1939. The microscopic fauna of the sandy beaches. In: *Problems of Lake Biology, Publ. Am. Assoc. Adv. Sci.* 10:94–106.

———. 1963. Ecological affinities and origins of free-living acelomate fresh-water invertebrates. In: *The Lower Metazoa,* pp. 435–551, ed. by E. C. Dougherty, 478 pp. Univ. California Press, Berkeley.

———. 1977. Trophic variables in Rocky Mountain trout streams. *Arch. Hydrobiol.* 80:253–285.

———. 1988. Ecology of the freshwater meiofauna. In: *Introduction to the Study of Meiofauna,* pp. 4.1–4.22, edited by R. P. Higgins and H. Thiel, Smithsonian Institution Press, Washington, D.C.

Prescott, G. W. 1969. *How to know the aquatic plants.* 171 pp. Brown, Dubuque, Iowa.

———. 1970. *How to know the freshwater algae.* 348 pp. Brown, Dubuque, Iowa.

Proctor, V. W. 1964. Viability of crustacean eggs recovered from ducks. *Ecology* 45:656–658.

Proctor, V. W. and C. R. Malone. 1965. Further evidence of the passive dispersal of small aquatic organisms via the intestinal tract of birds. *Ecology* 46:728–729.

Proctor, V. W., et al. 1967. Dispersal of aquatic organisms: viability of disseminules recovered from the intestinal tract of captive killdeer. *Ecology* 48:672–676.

Ramazotti, G. 1972. Il phylum Tardigrada. *Mem. Ist. Ital. Idrobiol. Dott. Marco de Marchi.* 28:1–732.

Remane, A., and C. Schlieper. 1971. Biology of brackish water. *Die Binnengewässer* 25:1–372.

Riser, D., and D. Morse. 1974. *Biology of the Turbellaria.* McGraw-Hill, New York.

Russell-Hunter, W. D. 1970. *Aquatic Productivity.* 306 pp. Macmillan, New York.

Scholander, P. F. et al. 1953. Studies on the physiology of frozen plants and animals in the Arctic. *J. Cell. Comp. Physiol.* **42** (Suppl. 1):1–56.

Schulz, E. 1935. *Actinarctus doryphorus* nov. gen. nov. spec., ein Markwürdiger Tardigrad aus der Nordsee. *Zool. Anz.* 111:285–288.

Schwoerbel, J. 1967. Das hyporheische Interstitial als Grenzbiotop zwischen oberirdischem und subterranem Ökosystem und seine Bedeutung für die Primar-Evolution von Kleinsthöhlenbewornern. *Arch. Hydrobiol. Suppl.* **33**:1–62.

Segerstråle, S. G. 1957. Baltic Sea. Pp. 751–800, in Treatise on marine ecology and paleoecology. *Mem. Geol. Soc. Am.* **67**:1–1296.

Sernov, S. A. 1927. Über die Überwinterung der Wasserorganismen im Eise und in der gefroren Erde nach dem Material von N. Boldyreva, P. P. Scharmina, und J. D. Schmeleva. *Atti Congr. Int. Limn. Teor. Appl.* **4**:555–563.

Steuer, H. 1910. *Planktonkunde.* 723 pp. Teubner, Leipzig and Berlin.

Stewart, K. W., L. E. Milliger, and B. M. Solon. 1970. Dispersal of algae, protozoans, and fungi by aquatic Hemiptera, Trichoptera, and other aquatic insects. *Ann. Entomol. Soc. Am.* **63**:139–144.

Thienemann, A. 1950. Verbreitungsgeschichte der Süsswassertierwelt Europas. *Die Binnengewässer* **18**:1–809.

Thorson, G. 1946. Reproduction and larval development of Danish marine bottom invertebrates. *Medd. Komm. Danmarks Fish. og Havunders. Ser. Plankton* **4**:1–523.

Valkanov, A. 1968. Das Neuston. *Limnologica* **6**:381–404.

Todd, D. K. (ed.). 1970. *The Water Encyclopedia.* 559 pp. Water Information Center, Port Washington, New York.

Vinogradov, A. P. 1953. The elementary chemical composition of marine organisms. *Mem. Sears Found. Mar. Res.* No. II, 647 pp.

Von Brand, T. 1946. *Anaerobiosis in invertebratess.* 328 pp. Normandy, Missouri.

Ward, H. B., and G. C. Whipple (ed. by W. T. Edmondson). 1959. *Fresh-water biology.* 1248 pp. Wiley, New York.

Warren, C. F. 1971. *Biology and water pollution.* 434 pp. Saunders, Philadelphia.

Wesenberg-Lund, C. 1939. *Biologie der Susswassertiere.* 817 pp. Vienna, Austria.

Wetzel, R. G. 1975. *Limnology.* 743 pp. Saunders, Philadelphia.

Whitton, B. A. 1975. *River ecology.* 725 pp. Univ. California Press, Berkeley.

Wiggins, G. B. et al. 1980. Evolutional and ecological strategies of animals in annual temporary pools. *Arch. Hydrobiol. Suppl.* **58**:97–206.

Williams, D. D. 1983. The natural history of a Nearctic temporary pond in Ontario with remarks on continental variation in such habitats. *Int. Rev. gesamten Hydrobiol.* **68**:239–253.

2

PROTOZOA

Beginning with the observations on *Vorticella*, reported in 1675 in a letter to the Royal Society by the pioneer Dutch microscopist Leeuwenhoek, an enormous literature on the Protozoa has accumulated. There are, for example, more than 4000 papers on the genus *Tetrahymena* alone. Unquestionably this phylum has received more attention by zoologists than has any other fresh-water group. Because of the fact that a protozoan cell embodies all the fundamental properties of living matter, it has long been a convenient object for the study of numerous phases of anatomy, cytology, physiology, behavior, and heredity; taxonomy has also been rather thoroughly investigated, but ecology is a relatively neglected phase of protozoan biology, and is only recently receiving emphasis.

Protozoa occur in an extraordinary variety of habitats. Free-living species are found wherever there is water, from an accumulation of a few drops to the largest lakes and seas, and over a wide range of chemical and physical conditions. Parasitic species live in certain organs of most of the larger Metazoa that have been carefully examined. It is estimated that 125,000 valid species of Protista have been described, of which more than 8000 are ciliates.

This chapter is based almost exclusively on free-living fresh-water Protozoa, and emphasis is placed on features that are utilized as key characters, but most of the generalizations apply to the phylum and classes as a whole. Physiology, morphology, and anatomy are discussed only briefly; these phases of protozoan biology are thoroughly covered in many texts and review articles.

Some protozoan genera are the subject of whole books, including *Amoeba, Paramecium, Stentor, Euglena,* and *Tetrahymena.*

The Protozoa literature is truly enormous, and the number of papers being published increases markedly every year. The situation has become even more complex during the past 25 years with increased interest in biochemical protozoology and the general use of electron microscopy, which is showing a host of structural details that were never before dreamed of. Our list of Protozoa references at the end of this chapter is only a small sampling, but we hope it is a fair sampling. Our conscience would not let us avoid citing some of the older classical books and papers. Since this book is primarily an identification manual, we have intentionally avoided references to the fine details of physiology, microanatomy, and biochemistry.

General characteristics. All Protozoa are single-celled organisms, though many species form colonies consisting of several to thousands of individuals. With the exception of certain reproductive individuals, however, all the cells in a colony are fundamentally independent and similar in structure and function.

The physical nature of protoplasm is perhaps best studied in Protozoa. Essentially it consists of water in which are dissolved a wide variety of organic and inorganic materials and in which is suspended a variety of minute droplets and solid particles, the whole constituting a complex colloidal system. Fibrils, networks, alveoli, Golgi bodies, chondriosomes, and other granules and droplets larger than colloidal size serve to increase further the physical complexity. Viscosity is more or less variable and reversible from sol to gel states. Typical protozoan protoplasm is translucent and colorless or family tinted, but many species are colored by ingested foods, stored materials, or pigments such as chlorophyll. Thus green, yellowish, brown, gray, bluish, and reddish are common colorations.

Size ranges from about 5 μm to 5 mm, but the great majority of species are more than 30 μm and less than 300 μm long. Body shape is highly variable, but spherical, oval, elongated, and more or less flattened are the most common body forms. In flattened species the ventral surface is arbitrarily defined as that surface nearest the substrate and usually bearing the cytostome. Symmetry includes spherical, radial, and bilateral patterns, and asymmetry.

Protozoologists are continuously embroiled in discussions of various classification and nomenclatorial schemes for protozoans, with no end in sight. Following is an outline of the major subdivisions used in this manual; it is modified from several sources and includes only those taxa that have free-living freshwater representatives. Parasitic and dominantly marine taxa are not included.

Subkingdom Protozoa

Phylum Sarcomastigophora. Flagella, pseudopodia, or both types of locomotor organelles; one type of nucleus; sexuality syngamic, if present.

    Subphylum Mastigophora (flagellates). One or more flagella; solitary or colonial; asexual reproduction by binary fission; nutrition phototrophic, heterotrophic, or both. Key on page 50.

        Class Phytomastigophorea. Chromatophores usually present; one or two flagella; amoeboid forms in some groups.

        Class Zoomastigophorea. No chromatophores; one to many flagella; amoeboid species in some groups.

    Subphylum Sarcodina (amoeboid forms). Pseudopodia typically present; flagella sometimes in developmental stages; cell naked or with external or internal supporting tests or skeletal elements;

asexual reproduction by fission. Key on page 65.

        Class Actinopoda. With spherical symmetry and axopodia.

        Class Rhizopoda. Without spherical symmetry; lobopodia or filopodia.

Phylum Ciliophora (ciliates and suctorians). With simple cilia or composite ciliary organelles; two types of nuclei; binary fission and several types of sexual reproduction. Key on page 71.

    Class Holotrichia. Ciliation simple and uniform.

    Class Spirotrichia. Ciliation sparse; cirri common; buccal ciliation conspicuous, with adoral zone of membranelles.

    Class Peritrichia. Somatic ciliation absent in mature forms, but oral ciliation conspicuous as rings around the cytostome; body often with a long contractile stalk; migratory ciliated larvae.

    Class Suctoria. No cilia in mature stage; sessile; attached to substrate by a noncontractile stalk; ingestion by means of suctorial tentacles; migratory larvae produced by budding.

The external surface of Sarcodina is usually a naked, thin, living plasmalemma, but in the ciliates and many flagellates there is also a protective pellicle of varied composition. In addition, both flagellates and Sarcodina often have a thicker shell, case, or armor composed of cellulose, jelly, tectin (pseudochitin), calcium carbonate, or silicon compounds. Frequently an outer protective sheath or lorica is distinctly separated from the periphery of the cell. Sometimes the outer case consists of foreign particles imbedded in a homogeneous matrix.

In ciliates and Sarcodina the layer of protoplasm just below the bounding membrane is usually a relatively clear ectoplasm in contrast to the more granular endoplasm internal to it.

The nucleus is nearly always difficult to

distinguish in living, unstained material. In the ciliates the macronucleus is granular, large, and of various shapes, ranging from spherical or oval to elongated, beaded, branched, or horseshoe shaped; the one or more micronuclei are very small. Sarcodina have a single type of nucleus; a few genera, such as *Pelomyxa* and *Actinosphaerium*, are multinucleate. The nucleus of free-living flagellates is single, vesicular, and usually more or less centrally located.

Locomotion. Typical Sarcodina move about by means of streaming movements and temporary extensions of the cell called pseudopodia. The mechanics of pseudopodial movement have been most extensively studied by Mast and his students, and by Jahn and Bovee. Adequate descriptions are included in most standard zoology texts and reference books.

Four types of pseudopodia are generally recognized, though there are intergradations and none of them can be clearly separated from the others. Lobopodia are fingerlike, tonguelike, rounded, or branched; they contain both ectoplasm and endoplasm (Figs. 1A, C). Filopodia are more or less filamentous,

usually pointed, and consist of ectoplasm only (Figs. 25E–G). Reticulopodia, or rhizopodia, are also filamentous and composed of ectoplasm but they are branching and anastamosing (Fig. 26D). Axopodia are semipermanent, owing to a relatively stiff axial rod, which in most cases ends internally in a small granule (Fig. 29).

In general, those Sarcodina having lobopodia move most rapidly, usually 0.5 to 3.0 μm/second, whereas those having other types of pseudopodia are more sluggish or do not move at all.

A flagellum is essentially a long, fine, whiplike protoplasmic thread that produces forward locomotion by spiral or undulating movements. Sometimes, however, the basal region is more or less rigid and only the distal half or tip moves about. Many Mastigophora have two flagella, both of which function in locomotion, but in some genera there are one to several accessory flagella that have little or no locomotor function. Accessory posteriorly directed flagella are thought to have a slight pushing or steering function in some species.

The axial filament, or axoneme, typically consists of nine fibrils arranged in two compact, parallel bundles. The axoneme is surrounded by a sheath and a limiting membrane.

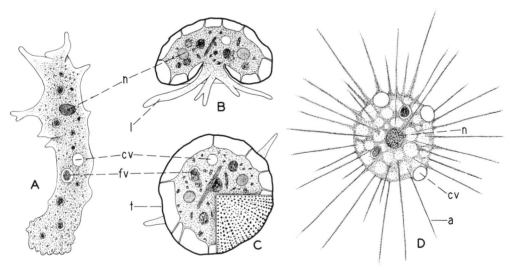

FIG. 1.–Typical Sarcodina. A, *Amoeba*; B, vertical optical section of *Arcella*; C, *Arcella,* showing partial horizontal section and partial external surface of test; D, *Actinophrys sol* Ehr. a, axopodium; cv, contractile vacuole; fv, food vacuole; n, nucleus; l, lobopodium; t, test.

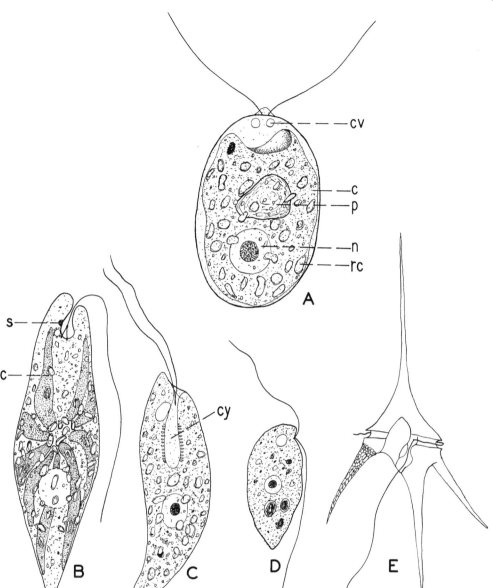

FIG. 2.–Typical Mastigophora. A, *Chlamydomonas*; B, *Euglena viridis,* Ehr. (chromatophores very difficult to distinguish in living animal); C, *Chilomonas paramecium* Ehr.; D, *Bodo*; E, *Ceratium hirundinella* (O.F.M.) with only a portion of shell sculpturing shown. c, chloroplast; cy, cytopharynx; cv, contractile vacuole; n, nucleus; p, pyrenoid; rc, reserve carbohydrate granules; s, stigma.

Flagella are commonly inserted at or near the anterior end of the cell, often in a groove, pit, or cytopharynx. The attachment is variable, complex, and sometimes double, but it is not visible in unstained specimens. Usually the flagellum originates in a basal granule just within the periphery of the cell. The basal granule may or may not be combined with another granule, the blepharoplast, just below. A fine fibril, or rhizoplast, sometimes connects the basal granule and blepharoplast with the edge of the nucleus.

Flagellates usually move at the rate of 15 to 300 μm/second. Under adverse environmental conditions the flagella may be lost, but regeneration occurs promptly when conditions are again favorable.

Fundamentally, cilia are similar to flagella, but there are several important differences. Cilia are much shorter, very abundant, and have only a single basal granule. They are arranged in longitudinal, diagonal, or oblique rows, and their movements are often coordinated so that waves of beats pass along the entire animal. In electron micrographs a cilium commonly appears to consist of a bundle of 11 (double) fibrils. A single cilium is thought to have a pendular or side-to-side movement, usually in one plane or with an oval pattern at the tip. Ciliates move more rapidly than other Protozoa, the usual speed being 200 to 1000 μm/second.

Cilia may also be fused to form more complex structures. A membranelle, for example, is a double lamella of cilia completely fused into a flexible plate and found in the region of the cytostome, oral groove, or peristome (Figs. 4F, 38–41). Adoral membranelles occur in all ciliates except the Holotrichia.

Undulating membranes of ciliates range in size from delicate aggregates no broader than the length of an ordinary cilium to enormous balloonlike structures (Figs. 4F, 36E). An undulating membrane is composed of a single longitudinal row of fused cilia, but there is often a group of from two to as many as five or six such membranes in the peristomial area near the cytostome or in the cytopharynx.

Cirri are the most specialized ciliary organs and are particularly characteristic of the Hypotrichida. They are highly variable in size and are located more or less definitely on the ventral surface (Fig. 4G). According to their location, they are known as frontal, marginal, ventral, anal, and caudal cirri. A cirrus has an elongated conelike shape and consists of a tuft of fused cilia. Unlike other locomotor organelles, cirri move about in any direction, and consequently they are used for walking, running, jerking, and jumping movements.

In addition to true locomotion, many Protozoa possess pronounced powers of contraction and extension. These movements are usually the result of contractility of unspecialized protoplasm, but in many ciliates there

are fine contractile fibrils, or myonemes, in the ectoplasm just below the surface of the cell.

Many genera of ciliates, flagellates, and amoeboid Protozoa are normally sessile and attached to the substrate, either with or without a stalk, but frequently these Protozoa metamorphose into contemporary "swarmer" stages that are capable of moving to new locations where they again attach to the substrate and assume the usual body form. A temporary, motile telotroch of a stalked, sessile ciliate is shown in Fig. 43C.

Feeding, nutrition. Those Protozoa that ingest solid particles of food, including bacteria, algae, other Protozoa, small Metazoa, or debris, are said to be holozoic. Many flagellates possess chromatophores and chlorophyll and are plantlike, or photosynthetic, in their nutrition; carbon dioxide and water are absorbed from the surroundings and synthesized into carbohydrates in the presence

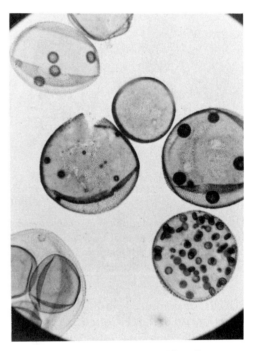

Fig. 3.–*Volvox,* a large colonial flagellate ($\times$100); note mature colonies, young and old daughter colonies, and clusters of microgametes.

of light and chlorophyll; this photosynthetic type of nutrition is called holophytic. From carbohydrates, and especially with the addition of absorbed nitrogen, phosphorus, and sulfur compounds, the other organic constituents of the cell are synthesized. Saprozoic nutrition involves absorbing dissolved salts and simple organic materials from the surrounding medium and synthesizing them into protoplasmic materials; this type of nutrition occurs in all classes of Protozoa. Parasitic nutrition is sometimes placed in a separate category, but since it usually involves the absorption of dissolved materials, it is essentially a specialized type of saprozoic nutrition. However, some parasites ingest particulate matter and might therefore be considered holozoic. Many Protozoa are mixotrophic; that is, they exhibit two of the preceding types of nutrition. Some ciliates, for example, are both holozoic and saprozoic, and some flagellates are both holophytic and saprozoic. *Ochromonas* is a flagellate in which holozoic, holophytic, and saprozoic nutrition all occur.

Some protozoans, such as *Chilomonas, Polytoma, Polytomella,* and *Astasia,* are entirely saprozoic and ingest no particulate food; they depend entirely on organic and inorganic nutrients in solution, usually in the vicinity of decaying organic matter.

Holozoic Protozoa may be conveniently classified according to their general food habits. *Spathidium, Dileptus, Didinium,* and *Actinobolina,* for example, are all carnivorous ciliates that feed on other Protozoa, usually only certain species. *Frontonia, Chilodon,* and most Sarcodina are herbivores that feed on algae; many other genera are bacterial feeders. *Blepharisma, Spirostomum,* and *Stentor* are examples of omnivores that feed on all kinds of particulate matter, both living and dead. There are convincing data to show "selection" of certain types of food by some omnivorous ciliates, but in general such observations are variable and inconclusive.

Lobopodia commonly advance in such a way as to surround and enclose a food particle. The latter is then within the cell in a droplet of water, the whole constituting a food vacuole. Sometimes amoeboid Protozoa ingest particles by an invagination process when the particles come in contact with the general body surface. Occasionally large particles, such as algal filaments, are seen to "glide" into the cell by an "import" process. Pseudopodia of many amoeboid groups are sticky so that a food particle sticks to the general body surface on making contact and is then drawn within to form a food vacuole.

With few exceptions, the chlorophyll of holophytic flagellates is localized in chromoplasts (chloroplasts), which may be disc, band, bowl, or cup shaped, but sometimes the green color is masked by accessory yellow, orange, or brown pigments. A red pigment, hematochrome, often appears in some green flagellates, but it is distributed throughout the cytoplasm and is thought to be formed as a response to low concentrations of nitrogen and phosphorus. Pyrenoids are small proteinaceous bodies that are imbedded in chloroplasts and around which the carbohydrate products of photosynthesis accumulate as laminae or granules. After they are formed, these particles (glycogen, paramylum, or starch) are evidently available to the general cytoplasm as reserve foods. Green flagellates vary widely with respect to their specific nitrogen requirements. Most species can use ammonium and nitrate compounds, others require amino acids, and some are even able to use peptones. These three groups of green flagellates are sometimes termed photoautotrophic, photomesotrophic, and photometatrophic, respectively.

Although many green flagellates are obligate phototrophs, there are numerous species that can also be grown in the dark, provided the correct dissolved nutrients are available in the culture medium. The Zoomastigophorea and many nonphotosynthetic Phytomastigophorea are holozoic, saprozoic, or mixotrophic. Like green flagellates, the colorless saprozoic species differ widely in their nitrogen requirements. In the holozoic species small food particles come in contact with the surface of the cell and sink or are drawn inward, sometimes by definite pseudopodial action. Many flagellates, both green and nongreen, possess an anterior cytostome, cytopharynx, and reservoir through which the flagellum projects, but particulate food is taken into the cell through this opening in only a very few holozoic forms.

Though many ciliates are saprozoic or

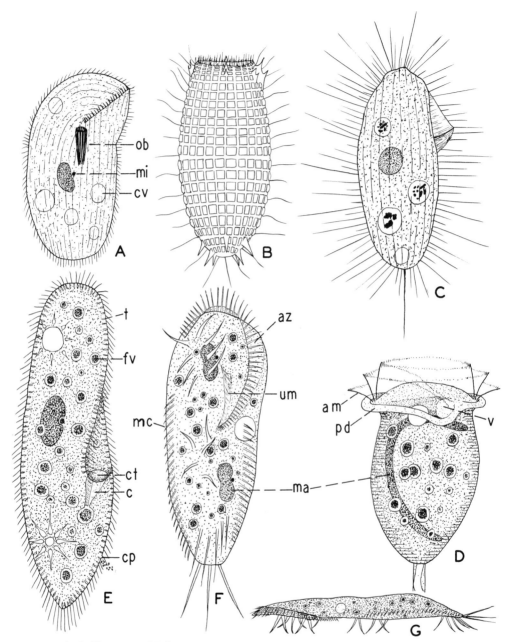

FIG. 4.–Typical ciliates. A, *Chilodonella cucullula* (O.F.M.); B, *Coleps octospinus* Noland; C, *Cyclidium*; D, *Vorticella*; E, *Paramecium*; F, ventral view of *Stylonychia mytilus* (O.F.M.), showing eight frontal, five ventral, five anal, and three caudal cirri; G, lateral view of *S. mytilus*. am, adoral membrane; az, adoral zone; c, cytopharynx; cp, cytopyge; ct, cytostome; cy, contractile vacuole; fv, food vacuole; ma, macronucleus; mc, marginal cirrus; mi, micronucleus; ob, oral basket; pd, peristomial disc; t, trichocyst; um, undulating membrane; v, vestibule. (B modified from Geiman, 1931.)

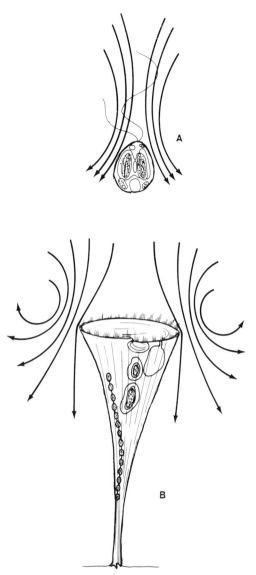

FIG. 5.–Water currents bringing in food. A, flagellate *Ochromonas*; B, ciliate *Stentor*. (Modified from Sleigh, 1973.)

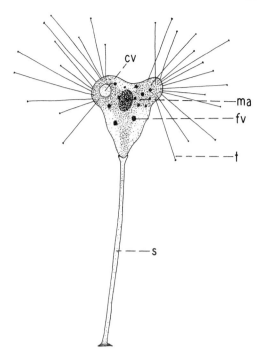

FIG. 6.–*Tokophrya lemnarum* Stein, a typical suctorian. cv, contractile vacuole; fv, food vacuole; ma, macronucleus; s, stalk; t, tentacle.

mixotrophic, the great majority are holozoic. Movement through the water as well as beating of cilia, membranelles of the peristome, and undulating membranes all serve to collect the food and bring it into the cytostome and cytopharynx. In raptorial genera, however, such as *Coleps, Didinium, Prorodon,* and *Lacrymaria,* the prey is relatively large, and current production has little to do with food getting.

At the lower end of the cytopharynx the particles enter the cell proper as food vacuoles. In some ciliates, especially gymnostomes, the wall of the cytopharynx is stiffened by minute rodlike trichites that collectively form the cytopharyngeal basket (Figs. 32, 33).

Though the Suctoria are all holozoic and feed mostly on living ciliates, much remains to be learned about the precise mechanisms of feeding. The struggling prey is held tightly at the tips of the tentacles, and contact with the tentacles often (but not always) kills the prey in a matter of seconds, the rigid tentacles being presumed to secrete a potent toxin. American freshwater species have only one type of tentacles; they are suctorial and often knobbed. Only a few of the tentacles attached to the prey serve as cytostomes and these enlarge to a diameter twice or more that of the nonsucking tentacles. The protoplasm flows through these hollow tentacles and is formed into food vacuoles as it enters the cell proper. The entire contents of even large ciliates may be sucked up in as little as 20 minutes.

Suction is said to approximate 0.2 atm, although some investigators postulate mechanisms other than suction (Rudzinska, 1973). Tentacles may be localized in bundles or more generally over the body surface. Figure 7 shows a large suctorian feeding on *Paramecium*.

Many ciliates regularly or occasionally harbor minute intracellular green zoochlorellae in a symbiotic relationship. In a few species, and especially in *Paramecium bursaria*, this has been shown to be an obligatory association. There are also numerous holozoic flagellates and rhizopods that contain intracellular green symbionts.

Digestion. In holozoic species the ingested food organisms may be killed in a few seconds after being taken into the food vacuole or they may stay alive for as long as an hour. Digestive enzymes are released into the food vacuoles from the surrounding cytoplasm. A variety of protein- and carbohydrate-digesting enzymes have been demonstrated, but the occurrence of fat-digesting enzymes is difficult to prove though they are undoubtedly present.

In most Protozoa that have been carefully studied the hydrogen-ion concentration of the food vacuoles is variable but usually within the range of pH 4.0 to 7.6, most digestion being under acid conditions. Depending on the food supply and the activity of a protozoan, there may be few to many contained food vacuoles. As digested materials are absorbed into the cytoplasm through the membrane of the food vacuole the contents of the vacuole become progressively smaller, and the indigestible residue is voided through a temporary rupture in the cell membrane, though in ciliates and flagellates there is often a permanent pore, the cytopyge, for this purpose.

Reserve foods. Depending on ecological and physiological conditions, a protozoan

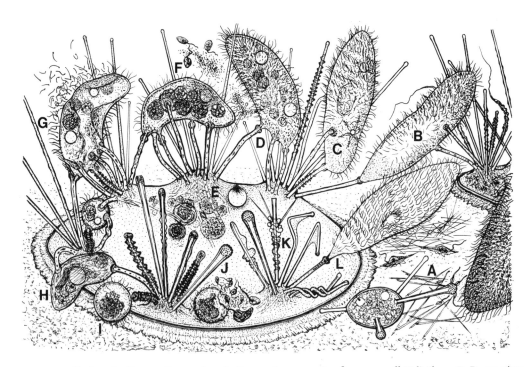

FIG. 7.–*Heliophrya* feeding on *Paramecium*. A, *Paramecium* escaping from a small *Heliophrya*; B, *Paramecium* caught by the tentacles of two different *Heliophrya*; C, D, F, and G, various feeding stages; G, deciliation of *Paramecium*; H, remnant of *Paramecium*; L, *Paramecium* about to escape from a tentacle. Note the wide variation in tentacular structure. (From Spoon et al., 1976.)

cell may contain little to much reserve food. The type of reserve food materials depends largely on the mode of nutrition. In holozoic and saprozoic species it is usually glycogen or paraglycogen. Holophytic species variously store starch, paramylum, and lipids. Paramylum can be easily distinguished from starch by the fact that it does not stain blue with iodine.

Respiration. The great majority of Protozoa are aerobic, the oxygen being absorbed from the surroundings through the cell membranes. Nevertheless, it is probable that more than 90% of our fresh-water species are capable of thriving in water that is less than 10% saturated with oxygen. Indeed, a great many species have been reported from anaerobic habitats such as sewage, the debris and mud of stagnant ponds, and the bottoms of lakes during periods of oxygen exhaustion, but these species are thought to be mostly facultative and temporary anaerobes capable of tolerating anaerobic conditions for variable periods but not able to maintain themselves indefinitely. The following genera, however, are known to have some obligatory anaerobic representatives: *Chlamydophrys, Vahlkampfia, Saprodinium, Mastigamoeba, Bodo, Holophrya, Trepomonas,* and *Metopus.* Under both permanent and temporary anaerobic conditions it is thought that energy is derived from reactions similar to fermentation processes.

Excretion. As in higher animals, the chief metabolic excretory products are water, carbon dioxide, and nitrogenous materials. Sometimes the latter are retained as granules or crystals for a variable time within the cell. It is now the general opinion that there are no special excretory organelles in Protozoa, but that dissolved excretory materials, notably urea, simply diffuse through the cell membranes into the surroundings.

Contractile vacuoles. These organelles are osmoregulatory devices that remove excess water from the cell. Since the internal salt concentration of fresh-water Protozoa is much higher than that of their surroundings, there is a continuous osmotic uptake of water through the cell membranes. The accumulation of water from the general cytoplasm and the release of water to the outside by the vacuoles therefore serve to maintain a constant internal salt-water balance.

Contractile vacuoles are found in all freshwater Sarcodina and Mastigophorea, and in all Ciliophora regardless of habitat. They do not occur in the Sporozoa or in most parasitic and marine Sarcodina and Mastigophora.

The Sarcodina have one or more contractile vacuoles of indefinite location. Each one starts out as a coalescence of several minute adjacent droplets. The small vacuole then grows, reaches a certain size, and bursts and releases its contents to the outside through a temporary rupture in the plasmalemma. New vacuoles appear in any part of the cell. In the Mastigophora there is a single small contractile vacuole. It is constant in location and usually anterior in the Phytomastigophorea and posterior in the Zoomastigophorea. Ciliophora have one to many contractile vacuoles, all constant in position and opening through permanent pores in the pellicle; sometimes these vacuoles are supplied by a complex system of collecting canals in the cytoplasm.

The rate of pulsation of a contractile vacuole is dependent on such factors as temperature, age, physiological state, food, salt concentration, and so on. In some Protozoa a volume of water equivalent to the volume of the entire cell may be voided in as short a time as 2 minutes; under other conditions and in other species an equivalent amount of water may be avoided in 24 to 48 hours.

It is probable that insignificant quantities of dissolved nitrogenous wastes leave the cell in the expelled fluid of the vacuole, but this is merely incidental to the main osmoregulatory function.

Irritability, behavior. Basic perception and response to contact, food, gravity, light, chemicals, and so on are probably all due to the general irritability of protozoan protoplasm. Nevertheless many Protozoa have special receptors. The stigma, or eyespot, of some flagellates is a minute clump of granules with an adjacent light-sensitive area of cytoplasm; such stigmata commonly produce a directional locomotion toward a light source, as shown, for example, by the accumulation of *Euglena* along the lighted side of a culture jar set in a north window. Special cilia or bristles are thought to be touch receptors in some ciliates.

A great many observations have been made on protozoan behavior, but only a few general examples can be cited here. Favorable environmental conditions elicit no visible responses, but under unfavorable circumstances, especially where there are gradients, Protozoa generally move away from the unfavorable and toward the favorable conditions. *Paramecium* and many other ciliates exhibit "trial and error" or "avoiding" reactions. On coming in contact with an unfavorable medium or object, they stop, retreat, turn slightly, and resume locomotion in a new direction; such movements may be repeated until the protozoan reaches favorable conditions. A few ciliates exhibit negative geotaxis.

As a group, only the stigma-bearing flagellates show reactions to light. They are generally photopositive to weak or medium light sources but photonegative to strong lights. Other Protozoa are mostly indifferent to usual light sources, though some are photonegative to exceptionally strong light.

Transmission of impulses is a regular function of the protoplasm, but in many ciliates a special neuromotor system of conducting fibrils has been demonstrated. Essentially, such a system consists of very fine longitudinal and transverse fibrils connecting the basal granules of the cilia. Many fibrils come together and are more or less centralized in the region of the cytopharynx. The neuromotor system functions as a coordinating apparatus for movements of the cilia, but elasticity, support, and contraction are also suggested functions.

In addition to (or in place of) the neuromotor system, many ciliates possess a silverline system in the ectoplasm. This is a latticework or irregular network of granules and fibrils that can be detected only when the protozoan is subjected to dry fixation and impregnation with silver nitrate.

Trichocysts are minute, cigar-shaped pockets of material, which are symmetrically arranged and abundant in the ectoplasm of numerous ciliates (about 4000 in *Paramecium*). When such ciliates are stimulated by pressure or various chemical substances, the trichocysts are discharged through minute surface pores, and on contact with water the substance of each trichocyst forms instantaneously into a long, rigid, needlelike projection from the cell. Little is known about the functions of trichocysts. In a few species they are used for anchoring to a substrate; perhaps they are useful in carnivorous species for capturing prey; possibly they function in warding off predators.

Reproduction. The usual method of reproduction involves mitosis and binary fission, resulting in two new individuals that quickly reconstitute themselves and form new organelles where necessary. Amitosis is probably limited to the macronucleus of Ciliophora. A few free-living fresh-water Protozoa exhibit multiple fission and plasmotomy.

Most protozoans have 4 to 12 chromosomes, but some have as few as 2 and *Amoeba proteus* has 500 to 600.

Syngamy is best developed in some of the colonial green flagellates where certain of the vegetative cells become specialized as gametes. In some genera the uniting gametes are

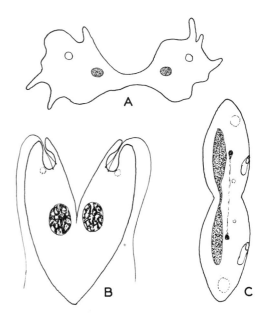

FIG. 8.–Binary fission in Protozoa, diagrammatic. A, *Amoeba*; B, *Euglena*; C, *Paramecium*.

essentially identical, and the condition is called isogamy. At the other extreme, such as in *Volvox*, the condition is clearly heterogamous, one of the gametes being small, motile, and spermlike and the other being very large, nonmotile, and egglike. In all cases, however, the zygote eventually gives rise to a new individual or colony.

Conditions most favorable for growth are also most favorable for rapid division, but under unfavorable, crowded conditions and in old cultures many ciliates commonly exhibit one of two complex phenomena, conjugation and autogamy.

Conjugating ciliates come together in pairs and unite temporarily along their lateral edges. Then follows a complex series of nuclear phenomena that differ from one species to another. In general, however, the process involves disintegration of the macronuclei, a series of mitotic and meiotic divisions of the micronuclei, and an exchange of micronuclear material. The migrating and stationary micronuclei then unite to form a single nucleus in each of the two conjugants. The animals separate, and there is a series of rapid nuclear and cell divisions accompanied by the reappearance of macronuclei derived from micronuclear material. Complex series of "mating types" have been found for several ciliates.

Autogamy involves nuclear changes within a single individual and includes segmentation and absorption of the macronucleus and a series of divisions of the micronuclei. Subsequent fusion of a pair of micronuclei is accompanied by reconstitution of the macronucleus. Like conjugation, the process of autogamy is completed within one to several days.

Basically, conjugation and autogamy seem to be rejuvenation phenomena by which there is a reshuffling of chromosomes and genes. General physiological vigor and fission rates appear to be increased in a population as a result of these processes. Figures of conjugation and autogamy are available in most elementary zoology textbooks.

Budding is almost restricted to the Suctoria, where it is the common method of reproduction. In exogenous budding one to several ciliated buds, each with a nucleus, are constricted off from the external surface of the

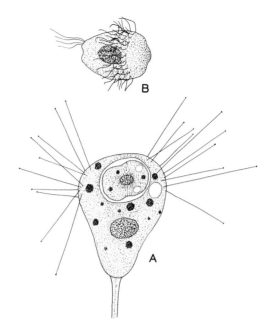

FIG. 9.–Reproduction in *Tokophrya lemnarum*. A, animal containing larva within brood pouch; B, released larva. (Modified from Noble, 1932.)

parent cell. Endogenous buds are liberated into internal spaces, or brood chambers, from which they soon escape to the outside. The ciliated embryos swim about for several hours, then become attached to the substrate, lose their cilia, and grow to mature size.

The capacity to regenerate lost parts of the cell is highly variable. In general, however, a bit of cytoplasm with no nuclear material soon dies, but if either a whole or part of a nucleus is present in a cell fragment, then that fragment often grows into a complete normal cell.

Encystment. The formation of resistant, protective cysts is rare among marine Protozoa but very common in fresh-water species. Cyst formation may be induced by any one of a variety of adverse environmental conditions, including drying, heat, cold, lack of food, and various chemical substances. The protozoan first rounds up and loses its flagella, cilia, and sometimes other organelles. The cyst wall that is then secreted around the periphery is usually double, consisting of a thin inner and a tough outer capsule. Sometimes a third

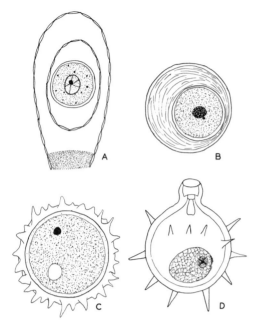

FIG. 10.–Typical protozoan cysts. A, *Euglypha* (amoeboid); B, *Colpoda* (ciliate); C, *Pleurotricha* (ciliata); D, *Ochromonas* (flagellate). (A modified from Kühn, 1926; B and D modified from Doflein, 1923; C modified from Ilowaisky.)

innermost albuminous membrane is also formed.

Such cysts are extremely resistant to desiccation, freezing, and high temperatures. Dry cysts commonly remain viable for several months to several years, though there are records of some that were kept for more than 40 years. Protozoans inhabiting damp soil are especially adapted for rapid encystment and excystment. Altogether, about 300 species have been reported from soil habitats.

Excystment of dry cysts and reconstitution of the cells are induced in a few minutes when the cysts come in contact with suitable natural waters or culture solutions will excyst when fresh water, salts, hay extracts, and various organic materials are added.

A few Protozoa are known to form reproductive cysts, in which fission occurs, or digestion cysts, which are formed after a large food intake.

The review of Corliss and Esser (1974) emphasizes our lack of information about the precise details of cyst biology, especially as exemplified by their comment "Recall that *Paramecium*, though widely dispersed, has no cysts of any kind."

Ecology. In general, ponds and pools contain the greatest numbers of species though not necessarily the greatest numbers of individuals. Such waters consist of a large number of microhabitats and afford a wide variety of niches. The open waters of lakes contain a characteristic assemblage but small number of species, including representatives of the following genera: *Dinobryon, Mallomonas, Peridinium, Ceratium, Gymnodinium, Trachelomonas,* and *Eudorina.* Sometimes summer populations of plankton species in lakes may reach more than 5 million individuals per liter.

Trickling filters, polluted waters, and Imhoff tanks have a rich and characteristic fauna. Certain species of *Amphileptus, Euglypha, Epistylis, Amoeba, Vorticella, Aspidisca, Trachellophyllum, Monas, Tokophrya, Opercularia, Litonotus, Difflugia, Paraglaucoma, Podophrya, Microthorax, Trepomonas, Hexamita, Anthophysis, Carchesium,* and *Bodo* are commonly known as "sewage Protozoa." Hänel (1979) found more than 50 species of colorless flagellates in the low-oxygen waters of sewage treatment plants. Colonial green flagellates, on the other hand, are usually most abundant in waters with a high oxygen content. Euglenoids are dominant in small ponds with a high organic content, while testate rhizopods are abundant and varied in sphagnum bogs. Even tree holes and the water held in large leaves and flowers have their characteristic faunas.

Chromatophore-bearing protozoans require light for photosynthesis and are therefore restricted to surface waters of lakes and to small bodies of water.

Owing to ease of transport, especially in the dry, cyst condition, most species are cosmopolitan, but local distribution seems to be governed by such factors as light, temperature, food, dissolved oxygen, acidity or alkalinity, and many chemical substances. Seldom can ecological generalizations be based on genera, only on species. Consequently an ecological classification of fresh-water Protozoa can be only of the most general sort. Following is such a classification suggested many years ago:

1. Katharobic Protozoa are found in springs, brooks, rivers, and ponds that are rich in oxygen and low in organic matter. Such habitats usually have a sparse fauna.

2. Oligosaprobic Protozoa occur in habitats that are relatively high in dissolved salts but low in organic matter; these habitats include the open waters of most lakes and reservoirs.

3. Mesosaprobic Protozoa occur in habitats where there is active oxidation and and decomposition. The majority of Protozoa thrive under these conditions.

4. Polysaprobic Protozoa live in habitats where there is much organic matter and carbon dioxide and often quantities of hydrogen sulfide and methane. Pollution, sewage, and anaerobic Protozoa belong in this category.

Coprozoic species grow abundantly in "cultures" or suspensions of fecal matter, which is rich in decomposing organic matter. Such species demonstrate a striking ability to adapt to unusual and abnormal conditions.

Most free-living species show wide ranges of tolerance to single environmental factors, but at the same time it is thought that their activities may be affected by relatively slight environmental changes. In general, flagellates react more directly to single factors than do ciliates, whereas amoebae are intermediate.

The abundance of particulate or dissolved nutrients is important for the distribution of most species, but food is restrictive only when feeding is selective, as it is for carnivorous species. Otherwise food is only a quantitative limiting factor. As might be expected, euryphagous species are most common and widely distributed.

Noland (1925) stressed the distribution of ciliates with respect to food supply. Bacterial growth is most vigorous in acid habitats and where there are small quantities of dissolved oxygen and large quantities of dissolved carbon dioxide; under such circumstances forms feeding on bacteria become abundant. On the other hand, in alkaline habitats where dissolved oxygen is high and dissolved carbon dioxide low, the growth of algae is favored; in such habitats ciliates that feed on algae are abundant. Although certain physical and chemical factors may not in themselves determine the presence or absence of specific ciliates, the sum of these interdependent factors produces bacterial and algal growths in varying degrees, and the precise make-up of the ciliate populations is correspondingly determined.

Pratt and Cairns (1985) recognize six freshwater protozoan functional groups with respect to food types, as follows, with common generic examples: dissolved mineral nutrients (*Chlamydomonas, Cryptomonas*); bacteria and detritus (*Bodo, Monas, Cyclidium, Vorticella*); algae, bacteria, and detritus (*Stentor*); algae, notably diatoms (*Stylonychia*); dissolved organic materials (*Astasia*); rotifers and protozoans (*Trachelius, Didinium, Actinophrys*, and *Acineta*). From the standpoint of number of genera, the bacteria and detritus category is by far the largest (228 genera), while the dissolved organic materials category is smallest (13 genera).

Picken (1937) has emphasized the fact that an assemblage of Protozoa is not a collection of individuals depending on one source of food, but a complex of herbivores, carnivores, omnivores, and detritus feeders, forming a closed social structure. He analyzed the food chains in several protozoan communities characterized by widely different proportions of diatoms, blue-green algae, and bacteria. *Stentor* and *Dileptus* were at the ends of these chains.

Little can be said about the geographic distribution of protozoans. Owing to their small size, production of cysts, and ease of distribution from one place to another, the great majority of species appear to be cosmopolitan so that they may be collected in similar habitats and microhabitats on a worldwide basis. Thus, one would expect that there is little opportunity for long-time isolation and speciation. Nevertheless, studies of mating types and isozymes suggest that incipient speciation with its geographic connotations may be more widespread than presently appreciated.

The most pronounced seasonal population changes are quantitative, and temperatures below 10 and above 28°C markedly affect the numbers of individuals present but not the numbers of species; summer and winter species lists from the same habitat are often strikingly similar. A few species have become adapted to hot springs where temperatures remain as high as 50°C, but most species have a thermal death point somewhere between 30 and 43°C. Optimum temperatures generally lie between 10 and 20°C. Under experimental conditions many protozoans become accustomed to very high temperatures if the transition is made gradually. Though trophic forms are killed by freezing, a great many species are active in water under ice throughout the winter.

The majority of species tolerate the wide range of ionic concentrations occurring in natural fresh waters, and although some species live in both fresh and marine waters, only a few can be transferred directly from one to the other. The transition is often successful, however, if it is carried out gradually in the laboratory. Seven species are known from the highly saline waters of Great Salt Lake.

Laboratory experiments have shown that many species can maintain themselves over a wide range in hydrogen-ion concentration. *Euglena viridis*, for example, has been cultivated in hydrogen-ion concentrations ranging from pH 2.3 to 11.0, *E. gracilis* in 3.0 to 9.9, *Glaucoma* sp. in 4.0 to 8.9, and *Chilomonas paramecium* in 4.1 to 8.4. *Stylonychia pustulata*, however, will grow only between pH 6.0 and 8.0, and *Spirostomum ambiguum* between 6.8 and 7.5. A few species are restricted to highly acid peat bogs in the range of pH 2.0 to 4.0; *Euglena mutabilis* is found in acid coalmine pits in the range of pH 1.8 to 3.9 and has been cultured in the range of pH 1.4 to 7.9. Nevertheless, the great majority of Protozoa have optimum conditions between pH 6.5 and 8.0. It should be borne in mind that pH is a gross measurement of a combination of many chemical factors, any of which, singly or collectively, may be limiting.

Noland and Gojdics (1967) is a summary of protozoan ecology; it contains an excellent list of older references.

Succession. Anyone who has kept hay infusions or other types of mixed cultures in the laboratory is aware that such cultures exhibit a rapid, complex, overlapping series of population growths and declines for the various species. Typically, the dominant hay infusion forms appear and become abundant in the following sequence: small flagellates, *Colpoda*, hypotrichs, *Paramecium*, *Vorticella*, and *Amoeba*. *Paramecium* does not usually become abundant until 8 to 14 days after the culture is begun, and *Amoeba* 2 to 6 weeks later when the culture is "dying out." No equilibrium is reached in such mixed cultures, and as yet sufficient information is not available to explain the sequence of species. Undoubtedly, such factors as food, temperature, light, dissolved oxygen, dissolved carbon dioxide, acidity, and the specific nature of the culture are important; and certainly the more intangible factors such as competition for food, available particulate and dissolved nutrients, and reproductive potential are also significant. Some investigators have presented evidence to indicate that a period of intensive growth of one species is brought to a halt by the accumulation of its own excretory products, and perhaps the growth cycle of one species may result in a favorable medium for another species. Seasonal and cyclic populations of Protozoa have also been studied in natural habitats, especially in the plankton, but these are likewise only poorly understood. The fundamental difficulty involved in protozoan succession studies is our inability to measure, control, and assess the relative significance of many interacting ecological factors.

Ectocommensal Protozoa. Many freshwater Metazoa have Protozoa living on their general external surfaces or gills, and some species are thought to be obligatory commensals, although it is sometimes difficult to decide just what special benefits the protozoan derives from such an association. Certain species of Suctoria, for example, are known only from the carapace of turtles; others are often abundant on the gills and appendages of Crustacea. Plankton cladocerans and copepods frequently carry epizoic protozoans, especially in eutrophic lakes and ponds. Common genera are *Vorticella, Scyphidia, Epistylis, Rhabdostyla,* and *Tokophrya*. The curiously modified ciliates *Kerona* and *Trichodina* are common on the external surface of hydras, where they run about rapidly in search of bits of food. However, some species of *Trichodina* are external parasites of fish, especially in hatcheries, but since they are frequently associated with other protozoan parasites such as *Chilodon, Costia,* and *Ichthyophthirius,* it is difficult to determine the relative importance of *Trichodina* as a true parasite.

Organisms living in and on Protozoa. Although there are a few Phycomycetes and many bacteria that are internal parasites of Protozoa, there are also some bacteria that are internal symbionts, or commensals, especially in amoeboid Protozoa, as well as some bacteria that are epiphytic on the external surface of Protozoa. Certain ciliates, including some species of *Paramecium, Stentor,* and others, commonly contain internal commensal green, blue-green, or yellow algae (therefore called zoochlorellae, zoocyanellae, and zooxanthellae, respectively). Such algae are ordinarily free living and autotrophic species, but when they occur as symbionts they are presumed to obtain carbon dioxide and nitrogen and phosphorus compounds resulting from the protozoan metabolism and to contribute oxygen and carbohydrates to the protozoan. *Proales parasita* is a common parasitic or predatory rotifer in the interior of *Volvox* colonies.

Economic importance. Although parasitic Protozoa are of inestimable importance because of their adverse effects on the well being and economic status of man, the free-living species are of little significance. During the summer months, however, some species become so abundant in municipal water supplies in rivers, lakes, and reservoirs as to impart an offensive odor to the water. Blooms of *Dinobryon, Chlamydomonas, Eudorina, Pandorina, Volvox, Glenodinium,* and *Peridinium* produce fishy odors. *Synura* smells like cucumber or muskmelon and has a bitter taste. *Mallomonas* has an aromatic, violet, or fishy odor. *Ceratium* produces a "vile stench." Such disagreeable odors and tastes are more or less eliminated from water supplies by copper sulfate treatment, filtration, or aeration.

Of considerable theoretical importance is the role played by Protozoa in aquatic food chains. Holophytic and saprozoic species are producers that utilize dissolved nutrients and serve as food for small Metazoa. Holozoic species are consumers that form an essential

link between particulate living and dead material, on the one hand, and rotifers, Crustacea, and other small Metazoa, on the other hand. Bird and Kalff (1986), for example, found that a large lake population of *Dinobryon* removed more bacteria from the water than all other plankters combined.

Collecting. Plankton and littoral species may be collected with a fine townet, but other free-living species are best obtained by collecting submerged green and rotting vegetation, bottom debris and ooze, and surface scum. A useful device for collecting on and above soft substrates is a long sucking tube attached to a collecting bottle. Materials should be brought back to the laboratory in a quantity of water in which they occur. If desirable, they may be concentrated in a smaller volume by pouring off a part of the water through bolting silk. Stacked laboratory finger bowls are convenient containers. Collections should be allowed to stand in moderate diffuse light where the temperature is not too high. Stigma-bearing species then migrate toward the side facing the strongest light; amoeboid forms remain in the debris, but ciliates often swim about in the water above. If such containers are kept in the laboratory for one to several weeks, additional species often appear.

Cultures. Mixed cultures for general study are easily made with hay, lettuce powder, wheat, and rice infusions that are inoculated with pond water. The most successful cultures are obtained by excluding small Metazoa such as oligochaetes and entomostraca, and by adjusting light, food, temperature, and pH to favorable conditions. Moldy cultures should be discarded. Detailed directions for mixed and pure cultures are given in Kudo (1966), Needham et al. (1937), and Page (1976). *Tetrahymena* can be raised in sterile cultures containing the proper nutrients, and for this reason it is probably now the most favored protozoan for laboratory experimentation.

Examination, preparation. There is no substitute for the study of *living* Protozoa;

killing and fixing agents almost invariably distort specimens badly and make them unfit for accurate identification. Furthermore, instead of using an ordinary glass slide, it is better to use one that has a small, shallow concavity. If a drop of culture solution is placed in the concavity, covered with a Number 1 cover slip, and sealed with vaseline, evaporation is prevented and the living specimens may be studied with the high power of the microscope for hours at a time. Sometimes similar hanging-drop preparations are advantageous. For making careful observations on cilia and flagella dark-field illumination is occasionally useful. It is always important that clean slides, cover slips, and pipettes be used for working with live Protozoa.

Living or preserved protozoans may be easily picked up, handled, and transferred with the smallest size Irwin loop—an indispensible item for the protozoologist.

Most ciliates move about so rapidly that they cannot be studied with the high powers of the microscope. Their movements may be impeded with cotton fibers or lens paper fibers. Narcotics, such as chloretone, isopropyl alcohol, nickel sulfate, or nickel chloride, are unsatisfactory because they usually alter body shape and physiology. Viscous agents such as agar, gelatin, tragacanth, or quince seed jelly are at best only moderately successful, but a viscous solution of methyl cellulose is excellent. This reagent is made up by dissolving 10 g of methyl cellulose (viscosity rating of 15 cP) in 90 ml of water. A drop of this syrupy solution is mixed with a drop of the protozoan culture on a slide before putting on the cover slip. A little experience will determine the relative amounts of culture solution and methyl cellulose that should be used. Movements of ciliates in this mixture are reduced to a minimum, and they may be carefully studied with the high powers. Furthermore, there is no distortion or detectable impairment of the activities of the organelles or physiological processes.

The nucleus and certain organelles can best be distinguished in living Protozoa with the aid of vital stains. Methylene blue is perhaps the best general vital stain; a small amount is placed on the slide and allowed to dry before water containing the Protozoa is placed on it. One percent aqueous solutions of Bismarck brown and neutral red are also good.

Noland's combined fixative and stain in excellent for demonstrating flagella and cilia in temporary mounts; it has the advantage of producing a minimum of distortion.

Phenol, saturated aqueous solution	80 ml
Formalin	20 ml
Glycerin	4 ml
Gentian violet	20 mg

The dye should be moistened with 1 ml of water before adding the other ingredients. A drop of this reagent should be mixed with a drop of the culture solution before putting on the cover slip.

Permanent, stained slides are mandatory for careful cytological work; a variety of fixatives, stains, and procedures are given in Kudo (1966) and in microtechnique handbooks. Schaudinn's fluid is perhaps the most generally used fixative. It is made as follows:

Saturated aqueous mercuric chloride	66 ml
Absolute or 95% alcohol	33 ml
Glacial acetic acid	1 ml

The first two ingredients can be kept mixed without deterioration, but the acid should be added just before using. After fixation, specimens can conveniently be brought through changes of 50, 70, 95%, and absolute alcohol in a centrifuge tube with gentle centrifuging between changes. A drop of the alcohol containing the concentrated Protozoa is then dropped with a pipette from a height of about 1 in. onto a slide coated with egg albumen. Following this, the slides are placed in absolute alcohol and thereafter treated as ordinary sections for staining.

Taxonomy. Unlike some metazoan phyla, the great majority of Protozoa die and disintegrate without leaving any fossil record, the only important exceptions being the skeleton-bearing Foraminifera and Radiolaria, both marine groups. Consequently, our ideas concerning the natural taxonomic relationships of the major groups must be based on existing morphological traits, physiology, and development.

The old idea of amoeboid Protozoa being ancestral to the other groups has been discarded, and most protozoologists believe that the flagellates, especially green flagellates, are most primitive. This conviction is based on the fact that holophytic organisms utilizing dissolved nutrients must necessarily have preceded holozoic forms feeding on particulate organic materials. Furthermore, it is comparatively easy to derive the Sarcodina from holozoic flagellates. The origin of the sporozoans (all parasitic), however, is obscure; they are probably polyphyletic, with flagellate and amoeboid ancestry. Both ciliates and suctorians are highly evolved and complex in structure. The former probably originated from some of the more advanced flagellates, whereas the Suctoria undoubtedly arose from ciliated ancestors by loss of cilia and development of tentacles during the adult stage.

Morphologically, the ciliates and suctorians are sharply defined groups, but many flagellates are known that are more or less amoeboid, and some amoeboid forms have flagella. The flagellate order Rhizomastigida, for example, has holozoic representatives with well-developed pseudopodia, including *Mastigamoeba* and *Mastigella*. Many representatives of other flagellate orders also have pseudopodia in varying degrees; some of these are *Rhizochrysis, Oikomonas, Chrysamoeba,* and *Bodo.* On the other hand, some Sarcodina such as *Pseudospora* and *Dimastigamoeba* are flagellated swarmers during a short period of their life history.

The criteria to be used for subdividing the Mastigophora and distinctions between "true flagellate algae" and "true flagellate protozoa" are kindred and highly controversial problems that have been discussed for many years. Many members of the Phytomastigophorea of the zoologist are classified by the botanist as members of the Chrysophyceae, Dinophyceae, Chlorophyceae, Euglenophyceae, and Cryptophyceae, and all of these classes of algae also include nonmotile genera. Most of the discussion has centered around the Phytomastigophorea, a group that contains pigmented and colorless forms, as well as holophytic, holozoic, saprozoic, and mixotrophic forms. At one extreme, a few investigators have advocated placing all chlorophyll-containing species with the algae and all nonpigmented forms with the Protozoa. Most protistologists, however, have a broader viewpoint of these questions; they maintain that a sharp division of flagellates into "plant" and "animal" species is purely an artificial device and serves no useful purpose. Indeed, there is much to be said for considering flagellates as a mutual and primitive stage of organization of both the plant and animal kingdoms, and there is no real reason why many flagellates cannot be claimed by both zoologists and botanists. The widely accepted opinion emphasizes the fact that the flagellates do not constitute a natural taxonomic group but represent diverse branches in a stage of cellular organization, that some orders are quite artificial, and that the majority of orders contain both pigmented and nonpigmented forms. Some holozoic and saprozoic colorless forms are easily linked with their pigmented relatives, but in other cases few indications of affinities can be found. The following respective pairs of pigmented and nonpigmented genera appear to be closely related: *Cryptomonas* and *Chilomonas, Euglena* and *Astasia, Chlamydomonas* and *Polytoma.*

About 500 genera of free-living Protozoa are known to occur in the fresh waters of the United States, but many of these are uncommon and have been reported only once or a few times. Such rare genera are mainly of interest only to the specialist. In order to determine just what genera should be included in this volume, however, the complete list of genera was sent to a group of American protozoologists along with a request that they strike from the list about 150 genera that they considered as being least likely to be encountered in field collections and laboratory cultures. For the first edition of this book, eight protozoologists collaborated in this effort, and as the result of their "votes" the complete list was cut to 303 genera, all of which are included in the following key. In this edition many additional genera are added to the keys.

With few exceptions this key is drawn up for the identification of *living* Protozoa. Measurements are given chiefly for general information and should not usually be used as key characters. In each case the greatest dimension is noted; it is given for the cell proper and does not include stalk, flagella, cilia, cirri, tentacles, or case. In colonial species the measurements are given for single individuals of the colony. Except where size ranges are given, lengths are rough averages and are subject to a plus or minus one-third variation. In general, a representative common species is figured for each genus. Most of these figures are diagrammatic optical sections or surface views intended to emphasize key characters. Nuclei are often shaded, but it should be remembered that they are generally translucent in living Protozoa. Some of the figures have been drawn especially for this manual, but a large number are modified, simplified, or composites from a great many literature sources. Almost without exception, these latter figures are so considerably modified from their antecedents as to make acknowledgments unnecessary and superfluous. All figures have been drawn free hand.

The numbers of species known for the various genera are not usually included in the following key, chiefly because few genera have been thoroughly monographed. A few genera are monospecific; examples are *Uroglena*, *Distigma*, *Teuthophrys*, *Urocentrum*, and *Bursaria*. The great majority of genera, however, are represented by more than three species, and some, such as the following, are represented by at least 20 species: *Euglena*, *Peridinium*, *Trachelomonas*, *Vorticella*, *Lionotus*, *Chilodonella*, *Cyclidium*, *Metopus*, *Euplotes*, *Epistylis*, *Spathidium*, and *Opercularia*.

Our main purpose in this key is to enable the student to identify to genus those protozoans likely to be encountered in fresh-water aquatic work. To keep key numbers within reason, we are subdividing the Protozoa into separately numbered keys. The flagellates begin on this page, the Sarcodina on page 65, and the Ciliophora on page 71.

In order to simplify the use of these keys, we have carried taxonomic terminology only to the order and genus levels. All family names have been dropped, especially since they are of limited value to the general student.

KEY TO COMMON GENERA OF FLAGELLATES

1. Usually with three chromatophores; one or two flagella (rarely three or four), but cell additionally amoeboid in some groups .... Class **PHYTOMASTIGOPHOREA**, 2
   No chromatophores; one to many flagella, but cell additionally amoeboid in some groups **86**
2. With two flagella, of which one extends transversely around the cell; body grooved; chromatophores variously colored; starch and lipid food reserves .... Order **DINOFLAGELLIDA**, 3
   With one to four flagella, directed anteriorly or trailing **11**
3. Naked or covered with a thin exoskeleton (Figs. 11A–E) **4**
   Covered with a thick exoskeleton (Figs. 11F–H) **8**
4. Covered with a thin exoskeleton; several yellow to brown chromatophores; annulus complete (Fig. 11A); 40 μm; common .... **Glenodinium**
   Naked; numerous chromatophores **5**
5. Annulus anterior (Fig. 11B) 30 μm .... **Amphidinium**
   Annulus not anterior **6**
6. Annulus spiral (Fig. 11J); 25 μm .... **Gyrodinium**
   Annulus not spiral **7**
7. Annulus complete (Fig. 11D); 40 μm; common .... **Gymnodonium**
   Annulus incomplete (Fig. 11E); 28 μm .... **Hemidinium**
8. Exoskeleton with transverse and longitudinal grooves and divided into plates (Figs. 11G, H) **9**
   Exoskeleton without grooves; plates absent; spherical or ellipsoidal (Fig. 11F); 45 μm.
                                                                                   Phytodinium
9. Flattened and with one anterior and three long posterior hornlike processes (Fig. 2E); shape highly variable; 100 to 500 μm; very common in lakes and ponds ... **Ceratium hirundinella** (Müller)
   Not flattened; without hornlike processes; 40 μm **10**
10. Annulus displaced (Fig. 11G) .... **Gonyaulax**
    Annulus not displaced (Fig. 11H); common in plankton .... **Peridinium**

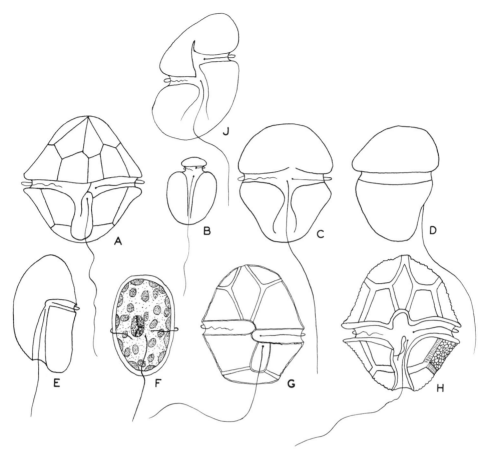

FIG. 11.–Representative Dinoflagellida. A, *Glenodinium*; B, *Amphidinium* (chromatophores present or absent); C, ventral view of *Gymnodinium*; D, lateral view of *Gymnodinium*; E, *Hemidinium*; F, *Phytodinium*; G, *Gonyaulax*; H, *Peridinium*; J, *Gyrodinium*. (A and H modified from Eddy, 1930.)

11. With yellow, brown, or orange chromatophores; usually quite small 12
    With green chromatophores, or chromatophores absent 37
12. With lipid and starch storage bodies; pellicle absent; body without a longitudinal furrow; one or two flagella inserted at or near the anterior end; many species form pseudopodia; most species both photosynthetic and phagotrophic . . . . Order **CHRYSOMONADIDA, 13**
    With stored carbohydrate granules; pellicle present; body with an oblique furrow; mostly holophytic, a few saprozoic or holozoic; two flagella, inserted anteriorly or laterally; no pseudopodia . . . . Order **CRYPTOMONADIDA, 32**
13. With one or two flagella; motile stage dominant; with a calcareous or siliceous shell or skeleton 15
    Flagella absent or transient; naked 14
14. Body amoeboid and solitary, with one or two chromatophores; not colonial (Fig. 15A).
                                                                                        Rhizochrysis
    Colonial; cells imbedded in a branching mass, 2 to 30 cm long; attached to substrate in gelatinous mosslike strands in cold mountain streams; with an acrid odor; recently divided cells often forming flagellated zoospores . . . . **Hydrurus foetidus** (Vill.)
15. With a single flagellum 16
    With two flagella 23
16. Cells in spherical colonies; each cell with two, long, siliceous rods (Fig. 12F); 15 μm; plankton.
                                                                           Chrysosphaerella
    Not colonial 17

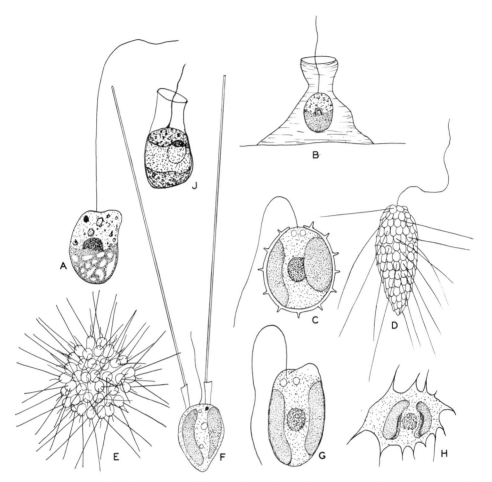

FIG. 12.–Representative Chrysomonadida. A, *Chrysapsis*; B, *Chrysopyxis*; C, *Chrysococcus*; D, *Mallomonas*; E, colony of *Chrysosphaerella*; F, individual *Chrysosphaerella*; G, *Chromulina*; H, *Chrysamoeba*; J, *Kephyrion*. (D modified from Smith, 1920.)

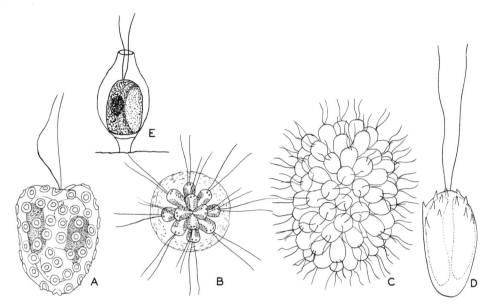

FIG. 13.–Representative Chrysomonadida. A, *Hymenomonas;* B, *Syncrypta* colony; C, *Synura* colony; D, individual *Synura;* E, *Derepyxis.* (A modified from Conrad; C and D modified from Smith, 1920.)

26. Cells imbedded in a gelatinous mass (Fig. 13 B); 11 μm .... **Syncrypta**
    Cells not imbedded in a gelatinous mass (Figs. 13 C, D); 35 μm; common .... **Synura**
27. Each cell enclosed within a delicate lorica (Figs. 14 A–C) **28**
    Each cell not enclosed within a lorica **29**
28. Lorica homogeneous, without growth rings (Figs. 14 A, B); both photosynthetic and phagotrophic; extremely common in plankton .... **Dinobryon**
    Lorica with growth rings (Fig. 14 C); epiphytic; sometimes solitary .... **Hyalobryon**
29. Cells arranged in a flat, wheellike colony composed of 10 to 20 wedge-shaped cells; cell 15 μm.
                                                                                Cyclonexis
    Cells arranged otherwise **30**
30. Solitary or colonial; individuals not arranged on periphery of a gelatinous mass (Figs. 14 D, E); 22 μm; photosynthetic and phagotrophic; more than 80 species .... **Ochromonas**
    Colonial; individuals arranged on periphery of gelatinous mass **31**
31. Center of colony without dichotomously branched strands (Figs. 14 F, G); 6 μm; when abundant imparts an offensive odor to water .... **Uroglenopsis**
    Center of colony with dichotomously branched strands, which are best seen in stained specimens (Figs. 14 H, J); 16 μm .... **Uroglena volvox** Ehr.
32. With two anterior flagella and an oblique furrow near anterior end (Figs. 2 C, 15 C–F).
                                                              Order **CRYPTOMONADIDA, 33**
    With two lateral flagella and an equatorial furrow (Fig. 15 B) .... **Nephroselmis**
33. With one or more chromatophores of various coloration **34**
    Without chromatophores; common in stagnant waters and infusions **36**
34. One red chromatophore; body often spindle-shaped; furrow strongly granulated (Fig. 15 E).
                                                                                Rhodomonas
    Two chromatophores; body not spindle-shaped; furrow lightly granulated **35**
35. Gullet present (Fig. 15 C) .... **Cryptomonas**
    Gullet absent (Fig. 15 F) .... **Cryptochrysis**
36. Body oval, flattened, and obliquely truncate at anterior end (Fig. 15 D); holozoic.
                                                              Cyathomonas truncata Ehr.
    Body cylindrical and elongated (Fig. 2 C); saprozoic .... **Chilomonas**

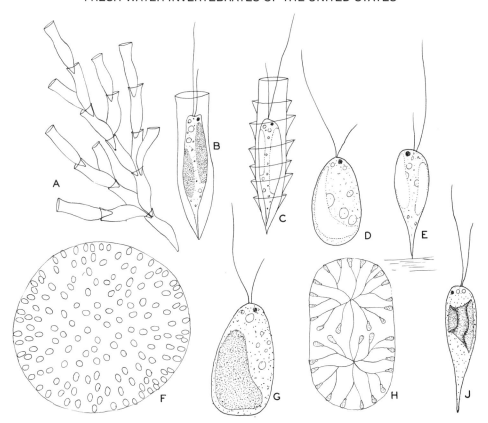

FIG. 14.–Representative Chrysomonadida, A, *Dinobryon* colony; B, *Dinobryon* individual; C, *Hyalobryon;* D, free-swimming *Ochromonas;* E, sessile *Ochromonas;* F, *Uroglenopsis* colony; G, *Uroglenopsis* individual; H, *Uroglena volvox* colony; J, *U. volvox* single individual. (A modified from Smith, 1920.)

37. Body more or less spherical; usually two or four apical flagella but three in one family and five in another family; glycogen reserve food; solitary or colonial; few chromatophores, usually shell or cup shaped; stigma present or absent .... Order **VOLVOCIDA, 39**
    Body more or less elongated; always solitary; many chromatophores, not cup shaped, sometimes absent; with another combination of other characters **38**
38. Uncommon genera; with one flagellum or with two flagella, of which one is trailing; flagella originating beside a superficial apical cavity or furrow; reserve foods lipids and glycogen; stigma absent .... Order **RAPHIDOMONADIDA, 83**
    Very common genera; with one, two, or three flagella, one trailing in one suborder; flagella originating in a definite reservoir; reserve food paramylum; stigma present.
                                                                          Order **EUGLENIDA, 65**
39. Solitary **40**
    Colonial **46**
40. With an external bivalve membrane (Figs. 16A, B); 16 μm **41**
    Without a bivalve membrane **42**
41. Bivalve membrane sculptured, halves evident in vegetative cell (Fig. 16A); stagnant water.
                                                                          Phacotus
    Bivalve membrane not sculptured, halves evident only in dividing cell (Fig. 16B) ...... **Pteromonas**
42. With two flagella **53**
    More than two flagella **43**
43. With three flagella; bean shaped (Fig. 16C); 14 μm .... **Trichloris**
    More than three flagella **44**
44. With four flagella **62**
    With more than four flagella **45**

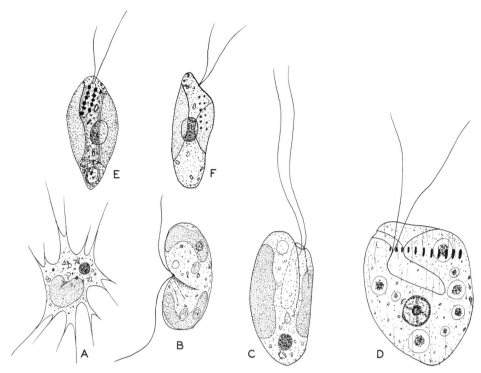

FIG. 15.–Representative Cryptomonadida and Chrysomonadida. A, *Rhizochrysis*; B, *Nephroselmis*; C, *Cryptomonas*; D, *Cyathomonas truncata* Ehr.; E, *Rhodomonas*; F, *Cryptochrysis*.

45. With five flagella; four anterior wings (Fig. 16D) .... **Chloraster**
    With six flagella (Fig. 16E); 13 μm .... **Pocillomonas**
46. Colony a flat plate (Figs. 17A, B); cell 13 μm **47**
    Colony spherical or subspherical **48**
47. Colony in a gelatinous envelope with anterior–posterior differentiation; 16 or 32 cells (Fig. 17A).
        Platydorina
    Colony in a gelatinous envelope without anterior-posterior differentiation; 4 to 16 cells (Fig. 17B).
        Gonium
48. Colony with at least 500 cells; colony up to 600 μm in diameter (Fig. 3) .... **Volvox**
    Colony with not more than 256 cells **49**
49. With cells of two different sizes; 32, 64, or 128 cells per colony (Fig. 17C) .... **Pleodorina**
    Cells all of same size **50**
50. Cells more or less spherical **51**
    Cells pear shaped; 8 to 16 cells per colony **52**
51. Cells close together; 4, 8, 16, or 32 cells per colony (Fig. 17D); cells 12 μm .... **Pandorina**
    Cells farther apart; 16, 32, or 64 cells per colony (Fig. 17E); cells 17 μm .... **Eudorina**
52. Cells with two flagella (Fig. 17F); cell 15 μm .... **Chlamydobotrys**
    Cells with four flagella (Fig. 17G); cell 19 μm .... **Spondylomorum**
53. With chromatophores **54**
    Without chromatophores; saprozoic **60**
54. Spindle shaped (Fig. 18A); 50 μm .... **Chlorogonium**
    Cell not spindle shaped **55**
55. Cell apparently naked (Fig. 2A); 8 to 22 μm .... **Chlamydomonas**
    Cell not naked **56**
56. Cell covered with thick gelatinous envelope (Fig. 18B); sometimes red pigmented; 10 to 50 μm;
      mostly in ephemeral ponds .... **Haematococcus**
    Cell not covered with a thick gelatinous sheath **57**

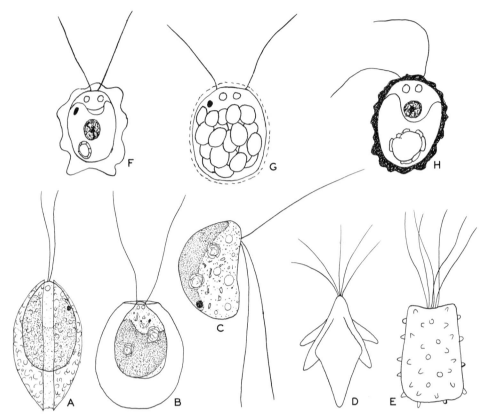

FIG. 16.—Representative Volvocida. A, *Phacotus*; B, *Pteromonas*; C, *Trichloris*; D, *Chloraster*; E, *Pocillomonas*; F, *Lobomonas*; G, *Gloeomonas*; H, *Thorakomonas*.

57. With blunt, irregular processes (Fig. 16F) .... **Lobomonas**
    Without blunt processes **58**
58. Cell not filling subcapsular space (Fig. 18C); 21 μm .... **Coccomonas**
    Cell filling subcapsular space **59**
59. Capsule thin and gelatinous (Fig. 16G) .... **Gloeomonas**
    Capsule crustose, brown to black (Fig. 16H) .... **Thorakomonas**
60. Spindle shaped (Fig. 18D); 30 to 80 μm .... **Hyalogonium**
    Not spindle shaped **61**
61. Ovoid (Fig. 18E); 23 μm .... **Polytoma**
    Anterior margin obliquely truncate (Fig. 18F); 15 μm .... **Parapolytoma**
62. Sickle-shaped (Fig. 18G); rarely biflagellated; 8 μm .... **Spermatozopsis**
    Not sickle shaped **63**
63. Pyramidal or heart shaped (Fig. 18H); 20 μm .... **Pyramimonas**
    Ovoid or ellipsoid **64**
64. Colorless (Fig. 18J); 14 μm .... **Polytomella**
    Green (Fig. 18K); 6 to 25 μm .... **Carteria**
65. With stigma; chromatophores almost invariably present (Figs. 2B, 19B–F) **67**
    Stigma and chromatophores absent **66**
66. With one flagellum **74**
    With two flagella **79**
67. Highly flattened, asymmetrical, and often longitudinally striated and twisted (Fig. 19B); 30 to 170
    μm .... **Phacus**
    Not highly flattened and asymmetrical **68**

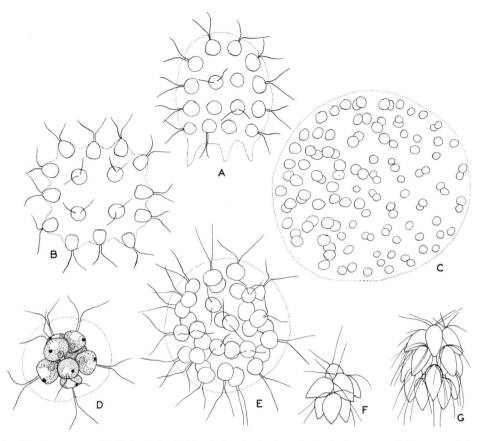

FIG. 17.–Representative Volvocida. A, *Platydorina*; B, *Gonium*; C, *Pleodorina*; D, *Pandorina*; E, *Eudorina*; F, *Chlamydobotrys*; G, *Spondylomorum*. (D–E modified from Smith, 1920.)

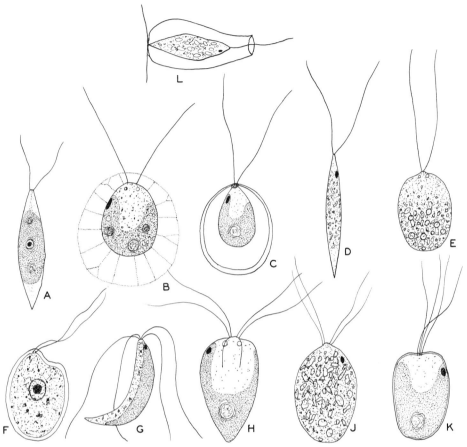

FIG. 18.–Representative Volvocida and Euglenida. A, *Chlorogonium;* B, *Haemotococcus;* C, *Coccomonas;* D, *Hyalogonium;* E, *Polytoma;* F, *Parapolytoma;* G, *Spermatozopsis;* H, *Pyramimonas;* J, *Polytomella;* K, *Carteria;* L, *Ascoglena.*

78. More or less spindle shaped; posterior end drawn out (Fig. 19L) .... **Astasia**
     Posterior end broad, rounded, or truncate during locomotion (Fig. 19M); very common.
                                                                              Peranema
79. Both flagella directed forward; plastic (Fig. 20A); holozoic; 40 to 120 µm; common in acid waters.
                                                                         Distigma proteus Ehr.
     One flagellum trailing **80**
80. With a long, slitlike furrow (Figs. 20B, C) **81**
     Without a furrow **82**
81. Trailing flagellum long (Fig. 20B); 15 to 60 µm .... **Anisonema**
     Trailing flagellum short (Fig. 20C); 8 to 25 µm; acid waters .... **Notosolenus**
82. Oval, flattened, and furrowed; with a long pharyngeal rod apparatus (Fig. 20E); 25 µm.
                                                                              Entosiphon
     Rounded to elongated; sometimes ridged or striated; pharyngeal rod apparatus short or absent (Fig.
     20D); 40 to 250 µm .... **Heteronema**
83. With many highly refractile trichocystlike bodies (Fig. 20F); 50 µm; acid waters .... **Gonyostomum**
     Without such bodies **84**
84. Cell with projecting hairlike setae, and sometimes pseudopodia (Fig. 20J) ... **Thaumatomastix**
     Without setae and pseudopodia **85**

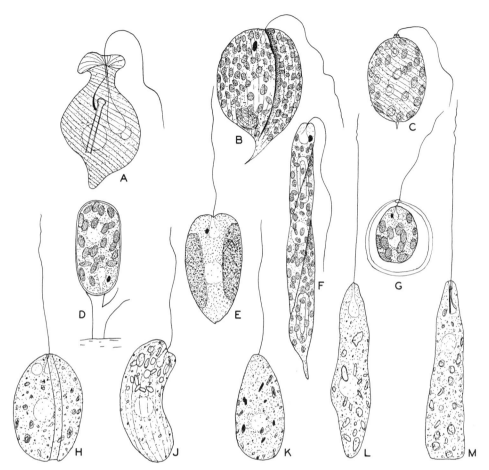

FIG. 19.–Representative Euglenida. A, *Urceolus;* B, *Phacus;* C, *Lepocinclis;* D, *Colacium;* E, *Cryptoglena;* F, *Euglena;* G, *Tracheolomonas;* H, *Petalomonas;* J, *Menoidium;* K, *Scytomonas;* L, *Astasia;* M, *Peranema.*

85. Anterior end narrow; body not flattened (Fig. 20H); 100 μm . . . . Vacuolaria
    Anterior end not especially narrow; body flattened (Fig. 20G); 60 μm . . . . Trentonia
86. Cell with annulus and sulcus (Figs. 11B, D, E, J) . . . . Class **PHYTOMASTIGOPHOREA,**
    Order **DINOFLAGELLIDA, 5**
    Cell without annulus and sulcus 87
87. Cell with cytostome and cytopharynx . . . . Class **PHYTOMASTIGOPHOREA, 88**
    Cell without cytostome and cytopharynx; holozoic or saprozoic 90
88. With stored carbohydrate granules; two flagella, both projecting forward.
    Order **CRYPTOMONADIDA, 33**
    With stored paramylum granules; sometimes oil droplets also; one flagellum, or two flagella with
    one usually trailing . . . . Order **EUGLENIDA, 89**
89. Stigma present; saprozoic; similar to *Euglena* but colorless; 45 μm . . . . Khawkinea
    Stigma absent 66
90. With well-defined pseudopodia in addition to flagella (Figs. 23G–L).
    Class **ZOOMASTIGOPHOREA,** Order **RHIZOMASTIGIDA, 123**
    With flagella only; only occasionally with one or two small blunt pseudopodia 91
91. With one or two flagella 92
    With three to eight flagella . . . . Class **ZOOMASTIGOPHOREA, 128**

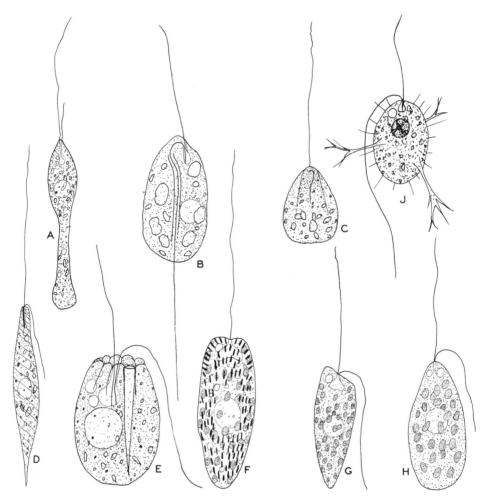

FIG. 20.–Representative Anisonemidae and Chloromonadina. A, *Distigma proteus*; B, *Anisonema*; C, *Notosolenus*; D, *Heteronema*; E, *Entosiphon*; F, *Gonyostomum*; G, *Trentonia*; H, *Vacuolaria*; J, *Thaumatomastix*.

92. With two flagella **93**
    With one flagellum .... Class **ZOOMASTIGOPHOREA, 101**
93. Flagella equally long **94**
    Flagella of unequal length .... Class **ZOOMASTIGOPHOREA, 113**
94. Naked or with gelatinous envelope; solitary or colonial; motile or sessile; sometimes with small
    pseudopodia .... Class **ZOOMASTIGOPHOREA, 95**
    Naked; solitary; motile; without pseudopodia .... Class **PHYTOMASTIGOPHOREA, 53**
95. Without lorica or gelatinous covering **96**
    With a lorica or gelatinous covering **98**
96. Oval or rounded amoeboid; free swimming or attached by a long stalk (Fig. 21A); 13 μm.
                                                                              Amphimonas
    Not oval or rounded amoeboid; stagnant waters **97**
97. Spirally twisted (Fig. 21B); 10 μm .... **Spiromonas**
    Ovate or pyriform plastic; anterior end pointed (Fig. 21C); 15 μm .... **Dinomonas**
98. Each cell with a stalk and individual lorica (Fig. 21D); 15 μm .... **Diplomita**
    Cells in gelatinous masses or tubes (Figs. 21E, F); 6 to 12 μm **99**
99. Individuals in dichotomously branching gelatinous tubes .... **Cladomonas**
    Tubes or mass not dichotomously branched **100**

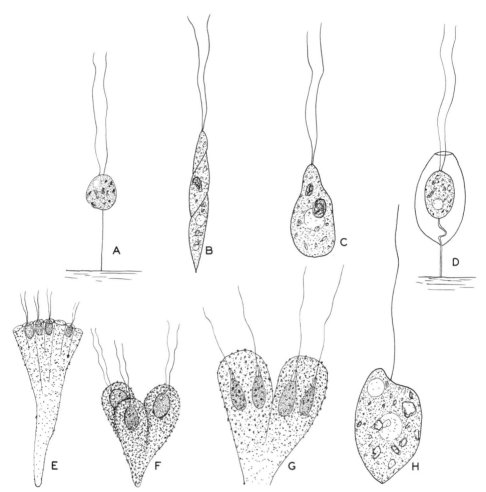

FIG. 21.–Representative Zoomastigophorea. A, *Amphimonas*; B, *Spiromonas*; C, *Dinomonas*; D, *Diplomita*; E, small part of colony of *Rhipidodendron*; F, small part of colony of *Spongomonas*; G, small part of colony of *Phalansterium*; H, *Oikomonas*.

100. Branched tubes united laterally (Fig. 21 E) .... **Rhipidodendron**
     Individuals imbedded in granular gelatinous masses that are not united laterally (Fig. 21 F).
                                                                       Spongomonas
101. Cell with collar (Fig. 22) .... Order **CHOANOFLAGELLIDA, 102**
     Cell without collar; flagellum anterior; spherical or oval (Fig. 21 H); 10 μm; stagnant water (usually
         included in the flagellated algaelike protozoans).
                                           Order **CHRYSOMONADIDA, Oikomonas**
102. Entire animal enclosed in a gelatinous mass (Fig. 21 G); 17 μm .... **Phalansterium**
     No gelatinous mass, or cell body only imbedded in gelatinous mass (Fig. 22); free swimming or
         sedentary; colonial or solitary 103
103. Without lorica (Figs. 22 A, B, F, M, N); stalked; holozoic or saprozoic; 5 to 15 μm 104
     With lorica (Figs. 26 C, G, H, J, O); stalked or unstalked; holozoic 109
104. With a double collar (Fig. 22 B); solitary or clustered .... **Diplosiga**
     With a single collar 105
105. Solitary (Fig. 22 F) .... **Monosiga**
     Colonial 106
106. Cells (except collar) imbedded in a gelatinous mass (Fig. 22 M) 107
     Cells not imbedded in a gelatinous mass (Figs. 22 A, N) 108

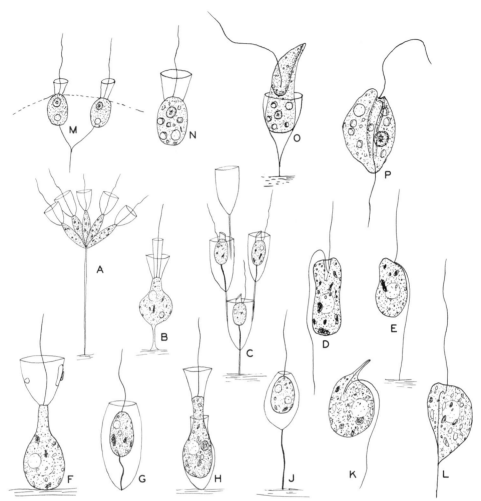

FIG. 22.–Representative Zoomastigophorea. A, colony of *Codosiga*; B, *Diplosiga*; C, colony of *Poteriodendron*; D, *Phyllomitus*; E, *Pleuromonas*; F, *Monosiga*; G, *Bicosoeca*; H, *Salpingoeca*; J, *Codonoeca*; K, *Rhynchomonas*; L, *Cercomonas*; M, two cells of a *Sephaeroeca* colony; N, one cell of a colony of *Desmarella*; O, *Histiona*; P, *Colponema*.

107. Colony spherical (Fig. 22M) .... **Sphaeroeca**
     Colony irregular and amorphous, usually flat .... Protospongia
108. Individuals clustered at end of a simple or branching stalk (Fig. 22A) .... **Codosiga**
     Cells united laterally (Fig. 22N) .... **Desmarella**
109. Lorica with a fine outer basal stalk (Figs. 22C, J) 112
     Lorica without a fine outer basal stalk (Figs. 22G, H, O) 110
110. Body small compared with lorica and attached to base of lorica with a fine stalk (Fig. 22G); sessile or
         free swimming; often in clusters; 23 μm .... **Bicosoeca**
     Body almost filling lorica 111
111. Distal end with lips and saillike projection (Fig. 22O) .... **Histiona**
     Distal end a cylindrical cone (Fig. 22H) .... **Salpingoeca**
112. Colonial (Fig. 22C); 35 μm .... **Poteriodendron**
     Solitary (Fig. 22J); 23 μm .... **Codonoeca**
113. One flagellum trailing (Figs. 22D, E, K, L, P); somewhat amoeboid; stagnant water **114**

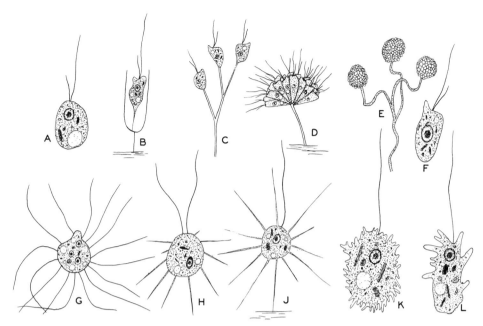

FIG. 23.–Representative Zoomastigophorea. A, *Monas*; B, *Stokesiella*; C, portion of a *Dendromonas* colony; D, portion of a *Cephalothamnium* colony; E, small portion of an *Anthophysis* colony; F, single individual of *Anthophysis*; G, *Multicilia*; H, *Dimorpha*, J. *Actinomonas*; K, *Mastigella*; L, *Mastigamoeba*.

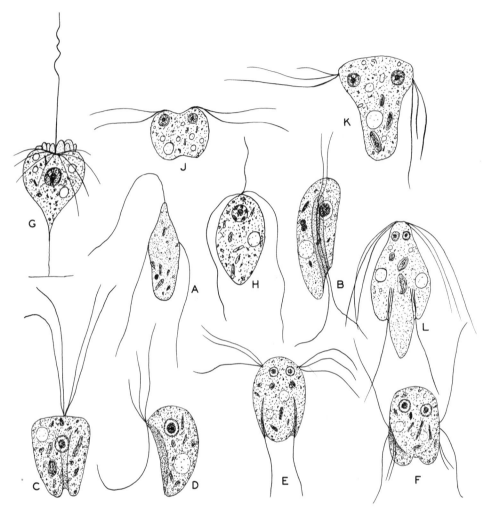

FIG. 24.–Representative Zoomastigophorea. A, *Dallingeria*; B, *Costia*; C, *Collodictyon*; D, *Tetramitus*; E, *Hexamita*; F, *Trepomonas*; G, *Pteridomonas*; H, *Macromastix*; J, *Gyromonas*; K, *Trigonomonas*; L, *Urophagus*.

126. With filopodia; usually spheroidal and attached (Fig. 23J); 10 μm .... **Actinomonas**
     With lobopodia 127
127. Pseudopodia very numerous; flagellum apparently not originating at nucleus (Fig. 23K); 130 μm.
                                                                               Mastigella
     Pseudopodia not so numerous; flagellum originating at nucleus (Fig. 23L); 20 to 200 μm.
                                                                               Mastigamoeba
128. With three flagella, two of which are trailing (Figs. 24A, H); free swimming or attached; 5 μm;
         stagnant water 129
     With four to eight flagella 130
129. Anterior end drawn out (Fig. 24A) .... **Dallingeria**
     Anterior end not drawn out (Fig. 24H) .... **Macromastix**
130. With four flagella originating in one area 131
     With four, six, or eight flagella, not all originating in one area; anaerobic in stagnant water 133
131. Flagella attached at or near anterior end; free swimming in stagnant water 132
     Flagella attached at base of a funnellike depression (Fig. 24B); ectoparasitic on fish; 10 μm ... **Costia**
132. Spherical, ovoid, or heart sharped (Fig. 24C); 45 μm .... **Collodictyon**
     Pyriform, with pointed posterior end (Fig. 24D); 12 μm .... **Tetramitus**

133. Posterior end truncate; six flagella near anterior end, two posterior (Fig. 24 E); 20 μm .... **Hexamita**
     Posterior end not truncate; with another arrangement of flagella 134
134. Body broadly oval and flattened (Figs. 24 F, J) 135
     Posterior end more or less tapered (Figs. 24 K, L) 136
135. With two anterior pairs of flagella (Fig. 24 J) .... **Gyromonas**
     With four flagella on each side (Fig. 24 F); 10 μm .... **Trepomonas**
136. With two cytostomes and six flagella (Fig. 24 K) .... **Trigonomonas**
     With one cytostome and eight flagella; two body furrows (Fig. 24 L) .... **Urophagus**

■ ■ ■

KEY TO COMMON GENERA OF AMOEBOID PROTOZOANS

1. Without chromatophores; with lobopodia, filopodia, or axopodia; pseudopodia the chief means of locomotion throughout the life history; the true amoeboid forms.
                                                            Subphylum **SARCODINA**, 3
   With one or two chromatophores and filopodia; 10 to 40 μm (usually classified with flagellates).
                          Subphylum **MASTIGOPHORA**, Class **PHYTOMASTIGOPHOREA**,
                                                            Order **CHRYSOMONADIDA**, 2
2. Filopodia branching; one or two chromatophores (Fig. 15A) .... **Rhizochrysis**
   Filopodia not branching; two chromatophores; usually solitary (Fig. 12H) ..... **Chrysamoeba**
3. With spherical symmetry and axopodia (Figs. 29A–E) .... Class **ACTINOPODA**, 47
   Without spherical symmetry; lobopodia or filopodia (Figs. 25, 27, 28) ...... Class **RHIZOPODA**, 4
4. Naked, without a test or shell 5
   With a sclerotized test or shell .... Order **TESTACIDA**, 16
5. With radiating reticulopodia (Fig. 25) .... Order **PROTEOMYXIDA**, 6
   With lobopodia (Figs. 1A, 25H, 26A, B) .... Order **AMOEBIDA**, 12
6. Small individuals grouped in a network of slightly branched and anastomosing reticulopodia (Fig. 25A); often colored by ingested algae; on *Vaucheria* .... **Labyrinthula**
   Cells solitary 7
7. With many reticulopodia that branch freely and anastomose; body shape and size inconstant; many small contractile vacuoles (Fig. 25C) .... **Biomyxa**
   Reticulopodia not anastomosing 8
8. Parasitic and completely within cells of algae and Volvocidae (Fig. 25B); 20 μm ...... **Pseudospora**
   Not completely within cells of algae and Volvocidae 9
9. Generally disc shaped, with reddish endoplasm and clear ectoplasm (Fig. 25D); 65 μm.
                                                                          Hyalodiscus
   Not generally disc shaped; endoplasm reddish or not 10
10. Heliozoa-like; when feeding on filamentous algae, body shape may change markedly (Fig. 25E); 50 to 700 μm .... **Vampyrella**
    Not Heliozoa-like; not feeding on filamentous algae 11
11. Stored food consisting of starch granules (Fig. 25F) .... **Protomonas**
    With other types of food inclusions; with or without mucous envelope (Fig. 25G); 45 μm.
                                                                            Nuclearia
12. With amoeboid and flagellated stages (Fig. 26A); stagnant water .... **Trimastigamoeba**
    Without a flagellate stage 13
13. Large, from 0.5 to 5.0 mm; many small contractile vacuoles and refringent bodies; many nuclei; on bottom in stagnant waters .... **Pelomyxa**
    Smaller, from 25 to 500 μm; a single nucleus 14
14. With one broad pseudopodium (Fig. 26B); 35 μm .... **Vahlkampfia**
    With numerous pseudopodia, never anastomosing (Figs. 1A, 25H); 25 to 500 μm 15
15. Temporary posterior end often with retractile papillae; body surface with many tapering pseudopodia and papillae (Fig. 25H) .... **Dinamoeba**
    Body with fewer typical pseudopodia (Fig. 1A) .... **Amoeba**

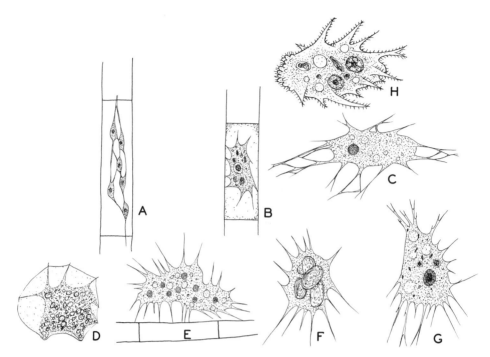

FIG. 25.–Representative Proteomyxida. A, *Labyrinthula* within an algal cell; B, *Pseudospora* within an algal cell; C, *Biomyxa*; D, *Hyalodiscus*; E, *Vampyrella* on a filamentous alga; F, *Protomonas*; G, *Nuclearia*; H, *Dinamoeba*.

16. Shell simple and membranous but sometimes with bits of adhering debris (Fig. 26) 17
    Shell composed of scales, plates, sand grains, or bits of debris Figs. 27, 28) 18
17. With numerous branching reticulopodia, usually anastomosing (Figs. 26C, D; 27K, L) 19
    With lobopodia or a few, simple, branched filopodia (Figs. 26F–M) 23
18. Shell composed of foreign materials (Figs. 27A–D; 28A–E) 32
    Shell composed of symmetrical and similar plates or scales (Figs. 27E–J; 28G, H) 40
19. Test with one aperture 20
    Test with two apertures, thin, spherical (Fig. 26C); 14 μm .... **Diplophrys**
20. Aperture lateral or subtermial (Fig. 26D); 110 μm .... **Lieberkühnia**
    Aperture terminal (Figs. 27K, L) 21
21. Filopodia branching but not anastomosing; test thin, flexible, colorless; diameter 20 to 45 μm (Fig. 27K) .... **Lecythium**
    Filopodia anastomosing (Fig. 27L) 22
22. Filopodia arising from a peduncle; up to 50 μm in diameter (Fig. 27L) .... **Microgromia**
    Filopodia not arising from a peduncle; test 100 to 400 μm long .... **Gromia**
23. Aperture circular, central, and inverted like a funnel (Fig. 1C); 30 to 250 μm; common in ponds; many species .... **Arcella**
    Aperture otherwise 24
24. Aperture very large (Figs. 26H, M) 25
    Aperture not especially large 26
25. Test patelliform and rigid (Fig. 26H); 20 μm .... **Pyxidicula**
    Test hemispherical or cup shaped, with an inner membranous sac with an elastic aperture (Fig. 27M).
    Diplochlamys
26. Test disclike, flexible, sometimes rolled up (Fig. 26G); 40 μm .... **Pseudochlamys**
    Test otherwise 27
27. Test semispiral, composed of curved rods or sand grains (Fig. 27N) .... **Lesquereusia**
    Test not semispiral 28

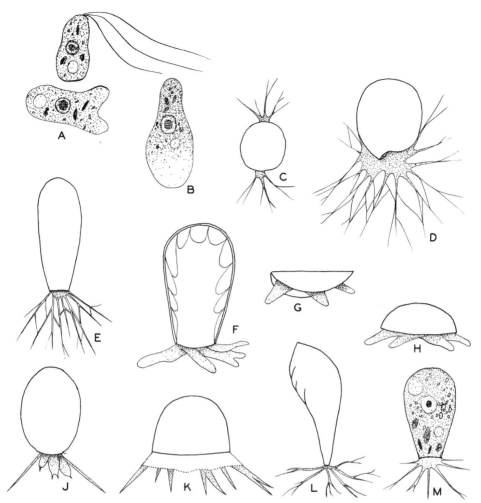

FIG. 26.–Representative Rhizopoda and Testacida. A, *Trimastigamoeba* (with and without flagella); B, *Vahlkampfia;* C, *Diplophrys;* D, *Lieberkühnia;* E, *Gromia;* F, *Hyalosphenia;* G, *Pseudochlamys;* H, *Pyxidicula;* J, *Difflugiella;* K, *Cochliopodium;* L, *Pamphagus;* M, *Chlamydophrys.*

28. With blunt pseudopodia only; test ovoid or pyriform; protoplasm partly filling test (Fig. 26F); 125
　　μm . . . . **Hyalosphenia**
　　Not with blunt pseudopodia only; protoplasm filling test 29
29. Median pseudopodia lobate or digitate and with pointed tips; lateral pseudopodia long, fine, and
　　tapering to a point; test ovoid and flexible (Fig. 26J); 40 μm . . . . . . **Difflugiella apiculata** Cash
　　All pseudopodia similar 30
30. Aperture large; test thin and flexible; pseudopodia blunt or pointed (Fig. 26K); 25 to 60 μm.
　　　　　　　　　　　　　　　　　　　　　　　　　　　　　　　　　　Cochliopodium
　　Aperture small (Figs. 26L, M); pseudopodia long and branching 31
31. Test hyaline and flexible; aperture very small (Fig. 26L); 40 to 100 μm . . . . **Pamphagus**
　　Test rigid; aperture not so small (Fig. 26M); 20 μm . . . . **Chlamydophrys**
32. Test with a curved neck, covered with small sand particles; filopodia (Fig. 27B); 25 μm . . . . **Campascus**
　　Test without a curved neck 33
33. Test a transparent membrane with two apertures at opposite ends (Fig. 28A) . . . . **Amphitrema**
　　Test with a single aperture 34

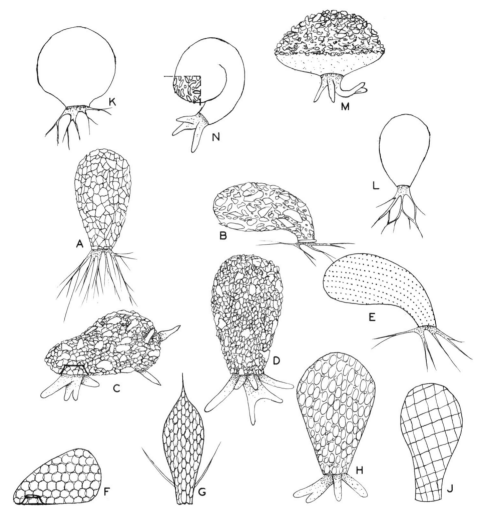

FIG. 27.–Representative Testacida. A, *Pseudodifflugia;* B, *Campascus;* C, a spiny *Centropyxis;* D, *Difflugia;* E, *Cyphoderia;* F, *Trinema;* G, a spiny *Euglypha;* H, *Nebela;* J, *Quadrulella;* K, *Lecythium;* L, *Microgromia;* M, *Diplochlamys;* N, *Lesquereusia.*

34. With long, straight or branching filopodia; test ovoid; aperture terminal (Fig. 27A); 25 to 70 μm.
    Pseudodifflugia
    Pseudopodia not filopodia 35
35. Aperture eccentric; test circular, discoid, or ovoid (Fig. 27C); 125 μm .... **Centropyxis**
    Aperture not eccentric (Fig. 28) **36**
36. Aperture surrounded by a four-lobed collar; test composed of sand grains (Fig. 28B).
    Cucurbitella
    Aperture without a four-lobed collar 37
37. Test with a necklike constriction; test composed of sand grains (Fig. 28C) .... **Pontigulasia**
    Test without a necklike constriction 38
38. Pseudopodia drawn out to points (Fig. 28D) .... **Phryganella**
    Pseudopodia not drawn out to points 39
39. Distal end of test covered with especially rough sand grains (Fig. 28E) .... **Heleopera**
    Distal end of test not covered with especially rough sand grains; test variable in shape; extremely common (Fig. 27D) .... **Difflugia**

FIG. 28.–Representative Testacida. A, *Amphitrema*; B, *Cucurbitella*; C, *Pontigulasia*; D, *Phryganella*; E, *Heleopera*; F, *Placocista*; G, *Assulina*; H, *Sphenoderia*.

40. Test with a curved neck, thin and covered with discs or scales; pseudopodia long and thin (Fig. 27 E); 60 to 200 μm .... **Cyphoderia**
    Test without a curved neck **41**
41. Test lenticular in cross section and with projecting spinous scales .... **Placocista**
    Test without projecting scales **42**
42. Test compressed anteriorly; circular siliceous scales (Fig. 27 F); 30 to 100 μm .... **Trinema**
    Test not compressed anteriorly **43**
43. With filopodia **45**
    With lobopodia **44**
44. Test composed of circular or oval plates (Fig. 27 H); 130 μm .... **Nebela**
    Test composed of quadrangular plates (Fig. 27 J); 80 to 140 μm .... **Quadrulella**
45. Aperture bordered by a dentate membrane; test colorless or brown, with elliptical scales arranged in diagonal rows; pseudopodia divergent, seldom branched (Fig. 28 G) .... **Assulina**
    Aperture not bordered by a dentate membrane; test hyaline, membranous; scales arranged in alternating series; pseudopodia often branched **46**
46. Test globular or oval, without spines; aperture relatively large (Fig. 28 H) .... **Sphenoderia**
    Test ovoid, often with scales; aperture relatively small (Fig. 27 G) .... **Euglypha**
47. Without scales, spicules, capsule, or an envelope of sand grains or diatoms **48**
    With scales, spicules, capsule, or an envelope of sand grains or diatoms **50**
48. Pseudopodia branched and with thickened bases (Fig. 29 A); 20 μm .... **Actinocoma**
    Pseudopodia unbranched and without thickened bases **49**

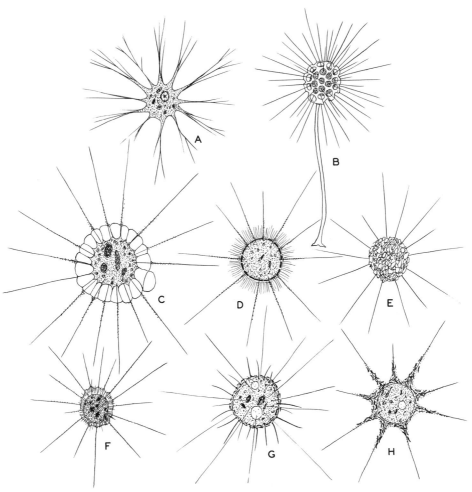

FIG. 29.–Representative Actinopoda. A, *Actinocoma*; B, *Clathrulina*; C, *Actinosphaerium*; D, *Myriophrys*; E, *Lithocolla*; F, *Heterophrys*; G, *Acanthocystis*; H, *Raphidiophrys*.

49. Granular endoplasm clearly set off from vacuolar ectoplasm (Fig. 29C); 70 to 300 μm.
                                                                        Actinosphaerium
    Endoplasm and ectoplasm not clearly divided (Fig. 1D); 40 μm .... **Actinophrys**
50. With a chitinoid, perforated capsule and a stalk (Fig. 29B); 75 μm .... **Clathrulina**
    With chitinoid or siliceous spicules or scales, or with an envelope of sand grains or diatoms 51
51. With numerous flagella among axopodia; siliceous scales (Fig. 29D); 40 μm ..... **Myriophrys**
    Without flagella **52**
52. With an outer envelope of sand grains, diatoms, or debris (Fig. 29E); 45 μm ...... **Lithocolla**
    With chitinous or siliceous scales or spicules **53**
53. Without siliceous scales; with indistinct radially arranged spicules projecting beyond the peripheral
       mucilaginous layer (Fig. 29F); 15 to 80 μm .... **Heterophrys**
    With siliceous scales **54**
54. With tangentially arranged siliceous scales and radiating siliceous spines (Fig. 29G); 40 μm.
                                                                       Acanthocystis
    With scales clustered around basal portions of pseudopodia; radiating siliceous spines absent (Fig.
       29H); 55 μm .... **Raphidiophrys**

KEY TO COMMON GENERA OF CILIATES

1. With cilia; free swimming or sessile 2
   Adult with suctorial tentacles (Fig. 44); cilia only in uncommon immature stages; usually sessile.
   Class **SUCTORIA, 156**
2. Without adoral zone of membranelles .... Class **HOLOTRICHIA, 4**
   With adoral zone of membranelles 3
3. With adoral zone winding clockwise to cytostome (Figs. 38–40) ... Class **SPIROTRICHIA, 96**
   With adoral zone winding counterclockwise to cytostome (Figs. 42, 43).
   Class **PERTRICHIA, 135**
4. Not commensal in mussels; without large ventral cilia 5
   Commensal in mantle cavity and gill chambers of mussels; flattened; with large ventral cilia for
     attachment (Fig. 31A); 60 to 250 μm .... Order **THIGMOTRICHIDA**, Conchophthirus
5. Cytostome on body surface or in peristome, without special cilia (Figs. 32, 33).
   Order **GYMNOSTOMATIDA, 7**
   Cytostome in peristome, with special cilia or membranelles 6
6. Peristome lined with rows of cilia (Fig. 34) .... Order **TRICHOSTOMATIDA, 59**
   Peristome with one undulating membrane and three membranelles; with or without cilia; cytostome
     ventral (Figs. 35, 36) .... Order **HYMENOSTOMATIDA, 71**
7. Cytostome at or near anterior end 8
   Cytostome not at or near anterior end 43
8. Region of cytostome more or less flattened; trichites present (Figs. 30A, B; 31B–F) 9
   Region of cytostome not flattened 15
9. With three curved anterior arms; containing zoochlorellae (Fig. 31B); 225 μm.
   Teuthophrys trisulca C. and B.
   Without such arms 10

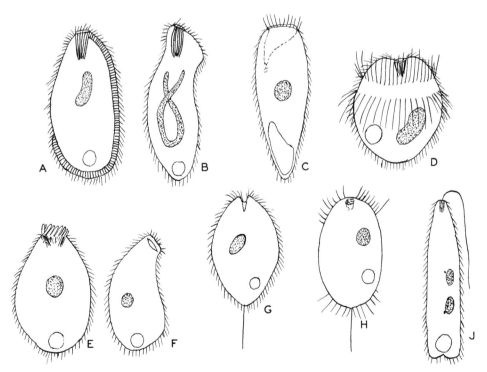

FIG. 30.–Representative Gymnostomatida. A, *Penardiella*; B, *Cranotheridium*; C, *Pelatractus*; D, *Askenasia*; E, *Spasmostoma*; F, *Platyophrya*; G, *Microregma*; H, *Pithothorax*; J, *Ileonema*.

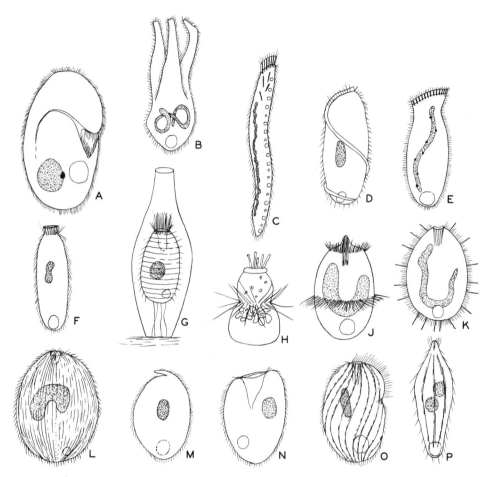

FIG. 31.–Representative Thigmotrichida and Gymnostomatida. A, *Conchophthirus*; B, *Teuthophrys*; C, *Homalozoon*; D, *Perispira*; E, *Spathidiium*; F, *Enchelydium*; G, *Vasicola*; H, *Mesodinium*; J, *Didinium*; K, *Actinobolina*; L, *Ichthyophthirius*; M, *Chilophyra*; N, *Bursella*; O, *Placus*; P, *Rhopalophrya*.

18. With an equatorial furrow and tentaclelike processes around cytostome (Fig. 31 H); 30 µm.
    Mesodinium
    Without an equatorial furrow and tentaclelike processes around cytostome (Figs. 30 D, 31 J) **19**
19. Two differing ciliary wreaths before midlength (Fig. 30 D) uncommon .... **Askenasia**
    With two or more similar ciliary wreaths, one beyond midlength (Fig. 31 J); 60 to 200 µm; feeds on
    other ciliates; common .... **Didinium**
20. Covered with regularly arranged, perforated plates; barrel shaped; often spinous (Fig. 4 B); 40 to 100
    µm .... **Coleps**
    Not covered with plates **21**
21. With tentacles scattered among cilia; ovate or spherical (Fig. 31 K); 150 µm ..... **Actinobolina**
    Without tentacles **22**
22. Parasitic on the integument of many fresh-water fishes; oval (Fig. 31 L); 100 to 1000 µm; become
    mature on host within small pustules .... **Ichthyophthirius**
    Not parasitic **23**
23. With a fingerlike process in front of cytostome; ovoid or ellipsoid (Fig. 31 M); 40 µm ... **Chilophrya**
    Without a fingerlike process in front of cytostome **24**
24. Cytostome with various special processes (Figs. 30 E, J) **25**
    Cytostome without special processes **26**
25. Cytostome surrounded by flaplike structures (Fig. 30 E) .... **Spasmostoma**
    Cytostome with a long flagellumlike device (Fig. 30 J) .... **Ileonema**
26. With a large anterior groovelike pit (Fig. 31 N); 250 to 550 µm .... **Bursella**
    Without a large anterior groovelike pit **27**
27. With prominent longitudinal grooves; 30 to 80 µm **28**
    Without prominent longitudinal grooves **29**
28. Grooves spiral; body ellipsoid or ovoid; cytostome a narrow slit (Fig. 31 O) .... **Placus**
    Grooves not spiral; body not ovoid or ellipsoid; with few cilia (Fig. 31 P) ...... **Rhopalophrya**
29. Cytopharynx terminating anteriorly in a small more or less distinct conelike process; macronucleus
    spherical to oval; body ovoid to short cylindrical; one side convex, the other somewhat flattened
    (Fig. 32 A); 80 µm .... **Lagynophrya**
    Cytopharynx not terminating anteriorly in a small cone; with other characteristics **30**
30. Pharyngeal trichites present (Figs. 32 B–D, K) **31**
    Pharyngeal trichites absent **34**
31. Body elongated (Fig. 32 K) .... **Enchelyodon**
    Body ovoid, globose, or ellipsoidal (Figs. 32 B–D) **32**
32. Trichites long (Fig. 32 B) .... **Pseudoprorodon**
    Trichites shorter (Figs. 32 C, D) **33**
33. Body 80 to 200 µm long (Fig. 37 B) .... **Prorodon**
    Body less than 60 µm long .... **Holophrya**
34. Body asymmetrical, compressed, and flasklike or ovoid (Fig. 30 F) .... **Platyophrya**
    Body radially symmetrical **35**
35. Body elongated (Figs. 32 G–J) **36**
    Body not especially elongated **38**
36. Greatly flattened, lancet-shaped or flask shaped; sometimes ribbonlike (Fig. 32 G).
    Trachelophyllum
    Not greatly flattened **37**
37. With a ringlike constriction around cytostome bearing long cilia; polymorphic and highly contractile
    (Fig. 32 H); 70 to 1000 µm .... **Lacrymaria**
    Without such a ringlike constriction; with a furrowed headlike structure (Fig. 32 J); 50 to 300 µm.
    Chaenea
38. Cytostome entirely or partially fringed with small tentaclelike processes (Figs. 32 E, F) **39**
    Cytostome without such processes **40**
39. Small tentaclelike processes surrounding cytostome; one or more long caudal cilia (Fig. 32 E); 15 to
    30 µm .... **Urotricha**
    Small tentaclelike processes only partially surrounding cytostome; with or without long caudal
    cilium (Fig. 32 F); 30 to 40 µm .... **Plagiocampa**
40. Flask shaped; anterior end obliquely truncate (Fig. 32 L); 40 to 200 µm .... **Enchelys**
    Not flask shaped **41**

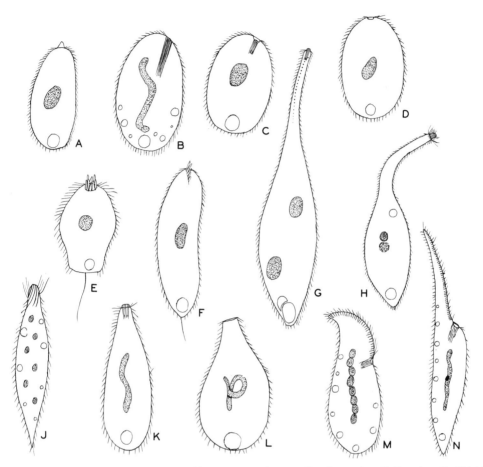

FIG. 32.–Representative Gymnostomatida. A, *Lagynophrya*; B, *Pseudoprorodon*; C, *Prorodon*; D, *Holophrya*; E, *Urotricha*; F, *Plagiocampa*; G, *Trachelophyllum*; H, *Lacrymaria*; J, *Chaenea*; K, *Enchelyodon*; L, *Enchelys*; M, *Branchiocoetes*; N, *Dileptus*.

41. Dorsal side convex; ventral side flat; cytostome small and slitlike (Fig. 30G); up to 60 μm
                                                                           Microregma
    Dorsal side not convex; ventral side not flat; no slitlike cytostome **42**
42. Ciliation uniform (Fig. 32D); 35 μm .... **Holophrya**
    Ciliation coarse and not uniform (Fig. 30H) .... **Pithothorax**
43. Cytostome lateral, slitlike or circular (Figs. 32M, N; 33A–G) **44**
    Cytostome on the flat ventral surface, in anterior half (Figs. 33H–N) **52**
44. Cytostome circular, at base of a long neck (Figs. 32M, N; 33A, B) **45**
    Cytostome a longitudinal cleft (Figs. 33C–G) **48**
45. Anterior end narrow and drawn out **46**
    Anterior end not narrow and drawn out (Fig. 32M); 150 μm; ectocommensals on aquatic isopods
        and amphipods .... **Branchiocoetes**
46. Elongated and narrow (Fig. 32N); 200 to 600 μm; often feeds on small and injured metazoans.
                                                                    Dileptus
    Not especially narrow (Figs. 33A, B) **47**
47. Anterior end in the form of a short fingerlike process; small peristomial field; cytostome in neck
      region (Fig. 33A); 300 μm .... **Trachelius**
    Anterior end drawn out into a long process; wide peristomial field; body broad at level of cytostome
      (Fig. 33B); 100 to 450 μm .... **Paradileptus**

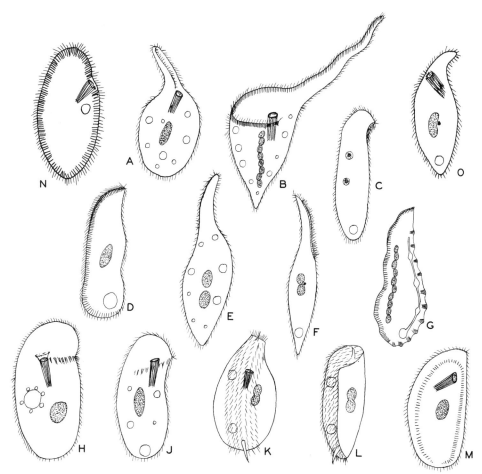

FIG. 33.–Representative Gymnostomatida. A, *Trachelius*; B, *Paradileptus*; C, *Loxodes*; D, *Bryophyllum*; E, *Amphileptus*; F, *Litonotus*; G, *Loxophyllum*; H, *Nassula*; J, *Chilodontopsis*; K, *Trochilia*; L, *Dysteria*; M, *Chlamydodon*; N, *Cyclogramma*; O, *Orthodonella*.

48. Cytostome on convex side of anterior portion of body (Figs. 33 D–G) **49**
    Cytostome on concave side of anterior end of body; lancet shaped; anterior end curved ventrally (Fig. 33 C); 100 to 700 μm .... **Loxodes**
49. Both left and right sides of body ciliated **50**
    Without cilia on the left side **51**
50. With a prominent ventral ridge bearing trichocysts (Fig. 33 D); 130 μm .... **Bryophyllum**
    Flask shaped, without such a ridge (Fig. 33 E); often on other colonial Protozoa; sometimes parasitic on gills of fishes and tadpoles; 40 to 135 μm .... **Amphileptus**
51. Without trichocyst borders; flask shaped, elongated, and flattened; cilia only on right side (Fig. 33 F); 80 to 500 μm .... **Litonotus**
    Ventral side with a trichocyst border; dorsal side with trichocyst border or clumps of trichocysts (Fig. 33 G); shape variable; 100 to 700 μm .... **Loxophyllum**
52. Ciliation complete; dorsal cilia less dense than those on flat ventral surface; 50 to 250 μm **53**
    Ciliation incomplete; dorsal surface without cilia or with only a few sensory bristles **56**
53. Brightly colored because of food vacuoles and symbiotic algae **54**
    Not brightly colored **55**
54. Trichocyst layer indistinct; often swims with a rolling motion; more than 70 μm long (Fig. 33 H).
                                                                                        Nassula
    Heavy trichocyst layer; glides on substrate; less than 70 μm long (Fig. 33 N) ... **Cyclogramma**

55. Cytopharynx nearly longitudinal (Fig. 33J) .... **Chilodontopsis**
    Cytopharynx at an angle (Fig. 33O) .... **Orthodonella**
56. Posterior ventral surface with a spinelike process (Figs. 33K, L) 57
    Posterior end without such a process 58
57. Ciliated portion of ventral surface free (Fig. 33K); 30 μm .... **Trochilia**
    Ciliated portion of ventral surface more or less covered by a fold of the unciliated portion (Fig. 33L);
       35 to 160 μm .... **Dysteria**
58. Variously shaped; dorsal surface without a transverse row of bristles; cytostome more or less
       elongated (Fig. 33M); 100 μm; rare .... **Chlamydodon**
    Ovoid; dorsal surface with a transverse row of bristles; cytostome round (Fig. 4A); very common; 50
       to 300 μm .... **Chilodonella**
59. With a gelatinous lorica (Fig. 34 A); 40 μm .... **Mycterothrix**
    Without a gelatinous lorica 60
60. Ciliation sparse; with two to nine keellike ridges; compressed (Figs. 34 B–D) 61
    Ciliation dense; without keellike ridges 63
61. Cytostome one third to one half of way from anterior end 62
    Cytostome near posterior end (Fig. 34B); 20 to 70 μm .... **Microthorax**
62. Cytopharynx long and tubular (Fig. 34C); 40 μm .... **Leptopharynx**
    Cytopharynx simple and short, near middle of body (Fig. 34D); 40 μm ...... **Drepanomonas**

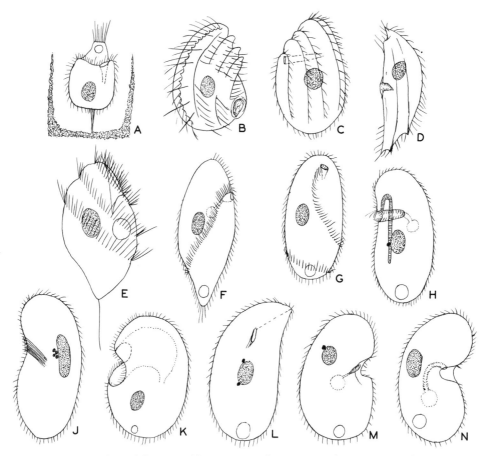

FIG. 34.–Representative Trichostomatida. A, *Mycterothrix*; B, *Microthorax*; C, *Leptopharynx*; D, view of *Drepanomonas*; E, *Trimyema*; F, *Spirozona*; G, *Trichospira*; H, *Plagiopyla*; J, lateral view of *Clathrostoma*; K, *Bresslaua*; L, *Bryophrya*; M, *Colpoda*; N, *Tillina*.

63. With a long cilium at the posterior end; cilia in a few spiral rows (Fig. 34 E); 65 μm; usually in sewage.
        Trimyema
     Without a single long posterior cilium; usually larger **64**
64. With a spiral zone of special cilia extending from cytostome to posterior end (Figs. 34 F, G); 90 μm.
        65
     Without such special cilia **66**
65. Spiral zone of special cilia extending from right anterior to left posterior (Fig. 34 F); spindle shaped;
        usually in sewage .... **Spirozona**
     Spiral zone of special cilia extending from left anterior to right posterior; cylindrical (Fig. 34 G).
        Trichospira
66. Peristome a ciliated cross furrow leading to cytostome; with a curved dorsal band originating in
        peristome (Fig. 34 H); 100 μm .... **Plagiopyla**
     Peristome otherwise **67**
67. Cytostome slitlike and in a small, flat, oval groove bearing a ciliated ridge (Fig. 34 J); 70 to 180 μm.
        Clathrostoma
     Cytostome funnellike and deep **68**
68. Cytopharynx very large, occupying anterior half of animal (Fig. 34 K); 80 to 250 μm; stagnant water.
        Bresslaua
     Cytopharynx not particularly large **69**
69. Ovoid to ellipsoid, anterior end slightly bent (Fig. 34 L); 50 to 120 μm .... **Bryophrya**
     Kidney shaped **70**
70. Cytopharynx short (Fig. 34 M); 20 to 120 μm .... **Colpoda**
     Cytopharynx long and curved (Fig. 34 N); 80 to 400 μm .... **Tillina**
71. Peristome lined with rows of free cilia; cigar or slipper shaped (Fig. 4 E); 60 to 300 μm.
        Paramecium
     Peristome with membrane; with or without free cilia **72**
72. Cytostome not connected with peristome **73**
     Cytostome at end or bottom of peristome **88**
73. Without long caudal cilia **80**
     With one or more long caudal cilia **74**
74. Body constricted in middle (Figs. 35 L, M) **75**
     Body not constricted in middle; with a single large posterior cilium **76**
75. Two ciliary bands; cluster of fused cilia at posterior end (Fig. 35 M); 75 μm.
        Urocentrum turbo (O.F.M.)
     Single ciliary band around anterior half of body; with a single posterior cilium (Fig. 35 L); 30 μm.
        Urozona
76. Body cylindrical, with pluglike ends .... **Balanonema**
     Body not cylindrical; no pluglike ends **77**
77. Body flat and with deep longitudinal furrows (Fig. 35 B) .... **Platynematum**
     Body without deep furrows **78**
78. Cytostome 0.7 to 0.8 as long as body; ventral side concave (Fig. 35 K); 140 μm ... **Lembadion**
     Cytostome shorter **79**
79. Body compressed; with cilia along anterior margin (Fig. 35 O); 40 μm; often in decaying animal
        matter .... **Saprophilus**
     Body not compressed; without cilia along anterior margin (Fig. 35 P); 50 to 200 μm.
        Loxocephalus
80. Body an oblique cone with rounded angles; 100 to 160 μm in diameter (Fig. 35 S) ... **Stokesia**
     Body shaped differently **81**
81. Cytostome opening pointed at anterior end (Fig. 35 A) .... **Frontonia**
     Cytostome opening more or less rounded at anterior end **82**
82. Posterior end bluntly pointed (Fig. 35 R) .... **Disematostoma**
     Posterior end not pointed **83**
83. Cytostome clearly oblique (Figs. 35 C, D) **84**
     Cytostome not clearly oblique (Figs. 35 E, F, H, J) **85**
84. Ovoid or ellipsoid; with seven postoral ciliary meridians (Fig. 35 D); 55 μm .... **Glaucoma**
     Elongate reniform; usually with a single postoral ciliary meridian (Fig. 35 D); 50 to 150 μm.
        Colpidium
85. Cytostome and cytopharynx with a single membranelle (Fig. 35 E); 75 μm ...... **Monochilum**
     Cytostome and cytopharynx with two to four membranelles (Figs. 35 F, H, J) **86**

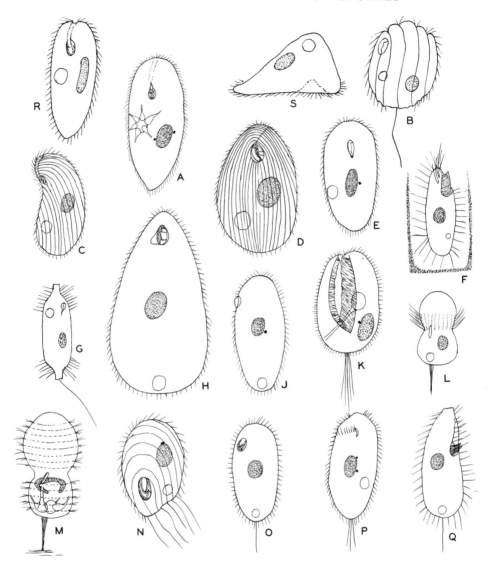

FIG. 35.–Representative Hymenostomatida. A, *Frontonia;* B, *Platynematum;* C, *Colpidium;* D, *Glaucoma;* E, *Monochilium;* F. *Cyrtolophosis;* G, *Balanonema;* H, *Tetrahymena;* J, *Dichilum;* K, *Lembadion;* L, *Urozona;* M, *Urocentrum;* N, *Cinetochilum;* O, *Saprophilus;* P, *Loxocephalus;* Q, *Uronema;* R, *Disematostoma;* S, *Stokesia.*

86. Within a mucilaginous envelope; cytostome with a pocket-forming membrane (Fig. 35F); 30 μm.
      Cyrtolophosis
   Without a mucilaginous envelope **87**
87. Cytostome about 0.1 or less the body length from the anterior end; pyriform or variable body shape; cytostome in median line (Figs. 37 D, 35 H); 40 to 200 μm; often a facultative parasite in aquatic invertebrates .... **Tetrahymena**
   Cytostome about 0.2 the body length from anterior end; ellipsoid; cytostome slightly to left of median line (Fig. 35J); 40 μm .... **Dichilum**
88. Peristome a small sickle-shaped slit; ellipsoid to cylindrical (Fig. 36A); some species parasitic in fresh-water invertebrates; 80 to 500 μm .... **Ophryoglena**
   Peristome long, beginning at anterior end of body **89**

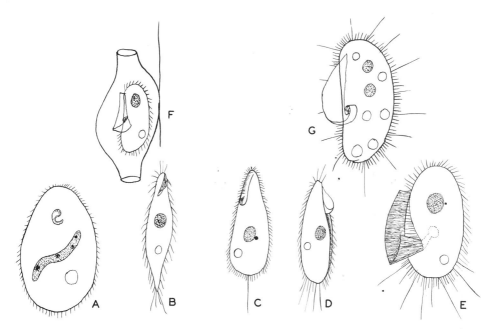

FIG. 36.–Representative Hymenostomatida. A, *Ophryoglena*; B, *Cohnilembus*; C, *Philaster*; D, *Cristigera*; E, *Pleuronema*; F, *Calyptotricha*; G, *Histiobalantium*.

89. Peristome with a prominent membrane forming a pocket surrounding cytostome; with one to
    several stiff posterior cilia (Figs. 36 D–G) 92
    Peristome without such a prominent membrane 90
90. With double undulating membrane on right side of peristomial furrow 91
    Membrane at right of oral furrow not double; 70 to 150 μm (Fig. 36 C) .... **Philaster**
91. Anterior end not ciliated (Fig. 35Q) .... **Uronema**
    Anterior end ciliated (Fig. 36B) .... **Cohnilembus**
92. With a postoral depression or groove; compressed (Fig. 36D); 50 μm .... **Cristigera**
    Without a postoral depression or groove 93
93. In a lorica that is open at both ends; body 50 μm long (Fig. 36F) .... **Calyptotricha**
    No lorica 94
94. Peristome extending for about two thirds of body length (Figs. 36E, G); 70 to 180 μm 95
    Peristome much shorter (Fig. 4C); with refractive pellicle; 20 to 50 μm .... **Cyclidium**
95. One contractile vacuole; long cilia only at posterior end (Fig. 36E) .... **Pleuronema**
    Many contractile vacuoles; long cilia interspersed all over body (Fig. 36G) ... **Histiobalantium**
96. With cilia only, sometimes reduced or absent; rarely with additional small groups of cirruslike
    structures 97
    With cirri only, confined to ventral surface; dorsal surface usually with rows of short bristles.
    Order **HYPOTRICHIDA, 115**
97. Body uniformly ciliated .... Order **HETEROTRICHIDA, 99**
    Ciliation reduced or absent 98
98. Circular in cross section; adoral zone consisting of cirri, bristles, or membranelles and enclosing a
    spiral peristomial field .... Order **OLIGOTRICHIDA, 108**
    Compressed and with a carapace; peristomial field consisting of eight membranelles; anaerobic.
    Order **ODONTOSTOMATIDA, 113**
99. Peristome sunken into the funnellike anterior end (Figs. 38A, B) 100
    Peristome exposed 101
100. Cytopharynx curved to the left (Fig. 38A); 500 to 1000 μm ...... **Bursaria** truncatella O.F.M.
    Cytopharynx curved to the right (Fig. 38B); 60 to 100 μm .... **Bursaridium**

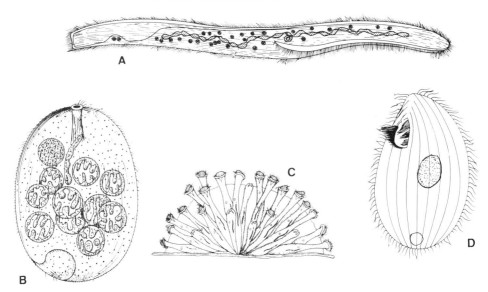

FIG. 37.–Representative ciliates. A, *Spirostomum*; B, *Prorodon,* showing uniform ciliation, cytopharynx, and food vacuoles; C, *Ophrydium,* a colonial peritrich with stalkless individuals imbedded in a secreted gelatinous mass; D, *Tetrahymena,* showing cytostomal organelles. (A and B modified from Grell, 1973; C from Sleigh, 1973; D modified from Corliss, 1959.)

101. With a narrow nonciliated zone to the right of the adoral zone (Figs. 38C–G) **102**
     Without such a nonciliated zone **106**
102. Adoral zone extending diagonally to right and posterior on ventral surface (Figs. 38C, E); anaerobic.
     103
     Adoral zone longitudinal on flat ventral surface and turning to right before cytostome (Figs. 38D, F,
     G) **104**
103. Body shape variable but more or less oblong or fusiform (Fig. 38C); 80 to 300 μm .... **Metopus**
     Bell shaped and with long, pointed posterior end (Fig. 38E); 185 μm .... **Caenomorpha**
104. Cylindrical and very much elongated; large posterior contractile vacuole (Figs. 37A, 38D); 150 to
     3000 μm .... **Spirostomum**
     Not especially elongated; not cylindrical; contractile vacuole smaller **105**
105. With marked grooves on body surface; cilia in cirruslike fused groups; oval (Fig. 38F); 100 μm.
     Phacodinium
     Without grooves; without fused cilia; spindle shaped or ellipsoid (Fig. 38G); 80 to 200 μm.
     Blepharisma
106. With a large undulating membrane; peristomial field not ciliated; ellipsoid (Fig. 38H); 100 to 400
     μm .... **Condylostoma**
     Without a large undulating membrane; peristomial field ciliated **107**
107. Highly contractile, trumpet shaped or cylindrical when extended but oval to pyriform while
     swimming; adoral zone encircling peristome in a spiral (Fig. 38K); 200 to 3000 μm .... **Stentor**
     Flattened and oval; adoral zone otherwise (Fig. 38J); 100 to 300 μm .... **Climacostomum**
108. Oral portion of peristome free on ventral surface; ovoid to spherical (Figs. 39A, B); 35 μm **109**
     Oral region enclosed by adoral zone (Figs. 39C, F) **110**
109. With body bristles or cirri (Fig. 39A) .... **Halteria**
     Without body bristles or cirri (Fig. 39B) .... **Strombidium**
110. Without lorica; pyriform (Fig. 39C); 55 μm .... **Strobilidium**
     With lorica or test, usually covered with debris **111**
111. Lorica divided into collar and basal portions (Fig. 39D); 65 μm .... **Codonella**
     Lorica not so divided **112**
112. Lorica viscous, irregular in form, usually elongated, and translucent (Fig. 39F); 40 to 200 μm.
     Tintinnidium
     Lorica not viscous, thin, bowl shaped (Fig. 39E); 50 μm .... **Tintinnopsis**

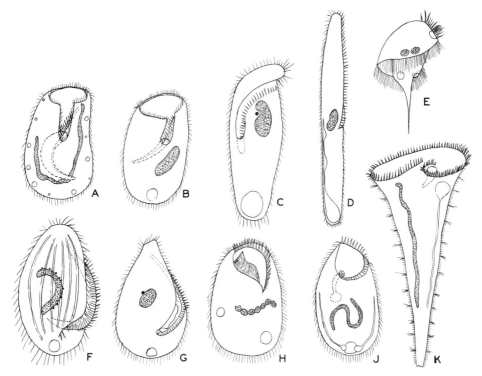

FIG. 38.–Representative Heterotrichida. A. *Bursaria*; B, *Bursaridium*; C, *Metopus*; D, *Spirostomum*; E, *Caenomorpha*; F, *Phacodinium*; G, *Blepharisma*; H, *Condylostoma*; J, *Climacostomum*; K, *Stentor*.

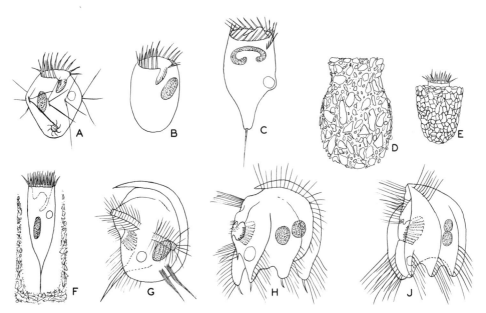

FIG. 39.–Representative Oligotrichida and Odontostomatida. A, *Halteria*; B, *Strombidium*; C, *Strobilidium*; D, lorica of *Codonella*; E, *Tintinnopsis*; F, *Tintinnidium*; G, *Discomorpha*; H, *Saprodinium*; J, *Epalxella*.

113. With frontal cilia; with posterior marginal teeth or ridges (Figs. 39H, J) 114
     Without frontal cilia; oval; without marginal posterior teeth or ridges (Fig. 39G); 80 μm.
     Discomorpha
114. Some posterior teeth spiny (Fig. 39H); 50 μm .... **Saprodinium**
     Posterior teeth not spiny (Fig. 39J); 25 to 80 μm .... **Epalxella**
115. Adoral zone reduced or rudimentary; seven frontoventral cirri (Fig. 40A); 35 μm ... **Aspidisca**
     Adoral zone well developed; usually larger 116
116. With two rows of marginal cirri 117
     Without marginal cirri; ovoid; macronucleus band shaped (Fig. 40 B); 80 to 200 μm ...... **Euplotes**
117. Anal cirri absent (Figs. 40C–E) 118
     Anal cirri present (Figs. 40, 41) 121
118. Ventral and marginal cirri spirally arranged (Figs. 40D, E); sometimes in gelatinous tubes and
     colonial 119
     Ventral and marginal cirri not spirally arranged; elongated, tail like region (Fig. 40C); 175 μm.
     Uroleptus
119. Peristome about one fourth of body length (Fig. 40D); 80 to 250 μm .... **Strongylidium**
     Peristome about one half of body length (Fig. 40E); 50 to 200 μm 120
120. Peristome-bearing part not flexible (Fig. 40E) .... **Stichotricha**
     Peristome-bearing part flexible .... **Chaetospira**

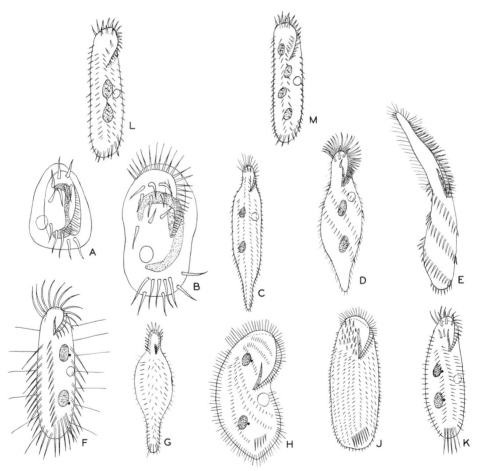

FIG. 40.–Representative Hypotrichida; ventral views. A, *Aspidisca*; B, *Euplotes*; C, *Uroleptus*; D, *Strongylidium*; E, *Stichotricha*; F, *Balladyna*; G, *Epiclintes*; H, *Kerona*; J, *Urostyla*; K, *Holosticha*; L, *Trichotaxis*; M, *Keronopsis*.

121. Frontal field without any special cirri (Figs. 40F, H, L, M) 122
       Frontal field with a few strong cirri (Figs. 40J, K; 41A–G) 126
122. Ellipsoid 123
       Not ellipsoid 124
123. One row of ventral cirri (Fig. 40F); 30 to 80 μm .... **Balladyna**
       Three rows of ventral cirri (Fig. 40L) .... **Trichotaxis**
124. Elongated and spoon shaped (Fig. 40G); free swimming; 375 μm .... **Epiclintes**
       Not spoon shaped 125
125. Reniform (Fig. 40H); commensal or parasitic on hydras; 160 μm ... **Kerona polyporum** Ehr.
       Elongated, with two ventral rows of setae reaching frontal field (Fig. 40M). .... **Keronopsis**
126. Ventral cirri numerous and always arranged in long longitudinal rows (Figs. 40J, K) 127
       Ventral cirri not particularly numerous and at least partly arranged in groups (Figs. 41A–G) 128
127. Ventral cirri in four or more rows (Fig. 40J); 200 to 800 μm .... **Urostyla**
       Ventral cirri in one to three rows (Fig. 40 K); 80 to 350 μm .... **Holosticha**
128. Ventral cirri arranged in one to three distinct longitudinal rows (Figs. 41 A–D) 129
       Ventral cirri not arranged in rows (Figs. 41 E–G) 132
129. With one to three longitudinal and parallel rows of cirri (Figs. 41A, B) 130
       With one or two oblique ventral rows of cirri (Figs. 41C, D) 131
130. With two of the seven anal cirri more posterior (Fig. 41A); 170 to 400 μm ...... **Pleurotricha**
       With a different arrangement of the anal cirri (Fig. 41B); 110 μm .... **Onychodromopsis**
131. With a long row of ventral cirri (Fig. 41C); 150 to 320 μm .... **Gastrostyla**
       With one or two short rows of ventral cirri (Fig. 41D); 120 μm .... **Gonostomum**
132. With 12 to 17 strong frontal cirri; four to eight macronuclei (Fig. 41E); 100 to 300 μm.
                                                                     Onychodromus grandis Stein
       With no more than 10 frontal cirri; 1 to 4 macronuclei (Figs. 41F, G) 133
133. Posterior end taillike (Fig. 41F); 110 to 250 μm .... **Urosoma**
       Posterior end not taillike; 50 to 300 μm 134

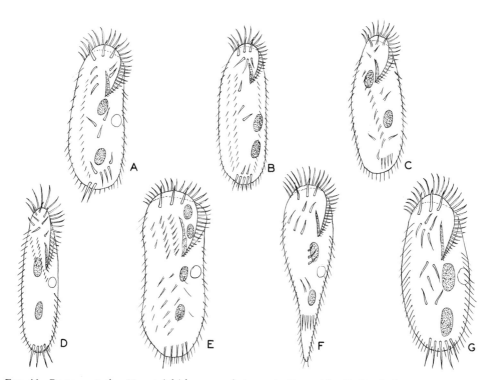

FIG. 41.–Representative Hypotrichida; ventral views. A, *Pleurotricha*; B, *Onychodromopsis*; C, *Gastrostyla*; D, *Gonostomum*; E, *Onychodromus grandis*; F, *Urosoma*; G, *Oxytricha*.

134. Body soft and flexible; caudal cirri short (Fig. 41 G) .... **Oxytricha**
     Body stiff; caudal cirri usually long (Fig. 4 F) .... **Stylonychia**
135. Usually attached; rarely with body cilia .... Order **SESSILIA, 136**
     Free swimming but with highly developed attachment disc on basal (aboral) surface.
                                                         Order **MOBILIA, 154**
136. Without a lorica but sometimes with a gelatinous envelope (Fig. 42) 137
     With a well-developed lorica 148
137. Posterior end usually with one or two short spines; free swimming (Figs. 42A, B) 138
     Posterior end without spines; sessile 139
138. Body with two to four rings of long, conical processes (Fig. 42A); 40 μm; common in rivers.
                                                      Hastatella
     Body without such processes; conical, ends broadly rounded; ring of cilia near aboral end (Fig. 42B);
        160 μm .... **Opisthonecta**
139. With a long, cylindrical contractile neck; colonial; individuals imbedded in common gelatinous
        mass (Figs. 37 C, 42 C, D); 200 to 500 μm; colonies up to 15 cm in diameter .... **Ophrydium**
     Without a necklike region 140
140. Without stalk, with posterior attachment disc (Fig. 42E); attached to debris, or epizoic; 40 to 100 μm.
                                    Scyphidia
     Stalk present 141
141. Stalk not retractile 142
     Stalk retractile; bell shaped; attached to plants and aquatic invertebrates 146
142. Stalk branched; colonial (Figs. 42F–H) 143
     Stalk not branched; attached to vegetation and a wide variety of aquatic invertebrates; 25 to 100 μm.
                                               145

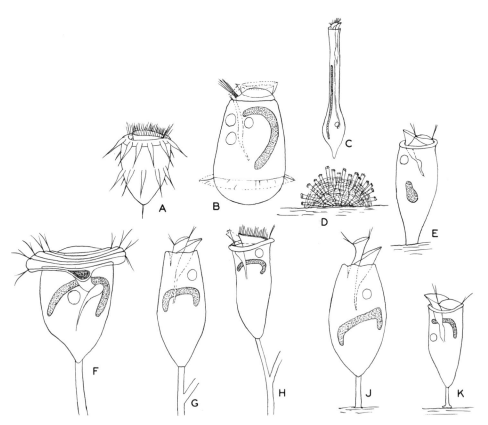

FIG. 42.–Representative Peritrichia. A, *Hastatella*; B, *Opisthonecta*; C, *Ophrydium*; D, very small colony of *Ophrydium*; E, *Scyphidia*; F, *Campanella*; G, *Opercularia*; H, *Epistylis*; J, *Pyxidium*; K, *Rhabdostyla*.

143. Adoral zone with four to six turns (Fig. 42F); 130 to 350 μm .... **Campanella**
    Adoral zone simpler; attached to vegetation and many aquatic invertebrates; 40 to 150 μm **144**
144. Peristome small, separated from peristome border and protrusible through it (Fig. 42G); often in
    polluted waters .... **Opercularia**
    Peristome nearly as wide as body, attached to peristome border at margin (Fig. 42H); colonies often
    macroscopic; common in polluted waters .... **Epistylis**
145. Frontal disc supported by stylelike process (Fig. 42J) .... **Pyxidium**
    Frontal disc not supported by stylelike process (Fig. 42K) .... **Rhabdostyla**
146. Solitary but often gregarious (Fig. 4D); 35 to 160 μm .... **Vorticella**
    Colonial; colonies up to 6 mm long **147**
147. Myonemes not continuous; individual stalks contract singly (Fig. 43A); 50 to 130 μm ... **Carchesium**
    Myonemes continuous; entire colony contracting or expanding simultaneously (Fig. 43B); 40 to 90
    μm .... **Zoothamnium**
148. Distal margin not attached to lorica; with a posterior stalk Figs. 43E–H) **149**
    Distal margin attached to aperture of lorica; unstalked; lorica with flattened attachment surface (Fig.
    43D); often ectocommensal on invertebrates; 75 μm .... **Lagenophrys**
149. Lorica with a stalk (Figs. 43E, F, N) **150**
    Lorica without a stalk (Figs. 43G, H, M) **152**
150. With a discoidal corneous operculum that closes lorica when animal contracts (Figs. 43F, N) ... **151**
    Without an operculum (Fig. 43E); 50 to 100 μm .... **Cothurnia**

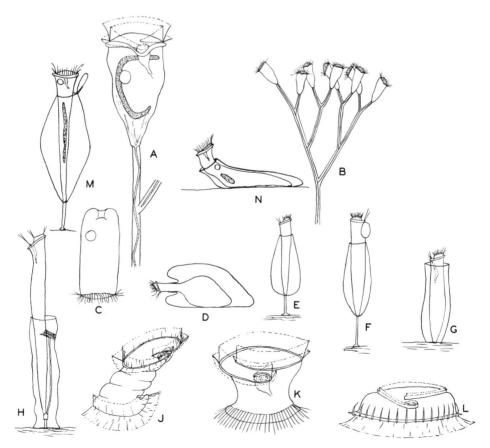

FIG. 43.–Representative Peritrichia. A, *Carchesium*; B, *Zoothamnium* colony; C, free-swimming telotroch stage of a vorticellid; D, *Lagenophrys*; E, *Cothurnia*; F, *Pyxicola*; G, *Vaginicola*; H, *Thuricola*; J, *Urceolaria*; K, *Trichodina*; L, *Cyclochaeta*; M, *Platycola*; N, *Caulicola*.

151. Operculum attached to lorica (Fig. 43N) .... **Caulicola**
    Operculum attached to body (Fig. 43F) .... **Pyxicola**
152. Lorica decumbent (Fig. 43M) .... **Platycola**
    Lorica upright 153
153. With a valvelike apparatus that closes the opening of the lorica when the animal is contracted (Fig.
    43H); 160 to 220 μm .... **Thuricola**
    Without such a valvelike apparatus (Fig. 43G); 50 to 200 μm .... **Vaginicola**
154. Peristome obliquely placed (Fig. 43J); 100 μm; commensal on Turbellaria; feeds on bacteria and
    unicellular forms .... **Urceolaria**
    Peristome not obliquely placed; barrel, bell, or saucer shaped 155
155. Outer row of cilia of attachment disc not bent upward (Fig. 43K); 55 μm; commensal on *Hydra*,
    fishes, and amphibians; sometimes thought to be true parasites on fishes .... **Trichodina**
    Outer row of cilia of attachment disc stiff and bent upward (Fig. 43L); 45 μm; commensal on sponges
    and fishes .... **Cyclochaeta**

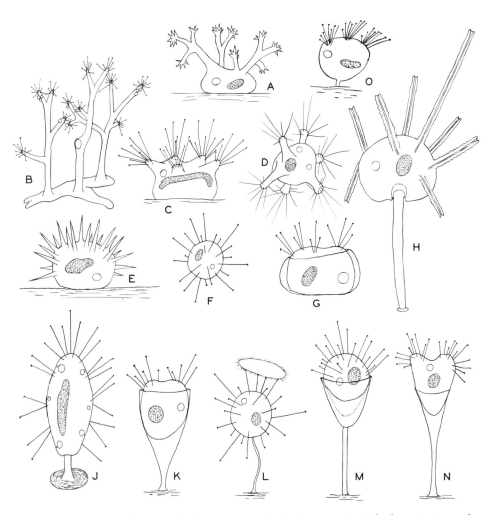

FIG. 44.–Representative Suctoria. A, *Dendrocometes*; B, *Dendrosoma*; C, *Trichophrya*; D, *Staurophrya*; E, *Stylocometes*; F, *Sphaerophrya*; G, *Solenophrya*; H, *Choanophrya*; J, *Discophrya*; K, *Metacineta*; L, *Podophrya* feeding on a ciliate; M, *Paracineta*; N, *Acineta*; O, *Hallezia*.

156. Body asymmetrical or branched 157
     Body more or less symmetrical 161
157. Without special arms 158
     With special arms (Figs. 44A, E); commensal on isopods, amphipods, and so on 160
158. Body dendritic, often large (Fig. 44B); up to 2.5 mm high; on vegetation .... Dendrosoma
     Body not dendritic 159
159. Body not rounded or elongated; tentacles knobbed; nucleus band-shaped (Fig. 44C); 30 to 240 μm; on vegetation and various invertebrates; also reported from gills of largemouth black bass.
     Trichophrya
     Body rounded and with six processes; tentacles not knobbed; nucleus round (Fig. 44D); 50 μm; often in suspension as a plankter .... Staurophrya
160. With branched arms (Fig. 44A); 75 μm .... Dendrocometes
     Arms not branched (Fig. 44E); 90 μm .... Stylocometes
161. Without a stalk 162
     With a stalk of variable length 163
162. Without a lorica; spherical (Fig. 44F); 75 μm; stagnant water .... Sphaerophrya
     With a lorica; tentacles in fascicles (Fig. 44G); 40 μm .... Solenophrya
163. With 5 to 12 tubular expansible tentacles used for engulfing large food particles (Fig. 44H); 65 μm.
     Choanophrya
     With normal type of tentacles 164
164. With a large attachment plate at the base of the stalk; tentacles evenly distributed or in fascicles (Fig. 44J); 70 μm .... Discophrya
     Without such a plate 165
165. Lorica drawn out, funnel shaped, and attached to stalklike lower end (Fig. 44K); 550 μm.
     Metacineta
     Lorica otherwise or absent 166
166. Body spherical to ellipsoidal (Figs. 44L, M) 167
     Body not spherical to ellipsoidal (Figs. 44N, O) 168
167. Without test or cup; tentacles distributed or in fascicles (Fig. 44L); 10 to 50 μm ... Podophrya
     With a close-fitting cup or gelatinous envelope without a visible rim (Fig. 44M); tentacles distributed; 40 μm .... Paracineta
168. Body completely or partly filling the delicate cup-shaped lorica; tentacles in one to three fascicles (Fig. 44N); 30 to 180 μm .... Acineta
     Lorica absent; tentacles in one to four fascicles 169
169. Stalk long (Fig. 6); 55 μm .... Tokophrya
     Stalk short (Fig. 44O) .... Hallezia

REFERENCES

Ahlstrom, E. H. 1937. Studies on variability in the genus Dinobryon. *Trans. Am. Microsc. Soc.* **56**:139–159.

Allegre, C. F., and T. L. Jahn. 1943. A survey of the genus Phacus (Protozoa: Euglenoidina). *Ibid.* **62**:233–244.

Balamuth, W. 1940. Regeneration in protozoa: a problem of morphogenesis. *Q. Rev. Biol.* **15**:290–337.

Bamforth, S. S. 1981. Protist biogeography. *J. Protozool.* **28**:2–9.

Bick, H. 1973. Population dynamics of Protozoa associated with the decay of organic materials in fresh waters. *Am. Zool.* **13**:149–160.

Bick, H., and S. Kunze. 1971. Eine Zusammenstellung von autökologischen und saprobiologischen Befunden an Süsswasserciliaten. *Int. Rev. gesamten. Hydrobiol.* **56**:337–384.

Bird, D., and J. Kalff. 1986. Bacterial grazing by planktonic lake algae. *Science* **231**:493–495.

Borror, A. C. 1972. Revision of the Order Hypotrichida (Ciliophora, Protozoa). *J. Protozool.* **19**:1–23.

Bovee, E. C., and T. L. Jahn. 1965. Mechanisms of movement in taxonomy of Sarcodina. II. The organization of subclasses and orders in relationship to the Classes Autotractea and Hydraulea. *Am. Midl. Nat.* **73**:293–298.

Brown, H. P., and M. M. Jenkins. 1962. A protozoan (Dileptus; Ciliata) predatory upon Metazoa. *Science* **136**:710.

Bütschli, O. 1887–1889. Protozoa. *Kl. Ord. Thier-Reichs* **1**:1–2035.

Cairns, J. 1964. The chemical environment of common fresh-water Protozoa. *Not. Nat.* **365**:1–6.

———. 1971. Factors affecting the number of species in fresh-water protozoan communities. In Cairns, *The structure and function of fresh-water microbial communities.* Virginia Polytechnical Institute State University Press, Blacksburg, Virginia, pp. 219–247.

Cairns, J., G. R. Lanza, and B. C. Parker, 1972. Structural and functional changes in algal and protozoan communities related to pollution. *Proc. Acad. Nat. Sci. Phila.* 124:79–127.

Cairns, J., and J. A. Ruthven. 1972. A test of the cosmopolitan distribution of fresh-water protozoans. *Hydrobiologia* 39:405–427.

Cairns, J., and W. H. Yongue. 1974. Protozoan colonization rates on artificial substrates suspended at different depths. *Trans. Am. Microsc. Soc.* 93:206–210.

Cairns, J., et al. 1969. The relationship of fresh-water protozoan communities to the Macarthur-Wilson equilibrium model. *Am. Nat.* 103:439–454.

Calaway, W. T., and J. B. Lackey. 1962. Waste treatment Protozoa. Flagellata. *Fla. Eng. Ser.* 3:1–140.

Calkins, G. N., and F. M. Summers (eds.). 1941. *Protozoa in biological research.* 1148 pp. Columbia University Press, New York.

Cash, J., J. Hopkinson, and G. H. Wailes. 1905–1921. *The British freshwater Rhizopoda and Heliozoa.* 5 vols. Ray Society, London.

Chen, T.-T. (ed.) 1967–1969. *Research in protozoology.* 3 vols. 1564 pp. Oxford University Press, Oxford, England.

Collin, B. 1912. Études monographique sur les Acinétiéns. *Arch. Zool. Exp. Gen.* 51:1–457.

Conrad, W. 1926. Recherches sur les flagellates de nos eaux saumâtres. II. *Arch. Protist.* 56:167–231.

Corliss, J. O. 1956. Evolution and systematics of ciliates. *Syst. Zool.* 5:68–91.

———. 1956a. On the evolution and systematics of ciliated Protozoa Part II. *Ibid.* 121–140.

———. 1959. An illustrated key to the higher groups of the ciliated Protozoa, with definition of terms. *J. Protozool.* 6:265–281.

———. 1959a. Comments on the systematics and phylogeny of the Protozoa. *Syst. Zool.* 8:169–190.

———. 1974. The changing world of ciliate systematics: historical analysis of past efforts and a newly proposed phylogenetic scheme of classification for the protistan Phylum Ciliophora. *Syst. Zool.* 23:91–137.

———. 1975. Taxonomic characterization of the suprafamilial groups in a revision of recently proposed schemes of classification for the Phylum Ciliophora. *Trans. Am. Microsc. Soc.* 94:224–267.

———. 1977. Annotated assignment of families and genera to the orders and classes currently comprising the Corlissian scheme of higher classification for the Phylum Ciliophora. *Trans. Am. Microsc. Soc.* 96:104–140.

———. 1979. *The ciliated protozoa: Characterization, Classification and Guide to the Literature.* 455 pp. Pergamon Press, New York.

———. 1984. The Kingdom Protista and its 45 phyla. *Biosystems* 17:87–126.

———. 1986. The kingdoms of organisms—from a microscopist's point of view. *Trans. Am. Microsc. Soc.* 105:1–10.

Corliss, J. O., and S. C. Esser. 1974. Comments on the role of the cyst in the life cycle and survival of free-living protozoa. *Ibid.* 93:578–593.

Curds, C. R. 1966. An ecological study of the ciliated Protozoa in activated sludge. *Oikos* 15:282–289.

———. 1969. An illustrated key to the British freshwater ciliated protozoa commonly found in activated sludge. *Water Pollut. Res. Tech. Pap.* 12:1–90.

———. 1973. The role of Protozoa in the activated-sludge process. *Am. Zool.* 13:161–169.

Deflandre, G. 1928. Le genre Arcella Ehrenberg. *Arch. Protistenk.* 64:152–287.

———. 1929. Le genre Centropyxis Sten. *Ibid.* 67:322–375.

Dillon, R. D., G. L. Walsh, and D. A. Bierle. 1968. A preliminary survey of Antarctic meltwater and soil amoebae. *Trans. Am. Microsc. Soc.* 87:486–492.

Doflein, F., and E. Reichenow. 1928–1929. *Lehrbuch der Protozoenkunde.* 5th ed. 1262 pp. Jena.

Doyle, W. L. 1943. The nutrition of the Protozoa. *Biol. Rev.* 18:119–136.

Eddy, S. 1930. The fresh-water armored or thecate dinoflagellates. *Trans. Am. Microsc. Soc.* 49:277–321.

Edmondson, C. H. 1906. The Protozoa of Iowa. *Proc. Davenport Acad. Sci.* 11:1–124.

Fenchel, T. 1986. *Ecology of Protozoa.* 208 pp. Blackwell, Palo Alto, California.

Finlay, B. J., and C. Ochsenbein-Gattlen. 1982. Ecology of free-living Protozoa. *Occas. Publ. Freshwater Biol. Assoc.* 17:1–167.

Finley, H. E. 1930. Toleration of freshwater Protozoa to increased salinity. *Ecology* 11:337–346.

Fjerdingstad, E. J. 1961. Ultrastructure of the collar of the choanoflagellate Codonosiga botrytis (Ehrenb.). *Z. Zellforsch.* 54:499–510.

Fritsch, F. E. 1935. *The structure and reproduction of the algae.* I. 791 pp. Cambridge University Press, Cambridge, England.

———. 1944. Present-day classification of algae. *Bot. Rev.* 10:233–277.

Geiman, Q. M. 1931. Morphological variations in Coleps octospinus. *Trans. Am. Microsc. Soc.* 50:136–143.

Gill, D. E. 1972. Density dependence and population regulation in laboratory cultures of Paramecium. *Ecology* 53:701–708.

Gojdics, M. 1953. *The genus Euglena.* 268 pp. University of Wisconsin Press, Madison.

Goulder, R. 1974. The seasonal and spatial distribution of some benthic ciliated Protozoa in Esthwaite Water. *Freshwater Biol.* 4:127–147.

Greiser, D. 1974. Ökologische Untersuchungen an Ciliaten in einer Modellselbstreinigungsstrecke. *Int. Rev. gesamten Hydrobiol.* 59:543–555.

Grell, K. G. 1973. *Protozoology.* 554 pp. Springer, Berlin.

Grimstone, A. V. 1961. Fine structure and morphogenesis in Protozoa. *Biol. Rev.* 36:97–150.

Gruchy, D. F. 1955. The breeding system and distribution of Tetrahymena pyriformis. *J. Protozool.* 2:178–185.

Hall, R. P. 1939. The trophic nature of the plantlike flagellates. *Q. Rev. Biol.* 14:1–12.

Hamilton, J. M. 1952. Studies on loricate Ciliophora. I. Cothurnia variabilis Kellicott. *Trans. Am. Microsc. Soc.* 71:382–392.

Hänel, K. 1979. Systematics and ecology of colourless flagellates in sewage. *Arch. Protistenk.* 121:73–137.

Henebry, M. S., and B. T. Ridgeway. 1979. Epizoic ciliated Protozoa of planktonic copepods and cladocerans and their possible use as indicators of organic water pollution. *Trans. Am. Microsc. Soc.* **98**:495–508.

Hirschfield, H. I. 1959. The Biology of the Amoeba. *Ann. N.Y. Acad. Sci.* **78**:1–303.

Hollande, A. 1942. Étude cytologique et biologique de quelques flagellés libres. *Arch. Zool. Exp. Gen.* **83** (Suppl.):1–268.

Honigberg, B. M., et al. 1964. A revised classification of the Phylum Protozoa. *J. Protozool.* **11**:7–20.

Ilowaisky, S. M. 1926. Material zum Studium der Cysten der Hypotrichen. *Arch. Protist.* **54**:92.

Jahn, T. L. 1934. Problems of population growth in the Protozoa. *Cold Spring Harbor Symp. Q. Biol.* **2**:167–180.

_____. 1946. The euglenoid flagellates. *Q. Rev. Biol.* **21**:246–274.

Jahn, T. L., and F. F. Jahn. 1949. *How to know the Protozoa.* 234 pp. Brown, Dubuque, Iowa.

Jakus, M. A. 1945. The structure and properties of the trichocysts of Paramecium. *J. Exp. Zool.* **100**:457–476.

Jennings, H. S. 1906. *Behavior of the lower organisms.* 366 pp. Columbia University Press, New York.

Johnson, L. P. 1944. Euglenae of Iowa. *Trans. Am. Microsc. Soc.* **63**:97–135.

Johnson, W. H. 1941. Nutrition in the Protozoa. *Q. Rev. Biol.* **16**:336–348.

_____. 1941a. Populations of ciliates. *Am. Nat.* **75**:438–457.

Kahl, A. 1930–1935. Urtiere oder Protozoa. I: Wimpertiere oder Ciliata (Infusoria). *Tierwelt Deutschlands* **18, 21, 25, 30**:1–886.

Kidder, G. W., and V. C. Dewey. 1945. Studies on the biochemistry of Tetrahymena. III. Strain differences. *Physiol. Zool.* **18**:136–157.

Kimball, R. F. 1943. Mating types in the ciliate Protozoa. *Q. Rev. Biol.* **18**:30–45.

King, R. L., and T. L. Jahn. 1948. Concerning the genera of amebas. *Science* **107**:293–294.

Kitching, J. A. 1938. Contractile vacuoles. *Biol. Rev.* **13**:403–444.

Kofoid, C. A., and O. Swezy. 1921. The free living, unarmored dinoflagellates. *Mem. Univ. Calif.* **5**:1–562.

Kudo, R. R. 1966. *Protozoology.* 5th ed. 1174 pp. Thomas, Springfield, Illinois.

Lackey, J. B. 1925. The fauna of Imhoff tanks. *Bull. N.J. Exp. Sta.* **417**:1–39.

_____. 1932. Oxygen deficiency and sewage protozoa: with descriptions of some new species. *Biol. Bull.* **63**:287–295.

_____. 1936. Some freshwater protozoa with blue chromatophores. *Ibid.* **71**:492–497.

_____. 1938. Protozoan plankton as indicators of pollution in a flowing stream. *Publ. Health Rep. Wash.* **53**:2037–2058.

_____. 1938a. A study of some ecologic factors affecting the distribution of Protozoa. *Ecol. Monogr.* **8**:501–527.

_____. 1940. The microscopic flora and fauna of tree holes. *Ohio J. Sci.* **40**:186–192.

Laybourn-Parry, J. 1984. *A Functional Biology of Free-Living Protozoa.* 224 pp. Croom Helm, London.

Lee, J. J., et al. (eds.). 1985. *An Illustrated Guide to the Protozoa.* 629 pp. Society of Protozoologists, Lawrence, Kansas.

Leedale, G. F. 1967. *Euglenoid flagellates.* 242 pp. Prentice-Hall, New York.

Leidy, J. 1879. Fresh-water rhizopods of North America. *Rep. U.S. Geol. Surv. Terr.* **12**:1–324.

Levine, N. D., et al. 1980. A newly revised classification of the Protozoa. *J. Protozool.* **32**:409–415.

Lowndes, A. G. 1944. The swimming of unicellular flagellate organisms. *Proc. Zool. Soc. London* **113A**:99–107.

Mackinnon, D. L., and R. S. J. Hayes. 1961. *An introduction to the study of Protozoa.* 506 pp. Clarendon Press, Oxford, England.

Manwell, R. D. 1961. *Protozoology.* 642 pp. St. Martin's Press, New York.

Marquardt, W. C., et al. 1966. Preservation of Colpoda steinii and Vahlkampfia sp. from an ice tunnel in Greenland. *Trans. Am. Microsc. Soc.* **85**:152–156.

Mast, S. O. 1925. Structure, movement, locomotion, and stimulation in Amoeba. *J. Morph.* **41**:347–426.

_____. 1947. The food-vacuole in Paramecium. *Biol. Bull.* **92**:31–72.

Mast, S. O., and W. J. Bowen. 1944. The food vacuole in the Peritricha, with special reference to the hydrogen-ion concentration of its contents and of the cytoplasm. *Biol. Bull.* **87**:188–222.

Milliger, L. E., K. W. Stewart, and J. K. G. Silvey, 1971. The passive dispersal of viable algae, protozoans, and fungi by aquatic and terrestrial Coleoptera. *Ann. Entomol. Soc. Am.* **64**:36–45.

Morishita, I. 1976. Protozoa in sewage and waste water treatment systems. *Trans. Am. Microsc. Soc.* **95**:373–377.

Noble, A. E. 1932. On Tokophrya lemnarum Stein (Suctoria) with an account of its budding and conjugation. *Univ. Calif. Publ. Zool.* **37**:477–520.

Noland, L. E. 1925. A review of the genus Coleps with descriptions of two new species. *Trans. Am. Microsc. Soc.* **44**:3–13.

_____. 1925a. Factors influencing the distribution of fresh water ciliates. *Ecology* **6**:437–452.

Noland, L. E., and H. E. Finley. 1931. Studies on the taxonomy of the genus Vorticella. *Trans. Am. Microsc. Soc.* **50**:81–123.

Noland, L. E., and M. Gojdics. 1967. Ecology of free-living Protozoa. *Res. Protozool.* **2**:215–266.

Page, F. C. 1976. An illustrated key to freshwater and soil amoebae. *Sci. Publ. Freshwater Biol. Assoc.* **34**:1–155.

Pascher, A. 1927. Volvocales. *Süsswasserfl. Dtschl.* **4**:1–506.

Penard, E. 1902. *Faune rhizopodique du bassin du Léman.* 714 pp. Geneva, Switzerland.

_____. 1904. *Les héliozoaires d'eau douce.* 341 pp. Geneva, Switzerland.

_____. 1922. *Études sur les infusoires d'eau douce.* 331 pp. Geneva, Switzerland.

Picken, L. E. R. 1937. The structure of some protozoan communities. *J. Ecol.* 25:368–384.

Pochmann, A. 1942. Synopsis der Gattung Phacus. *Arch. Protistenk.* 95:81–252.

Porter, K. G., et al. 1985. Protozoa in planktonic food webs. *J. Protozool.* 32:409–415.

Pratt, J. R., and J. Cairns. 1985. Functional groups in the Protozoa: roles in differing ecosystems. *J. Protozool.* 32:415–423.

Pringsheim, E. G. 1941. The interrelationships of pigmented and colourless Flagellata. *Biol. Rev.* 16:191–204.

_____. 1942. Contributions to our knowledge of saprophytic algae and flagellata. III. Astasia, Distigma, Menoidium, and Rhabdomonas. *New Phytol.* 41:171–205.

_____. 1948. Taxonomic problems in the Euglenineae. *Biol. Rev.* 23:46–61.

Rieder, J. 1936. Beitrag zur Kenntnis der Süsswasser-Suktorien und Revision der Schweizer Suktorien-Fauna. *Rev. Suisse Zool.* 43:359–395.

_____. 1936a. Biologische und ökologische Untersuchungen an Süsswasser-Suktorien. *Arch. Naturgesch.* 5:137–214.

Roux, J. 1901. *Fauna infusorienne des eaux stagnantes des environs de Génève.* 149 pp. Geneva, Switzerland.

Rudzinska, M. A. 1973. Do Suctoria really feed by suction? *BioScience* 23:87–94.

Schaeffer, A. A. 1926. Taxonomy of the amoebas with descriptions of thirty-nine new marine and fresh-water species. *Publ. Carnegie Inst. Wash.* 345:1–116.

Senn, G. 1899. Über einige Coloniebildende einzellige Algen. *Bot. Zeitgeb.* 57:39–104.

Slater, J. V. 1955. Some observations on the cultivation and sterilization of Protozoa. *Trans. Am. Microsc. Soc.* 74:80–85.

Sleigh, M. 1973. *The biology of Protozoa.* 315 pp. Arnold, London.

Small, E. B., et al. 1971. A survey of ciliate surface patterns and organelles as revealed with scanning electron microscopy. *Trans. Am. Microsc. Soc.* 90:283–294.

Smith, G. M. 1920. Phytoplankton of the inland lakes of Wisconsin. Part I. *Bull. Wis. Geol. Nat. Hist. Surv.* 57:1–243.

_____. 1950. *The fresh-water algae of the United States.* 2nd ed. 719 pp. McGraw-Hill, New York.

_____. 1944. A comparative study of the species of Volvox. *Trans. Am. Microsc. Soc.* 63:265–310.

Spoon, D. M., et al. 1976. Observations on the behavior and feeding mechanisms of the suctorian Heliophrya erhardi (Rieder) Matthes preying on Paramecium. *Trans. Am. Microsc. Soc.* 95:443–462.

Stout, J. D. 1956. Reaction of ciliates to environmental factors. *Ecology* 37:178–191.

Stump, A. B. 1935. Observations on the feeding of Difflugia, Pontigulasia, and Lesquereusia. *Biol. Bull.* 69:136–142.

Tartar, V. 1961. *The biology of Stentor.* 413 pp. Pergamon Press, New York.

Unger, W. B. 1931. The protozoan sequence in five plant infusions. *Trans. Am. Microsc. Soc.* 50:144–153.

Valkanov, A. 1940. Die Heliozoen und Proteomyxien. *Arch. Protistenk.* 93:225–254.

Von Brand, T. 1935. Der Stoffwechsel der Protozoen. *Ergeb. Biol.* 12:161–220.

Walton, L. B. 1915. A review of the described species of the Order Euglenoidina Bloch. Class Flagellata (Protozoa) with particular reference to those found in the city water supplies and in other localities in Ohio. *Bull. Ohio Biol. Surv.* 1:341–449.

Wang, C. C. 1928. Ecological studies of the seasonal distribution of Protozoa in a fresh-water pond. *J. Morphol. Physiol.* 46:431–478.

Webb, M. C. 1961. The effects of thermal stratification on the distribution of benthic Protozoa in Esthwaite Water. *J. Anim. Ecol.* 30:137–152.

Wichterman, R. 1953. *The biology of Paramecium.* 527 pp. Blakiston, New York.

Woodruff, L. L. 1912. The origin and sequence of the protozoan fauna of hay infusions. *J. Exp. Zool.* 12:205–264.

Yongue, W. H., and J. Cairns. 1971. Colonization and succession of fresh-water protozoans in polyurethane foam suspended in a small pond in North Carolina. *Not. Nat.* 443:1–13.

3

PORIFERA
(Sponges)

The great majority of the 4500 species of sponges are marine, only a single family, the Spongillidae, being represented in fresh waters. This family consists of about 170 species, of which about 30 have been reported from the United States.

Fresh-water sponges are common in clean ponds, lakes, streams, and rivers. Because they are sessile and because of their inconspicuous green, brown, gray, or yellowish coloration, however, they are frequently unnoticed. Indeed, until their morphology and physiology were first partially understood (beginning in 1825), sponges were often considered as plants. Syngamic reproduction has been known only since 1856. A suitable substrate for the matlike sponge growth may be provided by almost any stable submerged object, including rocks, pebbles, aquatic vegetation, logs, branches, and twigs. Sponges may be found on the upper surfaces, sides, or lower surfaces.

The Phylum Porifera may be divided into three classes, which are distinguished on the basis of their skeletal structure. The Class Calcarea consists of small marine sponges having a skeleton composed of calcium carbonate spicules, which may be one, three, or four rayed. Hexactinellida are the glass sponges, likewise all marine; their skeleton is an openwork structure of siliceous triaxon spicules or some modification of triaxon. In the Demospongea the skeleton may consist of horny fibers (spongin), siliceous spicules (not triaxon), or both. Sponges of commercial importance are in the Order Keratosa of the Demospongea; the skeleton consists entirely of spongin, spicules being absent. In four

other orders there is no spongin, the skeleton being composed entirely of tetraxon spicules. In the three remaining orders, however, the skeleton consists of both monaxon spicules and variable amounts of spongin. One of these orders, the Haplosclerina, contains the Spongillidae. In addition to the skeletal materials already mentioned, it should be emphasized that the noncellular matrix, or mesoglea, of all sponges also functions in support and cohesion of the living tissues.

General characteristics. The size of a single fresh-water sponge is enormously variable, depending on the species, age, and various ecological conditions. Some species, for example, even when mature may consist of a thin, slimy, delicate mat having a surface area of only a few square centimeters. Other species may cover an area of as much as 40 m^2 under highly favorable conditions. The thickness of a sponge growth is likewise variable; sometimes it may be only 1 or 2 mm thick, but under other circumstances it may be as much as 4 cm thick. The growth form may be typically encrusting and matlike, or there may be numerous papillae, fingerlike outgrowths, branches, or an extremely irregular lobed growth form (Figs. 1, 2).

Furthermore, few of our American species are limited to one type of growth form, so that field identification by this character is usually not feasible. *Eunapius fragilis,* for example, may be found as a fine-textured uniform crust; or it may be more robust, with an irregular surface and loose texture; or it may consist of luxuriant branching lobes. *Spongilla*

FIG. 1.–Colonies of *Spongilla lacustris* on twigs, ×0.3.

FIG. 2.–*Ephydatia muelleri.* A, growth form on a branched twig, ×0.3; B, proliferating growth on a single unbranched twig, ×0.8.

lacustris has every variation, from a simple encrustation to tufts of long fingerlike processes, and *Ephydatia muelleri* has a similar range. The more common growth forms for most of our species are indicated in the key that follows. In general, individuals of a particular species are usually more lobed and luxuriant in standing waters than in running waters.

Sponges are often collected in which the apparent thickness is as much as 30 to 60 mm, but if such growths are examined carefully it is usually found that the living portion of the mass forms only an outermost stratum 4 to 20 mm thick, all the underlying material being composed of the dead remains of the previous one or more years' growth.

Coloration of sponges may be produced by several factors. Most species found growing on the upper sides of objects are some shade of green, owing to contained algal cells (see page 96). If these same species occur in deep waters or on the undersides of objects, however, there is little chlorophyll, and the coloration may be some shade of dark brown, tan, gray, flesh color, or reddish (rarely). On the other hand, there are a few species that are seldom or never green, even when growing in bright light. Coloration may be modified by the ingestion of food and inorganic particles of a certain shade or by the presence of pigmented metabolic materials in the tissues.

The sponge body is simply constructed. There are no organs, and the tissues are relatively unspecialized and loosely organized.

FIG. 3.–Young spring sponges on bits of charred wood, ×1.6.

The external surface bears openings of two general sizes. The very abundant microscopic ostia are merely more or less circular openings in the epidermal membrane. The few oscular openings are scattered over the sponge surface. Usually they are 0.5 to 2.0 mm in diameter but may be as wide as 5 to 10 mm. An osculum is typically located at the distal end of a small, chimneylike projection. By means of a carmine suspension it can be easily shown that minute currents of water pass into the sponge through the ostia and out through the oscula.

Internally, the sponge is a maze of interconnected and ramifying spaces, channels, and chambers (Figs. 4, 5). In typical freshwater species the ostia open into an extensive subdermal cavity from which an abundance of minute narrow incurrent canals lead more deeply into the sponge. At their inner ends these incurrent canals open into small, more or less spherical cavities, the flagellated chambers. At the opposite side each flagellated chamber empties into another minute canal, the excurrent canal, which, along with many other excurrent canals, empties into a much larger irregular central cavity. Each central cavity opens to the outside through an osculum. The opening of the incurrent canal into the flagellated chamber is called the prosopyle; the larger opening from the chamber into the excurrent canal is called the apopyle. The entire arrangement of canals and chambers in the Spongillidae is the rhagon type.

Judging from the available literature, there is considerable variation in the structure of the thin, flat, outer epidermis, or pinacocyte layer. In some species it is apparently a syncitium with scattered nuclei, whereas in others there are definite cells (porocytes) with separating membranes. The ostia may be openings within special pore cells, but some workers maintain that the remnants of the pore cells eventually disappear, leaving the ostium as a simple opening in the epidermis.

In contrast to the situation in some marine sponges, the flagellated chambers of the Spongillidae are minute and lined with only a small number of collar cells, or choanocytes. Each choanocyte has a distal collar of microvilli surrounding a single flagellum. Except

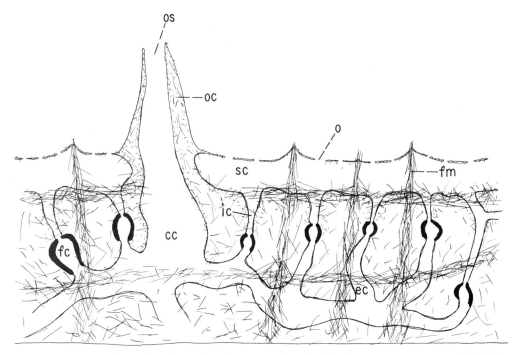

FIG. 4.–Diagrammatic section of peripheral portion of a typical fresh-water sponge. Heavy black lines represent layers of choanocytes lining flagellated chambers. Delicate matrix of spongin not shown. cc, central cavity; ec, excurrent canal; c, flagellated chamber; fm, fascicle of megascleres; ic, incurrent canal; o, ostium; oc, oscular chimney; os, osculum; sc, subdermal cavity.

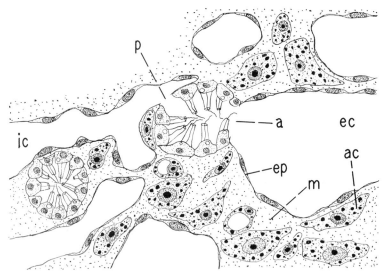

FIG. 5.–Detailed structure of Spongillidae, semidiagrammatic, showing two flagellated chambers. Spicules and delicate spongin matrix omitted. a, apopyle; ac, amoebocyte; ec, exhalent canal; ep, epithelial cell; ic, inhalent canal; m, mesoglea matrix; p, prosopyle. (Modified from Brien, 1936).

for the flagellated chambers, all internal chambers and canals of a sponge are lined with a flat layer variously characterized as a syncitium, semisyncitium, or layer of colloidal material with contained cells.

The space between layers of epithelium is filled with a gelatinous noncellular matrix (mesoglea or mesohyl) containing an abundance of several kinds of mesenchymal amoebocytes that have a wide variety of functions. Some secrete the skeleton; some function in nutrition, storage, transport, and excretion; others function in reproduction. Many amoebocytes are motile and move about slowly in the mesohyl.

Skeleton. Fresh-water sponges contain great numbers of needlelike microscopic spicules that aid in supporting the flimsy tissues. Such spicules are of two general types. The larger spicules, or megascleres (skeletal spicules), form a reticulate network through and in the tissues, and around and through the many cavities and channels. In addition to many megascleres arranged haphazardly, there are loose overlapping fascicles of megascleres radiating outward from the center or base of the sponge and frequently extending through the external epithelium to the outside. There are also transverse fascicles that run between the radiating fascicles. More or less minute amounts of fibrous and veil-like spongin bind the megascleres together.

The smaller type of spicules, or microscleres (flesh or dermal spicules), are most abundant in the peripheral and epidermal portion of the sponge body, although there are also some lining the deeper channels. Microscleres are arranged in a scattered fashion, rarely in fascicles. Microscleres are lacking in some species.

Megascleres and microscleres may be sharp pointed or blunt and with few to many spines of variable size, depending on the species. A few species have spherical, stellate, or birotulate microscleres.

Additional types of spicules are found associated with the special reproductive bodies, or gemmules, and these are discussed on page 97.

Unfortunately, the details of spicular structure are not always invariable taxonomic characters, especially because ecomorphic malformations are frequent in adverse habitats. Most commonly, the microspines of spicule shafts are reduced or absent in acid waters.

The siliceous material of sponge spicules is often said to have the composition of "hydrated silica" or "opal." Analyses show about 92% silicon dioxide, 7% water, and small amounts of magnesium, potassium, and sodium oxides.

Small and simple spicules are formed within single specialized amoebocytes, called silicoblasts, but large and complex spicules are formed by two or more silicoblasts. The first sign of a spicule is an extremely thin organic axial thread within the silicoblast, the siliceous material being deposited around it in every-increasing thickness.

Nutrition. Fundamentally, a sponge efficiently strains minute particles from the water in which it lives. Such particulates usually range from colloidal to 50 μm in diameter. The flagella of the choanocytes are directed more or less toward the apopyles of the flagellated chambers. Their spiral or undulating movements collectively produce weak water currents that enter the sponge through the ostia and leave through the oscula. As the water traverses the flagellated chambers, bacteria and bits of detritus are caught on the outer sticky surface of the collars of the collar cells as well as on the exposed surface of the cells proper. The collar has been likened to a thin pseudopodial membrane. All digestion is intracellular, but little or no digestion occurs within the choanocytes. Food particles are passed on to the adjacent mesoglea and into the amoebocytes where digestion is completed and where there may be considerable storage.

Prosopyles are so small that they frequently become plugged, and in such cases the particles are carried off to one side by slow "flowing" movements of the epidermal protoplasm. They may later sink below the surface and be taken up by amoebocytes. Various laboratory investigations have shown filtration rates of 0.01 to 0.28 ml of water per second per cubic centimeters of sponge tissue. Clearance rates are in the magnitude of 0.05 cm³ of food particles per second per gram of sponge tissue.

It is now thought that the inter- and intracellular "zoochlorellae" of Spongillidae are not essential to sponge metabolism, but there is some evidence to show that their presence results in better sponge growth. Such algae are certain ingested plankton species, including *Pleurococcus, Chlorella,* and other minute genera, that persist in the tissues, multiply, and are digested very slowly. In dim light or darkness they are digested much more rapidly than in bright light.

Circulation. There is no special circulatory mechanism in sponges. Materials are transferred by diffusion and amoebocytic transport.

Respiration. The exchange of oxygen and carbon dioxide occurs everywhere in exposed tissues of sponges, both externally and internally. Diffusion of these dissolved gases is probably rapid and efficient, since all sponge cells are either directly in contact with water or only a few cells distant from it.

Excretion. Dissolved metabolic waste materials simply diffuse into the water surrounding and passing through the sponge body. Granular metabolic materials are usually given off from amoebocytes into the excurrent canals but may also be retained by the cells for varying periods of time.

Choanocytes and many amoebocytes sometimes contain one or two contractile vacuoles, and it is probable that they function mainly in keeping the proper osmotic pressure in the tissues.

Contractile elements, impulse conduction. Although sponges are completely sessile and there is no definite muscle system, a few specialized amoebocytes, or myocytes, have the ability to contract or relax. A few bipolar and multipolar "nervous type" cells have been described by some investigators, but nothing is known about their role in the transmission of impulses.

The oscular chimneys are normally kept open by the excurrents, but if environmental conditions become unfavorable, the oscula will slowly close in several minutes. Such contraction is produced by the action of fusiform myocytes around the openings. In *Ephydatia fluviatilis* it has been shown that even under normal circumstances the oscular chimneys (1 to 3 mm high) continuously and slowly change in form, from high to low and wide to narrow. The osculum will close and the chimney will collapse in several seconds if the latter is struck a sharp blow with a needle. Complete recovery takes 20 to 30 seconds. Sticking a needle into the tissues 2 mm from the chimney base, however, has no effect on the chimney.

Some workers report that under unfavorable conditions ostia will slowly close as a result of contractility of the surrounding protoplasm, but there are no special sensory or nerve cells, and there is no coordination and no transmission of impulses (or perhaps transmission for 2 to 3 mm at most).

Reproduction, growth. Many details of syngamic reproduction are still obscure and confused. Some Spongillidae are hermaphroditic, and a single individual may produce eggs and sperm simultaneously, but in others the sexes are separate, or an individual may produce sperm at one time and eggs at another. Perhaps both conditions prevail, depending on the species. One investigator has found that *Spongilla lacustris* may exhibit alternative hermaphroditism, in which a sponge may be exclusively male one year and exclusively female the next year.

Sperm and egg production are thought to occur mostly during July and August, and to a lesser extent in June and September, but in a few species reproductive activity may be year round. The oogonia are amoebocytes that grow by absorbing and ingesting many adjacent cells. By the time the egg is mature it is large and lies loosely in one of the internal cavities of the sponge.

Although earlier investigators thought that sperm developed from amoebocytes, more recent work has shown that in at least two species they develop from choanocytes that lose their collars and flagella, become more or less amoeboid, migrate into the mesoglea, and function as spermatogonia. Mature flagellated sperm are formed in abundance in baglike "follicles." These sperm are released into the chambers of the sponge and out through the oscula into the surrounding water. Some of them are eventually drawn into the chambers of the same or other sponges where they fertilize the mature ova.

Several different and peculiar types of early embryology occur in the Spongillidae, but in all of them a flagellated embryo is formed that is released from the parent and swims about for a few hours or days before settling to the substrate and metamorphosing to form a minute young sponge (Fig. 3).

Growth is mainly a process of proliferation and accretion with accompanying development of additional orifices, channels, and chambers. Usually most of the sponge body dies or becmes dormant in the autumn and partly disintegrates during the winter, and although it sometimes does not revive in the following spring, it is more common for the surviving remnant to revive and continue growth during the following season. In this way a specimen may remain more or less intact from year to year, growing in thickness, but with only the outermost layer being alive and active. This persistence of sponges occurs especially on firm, broad substrates and where there is little ice action.

Many times, however, a sponge may remain active and green throughout the winter, even under ice.

Reduction bodies. Under highly unfavorable conditions a large sponge forms numerous "reduction bodies," and a small sponge may form a single reduction body. The tissues shrink into the skeleton and the canal system disappears, leaving the epidermis surrounding a mixture of mesenchyme cells, mesoglea, choanocytes, and spicules. Under favorable conditions such a reduction body may grow and regenerate a new sponge. Reduction bodies have been observed for only a few species, and it is not known whether they are produced regularly throughout the Spongillidae. Reduction bodies may be produced in the laboratory by placing the sponge in "tap water," where presumably there are toxic ions.

Regeneration. Like certain marine species, the Spongillidae have remarkable powers of regeneration. If bits of *Ephydatia* or *Spongilla* are macerated and dissociated by squeezing through a fine mesh net, small clumps of cells will coalesce within a few hours, slowly reconstitute, and within several days grow into a new minute sponge.

Gemmules. The formation of gemmules is one of the most characteristic features of fresh-water sponges. These are highly resistant resting stages similar in function to the statoblasts of fresh-water bryozoans. Briefly, a gemmule is a spherical structure with a multiple, dead, secreted outer layer, a covering of spicules, and an internal mass of undifferentiated mesenchymal amoebocytes. Gemmule diameter ranges from about 150 to 1000 μm. Coloration may be whitish or various shades of yellow or brown.

Gemmules are known to be produced by only about 10 marine species, but they are quite simple and have only a single thick enveloping membrane instead of the complex structure found in fresh-water species.

Formerly it was thought that gemmules appeared only at the close of the growing season, but it is now known that they may be formed at any time during the growing season. *Spongilla lacustris* is the only common species that regularly forms gemmules in September and October. In this species, and perhaps in others also, the control of gemmulation is influenced by endogenous factors during the summer and then triggered by environmental factors in the autumn. Because of autumn disintegration of superficial tissues, it is much easier to observe gemmules at that time.

Anheteromeyenia ryderi and *Heteromeyenia tubisperma* gemmules undergo diapause which is broken by low temperatures for one to several months.

Gemmules may be deposited in a compact basal layer in a sponge, or they may be scattered throughout the sponge body. The first sign of gemmule development is the grouping of many food-filled amoebocytes into discrete masses. A thin inner sclerotized membrane is formed around each mass of amoebocytes, followed by a heavier, middle

FIG. 6.–Disintegrated winter appearance of sponges. A, crust of gemmules on bit of bark, ×1; B, gemmules of *Spongilla lacustris* on rock, ×3. (B from K. Berg, 1948, by permission.)

(or outer) sclerotized membrane that has a single small foraminal area. The latter sometimes has a narrow, projecting foraminal tubule. Except for the location of the foraminal aperture or foraminal tubule, the gemmule is covered with a complete, crustlike, thick, honeycombed coating, or pneumatic layer. Sometimes a third thin, outer sclerotized membrane covers the pneumatic layer. *Spongilla lacustris* produces two distinctly different kinds of gemmules, one small, green, and thin walled, and the other large, brown, and thick walled. In some species many more or less cylindrical spicules are imbedded in the pneumatic layer and cover its outer surface. Such gemmule spicules are arranged tangen-

tially or in a heterogeneous manner; they differ markedly from the megascleres and microscleres. In other species, however, the pneumatic layer contains an abundance of radially arranged birotulate spicules, or amphidiscs (also called gemmoscleres). One of the wheellike ends of a birotulate is called a rotule.

Gemmules may remain attached to the substrate or become freed and rise to the surface or sink to the bottom. They winter over easily, withstand repeated freezings and thawings, and have been found to be viable after being kept dried for one to three years or more.

Germination may occur any time at water

temperatures of about 13 to 23°C, but the process requires 2 weeks or more in the autumn and only a few days in the spring. Some investigators maintain that gemmules require a quiescent dormant period of one or more cold months before they will germinate. When a gemmule germinates the membrane covering the foramen disappears, and the inner creamy mass of amoebocytes slowly flows out of the foraminal aperture and more or less covers the empty gemmule shell. This process is usually completed in 2 to 6 days. The amoebocytes undergo histological differentiation and form the first cells of the several tissue types. Some spicules, channels, and subdermal spaces may be formed as early as the third day, and choanocytes may be functioning as early as the fifth day. Further development and growth result in typical sponge structure.

Gemmules commonly contain zoochlorellae, and it has been found that such zoochlorellae, in both light and dark, initiate gemmule germination at lower temperatures and increase the speed of germination.

It is usually difficult to tell whether renewed growth of an old sponge in the spring is initiated by gemmule germination or by resumption of growth by dormant tissues, but probably both mechanisms are responsible.

Ecology. Sponges are most common in waters less than 2 m deep. They are seldom found as deep as 4 m although there are records of collections as deep as 50 m. In very favorable habitats sponge growths may almost obscure all suitable substrates. They are rarely found on a mud bottom, in silty water, or in small puddles. They are seldom found in extremely rapid waters. In general, individuals in running waters are smaller and more encrusting, and show little or no branching. *Spongilla lacustris* may often be found in both the branching and encrusting forms in the same body of water, but it is thought that the encrusting form is merely the earlier growth form that becomes branching as the sponge matures. Compared with most groups of fresh-water invertebrates, sponges are quite sensitive to variations in environmental conditions. *Ephydatia fluviatilis* will tolerate a minor

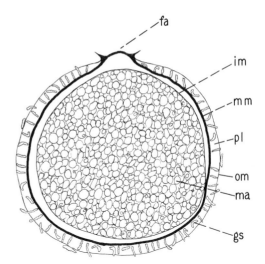

FIG. 7.–Diagrammatic section of a sponge gemmule, ×100, fa, foraminal area; gs, gemmule spicule; im, inner sclerotized membrane; ma, mesenchymal amoebocytes; mm, middle sclerotized membrane; om, outer sclerotized membrane (absent in some species); pl, pneumatic layer (containing spicules).

amount of pollution, and a few other species a lesser degree.

Most species are more or less photopositive. *Eunapius ingloviformis, Trochospongilla pennsylvanica,* and *Ephydatia fluviatilis,* however, prefer dim light and are usually on the undersides of objects in clear water.

Much of our detailed information about the effects of dissolved inorganic materials has been provided by Jewell (1935, 1939). The most important limiting factors seem to be calcium, silicon, bound carbon dioxide, and pH (the last as a rough measurement of the general suitability of water). The following species are acid (soft) water forms seldom occurring where the pH is above 7.0 or 7.3: *Heteromeyenia baileyi, Anheteromeyenia argyrosperma, Trochospongilla pennsylvanica, Eunapius ingloviformis,* and *Corvomeyenia everetti.* Alkaline species include *Heteromeyenia tubisperma, Ephydatia muelleri,* and *E. fluviatilis.* A few species, such as *Spongilla lacustris* and *Eunapius fragilis,* tolerate a wide range in bound carbon dioxide and hydrogen-ion concentration, although they fare better under alkaline conditions.

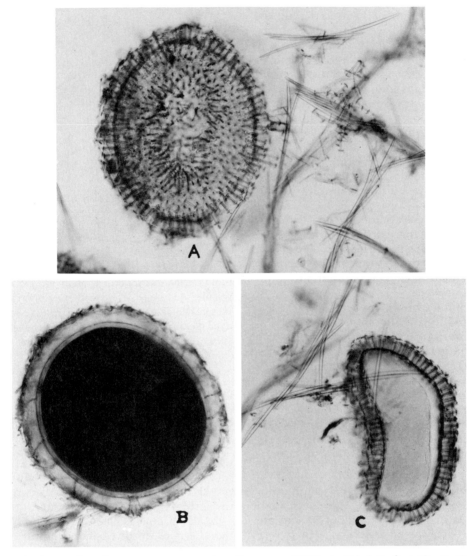

FIG. 8.–Gemmules of some common species of sponges, ×130. A, *Ephydatia fluviatilis*; B, *Spongilla lacustris*; C, *Heteromeyenia baileyi*.

All common species regularly occur in waters having less than 12.0 ppm of calcium. *Spongilla lacustris,* however, has been found where the calcium concentration was as high as 33.0 ppm, and *Heteromeyenia baileyi* where there was 53.4 ppm. The distribution of *Ephydatia muelleri* and *Eunapius fragilis* appears to be restricted by low calcium. The former has not been reported from waters containing less than 5.6 ppm and the latter 2.1 ppm. *Eunapius ingloviformis* and *Corvomeyenia everetti,*

on the other hand, appear to grow only where there are small quantities of calcium present; the former is restricted to waters having less than 3.2 ppm of calcium, and the latter 2.5 ppm. *Spongilla lacustris* is independent of ordinary calcium concentrations and grows luxuriantly in waters containing from 0.16 to 32.4 ppm.

In view of the fact that sponges require large amounts of silicon for spicule formation, it is surprising to find that most species attain good growth in waters that may contain only a trace of silicon. The only important exception is *Ephydatia muelleri,* which requires a minimum of 1.6 ppm of silicon dioxide. Nevertheless, some species show certain variations in spicule abundance and morphology, depending on the amount of silicon present. *Eunapius fragilis* has a poorly developed skeleton in water containing less than 1.0 ppm of silicon dioxide. In *Spongilla lacustris* both microscleres and megascleres become progressively attenuated in specimens collected in waters ranging from 13.0 ppm of silicon dioxide down to only a trace. In fact, a concentration of 0.56 ppm is the lowest value at which there is full microsclere development; there is no greater development above 0.56, but the microspines are absent below this concentration. In soft-water lakes this species has an abundance of spicules intermediate in size between megascleres and microscleres. *Trochospongilla pennsylvanica* has progressively heavier spicules and closer approximation in the size of the two rotules in lakes of increasing mineralization.

It is common to find three different species in the same lake, less common to find four, and rare to find five or six. Often there are definite associations of certain species, suggesting similar but obscure habitat requirements. The following pairs of species are commonly associated in the same lake or stream: *Ephydatia muelleri* and *Eunapius fragilis, Ephydatia muelleri* and *Spongilla lacustris, Eunapius fragilis* and *Spongilla lacustris, S. lacustris* and *Trochospongilla pennsylvanica.*

Larvae of spongilla flies (Sisyridae) pierce the cells and suck out the contents and are entirely dependent on sponges as a food source. A few additional insect larvae also feed on sponge tissues, and it is now known that crayfish (especially *Orconectes*) feed on sponges.

The irregular growth form of many sponges affords a favorable substrate for a wide variety of metazoans, including insects, crustaceans, water mites, annelids, nematodes, and even mollusks.

Spongillidae have occasionally been reported to clog water pipes and conduits, but otherwise they are of no economic importance.

Dispersal, geographic distribution. Sponges easily spread throughout single drainage systems, and by virtue of their resistant gemmules they are probably transported overland by wind, insects, and birds.

Spongilla lacustris, Eunapius fragilis, and *Ephydatia fluviatilis* are essentially cosmopolitan and have been found the world over wherever intensive collecting has been done in suitable areas. *Trochospongilla pennsylvanica, Anheteromeyenia ryderi, Radiospongilla crateriformis,* and probably several other species are widely distributed in the Northern Hemisphere. The four most common American species are *Spongilla lacustris, Eunapius fragilis, Ephydatia muelleri,* and *Heteromeyenia tubisperma.*

Certain other species, however, are rare and poorly known in the United States. *Spongilla wagneri, Ephydatia subdivisa, E. millsi,* and *E. robusta* are each known from only a single or several localities. Within the past 20 to 40 years several U.S. species have become extinct, chiefly because of pollution.

Collecting. Hand picking in the shallows and using a long-handled rake in deeper waters will produce an abundance of specimens. Sponges should be handled gingerly because of their fragility.

Culturing. Small *Ephydatia* and *Spongilla* have been kept alive in aquaria up to 5 or 6 months, but these are exceptionally long periods. Usually sponges remain in good condition for only 1 to 3 weeks, then begin to lose color and disintegrate. Deaths occurs in such short periods even if there is a continuous flow of water through the aquarium.

Some investigators have had success using a dead bacterial suspension for food.

If a small sponge or a bit of a larger one (about a cubic centimeter) is placed cut surface downward on a glass slide submerged under 5 to 10 cm of water, the sponge will often adhere and grow for a time, especially if the water is changed frequently. Never use tap water.

Germination of gemmules and early growth may be easily studied in the laboratory. Gemmules should be brought in and placed on slides or cover slips in shallow pans of water from the natural habitat. Germination will usually begin in 7 to 25 days, but sometimes the process may be hastened by first "vernalizing" the gemmules in water in a refrigerator at 5 to 10°C for a day or two. After germination is completed and the young sponge securely attached, the slide should be removed to an aquarium in which the water is frequently renewed. Under favorable conditions the sponge may grow and cover as much as half of the slide and live for a month or two.

Preserving, preparing. If a whole large specimen is to be preserved, it should be removed from the water with or without its substrate and placed in a warm, shaded place to dry out completely. Incidentally, anyone who works with fresh or drying sponges never forgets their peculiar odor, which is sometimes likened to that of garlic. Dried specimens must be stored and handled very carefully because of their fragility. Soft tissue paper, not cotton, should be used for shipping. Do not rub the eyes after handling dry sponges. Bits of sponge tissue and gemmules should be killed and preserved in 70% alcohol.

For rapid identification place a small bit of sponge on a slide, add two drops of concentrated nitric acid, heat over a flame until dry, and add mounting medium and cover slip. This treatment destroys essentially everything except the spicules. A fragment of dry epidermis mounted in balsam will usually show the presence or absence of microscleres.

Permanent mounts require more care, and the following method is one of several suitable ones. Place 4 or 5 cm³ of sponge in a small test tube, add 10 ml of concentrated nitric acid, and heat to boiling. Let stand for 1 day, remove most of the supernatant acid with a pipette, and *carefully* add water to half fill the test tube. Let stand for 1 hour with occasional agitation and draw off most of the supernatant. Repeat this washing process twice more. Replace the water with 95% alcohol and two changes of absolute alcohol (15 minutes each). Draw up some of the spicules and alcohol in a pipette and place on a slide, burn off the alcohol and add mounting medium and a cover slip.

Permanent mounts of gemmules are made by a slightly different technique. Place the gemmules in cold concentrated nitric acid in a test tube for 1 to 6 hours until they turn a translucent orange or yellow color, indicating that most of the pneumatic layer is dissolved. Then wash with water several times, dehydrate, and mount as indicated in the preceding paragraph.

Glycerin jelly mounts are not suitable because the refractive index of this substance does not result in sufficient differentiation of the sponge spicules.

Taxonomy. American fresh-water sponges have been studied by only a few investigators, some of the more important early contributors being Potts, Smith, Old, Penney, and Jewell. In recent years Harrison and his colleagues have been especially productive. Much remains to be done, however, especially from taxonomic, distributional, physiological, and ecological viewpoints.

Since general growth form is usually so variable within a species, it is a poor character for species identification. The detailed anatomy of gemmules, gemmule spicules, megascleres, and microscleres is essential for accurate identification.

There is still considerable disagreement among specialists concerning dividing lines between genera. Indeed, there is evidence of intergeneric or "parazoan" species mixtures in some forms. Such intergeneric hybrids have included a report of a sponge mass consisting of both *Anheteromeyenia* and *Eunapius* spicules, as well as the distinctive gemmules of both, with no intermediate structures. In this manual the nomenclature of Penney and Racek (1968) and others has been followed for most species.

Rough average or typical lengths of spicules are given in the key that follows, and it should be borne in mind that in most species these figures are subject to a variation of plus or minus 20 to 50%. It should also be noted that the various spicule types for a particular species are not drawn to the same scale in Figs. 9 to 13. Furthermore, the spicule figures represent only typical specimens, and it must be emphasized that there is considerable variation even in the same sponge or same gemmule.

KEY TO SPECIES OF PORIFERA

1. Gemmule spicules elongated and more or less cylindrical, or gemmule spicules absent 2
   Gemmule spicules always present and more or less birotulate (amphidisks) 9
2. Microscleres present 3
   Microscleres absent .... **Eunapius, 8**
3. Gemmules cylindrical; foramen only slightly elevated, not tubular; with a well-developed pneumatic layer in the gemmule coat; animal lead gray; known only from Osceola County, FL.
   Stratospongilla penneyi Harrison
   Gemmules not cylindrical; growth form a thin crust .... **Spongilla, 4**
4. Microscleres smooth and small (Fig. 9G); an encrusting species reported only from VA and NJ.
   Spongilla aspinosa Potts
   Microscleres spined 5
5. Gemmule spicules not curved; southeastern coastal areas, sometimes in brackish waters; growth form a thin crust .... **Spongilla wagneri** Potts
   Gemmule spicules slightly to strongly curved (Fig. 9F) 6
6. Microscleres with blunt tips (difficult to see with regular light microscope); rare; known only from slightly brackish waters in FL and LA .... **Spongilla alba** Carter
   Microscleres with sharp tips (difficult to see) 7

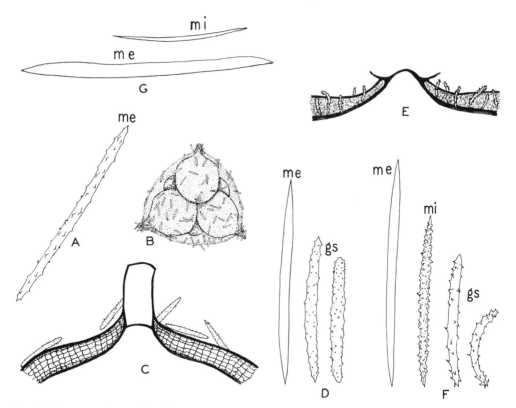

FIG. 9.—Structure of *Spongilla* and *Eunapius*. A, megasclere of *Eunapius ingloviformis*; B, group of gemmules of *E. fragilis* invested in a common pneumatic layer, ×50; C, section of foraminal region of gemmule of *E. fragilis*; D, typical spicules of *E. fragilis*; E, section of foraminal region of gemmule of *Spongilla lacustris*; F, typical spicules of *S. lacustris*; G, typical spicules of *S. aspinosa*. gs, gemmule spicules; me, megascleres; mi, microscleres. (B, C, and E modified from Sasaki, 1934.)

7. Microscleres with burrlike complex spines, especially in the middle third of the microscleres; rare in limestone habitats .... **Spongilla cenota** Penney and Racek
   Microscleres not burrlike; branched or unbranched growth form; in both lotic and lentic habitats; both green and white morphs, depending on exposure to light; common and very widely distributed .... **Spongilla lacustris** (L.)
8. Gemmules forming either a distinct pavement layer or present in clusters of two to eight scattered throughout the sponge body (Fig. 9B); growth form variable; common and widely distributed in running and standing waters .... **Eunapius fragilis** (Leidy)
   Gemmules not forming a distinct pavement layer, in compact hemispherical groups of 7 to 20; soft algalike encrusting growth form; often in highly colored standing waters; generally distributed but not common .... **Eunapius ingloviformis** (Potts)
9. Gemmules with one kind of birotulate **10**
   Gemmules with two types of birotulates, one long and the other short (Fig. 13) **21**
10. Margins of rotules entire, not serrate; birotulates short (Fig. 10); encrusting growth form; running or standing waters .... **Trochospongilla, 11**
    Margins of rotules serrated or incised; birotulates short to long (Figs. 11A–C) **13**
11. Rotules of gemmule spicules unequal in size, the proximal being much the larger; megascleres long, abundantly spined and ends rounded or pointed (Fig. 10A); slimy and thin growth form, seldom more than 5 mm thick; gray or flesh colored; generally distributed.
    Trochospongilla pennsylvanica (Potts)
    Rotules of gemmule spicules equal in size (Figs. 10B, C) **12**
12. Megascleres not spined (Fig. 10B); gemmules formed in a basal layer; color grayish or drab; rare, reported from NJ, PA, OH, KY, IL, LA, and WV ..... **Trochospongilla leidyi** (Bowerbank)
    Megascleres spined (Fig. 10C); gray, yellow, or brown color; cold-temperate states.
    Trochospongilla horrida Weltner
13. Birotulate microscleres present **14**
    Microscleres absent; if present then never birotulate **16**
14. With minute, slender, short, birotulate microscleres about 10 μm long (Fig. 11A) **15**
    Birotulate microscleres robust; growth form thin and encrusting, only 1 to 3 mm thick; dark gray to blackish coloration; known only from LA .... **Corvospongilla becki** Poirrier
15. With about 84% of microsclere shafts slightly to strongly curved; known only from a small pond in SC.
    Corvomeyenia carolinensis Harrison
    With about 34% of microsclere shafts slightly curved; skeleton poorly developed; finely branched growth form; common in bogs in northeastern states .... **Corvomeyenia everetti** (Mills)
16. Birotulates many times longer than diameter of rotules (Fig. 11B); gemmules small, white, numerous; thin, encrusting growth form, usually in running waters; generally distributed but not common.
    Radiospongilla crateriformis (Potts)
    Birotulates no more than three times as long as the diameter of rotules (Figs. 11C, E, F).
    Ephydatia, 17
17. Rays and spines of gemmule birotulates covered with microspines; megascleres smooth or microspined; massive, encrusting growth form; known only from FL and LA.
    Ephydatia subdivisa Potts
    Rays and spines of gemmule birotulates without microspines **18**
18. Margins of rotules finely serrate; megascleres entirely microspined; reported only from Sherwood Lake near De Land, FL .... **Ephydatia millsi** Potts
    Margins of rotules coarsely toothed (Figs. 11C, E, F) **19**

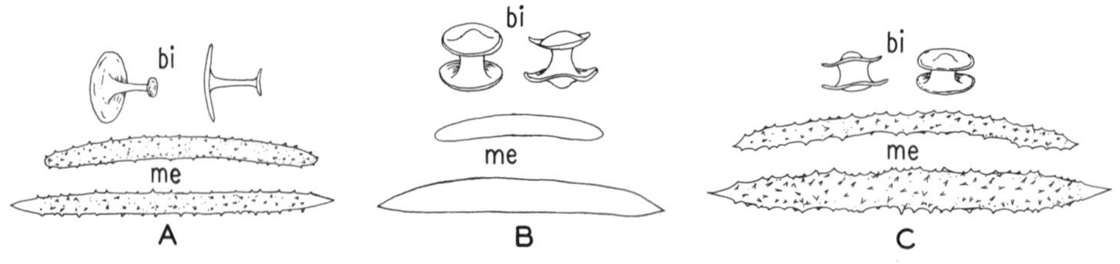

FIG. 10.–Spicules of *Trochospongilla*. A, spicules of *T. pennsylvanica*; B, spicules of *T. leidyi*; C, spicules of *T. horrida*. bi, gemmule birotulate; me, megasclere.

19. Megascleres smooth (Fig. 11E); color and growth form highly variable but usually flat and encrusting; more common in standing waters; generally distributed .... **Ephydatia fluviatilis** (L.)
    Megascleres with small and inconspicuous spines (Fig. 11F) 20
20. Birotulate length equal to or less than diameter of rotules, which are irregularly serrate or dentate, and with few rays (Fig. 11F); growth form highly variable but most luxuriant in standing waters where a single individual may cover as much as a square meter; usually in alkaline waters; common and generally distributed .... **Ephydatia muelleri** (Lieberkühn)
    Birotulates longer than diameter of rotule; rotules not split near to the center, with numerous spines, often malformed (Fig. 11F); reported only from eastern states and CA.

Ephydatia robusta (Potts)

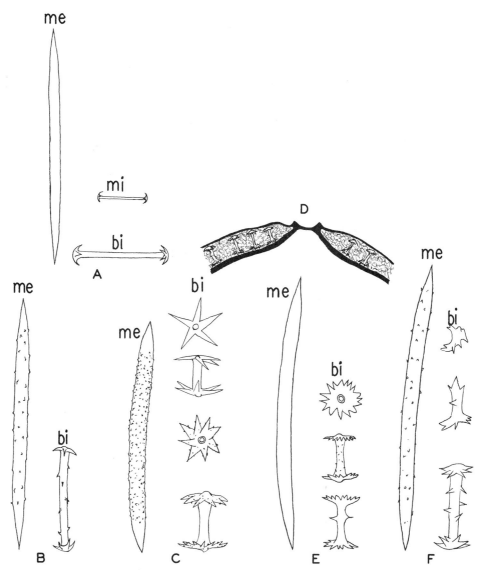

FIG. 11.–Structure of *Corvomeyenia, Radiospongilla,* and *Ephydatia.* A, spicules of *C. everetti*; B, spicules of *R. crateriformis*; C, spicules of *E. muelleri*; D, section of foraminal region of gemmule of *E. fluviatilis*; E, spicules of *E. fluviatilis*; F, spicules of *E. robusta.* bi, gemmule birotulates; me, megascleres; mi, microscleres.

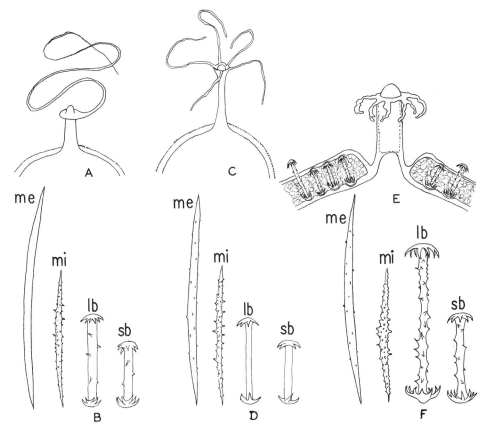

FIG. 12.–Structure of *Heteromeyenia*. A, foraminal region of gemmule of *H. latitenta*; B, spicules of *H. latitenta*; C, foraminal region of gemmule of *H. tubisperma*; D, spicules of *H. tubisperma*; E, foraminal region of gemmule of *H. baileyi*; F, spicules of *H. baileyi*. lb, longer birotulates; me, megascleres; mi, microscleres; sb, shorter birotulates. (A and C modified from Potts, 1887; E modified from Sasaki, 1934.)

21. Microscleres stellate (Fig. 13E); generally distributed but rare .... **Dosilia radiospiculata** (Mills)
    Microscleres absent, or acerate if present 22
22. Microscleres present, acerate .... **Heteromeyenia, 23**
    Microscleres absent .... **Anheteromeyenia, 26**
23. Foraminal tubule with one or two very long tendrils, often enveloping the tubule (Fig. 12A); shorter
    birotulates with rotules bearing deeply cut rays; longer birotulates with a smooth or sparsely
    macrospined shaft, rotules bearing deeply cut rays (Fig. 12B); usually thin and encrusting in
    running waters; northeastern quarter of the United States ..... **Heteromeyenia latitenta** (Potts)
    Foraminal tubule with three or more tendrils of varying length; megascleres sparsely microspined.
    24
24. Foraminal tubule one half to once the diameter of the gemmule body, with four to six long tendrils
    (Fig. 12C); shorter birotulates with smooth shaft and with rotules composed of three to five rays;
    longer birotulates usually with a thinner smooth shaft, and with rotules composed of three to five
    rays (Fig. 12D); encrusting form with soft papillate projections; usually in running waters; eastern
    half of the United States .... **Heteromeyenia tubisperma** (Mills)
    Foraminal tubule one fourth or less as long as diameter of gemmule body (Figs. 12A, C); both types of
    birotulates with macrospined shafts (Figs. 12F; 13A); encrusting growth form 25

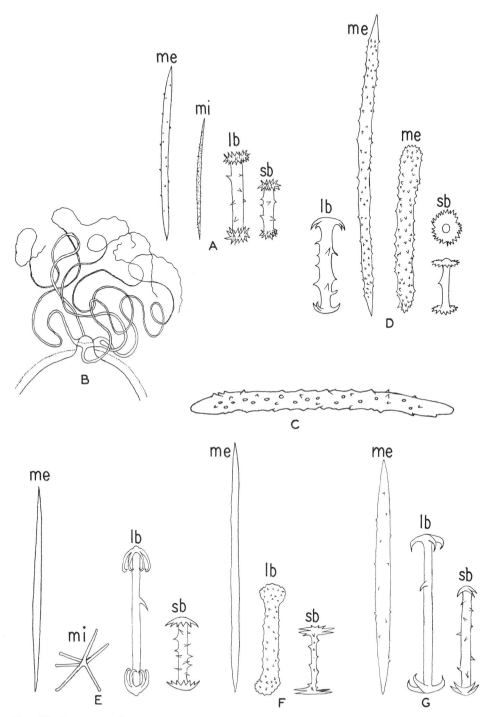

FIG. 13.–Structure of *Heteromeyenia, Dosilia,* and *Anheteromeyenia.* A, spicules of *H. tentasperma*; B, foraminal region of gemmule of *H. tentasperma*; C, megasclere of *A. pictovensis*; D, spicules of *A. ryderi*; E, spicules of *D. radiospiculata*; F, spicules of *A. biceps*; G, spicules of *A. argyrosperma.* lb, longer birotulates; me, megascleres; mi, microscleres; sb, shorter birotulates. (B modified from Potts, 1887.)

25. Rays of rotules of both types of birotulates deeply cleft and slightly incurved; longer birotulates 80 μm long; shorter birotulates 60 μm long (Fig. 12F); foraminal tubule variable, usually with 6 to 10 irregular and sometimes branched tendrils (Fig. 12E); but tendrils sometimes fewer, reduced, or absent; widely scattered in eastern states .... **Heteromeyenia baileyi** (Bowerbank)
    Rays of rotules of both types of birotulates less deeply cleft and arranged so as to give the rotule a burrlike appearance (Fig. 13A); foraminal tubule with about three to five long, irregular tendrils (Fig. 13B); reported from northeastern states .... **Heteromeyenia tentasperma** (Potts)
26. Rotules of long birotulates in the form of small knobs; megascleres smooth (Fig. 13F); encrusting growth form, seldom more than 5 cm in diameter; reported only from two streams near Douglas Lake, MI; perhaps only a variety of *Ephydatia muelleri*.

    Anheteromeyenia biceps Lindenschmidt
    Rotules of long birotulates not in the form of knobs; megascleres spined (Figs. 13D, G) **27**
27. Rotules of short birotulates with a few large rays (Fig. 13G); encrusting growth form; eastern United States .... **Anheteromeyenia argyrosperma** (Potts)
    Rotules of short birotulates disclike, with numerous small teeth (Fig. 13D) **28**
28. Megascleres finely spined (Fig. 13D); averaging 200 μm long; standing and running waters in eastern half of the United States .... **Anheteromeyenia ryderi** (Potts)
    Megascleres coarsely spined (Fig. 13C); averaging 165 to 210 μm long; extreme northeastern United States .... **Anheteromeyenia pictovensis** (Potts)

REFERENCES

Arndt, W. 1928. Porifera, Schwämme, Spongien. *Tierwelt Dtschl.* 4:1–94.

Berg, K. 1948. Studies on the bottom animals of Esrom Lake. *Mem. Acad. Roy. Soc. Lett. Danemark,* Ser. 9, 8:1–255.

Bowerbank, J. S. 1863. A monograph of the Spongillidae. *Proc. Zool. Soc. London* (1863):440–472.

Brien, P. 1932. Contribution à l'étude de la régénération naturelle chez les Spongillidae. *Arch. Zool. Exp. Gén.* 74:461–506.

_____. 1936. La réorganisation de l'éponge après dissociation par filtration et phénomènes d'involution chez Ephydatia fluviatilis. *Arch. Biol.* 48:185–268.

Brønsted, A., and H. V. Brønsted. 1953. The effects of symbiotic zoochlorellae on the germination rate of gemmules of Spongilla lacustris (L.). *Vidensk. Medd. Dan Naturhist. Foren. København* 115:133–144.

Eshleman, S. K. 1949. A key to Florida's fresh-water sponges, with descriptive notes. *Q. J. Fla. Acad. Sci.* 12:35–44.

Frost, T. M., and C. E. Williamson, 1980. *In situ* determination of the effect of symbiotic algae on the growth of the freshwater sponge Spongilla lacustris. *Ecology* 61:1361–1370.

Gee, N. G. 1932. Genus Trochospongilla of the fresh water sponges. *Peking Nat. Hist. Bull.* 6:1–32.

_____. 1932a. The known fresh-water sponges. *Ibid.* 25–51.

Gilbert, J. J. 1975. Field experiments on gemmulation in the fresh-water sponge Spongilla lacustris. *Trans. Am. Microsc. Soc.* 94:347–356.

Gilbert, J. J., and T. L. Simpson. 1976. Sex reversal in a freshwater sponge. *J. Exp. Zool.* 195:145–151.

Harrison, F. W. 1982. Developmental biology of fresh-water sponges. Pp. 1–67 in *Developmental Biology of Freshwater Invertebrates*, 588 pp. (F. W. Harrison and R. R. Cowden, eds.), Liss, New York.

Harrison, F. W., and R. R. Cowden (eds.). 1976. *Aspects of Sponge Biology,* 354 pp. Academic Press, New York.

Harrison, F. W., and M. B. Harrison. 1977. The taxonomic and ecological status of the environmentally restricted spongillid species of North America. II. Anheteromeyenia biceps (Lindenschmidt). *Hydrobiologia* 55:167–169.

Jewell, M. E. 1935. An ecological study of the fresh-water sponges of northern Wisconsin. *Ecol. Monogr.* 5:461–504.

_____. 1939. An ecological study of the fresh-water sponges of Wisconsin, II. The influence of calcium. *Ecology* 20:11–28.

_____. 1952. The genera of North American fresh-water sponges. Parameyenia, new genus. *Trans. Kans. Acad. Sci.* 55:445–457.

Jorgensen, C. B. 1944. On the spicule-formation of Spongilla lacustris (L.). I. The dependence of the spicule formation on the content of dissolved and solid silicic acid on the milieu. *Biol. Medd. Kjøbenhavn* 19:1–45.

Leveaux, M. 1939. La formation des gemmules chez les Spongillidae. *Ann. Soc. R. Zool. Belg.* 70:53–96.

_____. 1941, 1942. Contribution à l'étude histologique de l'ovogénèse et de la spermatogénèse des Spongillidae. *Ibid.* 72:251–269; 73:33–50.

Lindenschmidt, M. J. 1950. A new species of fresh-water sponge. *Trans. Am. Microsc. Soc.* 69:214–216.

McNair, G. T. 1923. Motor reactions of the fresh-water sponge, Ephydatia fluviatilis. *Biol. Bull.* 44:153–166.

Neidhofer, J. R. 1940. The fresh-water sponges of Wisconsin. *Trans. Wis. Acad. Sci. Arts Lett.* 32:177–197.

Old, M. C. 1931. A new species of fresh-water sponge. *Trans. Am. Microsc. Soc.* 50:298–299.

_____. 1931a. Taxonomy and distribution of the fresh-water sponges (Spongillidae) of Michigan. *Pap. Mich. Acad. Sci. Arts Lett.* 15:439–476.

———. 1932. Environmental selection of the freshwater sponges (Spongillidae) of Michigan. *Trans. Am. Microsc. Soc.* 51:129–137.

———. 1932a. Contribution to the biology of freshwater sponges (Spongillidae). *Papers Mich. Acad. Sci. Arts Lett.* 17:663–679.

Penney, J. T. 1933. Reduction and regeneration in fresh water sponges (Spongilla discoides). *J. Exp. Zool.* 65:475–495.

———. 1933a. A new fresh-water sponge from South Carolina. *Proc. U.S. Natl. Mus.* 82:1–5.

———. 1954. Ecological observations on the fresh-water sponges of the Savannah River Project area. *Univ. S.C. Publ. Ser.* 3, 1:156–192.

———. 1960. Distribution and bibliography (1892–1957) of the freshwater sponges. *Ibid.* 3:1–97.

Penney, J. T., and A. A. Racek. 1968. Comprehensive revision of a world-wide collection of fresh-water sponges (Porifera: Spongillidae). *Bull. U.S. Nat. Mus.* 272:1–184.

Poirrer, M. A. and others. 1987. Comparative morphology of microsclere structure in Spongilla alba, S. cenota, and S. lacustris (Porifera, Spongillidae). *Trans. Am. Micros. Soc.* 106:302–310.

Potts, E. 1887. Fresh water sponges. A monograph, *Proc. Acad. Nat. Sci. Phila.* (1887):157–279.

Sasaki, N. 1934. Report on the fresh-water sponges obtained from Hokkaido. *Sci. Rep. Tohoku Imp. Univ. (Ser. 4)* 9:219–248.

Schröder, K. 1936. Beiträge zur Kenntnis der Spicula-bildung der Larvenspiculation und der Variationsbreite der Gerüstnadeln von Süsswasserschwämmen. *Z. Morphol. Oekol. Tiere* 31:245–267.

Simon, L. 1953. Uber die Spezifität der Nadeln und die Variabilität der Arten bei den Spongilliden. *Zool. Jahrb. Abt. Allg. Zool. Physiol. Tiere* 64:207–234.

Simpson, T. L., and P. E. Fell. 1974. Dormancy among the Porifera: gemmule formation and germination in fresh water and marine sponges. *Trans. Am. Microsc. Soc.* 93:544–577.

Smith, D. G. 1976. An intergeneric fresh-water sponge mixture (Demospongiae: Spongillidae). *Ibid.* 95:235–236.

Smith, F. 1918. A new species of Spongilla from Oneida Lake, New York. *N.Y. State Coll. For. Syracuse Univ., Tech. Publ.* 9:239–243.

———. 1921. Data on the distribution of Michigan fresh-water sponges. *Pap. Mich. Acad. Sci. Arts Lett.* 1:418–421.

———. 1921a. Distribution of the fresh-water sponges of North America. *Bull. Ill. Nat. Hist. Surv.* 14:9–22.

Van Trigt, H. 1919. A contribution to the physiology of the fresh-water sponges. (Spongillidae). *Tijd. Ned. Dierkd. Vereen. (II.)* 17:1–220.

Wierzejski, A. 1935. Süsswasserspongien. Monographische Bearbeitung. *Mem. Acad. Polon. Sci. Lett., Ser. B,* 9:1–242.

4

COELENTERATA
(Hydroids, Jellyfish)

Although the Phylum Coelenterata contains in excess of 9000 species, the vast majority are marine. Two classes, the Scyphozoa (true jellyfish) and the Anthozoa (corals, sea anemones, etc.), are exclusively marine, and in the third class, the Hydrozoa, only about 20 species are known from the fresh waters of the United States. These include 16 hydras, one uncommon "fresh-water jellyfish," one uncommon species of colonial polyp, and one rare protohydroid from brackish coastal waters.

Although not conspicuous, hydras are typical representatives of the fauna of ponds, spring brooks, unpolluted streams and rivers, and the littoral zone of lakes. They are all sessile and may be found attached to stones, twigs, vegetation, or debris, occasionally in enormous numbers, but rarely at depths exceeding 1.5 m. Populations are most dense during the summer months. When there is a food shortage, a hydra often secretes a gas bubble beneath the pedal disc and floats to the surface film, where the bubble bursts, leaving the hydra hanging downward from the surface film.

General characteristics. Hydras should be familiar to all who have had an elementary biology or zoology course. Like other coelenterates, they are radially symmetrical, the main body, or column, being an elongated cylinder from 1 to 25 mm long. Attachment to the substrate is effected by a pedal disc that consists largely of special secretory cells. The distal end of the column has a circlet of tentacles whose length varies from one half to five times the length of the column. Usually there are 5 or 6 tentacles, sometimes 3, 4, 7, or 8, and rarely up to 12. The end of the column in the center of the circlet of tentacles is raised and domelike to form a hypostome that bears a single opening, the combined mouth and anus. The single, continuous, internal body cavity is a gastrovascular cavity; it continues out into the hollow tentacles.

Although some hydras have a characteristic green or brown color, others have a variable coloration depending on age and the kind and amount of ingested food. Thus there may also be various shades of translucent whitish, grayish, tan, or red.

The body wall is simply constructed. Externally it consists of a layer of epidermis, and internally there is a layer of gastrodermis lining the gastrovascular cavity. Several different histological types of epidermal and gastrodermal cells can be distinguished. Between these two epithelia is a very thin, secreted, noncellular, cementlike mesoglea.

All coelenterates have minute stinging capsules, or nematocysts, imbedded in certain epidermal cells. Each nematocyst contains a coiled capillary thread, which, on proper stimulation, can be "exploded" and extruded. Nematocysts function in food getting and protection by means of their entangling, adhesive, and paralyzing properties. Nematocysts are especially abundant on the tentacles and distal part of the column.

Although metagenesis and the alternation of medusoid and polyp stages are typical of marine hydrozoans, the hydras have only the polyp stage in their life history.

Locomotion. Hydras are usually considered sessile creatures; when they are undisturbed the column is extended, the tenta-

110

cles are spread out, and there is no movement visible to the naked eye. Occasionally, however, the tentacles wave about or the whole animal contracts and expands unaccountably. But hydras are also capable of moving slowly on the substrate. Most commonly they have a very slow, gliding locomotion on the pedal disc. Occasionally they exhibit somersaulting or inchwormlike looping movements. More rarely they attach the distal end of the body to the substrate and pull themselves along by contractions and elongations. The central part of the basal disc may secrete a small gas bubble that enables the hydra to rise to the surface and remain suspended at the surface film in quiet waters.

Feeding, digestion. Hydrozoans are strictly carnivorous and eat many kinds of small metazoans, including cladocerans, copepods, insects, and annelids (but not ostracods). Occasionally hydras become a hatchery nuisance by killing fish fry, but *Hydra oligactis* is probably the only species large enough to be of any real importance. When a food organism is swimming or creeping about and by chance comes in contact with the tentacles or column, it is paralyzed and killed in a few seconds by the nematocysts that are discharged from those parts of the column or tentacles contacted. The organism is then manipulated so that it is taken into the mouth. The latter is variously described as star shaped or circular; both it and the column are capable of great distension, and a cladoceran four times the diameter of the column may be ingested. When the usual food supply is insufficient, hydras may feed on the organic material of the substrate, but this is certainly not a normal means of feeding. The feeding reaction of hydras (writhing and twisting of the tentacles toward the proctostome) is induced by reduced glutathione (and perhaps by other substances). Presumably this is the key mechanism in the feeding process of hydras.

Digestion is both extra- and intracellular. Preliminary extracellular digestion occurs in the gastrovascular cavity, resulting in the maceration of the food organisms into a mush within several hours. These small particles are then taken up by pseudopodial action

FIG. 1.—*Hydra* with two buds (partially contracted, ×12. (Courtesy General Biological Supply House, Inc., Chicago, Ill.)

of many of the gastrodermal cells. Further digestion is intracellular. Only the gastrodermal cells in the upper part of the column appear to function in digestion; those in the more basal part of the column (stalk) are highly vacuolated and have no digestive power.

Excess food is stored in the gastrodermal cells as fat and glycogen. Indigestible material is voided from the mouth.

Chlorohydra viridissima is the so-called green hydra because of the abundance of algal cells (probably mostly *Chlorella*) in the gastrodermis. Such living algae probably have a true symbiotic relationship with the hydra, but dead and disintegrating cells are digested and used as food. The algae are passed from one generation to the next in the egg cytoplasm.

Nematocysts. A nematocyst is a spherical to elongated, capsular, sclerotized organoid ranging from 5 to 25 μm in length in fresh-

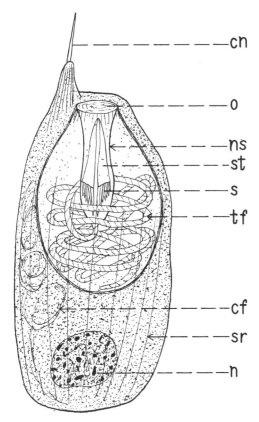

FIG. 2.–Undischarged stenotele nematocyst within its cnidoblast. cf, cnidoblast filament; ns, nematocyst shaft (butt); o, operculum; s, spines; sr, supporting rods of cnidoblast; st, stylet; tf, tubular filament (actually much long than shown.)

fine spinules (Figs. 3D, E). Desmonemes appear to function mainly in entangling and wrapping around bristles of the prey.

Stenoteles (penetrants) are large, spherical nematocysts. The coiled tubule is open at the tip and has a definite butt, or enlarged basal portion, which bears three spines and three spiral rows of minute spinules (Figs. 2, 3F). Exploding stenoteles pierce the prey and inject a paralyzing material.

There are two types of isorhiza (glutinant) nematocysts, but both are characterized by a tubule that is open at the tip, by the absence of a butt, and by the secretion of a paralyzing material. Holotrichous isorhizas are larger and have the tubule covered with minute spinules (Fig. 3C); they function mainly in defense. Atrichous isorhizas are smaller and have a smooth tubule (Figs. 3A, B); they appear to function mainly in locomotion by discharging and sticking when the tentacles are placed in contact with a substrate.

Nematocysts have an unusual mode of origin and migration. They are formed within special interstitial cnidoblast cells of the epidermis. Most cnidoblasts first appear in the distal third of the column. Before their contained nematocysts are fully formed, many cnidoblasts pass through the mesoglea and gastrodermis into the gastrovascular cavity. Here they circulate about briefly before being taken up again by gastrodermal cells, especially in the tentacles. Then they are transported back through the mesoglea to their final location in the epidermis.

According to one theory, when the projecting cnidocil is touched, the operculum springs open and the nematocyst tubule instantaneously turns inside out as it whips out of the capsule. The base of the tubule emerges first and the rest follows from base to tip. According to a more recently elaborated theory, the external tubule is formed by the rapid extrusion of a fluid that hardens on contact with water. At the same time the internal coiled tubule disintegrates. Discharged nematocysts are lost and are of no further use to the hydra.

It is thought that stenotele and desmoneme discharge is usually dependent on a combined chemical and mechanical stimulus. The juice and body fluids of the normal prey will not in themselves discharge nematocysts,

water species. It is contained within a special epidermal cell, the cnidoblast. Inside of the nematocyst is a minute, coiled, hollow thread, or tubule, whose base is attached to the inner distal wall of the nematocyst. A lid, or operculum, covers the distal opening of the capsule, and a tiny, triggerlike bristle, or cnidocil, projects into the surrounding water from the surface of the cnidoblast near the edge of the operculum. In the Phylum Coelenterata as a whole there are many different types of nematocysts, but in the fresh-water hydras there are only four types as follows:

Desmonemes (volvents) are small, pyriform, or spherical nematocysts. The tubule forms only one loop, is closed at the distal end, and is usually unarmed but many have extremely

nor will a fine glass rod touched to the cnidocils. In addition, these nematocysts of satiated animals do not react when food organisms touch the tentacles.

The physiological mechanisms of nematocyst action are poorly understood, but it is thought that the discharge is produced by an increased pressure within the capsule caused by increased permeability of the capsule wall. Perhaps a chemical stimulus acts first by reducing surface tension and lowering the resistance of the cnidoblast and cnidocil to mechanical stimuli.

One of the most unusual phenomena centering around nematocyst biology is the use of nematocysts by certain species of *Microstomum* for their own protection and food getting. When one of these turbellarians feeds on a hydra only the desmonemes and atrichous isorhizas are digested, and the other two types of nematocysts are passed through the mesenchyme of the *Microstomum* and into the epidermis. Here they are correctly oriented by associated phagocytes and discharged when touched by other animals.

Physiology. Circulation of materials within the gastrovascular cavity is effected by movements of the whole animal and by beating of the flagella of some of the gastrodermal cells. Such cells usually have one or two flagella.

Respiration occurs through the general body surface. None of the hydrozoans are active in the presence of reduced quantities of dissolved oxygen.

There seems to be no special excretory mechanism, and it is presumed that metabolic wastes simply diffuse out into the surrounding water. A few workers, however, have reported a small "aboral" or "excretory" pore in the center of the basal disc. Osmoregulation seems to be a function of the individual cells.

When strongly stimulated, and under adverse environmental conditions, hydras contract markedly and assume a barrellike shape with small, budlike contracted tentacles. Though there are no special bands or sheets of muscle tissue to produce such contractions and elongations, many of the epidermal cells and some gastrodermal cells have one or a few threadlike myonemes associated with

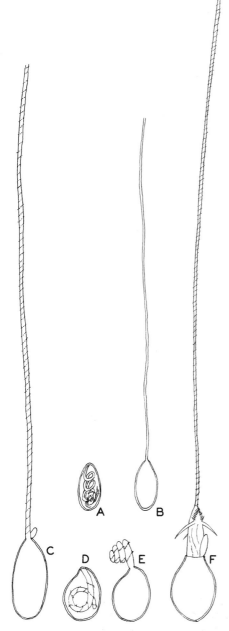

FIG. 3.–Typical hydra nematocysts. A, undischarged atrichous isorhiza; B, discharged atrichous isorhiza; C, discharged holotrichous isorhiza; D, undischarged desmoneme; E, discharged desmoneme; F, discharged stenotele. A to E, ×2000; F, ×1000. (Cnidoblasts not shown.)

their basal portions. Most of the longitudinal fibers are restricted to the epidermis and the circular fibers to the gastrodermis; the former are more abundant.

Elongated sensory cells are numerous in the epidermis of the tentacles and oral region. They are generalized receptors for touch, temperature, and substances in solution, and pass basally into one or more processes of the nervous system, which is a nerve net ramifying everywhere at the base of the epidermis. The neurones are bi- and multipolar ganglion cells and their neurites, and the entire system has the appearance of a continuous network. There are no coordinating or relay centers, and stimuli simply diffuse throughout the nerve net in all directions. The nervous system has no connections with nematocysts.

Reproduction, development. Budding is the common asexual method of reproduction. Such buds usually originate at the junction of the gastric and stalk regions. A bud first appears as a hemispherical outpouching that later elongates, becomes cylindrical, and develops tentacles. Subsequent buds appear spirally. The gastrovascular cavity of the budding individual and parent are continuous, but soon the base of the bud pinches off completely and the new individual becomes separated and independent. Under favorable conditions a new bud may be produced every two or three days.

Both transverse and longitudinal fission are occasionally observed in hydras, but these methods of reproduction are not normal and usually follow injuries or depression periods (page 118). Powers of regeneration and susceptibility to grafting are well developed, and a large amount of experimental work has been done on these phases of hydra biology.

Syngamic reproduction is usually limited to the autumn months when water temperatures are dropping, but in a few species it may appear in spring or summer when temperatures are highest or when the habitat is drying up. In hermaphroditic species the testes are distal and the ovaries proximal; sometimes ovaries and testes mature simultaneously, sometimes at different periods. In dioecious species the ovaries or testes may be distributed throughout most of the distal half or three quarters of the column. In general, males far outnumber females, in striking contrast to

the situation in most fresh-water invertebrate groups.

A gonad consists of an accumulation of epidermal interstitial cells between mesoglea and epidermis, but in a developing ovary these cells fuse or engulf each other, leaving a single large food-filled ovum. A mature testis is usually hemispherical or helmet shaped. It usually has a "nipple"; if present, many of the sperm escape to the outside through this small area. Ovaries are hemispherical or bulbous swellings.

When an egg is mature, the covering epidermis ruptures and withdraws, forming a small cup or cushion under the egg. If not fertilized promptly, the egg dies. Sperm may come from the same or a different individual.

During late cleavage the embryo secretes a sclerotized yellowish shell, the embryonic theca. When the theca is being formed it is somewhat sticky, and at this stage the whole embryo drops off the parent or is fastened to the substrate by bending movements of the parent. It sticks to the substrate while the theca hardens. Such thecated embryos may be spherical or planoconvex, spinous or smooth. The diameter ranges from 0.40 to 1.00 mm. They are highly resistant to adverse environmental conditions and endure drying and freezing temperatures. After a dormant period of 3 to 10 weeks or more, the theca softens and splits, permitting the exit of an embryo that already has a gastrovascular cavity and tentacles.

Little is known about the length of life under normal conditions. Under favorable conditions in the laboratory, however, hydras live for 3 to 12 months or more. A few specimens have been kept alive for more than 2 years, during which time they produced scores of buds.

Fresh-water jellyfish. Although hydras are by far the most familiar fresh-water hydrozoans, there is one American species of "jellyfish," *Craspedacusta sowerbyi.* Unlike large marine jellyfish, however, it has a velum and its life history includes a colonial (but simple) polyp.

C. sowerbyi was first found in England in 1880; more recently it has been noted many times in Europe and on other continents but only rarely in Africa. It has been reported from about 60 localities in the United States

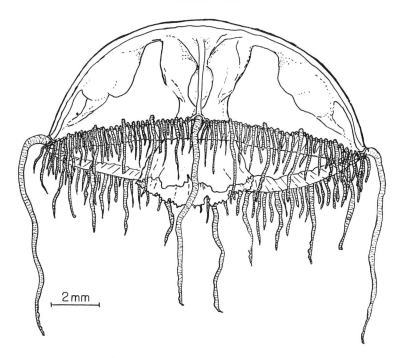

2 mm

F_IG. 4.–*Craspedacusta sowerbyi.* (From Belk and Hotaling, 1971.)

since 1908, when it was discovered in this country. It has been found mostly in small lakes, ponds, fishponds, and old water-filled quarries, especially between July and October. In addition, there are a few reports for rivers, sluggish streams, and large reservoirs (e.g., Lake Mead).

A mature medusa stage of *C. sowerbyi* (Fig. 4) has the typical umbrella or bell shape, manubrium, and four radial canals and one circular canal forming the gastrovascular cavity. The diameter of the bell is about 5 to 22 mm, and the circumference bears 50 to 500 solid tentacles of varying length and arranged in three to seven series, depending on the medusa size. Only short tentacles are involved in feeding; long tentacles are presumably stabilizers for swimming. The bell margin also bears many statocysts, often 60 to 100 or more. The velum is rather thick. Nematocysts are especially abundant around the mouth, at the edge of the bell, and on the tentacles. *Craspedacusta* actively swims up, down, or sideways in a dancing sort of movement produced by varying contractions of the bell. Food consists of zooplankters, chiefly in the 0.2 to 2.0 mm range. It is not

thought to be an important predator on very small fish. Crayfish are the only important group that feed on the medusae, especially when the latter are at rest on the substrate.

Most populations appear to consist of all males or all females, the occurrence of mixed populations having been observed only a few times. The four gonads consist of convoluted or flaplike masses on the subumbrellar surface of the bell in the vicinity of the radial canals. Both eggs and sperm are released into the surrounding water, where fertilization occurs.

The zygote develops into a simple, branching, colonial hydroid (Fig. 5) 2 to 8 mm long, usually consisting of two to four individuals (seldom up to 10). Single individuals are usually less than 2 mm long, and tentacles are absent. The hydroid creeps about slowly on the substrate, feeding on small metazoans. The region around the mouth bears clumps of nematocysts. The body secretes a sticky, mucous material to which debris adheres and more or less covers the body. During the winter months the hydroids contract and secrete a sclerotized protective coat.

The hydroid stage has three methods of

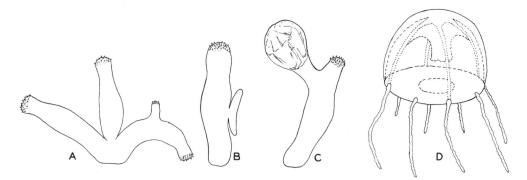

FIG. 5.–Stages in the life history of *Craspedacusta sowerbyi*. A, typical colony of hydroids, ×12; B, hydroid with a maturing planulalike bud, ×15; C, hydroid with a maturing medusa bud, ×15; D, newly released medusa with eight tentacles, ×100, (A and D greatly modified from Payne, 1924.)

asexual reproduction. Ordinary budding is similar to that of the hydras, and new individuals may remain attached or constrict off from the parent. Nonciliated planulalike buds (frustules) may also be produced. These constrict off, elongate, and creep about slowly on the substrate for a time, and then develop into new polyps. The third method involves the production of medusa buds. When these break free from the parent polyp the bell diameter is about 0.4 mm, and there are only eight tentacles, but rapid growth and development follow, resulting in the large, sexually mature medusae. Both polyps and medusae feed on a wide variety of small invertebrates.

The occurrence of the medusae is highly sporadic and unpredictable. In some years they may be found in abundance in certain ponds, but in subsequent years they may be absent, only to reappear several years later or in several successive years. In other ponds they have been found only once and have never reappeared, in spite of careful searching. There is some evidence to show that the medusae are positively phototropic.

The hydroid generation is only occasionally noticed, probably chiefly because of its small size and inconspicuous nature. It is possible that the hydroids live undetected for long periods, only occasionally giving rise to the medusoid generation. For many years the hydroid was called *Microhydra ryderi,* and it was not until 1924 that it was definitely associated with *Craspedacusta sowerbyi*.

The fresh-water jellyfish has been reported from nearly all states, but is not known from northern New England. On a worldwide basis, *C. sowerbyi* appears to be chiefly confined to a broad band between 45° north and 45° south latitude.

Several related species of fresh-water medusae have been reported from other continents, including one species from China, two or three in Africa, seven in India, and one in Trinidad. All of these species have their closest relatives in marine situations, and most invasions of fresh waters appear to transpire along tropical shores.

Colonial polyps. Another unusual fresh-water coelenterate is *Cordylophora lacustris*. It is a profusely branching, mosslike colonial hydrozoan superficially similar to the marine *Obelia* and its many relatives. It does best at 15⁰/₀₀ salinity and occurs in brackish inlets and estuaries from New Jersey to the Gulf of the St. Lawrence, but has also been found in inland rivers in Illinois, Arkansas, Arizona, Kansas, Tennessee, and Louisiana. This species is probably cosmopolitan. It is widely distributed in northern Europe and has been found in numerous other parts of the world. Some taxonomists believe this species should be called *Cordylophora caspia*.

During the summer months brackish water colonies attain a height of 20 to 100 mm from the substrate, but in fresh waters growths seldom exceed 30 mm in height. There are two kinds of polyps making up such colonies; hydranths are spindle shaped feeding individuals with a prominent manubrium and 10 to 20 thin scattered tentacles, and gonophores are ovoid reproductive polyps. It

FIG. 6.–*Cordylophora lacustris,* ×5. A, small portions of summer growth of colony; B, colony in winter condition (lower portion of colony covered with detritus). (B from K. Berg, 1948, by permission.)

is notable that only a few gonophores are produced on colonies growing in strictly fresh waters. A single female gonophore usually produces 6 to 20 ova in brackish waters, but only three to six in fresh waters. Mature ova are retained on the gonophores, but the sperm from male gonophores are released into the water, and after swimming about for a short time they penetrate through the covering of the female gonophores and fertilize the eggs. The embryo develops on the gonophore and forms a free-swimming ciliated planula that is released from the colony. The planula attaches to the substrate and by growth and differentiation gives rise to a new colony. Thus there is no true medusoid generation in *Cordylophora*.

In the autumn the hydranths disintegrate and only a few horizontal and more basal upright portions of the colony winter over (Fig. 6B).

Protohydroids. A third unusual type of hydroid is *Protohydra leuckarti* (Fig. 8). Although not a true fresh-water species, it occurs in brackish estuaries, swamps, and creeks, some of which are almost fresh. It has the general appearance of a hydra but has no tentacles. The body is usually less than 2 mm long but may attain a maximum of 5 mm. The species exhibits dioecious syngamic reproduction and also reproduces by transverse and longitudinal fission. *Protohydra* has been reported

several times along the New England coast and from numerous places in Great Britain and northwestern Europe.

The writer has seen several related minute forms in interstitial psammolittoral and stream

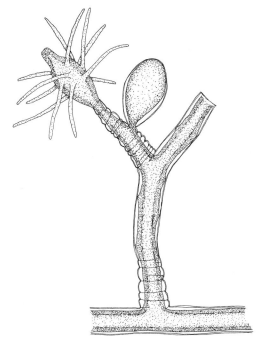

FIG. 7.–Small portion of colony of *Cordylophora lacustris,* ×15, showing one hydranth and one immature gonophore.

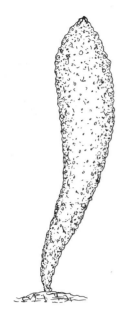

FIG. 8.–*Protohydra leuckarti*, ×35.

bottom waters. These were very short and had only three or four blunt tentacles.

Ecology. Although hydroids are typical members of the littoral and shallow stream associations, they have occasionally been reported from waters as deep as 40 to 350 m. It is not unusual to find them in swift streams or in the wave-swept rocky littoral of large lakes. Unless there are large pieces of debris present, however, they do not occur on fine, muddy bottoms. Greatest populations are usually found in late spring and early summer, followed by a marked decrease in July or August. Sometimes there is also an autumn maximum in September or October. Populations may be vigorous in midwinter but are usually scanty by the time of the ice melt in late winter.

Gonad production is often induced in 2 or 3 weeks by lowered laboratory temperatures, usually 10 to 15°C, but often a culture becomes sexual without any apparent stimulus. Medium hard waters with pH readings of 7.6 to 8.4 seem to be especially favorable for most species. Although little is known about minimal oxygen concentrations, about 3.0 ml/l seems to be a typical value. Hydras do not occur in oxygen-deficient depths of lakes. In general, any lasting contraction of the tentacles is an indication of unfavorable environmental conditions.

Periods of "depression" are often observed in laboratory cultures and are presumably induced by a variety of unfavorable conditions, including temperatures above 20°C, lack of sufficient dissolved oxygen, toxic salts, transfer to clean water, and accumulation of metabolic materials. The bodies of hydras undergoing depression rapidly shorten and the tentacles contract and then disappear. The column then further shortens into a small brownish mass that soon disintegrates. If not too far gone, hydras may recover on being placed in more favorable conditions.

Commensals, enemies. Commensal or parasitic ciliates, such as *Kerona* (page 82) and *Trichodina* (page 85), are common on the external surface of hydras where they feed on small bits of food. They do not discharge the nematocysts, although most free-living ciliates, both smaller and larger than *Kerona* and *Trichodina,* do easily discharge them.

Carnivorous enemies are not abundant or important but include a few turbellarians, nemertines, crustaceans, snails, and aquatic insects. *Anchistropus minor* is a cladoceran that feeds exclusively on hydras. Epidemics of an external amoeboid parasite, *Hydramoeba hydroxena,* sometimes decimate vigorous hydra growths.

Geographic distribution, dispersal. *Chlorohydra viridissima* is known from North America, Europe, Greenland, South America, and New Zealand. *Hydra oligactis* is widely distributed in the Northern Hemisphere and has been recorded in Australia. The other species of hydras occurring in the United States, however, have not been reported from continents other than North America. About 25 species are recorded in the literature as occurring outside of the continent of North America, but about half of these are so sketchily described as to be unrecognizable. Europe has eight species. Six American hydras are known from single localities, and several others have been reported from a few scattered areas. Undoubtedly when more intensive collecting is done, especially in the western half of the country, the geographical ranges of most species will be greatly extended.

Hydras attached to mollusk shells may be easily distributed through single drainage systems, and thecated embryos are probably easily blown or transported overland.

Collecting. Unless hydras are exceedingly abundant, they may be collected most easily by bringing small objects and aquatic plants into the laboratory where they may be picked off. If the containers are placed in semidarkness, the hydras will migrate to the surface film as the dissolved oxygen in the water decreases. The grayish or brownish thecated embryos should be sought for on aquatic vegetation or debris with the low powers of a binocular dissecting microscope.

Culturing. Specimens keep well in aquaria in moderate light if the temperature is below 20°C and if they are well fed with cladocerans, copepods, or brine shrimp. Three or four daphnids per day per hydra will produce prolific budding. A fraction of the aquarium water should be changed frequently, and only pond water should be used, not tap water.

Periods of depression may be overcome by changing the culture water. If necessary, hand feed depressed individuals by holding (with fine forceps) a freshly killed daphnid in contact with the hydra.

Preserving, preparing. In order to kill hydras in an expanded condition they should first be squirted off their normal substrate into a dish. After they are fully expanded in about 2 ml of water, dash a copious quantity of hot Bouin's fixative into the dish. After a few minutes place the specimens in cold Bouin's fluid for 10 hours. Then wash in several changes of alkaline 30% alcohol, stain, dehydrate, and mount permanently. Hance (1957) gives an alternate excellent method for killing hydras in an expanded condition.

Dissociation of the tissues to permit an examination of the nematocysts may be accomplished by placing hydras in a few drops of Hertwig-Schneider's fluid on a slide for 15 to 20 minutes. This solution consists of one part 0.02% osmic acid and four parts 5.0% glacial acetic acid. Dilute Bouin's fluid and 1% nitric acid are also fairly useful for dissociation.

Taxonomy. Modern taxonomic concepts and generic designations date back to the important work of Schulze (1917). Essential characteristics for identification include shape of testes, form of the theca, relative body and tentacle length, hermaphroditism or the dioecious condition, and the shape, size, and structure of the nematocysts. The detailed anatomy of nematocysts can best be distinguished on macerated or dissociated specimens with an oil immersion objective. Preserved nematocysts are of little value.

Most of the basic information on our American hydras has been gained through the extensive and careful investigations of Libbie H. Hyman.

The following key includes all American species that have been adequately described in the literature. All lengths and proportions are based on measurements of the living, expanded animal. Ordinarily only living hydras can be accurately identified.

KEY TO SPECIES OF COELENTERATA

1. Solitary polyps with tentacles; without a medusa stage in the life history.
   Order **HYDROIDA**, HYDRIDAE, 2
   Colonial polyps, free-swimming medusae, or solitary polyps without tentacles 16
2. Gastrodermis containing green algae; hermaphroditic; holotrichous isorhizas narrowed as base (Fig. 9L); tentacles always shorter than the column; green hydras (the taxonomic status of this genus is questionable; possibly it should lapse, with the species being included in the genus *Hydra.*).
   Chlorohydra, 3
   Gastrodermis without green algae; hermaphroditic or dioecious 4

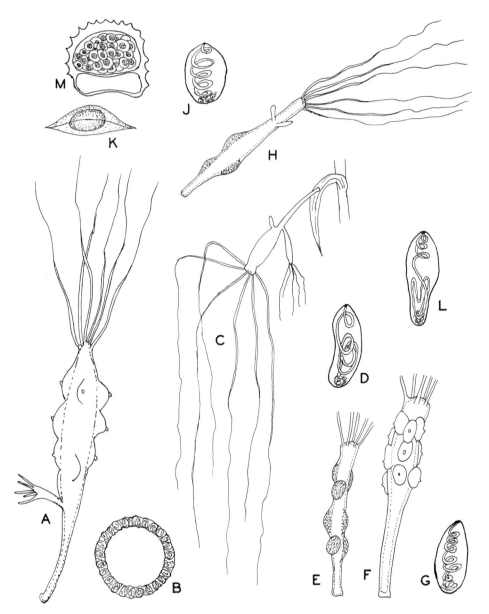

FIG. 9.–Structure of hydras. A, male *Hydra cauliculata,* ×4; B, optical section of thecated embryo of *H. cauliculata*; C, *H. oligactis,* ×2; D, holotrichous isorhiza of *H. oligactis*; E, male *H. oligactis* with testes, ×3; F, male *H. pseudoligactis* with testes, ×3; G, holotrichous isorhiza of *H. pseudoligactis*; H, sexually mature *H. utahensis,* showing basal ovaries and distal testes, ×7; J, holotrichous isorhiza of *H. utahensis*; K, theca and contained zygote of *H. utahensis*; L, holotrichous isorhiza of *Chlorohydra viridissima*; M, cross section of theca of *Chlorohydra hadleyi.* (A and B modified from Hyman, 1938; C–E modified from Hyman, 1930; F–K modified from Hyman, 1931a.)

3. Embryonic theca two chambered (Fig. 9M); column of mature individuals less than 15 mm long; usually 6 to 10 tentacles; ovaries and testes develop simultaneously; reported only from PA and NJ but undoubtedly common elsewhere in the eastern states . . . . . **Chlorohydra hadleyi** Forrest

Embryonic theca with a single chamber; column up to 30 mm. long but usually less than 15 mm; usually from 4 to 12 tentacles; ovaries and testes do not develop simultaneously; widely distributed but not common . . . . **Chlorohydra viridissima** (Pallas)

4. Tentacles exceptionally long, up to five times column length; column up to 17 mm long; reported only from lakes near Edmonton, Alberta, Canada; a questionable species.
Hydra canadensis Rowan

Tentacles shorter **5**

5. Tentacles shorter than column length **6**

Tentacles longer than column length **8**

6. Embryotheca not spined; embryo spherical; pale whitish coloration; column 1.5 to 7.0 mm long; hermaphroditic; known only from GA . . . . **Hydra lirosoma** Campbell

Embryotheca spined; embryo not spherical; column more than 5 mm long; hermaphroditic **7**

7. Tentacles one-half column or more; theca spined and helmet shaped (Fig. 10); hermaphroditic; in ponds and brooks; variable coloration, depending on contents of gastrovascular cavity; reported only from NJ . . . . **Hydra hymanae** Hadley and Forrest

Tentacles one-half column or less; theca spined and spherical (Fig. 10); dioecious, rarely hermaphroditic; in standing or slowly flowing waters; whitish or tan coloration; the white hydra; probably widely distributed throughout the eastern states . . . . . . **Hydra americana** Hyman

8. Holotrichous isorhizas broadly oval (Fig. 9) **9**

Holotrichous isorhizas narrowly oval or cylindrical (Figs. 9G, L) **10**

9. Column 2 to 6 mm long, usually less than 5 mm; embryotheca slightly spined; known only from a brook in Milburn Township, NJ . . . . **Hydra minima** Forrest

Column 10 mm or more long; embryotheca spineless; known only from a pond near Salem, UT.
Hydra utahensis Hyman

10. Tentacles more than three times column length **11**

Tentacles less than three times column length **13**

11. Theca spherical and spineless; testes without nipples, or pumpkin shaped and with nipples (Figs. 9E, F); brownish; common **12**

Theca spherical and spined; testes pumpkin shaped, with nipples; a questionable species, reported only from ponds near Portland, OR . . . . **Hydra oregona** Griffin and Peters

12. Holotrichous isorhizas with lengthwise loops (Fig. 9D); testes without nipples (Fig. 9E); column up to 20 mm long; the true brown hydra; common in northern states from MT, WY, and CO east to New England, north of the Ohio River . . . . **Hydra oligactis** (Pallas)

Holotrichous isorhizas with transverse loops (Fig. 9G); testes pumpkin shaped, with nipples (Fig. 9F); column up to 25 mm long; the false brown hydra; most common in central states, scattered elsewhere . . . . **Hydra pseudoligactis** (Hyman)

13. Column slightly stalked; brownish; in standing waters; desmonemes as long as atrichous hydrorhizas; probably common in the Atlantic Coast states . . . . **Hydra cauliculata** Hyman

Column not stalked; desmonemes shorter than atrichous hydrorhizas **14**

14. Buds usually in opposite pairs; column to 18 mm; known only from brooks in NJ.
Hydra rutgersensis Forrest

Buds not usually in opposite pairs **15**

15. Column to 10 mm; hermaphroditic; theca with short, thick spines; in standing waters as far west as NE.
Hydra carnea Agassiz

Column to 15 mm; dioecious; theca with long branching spines; in swift waters or on wave-swept shores; reported from scattered localities as far west as OK . . . . **Hydra littoralis** Hyman

16. Macroscopic colonial polyps 10 to 100 mm high and consisting of feeding hydranths and reproductive gonophores (Figs. 6, 7); uncommon in brackish waters of Atlantic and Pacific coasts; also reported from rivers in IL, AR, OH, TN, and LA.
Order **HYDROIDA**, CLAVIDAE, **Cordylophora lacustris** Allman

Not macroscopic colonial polyps **17**

17. Free-swimming bell-shaped medusae with numerous peripheral tentacles; bell diameter ranging from 0.4 to 22 mm (Fig. 4); sporadic and uncommon, especially in man-made ponds, small lakes, fish ponds, and old flooded quarries.
Order **TRACHYLINA**, PETASIDAE, **Craspedacusta sowerbyi** Lankester

Sessile solitary or simple colonial hydroids without tentacles **18**

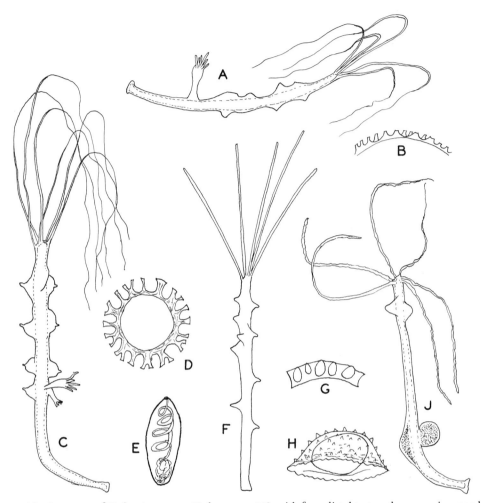

FIG. 10.–Structure of *Hydra*. A, mature *Hydra carnea,* ×8, with four distal testes, three ovaries, one basal testis, and one bud; B, spines of embryonic theca of *H. carnea* ; C, mature male *H. littoralis,* with eight testes and two buds, × 5; D, optical section of thecated embryo of *H. littoralis* ; E, holotrichous isorhiza of *H. littoralis*; F, mature male *H. americana,* ×6; G, portion of embryonic theca of *H. americana* ; H, thecated embryo of *H. hymanae* ; J, sexually mature *H. hymanae* with two testes and two ovaries, ×6. (A and B modified from Hyman, 1931; C and D modified from Hyman, 1938; E modified from Hyman, 1931b; F and G modified from Hyman, 1929; H and J modified from Hadley and Forrest, 1949.)

18. Simple colonial hydroids, 2 to 8 mm long, usually consisting of 2 to 4 individuals and seldom up to 12 (Fig. 5); rare in ponds, small lakes, fish ponds, and old flooded quarries.
     Order **TRACHYLINA**, PETASIDAE, **Craspedacusta sowerbyi** Lankester
   Solitary hydroid up to 3 mm long (Fig. 8); in brackish estuaries, swamps, and creeks, some of which are almost fresh; rare along New England coast.
     Order **HYDROIDA**, HYDRIDAE, **Protohydra leuckarti** Greeff

REFERENCES

Acker, T. S., and A. M. Muscat. 1976. The ecology of Craspedacusta sowerbii Lankester, a freshwater hydrozoan. *Am. Midl. Nat.* **95**:323–336.

Ahmad, M. F. et al. 1987. A new species of Limnocnida (Limnomedusae, Coelenterata) from a freshwater aquarium in India. *Hydrobiologia* **144**:33–36.

Belk, D., and D. Hotaling. 1971. Guam record of the freshwater medusa Craspedacusta sowerbyi. *Micronesia* **7**:229–230.

Berg, K. 1948. Studies on the bottom animals of Esrom Lake. *Mem. Acad. Roy. Soc. Lett. Danemark,* Ser. 9, **8**:1–255.

Bryden, R. R. 1952. Ecology of Pelmatohydra oligactis in Kirkpatricks Lake, Tennessee, *Ecol. Monogr.* **22**:45–68.

Campbell, R. D. 1983. Identifying hydra species. Pp. 19–28 in H. M. Lenhoff (ed.) *Hydra: Research Methods,* 463 pp. Plenum, New York.

———. 1987. A new species of Hydra (Cnidaria: Hydrozoa) from North America with comments on species clusters within the genus. *Zool. J. Linn. Soc.* **91**:253–263.

Dejdar, E. 1934. Die Süsswassermeduse Craspedacusta sowerbii Lankester in monographischer Darstellung. *Z. Morphol. Oekol. Tiere* **28**:595–691.

Dodson, S. I., and S. D. Cooper. 1983. Trophic relationships of the freshwater jellyfish Craspedacusta sowerbyi Lankester 1880. *Limnol. Oceanogr.* **28**:345–351.

Ewer, R. F. 1947. On the function and mode of action of the nematocysts of Hydra. *Proc. Zool. Soc. London* **117**:365–376.

———. 1948. A review of the Hydridae and two new species of Hydra from Natal. *Ibid.* **118**:226–244.

Forrest, H. 1963. Taxonomic studies on the hydras of North America. VIII. Description of two new species, with records and a key to the North American hydras. *Trans. Am. Microsc. Soc.* **82**:6–17.

Grayson, R. F. 1971. The freshwater hydras of Europe. I. A review of the European species. *Arch. Hydrobiol.* **68**:436–449.

Hadley, C. E., and H. Forrest. 1949. Taxonomic studies on the hydras of North America. 6. Description of Hydra hymanae, new species. *Am. Mus. Novit.* **1423**:1–14.

Hance, R. T. 1957. Mass fixation of Hydra. *Trans. Am. Microsc. Soc.* **76**:387–389.

Hyman, L. H. 1929. Taxonomic studies on the hydras of North America, I. General remarks and description of Hydra americana, new species. *Trans. Am. Microsc. Soc.* **48**:242–255.

———. 1930. Studies on hydras, II. The characters of Pelmatohydra oligactis (Pallas). *Ibid.* **49**:322–333.

———. 1931. Studies on hydras, III. Rediscovery of Hydra carnea L. Agassiz (1850) with a description of its characters. *Ibid.* **50**:20–29.

———. 1931a. Studies on hydras, IV. Description of three new species with a key to the known species. *Ibid.* 302–315.

———. 1938. Taxonomic studies on the hydras of North America. V. Description of Hydra cauliculata, n. sp., with notes on other species, especially Hydra littoralis. *Am. Mus. Novit.* **1003**:1–9.

Jones, C. S. 1941. The place of origin and the transportation of cnidoblasts in Pelmatohydra oligactis (Pallas). *J. Exp. Zool.* **87**:457–476.

———. 1947. The control and discharge of nematocysts in Hydra. *Ibid.* **105**:25–60.

Kepner, W. A., W. C. Gregory, and R. J. Porter. 1935. The manipulation of Hydra's nematocysts by Microstomum. *Science* **82**:621.

Kepner, W. A., B. D. Reynolds, L. Goldstein, and J. H. Taylor. 1943. The structure, development and discharge of the penetrant of Pelmatohydra oligactis (Pall.). *J. Morphol.* **72**:561–587.

Kepner, W. A., et al. 1951. The discharge of nematocysts of the hydra, with special reference to the penetrant. *J. Morphol.* **88**:23–48.

Lenhoff, H. M. 1968. Behavior, hormones, and Hydra. *Science* **161**:434–442.

Lytle, C. F. 1982. Development of the freshwater medusa Craspedacusta sowerbii. Pp. 129–150 in F. W. Harrison and R. R. Cowden (eds.) *Developmental Biology of Freshwater Invertebrates,* 588 pp. Liss, New York.

Miller, D. E. 1936. A limnological study of Pelmatohydra with special reference to their quantitative seasonal distribution. *Trans. Am. Microsc. Soc.* **55**:123–193.

Mueller, J. F. 1950. Some observations on the structure of Hydra, with particular reference to the muscular system. *Trans. Am. Microsc. Soc.* **69**:133–147.

Pauly, R. 1902. Untersuchungen über den Bau und die Lebensweise der Cordylophora lacustris Allman. *Jena. Z. Naturwiss.* **36**:737–780.

Payne, F. 1924. A study of the fresh-water medusa, Craspedacusta ryderi. *J. Morphol.* **38**:387–430.

Pennak, R. W. 1956. The fresh-water jellyfish Craspedacusta in Colorado with some remarks on its ecology and morphological degeneration. *Trans. Am. Microsc. Soc.* **75**:324–331.

Rowan, W. 1930. On a new Hydra from Alberta. *Trans. R. Soc. Can., Sect. 5, Biol. Sci.* **24**:165–170.

Schulze, P. 1917. Neue Beiträge zu einer Monographie der Gattung Hydra. *Arch. Biontologie* **4**:33–119.

Welch, P. S., and H. A. Loomis. 1924. A limnological study of Hydra oligactis in Douglas Lake, Michigan. *Trans. Am. Microsc. Soc.* **43**:203–235.

5

TURBELLARIA
(Flatworms)

Considering the Phylum Platyhelminthes as a whole, only a small fraction of the described species are free-living in fresh-water habitats. The classes Cestoidea (tapeworms) and Trematoda (flukes) are entirely parasitic, but the Class Turbellaria is almost exclusively a free-living group. Many of the species of turbellarians are marine and a few (about 10 in the United States) are terrestrial in warm, damp habitats or parasitic on marine invertebrates; the remainder are fresh-water forms.

Three land planarians have been reported from the United States, all belonging to the genus *Bipalium,* and all are introduced from their original extensive ranges in southeastern Asia and Japan. They are large planarians, usually 60 to 350 mm long, customarily found in the warm damp soil of greenhouses.

The Class Turbellaria is customarily divided into seven orders. Two of these orders, the Acoela and Polycladida, are essentially marine. The Alloeocoela is a marine and fresh-water order, but the Tricladida, Catenulida, Macrostomida, and Neorhabdocoela have fresh-water, marine, and terrestrial representatives.

There is one record of a microscopic acoel flatworm being found in U.S. fresh water (Mirror Lake, New Hampshire), and two other rare species are known from Germany and Poland (see figure on page 128). These worms are only about 0.5 mm long and lack a digestive cavity. Probably all are recent immigrants from ancestors in the marine environment. Careful collecting should reveal additional fresh-water Acoela, especially in rivers aand lakes not far from the seacoast.

Fresh-water turbellarians are found everywhere, usually on or closely associated with a substrate. The larger species are sometimes confused with small leeches, which they resemble somewhat in shape and color; the microscopic species resemble large ciliates in size, shape, and general habits.

General characteristics. All fresh-water flatworms are more or less elongated. They may be highly flattened and leaflike, cylindrical, or spingle shaped. Almost invariably, however, there is some evidence of flattening on the ventral surface. In many species the anterior end is differentiated and sufficiently specialized to resemble a "head."

Tricladida are usually 5 to 30 mm long, but the other orders are seldom more than 4 mm long, the majority being microscopic and 300 to 1200 μm long. Because of their habit of reproducing by fission, some small species customarily occur in chains of several individuals, or zooids. The general surface of the body is usually more or less covered with cilia, and the term "Turbellaria" was coined by Ehrenberg in 1831, when he detected the minute vortical currents of water generated at the anterior end by ciliary action. In the familiar species of planarians, to which every zoology student is exposed in college, cilia are absent from the dorsal surface and edges of the body.

The body cavity of flatworms is a typical gastrovascular cavity having but a single external opening, which functions as both mouth and anus. In the Tricladida the gastrovascular cavity has three main irregular branches, one median and anterior and the others posterior and lateral (Fig. 1). The other orders have a single median gastrovascular chamber, but in a few representatives its outline is more irregular, owing to small lateral caeca.

In a few Turbellaria the mouth is terminal or subterminal. More characteristically it is ventral and somewhere between one fifth and three fifths of the way back from the anterior end. A protrusible or nonprotrusible muscular pharynx surrounds that portion of the gastrovascular cavity just inside the mouth. If protrusible, the pharynx often lies in a separate pharyngeal chamber.

Almost all species are hermaphroditic and

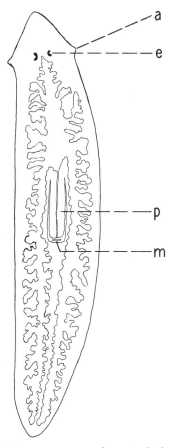

FIG. 1.–Gross anatomy of a typical planarian, showing the three divisions of the gastrovascular cavity, ×15. a, auricle; e, eyespot; m, mouth (on ventral surface of body); p, pharynx (lying in pharyngeal chamber).

reproduce sexually. One or two genital pores are found on the ventral surface, usually posterior to the mouth.

A wide range of coloration occurs in freshwater Turbellaria, although the brilliant patterns of marine species are never approached. The more familiar planarians (Tricladida) are various shades of gray, brown, or black; often there are dorsal stripes, spots, or mottlings. The ventral surface, however, is usually lighter in color or gray and without a color pattern. Some species have a highly variable coloration. *Dugesia tigrina,* for example, may be dark yellow, olive, brown, or brownish black, and the pattern may range from lightly mottled to dense coloration (Fig. 2). The lighter shades usually develop in laboratory cultures. Cave species and a few noncave species are whitish.

The other four orders are characteristically grayish or colorless, but a few are brilliant yellow, gold, orange, red, or rose. Other species are a bright green, owing to internal symbiotic zoochlorellae. In all naturally light-colored Turbellaria the body may take on a variety of darker shades, depending on the particular contents of the gastrovascular cavity.

Two darkly pigmented eyespots are usually present near the anterior end, but in some species eyespots are absent, and in a few others they are numerous.

Epidermal and subepidermal glands secrete an abundance of mucus that usually covers the body. Minute rodlike rhabdites are produced in the epidermis and subepidermal tissues of many genera, and these, on being extruded to the surface of the body, disintegrate, and produce an additional adhesive mucus.

Thin layers of circular and oblique, and a thicker layer of longitudinal muscle fibers lie below the basement membrane of the epidermis. The cells that line the gastrovascular cavity are large and phagocytic. Between the inner gastrodermis and the outer muscle fibers and epidermis the body consists largely of an undifferentiated mass of parenchyma cells. A few muscle fibers are found in the parenchyma, but otherwise it appears to function as a loose packing material between the excretory and reproductive organs. In the smaller turbellarians the parenchyma is often highly vacuolated, glandular, and watery.

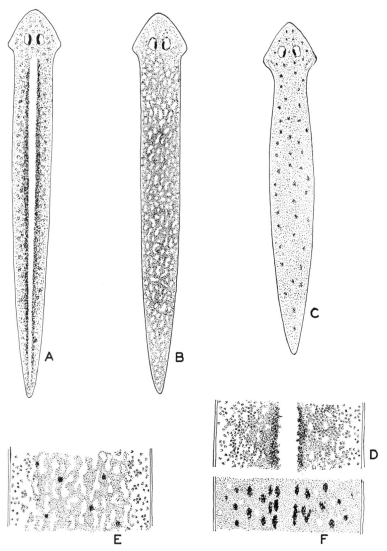

FIG. 2.–Variations in color pattern of *Dugesia tigrina*. A, striped form; B, spotted form; C, dark spots on uniform brown background; D to F, enlarged views of similar patterns. (From Hyman, 1939.)

Locomotion. Triclad flatworms cannot swim but are only capable of locomotion on a substrate or on the underside of the surface film of water. Smooth gliding movements are produced by the action of cilia on the thin coat of mucus secreted onto the substrate by cyanophilous glands in the ventral parenchyma. Faint waves of muscular contractions pass longitudinally down the body and supplement ciliary action. On strong stimulation, planarians may crawl actively by means of pronounced waves of muscular contraction that pass from anterior to posterior. A few species have anterior adhesive organs, which may be used as a leech uses its anterior sucker for locomotion.

Catenulids, macrostomids, neorhabdocoels, and alloeocoels move on a substrate as do planarians, but many are also capable of swimming about freely in the water, owing to their abundant cilia.

When conditions are unfavorable, when

the water is too warm, or when fully fed, planarians are not active but remain more or less contracted for days at a time.

Feeding, digestion. The most familiar type of feeding occurs in the planarians. Here the pharynx is long, highly muscular, and protrusible through the mouth. When the presence of food (usually living, dead, or crushed animal matter) is detected in the water, the animals quickly move toward it and place the ventral surface of the body in contact with it. The pharynx is then extruded through the mouth, and its tip is placed against the food (Fig. 3). Sometimes the pharynx may be extended to as much as one half of the body length. There appears to be no appreciable secretion of digestive juices from the tip of the pharynx, and apparently only soft or disintegrating tissues are capable of being sucked up into the main gastrovascular cavity by the muscular action of the pharynx. The gastrovascular cavity becomes filled with fluid and small bits of tissues after 30 to 80 minutes of feeding. There is no good evidence for extracellular digestion in planarians. Instead, individual particles are ingested by pseudopodial action of the large gastrodermal cells, and digestion is intracellular. Evidence based on the small amount of work that has been done on microturbellarians, however, indicates that it is probable that they have considerable extracellular digestion.

Some species of *Phagocata* (Tricladida) are interesting in that they are polypharyngeal (Fig. 14F). In microturbellarians the pharynx is short and generally not greatly protrusible, but the normally small mouth may be expanded to a marked degree during feeding.

Although some flatworms are sometimes said to be vegetarians, the great majority of Turbellaria are zoophagous and feed on small living invertebrates. Nevertheless, dead

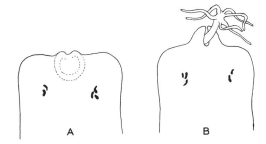

FIG. 4.–Dorsal views of head of *Procotyla fluviatilis*. A, with adhesive organ retracted; B, with adhesive organ grasping a hydra. (Modified from Redfield, 1915.)

animal matter may often be ingested, and the microscopic species subsist on fresh corpses of larger metazoans as well as on whole living or dead Protozoa, rotifers, nematodes, and gastrotrichs. Small individuals, such as *Catenula* and *Macrostomum,* are especially abundant in decaying organic matter and perhaps are essentially omnivorous. Some species of *Phaenocora* and *Macrostomum,* however, prefer animal tissues and eject plant material from the pharynx. Many substrate species prefer a diatom diet.

Some few species are so voracious that they will exterminate other species in a mixed culture. Cannibalism has occasionally been observed, especially in crowded cultures where there are injured specimens.

A few microturbellarians have been found to form an adhesive layer or mass on the substrate by the discharge of rhabdites at the anterior end. Such mucus with its entangled detritus, bacteria, algae, and protozoans is then gathered into a ball and ingested. Some investigators suggest that this method of feeding may be more common than is suspected. Planarians also exude rhabdites, which form a mucus on coming in contact with water. Small crustaceans get caught in this material, as well as in the usual slime track, and are thus easy prey.

Some Dendrocoelidae have an interior, adhesive, suckerlike organ that is used in feeding. Such species pounce on small crustaceans and hydras and grasp them firmly by bringing the lateral edges of the sucker together. The prey is then held against the substrate in the pharyngeal region while

FIG. 3.–Planarian with extruded pharynx, just beginning to feed on a bit of food.

feeding proceeds. *Procotyla fluviatilis* feeds chiefly on living crustaceans; it cannot be cultured on raw meat.

In general, the larger species of microturbellarians feed on living plankton and substrate crustaceans (up to 10/day) and annelids, but dead asellids and gastropods are common foods also.

When reared in the laboratory, planarians should be fed one to three times per week. They may be kept for months without any food, however, while the body gets progressively smaller and more simplified in structure. No reproduction occurs during such periods of starvation.

Microstomum and a few other genera have become curiously adapted for retaining the undischarged nematocysts of ingested hydras. Within 10 to 20 hours after ingestion, the nematocysts are found distributed over the body surface between the epidermal cells. Such nematocysts are retained as defense devices for the remainder of the life of the turbellarian.

There is no special circulatory system in flatworms. Materials are distributed throughout the gastrovascular cavity by means of general muscular and body movements. The tissues are nowhere thick or bulky, and diffusion of materials from cell to cell is easily facilitated.

Respiration. The respiratory exchange of oxygen and carbon dioxide occurs through the general epidermis and perhaps to a slight extent through the gastrodermis also. Many microturbellarians are characteristic of habitats where there is much decay and low concentrations of dissolved oxygen; some are able to withstand anaerobic conditions for variable lengths of time, although there is little definite information on this phase of flatworm biology in the literature. Planarians, however, all require high concentrations of oxygen and occur only in well-oxygenated habitats.

Symbiotic zoochlorellae occur in the parenchyma and gastrodermis of many microturbellarians, although they are not necessarily consistently present in a particular species. Photosynthesis releases oxygen to the turbellarian tissues, while the worm contributes carbon dioxide and nitrogenous

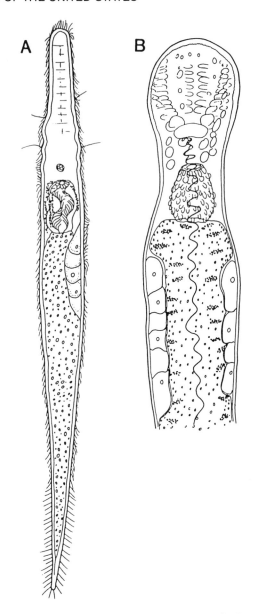

FIG. 5.–Two undescribed species of *Rhynchoscolex,* showing the range of shape in the proboscis. (From Kolasa et al. 1987, with permission.)

compounds for algal metabolism. Algae may be passed on to the succeeding generation in the turbellarian egg.

Excretion. The flame bulb, or protonephridial, system of flatworms has a variable pattern. There may be few to many flame bulbs, but their ducts usually empty into two longitudinal collecting ducts (Fig. 6). *Stenostomum* and *Catenula* are exceptional in having a single median collecting duct. Collecting ducts open to the ventral surface by means of one or two pores, which are variable in position.

As in certain other phyla, the excretory process is a function of the ducts rather than the flame bulbs themselves. Nevertheless the physiological significance of the whole flame bulb system appears to vary from one species to another. Sometimes the system has an important excretory function; sometimes it is apparently unimportant. Often it is of great importance in osmoregulation; in other species it appears to be relatively unimportant.

Gyratrix hermaphroditus occurs in fresh-water, brackish, and marine habitats. In the former it has a highly developed flame bulb system. In brackish habitats the system is less highly developed, with the bladder and ampullae

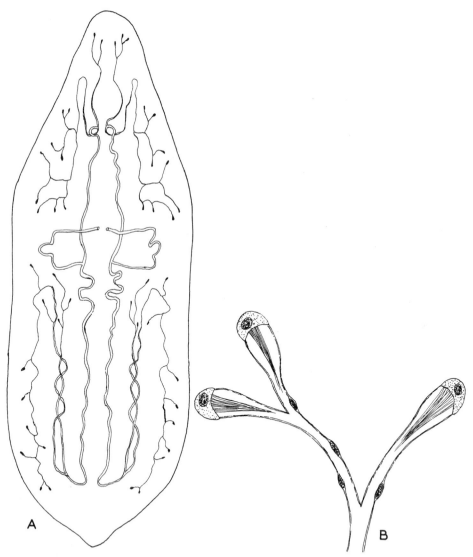

FIG. 6.–Diagrams of protonephridial structures. A, protonephridial system of *Mesostoma* (the two proximal pores are located just inside the mouth); B, three flame bulbs at terminal endings of tubules.

lacking. In marine habitats, however, it is poorly developed or absent.

It is probable that a large fraction of the excretory materials are disposed of through the general epidermal surface and also perhaps to a slight extent through the gastrodermis into the gastrovascular cavity. Some excretory granules are commonly retained and stored in the tissues until the animal dies.

Nervous system. The brain consists of two more or less well-defined lobes, or ganglia, near the anterior end. Usually there are two prominent longitudinal nerve trunks, one originating in each ganglion, but often there are one to several additional, smaller trunks. Many cross commissures and small lateral nerves originate from the longitudinal trunks and innervate all of the organs and tissues. An abundance of sensory fibers are given off from the brain and innervate the anterior end (Fig. 7).

Sensory receptors. Eyespots are the most highly developed sense organs, especially in planarians. Each eyespot is shaped like a bean or a pigmented cup containing translucent light-sensitive receptor cells, which are attached to nerve fibers at the outer portion of the cup. Some spring and most cave species of Tricladida have no eyespots. The majority of microturbellarians have one pair, but some are blind, and a few have multiple eyespots (Figs. 17–19).

Such eyespots are relatively simple in structure and are sometimes unpigmented and called light-refracting bodies rather than eyes. In *Stenostomum,* for example, the so-called eyespot is colorless and consists of a few to many basal refractive spherules and a distal low cone-shaped vesicle. Actually a visual function has not yet been definitely established for this structure.

Although eyespots may be well developed, most turbellarians are photonegative, especially in moderate to strong light, and as a consequence are usually found in shaded areas or under objects.

Some species have a pair of anterior sensory pits, often ciliated, which are presumably chemoreceptors. They may be rounded, oblong, slitlike, and shallow or deep (Fig. 16D).

Larval *Rhynchoscolex* and adult *Catenula* and *Otomesostoma* have a median anterior statocyst (Figs. 16A, 20B), but none of the other American fresh-water genera have such a structure.

In addition to the general coat of short cilia, many have a few long sensory cilia, especially at the anterior and posterior ends (Figs. 9, 10). These move independently of each other and of the short cilia. In some species they are relatively stiff and are called "spines."

Other receptors, mostly in the epidermis, function in the detection of currents, in olfaction, and in response to contact. The pharynx and anterior end are especially sensitive.

Reproduction, development. With reference to their reproductive habits, there are three general groups of turbellarians: (1) species that reproduce only asexually, by budding or fission; (2) species that reproduce solely by sexual methods; and (3) species that are capable of reproducing sexually or asexually, depending on the genetic strain, physiological strain, and prevailing ecological conditions.

Fission does not occur in the majority of triclads, though certain common species are in the third category. Among these latter species the relative frequency of sexual and asexual reproduction varies greatly from one species to another. In *Dugesia tigrina,* for example, there appear to be at least two physiological varieties. One variety reproduces only asexually. Under appropriate conditions of temperature and food supply, individuals pinch in at about midlength on each edge. The pinching in proceeds rapidly until the two halves are completely separated. Then the anterior end of the posterior individual reconstitutes a head, and the posterior end of the anterior individual forms a tail. External factors may change the rate of fission or inhibit it, but they do not change the reproduction to the sexual type. Fission occurs only above 10°C, with frequency increasing to a maximum at 25 to 28°C. Such fission strains have been kept as long as 5 years in the laboratory without development of sex organs.

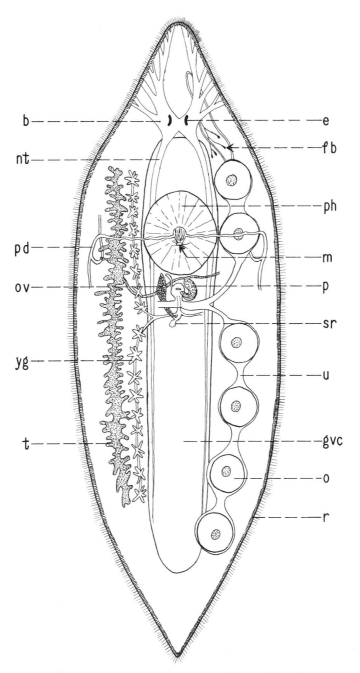

FIG. 7.–Structure of *Mesostoma ehrenbergi* Focke, ×20, ventral view, b, brain; e, eyespot; fb, flame bulb (distributed over entire body); gvc, gastrovascular cavity; m, mouth; nt, longitudinal nerve trunk; o, ovum; ov, ovary; p, penis (projected into genital atrium); pd, protonephridial duct; ph, pharynx; r, rhabdites in epidermis; sr, seminal receptacle; t, testes (shown only on left side of body); yg, yolk gland. (Modified from Ruebush, 1940.)

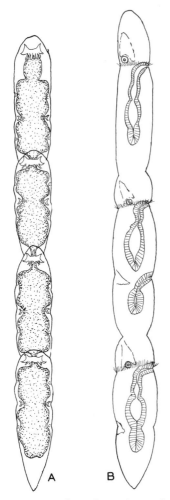

FIG. 8.–Chains of zooids in microturbellarians. A, dorsal view of four zooids of *Stenostomum*, ×24 (gastrovascular cavity stippled); B, lateral view of three zooids of *Catenula*, ×85. (Modified from Marcus, 1945.)

In other physiological races of *D. tigrina* there may be an alternation of reproductive methods according to water temperature. Sex organs may develop during the winter and early spring, and egg capsules are deposited in May and June. As the water temperature increases, however, the reproductive organs degenerate, and in July, August, and September reproduction may be entirely by fission. By late autumn the sex organs are again beginning to develop. If the water temperature remains low throughout the year, however, sexual reproduction may be continuous.

It has also been suggested that there are still other strains of *D. tigrina* that reproduce sexually regardless of normal seasonal temperature variations.

Dugesia dorotocephala and a few other triclads also appear to exist in several physiological varieties comparable to the situation in *D. tigrina*. Nevertheless *D. dorotocephala* is rarely observed to reproduce sexually under natural conditions.

Procotyla fluviatilis is incapable of asexual reproduction; it is sexual and deposits egg cocoons during the winter months. Asexual reproduction is also unknown in *Cura foremanii*, but it reproduces sexually at all seasons. In *Phagocata vernalis* and *P. velata*, however, sexual reproduction is rare and apparently occurs only during the winter months in permanent bodies of water.

The seasonal reproductive habits of microturbellarians are poorly known, but most of the genera reproduce solely by the sexual method. *Catenula, Rhynchoscolex, Suomina, Microstomum,* and *Stenostomum* are the only genera capable of reproducing asexually. In these forms fission proceeds rapidly and often along several transverse planes so that there may be a series of two to eight zooids in a chain (Fig. 8). Occasionally a few species in this group of five genera have been observed to develop sex organs and reproduce sexually. At least one species of *Phaenocora* spends about 7 months in a diapause egg stage.

Asexual reproduction in triclads involves either fission or fragmentation. In fission the posterior part of the body tears itself away; each of the two resulting individuals then grows and regenerates the lost parts. In fragmentation the animal breaks into several parts, each of which encysts in a slime mass. After an inactive period, a small individual emerges from the cyst, becomes active, and grows to normal size. Triclads have seldom been observed dividing, and there is some evidence to show that the process occurs at night and is completed within 30 minutes to several hours. In microturbellarian fission a full complement of organs is usually developed before separation. Under favorable conditions divisions may occur as frequently

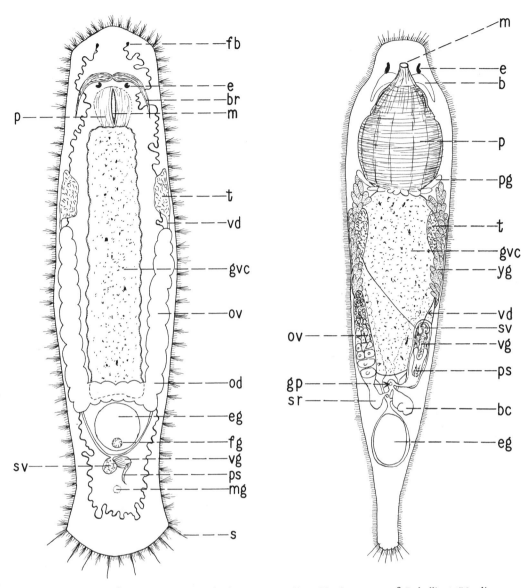

FIG. 9.–Structure of *Macrostomum appendiculatum* (Fabr.), ×140, a species common everywhere in fresh, brackish, and salt waters. br, brain; e, eyespot; eg, egg; fb, flame bulb; fg, female genital pore; gvc, gastrovascular cavity; m, mouth; mg, male genital pore; od, oviduct; ov, ovary; p, pharynx; ps, penis stylet; s, spine; sv, spermiducal vesicle; t, testes; vd, vas deferens; vg, prostatic vesicle (Modified from Ferguson, 1939.)

FIG. 10.–Structure of *Dalyellia*, ×70, diagrammatic. b, brain; bc, bursa copulatrix; e, eyespot; eg, egg; gp, genital pore; gvc, gastrovascular cavity; m, mouth; ov, ovary; p, pharynx; pg, pharyngeal gland; ps, penis stylet; sr, seminal receptacle; sv, spermiducal vesicle; t, testes; yg, yolk glands; vd, vas deferens; vg, prostatic vesicle. (Modified from Ruebush and Hayes, 1939.)

as every 5 to 10 days in triclads, but in micro-turbellarians the time interval is much shorter.

All American fresh-water turbellarians are hermaphroditic, and the male and female organ systems are unusually complex in many genera. They are not present during periods of fission but develop later, presumably from undifferentiated parenchyma cells. The male organs sometimes mature before the female organs. Various types of reproductive systems are diagrammed in Figs. 7 and 9–11, and only a typical planarian arrangement will be described here (Fig. 11).

In the female reproductive system the two ovaries are on either side of the body near the anterior end. A longitudinal oviduct (or ovo-vitelline duct) proceeds posteriorly from each ovary and collects yolk cells from numerous lateral yolk glands. The two ovovitelline ducts unite near the midline, posterior to the mouth, and enter a common cavity, the female atrium. A saclike bursa copulatrix connects with the female atrium by means of a bursal canal.

The testes are small spherical bodies arranged roughly in two longitudinal rows. Each one connects with one of the two vasa deferentia by means of a minute duct. As the vasa deferentia proceed posteriorly they are widened and convoluted to form spermiducal vesicles where ripe sperm are stored. The spermiducal vesicles, either with or without uniting into a common duct, enter the basal portion of the highly muscular penis, and the tip of the latter projects into a cavity, the male atrium. Male and female atria unite ventrally to form a common atrium, which opens externally as a common genital pore.

Additional cement glands, copulatory glands, and other accessory reproductive structures occur in many turbellarians. Other orders often have the distal end of the penis modified to form a stylet, and sometimes the basal part is a prostatic vesicle, which secretes a granular material vital to the sperm. Many microturbellarians have two genital pores.

As is the case in most hermaphroditic animals, two turbellarians usually copulate and exchange sperm. The penis of each animal is elongated and protruded from its genital pore and through the genital pore of the other animal, the sperm being discharged into the bursal canal in planarians. Both

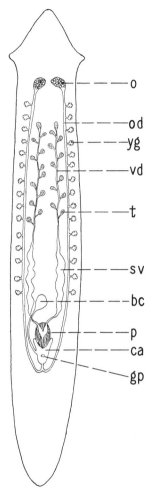

FIG. 11.–Diagrammatic ventral view of a typical triclad reproductive system, ×16. (The relationships of the male and female atria are distinct only in a lateral view.) bc, bursa copulatrix; ca, common atrium; gp, genital pore; o, ovary; od, ovovitelline duct; p, penis; sv, spermiducal vesicle; t, testis, vd, vas deferens; yg, yolk gland.

animals exude quantities of mucus from the genital pore region during copulation. On completion of copulation, sperm are stored in the bursa copulatrix for a short time and then leave and migrate up the oviducts to the region of the ovaries. After a variable interval, the eggs are fertilized as they leave the ovaries.

As the zygote moves down the oviduct it accumulates yolk cells, and in the male atrium aggregates of zygotes and yolk cells become surrounded by a proteinaceous capsule. Such a cocoon usually contains 2 to 20 zygotes and thousands of yolk cells. It leaves the body through the genital pore, and in *Dugesia, Cura,* and a few other species it is attached to the substrate by means of a stalk (Fig. 12A). Each worm may deposit a series of cocoons during the breeding season.

Hypodermic impregnation appears to be the rule in *Stenostomum.* A single egg is ripened in each ovary during the reproductive season. It is fertilized inside the body and is liberated to the outside by a rupturing of the body wall.

More commonly, however, microturbellarians copulate and exchange sperm in the usual fashion. Fertilized eggs are usually laid singly, each enclosed in a capsule or shell that is sometimes stalked (Fig. 12B). Many of the Typhloplanidae produce two kinds of eggs. Thin-shelled "summer" eggs hatch promptly after deposition, but the larger, thick-shelled "winter" eggs have delayed hatching and may be dormant during the winter; they are capable of withstanding unfavorable environmental conditions. Winter eggs may be extruded from the parent body in the customary way or they may be liberated on the death and disintegration of the parent. Some species are regularly viviparous; the summer eggs hatch within the body and the young break out of the posterior region of the parent to the outside. In microturbellarians other than the Typhloplanidae only eggs of the winter type are produced.

Self-fertilization is rare in the Tricladida and has been definitely demonstrated only for *Cura foremanii.* Certain species sometimes produce sterile cocoons. Some investigators think that summer eggs of typhlopanids are self-fertilized.

Cocoons of triclad turbellarians are 2 to 4 mm in diameter and light colored when recently deposited, but soon become dark brown or blackish. Although they are resistant to low temperatures, they cannot withstand isolated drying. Depending on the season, a cocoon may winter over or hatch in about 2 weeks. The several worms that emerge are only 1 to 3 mm long. Development is direct and there are no special larvae stages in the American fresh-water species. With the exception of reproductive structures, the newly

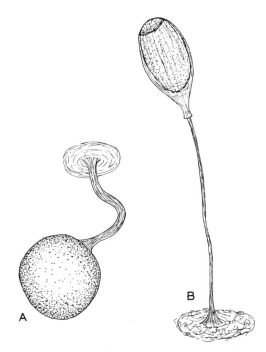

FIG. 12.–Turbellarian cocoon and microturbellarian encapsulated egg. A, stalked planarian cocoon, ×7; B, stalked encapsulated egg of *Gyratrix,* ×100 (note operculum).

hatched worm often has the full complement of organ systems.

Phagocata velata and *P. vernalis* have an unusual habit of fragmenting, especially when their habitat becomes warm in late spring and summer. When a worm reaches a length of 12 to 15 mm, the internal organs degenerate and the whole worm fragments into numerous small bits. Each piece secretes mucus, which hardens to form a small cyst, and a whole miniature individual is reconstituted within the cyst. Encystment of a whole mature animal has also been reported. Such cysts may hatch in a few weeks or may winter over.

Length of life is known chiefly from laboratory cultures. Sexually reproducing species live for only a few weeks or months, usually dying in summer or autumn after producing winter eggs. A few species, however, may survive a year or more by encysting during unfavorable conditions. Species capable of fission can presumably continue indefinitely under appropriate conditions. Such laboratory cultures have been kept going for 2 to 11 years.

The larger species of planarians, especially *Dugesia,* have long been classical objects for regeneration and grafting experiments. A great deal of such work was done in the United States, especially between 1910 and 1930. If extended specimens are carefully cut transversely into two or several pieces with a razor blade, a large percentage of the pieces will regenerate into whole new worms, the original longitudinal axis being maintained. Worms cut into two along the median line regenerate respective right and left halves to form two new worms. Partial longitudinal cuts at the anterior end often produce worms with two or more complete heads. Regeneration potentialities are most highly developed in the anterior half or three quarters of the body.

In spite of this regeneration capacity, planarians are rather delicate, and when handled roughly in the laboratory they easily develop breaks and irregularities in the body.

Ecology. Springs, spring brooks, ditches, marshes, pools, ponds, lakes, and caves all have their characteristic turbellarian populations. Most species, and especially triclads, are photonegative and are found under objects or in debris during the daytime. Flatworms thrive on any kind of substrate where there is an appropriate food supply. Although they are most characteristic of the shallows, a few species have been collected from lake bottoms as deep as 100 m.

Triclads are typical of brooks and streams; the other four orders are never abundant in such habitats. Rheophile forms are also usually cold stenotherms; common species are *Cura foremanii, Dugesia dorotocephala, Polycelis coronata, Phagocata velata,* and *P. morgani.* The last-named species is restricted to water between 4 and 14°C. *Procotyla fluviatilis* occurs in all kinds of habitats, including running waters, standing waters, and even brackish areas. In general, specimens from running waters attain greater size than those from standing waters. *Dugesia tigrina* and *Phagocata vernalis* are both standing-water triclads, but whereas the former species is a eurytherm, the latter is active only during the cold months in temperate zone ponds and ditches.

Small colonies of a well-known European species, *Crenobia alpina,* are restricted to cold brooks of the Alps, the Scandinavian countries, and Great Britain. The species is seldom found in water that becomes warmer than 12, and 7°C is the optimum temperature. Sexual reproduction is most common in the winter, but fission may occur in spring and early summer. Encystment is unknown. Because of its peculiar stenothermal distribution, this species is thought to be a glacial relict. No species of comparable habits are known from the United States.

The nontriclads are especially abundant during the warm months in pools and ponds containing quantities of algae, other vegetation, and organic debris. In a single lake, where many microhabitats or niches exist, as many as 30 species have been recorded, especially in hard-water lakes and ponds. Almost any stagnant pool will contain one or more species of *Dalyellia,* and usually one or more additional species in each of *Stenostomum, Macrostomum, Catenula, Microstomum,* and *Mesostoma.* Such species are eurythermal since they occur in habitats having a wide daily and seasonal temperature range. *Gyratrix, Phaenocora,* and *Otomesostoma* are also chiefly eurythermal. *Rhynchomesostoma rostratum* and certain species of *Macrostomum, Dalyellia,* and *Castrada,* however, are some of the few genera known to be cold stenotherms.

Some species spend as much as 8 months per year in diapause.

Kolasa et al. (1987) have recently demonstrated the presence of a rich microturbellarian fauna on the bottom and in the interstitial waters of streams and springs in New York state. Although the worms were never abundant, these investigators found 33 species of which 5 were undescribed and 8 were new to North America. Twelve species of *Stenostomum* accounted for the great majority of individuals collected.

The great majority of turbellarian genera occurring in fresh waters are restricted to that environment. *Macrostomum,* however, is the most distinct exception, with many species in brackish and marine environments. As implied in a previous section, *Gyratrix hermaphroditus* is one of the very few euryhaline species; it is found in fresh-water, brackish, and marine environments. There is some evidence to show that this taxon actually consists of a large cluster of incipient species.

In those microturbellarian genera occurring in both standing and running waters there is often a correlation of size and activity with habitat. Still waters are characterized by larger, more sedentary species, whereas the small, faster-moving species are more abundant in waters where there is a current, however slow.

Some species of *Pseudophaenocora* thrive in the presence of only a trace of oxygen, and many other small species occur in water where the concentrations are as low as 5 to 40% saturation. Most triclads are restricted to water whose oxygen concentration does not fall below 70% saturation.

Hymanella retenuova is the only planarian known to produce a thick-walled cocoon capable of withstanding drought. In addition, however, *Phagocata velata* and *P. fawsetti* undergo multiple fission during drought; each fragment secretes a slime layer which hardens into a resistant cyst.

Enemies, parasites. Turbellaria are seldom an important element in the diet of other animals. Odonata nymphs consume a few planarians, and members of the other orders appear to be taken occasionally by nematodes, annelids, and a few crustaceans and aquatic insects.

Internal parasites of turbellarians include ciliates, gregarines, and (rarely) larval mermithid nematodes. Some turbellarians may be found in the brood pouch of isopods and amphipods, but they are probably commensals rather than true parasites.

Geographical distribution. So far as flatworms are concerned, that portion of the United States west of the Great Plains is almost a *terra incognita*. A few triclad records have been published for this region, but microturbellarians have been even more greatly neglected. East of the Mississippi the triclads have been extensively studied, especially by Hyman, and although the nontriclads have been studied to a considerable degree by Kepner, Ferguson, and Ruebush, an enormous amount of work yet remains to be done. The number of microscopic species

appears to be very large, and each new investigation turns up new species. As shown in the key beginning on page 138, some triclads are common and widely distributed, whereas others are known from only one or two localities. Cave species are highly endemic, each form being restricted to a single cave or cave system, or to a group of caves in the same general area.

Although many microturbellarians have thus far been reported from only one or several collecting areas, a few appear to be common and essentially cosmopolitan in the Holarctic region. Typical examples are *Macrostomum appendiculatum, Stenostomum leucops* (Dugès), *Castrella truncata* (Abildgaard), *Phaenocora unipunctata* (Ørsted), and *Bothromesostoma personatum* (Schmidt).

Little definite information is available concerning the dissemination of turbellarians. *Dugesia dorotocephala* is generally distributed in spring brooks and marshy springs, yet it probably is unable to migrate actively throughout a single drainage system because of its ecological preference for headwaters. The same is true of *Phagocata morgani,* which is widely distributed east of the Mississippi. *Dugesia tigrina,* however, is never found in springs or spring brooks, but it does occur in a wide variety of other lotic and lentic habitats and presumably can migrate actively through drainage systems over wide areas. *Procotyla fluviatilis* and *Cura foremanii* also probably spread by active migration. Species such as *Phagocata vernalis,* however, which occur in temporary ponds and ditches, must rely entirely on passive means of distribution.

Even less is known about the dissemination of microscopic species, but the mechanisms of distribution must be fairly efficient, since the same species may be found everywhere in suitable habitats, even in temporary ponds and puddles that are dry during the late summer and early autumn.

Undoubtedly the cyst stages, winter eggs, and perhaps even cocoons are transported from place to place by natural agencies and larger animals that are capable of moving or flying overland from one pond, lake, or stream to another.

Oligochoerus, a fresh-water acoel turbellarian, has recently been found in several European fresh-water canals, rivers, and lakes. We predict that this group will also eventually be found in U.S. fresh waters.

Collecting. Spring brooks may be "baited" for planarians with cubes of raw, lean beef. After 15 minutes the meat should be examined and the worms shaken off into a container of water. Otherwise they may be washed or picked off the undersides of objects with a large pipette. If planarians are abundant in water cress, the plants should be placed in containers in the laboratory. As oxygen decreases, the worms will migrate to the top, where they may be picked off with a glass rod or pipette.

Planarians may be taken from pond and lake substrates by placing the bait in a glass jar with a well-perforated lid. Retrieve the jar after 24 hours.

Because of their small size, the other orders are more difficult to obtain and concentrate in numbers. The uppermost 1 or 2 cm of bottom debris as well as algae and rooted vegetation from any body of water will yield these forms if such material is placed in wide-mouthed jars or aquaria in the laboratory. In 6 to 12 hours many of the specimens will migrate to the surface film and may be picked up with a pipette. Such containers should also be allowed to stand undisturbed for 2 to 20 days. As decomposition increases, more individuals and additional species will be found.

Sometimes a Birge cone net drawn through dense vegetation will take large numbers. Small species are common inhabitants of the under surface of lily pads.

The Bou-Rouch or comparable sampler is useful for collecting interstitial species (see page 610).

Culturing. Planarians may be kept in standing water, provided it does not become too warm and is changed every other day. Tap water is often toxic. Enameled pans and glass or crockery containers are all suitable. Since they are photonegative animals, it is best to put objects into the container, such as pebbles or bits of broken flowerpots, in order to provide shaded places. Also, the whole container may be covered. Cubes of fresh meat (especially beef liver), earthworm fragments, or chopped meal worms are suitable foods

and should be supplied once or twice a week. *Procotyla fluviatilis,* however, requires small living crustaceans. After 2 or 3 hours any remaining food fragments should be removed, and if the cultures are in standing water, the water should be changed. Planarians will live for 3 to 12 weeks without any food but they get progressively smaller.

Most planarians are seldom found reproducing sexually, but if sexual specimens of *Dugesia tigrina* are collected in the field, they will usually continue producing egg capsules in the laboratory. *Procotyla fluviatilis* is especially useful for a study of sexual reproduction because of its translucency. Its sexual period extends from autumn through winter and sometimes into early spring. Nevertheless, *Cura foremanii* is about the only species that can really be depended on to produce egg capsules regularly under laboratory conditions.

Temporary cultures of microturbellarians may develop and be kept for several weeks in the original containers of decaying aquatic vegetation and debris in the laboratory. More or less permanent cultures may be maintained in Petri dishes or finger bowls in a variety of ways. *Stenostomum* and certain other genera may be cultured in boiled wheat, rice, or rye grain infusions similar to those used for other micrometazoans and Protozoa. Subcultures should be started every 5 or 6 weeks. Pablum or cultures of similar foods are often successful. Some cold stenothermal species of *Macrostomum* can be cultured only in a refrigerator. Food may be crushed entomostracans, chopped meal worms, or chopped aquatic oligochaetes.

Preparing, preserving. There are several suitable methods of killing and fixing planarians, but all methods should begin with the animals extended in a very small amount of water. Hot saturated mercuric chloride in 0.9% sodium chloride is good. Two percent nitric acid for 1 minute followed by 70% alcohol is also fairly good. Two percent nitric acid may be dropped on the planarians, and then they should be flooded with hot saturated mercuric chloride in 0.9% sodium chloride. A biological supply company advocates narcotizing in strychnine water for a few minutes and then flooding with hot Gilson's fluid. The latter solution consists of the following: 5 g of mercuric chloride, 5 ml 80% nitric acid, 1 ml of glacial acetic acid, 25 ml 70% alcohol, and 220 ml of water; filter after 3 days.

Because of their greater translucency, microturbellarians are advantageously studied alive with the 4, 8, or 16 mm objective in a hollow-ground slide or by placing them between two square cover slips sealed at the edges with petroleum jelly. Such a mount may be turned over on a slide and both sides of the specimens examined. Slight pressure compresses the worms so that they move about only slowly. If the preparation is carefully made, even an oil immersion lens may be used.

Nontriclads should usually be anesthetized with 0.1% chloretone, 10% alcohol, 1% hydroxylamine hydrochloride, or near-freezing temperatures. Fixation with Helly's fluid produces little shrinkage. This reagent consists of the following: 2.5 g of potassium dichromate, 5 g of mercuric chloride, 1 g of sodium sulfate, 100 ml of water, and 5 ml of formalin. Hayes (1941) recommends Harper's modification of Allen's B-15 fixative. It consists of two solutions, which should be mixed immediately before use. Solution A contains 380 ml saturated picric acid, 40 ml of glacial acetic acid, and 8 g of chromic acid. Solution B contains 220 ml of saturated picric acid, 200 ml of formalin, and 8 g of urea.

Flatworm fixation usually takes 30 to 60 minutes. If the fixative contains mercuric chloride, the specimens should be washed several times in 50% alcohol containing a little tincture of iodine. Pigmented specimens may be bleached in undiluted hydrogen peroxide. Seventy percent alcohol and dioxan are suitable preservatives. Stained whole mounts of rhabdocoels are sometimes useful, orange G and Delafield's hematoxylin being suitable stains.

Taxonomy. If serial sections are not available, identification of triclads to genus or species should be made on living specimens whenever possible. Shape, and especially the shape of the anterior end, is highly variable in preserved specimens, even though carefully prepared.

Generic identification of nontriclads is usually possible with live specimens, but they must be mature.

On the other hand, species identification of almost all nontriclads and species identification of almost all nontriclads and species identification of some triclads are entirely dependent on serial sections, and especially on the detailed anatomy of the reproductive system of the mature individual. It is unfortunate that such work can be done only by the experienced specialist, and the considerable effort involved in making species identifications will probably always maintain a dearth of investigators.

The turbellarian fauna of the United States is poorly known, especially in the western half of the country, and therefore a species name should never be given to a specimen simply because it is the only one previously reported from a particular region.

One of the most perplexing and variable genera is *Phagocata*. Most of the species are known only from restricted localities in widely scattered areas of the United States. Some are epigean; some are hypogean. The group is sorely in need of revision and should perhaps be split into at least three genera.

Many of our species have a long history of taxonomic confusion and synonymy, chiefly as the result of variable characters, poor fixation, and identification by inexperienced workers. The section of the key that includes the microturbellarian genera has been greatly modified from Ruebush (1941). There are so many undescribed and unreported species of microturbellarians that it is fruitless to carry this part of the key beyond the generic level. The triclad section is greatly modified from Kenk (1972). Both the triclad and microturbellarian sections are constructed with practicability in mind, without strict attention to fundamental generic criteria. Several rare genera are omitted. Following is an outline of the taxonomic relationships recognized in this manual.

Order Tricladida
  Planariidae
    Cura
    Dugesia
    Hymanella
    Kenkia
    Phagocata
    Planaria
    Polycelis
    Sphalloplana
  Dendrocoelidae
    Dendrocoelopsis

Macrocotyla
Procotyla
Rectocephala
Order Catenulida
  Catenulidae
    Catenula
    Rhynchoscolex
    Suomina
  Stenostomidae
    Stenostomum
Order Macrostomida
  Macrostomidae
    Macrostomum
  Microstomidae
    Microstomum
Order Neorhabdocoela
  Provorticidae
    Provortex
  Dalyelliidae
    Castrella
    Dalyellia
    Gieysztoria
    Microdalyellia
  Typhloplanidae
    Amphibolella
    Bothromesostoma
    Castrada
    Krumbachia
    Mesostoma
    Olisthanella
    Opistomum

Phaenocora
Prorhynchella
Protoascus
Pseudophaenocora
Rhynchomesostoma
Strongylostoma
Typhloplana
  Gyratricidae
    Gyratrix
  Polycystidae
    Klattia
    Polycystis
  Koinocystidae
    Koinocystis
    Microkalyptorhynchus
Order Alloeocoela
  Prorhynchidae
    Geocentrophora
    Prorhynchus
  Otomesostomidae
    Otomesostoma
  Bothrioplanidae
    Bothrioplana
  Plagiostomidae
    Hydrolimax

The figures associated with this key are all diagrammatic, and usually all anatomical details except significant key characters are omitted. Wherever possible, characters are used that do not require serial sections.

KEY TO TURBELLARIA

1. Large species, living animal longer than 5 mm; gastrovascular cavity consisting of three main branches, one anterior and two posterior and lateral (Fig. 1) ..... Order **TRICLADIDA, 2**
Small species, rarely longer than 4 mm; gastrovascular cavity almost invariably a single, median, longitudinal cavity (Figs. 7, 9, 10); collectively, the "microturbellarians".
    Orders **CATENULIDA, MACROSTOMIDA, NEORHABDOCOELA, ALLOEOCOELA, 37**
2. Dorsal surface more or less opaquely pigmented, usually grayish to brownish or black; eyes present.
    3
Dorsal surface translucent, whitish, or creamy; eyes present or absent **18**
3. With many eyespots forming an anterior band (Figs. 13 D, E); up to 19 mm long; cold streams in Rocky Mountain area; 3 species, of which only one is common ...... **Polycelis coronata** (Girard)
With two eyespots **4**
4. With multiple pharynges (Fig. 14 F); several polytypic species, very difficult to distinguish.
    Phagocata, 5
With a single pharynx **6**
5. Poorly defined auricles (Fig. 14 F); penis papilla long, conical; northeastern states.
    Phagocata gracilis (Haldeman)
Without auricles (Fig. 14 K); penis papilla short, truncate; southeastern states.
    Phagocata woodworthi Hyman

6. Anterior end with adhesive organ (Fig. 4) 7
   Anterior end without adhesive organ 8
7. In springs, streams, and lakes west of the Continental Divide; 14 to 22 mm long.
   Dendrocoelopsis vaginata Hyman
   Known only from one lily pond in Washington, D.C., apparently introduced; 14 mm long.
   Rectocephala exotica Hyman
8. Auricles elongated, sharply pointed (Fig. 13A); up to 30 mm long; in running and standing waters
   everywhere .... **Dugesia dorotocephala** (Woodworth)
   Auricles not especially elongated and sharply pointed 9
9. Anterior end more or less triangular (Fig. 2) 10
   Anterior end truncated (Fig. 13G) 12
10. Eyes close to anterior end; up to 20 mm long; a European species now found in the St. Lawrence River
    system and NY .... **Dugesia polychroa** (Schmidt)
    Eyes not especially close to anterior end 11
11. Coloration highly variable but patterned (Fig. 2); pharynx with white tip; 6 to 18 mm long; generally
    distributed and very common .... **Dugesia tigrina** (Girard)
    Color uniformly gray, brown, or blackish; pharynx all white; 7 to 15 mm long (Fig. 13F); eastern half
    of the United States .... **Cura foremanii** (Girard)
12. Dorsal surface brown, but faded marginally; up to 12 mm long; CA .... **Phagocata**
    Coloration extending to margin of body 13

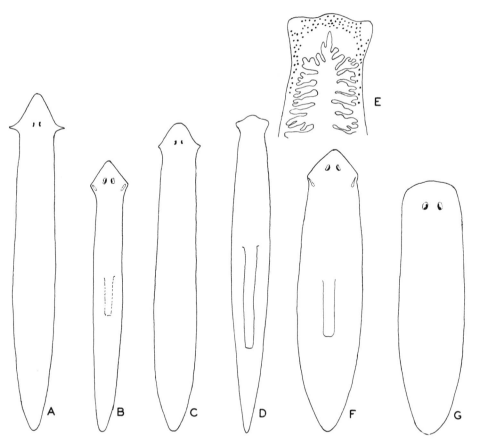

FIG. 13.–Structure of Planariidae. A, *Dugesia dorotocephala,* ×6; B and C, *D. tigrina,* ×7; D, *Polycelis coronata,* ×5; E, anterior end of *P. coronata;* F, *Cura foremanii,* ×6; G, *Hymanella retenuova,* ×6, (B redrawn from Hyman, 1931b; D and E modified from Hyman, 1931; F modified from Kenk, 1935.)

13. Lateral corners of head rounded, no distinct necklike constriction behind them (Fig. 13G); up to 14 mm long; vernal pools and seepages in eastern states .... **Hymanella retenuova** Castle
Lateral corners of head not rounded **14**

14. Darkly pigmented gray, brown, or black; with an adenodactyl (a dense prostate gland behind the gonopore); up to 13 mm long; rare in springs and streams of southeastern states.
Planaria dactyligera Kenk
Pigment light to dark; up to 22 mm long; adenodactyl absent **15**

15. A high-altitude species, restricted to cold streams and springs in western states; up to 22 mm long; gray to black coloration .... **Phagocata**
Not a high-altitude species **16**

16. Up to 20 mm long; coloration dark gray to almost black; widely distributed and common in vernal ponds and springs in eastern states .... **Phagocata**
Up to 12 mm long; coloration gray; distribution more restricted **17**

17. Known only from seepage springs in NC .... **Phagocata bulbosa** Kenk
In temporary ponds and ditches in early spring in midwestern states.
Hymanella retenuova Castle

18. Auricles long, thin, and pointed; body up to 27 mm long and 2 mm wide; two species in caves and seeps in GA, AL, and TN .... **Phagocata**
Auricles absent or poorly developed **19**

19. Adhesive organ in the form of a large, permanent, anterior snout (Fig. 15A); known only from Malheur Cave, Harney County, OR .... **Kenkia rhynchida** Hyman
Snout absent **20**

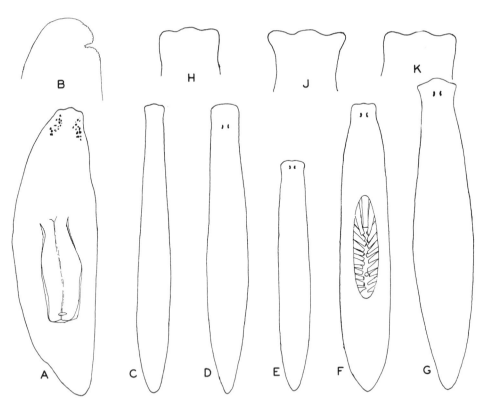

FIG. 14.–Structure of Planariidae and Dendrocoelidae. A, *Dendrocoelopsis americana*, ×18; B, longitudinal section of anterior end of *D. americana*, showing adhesive organ; C, *Procotyla typhlops*, ×7; D, *Phagocata morgani*, ×7; E, *P. velata*, ×5; F, *P. gracilis*, ×5; G, *P. woodworthi*, ×7; H, anterior end of *Planaria dactyligera*; J, anterior end of *Phagocata gracilis*; K, anterior end of *P. woodworthi*.

20. Eyes present 21
    Eyes absent 23
21. One pair of eyes; adhesive organ absent (Fig. 14D); up to 14 mm long; about 10 southeastern species found mostly in subterranean waters .... **Phagocata**
    Usually more than one pair of eyes; adhesive organ present (Fig. 4) 22
22. One to eight eyes on each side (Fig. 4); up to 20 mm long; widely distributed and common in eastern half of the United States .... **Procotyla fluviatilis** Leidy
    Eyes numerous (Fig. 14A); up to 18 mm long; caves and springs of OK and AR.
    .... **Dendrocoelopsis americana** (Hyman)
23. Restricted to surface waters 24
    Restricted to caves, or caves and springs; (about eight recently described rare species are not keyed out here) 25
24. Known only from Lake Tahoe deep waters; up to 14 mm long.
    .... **Dendrocoelopsis hymanae** Kawaktsu
    Known only from a spring in KS; up to 20 mm long .... **Sphalloplana kansensis** Hyman
25. Adhesive organ absent; up to 12 mm long; subterranean in VA ...... **Procotyla typhlops** Kenk
    Adhesive organ present 26
26. Coloration a very light reddish brown or orange; up to 30 mm long; caves and springs in MO and IA.
    .... **Macrocotyla glandulosa** Hyman
    Coloration creamy or white; a group of cave species, each of which is known from a single cave or cave system; several rare TX and VA species are not included in this key 27
27. Anterior end bluntly rounded; body margin with a row of large rhabdites (Fig. 15B) 28
    Anterior end truncate; with large rhabdites (Figs. 15F, J) 30
28. Caves and springs in IL and MO; up to 17 mm long (Fig. 15E).
    .... **Sphalloplana hubrichti** Hyman
    With another distribution; smaller 29

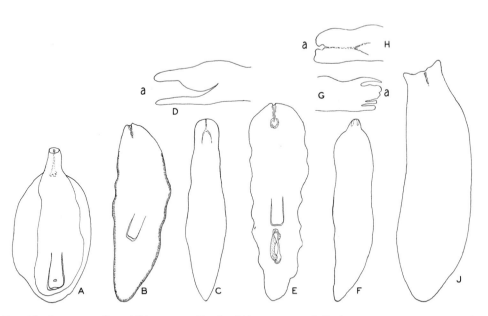

FIG. 15.–Structure of Kenkiidae. A, *Kenkia rhynchida,* ×11; B, *Sphalloplana virginiana,* ×4; C, *S. buchanani,* ×4; D, longitudinal section of anterior end of *S. buchanani,* showing the invaginated adhesive gland; E, *S. hubrichti,* ×3 (penis stippled); F, *S. pricei,* ×5; G, longitudinal section of anterior end of *S. pricei,* showing partly protruded adhesive gland; H, same, showing retracted adhesive gland; J, *S. percoeca,* ×6, a, anterior end. (A, C, D, and F–H modified from Hyman, 1937a; B and E modified from Hyman, 1945; J modified from Buchanan.)

29. Known only from a VA cave; up to 12 mm long (Fig. 15B) .... **Sphalloplana virginiana** Hyman
    Known only from WV caves; up to 11 mm long .... **Macrocotyla hoffmasteri** Hyman
30. Maximum length of live worm less than 10 mm **31**
    Maximum length of live worm more than 10 mm **33**
31. Known only from an IN cave; up to 9 mm long .... **Sphalloplana weingartneri** Kenk
    With another distribution **32**
32. Known only from a GA cave; up to 8 mm long .... **Sphalloplana georgiana** Hyman
    Known only from an AL cave; up to 6 mm long .... **Sphalloplana alabamensis** Hyman
33. Known only from TX caves; up to 35 mm long; variable, may be a group of species.
                                                           Sphalloplana mohri Hyman
    Known only from other states **34**
34. Found only in PA caves; up to 28 mm long (Fig. 15F) .... **Sphalloplana pricei** (Hyman)
    Found in other cave systems **35**
35. Found only in KY caves **36**
    Found only in MO caves; up to 30 mm long .... **Macrocotyla lewisi** Kenk
36. Head with rounded auricles; up to 16 mm long (Fig. 15J) .... **Sphalloplana percoeca** (Packard)
    Anterior end truncate, no auricles; up to 15 mm long ...... **Sphalloplana buchanani** (Hyman)
37. Body usually composed of a series of two to eight zooids whose anterior and posterior extremities
    appear as paired lateral indentations; eyes usually absent (Fig. 8) **38**

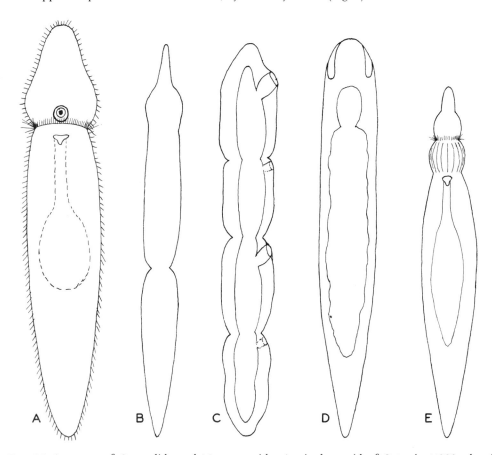

FIG. 16.–Structure of Catenulida and Macrostomida. A, single zooid of *Catenula,* ×230, showing statocyst, mouth, and gastrovascular cavity; B, two zooids of *Rhynchoscolex,* ×55; C, lateral view of four zooids of *Microstomum,* ×35; D, dorsal view of single zooid of *Stenostomum,* ×220; E, ventral view of *Suomina,* ×200, showing mouth, gastrovascular cavity, and ciliated anterior transverse groove. (A modified from Heinlein and Wachowski.)

Body composed of a single individual; eyes present or absent **41**

38. With a statocyst at the anterior end of each zooid; each zooid composed of a cephalic lobe and body (Fig. 16A); 2 species .... **Catenula**
    Without a statocyst **39**
39. With a narrow or swollen proboscislike structure at the anterior end (Figs. 5, 16B) ... **Rhynchoscolex**
    Without such a structure **40**
40. With a preoral blind pouch at the anterior end of the gastrovascular cavity (Fig. 16C); common; many species .... **Microstomum**
    Without such a pouch; with a ciliated pit on either side of the head (Fig. 16D); the most common of all microturbellarians; generally distributed in a wide variety of habitats; many species.

    Stenostomum
41. With a simple pharynx (Figs. 9, 16E) **42**
    With a complex pharynx (Figs. 17, 20) **43**
42. With a ciliated groove encircling the anterior part of the body (Fig. 16E) .... **Suomina**
    Without such a groove; dorsoventrally flattened and posterior end somewhat broadened and with adhesive papillae; spatulate (Fig. 9); 0.7 to 3.5 mm long; many common species.

    Macrostomum

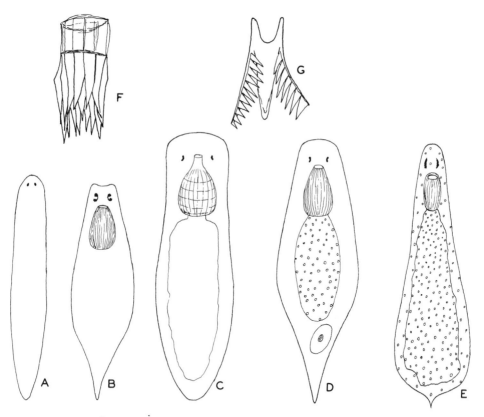

FIG. 17.–Structure of Neorhabdocoela and Alloeocoela. A, *Hydrolimax grisea,* ×4.5; B, *Castrella,* ×60, showing pharynx and eyespots; C, *Provortex,* ×125, showing eyespots, pharynx, and gastrovascular cavity; D, *Microdalyellia,* ×50, showing eyespots, pharynx, gastrovascular cavity, zoochlorellae, and single egg; E, *Phaenocora,* ×37, showing eyespots, pharynx, gastrovascular cavity, and zoochlorellae; F, cuticular apparatus of male copulatory organ of *Gieysztoria;* G, Cuticular apparatus of male copulatory organ of *Dalyellia* and *Microdalyellia.* (A modified from Hyman, 1938; B, D, and E modified from Ruebush, 1941; C modified from Ruebush, 1935a.)

43. Pharynx more or less bulbous (Fig. 17) **44**
    Pharynx not bulbous (Figs. 20 C, D) **66**
44. Pharynx at anterior end of gastrovascular cavity, directed anteroventrally, and caskshaped (Figs. 17B, C) **45**
    Pharynx not at anterior end of gastrovascular cavity, directed ventrally, and rosulate in shape (Figs. 18A, E) **52**
45. A large species, up to 15 mm long; dorsal surface dark gray, ventral surface whitish; body plump and cylindroid; reported from NJ and eastern PA (Fig. 17A) .... **Hydrolimax grisea** Haldeman
    Usually no more than 5 mm long **46**
46. Body nearly cylindrical and more or less tapering to a pointed tail (Fig. 17B); usually free swimming. **47**
    Body triangular in section; posterior end truncate, with a small tail (Fig. 17E); usually creeping **51**
47. Each eye consisting of two pigmented spots connected by a pigmented band; anterior end truncate (Fig. 17B); body usually darkly pigmented with blue or black .... **Castrella**
    Each eye a single concave mass; anterior end more rounded (Figs. 10, 17C); brownish or green pigmentation **48**
48. Mature animals about 0.5 mm long; not pigmented; posterior end slightly rounded (Fig. 17C); each egg containing two embryos; only one or two species reported from the United States.
    Provortex

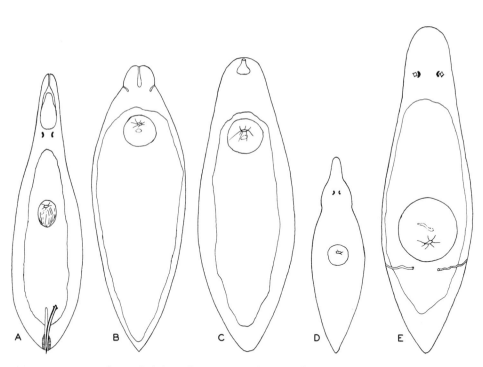

FIG. 18.–Structure of Neorhabdocoela. A, ventral view of *Gyratrix hermaphroditus,* ×40, showing proboscis, eyespots, mouth, pharynx, gastrovascular cavity, and penis stylet; B, ventral view of *Prorhynchella,* ×100, showing anterior sensory pit, anterolateral ciliated pits, mouth, pharynx, and gastrovascular cavity; C, ventral view of *Microkalyptorhynchus,* ×80, showing anterior sensory pit, mouth, pharynx, and gastrovascular cavity; D, ventral view of *Rhynchomesostoma,* ×70, showing eyespots, mouth, and pharynx; E, ventral view of *Krumbachia minuta,* ×180, showing eyespots, mouth, pharynx, gastrovascular cavity, protonephridial tubules, and protonephridial pores. (B modified from Ruebush, 1939; C modified from Ruebush, 1935; D modified from Ruebush, 1941; E modified from Ruebush, 1938.)

Mature animals 0.5 to 3 mm long; usually brown or green coloration; posterior end more or less pointed; each egg with a single embryo **49**

49. Cuticular apparatus of male copulatory organ consisting of a collar with stout spines at the distal end (Fig. 17F); one pair of dark eyes .... **Gieysztoria**
    Cuticular apparatus of male copulatory organ consisting of a long tongue and two spinous processes (Fig. 17G) **50**

50. Mature animals 2 to 3 mm long (Fig. 10); one to many eggs present and scattered throughout the body; with zoochlorellae and brownish pigment in parenchyma; several species in the United States .... **Dalyellia**
    Mature animals only 1 to 2 mm long; only one egg carried at a time (Fig. 17D), rarely two to four; zoochlorellae, if present, restricted to the gastrovascular cavity; about 30 species.
    Microdalyellia

51. Mature animals up to 4.8 mm long; without zoochlorellae; in sulfur springs.
    Pseudophaenocora sulfophila Gilbert
    Mature animals 1 to 3 mm long; often with zoochlorellae in gastrovascular cavity and parenchyma (Fig. 17E); about 5 species .... **Phaenocora**

52. Anterior end with a heavy, muscular proboscis contained in a sheath; body usually cylindrical (Fig. 18A); free swimming **53**
    Without such a proboscis; body usually flattened; usually creeping **55**

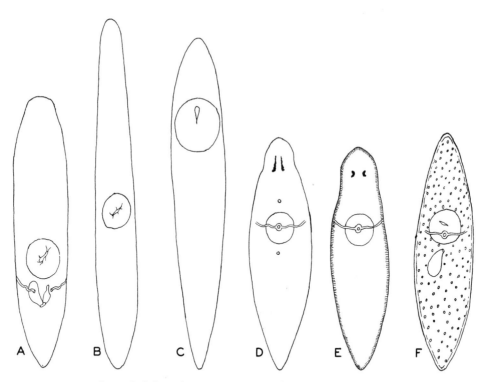

FIG. 19.–Structure of Neorhabdocoela. A, ventral view of *Krumbachia virginiana,* ×30, showing mouth, pharynx, penis, bursa copulatrix, protonephridial tubules, and protonephridial pores; B, ventral view of *Amphibolella,* ×125; C, *Protoascus wisconsinensis,* ×95; D, ventral view of *Bothromesostoma,* ×11, showing eyespots, pore of sensory pouch, protonephridial tubules, mouth, pharynx, and common genital pore; E, ventral view of *Strongylostoma,* ×50, showing eyespots, protonephridial tubules, mouth, pharynx, and epidermal rhabdites; F, ventral view of *Typhloplana,* ×65, showing zoochlorellae in parenchyma, mouth, pharynx, protonephridial tubules, protonephridial pore, and penis. (A, D, E, and F modified from Ruebush, 1941; B modified from Kepner and Ruebush, 1937.)

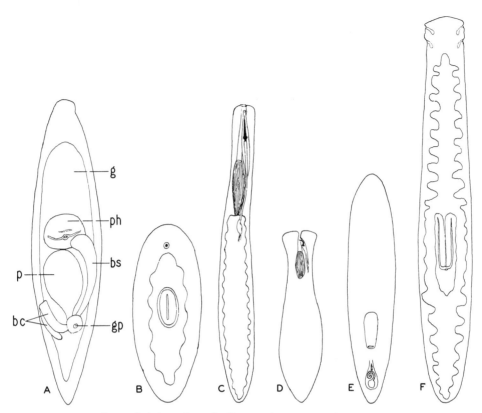

FIG. 20.–Structure of Neorhabdocoela and Alloeocoela. A, *Castrada,* ×48; B, *Otomesostoma,* ×35; C, *Prorhynchus,* ×17; D, *Geocentrophora,* ×25; E, *Opistomum,* ×16; F, *Bothrioplana,* ×16. (A modified from Kepner, Ruebush, and Ferguson, 1937; D modified from Ruebush, 1941.)

53. With a posterior penis stylet; eyes black (Fig. 18A); body coloration whitish; 2 mm long; common; actually a difficult complex of (incipient) sibling species with close marine relatives.

          Gyratrix hermaphroditus Ehr.

    Without a posterior penis stylet **54**

54. Up to 2 mm long; colorless anteriorly, reddish posteriorly; proboscis opening terminal; 1 rare species

          Polycystis

    Up to 1.5 mm long; white to gray coloration; proboscis opening subterminal; 1 rare species.

          Klattia

55. With a median anterior sensory pit (Figs. 18B, C) **56**

    Without a sensory pit **57**

56. With a pair of anterolateral ciliated pits (Fig. 18B) .... **Prorhynchella**

    Without a pair of anterolateral ciliated pits (Fig. 18C) .... **Microkalyptorhynchus**

57. With a retractile and sensory anterior end (Fig. 18D) .... **Rhynchomesostoma**

    Without such a structure at the anterior end **58**

58. Protonephridial tubules opening separately on the ventral surface of the body (Fig. 18E) **59**

    Protonephridial tubules opening into the mouth cavity (Figs. 19D, E) **62**

59. With a large refractive rectangle in each eye (Fig. 18E); with a bursa copulatrix; 0.5 mm long.

          Krumbachia minuta Ruebush

    Eyes absent; without a bursa copulatrix **60**

60. Body 2 to 3 mm long, opaque (Fig. 19A) .... **Krumbachia virginiana** (Kepner and Carter)

    Body up to 1.5 mm long, more or less translucent; rare **61**

61. Pharynx near middle of body (Fig. 19B) .... **Amphibolella**
Pharynx near anterior end (Fig. 19C); reported only from WI.

Protoascus wisconsinensis Hayes

62. With a ventral pore leading into the sensory pouch anterior to mouth; common genital pore present (Fig. 19D); 5 to 7 mm long .... **Bothromesostomum**
Without a ventral sensory pouch; common genital pore present or combined with the mouth (Fig. 19E) 63
63. With rhabdites in the epidermis (Figs. 7, 19E) 64
Without rhabdites in the epidermis 65
64. Large species, usually 2 to 4 mm long, but a few species as long as 10 mm; with long parenchymal rhabdites; eyes, when present, black (Fig. 7); coloration pale brown, gray, or gold; very common in temporary ponds and puddles; about 10 species .... **Mesostoma**
Small species, only 1 to 1.5 mm long; without parenchymal rhabdites; eyes red (Fig. 19E); several species .... **Strongylostoma**
65. About 1 mm long; without bursa copulatrix; with abundant zoochlorellae in parenchyma (Fig. 19F).

Typhloplana

About 1 to 2 mm long; with bursa copulatrix (Fig. 20A); body whitish or yellowish; with or without zoochlorellae; several species .... **Castrada**
66. With an anterior statocyst (Fig. 20B) .... **Otomesostoma**
Without a statocyst 67
67. Pharynx located at anterior end of the gastrovascular cavity (Figs. 20C, D) 68
Pharynx behind middle of body and directed posteriorly (Figs. 20E, F) 69
68. Penis stylet straight or slightly bent (Fig. 20C) .... **Prorhynchus**
Penis stylet bent almost at a right angle (Fig. 20D); several species up to 8 mm long, including one from KY caves .... **Geocentrophora**
69. Pharynx not connected with gastrovascular cavity; penis near posterior end of body and heavily spined; without ciliated pits (Fig. 20E) .... **Opistomum**
Pharynx connected with gastrovascular cavity; with two pairs of anterior ciliated pits (Fig. 20F).

Bothrioplana

REFERENCES

Ax, P., and J. Dörjes. 1966. Oligochoerus limnophilus, nov. spec., ein kaspisches Faunenelement als erster Süsswasservertreter der Turbellaria Acoela in Flussen Mitteleuropas. *Int. Rev. gesamten Hydrobiol.* 51:15–44.

Ball, I. R. et al. 1981. The planarians (Turbellaria) of temporary waters in eastern North America. *Life Sci. Contr. R. Ont. Mus.* 127:1–27.

Buchanan, J. W. 1936. Notes on an American cave flatworm, Sphalloplana percoeca (Packard). *Ecology* 17:194–211.

Bauchhenss, J. 1971. Die Kleinturbellarien Frankens. Ein Beitrag zur Systematik und Ökologie der Turbellaria excl. Tricladida in Süddeutschland. *Ibid.* 56:609–666.

Brønsted, H. V. 1955. Planarian regeneration. *Biol. Rev.* 30:65–126.

Cannon, L. R. G. 1986. *Turbellaria of the World: A Guide to Families and Genera.* 136 pp. Brisbane, Queensland Museum.

Carpenter, J. H. 1970. Geocentrophora cavernicola n. sp. (Turbellaria, Alloeocoela): first cave alloeocoel. *Trans. Am. Microsc. Soc.* 89:124–133.

Castle, W. A., and L. H. Hyman. 1934. Observations on Fonticola velata (Stringer), including a description of the anatomy of the reproductive system. *Trans. Am. Microsc. Soc.* 53:154–171.

Chandler, C. M. 1966. Environmental factors affecting the local distribution and abundance of four species of stream-dwelling triclads. *Invest. Ind. Lakes Streams* 7:1–56.

Chodorowski, A. 1959. Ecological differentiation of turbellarians in Harsz-Lake. *Pol. Arch. Hydrobiol.* 6:33–73.

———. 1960. Vertical stratification of Turbellaria species in some littoral habitats of Harsz Lake. *Ibid.* 8:153–163.

Curtis, W. C. 1902. The life history, the normal fission, and the reproductive organs of Planaria maculata. *Proc. Boston Soc. Nat. Hist.* 31:515–559.

Faubel, A., and J. Kolasa. 1978. On the anatomy and morphology of the freshwater species of Acoela (Turbellaria): Limnoposthia polonica (Kolasa et Faubel, 1974). *Bull. Acad. Pol. Sci. Ser. Sci. Biol. Classe II,* 26:393–397.

Ferguson, F. F. 1939–1940. A monograph of the genus Macrostomum O. Schmidt 1848. Parts I to VIII. *Zool. Anz.* 126:7–20; 127:131–144; 128:49–68; 188–205, 274–291; 129:21–48, 120–146, 244–266.

———. 1954. Monograph of the macrostomine worms of Turbellaria. *Trans. Am. Microsc. Soc.* 73:137–164.

Ferguson, F. F., and W. J. Hayes, Jr. 1941. A synopsis of the genus Mesostoma Ehrenberg 1835. *J. Elisha Mitchell Sci. Soc.* 57:1–52.

Folsom, T. C., and H. F. Clifford. 1978. The population

biology of Dugesia tigrina (Platyhelminthes: Turbellaria) in a thermally enriched Alberta, Canada lake. *Ecology* 59:966–975.

Hayes, W. J., Jr. 1941. Rhabdocoela of Wisconsin. I. Morphology and taxonomy of Protoascus wisconsinensis n. g., n. sp. *Am. Midl. Nat.* 25:388–401.

Heinlein, E., and H. E. Wachowski. 1944. Studies on the flatworm Catenula virginia. *Amer. Midl. Nat.* 31:150–158.

Heitkamp, U. 1978. Speziationsprozesses bei Gyratrix hermaphroditus Ehrenberg, 1831 (Turbellaria, Kalyptorhynchia). *Zoomorphologie* 90:227–251.

Hyman, L. H. 1925. The reproductive system and other characters of Planaria dorotocephala. *Trans. Am. Microsc. Soc.* 44:51–89.

_____. 1928. Studies on the morphology, taxonomy, and distribution of North American triclad Turbellaria, I. Procotyla fluviatilis, commonly but erroneously known as Dendrocoelum lacteum. *Ibid.* 47:222–255.

_____. 1931. Studies on triclad Turbellaria, III. On Polycelis coronata (Girard). *Ibid.* 50:124–135.

_____. 1931a. Studies on triclad Turbellaria, IV. Recent European revisions of the triclads, and their application to the American forms, with a key to the latter and new notes on distribution. *Ibid.* 50:316–335.

_____. 1931b. Studies on triclad Turbellaria. V. Descriptions of two new species. *Ibid.* 50:336–343.

_____. 1937. Studies on triclad Turbellaria, VIII. Some cave planarians of the United States. *Ibid.* 36:457–477.

_____. 1938. North American Rhabdocoela and Alloeocoela. II. Rediscovery of Hydrolimax grisea Haldeman. *Am. Mus. Novit.* 1004:1–19.

_____. 1939. North American triclad Turbellaria. IX. The priority of Dugesia Girard 1850 over Euplanaria Hesse 1897 with notes on American species of Dugesia. *Trans. Am. Microsc. Soc.* 58:264–275.

_____. 1939a. North American triclad Turbellaria, X. Additional species of cave planarians. *Ibid.* 276–284.

_____. 1945. North American triclad Turbellaria, XI. New, chiefly cavernicolous, planarians. *Am. Midl. Nat.* 34:475–484.

_____. 1951. North American triclad Turbellaria, XII. Synopsis of the known species of freshwater planarians of North America. *Trans. Am. Microsc. Soc.* 70:154–167.

Jenkins, M. M., and H. P. Brown. 1963. Cocoon-production in Dugesia dorotocephala (Woodworth) 1897. *Trans. Am. Microsc. Soc.* 82:167–177.

Karling, T. G. 1963. Die Turbellarien Ostfennoskandiens. V. Neorhabdocoela. 3. Kalyptorhynchia. *Fauna Fenn.* 17:1–59.

Kenk, R. 1935. Studies on Virginian triclads. *J. Elisha Mitchell Sci. Soc.* 51:79–126.

_____. 1941. Induction of sexuality in the asexual form of Dugesia tigrina (Girard). *J. Exp. Zool.* 87:55–70.

_____. 1970. Freshwater triclads (Turbellaria) of North America. II. New or little known species of Phagocata. *Ibid.* 83:13–34.

_____. 1972. Freshwater planarians (Turbellaria) of North America. *Biota Freshwater Ecosystems, E.P.A. Identification Manual No.* 1:1–81.

_____. 1973. Freshwater triclads (Turbellaria) of North America, V: The genus Polycelis. *Smithson. Contrib. Zool.* 135:1–15.

_____. 1973a. Freshwater triclads (Turbellaria) of North America, VI: the genus Dendrocoelopsis. *Ibid.* 138:1–16.

_____. 1974. Index of the genera and species of the freshwater triclads (Turbellaria) of the world. *Ibid.* 183:1–90.

_____. 1975. Fresh-water triclads (Turbellaria) of North America. VII. The genus Macrocotyla. *Trans. Am. Microsc. Soc.* 94:324–339.

_____. 1975a. Freshwater triclads (Turbellaria) of North America. VIII. Dugesia arizonensis, new species. *Proc. Biol. Soc. Wash.* 88:113–120.

_____. 1977. Freshwater triclads (Turbellaria) of North America. X. Three species of Phagocata from the Eastern United States. *Proc. Biol. Soc. Wash.* 89:645–652.

_____. 1977a. Freshwater triclads (Turbellaria) of North America. IX. The genus Sphalloplana. *Smithson. Contrib. Zool.* 246:1–38.

_____. 1982. Freshwater triclads (Turbellaria) of North America. XIII. Phagocata hamptonae, new species, from Nevada. *Proc. Biol. Soc. Wash.* 95:161–166.

_____. 1984. Freshwater triclads (Turbellaria) of North America. XV. Two new subterranean species from the Appalachian region. *Proc. Biol. Soc. Wash.* 97:209–216.

_____. 1987. Freshwater triclads (Turbellaria) of North America. XVI. More on subterranean species of Phagocata of the eastern United States. *Proc. Biol. Soc. Wash.* 100:664–673.

Kepner, W. A., E. D. Miller, and A. W. Jones. 1934. Observations upon Rhynchomesostomum rostratum. *Zool. Anz.* 107:188–192.

Kepner, W. A., and T. K. Ruebush. 1935. Microrhynchus virginianus n. gen. n. sp. *Zool. Anz.* 111:257–261.

_____. 1937. Amphibolella Virginiana n. sp. *Zool. Anz.* 118:103–107.

Kepner, W. A., T. K. Ruebush, and F. F. Ferguson. 1937. Castrada virginiana n. sp. *Zool. Anz.* 119:307–314.

Kolasa, J., and A. Faubel. 1974. A preliminary description of a freshwater Acoela (Turbellaria): Oligochoerus polonicus nov. spec. *Boll. Zool.* 41:81–85.

Kolasa, J. et al. 1987. Microturbellarians from interstitial waters, streams, and springs in southeastern New York. *J. N. A. Benthol. Soc.* 6:125–132.

Kromhout, G. A. 1943. A comparison of the protonephridia in fresh-water, brackish-water, and marine specimens of Gyratrix hermaphroditus. *J. Morphol.* 72:167–169.

Luther, A. 1955. Die Dalyelliiden (Turbellaria Neorhabdocoela). Eine Monographie. *Acta Zool. Fenn.* 87:1–337.

_____. 1960. Die Turbellarien Ostfennoskandiens. I. Acoela, Catenulida, Macrostomida, Lecithoepitheliata, Prolecithophora und Proseriata. *Fauna Fenn.* **7**:1–154.

_____. 1962. Die Turbellarien Ostfennoskandiens. IV. Neorhabdocoela I. *Ibid.* **12**:1–71.

_____. 1963. Die Turbellarien Ostfennoskandiens. IV. Neorhabdocoela 2. *Ibid.* **16**:1–163.

Maguire, B. 1963. Population regulation by ectocrines in Cura foremanii (Turbellaria). *Proc. XVI Int. Cong. Zool.* **1**:64.

Marcus, E. 1945. Sôbre Catenulida brasileiros. *Univ. Sao Paulo, Bol. Fac. Filos. Cienc. Lett., Zool. No.* **10**:8–100.

_____. 1946. Sôbre Turbellaria limnicos brasileiros. *Ibid.* **11**:5–253.

Mitchell, R. W. 1968. New species of Sphalloplana (Turbellaria; Paludicola) from the caves of Texas and a reexamination of the genus Speophila and the Family Kenkiidae. *Ann. Speleol.* **23**:597–620.

Nuttycombe, J. W. 1956. The Catenula of the eastern United States. *Am. Midl. Nat.* **55**:419–433.

Nuttycombe, J. W., and A. J. Waters. 1938. The American species of the genus Stenostomum. *Proc. Am. Philos. Soc.* **79**:213–284.

Pörner, H. 1966. Die rhabdocoeliden Turbellarien der Gewässer von Jena und Umgebung. *Limnologica* **4**:27–44.

Redfield, E. S. P. 1915. The grasping organ of Dendrocoelum lacteum. *J. Anim. Behav.* **5**:375–380.

Reynoldson, T. B. 1958. The quantitative ecology of lake-dwelling triclads in northern Britain. *Oikos* **9**:94–138.

_____. 1964. Evidence for intra-specific competition in field populations of triclads. *J. Anim. Ecol.* **33**:187–201.

_____. 1967. A key to the British species of fresh-water triclads. *Sci. Publ. Freshwater Biol. Assoc.* **23**:1–28.

Reynoldson, T. B., and L. S. Bellamy. 1971. Intraspecific competition in lake-dwelling triclads. *Oikos* **22**:315–328.

_____. 1973. Interspecific competition in lake-dwelling triclads. *Ibid.* **24**:301–313.

Reynoldson, T. B., and A. D. Sefton. 1976. The food of Planaria torva (Müller) (Turbellaria–Tricladida), a laboratory and field study. *Freshwater Biol.* **6**:383–393.

Riser, N. W., and M. P. Morse. 1974. *Biology of the Turbellaria.* 530 p. McGraw-Hill, New York.

Rixen, J.-U. 1961. Kleinturbellarien aus dem Litoral der Binnengewässer Schleswig-Holsteins. *Arch. Hydrobiol.* **57**:464–538.

Ruebush, T. K. 1935. The genus Olisthanella in the United States. *Zool. Anz.* **112**:129–136.

_____. 1935a. The occurrence of Provortex affinis Jensen in the United States. *Ibid.* **111**:305–308.

_____. 1937. The genus Dalyellia in America. *Ibid.* **119**:237–256.

_____. 1938. Krumbachia minuta n. sp. (Turbellaria Rhabdocoela). *Zool. Anz.* **122**:260–265.

_____. 1939. A new North American rhabdocoel turbellarian, Prorhynchella minuta n. gen., n. sp. *Ibid.* **127**:204–209.

_____. 1940. Mesostoma ehrenbergii wardi for the study of the turbellarian type. *Science* **91**:531–532.

_____. 1941. A key to the American freshwater turbellarian genera, exclusive of the Tricladida. *Trans. Am. Microsc. Soc.* **60**:29–40.

Ruebush, T. K., and W. J. Hayes, Jr. 1939. The genus Dalyellia in America. II. A new form from Tennessee and a discussion of the relationships within the genus. *Zool. Anz.* **128**:136–152.

Sanders, O., and R. M. Sanders. 1970. Webs secreted by planarians. *J. Biol. Psych.* **12**:65–70.

Schwank, P. 1980. Neue limnische Turbellarien. *Arch. Hydrobiol.* **88**:463–490.

Schwartz, S. S., and P. D. N. Hebert. 1982. A laboratory study of the feeding behavior of the rhabdocoel Mesostoma ehrenbergii on pond Cladocera. *Can. J. Zool.* **60**:1305–1307.

Von Graf, L. 1911. Acoela, Rhabdocoela, und Alloeocoela des Ostens der Vereinigten Staaten von Amerika. *Z. Wiss. Zool.* **99**:1–108.

_____. 1913. Turbellaria. II. Rhabdocoelida. *Das Tierreich* **35**:1–484.

Young, J. O. 1970. British and Irish freshwater microturbellaria: historical records, new records and a key for their identification. *Arch. Hydrobiol.* **67**:210–241.

_____. 1975. The population dynamics of Phaenocora typhlops (Vejdovsky) (Turbellaria: Neorhabdocoela) living in a pond. *J. Anim. Ecol.* **44**:251–262.

Young, J. O., and T. B. Reynoldson. 1965. A laboratory study of predation on lake-dwelling triclads. *Hydrobiologia* **26**:307–313.

6

NEMERTEA
(Proboscis Worms)

The majority of fresh-water biologists never have the experience of observing living fresh-water nemerteans. One reason is they are seldom sought intentionally and are usually overlooked or disregarded in collections made with other taxa uppermost in mind. Another reason is their extremely "spotty" and local abundance. Weedy ponds, masses of filamentous algae, the undersides of lily pads, and the general substrate of littoral areas are preferred habitats, but often nemerteans may be taken from only one restricted area in a pond or stream and are apparently absent elsewhere. A third reason they are often overlooked is their superficial resemblance to oligochaetes because of their pseudosegmental appearance owing to the many gonads and intestinal diverticula, especially when these diverticula are full of food.

Nemerteans have been badly neglected in the United States, but presumably there is only one well-known fresh-water species. It was originally described in 1850 as *Emea rubrum*, but has a long history of taxonomic and nomenclatorial confusion. It is now called *Prostoma graecense* (Böhmig). This species is widely distributed in the United States, as well as worldwide. A second species, rarely found in the United States, is *Prostoma eilhardi* (Montgomery). It is likewise worldwide in its distribution. Unfortunately, serial sections are necessary for the separation of these two species. *P. graecense* has a well-developed ciliated esophagus; *P. eilhardi* has an indistinct, unciliated esophagus. Five other species of *Prostoma* are known from other countries, chiefly from single localities. In addition, there are six other monospecific fresh-water genera, each known from a single locality, including Java, Germany, Brazil, Chile, and New Zealand. This situation is in striking contrast to that prevailing in the marine littoral where many species are found. Apparently nemerteans have not generally found the means necessary for the invasion of fresh waters.

General characteristics. *Prostoma graecense* is elliptical in cross section, somewhat flattened ventrally, up to 30 mm long, and 0.6 to 2.0 mm in diameter. Coloration is highly variable. Young and well-fed specimens are whitish or pale yellow, but mature specimens are usually yellowish red, orange, or deep red. There is also a rare green variety. The anterior end is rounded and without a definite head; the posterior end is tapered to a point. Three pairs of ocelli are usually present, but occasionally there are two or four pairs. They are arranged in two longitudinal rows on the dorsal surface of the anterior end and are sometimes placed asymmetrically. The body is smooth, unsegmented, and covered with a columnar ciliated epithelium.

Owing to a well-developed longitudinal and circular muscle system, these worms are highly contractile. A smooth, gliding type of locomotion is produced by cilia acting on a copious slime track secreted by the epidermis.

Digestive system, feeding. The subterminal mouth opens into a buccal cavity, which is followed by an esophagus, stomach, intestine, and terminal anus. The entire length of the long intestine has pairs of lateral diverticula, and from the anterior end of the intestine a pair of pyloric caeca extend forward. The digestive tract has no musculature of its own.

The proboscis apparatus is a unique feature of nemerteans. A fluid-filled longitudinal cavity, the rhynchocoel, runs throughout most of the body length dorsal to the digestive tract. The muscular wall of the rhynchocoel is called the proboscis sheath. Lying within the rhynchocoel is a tubular muscular proboscis, which can be protruded anteriorly and withdrawn back into the rhynchocoel by a long retractor muscle. The proboscis is very long, sometimes two or three times the body length, and when not in use it is strongly contracted and coiled within the rhynchocoel. The rhynchodaeum is the cavity of the proboscis and it has an opening to the outside in common with the mouth. There are actually three parts to the proboscis. Anteriorly it is a thick-walled tube. Farther back is a middle bulbous portion bearing a functional sclerotized stylet set in a heavy base along with two lateral pockets containing accessory stylets in various stages of development (Figs. 1, 3). The functional stylet is frequently lost and is then replaced by another from the lateral pockets. Posterior to the bulb is a blind tube. When the proboscis is protruded, the bulb and posterior blind tube are everted through the anterior thick-walled tube and out of the mouth in such a fashion that the stylet is at the tip of the extended proboscis. Proboscis eversion is effected by muscle contraction exerting pressure on the fluid in the rhynchocoel.

The proboscis functions in both defense and food getting. While *Prostoma* creeps about on the substrate its extended proboscis probes and darts constantly. If the proboscis comes in contact with a small metazoan, the stylet may be used to stab the animal repeatedly. A toxic material is thought to be secreted from the posterior chamber of the proboscis and discharged through the stylet apparatus. The proboscis is also coiled around the prey and quantities of sticky mucus are secreted, thus further immobilizing the prey. The mouth and buccal cavity are capable of great distension, and most food organisms are ingested intact. If the prey is too large, however, only the body juices are sucked out. *Prostoma* is chiefly carnivorous and feeds on all kinds of small metazoans, and especially on small oligochaetes. It may also feed on recently dead animals, and a few times it has been observed ingesting organic detritus.

Preliminary digestion occurs in the stomach and further digestion in the intestine and its diverticula. Specimens may be kept for 6 months or more without food, but the metabolic requirements during this period may produce a shrinkage to one tenth or less of normal size.

General anatomy, physiology. Below the epidermis *Prostoma* has a thick connective tissue dermis and well-defined layers of longitudinal and circular muscle. Visceral organs are separated by mesenchymal connective tissue.

A closed circulatory system is present, the main vessels being a dorsal longitudinal contractile vessel and two lateral longitudinal contractile vessels. The blood is clear, sometimes faintly yellowish reddish. Possibly it contains hemoglobin.

Excretion is effected by a protonephridial system consisting of many branched tubules with clusters of terminal flame bulbs and several nephridiopores along each side of the body. Presumably this apparatus also functions in osmoregulation. The protonephridial system is much more extensive than in marine nemertines.

The brain is dark colored and four lobed and surrounds the esophagus. Two prominent lateroventral nerve cords pass from it and extend to the posterior end. Smaller paired nerves supply the head, proboscis, and proboscis sheath.

The ocelli lie beneath the epidermis and are irregular cup-shaped masses of black, brown, or reddish pigment granules. Anterior sensory grooves are presumably chemotactile, and sensory cells with long cilia are especially abundant at the extremities.

There are no special respiratory structures or adaptations. It is possible that some respiratory exchange occurs through the wall of the digestive tract in addition to the general body surface.

Reproduction, life history. *Prostoma graecense* is hermaphroditic and generally protandric; saclike ovotestes alternate with the intestinal diverticula. Each ovotestis opens to

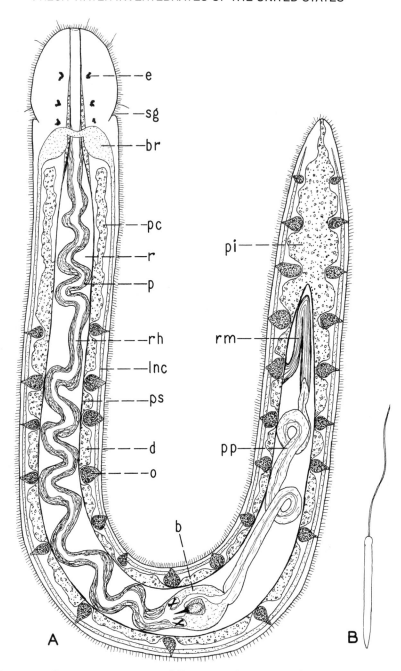

FIG. 1.–Structure of *Prostoma graecense,* semidiagrammatic. A, general anatomy, dorsal view, ×16; B, with protruded proboscis, ×1.5. b, bulbous portion of proboscis; br, brain; d, diverticulum of intestine; e, ocellus; lnc, lateral nerve cord; o, ovotestis; p, anterior portion of proboscis; pc, pyloric caecum; pi, posterior portion of intestine; pp, posterior portion of proboscis; ps, proboscis sheath; r, rhynchocoel; rh, rhynchodaeum; rm, retractor muscle of proboscis; sg, sensory groove.

the outside by a lateral genital pore and produces many sperm but only a single ripe egg. The additional oocytes degenerate. Breeding may occur at any time, but most often between May and December, provided the water temperature is above 10°C. A sexually mature individual secretes a mucous sheath around itself, releases sperm and a double row of eggs into the sheath, and then crawls out of the sheath. Not infrequently two individuals, one of which is at the time functionally male and the other female, occupy the same mucous sheath, and cross-fertilization occurs. Contact with water causes dissolution of the germinal vesicles around the eggs and they become fertilized. Development is direct, and there is no special larval stage. Young worms leave the mucous mass and become independent. Newly hatched worms frequently swim about freely above the substrate.

Prostoma has considerable powers of regeneration. The whole posterior portion of the body may be regenerated provided the remaining anterior portion contains at least half of the esophageal region. A new proboscis may be regenerated if the old one is lost.

Ecology. *Prostoma graecense* can be most easily found in autumn when it is relatively abundant. It occurs only in well-oxygenated shallow and littoral standing waters and sluggish streams, especially in masses of filamentous algae, on rooted aquatics, or in the debris of the general substrate. No explanation is known for its localized distribution in such places. It is especially active at night, when it does most of its feeding and egg deposition. It is highly thigmotactic and reacts positively to water currents. With the approach of autumn and freezing temperatures *Prostoma* migrates to deeper waters and returns to the shallows in spring.

Under adverse environmental conditions, such as insufficient oxygen, high temperature, abnormal substrate, and lack of food, *Prostoma* forms resting cysts by rounding up within a secreted outer layer of mucus, which soon hardens and becomes covered with detritus. The worm may remain alive within such a cyst for a few days to a few weeks, emerging when ecological conditions are again appropriate.

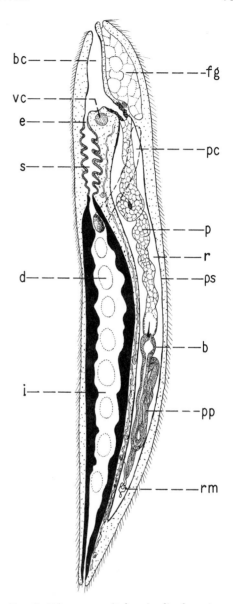

Fig. 2.–Diagrammatic longitudinal section of *Prostoma*, ×7.5. b, bulbous portion of proboscis; bc, buccal cavity; d, diverticulum of intestine; e, esophagus; fg, frontal gland; i, intestine; p, anterior portion of proboscis; pc, pyloric caecum; pp, posterior portion of proboscis; ps, proboscis sheath; r, rhynchocoel; rm, retractor muscle of proboscis; s, stomach; vc, ventral commissure of brain. (Modified from Reisinger, 1926.)

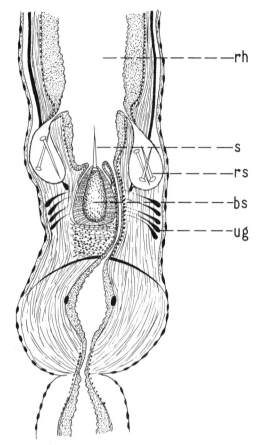

FIG. 3.—Bulbous portion of proboscis and adjacent structures of *Prostoma*. bs, base of stylet; rh, rhynchodaeum; rs, accessory stylet; s, functional stylet; ug, unicellular glands. (Greatly modified from Reisinger, 1926.)

Cysts cannot withstand complete drying and are of limited value in dissemination.

Prostoma is preyed upon by larger worms and carnivorous insects and crustaceans.

Culturing, preserving. *Prostoma* may be cultured in an aquarium that has a thin layer of soil on the bottom and plenty of rooted aquatics. Excessive bacterial decomposition inhibits nemerteans and should be avoided. Feed the worms bits of liver, earthworms, or small metazoans.

Because it is so highly contractile, *Prostoma* should always be anesthetized before killing. Chloretone and other solutions may be used with varying success. Kill in 80% alcohol or hot FAA. Preserve in 70% alcohol.

Taxonomy. About 650 valid species of nemerteans are known. The great majority are marine, but a few occur on land in fresh waters. The genus *Prostoma* is essentially cosmopolitan.

Because of wide variations in color, number of reserve stylets, length of rhynchocoel, number of ocelli and size, the taxonomy of *Prostoma* is confused and difficult. The situation has become especially involved because some of the descriptions and identifications have been based on contracted and faded alcoholic specimens. *Prostoma graecense* has also gone under the generic designations of *Emea, Tetrastemma,* and *Stichostemma* and under the species designations of *asensoriatum, aquarium,* and *rubrum.*

Although it is probably true that European species have been introduced into this country on imported aquatic plants, it is not known how widespread they have become. Furthermore, it is quite possible that additional native species await discovery. Indeed, the present writer has collected several nemerteans that are obviously undescribed species.

REFERENCES

Böhmig, L. 1898. Beiträge zur Anatomie und Histologie der Nemertinen. *Z. Wiss. Zool.* 64:479–564.

Child, C. M. 1901. The habits and natural history of Stichostemma. *Am. Nat.* 35:975–1006.

Coe, W. R. 1943. Biology of the nemerteans of the Atlantic coast of North America. *Trans. Conn. Acad. Arts Sci.* 35:129–328.

Gibson, R., and J. Moore. 1978. Freshwater nemerteans: new records of Prostoma and a description of Prostoma canadiensis sp. nov. *Zool. Anz.* 201:77–85.

Gibson, R., and J. O. Young. 1974. First records of Prostoma (Hoplonemertea) from East Africa. *Ibid.* 193:103–109.

_____. 1976. Freshwater nemerteans. *Zool. J. Linn. Soc.* 58:177–218.

Montgomery, T. 1896. Stichostemma aesensoriatum n. sp., a freshwater nemertean from Pennsylvania. *Zool. Anz.* 19:436–438.

Moore, J., and R. Gibson. 1985. The evolution and comparative physiology of terrestrial and freshwater nemerteans. *Biol. Rev.* 60:257–312.

Reisinger, E. 1926. Nemertini. *Biol. Tiere Deutschl.* 7:1–24.

Stiasny-Wijnhoff, G. 1938. Das Prostoma Dugés, eine Gattung von Süsswasser-Nemertinen. *Arch. Neerl. Zool. Suppl.* 3:219–230.

7

GASTROTRICHA

In company with protozoans, rotifers, nematodes, and small oligochaetes, the Gastrotricha are a part of the characteristic assemblage of microorganisms on aquatic vegetation and on the debris (periphyton) forming the substrates of standing waters. They are also commonly a part of the upper interstitial, or meiobenthos.

Formerly it was thought that gastrotrichs were almost restricted to fresh waters, but they are now known to be abundant in marine habitats. The phylum consists of only two orders; the Order Macrodasyoidea is strictly marine, and the Order Chaetonotoidea is predominantly freshwater.

General characteristics. Most fresh-water gastrotrichs range from 100 to 300 μm in length, although a few are as small as 70 μm or as large as 600 μm. The body is short to long and wormlike, with a flat ventral surface. Head and arched trunk regions are obvious, and often there is a more or less distinct neck. Two toelike projections, forming the furca, are commonly present at the posterior end.

The margin of the head is variable, even within the same genus. Sometimes the entire margin is smooth, sometimes it has a median anterior lobe and one or two pairs of variously developed lateral lobes. Tufts of sensory cilia are customarily situated in the depressions between lobes. The mouth is terminal, subterminal, or ventral, and is often surrounded by short delicate bristles. In fresh-water species the anus lies in a dorsal position at the base of the furca; rarely is it terminal or subterminal.

Locomotor cilia occur chiefly on the ventral surface of the head and trunk. Most commonly there are several patches on the head and two longitudinal bands running throughout most of the body length (Figs. 3 H, J). Unlike turbellarians, gastrotrichs have only monociliated epithelial cells.

Coloration is variable, usually light gray to reddish brown, and mostly determined by the contents of the digestive tract. A starved gastrotrich is grayish and translucent.

Internal organization is relatively uncomplicated. The digestive tract is tubular and without diverticula. The space between the digestive tract and thin body wall is a pseudocoel and it contains the simple muscle, protonephridial, nervous, and reproductive systems (Fig. 1). There are no special circulatory or respiratory structures.

Many workers have discussed the phylogenetic relationships with the Nematoda, Rotatoria, Kinorhyncha, Turbellaria, and other groups, but the Gastrotricha are here considered sufficiently distinctive to constitute a separate phylum.

Cuticular structures. The general body surface is gastrotrichs is covered with a cuticle secreted by the underlying syncytial hypodermis. In the great majority of species, however, the cuticle is not simple and thin but patterned and with a variety of outgrowths, especially dorsally and laterally (Figs. 2, 7, 8).

In a few species the cuticle is locally thickened to form variously arranged plates, but more commonly there are series of cuticular scales that usually overlap more or less. Both plates and scales may have small keels or spines; the latter range from very short to exceedingly long and robust. In the absence of plates and scales, spines are often attached directly to the thin cuticle. Sometimes spines are barbed and have a triradiate or platelike base. A few species have midventral scales or

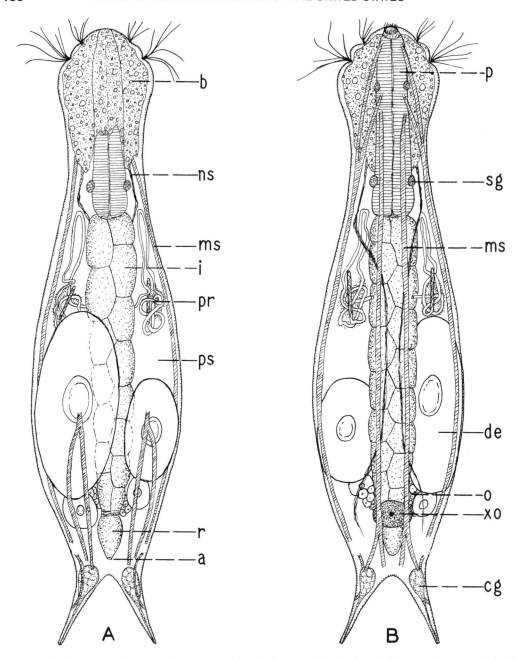

FIG. 1.–Diagrammatic internal structure of a typical gastrotrich. A, dorsal; B, ventral. a, anus; b, brain; cg, cement gland; de, developing egg; i, intestine; ms, muscle strand; ns, longitudinal nerve strand; o, ovary; p, pharynx; pr, protonephridium; ps, pseudocoel; r, rectum; sg, salivary gland; xo, "X organ." (Greatly modified from Zelinka, 1889.)

plates. Size and shape of plates, scales, and spines commonly vary within the same individual, depending on the location on the body. A few simple setae are often present, especially near the anterior and posterior ends of the body. The cephalic shield is a thickened cuticular plate covering the anterior end of the head above the mouth (Fig. 3 G). Sometimes there are less distinct special plates posterior to the mouth and laterally on the head.

Locomotion. Beating movements of the ventral cilia against the substrate produce a characteristic smooth, graceful, gliding type of location that is sometimes quite rapid; presumably the lateral tufts of head cilia are also of some importance. Such gliding locomotion is especially notable in the Chaetonotidae.

The long spines of the Dasydytidae are used in springing and leaping. Although the spines are not supplied with special muscles, they are moved vigorously toward the plane of the body when the arched body is straightened.

Most gastrotrichs may temporarily leave the substrate and swim about in the water with the ventral and head cilia, but they seldom go far above the substrate.

Polymerurus is said to have a sinuous type of locomotion characteristic of nematodes.

A cement gland is usually found in the base of each process of the furca, with its duct opening at or near the tip. The adhesive secreted material is used for temporary attachment to objects. A similar arrangement is found in rotifers. In the marine order Macrodasyoidea there is often an abundance of small lateral and posterior adhesive tubules.

Feeding, digestive system. Gastrotrichs browse about on the substrate, ingesting bacteria, algae, small protozoans, and organic detritus. Currents induced by head cilia aid in concentrating and bringing food particles to the mouth.

Often the mouth opening is surrounded by a projecting cuticular collarlike lip bearing the oral bristles. The mouth cavity opens

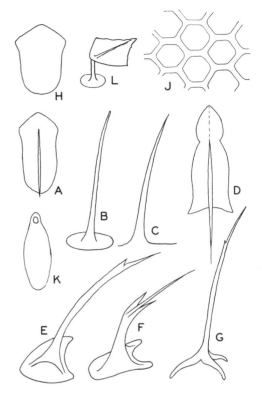

FIG. 2.–Cuticular structures of Gastrotricha. A to G, *Chaetonotus;* H and J, *Lepidodermella;* K, *Polymerurus;* L, *Aspidiophorus.* H, J, and K, simple cuticular plates; A, keeled plate; B, D, E, and F, spined plates; C and G, cuticular spines. (J from Brunson, 1950.)

directly into the pharynx, which is remarkably similar to the nematode pharynx. It is a long, cylindrical, muscular organ, usually slightly larger toward the inner end and sometimes with one or two indistinct bulbs. Its cavity is more or less triquetrous and lined with thin cuticle. Four small unicellular salivary glands are closely associated with the pharynx, one pair anteriorly and one posteriorly.

The intestinal wall consists of only four longitudinal rows of cells. There are no caeca, although a short rectum is distinguishable just before the anus.

One investigator has reported that reserve food materials are stored in special subepidermal cells, which break loose and float about in the pseudocoel.

Excretion. The Chaetonotidae, including the great majority of fresh-water species, all have a single pair of flame bulbs at about midlength. The two protonephridial tubules are considerably coiled in the pseudocoel and open ventrally near the midline. Undoubtedly the protonephridial system functions in both excretion and osmoregulation. Although there is some disagreement, it is probable that the other families of the Chaetonotoidea also have a protonephridial system. Some investigators claim that the marine species of Chaetonotidae have a protonephridial system; others maintain that it is absent. In the marine order Macrodasyoidea, however, the system is lacking. It is possible that the intestine of all gastrotrichs has an accessory excretory function.

Muscle system. The longitudinal muscle strands are quite prominent, especially ventrally. Usually there are about six pairs, with origins and insertions on the body wall. Their contraction produces shortening, curving, and partial rolling up of the body. A few delicate circular fibers in the body wall have recently been demonstrated for a few species, but they are thought to be of little importance in movement. In addition to the radial pharyngeal musculature, there are a few circular fibers on the outside of the pharynx and in the region of the rectum.

Nervous system. A large, dorsal, lobed, saddled-shaped brain covers the anterior portion of the pharynx, and two lateral nerve strands extend almost to the posterior end of the body. Fine branches innervate the body wall and viscera.

Sensory receptors. One or two pairs of pigmented "eyespots" have been seen on the lateral surfaces of the brain in a few species, but their function is questionable.

The tufts of cilia on the head are thought to function in touch, current detection, and the detection of dissolved materials. They are inserted in special thick glandular cells of the hypodermis.

Pairs of anterior and posterior tactile bristles are of common occurrence.

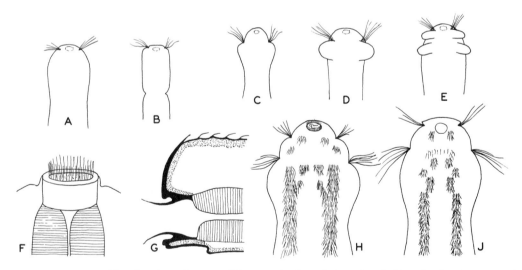

FIG. 3.–Structure of head region of Gastrotricha. A to E, dorsal view of typical heads, showing general shape and head lobes; F, semidiagrammatic ventral view of mouth region, showing circular collar, oral bristles, and anterior portion of pharynx; G, diagrammatic longitudinal section of anterior end, showing oral bristles, oral cavity, anterior portion of pharynx, and the cephalic shield; H, ventral view of anterior end of a typical *Chaetonotus,* showing arrangement of cilia; J, ventral view of anterior end of a typical *Ichthydium,* showing arrangement of cilia. (A–E modified from Brunson, 1950; F–J modified from Remane, 1935–1936.)

Reproduction, life history. Fresh-water gastrotrochs are all parthenogenetic females, males being unknown. Recently, however, atypical sperm have been seen in the genital area of a few of these "parthenogenetic" species (Weiss and Levy 1979). The two parts of the ovary are lateral to the intestine. Each portion consists of a few oocytes and eggs in various stages of development. The oviduct is so skimpy that some investigators doubt its existence as a definite duct. The genital pore is ventral. A single ventral saclike "X organ" lies just inside the genital pore; possibly it represents a vestigial copulatory bursa.

A gastrotrich usually produces one to five eggs during its lifetime. A mature egg is quite large, and before leaving the parent it distends the body greatly. Nevertheless, it is plastic, and as it emerges from the genital pore it is momentarily constricted into a dumbbell shape. On coming in contact with the water an outer shell hardens. Sometimes the shell bears spines or other protuberances. Eggs are usually attached to some small object on the substrate.

Brunson (1949) distinguishes two definite types of eggs, differing slightly in size. Tachyblastic eggs begin cleavage as soon as deposited and hatch 12 to 70 hours later. Opsiblastic eggs are slightly larger and are produced in old cultures and when environmental conditions become unfavorable. They have a heavier shell and survive desiccation, freezing, and unusually high temperatures. The duration of dormancy is variable, depending on ecological conditions, but it is known to last as long as 2 years. Opsiblastic eggs are usually

the last ones laid by a particular gastrotrich, but there are some exceptions.

Gastrulation begins at the 32-cell stage. Development is direct. On emergence from the egg a gastrotrich is quite large, and during the ensuing several hours it grows by 20 to 30% to attain mature size. As compared with the Macrodasyoidea, the Chaetonotoidea are composed of relatively few cells; perhaps the number is constant within each species. Little is known about the normal length of life; individuals in cultures live for 3 to 22 days, depending on the species and ecological conditions. Like the situation in rotifers, the cuticle is never shed during the life history.

Ecology. Only seven genera of gastrotrichs are restricted to fresh waters, including *Dasydytes, Dichaetura, Heterolepidoderma, Kijanebalola, Lepidodermella, Neogossea,* and *Proichthydium.* Four genera occur in both marine and fresh waters, or brackish and fresh waters. These are *Aspidiophorus, Chaetonotus, Ichthydium,* and *Polymerurus.* Only a very few species are known to occur in both types of environments, and some investigators believe that these are all actually pairs of varieties. About 35 genera are restricted to marine habitats.

Gastrotrichs are typical of puddles, marshes, wet bogs, and the shallow litoral, especially where there is detritus and decaying material. Old laboratory protozoan cultures or aquaria with disintegrating aquatic plants commonly have gastrotrich populations. They are uncommon in running waters generally and perhaps absent from swift streams. For some unknown reason, tropical gastrotrichs are often found swimming about above the substrate.

Most species remain closely associated with the substrate, although the Neogosseidae and Dasydytidae may swim about above it. Their occurrence in the plankton of the shallows, however, is atypical and fortuitous. Little is known about the depth distribution and concentrations of fresh-water species, although *Chaetonotus* has been found in bottom deposits as deep as 36 m. Recently it has been found that the capillary water of sandy beaches a meter or two from the water's edge supports a considerable population of gastrotrichs.

Although maximum numbers are usually

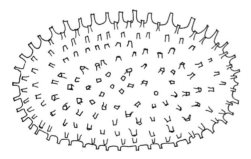

FIG. 4.–Opsiblastic egg of *Chaetonotus.* (Redrawn from Zelinka, 1889.)

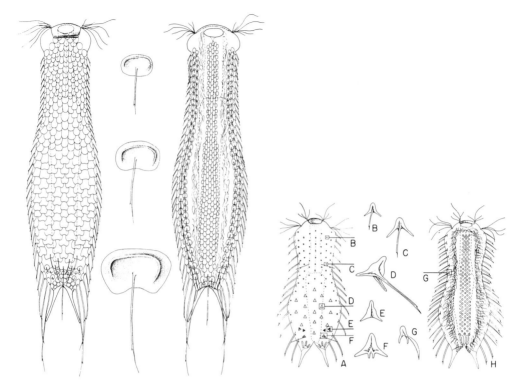

Fig. 5.–Detailed structure of two species of *Chaetonotus.* Note ventral ciliation patterns. (From Balsamo, 1983.)

found in summer and autumn, gastrotrichs may be collected at any time of the year. Other than opsiblastic eggs, no special desiccation or overwintering mechanisms are known.

Many species may be collected from habitats where there is much decay, and where the dissolved oxygen content consequently may be less than 1 ppm. Minimal oxygen concentrations are not definitely known, however, although it is probable that gastrotrichs may withstand at least temporary anaerobiosis.

Geographical distribution, dispersal. Except for some areas in Europe and three or four in the United States, there has been almost no intensive, systematic collection of gastrotrichs. Most records are the result of incidental or superficial studies, often associated with collections of other groups of fresh-water invertebrates. As a result, we know little about the geographical distribution

of individual species. Most fresh-water genera, however, appear to be cosmopolitan and occur the world over in suitable habitats; these are *Aspidiophorus, Chaetonotus, Dasydytes, Ichthydium, Lepidodermella, Neogossea,* and *Polymerurus.* Overland dispersal is undoubtedly effected by the resistant opsiblastic eggs.

Enemies. In addition to the toll taken by browsing insects and crustaceans, gastrotrichs are also devoured by amoeboid protozoans, hydras, oligochaetes, turbellarians, and perhaps by tardigrades and nematodes.

Collecting. Sandy beach washings, and rinsings of aquatic vegetation often yield good collections. Specimens may be concentrated by straining the wash water through fine bolting silk. For bottom species the thin layer of surface mud and debris should be carefully scooped up and placed in a suitable container with water. After standing for a time in the laboratory, material at the debris and water interface should be sucked up with a long pipette and examined under the high power of the binoculars.

Culturing. Old protozoan cultures are sometimes suitable if gastrotrichs are seeded into them, and one investigator recommends using two drops of uncooked egg yolk in 100-ml of spring or pond water. In either case it is best to add a little bottom debris to the culture.

One-tenth percent malted milk in well water is especially useful for pure cultures. After the water is brought to a boil, malted milk powder is added, and the solution is then further boiled for about a minute. Let stand a week before inoculating with gastrotrichs.

Regardless of the culture method used, about one half the volume of fluid should be renewed every 2 to 3 weeks.

Preserving, preparing. Two percent osmic acid is an excellent fixative. Place the living specimens in a drop of water on a slide and invert for 5 to 10 seconds over the mouth of the bottle of osmic acid. Saturated mercuric chloride, 10% formalin, and Bouin's fluid are alternative fixatives that often give good results. Specimens may be stored in 70% alcohol or 5% formalin. If glycerin is to be used for mounting, transfer the gastrotrichs to a concave watch glass containing 5% glycerin in 50% alcohol. Cover loosely to keep out dust and permit evaporation. After about a week the water and alcohol will all have evaporated and the specimens will be in concentrated glycerin. Any of the several mounting methods for rotifers may be used (page 195). Glycerin is excellent for permanent storage.

Some species require narcotization before fixing, and a wide variety of substances may be tried with varying success, including a concentrated aqueous menthol solution, chloretone, 2% benzamine lactate, 2% butyn, and 2% hydroxylamine hydrochloride.

Scales, plates, and spines may be isolated by the judicious use of dilute acetic acid. Stain with fuchsin. Oil immersion is essential for critical work and species identification. An Irwin loop is ideal for sorting and handling individual gastrotrichs.

Taxonomy. Only two investigators have made important taxonomic contributions to the biology of American gastrotrichs. Stokes published the results of his New Jersey observations in 1887 and 1888, and Brunson began his series of papers in 1947. Obviously, therefore, an enormous amount of work remains to be done in this country. Some of the most important European investigations are those of Zelinka (1889), Grünspan (1910), and Remane (1935–1936). By way of contrast, a great deal of research is being done on marine gastrotrichs.

The most reliable and significant characteristics for distinguishing genera are the nature of the furca, and the presence or absence and detailed structure and arrangement of cuticular plates, scales, and spines.

The majority of American species are unreported or undescribed. As a consequence, almost every study, however superficial, turns up new species. It is therefore fruitless to include a key to the approximately 45 American species known to date.

About 250 species have been described from the fresh waters of the world, but they may all be grouped into only 14 genera. The key that follows includes only nine genera; two of these have not yet been reported from the United States (*Aspidiophorus* and *Dichaetura*), but they are widely distributed and will undoubtedly be found here eventually. Five other rare fresh-water genera are not included in the key; *Proichthyidium* has been collected only once in the Argentine, *Kijanebalola* is known only from central Africa, *Metadasydytes* has been collected only in Poland, and one species each of *Redudasys* and *Arenotus* have been found in Brazilian psammon.

Remane (1935–1936) gives good evidence for grouping *Dasydytes, Setopus, Anacanthoderma, Stylochaeta,* and so on, into a single genus, *Dasydytes,* and this practice is followed in the present manual. *Lepidoderma* is preoccupied and *Lepidodermella* is used instead.

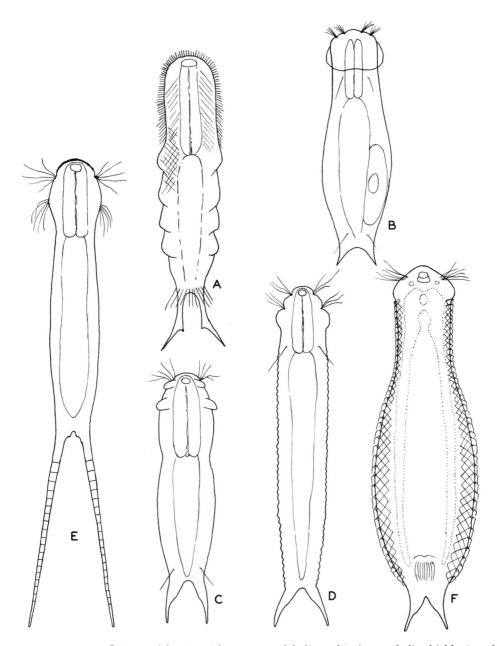

FIG. 6.–Structure of Gastrotricha. A, *Dichaetura;* B, *Ichthydium,* showing cephalic shield; C and D, *Ichthydium;* E, *Polymerurus;* F, ventral view of *Aspidiophorus,* showing arrangement of cilia. (A modified from Metschnikoff; B modified from Konsuloff; C modified from Brunson, 1949; D and E modified from Brunson, 1950; F modified from Saito, 1937.)

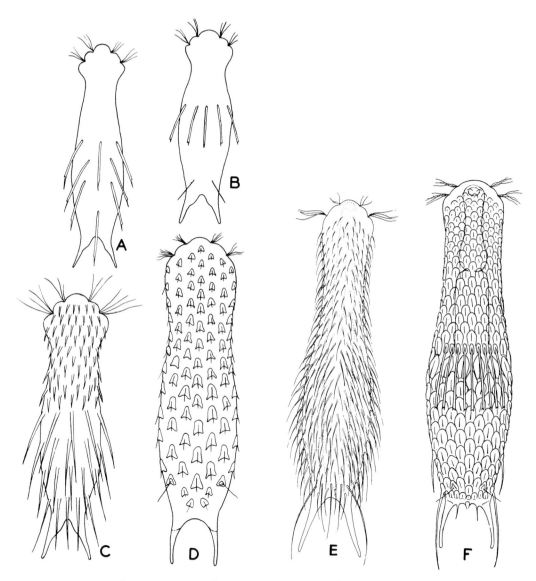

FIG. 7.–Structure of various species of *Chaetonotus*. (A–C modified from Brunson, 1950; D modified from Saito, 1937; E from Brunson, 1950; F from Saito, 1937.)

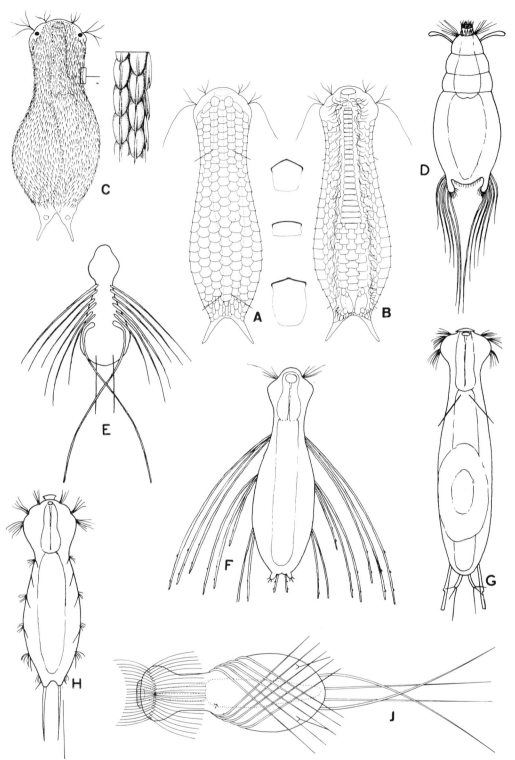

FIG. 8.–Structure of Gastrotricha. A, *Lepidodermella,* dorsal; B, *Lepidodermella,* ventral; C, *Heterolepidoderma* with detail; D, *Neogossea;* E–J, *Dasydytes.* (A–C, and J from Balsamo, 1983; D modified from Daday; E and F modified from Brunson, 1950; G modified from Voigt, 1960; H modified from Greuter.)

KEY TO GENERA OF GASTROTRICHA

1. Caudal furca present (Figs. 6, 7) 2
   True caudal furca absent, but sometimes with rudimentary protuberances, long spines, or setose styloid processes (Figs. 8 D–J) 8
2. Caudal furca not forked; cuticle smooth, or with spines, plates, simple scales, or spined scales or plates; head usually with two to four tufts of cilia; cephalic shield present or absent; usually with one pair of posterior bristles (Figs. 6 B–F; 8A, B); two longitudinal bands of cilia; common and widely distributed; includes about 90% of the known fresh-water species ... CHAETONOTIDAE, 3
   Caudal furca forked; cuticle covered with spined plates; head without tufts of cilia but with many single cilia; cephalic shield absent; four to many posterior bristles (Fig. 6A); several rare and poorly known species reported from Europe; not known in the United States.
   DICHAETURIDAE, **Dichaetura**
3. Caudal furca very long and jointed; body usually with spines or spines and scales (Fig. 6E); several species .... **Polymerurus**
   Caudal furca not especially long, not jointed 4
4. Cuticle smooth, without plates, scales, or spines, but sometimes with setae (Figs. 6C, D); about 25 species some of which are marine .... **Ichthydium**
   Cuticle not smooth 5
5. Body covered with complex stalked plates, each consisting of a circular basal plate, a short stalk, and a rhombic end plate (Fig. 6F); about 15 species; not yet reported from the United States.
   Aspidiophorus
   Body with simple plates, scales, or spines 6
6. Body with spines or spined scales (Fig. 7); very common; about 110 species .... **Chaetonotus**
   Body without spines or scales but covered with plates 7
7. With flat broad plates or scales (Figs. 8A, B); about 15 species .... **Lepidodermella**
   With small, keeled plates (Fig. 8C); several uncommon species; rare in the United States.
   Heterolepidoderma
8. Head with two club-shaped tentacles; body with long spines, with or without basal plates; two long spines or two styloid processes at posterior end of body; several uncommon species (Fig. 8D); rare in the United States .... NEOGOSSEIDAE, **Neogossea**
   Head without tentacles; body with long spines, arranged singly or grouped; no plates or scales; ventral ciliation in two longitudinal bands or two longitudinal rows of tufts; head ciliation usually with three ventral transverse bands of cilia (Figs. 8E–J); a highly variable genus in need of taxonomic revision; about 25 species .... DASYDYTIDAE, **Dasydytes**

REFERENCES

Balsamo, M. 1977. Prime recerche sui Gastrotrichi dulciacquicoli Italiani. *Att. Soc. Toscana Sci. Nat.* **84**:87–150.

———. 1983. Gastrotrichi (Gastrotricha). *Guide Riconosc. Spec. Anim. Acque Int. Ital.* **20**:1–92.

Bennett, L. W. 1979. Experimental analysis of the trophic ecology of Lepidodermella squammata (Gastrotricha: Chaetonotoidea). *Trans. Am. Microsc. Soc.* **98**:254–260.

Brunson, R. B. 1947. Gastrotricha of North America. II. Four new species of Ichthyidium from Michigan. *Pap. Mich. Acad. Sci. Arts Lett.* **33**:59–62.

———. 1948. Chaetonotus tachyneusticus. A new species of gastrotrich from Michigan. *Trans. Am. Microsc. Soc.* **67**:350–351.

———. 1949. The life history and ecology of two North American gastrotrichs. *Ibid.* **68**:1–20.

———. 1950. An introduction to the taxonomy of the Gastrotricha with a study of eighteen species from Michigan. *Ibid.* **69**:325–352.

Daday, E. 1905. Untersuchungen über die Süsswasser-Mikrofauna Paraguays. *Zoologica* **18**:72–86.

De Beauchamp, P. M. 1934. Sur la morphologie et l'ethologie des Neogossea (gastrotriches). *Bull. Soc. Zool. Fr.* **58**:331–342.

Faucon, A. S., and W. D. Hummon. 1976. Effects of mine acid on the longevity and reproductive rate of the Gastrotricha Lepidodermella squammata (Dujardin). *Hydrobiologia* **50**:265–269.

Greuter, R. 1917. Beiträge zur Systematik der Gastrotricha in der Schweiz. *Rev. Suisse Zool.* **25**:35–70.

Grünspan, T. 1910. Die Süsswassergastrotrichen Europas. Eine zusammenfassende Darstellung ihrer Anatomie, Biologie und Systematik. *Ann. Biol. Lacustre* **4**:211–365.

Kisielewski, J. 1986. Freshwater Gastrotricha of Poland. *Fragmenta Faunistica* **30** (9–16):139–295.

———. 1987. Two new interesting genera of Gastrotricha (Macrodasyida and Chaetonotida) from the Brazilian freshwater psammon. *Hydrobiologia* **153**:23–30.

Konsuloff, S. 1913. Notizen über die Gastrotrichen Bulgariens. *Zool. Anz.* **43**:255–260.

Metschnikoff, E. 1865. Über einige wenig bekannte niedere Tierformen. *Zeitsch. Wiss. Zool.* **4**:450–465.

Mola, P. 1932. Gastrotricha delle acque dolci italiane. *Int. Rev.* **26**:397–423.

Packard, C. E. 1936. Observations on the Gastrotricha indigenous to New Hampshire. *Trans. Am. Microsc. Soc.* **55**:422–427.

Remane, A. 1927. Beiträge zur Systematik der Süsswassergastrotrichen. *Zool. Jahrb. Abt. Syst.* **53**:269–320.

———. 1935–1936. Gastrotricha. *Klassen u. Ordnungen des Tierreich* **4** (Abt. 2, Buch 1, Teil 2, Lfg. 1–2):1–242.

Saito, I. 1937. Neue und bekannte Gastrotrichen der Umgebung von Hiroshima (Japan). *J. Sci. Hiroshima Univ., Ser. B, Div. 1*, **5**:245–265.

Stokes, A. C. 1887. Observations on Chaetonotus. *The Microscope.* **7**:1–9, 33–43.

———. 1888. Observations on a new Dasydytes and a new Chaetonotus. *Ibid.* **8**:261–265.

Voigt, M. 1960. Gastrotricha. *Die Tierwelt Mitteleur.* **1**:4a:1–74.

Weiss, M. J., and D. P. Levy. 1979. Sperm in "parthenogenetic" freshwater gastrotrichs. *Science* **205**:302–303.

Zelinka, C. 1889. Die Gastrotrichen. Eine monographische Darstellung ihrer Anatomie, Biologie und Systematik. *Z. Wiss. Zool.* **49**:209–384.

8

ROTIFERA
(Rotifers)

If we were to designate a single major taxonomic category that is most characteristic of fresh waters, it could only be the Phylum Rotifera. For the rotifers are one of the few groups that have unquestionably originated in fresh waters, and it is here that they have attained their greatest abundance and diversity. For this reason we have chosen to give more detailed consideration to rotifers than to any other fresh-water taxon. Probably 2500 species have been described, but less than 5% of these are found in marine and brackish waters. Only two plankton species occur in the mid-Atlantic.

Except for occasional new species descriptions, the basic taxonomy of the Rotifera is well established. Recent emphasis has centered on experimental work, especially biochemical phenomena, clone genetics, species competition, selective feeding, and sensory physiology.

Rotifers were first studied and described by Leeuwenhoek in 1703, and since his time they have become classical objects for study by the amateur microscopist and professional hydrobiologist alike. They are essentially all microscopic, the length range for the phylum being about 40 μm to 2.5 mm, but the great majority are between 100 μm and 500 μm long. These fascinating creatures have long been called Rotifera, or rotifers, because in some species the disclike ciliated anterior end, or corona, has a fancied and illusory resemblance to a pair of revolving wheels owing to the synchronized beating of the coronal cilia. Rotifers occur in an endless variety of aquatic and semiaquatic habitats, including the limnetic and deepest regions of the largest lakes and the smallest puddles.

They are found in damp soil and vegetable debris, in mosses that may be wetted or dampened only occasionally, in the interstices between the sand grains of lake beaches, from the Arctic and Antarctic to the tropics, in small rock depressions, and even in cemetery flower urns and eaves troughs.

About 20 species of bdelloids have been found in Antarctica, but the total bdelloid fauna there is probably much larger.

The vast majority of rotifers encountered under natural conditions are females. Males are definitely known for relatively few species; they are much smaller than the females, degenerate, and seldom live for more than 2 or 3 days. Most of the remarks in this chapter concern only female rotifers; the biology of males is discussed in a special section on page 185.

General characteristics. The Phylum Rotifera is commonly divided into two classes, the Digononta and the Monogonta. The female members of the former class are characterized by paired ovaries, a ramate mastax, and the absence of a secreted tube or lorica. The Class Digononta is subdivided into two orders, the Seisonidea and the Bdelloidea. The Seisonidea includes only a single genus, *Seison,* occurring as commensals on marine Crustacea; the ovaries have no vitellaria, males are well developed, and the corona is rudimentary. Almost all Bdelloidea, however, occur in fresh waters; the ovaries have vitellaria, males are unknown, reproduction is exclusively by parthenogenesis, and the corona is well developed. The Bdelloidea are appropriately named since the body is highly contractile and they creep about on the substrate in a leechlike manner.

The members of the Class Monogononta constitute about 90% of the known species of rotifers and are characterized by a single ovary, by a mastax that is not ramate, and by the presence or absence of a lorica or a secreted tube. Males are known for some species and not for others; they are small and degenerate. This class is divided into three orders: Ploima, Flosculariacea, and Collothecacea.

The Order Ploima includes most of the free swimming, limnetic plankton, and lit-toral species. These rotifers typically have a posterior "foot" and two "toes." The corona is not especially large.

The Order Flosculariacea includes some species that are free swimming and many more that are sessile as adults. The foot, when present, has no toes. Often there is a secreted gelatinous envelope or case. The corona is not especially large.

The Order Collothecacea includes those rotifers with a very large, lobed corona and a mouth that is centrally located at the lower end of a funnellike infundibulum; they are usually solitary and sessile.

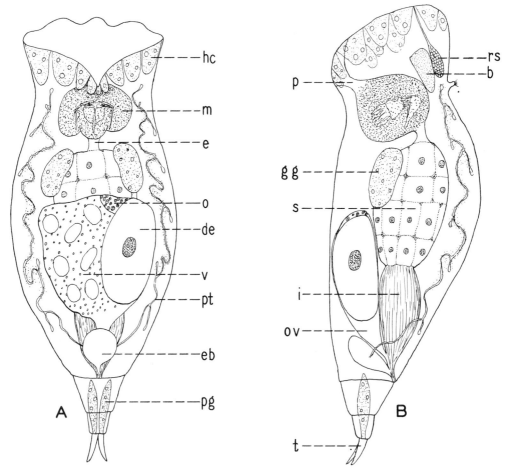

FIG. 1.–Diagrammatic structure of *Epiphanes senta* (Müller), ×400. A, ventral view; B, lateral view. (Musculature and coronal ciliation not shown; see Fig. 102.) b, brain; de, developing egg; e, esophagus; eb, excretory bladder; gg, gastric gland; hc, hypodermal cell of head; i, intestine; m, mastax; o, ovary; ov, oviduct; p, pharynx; pg, pedal gland; pt, protonephridial tubule; rs, retrocerebral sac; s, stomach; t, toe; v, vitellarium.

As a group, the rotifers display an amazing range of morphological variations and adaptations. Yet the great majority have several fundamental features in common. The body is usually elongated and cylindrical, though the complete range of shape extends from linear attenuation to spherical. Usually three main body regions may be distinguished; head, trunk, and foot. The head is more or less distinctly set off from the large trunk region, although there is rarely an elongated neck region. In the bdelloid rotifers the body

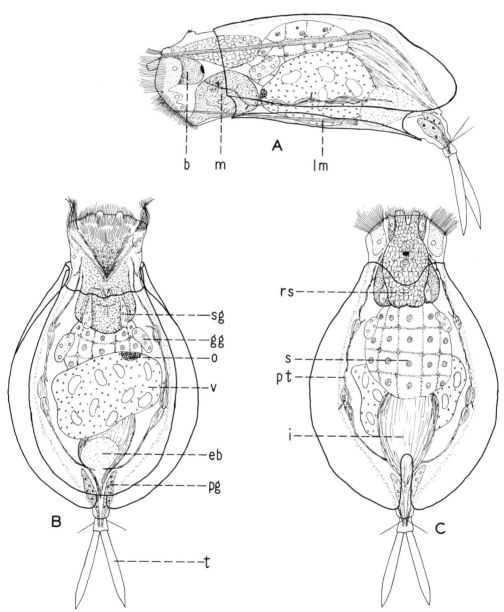

FIG. 2.–Structure of *Euchlanis dilatata* Ehr., ×320, a typical loricate rotifer, semidiagrammatic. A, lateral; B, ventral; C, dorsal. b, brain; eb, excretory bladder; gg, gastric gland; i, intestine; lm, longitudinal muscle; m, mastax; o, ovary; pg, pedal gland; pt, protonephridial tubule; rs, retrocerebral sac; s, stomach; sg, salivary gland; t, toe; v, vitellarium. (Modified from Myers, 1930.)

appears to be composed of segments, but these are only superficial annuli indicating the zones of folding and telescoping of the cuticle when the animal contracts; typically there are 15 to 18 such annuli. There is no true segmentation in rotifers. The posterior tapering foot usually consists of two or more superficial segmentlike parts, or it may be highly retractile and annulated or creased; sometimes it is absent. When present, the foot usually has two (sometimes none, one, three, or four) terminal "toes."

The general body surface of a rotifer is covered with a cuticle that is secreted by a syncytial hypodermis. The hypodermis forms a thin layer under most of the cuticle, but at the anterior end it is thick, cushionlike, and inwardly lobed. In some rotifers the cuticle is very thin and flexible, but in many genera a portion of the cuticular surface is thickened, more or less rigid, and is termed a lorica. The lorica may be poorly developed and consists of several thin, elastic, platelike sections of the trunk cuticle, or it may be thick, rigid, box-like, sculptured, and inelastic and may in-volve the entire surface of the trunk, much of the foot, and even a part of the head. Between these two extremes, every gradation in lorica development is to be found, sometimes even within a single genus (e.g., *Cephalodella*).

The periphery and more or less of the corona surface are ciliated, but the density and pattern of ciliation are highly variable. In general, these cilia serve the double function of locomotion and feeding.

The mouth may be anterior and in the center of the corona, peripheral, subterminal, or ventral. The anus, or cloacal aperture, is typically located dorsally at the base of the foot.

The mastax is a structure peculiar to the digestive system of rotifers, and no com-parable device is known elsewhere in the animal kingdom, with the possible exception of the marine Gnathostomulida. Superficially, it has the appearance of a bulbous swelling lying between the pharynx proper and the esophagus. It consists of a complicated ar-rangement of muscles that activate a set of translucent sclerotized jaws (collectively called the trophi), which are used to seize, tear, grind, or macerate the food (Figs. 8–10).

True coloration of the tissues is usually

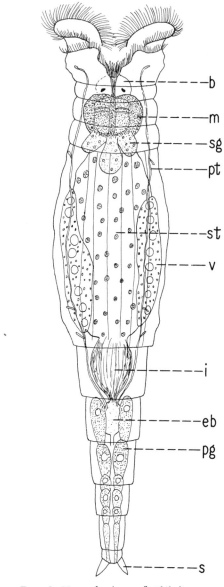

Fig. 3.–Ventral view of *Philodina*, a typical bdelloid rotifer, ×375, semidiagrammatic. b, brain; eb, excretory bladder; i, intestine; m, mastax; pg, pedal gland; pt, protonephridial tu-bule; s, spur (toes are completely retracted); sg, salivary gland; st, stomach; v, vitellarium. (Modified from Hickernell, 1917.)

grayish, yellowish, or sometimes a pink or bluish cast. The apparent coloration, however, is more often tan, green, or brownish, owing to the varied contents of the digestive tract and stored excretory granules.

Corona and its ciliation. It is probable that the most primitive type of corona is that of the creeping and browsing notommatid Ploima. It consists merely of an anterior, ventral, or oblique surface more or less densely covered with cilia; the buccal area, which surrounds the mouth, is always densely ciliated, but this buccal field may be extended so that it covers some or all of the corona (Fig. 4A). From this simple type, all the more complicated and intergrading arrangements are presumed to be derived. In addition to the dense, short cilia of the coronal surface, for example, there may be a peripheral wreath of large, strong cilia (Figs. 4B–E). On the other hand, with the exception of the buccal and peripheral areas, some genera, such as *Eosphora* (Fig. 39K), have little or no ciliation on the coronal surface. *Epiphanes* and many

other genera have a small buccal field, a strong peripheral wreath, and several transverse rows of long cilia and clumps of cilia located on protuberances on the corona. *Synchaeta* has several long sensory coronal bristles in addition to ordinary cilia (Fig. 4F). In many Notommatidae, *Synchaeta*, and a few other genera the corona bears a pair of lateral ciliated earlike processes, or auricles (Fig. 39D). In the Bdelloidea and Flosculariacea the surface of the corona is usually bare, but it has two peripheral concentric wreaths of long, strong cilia with a densely ciliated furrow between (Figs. 42–46). In *Stephanoceros* and *Collotheca* the corona is a greatly concave infundibulum and its upper edge is folded and developed into prominent lobes or arms. These protuberances bear long, stiff setae. The mouth, at the bottom of the infundibulum, is surrounded by bands of cilia. In *Acyclus, Cupelopagis,* and *Atrochus* there are no coronal cilia.

In addition to these several types of corona, there are other arrangements and intergradations involving sparse to dense ciliation.

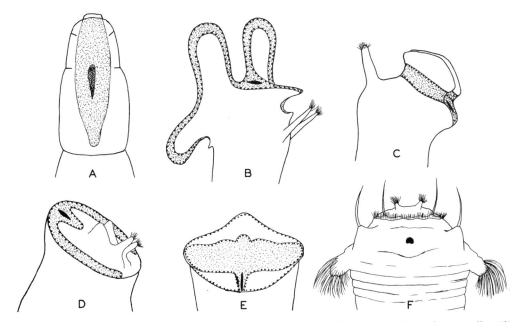

FIG. 4.–Examples of coronal ciliation. Membranelles and large cilia shown by large dots, smaller cilia by small dots. A, *Dicranophorus,* ventral; B, *Floscularia,* lateral, showing the ventral pellet-making concavity; C, *Philodina,* lateral; D, *Conochilus,* lateral; E, *Cyrtonia,* ventral; F, *Synchaeta,* dorsal.

Locomotion. Movements through the water are mostly dependent on the beating of the peripheral cilia, and many plankton and limnetic species remain in permanent suspension throughout life without ever coming in contact with a substrate. Such locomotion is often a combination of twisting on the longitudinal axis and spiral movements of the whole animal. A few plankton genera, such as *Filinia, Hexarthra,* and *Polyarthra,* also move by sudden jerks and leaps, owing to sudden beating movements of their long appendages. Most of the free swimming, nonplankton species having a foot and toes are able to creep about and browse as a result of the beating of the coronal cilia against the substrate combined with a pushing action of the toes. Toes are useful in steering during both creeping and swimming.

In the Bdelloidea, however, the corona is retracted during creeping and in such a manner as to produce an anterior terminal suckerlike proboscis that functions in the leechlike movements. The toes of bdelloids are usually everted only during active creeping (Figs. 44C, F; 45K; 46C).

Most rotifers with a foot and toes have pedal glands. The Ploima usually have two pedal glands in the base of the foot or in the posterior part of the trunk, whereas the Bdelloidea have from 4 to 20 such glands. Pedal glands open via ducts at the tips of the toes, at the base of the toes, or, in some Bdelloidea, also into the spurs. The secretion is used for anchoring the animal to the substrate temporarily by the tips of the toes or spurs.

Most of the Collothecacea and Flosculariacea are sessile as adults, but their newly hatched young swim about actively for a short time before attaching to the substrate.

Tube construction. A wide variety of protective tubes are constructed or secreted by many genera of sessile rotifers, and even by a few free-swimming species. In *Ptygura* the tube is usually gelatinous, sometimes flocculose, colorless to brown, and it appears to be secreted by the general body surface; in many species it is more or less covered with bits of adherent debris, and in one species the tube is supplemented with fecal pellets. *Stephanoceros* secretes a thin, hyaline tube, and most species of *Limnias* form a thin, firm, and sometimes annulated tube.

Several species of *Floscularia* construct their tubes of small pellets of debris, and needless to say, it is a fascinating process to watch. The newly settled young *Floscularia* secretes a rough gelatinous tube that is to form the base of the main tube. On the ventral side of the body,

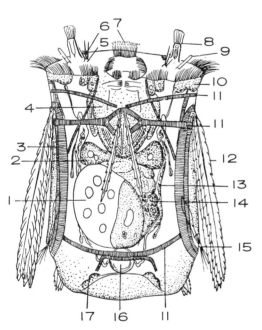

FIG. 5.–*Polyarthra vulgaris,* showing lateral locomotor appendages. (Modified from various sources.)

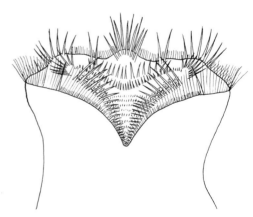

FIG. 6.–Coronal ciliation of *Epiphanes senta.* (Modified from de Beauchamp, 1909.)

just posterior to the mouth, there is a conical, ciliated depression into which a large gland secretes gelatinous material. As an accessory part of the coronal cilia action, small bits of detritus are collected and concentrated in the ciliated cup. This detritus is thoroughly mixed with the gelatinous secretion, and by means of ciliary action in the cup the mass is rotated on its axis and is formed into a hard symmetrical pellet having the shape of a pistol bullet. By appropriate movements of the head and trunk, the pellet is pushed into place at the top of the tube with the tip pointed outward. The upper edge of the tube is always level with the pellet-making cup, and as the animal elongates, it continuously manufactures pellets so that growth closely corresponds with the length of the tube (Edmondson, 1945).

According to the specific constituents of the pellets, the tubes are usually yellow, green, or brown, although in one species they are almost colorless and composed largely of the gelatinous secretion. Another species of *Floscularia* constructs its tube out of fecal pellets.

Mastax and trophi. The anatomy and action of the mastax and its contained trophi are so variously modified for special feeding habits that it is best to consider these structures before discussing feeding and the digestive system.

Food is brought into the mouth by ciliary

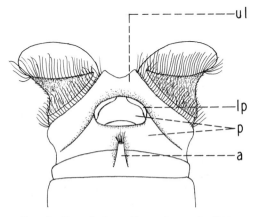

FIG. 7.–Dorsal view of anterior end of *Macrotrachela,* a typical bdelloid rotifer. a, antenna; lp, lamella of proboscis; p, proboscis; ul, upper lip.

action, down the short pharynx, and into the mastax. Only the anterior, dorsal, or anterodorsal part of the mastax contains a cavity through which the food passes. The greater, basal part of this organ consists of minute muscles and the sclerotized trophi, whose movements the muscles bring about. In traversing the mastax, however, the food must pass through the moving, distal, toothlike portions of the trophi.

The trophi consist of one median piece and three paired lateral pieces, all of which are subject to morphological variation and specialization. The fulcrum is basal and median and serves as an attachment for the two rami; these three pieces collectively form the incus. The two unci are usually toothed to a varying degree, and each of these pieces is attached laterally to a manubrium. An uncus and its associated manubrium collectively form a malleus. Thus the trophi consist of one incus (composed of one fulcrum and two rami) plus two mallei (each composed of one manubrium and one uncus) (Figs. 8A, B). Eight basic, but variable, types of mastax are recognized according to the relative development and specialization of their parts. These are characterized briefly below, and their occurrence in the various families of rotifers is given in the taxonomic outline on pages 196–197.

Malleate (Fig. 8B). Specialized for horizontal grinding of plankton and particulate detritus. Feebly prehensile.

Virgate (Fig. 10). The fulcrum is long and expanded at the base for the attachment of the hypopharynx, a powerful domelike mastax muscle; the upper (dorsal) end of the hypopharynx is free and covered with cuticle, and when it contracts it acts as a piston and sucks food into the mouth and mastax cavity. Rotifers with a virgate mastax commonly extract the fluids of plant cells and the body fluids of microscopic animals by this suction mechanism. In addition to suction, the trophi of the virgate mastax may be used for biting or nibbling preliminary to ingestion and the grinding action. Plankton, periphyton, and particulate detritus are also used as the chief food by many rotifers with a virgate mastax.

Cardate (Figs. 9A, B). This type of mastax occurs only in the Subfamily Lindiinae. Like the virgate type, the cardate mastax functions

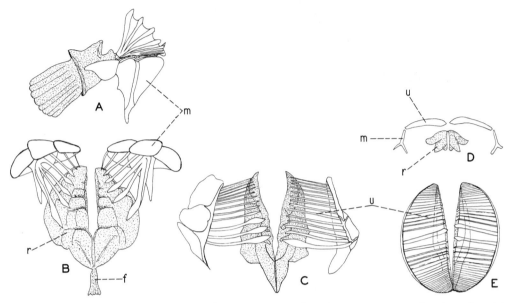

FIG. 8.–Typical trophi. A, left side view of malleate trophi of *Epiphanes senta*; B, anterior view of trophi of *E. senta*; C, anterior view of malleoramate trophi of *Floscularia*; D, optical section of ramate trophi of a bdelloid rotifer; E, anterior view of ramate trophi of a bdelloid rotifer. Inci are stippled. f, fulcrum; m, manubrium; r, ramus; u, uncus. (Modified from de Beauchamp, 1909.)

by suction, but the whole mastax is able to oscillate on a transverse axis, and a complicated epipharynx supports the mouth. The Lindiinae are only incidentally predatory, the chief food being periphyton and particulate detritus.

Forcipate (Fig. 9E). The trophi of this type of mastax are elongated, strongly compressed dorsoventrally, and adapted for protrusion through the mouth and the capture and tearing of the prey (Protozoa and micrometazoa). The trophi can sometimes be protruded for half their length.

Incudate (Figs. 9C, D). The trophi are prehensile and all parts are reduced except the pincerlike protrusile rami. The mastax musculature is likewise reduced. Food consists mostly of zooplankton.

Ramate (Figs. 8D, E). This mastax is specialized for the grinding of plankton, periphyton, and detritus. All parts of the trophi are reduced or vestigial except the semicircular unci, which have opposing linear teeth. The fulcrum is usually absent.

Malleoramate (Fig. 8C). This type is much like the ramate mastax, but the first few ventral teeth are larger, more highly developed, and detached, while the remaining teeth are progressively smaller striations. The fulcrum is present but small.

Uncinate (Fig. 9F). Specialized for laceration of ingested plankton, periphyton, and detritus. An intermediate piece, or subuncus, facilitates greater movement of the unci on the rami.

The trophi of all types of mastax are minutely sculptured, ridged, and barred. These complex markings can be discerned to advantage only with the oil immersion lens and they are mostly omitted in the trophi figures in this chapter.

Food, feeding. In both sessile and free swimming nonpredatory rotifers the coronal cilia are of primary importance in creating localized currents that guide and concentrate the loose periphyton, small plankton organisms, and detritus at the region of the mouth. Indeed, the great majority of species are omnivorous and ingest all organic particles of the appropriate size. Common ex-

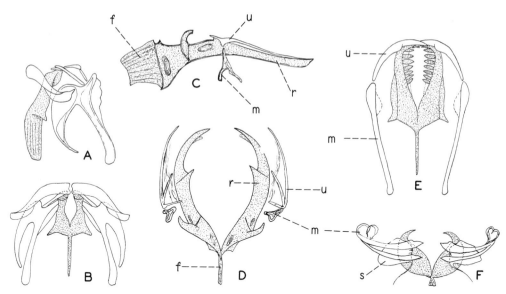

FIG. 9.–Typical trophi. A, lateral view of cardate trophi of *Lindia*; B, ventral view of trophi of *Lindia*; C, lateral view of incudate trophi of *Asplanchna*; D, anterior view of trophi of *Asplanchna*; E, anterior view of forcipate trophi of *Dicranophorus*; F, anterior view of uncinate trophi of *Stephanoceros*. Inci are stippled. f, fulcrum; m, manubrium; r, ramus; s, subuncus; u, uncus. (A and B modified from Harring and Myers, 1922; C to F modified from de Beauchamp, 1909; E modified from Harring and Myers, 1928.)

amples are *Cephalodella, Filinia, Keratella, Lecane, Proales, Euchlanis, Epiphanes, Brachionus,* the Bdelloidea, and most sessile species.

Predatory species probably detect their prey by touch or by biochemical stimuli. The following common genera feed on other rotifers and all kinds of small Metazoa, either in the plankton or on a substrate: *Asplanchna, Dicranophorus, Ploesoma, Synchaeta,* and *Trichocerca.* A careful investigation has shown that *Asplanchna* will usually ingest all plankton organisms larger than 15 μm.

Many notommatids, as well as a few genera in other families, feed mostly on the fluid contents of filamentous algal cells.

Cupelopagis, Acyclus, and *Atrochus* are all sessile and without coronal cilia. They have a large funnellike corona, or infundibulum, however, and when the prey wander into this cavity, they are quickly enclosed and ingested.

A few free-living rotifers have highly specialized food habits. *Acyclus inquietus,* for example, lives among *Sinantherina* colonies and eats the motile young. *Dicranophorus isothes* is thought to subsist entirely on small Cladocera, and *D. thysanus* feeds on dead copepods, Cladocera, and oligochaetes.

Digestive system. Most Flosculariacea, Bdelloidea, and Ploima have similar digestive systems (Figs. 13A, B, D–F). The mouth opens into a narrow, ciliated pharynx which leads to the cavity of the mastax. A variable number of small salivary glands are closely associated with the mastax. After the food has been macerated in the mastax, it passes into the short to long esophagus, which leaves the dorsal or posterodorsal surface of the mastax. The esophagus extends to the large, thick-walled, ciliated stomach, where most of the digestion and absorption are thought to occur. Usually there is a pair of oval or bean-shaped gastric glands on the anterior margin of the stomach. The intestine may be distinctly or indistinctly set off from the stomach. It is much smaller, sometimes very narrow, thin walled, and often ciliated. The cloaca is short and seldom ciliated; it opens to the outside dorsally at the base of the foot.

The anterior end of the Collothecacea

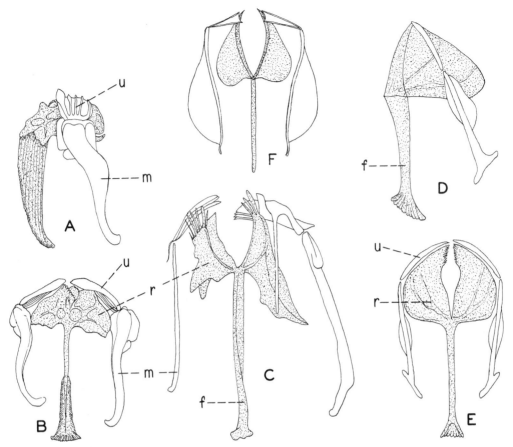

FIG. 10.–Typical virgate trophi. A, lateral view of trophi of *Notommata*; B, ventral view of trophi of *Notommata*; C, ventral view of trophi of *Trichocerca*; D, lateral view of trophi of *Cephalodella*; E, ventral view of trophi of *Cephalodella*; F, ventral view of trophi of *Synchaeta*. Inci are stippled. f, fulcrum; m, manubrium; u, uncus; r, ramus. (A and B modified from Harring and Myers, 1922; C modified from de Beauchamp, 1909; D and E modified from Harring and Myers, 1924.)

digestive tract is much different (Fig. 13C). The buccal area is situated at the base of the large funnellike infundibulum. Food taken into the mouth at the bottom of the infundibulum passes through a narrow esophageal tube, which hangs freely in a very large cavity, the proventriculus. The small mastax is situated at the bottom of the proventriculus.

Digestion is regularly extracellular, but in a few genera, such as *Chromogaster, Ascomorpha,* and a few others, it is clearly intracellular. No gastric glands are present in these rotifers, but the stomach has large gastric caeca that fill much of the pseudocoel.

Although many rotifers have true symbiotic zoochlorellae in the stomach wall, it is now believed that some instances previously reported as symbiosis are in reality ingested algal cells that persist for several days during their very slow intracellular digestion in the cells of the stomach wall.

In members of the bdelloid family Habrotrochidae there is no stomach cavity and the stomach is a continuous syncytial mass. As food leaves the esophagus it enters the stomach protoplasm in the form of food pellets, or food vacuoles, which circulate about and become digested.

Respiration. Most littoral and plankton rotifers normally have rather high oxygen

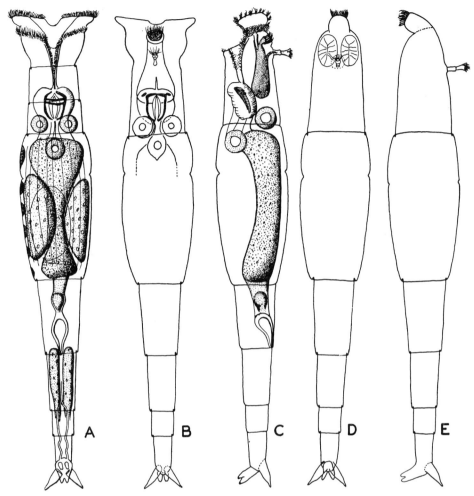

FIG. 11.–Structure of *Philodina,* diagrammatic. A, ventral view; B, dorsal view with most internal organs not shown; C, lateral view with reproductive system not shown; D, dorsal view with corona retracted; E, lateral view with corona retracted. (From Calaway, 1968, by permission.)

requirements, but it is well known that certain genera are capable of withstanding anaerobic conditions for short periods and very low concentrations of dissolved oxygen (0.1–1.0 ppm, for example) for extended periods.

Limnetic plankton genera, such as *Asplanchna, Filinia, Polyarthra,* and *Keratella,* commonly occur in the oxygen-poor hypolimnion of lakes during midsummer and midwinter. Rotifers living at a depth of a few centimeters in the interstices of sandy beaches likewise are in a region of low oxygen. Trickling filters of sewage plants and oxygen-deficient bottom muds and hay infusions characteristically contain species of *Lecane, Lepadella,* and many bdelloids. Undoubtedly the creation of water

currents by ciliary action ensures an adequate supply of oxygen to rotifers under most circumstances.

Osmotic control, excretion. Like most small fresh-water organisms, rotifers are readily permeable to water, but the internal osmotic pressure is kept constant by the action of the protonephridial system (Fig. 14). There are from four to 50 symmetrically arranged flame bulbs throughout much of the body. These are connected to a long, convoluted collecting tubule on each side, a part of the tubule being thin walled, another part thick walled and glandular. The two collecting

FIG. 12.–*Asplanchna* about to ingest a *Brachionus*. (From Halbach, 1971.)

tubules emptly into an excretory bladder that has a very short duct leading to the ventral side of the cloaca. Excess water and probably a certain amount of metabolic waste are absorbed from the fluid of the pseudocoel by the flame bulbs and wall of the collecting ducts and are excreted into the cavity of the ducts. The weak pressure produced by the beating of the tufts of cilia in the flame bulbs creates a slight current, which aids in carrying the excretory fluid through the tubules to the bladder. Here it is stored only temporarily, since the bladder normally contracts and empties its contents to the outside through the cloacal aperture (anus) one to six times per minute. In a few Ploima and in typical Bdelloidea there is no separate bladder; the collecting tubules merely empty into the thin-walled posterior part of the cloaca, which is modified into a bladder.

Old rotifers are usually distinguished by their darker pigmentation, most of which consists of accumulated excretory granules retained in the tissues.

Muscles. Both smooth and striated muscle fibers occur in rotifers, the latter often as-

sociated with the rapid movement of long spines and special appendages, as in *Polyarthra* and *Hexarthra*.

Muscles are always arranged in small, bandlike groups of fibers, never in flat sheets. The circular muscle system consists of 4 to 15 bands attached just beneath the hypodermis. These bands are typically complete in the Monogononta, but in the Bdelloidea the bands usually havve a ventral gap. One set of posterior longitudinal muscle bands are inserted in the foot or posterior of the trunk. An anterior set of longitudinal bands have their insertions in the coronal region and their origins near the middle of the trunk. *Testudinella* has short dorsoventral muscles, no circular ones. *Cupelopagis* has a remarkably complex system of muscles.

A few delicate visceral and cutaneovisceral muscles help move and hold the internal organs in place.

Nervous system. The arrangement of the rotifer nervous system is nonlinear. The largest mass of nervous tissue is the sacshaped cerebral ganglion, or "brain," lying on the dorsal surface of the mastax and esophagus and sometimes more or less obscured by the hypodermal syncytium of the corona. By means of fine paired nerve fibers it is connected with the mastax ganglion (sometimes absent) on the ventral surface of the mastax and with the caudal ganglion in the foot region. A few additional fine connecting fibers and associated nerve cell bodies innervate the sensory areas, muscles, and viscera (Fig. 16).

Sense organs. Perhaps the most conspicuous sensory area is the cervical eyespot. Characteristically it lies at the lower end of the brain or is imbedded in the brain and consists of a cup- or bowl-shaped mass of red pigment granules containing a refractile globule. Occasionally it is double. Many rotifers have two additional frontal eyespots located on the corona and widely separated. Eyespots are lacking in the adults of most sessile species, but they are present in the active immature stages.

A small, median, dorsal, setose antenna is usually present; rarely it is paired. In addition,

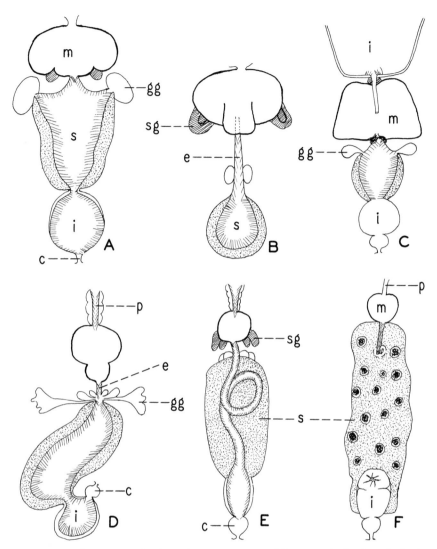

FIG. 13.–Diagrams of typical rotifer digestive systems. A, *Epiphanes*; B, *Asplanchna*; C, *Stephanoceros*; D, *Testudinella*; E, *Mniobia*; F, *Habrotrocha*. c, cloaca; e, esophagus; gg, gastric gland; k, intestine; inf, infundibulum; m, mastax; p, pharynx; s, stomach; sg, salivary gland. (A–E greatly modified from de Beauchamp, 1909.)

the Monogononta have a pair of lateral antennae in the posterior third of the body and often a minute caudal antenna on the dorsal surface of the foot. Special tufts of cilia, membranelles, setae, and sensory protuberances or papillae are of common occurrence on the corona of many Monogononta.

For many years the retrocerebral organ of rotifers went almost unnoticed. It consists of two parts: a pigmented retrocerebral sac situated dorsal or posterior to the brain and opening on two coronal papillae by a paired duct, and a small paired subcerebral gland on either side of the sac with ducts alongside those of the sac (Fig. 17). Either of these two components may be lacking, but usually it is the latter. "Bacteroids" often occur in the retrocerebral sac and occasionally in the subcerebral glands. A variety of functions have been postulated for the retrocerebral organ, none of them conclusive. The most plausible suggestion is a special sense organ.

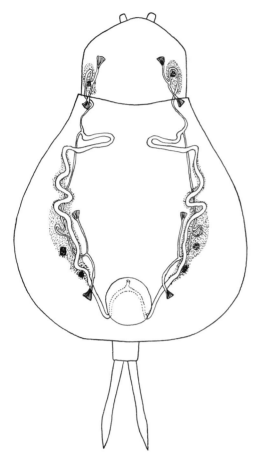

FIG. 14.–Protonephridial system of *Euchlanis*. (Modified from Stossberg, 1932.)

Reproduction. The Monogononta have a simple, ventral, elongated, saclike reproductive system (Figs. 1, 2). Just within the distal end of the sac is a small ovary consisting of a clump of developing oocytes. The vitellarium is a syncytium with a few large nuclei; it occupies half or more of the capacity of the sac, and as eggs in the ovary mature one at a time and move downward slightly, the vitellarium contributes a large mass of yolk to the egg so that it becomes relatively large. On maturing, the egg passes down the short oviduct, into the cloaca, and out the cloacal aperture. The egg is quite elastic and is momentarily constricted and elongated as it passes through these small ducts.

The Bdelloidea reproductive system is similar to that of the Monogononta except that it is V or Y shaped and has two ovaries and two vitellaria (Fig. 18). The two oviducts are very delicate or rudimentary.

Males are unknown in the bdelloids, and reproduction is solely by parthenogenesis. In the Monogononta, however, males have been reported or described for those species whose biology has been carefully worked out, and it is thought that males occur regularly in this group, although there are many genera in which males have never been seen, and perhaps in these groups reproduction is also exclusively parthenogenetic. It is commonly stated that males are rare, but this belief has arisen simply because the occurrence of males in most species is restricted to only one to three weeks during the year. If one collects during the brief time of male occurrence, they can be found in abundance.

Ploimate rotifers have two distinct types of females, which, with the exception of two or three species, are morphologically indistinguishable. During the greater part of the year the females reproduce exclusively by parthenogenesis. These females are said to be amictic, and both their body cells and their amictic eggs all have the diploid number of chromosomes. Their eggs undergo only one, nonreductional division during maturation in the ovary. The second type of female is called mictic. Such females appear only at critical times of the year, especially when there are marked changes in the factors of the environment (see page 183). The eggs of a mictic female undergo the usual double meiotic division and are therefore haploid. In contrast to amictic females, the mictic females are capable of being fecundated by the males. If fecundation does occur, the fertilized eggs (also called winter eggs, or resting eggs) they deposit are thick walled and highly resistant to adverse environmental conditions. If, however, a mictic female is not impregnated, she deposits eggs that promptly hatch into males. Fundamentally, a resting egg is a potential male-producing egg that has been fertilized and that, on hatching, produces an amictic female.

One mictic female may produce both fertilized eggs and male eggs, but males and females, or fertilized eggs and females, are never produced by the same parent. In fact mictic females are commonly seen carrying both male eggs and fertilized eggs at the posterior end of the body. Fecundation of an

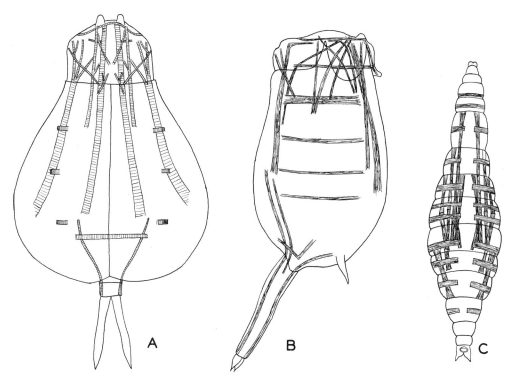

FIG. 15.–Rotifer musculature, semidiagrammatic. A, *Euchlanis,* dorsal; B, *Brachionus,* lateral; C, typical bdelloid, ventral. (A and B modified from Stossberg, 1932.)

amictic female by a male has no effect. The two kinds of female are always physiologically distinct, not interchangeable. A typical sequence of generations is shown in Fig. 20. Under natural conditions, there are usually only one or two mictic generations per year, whereas there may be up to 20 or 40 amictic generations. The female that hatches from a resting egg is always amictic, but the following generation may be amictic or mictic.

The number of amictic eggs that a female may produce during her lifetime is highly variable and depends on both the species and environmental conditions. Specific laboratory cultures have shown that *Brachionus calyciflorus* averages only 3.6 eggs and *Testudinella elliptica* only 5.0, whereas *Epiphanes senta* averages 45.4 and *Proales sordida* 24.3. Little is known about the number of eggs that a mictic female produces, but presumably it does not differ greatly from the number produced by an amictic female.

Some rotifers are ovoviviparous, the eggs being retained in the oviduct or pseudocoel until hatched. These include two species of Notommatidae, many bdelloids, *Asplanchna, Conochilus, Rhinoglena,* and a few others. Their young are usually released through the cloaca, but many bdelloids are known to release the young through ruptures in the body wall. In *Asplanchna* the male may fecundate its own mictic mother before it is released from the body of the mother.

Amictic, or summer, eggs are usually released from the body of the female and hatch within a day or two. The summer eggs of some plankton species contain large oil droplets, which facilitate suspension and flotation. In the following genera, however, the summer eggs are usually carried loosely attached to the posterior end of the body until hatching: *Polyarthra, Pompholyx, Keratella, Hexarthra, Brachionus,* and *Filina.*

Resting (or winter, or fertilized) eggs are heavy, thick shelled, and often sculptured. On release from the female they usually sink to the bottom, but in the autumn they may sometimes be found floating on the surface

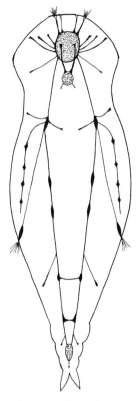

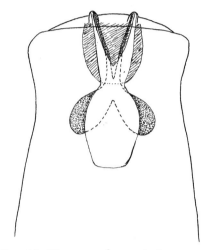

FIG. 16.–Diagram of rotifer nervous system. (Greatly modified from Remane, 1932.)

FIG. 17.–Diagram of a typical retrocerebral organ (dorsal view). Retrocerebral sac clear, subcerebral glands stippled, and brain with parallel diagonal lines.

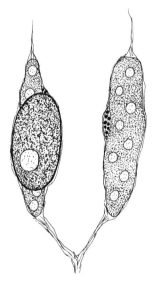

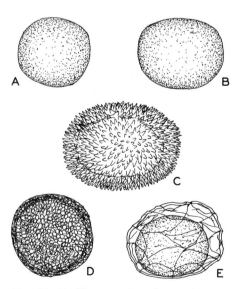

FIG. 18.–Diagram of typical bdelloid reproductive system, showing two vitellaria, two ovaries, two vestigial oviducts, and one developing egg.

FIG. 19.–Rotifer eggs. A and B, amictic eggs; C to E, fertilized (resting) eggs.

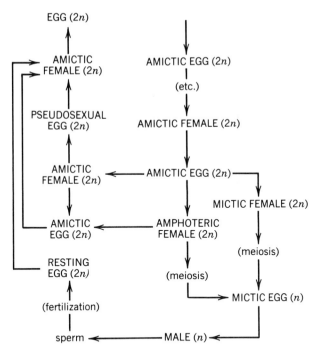

Fig. 20.–Diagrammatic sequence of generations and chromosome complements in a typical ploimate rotifer.

or blown ashore. They are remarkably resistant to desiccation, high temperatures, low temperatures, and adverse chemical conditions for long periods. Many species winter over in the resting egg stage. Unlike summer eggs, the resting eggs require a latent period of several weeks to several months before they can hatch. Hatching is stimulated by combinations of changes in temperature, osmotic pressure, water chemistry, aeration, and even algae. In spite of some textbook statements to the contrary, resting eggs are known for less than 1% of all rotifer species.

A curious fourth type of rotifer egg has recently been reported from cultures of *Keratella quadrata* and *K. hiemalis*. It is an unfertilized "pseudosexual resting egg" produced by parthenogenesis in the absence of males. The morphology and development are similar to those of ordinary resting eggs.

Figure 20 summarizes the complicated and interrelated reproductive events as they have been worked out for a few species of ploimate rotifers.

Biology of males. With few exceptions, male rotifers are minute, degenerate, and short lived. Commonly they are about a third as long as the female. The digestive tract is absent or vestigial, although a few species have a mastax and stomach. Mouth and anus are never present. The males have no well-developed lorica or spines, and the corona is always anterior and well supplied with cilia. They are very rapid swimmers, never sessile.

The simple reproductive system occupies most of the pseudocoel. The testis is pear shaped or globular and is held in place by a small strand supposed to be a remnant of the digestive tract. The vas deferens is short to long and sometimes has two or four associated prostate glands. Some species have no true penis, the vas deferens being everted through a small ciliated orifice during copulation. Other species have a well developed penis that is everted continuously; sometimes it is larger than the foot. In a few males the whole posterior end of the body is tubelike and specialized for copulation.

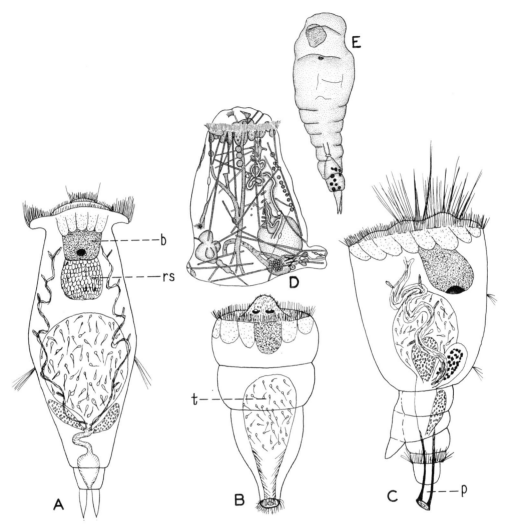

F I G. 21.–Typical male rotifers, semidiagrammatic and with musculature omitted (except in D). A, dorsal view of *Notommata*; B, dorsal view of *Filinia*; C, lateral view of *Brachionus*; D, *Asplanchna*; E, *Lecane*. (A–C greatly modified from Wesenberg-Lund, 1923; D from Sudzuki, 1956; E from Sudzuki, 1959.)

Males of plankton species are the most strongly reduced. Some are not much more than a sperm sac surrounded by protoplasm and a cuticle, and having an anterior tuft of cilia.

Males are ready for copulation within an hour after hatching. They are extremely active, especially in the presence of females, and swim about erratically until they come in contact with one. Copulation and sperm transfer occur either through the cloacal aperture of the female or through the body wall (*Asplanchna*), and the male dies immedi-

ately thereafter. If a female is not located, the male lives for only 2 or 3 days; *Asplanchna* males, however, are known to live for as long as 4 to 7 days.

Males have been reported for only about 10% of the species of Monogononta. Few have been carefully studied and figured. Males have been reported for only three or four species in the large genus *Lecane,* and there are only three or four records for the family Trichocercidae.

In the so-called perennial species of rotifers, males usually appear in spring and autumn; in "summer" species they appear in autumn; and in "winter" species they ordinarily occur in spring.

Development and longevity. The rotifer zygote undergoes unequal and regular cleavage, and the gastrula consists of an outer layer of cells (ectoderm) enclosing an inner cell mass (mesoderm and endoderm). Cell division and differentiation proceed rapidly, with each organ developing from a definite cell complex. When all organs have received their full complement of cells, cell division ceases.

Most plankton and littoral species grow rapidly during the several hours after hatching, but thereafter very slowly if at all. Many sessile species, however, grow throughout life, chiefly as a result of the elongation of the posterior end. Although rotifers have no ecdyses, the mature adult may be three to five times as large as the newly hatched young.

Although there may be considerable size variations among the individuals of a single species, it has been found, for the several species carefully studied, that there is an approximately constant number of cells in all adults of the same species. In *Epiphanes senta,* for example, there are about 959 nuclei. Each organ in the great majority of specimens has the same number of nuclei, though a few tissues, especially the vitellarium, show slight variations in the number of nuclei.

Length of life, from hatching until death, is highly variable. *Epiphanes senta* averages about 8 days, *Lecane inermis* 7.4 days, *Brachionus calyciflorus* 6 days, and *Proales decipiens* only 5.5 days. At the other extreme, *Euchlanis incisa* averages 21 days, *Keratella quadrata* 22 days, *Cupelopagis vorax* 6 weeks, and many bdelloids are active for 3 to 6 weeks.

As might be expected, the ability to regenerate lost or damaged organs is poorly developed. Amputation of the coronal lobes of *Stephanoceros* results in regeneration only in immature specimens; regeneration is lacking or very slow in full-grown individuals.

Syngamic reproduction. Since about 1904 there has been a vigorous and many-sided controversy over the relative significance of physiological, ecological, and genetic factors in the production of mictic females and males, and bisexual reproduction. Scores of papers dealing with a wide variety of field observations and controlled laboratory experiments have appeared. Reduced to simplest terms, the chief problem may be stated as follows: most ploimate rotifers reproduce by a series of parthenogenetic generations throughout most of the year; yet suddenly mictic females and males may appear in the population, and resting eggs are produced, usually in the autumn, but sometimes in the spring, or both. How may this phenomenon be explained? What factors are responsible for male production?

Most of the important observations and experimental work centering around this involved problem have stemmed from the laboratories of Shull, Whitney, Tauson, Wesenberg-Lund, and (recently) Gilbert. Their many papers should be consulted for a complete account. For our purposes, however, it is sufficient to summarize the present status of the matter and to mention some of the more significant items relating to sexual periodicity that have appeared as side issues in these studies.

To begin, it is obvious that species differ considerably from one another with respect to the nature and degree of ecological factors that instigate their bisexual periods. Furthermore, some species behave differently from one body of water to another and from one year to another. In one lake, for example, a species may regularly produce males in the autumn; in another lake it may have both spring and autumn bisexual periods; in a third lake the occurrence of bisexual periods may be sporadic. *Polyarthra,* for example, sometimes produces males at almost any time of the year, in both ponds and lakes.

In general, it has been shown that small populations of rotifers usually remain parthenogenetic, and that bisexuality appears during population increases. External factors, rather than genetic makeup, control sexual periodicity, but these factors are nonspecific. Careful experimental work indicates that the chief factors responsible for the appearance of males in a population are (1) a change in the type of food (green to nongreen, and vice versa), (2) an increase in food supply, and (3) a decrease in food supply. Other factors that have been shown to be of importance in instigating male production are crowding, addition of fresh culture medium, addition of distilled water, making the culture medium more alkaline, changing the temperature (rarely), and starving (rarely). Sometimes, however, even if measurable conditions remain unaltered, males unaccountably appear in a population.

The solution of the problem must, of course, be sought in a study of the amictic female, for whether she produces eggs that will hatch into more amictic females or eggs that will hatch into mictic females is a matter determined by her physiology. The decision as to which type of egg she will deposit is made in the 3-hour period before she releases the egg from her oviduct. Eggs hatching into mictic females are slightly smaller than eggs hatching into amictic females, and several investigators contend that in this difference lies the crux of the whole problem. The production of a large or small egg might very well depend on the available amount of yolk in the ovary, which in turn, might indirectly depend on environmental conditions. It is known, for example, that the appearance of mictic females is in part brought on by a change in the photoperiod, by greater population densities, and by minute quantities of vitamin C in the water.

This matter is further complicated by the fact that some species produce polymorphic resting eggs that may show a considerable range of size, shell sculpture, and color.

Epiphanes senta invariably has a sexual period under natural conditions, but experimentally this species has been carried through more than 500 successive parthenogenetic generations; such old cultures are characterized by a considerably lowered vigor and reproductive rate. *Proales decipiens,* however, has been carried through more than 250 successive parthenogenetic generations with no changes in vigor or fecundity, and it is interesting to note that the male is unknown in this species. Such facts appear to constitute some evidence that periodic bisexuality may have a rejuvenating effect on rotifer populations through the associated meiotic and fertilization processes.

Periodicity. Like so many other plankton and littoral organisms, rotifers exhibit periodic and sometimes quite striking cycles of abundance during the year, a fact familiar to anyone who has done seasonal plankton work. Various investigators have designated certain species as monocyclic, dicyclic, polycyclic, or acyclic and perennial, according to whether their annual population curves have one, two, several, or no pronounced peaks. *Kellicottia longispina* and *Conochilus unicornis* are usually considered monocyclic, although they are both perennial in certain lakes. *Euchlanis dilatata* commonly has an autumn maximum. *Brachionus angularis* and *Keratella cochlearis* are often considered dicyclic, with spring and autumn maxima, but are sometimes perennial. Other species are more variable. *Polyarthra* is dicyclic (summer and autumn), polycyclic, or perennial. *Keratella quadrata* may be most abundant in spring, summer, or autumn, or it may be perennial. *Asplanchna priodonta* is variously considered monocyclic, dicyclic, or perennial.

There has been an unfortunate tendency on the part of some investigators to generalize on the periodicity of a species on the basis of their observations of a few large lakes during only 1 or 2 years. As emphasized by Pennak (1949), however, the cycles of abundance for plankton species are highly variable within each species, variable from year to year within a single lake, and especially variable from one small lake to another.

Variation, cyclomorphosis. Like the cladocerans, many species of plankton rotifers exhibit a remarkable range of variation in size and lorica development from one habitat to another. The most striking examples of these local "races" occur in *Keratella* and *Brachionus*. In *Brachionus calyciflorus,* for instance, all degrees of development in the four posterior spines have been reported, ranging from long and well developed to very short or absent (Figs. 32B–D). At least in part, the production of spines in this species is induced by starvation, low temperatures, and substance(s) elaborated into the water by the predatory *Asplanchna*. Comparable variations in posterior spine development occur in *B. quadridentata* (Fig. 32H), *B. caudatus* (Fig. 32F), and *Platyias patulus*; to some extent the anterior spines are also variable. Until about 1915 these variations were often described as distinct species, and taxonomists have had to contend with the resultant confusing synonymy. Now, however, they are recognized as forms or races comprising relatively few species. An extreme example is shown by *Keratella cochlearis,* which supposedly exhibits at least 13 "formes" (see Fig. 26).

Variations in the size of races from different localities are also striking. Ahlstrom (1940) gives the average width of the lorica of *Brachionus calyciflorus* for 21 localities all over the world, and the variation is 124 to 300 μm. This is a remarkable range of average size for a species having determinate cell division.

Superimposed on the confusion of varieties in some loricate plankton species is an annual cyclomorphic series, that is, a sequence of generations in the same strain marked by progressive modifications in morphology through a long parthenogenetic series. In *Keratella cochlearis* (Figs. 25A–D, 26), for example, the lorica has a long posterior spine during the cold months followed by a gradual reduction in length until July and August when there is a very short spine or no spine. At this point syngamic reproduction often occurs, followed by a gradual reversal in spine length during the remainder of the year. Such well-defined cycles usually occur in small lakes that become quite warm in summer. In Lake Michigan there is no cycle, and in other bodies of water only a partial cycle. Sometimes the two extreme lorica forms have no transitional forms. The litera-

ture contains scores of papers dealing with *K. cochlearis* cyclomorphosis and each study seems to show results that are more or less different from others.

In *Keratella quadrata* (Figs. 25H–L) a form with long posterior spines initiates the cycle. The spines are progressively reduced in successive generations until a form with very short spines or no spines is dominant. Syngamic reproduction may then occur, and the cycle starts all over again with a long-spined form. But often this species is long spined the year round even in small ponds.

From these examples, and judging from the occurrence of two or three widely differing varieties in the same population, it is obvious that cyclomorphosis is a variable phenomenon within a single species. Certainly it cannot be ascribed to an inherent mechanism or simply to changes in environmental factors, but rather to a complicated interaction between organism and environment. In fact, the problems of cyclomorphosis, population fluctuations, local races, and bisexual reproduction all seem to be inextricably bound up with each other and with a multitude of ecological factors. Although this complex is a most difficult one, it is probably capable of solution by means of careful field and laboratory observation and experimentation. Gilbert (1973), for example, has shown that the lateral humps of male and female *Asplanchna sieboldi* protect them from cannibalism.

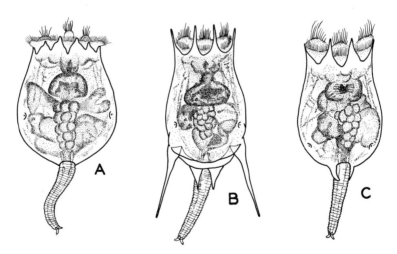

FIG. 22.–A, *Brachionus rubens*; B, *B. calyciflorus* with large posterolateral spines; C, *B. calyciflorus* without posterolateral spines. (From Halbach, 1971.)

The species problem. Without detailing the evidence here, information is accumulating at an increasing rate that the genetics of single species and related subspecies is a complex matter. For example, some species show evidence of two or more genetic "clones" living in a single body of water. While these clones differ imperceptably in their detailed morphology, they do show clear distinctions in such items as their feeding rates, reproductive habits, geographic distribution, seasonal periodicity, longevity, and biochemistry. Thus, for all practical purposes, they should be classified as good and separate species. But most rotiferologists have long been staunch "lumpers," and it will take a long time to generalize and clarify most species and subspecies problems.

General ecology. As emphasized earlier, rotifers occur in an extreme variety of habitats. They are poorly represented only in hot and cold springs, rushing streams, brackish waters, and extremely saline waters. Some common species are usually present in the same pond year after year, but sometimes unaccountably disappear for one or more years, then suddenly reappear.

Approximately 75% of the known species occur in the littoral areas of lakes and ponds, but only about 100 species are limnetic or planktonic enough in their habits to be completely emancipated from a substrate. The Bdelloidea are most characteristic of mosses and sphagnum. These habitats may be submerged, emergent, or terrestrial and only occasionally wet. Only a few species of the following genera of bdelloids regularly occur in the littoral zone: *Rotaria, Embata, Philodina,* and *Dissotrocha*; two or three other genera are uncommon in littoral waters. On the other hand, only a few ploimate species occur commonly in moss and only if it is kept wet.

A few free-living species are highly specific in their habitat preferences. *Acyclus inquietus* is invariably found among colonies of *Sinantherina*. *Collotheca algicola* appears to be restricted to clusters of *Gloeotrichia* filaments, and one or two other sessile species are also common on this alga. *Brachionus plicatilis* and *B. pterodinoides* are confined to extremely alkaline ponds and lakes, especially in the western states. *Proales*

rheinardti is unusual in that it occurs in mountain springs and among *Fucus* in marine bays and inlets! Only two genera, *Seison* and *Zelinkiella*, are known to be restricted to marine habitats.

Brackish waters contain many euryokous fresh-water species, as well as some species that are restricted to this habitat. The genus *Synchaeta* is remarkable for its many brackish and estuarine species.

Rotifers serve as food to a host of predators, including a few protozoans, *Craspedacusta*, the larger cladocerans, many cyclopoid and calanoid copepods, *Chaoborus*, and *Mysis*. Even very small fish may capture rotifers.

Although the Rotifera are primarily free-living organisms, a few species have become specialized for a parasitic existence. *Notommata trypeta* is parasitic in colonies of the alga *Gomphosphaeria*. *Proales parasitica* is a common parasite inside *Volvox* colonies. A few species of *Albertia* are ecto- and endoparasites of aquatic oligochaetes. *Drilophaga judayi* has been found only in the free-living state, but a second species in this genus is parasitic on *Lumbriculus*, and a third on *Erpobdella*.

Aside from such true parasites, many free-living rotifers live as commensals or epizoics on a wide variety of fresh-water invertebrates, especially crustaceans and insects. Common examples include certain species of Bdelloidea, *Pleurotrocha, Lepadella,* and *Testudinella*.

On the other hand, some of the larger plankton species often have the pseudocoel packed with ramifying filaments or sausage-shaped bodies, especially during the warm months. These are parasitic fungi that occasionally become epidemic.

Judging from the large plankton literature, it is obvious that most plankton communities average between 40 and 500 rotifers per liter, with populations in excess of 1000 per liter being unusual. The waters of mountain lakes and large oligotrophic lakes, however, sometimes contain less than 20 rotifers per liter, especially during the cold months. The most dense plankton rotifer population ever recorded from unpolluted waters is 5800 specimens per liter, cited by Pennak (1949).

Within ordinary limits, the density of natural rotifer populations appears to be governed by the amount of available food.

The maximum density of rotifer populations appears to be closely correlated with the relative amount of available substrate and exposed surfaces. Using 5800 rotifers per liter as a maximum for plankton, the corresponding maximum for sessile species inhabiting finely divided littoral plants is about 25,000 rotifers per liter (Edmondson, 1946). Furthermore, rotifers inhabiting the wet interstices of sandy beaches are reported to attain a maximum density of 1.155 million individuals per liter of damp sand (Pennak, 1940). The greater the surface, therefore, the greater the population. Presumably this phenomenon is associated with more food, more attachment space, more nooks for protection, and reduced predation loss. Below the water's edge the sand contains a negligible rotifer population, but the top 3 or 4 cm of sand extending from the water's edge as far as 3- or 4-m landward is a favorable habitat. Although many of the sandy beach rotifers occur regularly in the littoral, about 45 psammobiotic species are known to be restricted to the sand habitat, especially species of *Trichocerca, Encentrum, Lecane,* and *Wierzejskiella.*

In contrast to littoral forms, the limnetic, or open water, plankton genera occur over a wide depth range, even as deep as 100 or 200 m. *Asplanchna, Filinia, Polyarthra,* and *Keratella* are all good examples.

If open-water limnetic plankton samples from many lakes are analyzed statistically, the following rotifers will be found to be most abundant and generally distributed: *Keratella cochlearis, K. quadrata, Kellicottia longispina, Filinia longiseta, Asplanchna* spp., and *Conochilus unicornis. Brachionus* spp., *Synchaeta* spp., and *Conochiloides* spp. will be next most common. As shown by Halbach (1973), there are positive association coefficients involved between most of these species cooccurrences.

Those pond and lake plankters having eyespots exhibit, in varying degrees, a 24-hour cycle of vertical movements in accordance with the variations in subsurface illumination. Beginning late in the afternoon a slow drift upward begins, and maximum abundance occurs near the surface in the early morning. Around dawn a reverse drift downward begins, until maximum concentrations in the deep waters are attained about noon. Such movements are also of general occurrence among copepods and cladocerans,

but in rotifers they are of course much less pronounced, average amplitudes of only 1 to 3 m being common and amplitudes as high as 8 or 10 m being very unusual.

Unfortunately, many papers contain generalizations on rotifer ecology that are based on "controlled" conditions in the laboratory, and do not apply to natural field conditions. Experimental work on rotifers has burgeoned during the past 20 years, and it is difficult to generalize since results are often contradictory and confusing because of reliance on manipulating a single ecological factor at a time. Such studies as food clearance rates, competition with other rotifer species and with cladocerans, biochemical control of polymorphism, cannibalism phenomena, selective feeding, food selection, predation effects, induction of parthenogenetic and bisexual reproduction, and sensory physiology—these are all complicated phenomena that await additional critical research.

Ecology of sessile rotifers. The work of Edmondson (1944, 1945) has contributed much information to the ecology of sessile rotifers (Flosculariidae and Collothecidae). For the most part, sessile species are never present in quantities below a temperature of 15°C, the largest populations always being present at temperatures about 20°C. They are usually most plentiful in late spring, especially in small bays and littoral nooks in ponds and lakes. Although any substrate may be used, sessile species are by far the most abundant on submerged aquatic vegetation. *Utricularia vulgaris* var. *americana* has the most extensive, dense, and varied rotifer fauna of all species. Other favored plants are *Myriophyllum,* sphagnum, potamogetons, *Anacharis,* and *Vallisneria.*

Some species occur on a wide variety of plants, but others exhibit striking preferences. Thus *Collotheca gracilipes* is almost invariably on *Utricularia vulgaris americana, Ptygura melicerta* inhabits *Gloeotrichia* and *Coleochaete,* whereas *Collotheca campanulata* and *Ptygura barbata* are usually found on epiphytic algae. *Cupelopagis vorax,* however, is restricted to broad, flat leaf surfaces. Edmondson concludes that species distribution is determined mostly by water chemistry and available plant substrates.

Hydrogen-ion effect. Harring and Myers (1928) first pointed out that there are some striking correlations between the pH of a body of water and the composition of its ploimate rotifer fauna. In general, alkaline waters (above pH 7.0) contain relatively few species but large numbers of individuals, whereas acid waters (below pH 7.0) contain large numbers of species and few individuals. In the hard-water lakes near Madison, WI, Myers identified 100 species in a week of collecting, but he found the same number of species in only 1 hour of collecting in a small acid lake in northern Wisconsin (personal communication). In the alkaline Yahara basin of southern Wisconsin he collected 138 species, but in the acid waters of Mount Desert Island in Maine he collected 497 species (Myers, 1931, 1933, 1934).

As the result of his extensive field collecting, Myers distinguished three ecological groups of species with reference to their occurrence in waters of various hydrogen-ion concentrations. (1) Alkaline water species are those that are confined to waters having pH readings above 7.0. *Asplanchna, Asplanchnopus, Mytilina, Brachionus, Filinia, Lacinularia, Sinanthernia, Eosphora,* and *Notholca* are all alkaline genera, but there are also species in other genera that are restricted to alkaline waters. (2) Transcursion species are those that occur in both alkaline and acid waters. By far the large majority of rotifers belong in this category. (3) Acid water species are those that are confined to lakes, ponds, and bogs having an acid reaction. There are no genera restricted to acid waters, only certain species in genera that also have transcursion and alkaline representatives. The following genera contain numerous acid water species: *Cephalodella, Lepadella, Lecane, Monostyla, Trichocera,* and *Dicranophorus.*

Most genera contain representative species in each of these three groups, but a specialist can tell at a glance whether a particular collection came from an acid or alkaline body of water. Even though an acid lake and an alkaline lake may be separated by a ridge of land only 50 m wide, the rotifer faunas are completely different.

In a study of many areas in Wisconsin and New England, Edmondson (1944) lists 8 alkaline, 36 transcursion, and 12 acid species of sessile rotifers. In an investigation of many ponds and lakes (mostly acid) of the Pocono Plateau in Pennsylvania, Myers (1942) found 4 alkaline species, 313 transcursion, and 75 acid water species. In a study of the acid waters of the Adirondacks, Myers (1937) found 42 acid water species and 171 transcursion species. On the other hand, he found (1936) a total of only 106 alkaline and transcursion species in the alkaline waters of southern California, even though he collected almost daily for 3 months.

To be sure, pH 7.0 is by no means a sharp dividing line between acid or alkaline water species; it is only an approximation. Some acid species, for example, are found in waters giving readings as high as pH 7.1 or 7.2, and alkaline species likewise occur as low as pH 6.9 or 6.8. It is not meant to imply that the hydrogen-ion concentration itself is a governing factor in ecological distribution, since pH is a reflection of a whole complex of physical and chemical variables, and actually the limiting factor may be one or more of these ecological variables. Total dissolved salts, calcium, and bicarbonate, for example, have been suggested as being fundamentally important. Acid water rotifers can easily be transferred to alkaline waters where they will live, and vice versa, but their parthenogenetic eggs will not hatch in the unfavorable environment.

The preceding remarks relating to pH refer only to ploimate rotifers; they do not apply to the Bdelloidea, which are apparently all transcursional and capable of tolerating a relatively wide range of ecological conditions.

Desiccation. Even certain modern textbooks include the statement or inference that all rotifers can be dried to form a resistant cyst, which when returned to water will excyst and become active again. In the first place, true desiccation is restricted to the majority of Bdelloidea and does not occur at all in the much larger ploimate group. Second, in the great majority of bdelloids desiccation involves drying and shrinking of the body contents and cuticle without the formation of a true cyst. Only a few bdelloids are known to form true secreted cysts; when subjected to drying, these species secrete a surrounding

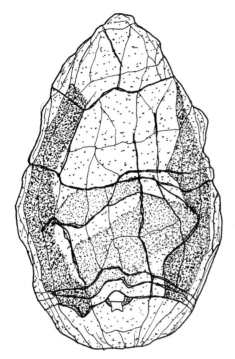

FIG. 23.–Desiccated *Philodina*. (Modified from Hickernell, 1917.)

gelatinous material that quickly hardens and becomes resistant.

Bdelloids typically inhabit moss on old walls, bark, or ground; they are much more abundant on emergent sphagnum than on submerged sphagnum. Such habitats are wet only intermittently from rains, wave action, or high water, and the process of desiccation allows the rotifer to tide over the frequent unfavorable periods of dryness. Very few truly aquatic bdelloids are able to undergo desiccation.

Bdelloids that are isolated in a little water on a smooth, bare surface are unable to undergo desiccation and they die. They must be associated with irregular surfaces and interstices where drying occurs more gradually and where there is a protective effect, such as in debris, sand, or among moss or sphagnum. When drying begins, the bdelloid draws both ends of the body into the central portion of the trunk and then puckers the two ends as though they were drawn up by purse strings (Fig. 23). This globular body mass then becomes wrinkled and loses much water.

The pseudocoel is almost nonexistent, and the volume of the animal is only one third or one fourth of its normal size.

A remarkable rearrangement of chromatin occurs during desiccation. Instead of remaining in its usual large karyosome mass in the center of the cell, it fragments, becomes highly dispersed, and adheres to the inner surface of the cell membrane. Presumably this adjustment facilitates suspended animation and the utilization of oxygen at a greatly retarded rate.

Desiccated rotifers may be kept dry as dust for months or even years. There is one record of specimens being revived from moss kept dry for 27 years. When the dried rotifers are placed in water, they quickly absorb water and become extended and normally active. Sometimes complete recovery occurs in only 10 minutes, but under other circumstances it may take a whole day.

Desiccation always produces some mortality, even under favorable conditions, and only a fraction of a group of desiccated rotifers are viable and can be revived. Rapid drying is injurious, and, in general, the longer the interval of desiccation, the lower will be the percentage of recovery. Nevertheless, even under natural conditions, bdelloids are able to survive several alternating periods of activity and desiccation.

Dispersal, geographic distribution. Overland transport of resting eggs or desiccated bdelloids is easily effected by winds and animals, and many rotifers are potentially cosmopolitan. By "cosmopolitan" we mean that similar habitats all over the world have similar rotifer faunas, and whether a given species is present in a particular body of water depends partially on geographic location, and partially on the precise complex of ecological conditions prevailing there. As Jennings (1903) has said, "The problem of the distribution of the Rotifera is, then, a the distribution of the Rotifera is, then, a problem of the means of distribution." The same species of rotifers are found in limnetic areas, in moss, in sandy beaches, in sphagnum bogs, in comparable farm ponds, and so on, the world over. Even the "rare" species are being reported from localities everywhere that appropriate habitats prevail and where the collector is diligent. Nevertheless there are many species with restricted distributions, for example, the species of *Notholca* endemic to Lake Baikal in the USSR. Similarly, there are 46 described species of *Brachionus*; only 22 are cosmopolitan, and the others are confined to single continents, especially in the Southern Hemisphere. None are endemic to Europe.

Pseudoploesoma formosum and *Trichocerca platessa* are common in the acid waters of the northeastern states but have never been collected elsewhere in the world. Many such similar instances are difficult to explain on the basis of our present knowledge of rotifer ecology. Presumably, however, such species do not have the proper distribution mechanisms to overcome their restricted ranges.

In spite of rigorous climatic isolation, rotifers have been remarkably successful in colonizing Antarctic and Subantarctic areas. DePaggi and Koste (1984), for example, list 27 Monogononta and 16 Bdelloidea as occurring south of 60° south latitude.

Collecting. Plankton species are easily taken with a townet, thrownet, or a fine dipnet with a vial or centrifuge tube fastened to the lower open end.

Vegetated littoral areas are much richer collecting grounds, especially for browsing species, but nets are of little use there. Instead, masses of the plant material should be carefully placed in large screwtop jars along with water from the same habitat. The jars should be no more than half filled with plant material. When the jars are brought to the laboratory, they should be uncapped and placed in a north window. As the oxygen slowly decreases the rotifers will gather at the top of the water and usually toward the lighted side, where they may be pipetted out. Sessile species may be found with the dissecting binoculars on bits of plants. Such temporary culture jars will keep well for a week or more if the lowermost third of the water is changed daily.

Bdelloids may be brought to the laboratory in handfuls of wet or dry moss, which may be stored dry. When the material is to be studied it should be soaked for a few hours in water, then washed vigorously several times. The wash water and sediment should be kept in covered watch glasses or petri dishes for weeks or months and examined periodically under the microscope. Some bdelloids appear promptly, others not for 2 or 3 weeks.

Interstitial species may be washed out of streamside and lakeside sand and gravel.

Culturing. Many rotifer culture methods are suggested in the literature. Some methods are better for certain species than others, and if an investigator wishes to culture a previously uncultured species, he should try a variety of methods and modifications in order to get best results. Species of *Lecane* have been raised in 0.1% malted milk in pond water, the solution being changed daily. The same genus has been cultured in rye extract prepared by grinding 20 grains of rye in a mortar and boiling in 100 ml of water for 20 minutes; the rotifers should be kept in drops of this solution and changed daily. Beef cube extract prepared by boiling one cube in 400 ml of tap water is sometimes good. Dried breakfast oats may be used by boiling 8 to 30 flakes for 3 minutes in 100 ml of water, filtering, and using after 24 hours. *Epiphanes* and certain other pond genera may be cultured in hay infusions that have been inoculated with a suitable protozoan or algal food organism. "Stable tea" is prepared by boiling 800 ml of horse manure in 1000 ml of water for 1 hour; cool, strain, dilute with two parts boiled rain water, inoculate with a food organism, and age a week or 10 days before inoculating with the rotifer. Bdelloids may be cultured easily in 0.1% nonfat dried milk in tap or pond water. Sessile species are extremely difficult to culture.

Aloia and Moretti (1973) describe a pure culture sequence of *Aerobacter-Paramecium-Asplanchna*. Many modern laboratory cultures of rotifers are maintained with pure cultures of algae used for food.

The most important precaution to be observed in culturing rotifers is to change the culture solution frequently and prevent excessive bacterial activity.

Several species, especially *Lecane,* have been successfully cultured in sterile media containing various mixtures of liver extract, vitamins, salts, and peptones.

Preparing, preserving. Since the identification of loricate rotifers depends largely on the details of the lorica, they are conveniently killed and fixed by placing directly in 10% formalin. This causes the soft parts of the animal to contract strongly, leaving the lorica well exposed for easy study.

The illoricate, free-swimming Ploima, however, require a much different technique. The identification of these forms depends largely on anatomical details that can be studied only in an uncontracted, natural state. As a consequence, these rotifers must be narcotized before killing and fixing. The following method is adapted from Myers (1942) and should be performed under dissecting binoculars. With a fine pipette place the rotifers in about 3 ml of water in a concave watch glass. If the specimens are from acid waters, add three or four drops of 1% neosynephrin hydrochloride; repeat at 5-minute intervals until the rotifers are all lying on the bottom and ciliary action has about stopped. Kill and fix by adding three drops of 0.5 or 1.0% osmic acid from a dropping bottle and quickly stirring into the water. Pour the contents of the watch glass into a small vial, and after 5 minutes decant or pipette off most of the fluid; then add water. Repeat the decantation and addition of wash water three times at 15-minute intervals.

Another useful method for illoricate rotifers involves narcotizing and killing with 0.04% procaine hydrochloride, although it takes 10 to 15 hours of contact!

If the rotifers are from alkaline waters, a 5% solution of novocain or cocaine hydrochlorate in 50% methyl alcohol gives better results than the neosynephrin. Usually, however, these narcotics are very difficult to obtain, even for such scientific purposes, and other more readily available substitutes must be used. Chloretone and 2% benzamine lactate are useful, and the writer has had considerable success with 2% butyn and 2% hydroxylamine hydrochloride.

The osmic acid solution will last much longer without deteriorating if it is dissolved in 1% chloroplatinic acid (platinic chloride) and kept in a dropping bottle that is painted black. If rotifers are exposed to osmic acid too long, they will blacken but they can be bleached in weak potassium hydroxide.

Many illoricate species, especially Notommatidae and bdelloids, are so delicate and sensitive that narcotization is not successful, and in such cases it is advisable to flood the specimens with an equivalent amount of *boiling* water in a watch glass; this treatment kills a large fraction of the animals in a well-expanded and lifelike condition. Formalin or other fixative should then be added.

Hanley (1949) strongly recommends the following procedure. Narcotize by adding several drops of 2% benzamine hydrochloride to a small amount of water containing the rotifers, the time being variable from one species to another. Kill in 10% formalin. Wash at least six times in 5% formalin to remove all the benzamine hydrochloride, which might later crystallize out.

After killing and fixing, all rotifers can be stored in 2 to 5% formalin containing 2% glycerin. Many workers add a little eosin as this colors the rotifers and makes them easier to study and handle.

Preserved specimens can be studied and kept for several days in a shallow, hollow-ground slide, provided the cover slip is sealed with petroleum jelly.

Permanent mounts are best made with glycerin or glycerin jelly. In either case the rotifers should be slowly run up to pure glycerin. This is done by placing the preserved specimens in 5% glycerin in a depression embryology slide or concave watch glass and leaving it loosely covered for about a week so that all the water evaporates. In this way the specimens are penetrated slowly and partially cleared, and they retain their turgidity and natural shape.

Individual specimens in glycerin should never be picked up with a pipette since they usually stick to the inner surface of the glass and are easily lost. Instead, they may be readily picked out of glycerin under the dissecting binoculars with an Irwin loop. Specimens handled in this way may be conveniently set aside while other preparations are being made. Glycerin jelly slides are made by placing a melted drop on a slide, touching the drop with the tip of the bristle or loop holding the rotifer, and quickly orienting the specimen under the binoculars before putting on the round cover slip. The glycerin jelly should be allowed to set for several hours.

Then the excess may be trimmed from the edge of the cover slip and several coats of Murrayite applied to its edge with a camel's hair brush and turntable at 24-hour intervals.

Fluid glycerin mounts are much more difficult to prepare but are superior for some purposes. Detailed directions for one method are given in Harring and Myers (1928). An alternative method is as follows: Three or four bits of *passe partout* or gummed paper are fixed symmetrically to a 1 × 3 slide and spaced about 10 or 12 mm apart to act as supports for the cover slip and thus prevent crushing of the specimen. The rotifer is then placed in a droplet of glycerin in the center of the slide and midway between the supports. The cover slip is next lowered into place very carefully, and by touching a camel's hair brush of Murrayite to the edge of the cover slip, capillarity will draw the Murrayite between the cover slip and slide and it will surround the glycerin droplet and completely fill the capillary space. Such slides should be set aside to dry for several hours and then given several rings of Murrayite at 24-hour intervals for greater permanence. Incidentally, as of this writing, Murrayite is almost impossible to obtain. Alternatives, such as various types of fingernail polish, are usually suitable.

Trophi mounting. Since genus and species identifications are often dependent on the anatomical details of the trophi, it is sometimes essential that permanent or semi-permanent mounts of these jaws be made. The following method, requiring some skill and practice and much patience, is modified from Myers (1937). Place a drop of 1:10 Clorox, or similar caustic alkali, just inside the concavity of a shallow concavity slide, and place a similar drop just outside the concavity. Then place a 22-mm square cover slip on the outside drop and push it over the concavity until it is almost in contact with the inside drop. Next place the rotifer in the inner drop by means of a bristle or Irwin loop. If the cover slip is pushed slowly over the inner drop, the rotifer will be drawn under. And by working the cover slip over the concavity by short pushes, and adding small quantities of solution after each advance, the rotifer will be forced into the acute angle formed by the edge of the concavity and the under surface of the cover slip. The edges of the cover slip should then be carefully dried and painted with petroleum jelly or Murrayite. The position of the rotifer should be noted, and in about a half hour it will all have dissolved except the trophi. Such trophi slides will last for several to many months, and if a petroleum jelly seal is used, the trophi may be oriented by tapping or moving the cover slip slightly.

Taxonomy. Rotifer taxonomy owes much to Harring's painstaking *Synopsis of the Rotatoria,* which appeared in 1913, bringing order out of chaotic nomenclature. Nevertheless, with the exception of specialists, many aquatic biologists still clung to old, ambiguous, and confused nomenclature. Even to this day, some outdated terminology still persists.

The arrangement of genera and families used here is greatly modified the plan given by Remane in *Das Tierreich* (1929–1933). It is based primarily on the fundamental structure and modifications of the mastax, in addition to other obvious morphological features. The outline presented below includes the genera given in the key that follows:

Class Digononta
  Order Seisonidea
  Order Bdelloidea (ramate mastax)
    Philodinidae
      Dissotrocha, Embata, Macrotrachela, Mniobia, Philodina, Pleuretra, Rotaria
    Habrotrochidae
      Ceratotrocha, Habrotrocha, Scepanotrocha
    Philodinavidae
      Philodinavus
    Adinetidae
      Adineta, Bradysclea
Class Monogononta
  Order Flosculariacea (malleoramate mastax)
    Flosculariidae
      Beauchampia, Floscularia, Lacinularia, Limnias, Octotrocha, Pseudoecistes, Ptygura, Sinantherina
    Conochilidae
      Conochilus, Conochiloides
    Hexarthridae
      Hexarthra
    Testudinellidae
      Pompholyx, Testudinella

Filiniidae
  Finilia, Tetramastix
Trochosphaeridae
  Trochosphaera
Order Collothecacea (uncinate mastax)
  Collothecidae
    Collotheca, Stephanoceros
  Atrochidae
    Acyclus, Atrochus, Cupelopagis
Order Ploima
  Notommatidae
    Notommatinae (virgate or virgate-forcipate mastax)
      Cephalodella, Drilophaga, Enteroplea, Eosphora, Eothinia, Itura, Monommata, Notommata, Pleurotrocha, Resticula, Rousseletia, Scaridium, Sphyrias, Taphrocampa, Tylotrocha
    Tetrasiphoninae (modified virgate mastax)
      Tetrasiphon
  Proalidae (modified malleate mastax)
    Bryceella, Proales, Proalinopsis
  Lindiidae (cardate mastax)
    Lindia
  Birgeidae (aberrant mastax, with a pair of pseudunci)
    Birgea
  Dicranophoridae (forcipate mastax)
    Albertia, Aspelta, Dicranophorus, Dorria, Encentrum, Erignatha, Myersinella, Pedipartia, Streptognatha, Wierzejskiella
  Synchaetidae (virgate or virgate-forcipate mastax)
    Polyarthra, Ploesoma, Pseudoploesoma, Synchaeta

Microcodonidae (forcipate mastax)
  Microcodon
Gastropodidae (virgate mastax)
  Ascomorpha, Chromogaster, Gastropus
Trichocercidae (virgate mastax)
  Ascomorphella, Elosa, Trichocerca
Asplanchidae
  Asplanchna, Asplanchnopus, Harringia
Epiphanidae (malleate mastax)
  Cyrtonia, Epiphanes, Mikrocodides, Rhinoglena
Brachionidae (malleate mastax)
  Anuraeopsis, Argonotholca, Brachionus, Kellicottia, Keratella, Notholca, Platyias
Euchlanidae (malleate mastax)
  Beauchampiella, Dipleuchlanis, Euchlanis, Tripleuchlanis
Mytilinidae (malleate mastax)
  Lophocharis, Mytilina
Trichotriidae (malleate mastax)
  Macrochaetus, Trichotria
Colurellidae (malleate mastax)
  Colurella, Lepadella, Paracolurella, Squatinella
Lecanidae (malleate mastax)
  Lecane (Monostyla)

The following key includes all genera likely to be encountered in the United States. It does not include about 30 rare fresh-water genera that have been found only on other continents and that may eventually be found in the United States. Numbers of species are only approximations for the numbers reported or likely to be reported for the United States.

KEY TO GENERA OF ROTIFERA

1. With a single ovary (Figs. 1, 2); mastax not ramate; lorica or secreted tube present or absent.
                                                                Class **MONOGONONTA, 3**
   With paired ovaries (Figs. 3, 11); mastax ramate (Fig. 8E); without a lorica or secreted tube.
                                                                Class **DIGONONTA, 2**
2. Ovaries without a vitellarium; male well developed; corona disc rudimentary; one genus and several species commensal on marine Crustacea . . . . Order **SEISONIDEA**
   Each ovary with a vitellarium (Fig. 18); males absent; reproduction entirely by parthenogenesis; corona usually well developed; body cylindrical, highly contractile, and telescopic; free swimming or with leechlike creeping movements; with a dorsal proboscis; toes, when present, usually retracted within foot except when creeping; two nonretractile spurs usually present; about 380 species worldwide . . . . Order **BDELLOIDEA, 94**
3. Female almost invariably free swimming or creeping; lorica present or absent; without a secreted tube; corona not especially large; foot, when present, typically with two toes; solitary . . . **5**

Female usually sessile; without a lorica but often in a secreted tube; corona large; foot, when present, long and annulated and without toes but with a terminal attachment disc; solitary or colonial.

4

4. Corona surrounded by two concentric wreaths of cilia with a ciliated furrow between (Figs. 4B, D); cilia of outer wreath always shorter than inner; mouth not central; usually with one or two well-developed antennae; mastax malleoramate (Fig. 8C); solitary or colonial.

Order **FLOSCULARIACEA, 81**

Corona very large, without such a double wreath of cilia; without well-developed antennae; mouth central; mastax uncinate or poorly developed; buccal field horseshoe shaped and located at the bottom of a large concave infundibulum (Fig. 43); solitary, often sessile.

Order **COLLOTHECACEA, 90**

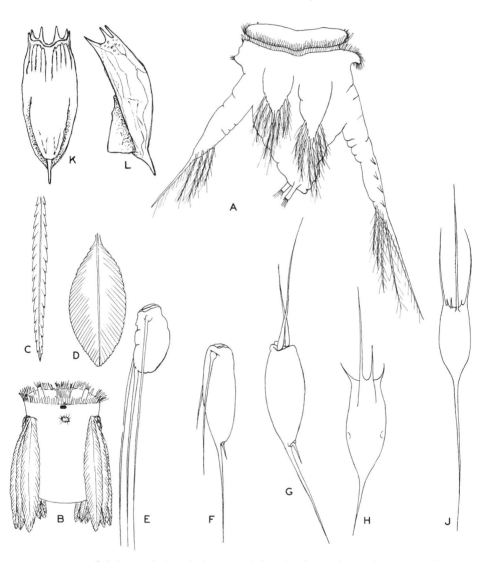

FIG. 24.–Structure of Ploima and Flosculariacea. A, right side of *Hexarthra*; B, dorsal view of *Polyarthra*; C and D, single appendages of *Polyarthra*; E, lateral view of *Filinia*; F, *Tetramastix opoliensis* with anterior spines in normal position; G, *T. opoliensis* with anterior spines retracted forward; H, dorsal view of lorica of *Kellicottia bostoniensis*; J, dorsal view of *K. longispina*; K, ventral view of *Argonotholca*; L, side view of *Argonotholca*. (K and L modified from Donner, 1966.)

5. Foot and toes always absent 6
    Foot always present; toes usually present 20
6. With 12 transparent, movable, sword- or blade-shaped lateral appendages (Fig. 24B); body short and more or less cylindrical; about 10 variable species, of which several are common plankton forms .... Order **PLOIMA**, Polyarthra
    Without 12 such appendages 7
7. With six stout, muscular, setose appendages; body conical; with a double ciliary wreath and a ciliated groove between (Fig. 24A); the "jumping rotifer"; usually in small alkaline lakes and ponds during the summer months; several species ..... Order **FLOSCULARIACEA**, Hexarthra
    Without such appendages 8
8. Without a lorica but with three or four long movable spines (Figs. 24E, F).
                                                                Order **FLOSCULARIACEA**, 9
    Lorica present or absent; spines, if present, rigid and confined to anterior and posterior margins of lorica 10
9. With two lateral and one posterior spine (Fig. 24E); several common limnetic species, one very common .... **Filinia**
    With two unequal lateral and two unequal posterior spines (Fig. 24F); one uncommon plankton species; sometimes included in *Filinia* .... **Tetramastix opoliensis** Zach.
10. With a spinous lorica composed of two plates immovably fused laterally ...... Order **PLOIMA**, 11
    Lorica absent, or present and not spinous 14
11. With one very long posterior spine and four or six anterior spines, of which three are very long (Figs. 24H, J); two limnetic species, one of which is very common .... **Kellicottia**
    With one, two, or no posterior spines, and with six short to medium anterior spines; highly variable plankton and limnetic species 12
12. Dorsal surface of lorica with a pattern of polygonal facets (Figs. 25A–L, 26, 27); cyclomorphic; about 20 species, of which several are very common .... **Keratella**
    Dorsal surface of lorica plain, pustulose, or longitudinally striated; usually in ponds 13
13. With a broad ventral bulbous keel (Figs. 24K, L); rare in coastal waters, common in brackish areas.
                                                                Argonotholca
    Without such a keel (Figs. 25M, N); about 10 species .... **Notholca**
14. Large transparent species, usually 400 to 2000 μm long 15
    Small, more opaque species, less than 200 μm long 16
15. Sac shaped, with a well-developed corona; intestine and anus lacking; mastax incudate; vitellarium horseshoe shaped or globose (Figs. 28A, B); often viviparous; predatory; morphology variable; body sometimes with humps or winglike processes; about 6 very common plankton species.
                                                                Order **PLOIMA**, Asplanchna
    Spherical and without a true corona, the body being divided into unequal hemispheres by a ciliary band; intestine and anus present; mastax malleoramate (Fig. 28C); up to 750 μm in diameter; very rare in small polluted ponds; more likely to be found in southern states; 2 species.
                                                                Order **FLOSCULARIACEA**, Trochosphaera
16. With a thin lorica, or lorica absent; sac shaped; with a very large lobed stomach filling much of the body cavity; anus absent; usually densely colored or opaque (Fig. 28D); mastax virgate; several plankton species .... Order **PLOIMA**, Ascomorpha
    With a thick lorica; not sac shaped 17
17. Lorica composed of a single cylindrical piece, trilobate in section and with a crescentic opening posteriorly on the left side (Fig. 28E); trophi asymmetrical; one uncommon species.
                                                                Order **PLOIMA**, Elosa worrali Lord
    Lorica composed of a dorsal and a ventral plate (Figs. 28F–K); trophi symmetrical 18
18. Corona with one to three fingerlike processes in addition to ciliated papillae; lorica composed of two convex plates fused laterally (Figs. 28F, G); mastax virgate; probably two species.
                                                                Order **PLOIMA**, Chromogaster
    Corona without fingerlike processes 19
19. Lateral edges of dorsal and ventral plates of lorica closely confluent; flat or four lobed in cross section; two frontal eyespots (Figs. 28H, J); mastax malleoramate; several species, including only two in the United States .... Order **FLOSCULARIACEA**, Pompholyx
    Lateral edges of dorsal and ventral plates of lorica connected by infolded cuticle; ventral plate flat, dorsal plate arched (Fig. 28K); mastax malleate; 1 common and variable species; several very rare species .... Order **PLOIMA**, Anuraeopsis fissa (Gosse)
20. With a well-developed rigid lorica 21
    Lorica absent, or present and poorly developed and flexible .... Order **PLOIMA**, 43

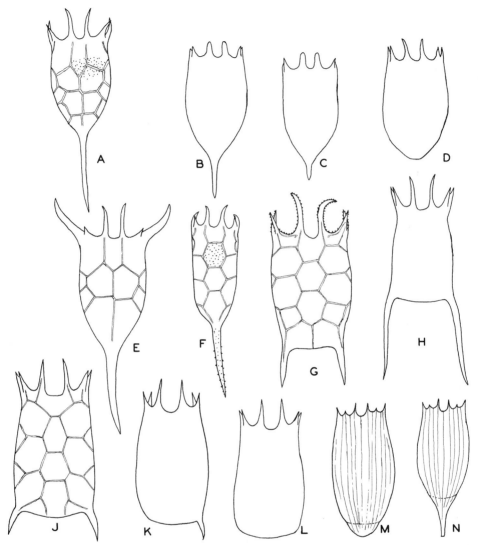

FIG. 25.–Lorica of typical species of *Keratella* and *Notholca* (dorsal views). A to D, *Keratella cochlearis*; E, *K. taurocephala* (an acid water species); F, *K. gracilenta*; G, *K. serrulata* (common in acid waters); H to L, *K. quadrata*; M and N, *Notholca*. Dorsal plates shown only for A, E–G, and J; small amount of pustulation shown only in A and F. (A–L modified from Ahlstrom, 1943.)

21. Foot attached to the ventral surface; lorica composed of one rigid piece, never dorsoventrally flattened; foot usually annulated (Figs. 30A, B); mastax virgate ...... Order **PLOIMA, 22**
    Foot terminal or subterminal **24**
22. Surface of lorica plain; body laterally compressed; lorica with a small ventral opening for the foot; foot with one or two toes (Fig. 30A); 3 species .... **Gastropus**
    Surface of lorica marked with ridges or vesicles; body not laterally compressed; lorica usually open along the midventral line; foot with two toes (Fig. 30B, C) **23**
23. Corona with frontal palps; foot annulated; lorica open along the midventral line (Fig. 30B); mastax virgate and adapted for prehension; about 4 species .... **Ploesoma**

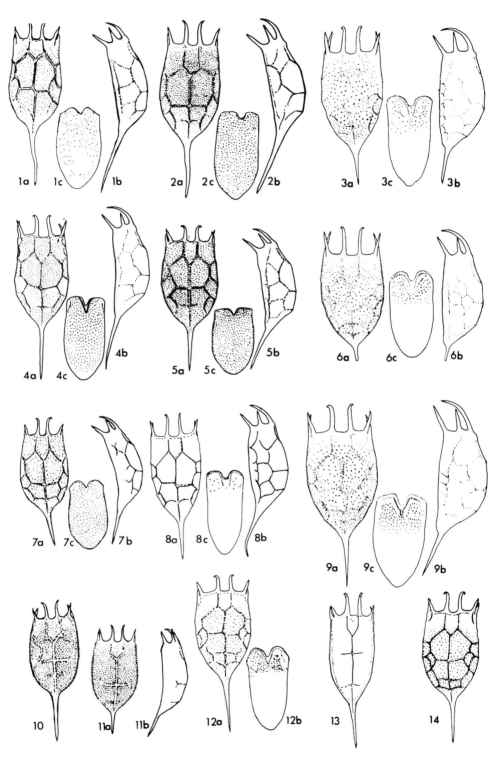

FIG. 26.–Forms and varieties of *Keratella cochlearis*. (From Koste, 1978, from various sources.)

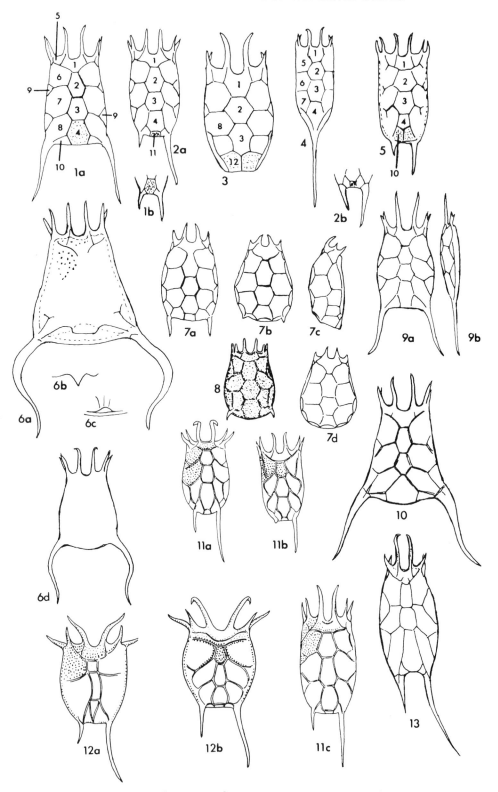

FIG. 27.–Various species and varieties of *Keratella*. (From Koste, 1978, from various sources.)

Corona without frontal palps but with two unique juxta-buccal protuberances; foot composed of two telescoping segments and emerging from an oval opening in the lorica (Fig. 30C); mastax virgate but not adapted for prehension; 1 rare species in acid waters.
Pseudoploesoma formosum (Myers)

24. Body cylindrical, more or less curved, and asymmetrical; lorica a single cylindrical piece, often with teeth and longitudinal grooves or ridges; toes spinelike, of unequal length, and with several small spinules at their base (Figs. 30D–J); creeping or free swimming; mastax virgate; trophi asymmetrical (Fig. 10C); common; many species .... Order **PLOIMA**, Trichocerca
    With another combination of characters; rarely asymmetrical; toes not spinelike 25
25. Foot long, retractile, annulated, and terminating in a tuft of cilia; dorsal and ventral plates of lorica completely fused laterally; greatly flattened dorsoventrally and sometimes nearly circular (Fig. 31B); mastax malleoramate; 120 to 300 μm long; about 15 littoral species.
Order **FLOSCULARIACEA**, Testudinella

    Foot ending in one or two toes .... Order **PLOIMA**, 26
26. Entire margin spiny; lorica heavy, with 4 to 10 long dorsal spines and six posterior spines; foot short (Fig. 31A); mastax malleate; about 6 littoral species .... **Macrochaetus**
    Entire margin of body not spiny 27
27. Head bearing a wide, circular shield; lorica cylindrical or pyriform, often with one or two median dorsal spines, or several spines on posterior margin of lorica; sometimes with a spine at the base of the toes (Figs. 31C, D); mastax malleate; about 10 uncommon species .... **Squatinella**
    Head without a wide circular shield 28

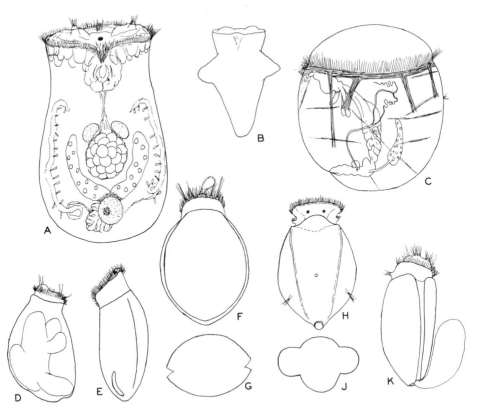

FIG. 28.–Structure of Ploima and Flosculariacea. A, ventral view of *Aplanchna* (muscle fibers not shown); B, *Asplanchna* with winglike processes; C, lateral view of *Trochosphaera*; D, lateral view of *Ascomorpha* showing lobed stomach; E, lateral view of *Elosa worralli*; F, dorsal view of *Chromogaster*; G, cross section of *Chromogaster*; H, dorsal view of *Pompholyx*; J, cross section of *Pompholyx*; K, lateral view of *Anuraeopsis fissa* carrying egg.

28. Usually with two spines at base of foot, or rarely with two posterior dorsolateral spines on the lorica; lorica composed of one boxlike piece, usually thick and with large facets (Figs. 31E, K); mastax malleate; about 8 species .... **Trichotria**
Without two such spines **29**
29. Body moderately flattened dorsoventrally; dorsal and ventral plates of lorica completely fused laterally; anterior dorsal margin of lorica usually with four or six spines; posterior margin with or without spines; two toes; mastax malleate **30**
With another combination of characters; mastax usually malleate or submalleate; never dorsoventrally flattened and never with four or six spines on anterior dorsal margin of lorica **31**
30. Foot segmented and retractile (Fig. 32N); three species .... **Platyias**
Foot long, annulated, retractile, not segmented (Fig. 32K); about 20 plankton species, some of which are common and highly variable, especially in hard waters; sometimes in dense populations.
Brachionus
31. Strongly compressed laterally; lorica composed of two lateral plates; frontal head hood present; toes long and tapering; lorica open along anterior, ventral, and posterior margins (Figs. 31F, G); small species **32**
Not strongly compressed laterally; lorica of different construction; frontal hood absent **33**
32. Terminal foot segment long (Fig. 31F); three rare species .... **Paracolurella**
Terminal foot segment short; about 20 species, some common (Figs. 31G, H) ...... **Colurella**
33. Foot long; toes long; lorica without spines or plates **34**
Foot short; toes short to long **35**
34. Lorica vase shaped and thin; toes shorter than rest of body (Fig. 33B); mastax virgate; 1 species.
Scaridium longicaudum (Müller)
Lorica pear shaped; toes about as long as rest of body (Fig. 33A); mastax malleate; 1 rare species.
Beauchampiella eudactylota (Gosse)
35. Lorica composed of a ventral plate and two lateral plates, the edges of the latter forming two dorsal ridges; often with anterior and posterior spines (Figs. 33C–E); about 10 littoral species, usually in hard waters .... **Mytilina**
Lorica constructed differently **36**
36. Lorica composed of one piece, capacious, rigid, boxlike, and with a prominent dorsal keel or low ridge extending the entire length of the lorica (Figs. 33F, G); several littoral species.
Lophocharis
Lorica usually composed of a dorsal and a ventral plate **37**
37. Dorsal and ventral plates rigidly united at the edges; with an anterior opening for the protrusion of the head and a posterior opening through which the foot projects; foot well-developed (Figs. 33H–K); about 35 very small species, 60 to 180 μm long; littoral .... **Lepadella**
Dorsal and ventral plates not united at the edges; foot not well developed; associated with the substrate but occasionally found in the plankton **38**
38. With a single long toe; strongly compressed dorsoventrally and oval to ovate in outline (Figs. 33L–N); about 25 small littoral species; usually united with *Lecane* .... **Monostyla**

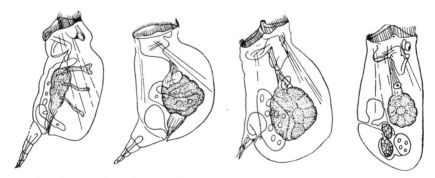

FIG. 29.–Series of saccate body forms; left to right: *Enteroplea, Harringia, Asplanchnopus,* and *Asplanchna.* (Modified from various sources.)

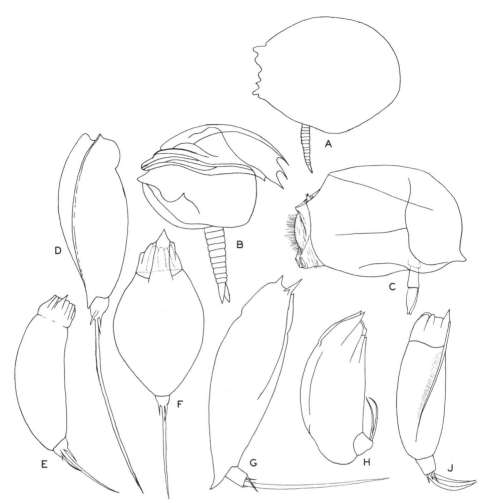

FIG. 30.–Structure of ploimate rotifers. A, lateral view of *Gastropus*; B, lateral view of *Ploesoma*; C, lateral view of *Pseudoploesoma formosum*; D to J, common species of *Trichocerca* (contracted). (B modified from Wesenberg-Lund, 1930; C modified from Myers, 1938.)

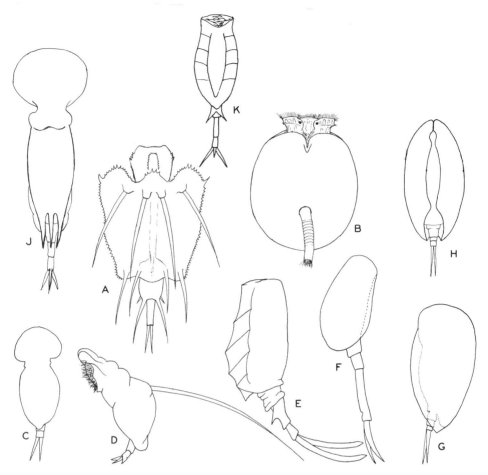

FIG. 31.–Structure of Colurellidae, Trichotriidae, Testudinellidae, and Brachionidae. A, dorsal view of contracted *Macrochaetus*; B, ventral view of *Testudinella*; C, dorsal view of *Squatinella*; D, lateral view of a spined species of *Squatinella*; E, lateral view of contracted *Trichotria*; F, lateral view of contracted *Paracolurella*; G, lateral view of contracted *Colurella*; H, ventral view of contracted *Colurella*; J, dorsal view of *Squatinella*; K, dorsal view of *Trichotria*. (D modified from Hauer, 1935; F modified from Myers, 1934.)

Lateral sulci without a flange (Figs. 2; 34F, G); about 15 large species, some very common.

Euchlanis

43. Foot rudimentary; body saccate (Fig. 37G); one species parasitic in *Volvox* colonies.

Ascomorphella volvocicola (Plate)

Foot not rudimentary; body not saccate; free living **44**

44. With four long, prominent sensory bristles on the corona; body conical; toes small (Fig. 34L); mastax virgate; about 15 plankton species, several of which are common; some species occur in salt and brackish water .... **Synchaeta**

Without four such bristles **45**

45. Foot very long, about half of total length; a single toe; corona flat and circular (Fig. 37B); mastax forcipate; a single uncommon species, 200 μm long .... **Microcodon clavus** Ehr.

Foot much shorter, almost invariably with two toes **46**

46. Corona complex, with an outer band of cilia and an inner band of cilia, sometimes also with accessory rows of cilia and ciliated protuberances (Figs. 1; 37A, C, D); mastax malleate; large species **47**

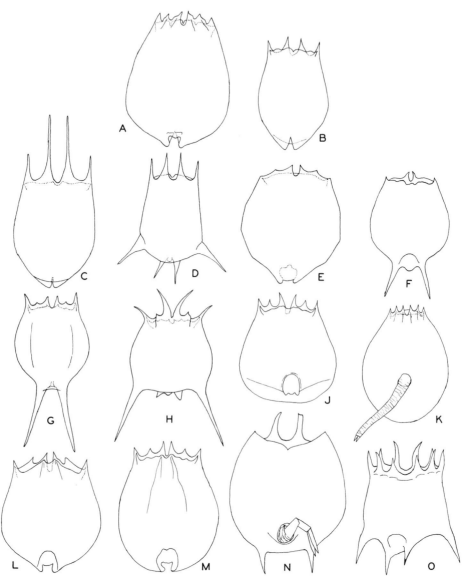

FIG. 32.–Structure of lorica of some common species of *Brachionus* and *Platyias*. A dorsal view of
Brachionus plicatilis, a species characteristic of highly alkaline waters; B, ventral view of *B. calyciflorus*; C and
D, dorsal views of *B. calyciflorus*; E, dorsal view of *B. angularis*; F, ventral view of *B. caudatus*; G, dorsal view of
B. havanaensis; H, dorsal view of *B. quadridentata*; J, ventral view of *B. quadridentata*; K, ventral view of *B.
pterodinoides,* a species characteristic of highly alkaline waters; L, dorsal view of *B. rubens*; M, ventral view of
B. urceolaris; N, ventral view of *Platyias quadricornis*; O, ventral view of *P. patulus.* (Redrawn and modified
from Ahlstrom, 1940.)

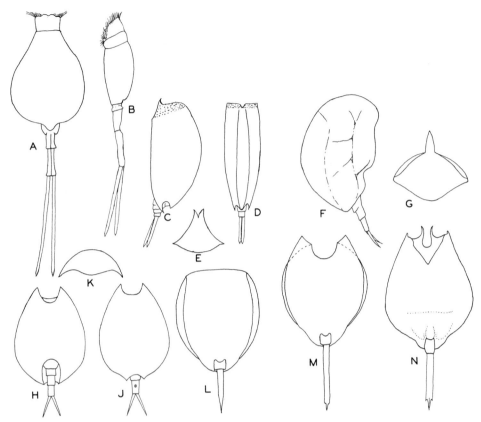

FIG. 33.–Structure of ploimate rotifers. A, contracted *Beauchampiella eudactylota*; B, lateral view of *Scaridium longicaudum*; C, lateral view of contracted *Mytilina*; D, dorsal view of contracted *Mytilina*; E, cross section of *Mytilina*; F, lateral view of contracted *Lophocharis*; G, cross section of *Lophocharis*; H, ventral view of contracted *Lepadella*; J, dorsal view of contracted *Lepadella*; K, cross section; L to N, ventral views of three contracted species of *Monostyla* (=*Lecane*).

Corona simple, with a single peripheral band of cilia; general surface of corona without cilia, with few scattered cilia, or with cilia more or less covering the corona; corona occasionally with two slight sensory protuberances **50**

47. With a single toe; sometimes a small spur at the base of the toe; foot thick (Fig. 37A); 1 U.S. species.
Mikrocodides chlaena (Gosse)

With two toes **48**

48. With a large dorsal proboscis at the anterior end bearing two eyespots (Fig. 37C); one littoral species, 270 to 400 μm long . . . . **Rhinoglena frontalis** Ehr.

Without such a proboscis **49**

49. Eyespot absent; body not strongly tapered posteriorly (Fig. 1); common during spring in stock ponds and puddles containing much organic matter; probably 5 species; 300 to 600 μm long.
Epiphanes

Eyespot present; body strongly tapered posteriorly (Fig. 37D); one littoral species; 250 to 300 μm long . . . . **Cyrtonia tuba** Ehr.

50. Large, globular species; mastax incudate (Figs. 37E, F); carnivorous **51**

Smaller species, more or less opaque; mastax not incudate; mostly illoricate or weakly loricate (see couplet 61) **52**

51. Intestine and anus absent (Fig. 37F); three rare littoral species; up to 1 mm long.
Asplanchnopus

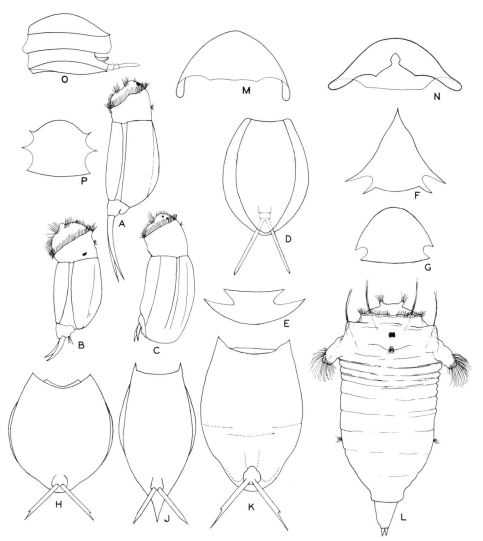

FIG. 34.–Structure of ploimate rotifers. A to C, three typical species of *Cephalodella*; D, dorsal view of retracted *Dipeuchlanis*; E, cross section of *Dipeuchlanis*; F and G, cross sections of *Euchlanis*; H to K, ventral views of typical species of *Lecane* (contracted); L, dorsal view of *Synchaeta*; M, cross section of *Euchlanis*; N, anterior edge of *Euchlanis*; O, lateral view of lorica of *Tripleuchlanis*; P, cross section of *Tripleuchlanis*. (D–G modified from Myers, 1930; H–K modified from Harring and Myers, 1926; M–P from Myers, 1930.)

54. Lateral antennae located near base of foot, long, knobbed, and with excessively long setae; ovary long, slender, and ribbonlike; auricles absent; dorsal antenna double and very long (Fig. 38C); corona oblique; total length 750 to 1000 μm; 1 rare species in acid waters.

Tetrasiphon hydrocora Ehr.

Lateral antennae not located near base of foot; ovary not long, slender, and ribbonlike; auricles present or absent; dorsal antenna not especially long; corona frontal or oblique **57**

55. Mastax modified malleate; unci adapted for crushing and grinding; corona oblique, with two lateral tufts of cilia but no auricles (Figs. 40D–F); total length 90 to 400 μm **71**

Mastax not modified malleate or malleate-virgate; corona oblique or frontal; auricles present or absent **56**

56. Body cylindrical and tapering posteriorly, sometimes indistinctly annulated; foot not set off sharply from body (Fig. 38B); mastax cardate (Fig. 9A); total length 250 to 1200 μm; about 10 species.

Lindia

Foot slender and sharply set off from the wide body (Fig. 38A); trophi aberrant, with a pair of pseudunci; 240 to 275 μm long; 1 rare littoral species ...... **Birgea enantia** Harring and Myers

57. Toes longer than rest of body, usually unequal in length (Fig. 38D); about 20 species, usually in acid waters .... **Monommata**

Toes short **58**

58. Gastric glands long, ribbonlike, and bifurcated; with four slender accessory gastric appendages; body short and saclike (Fig. 38E); one uncommon littoral species; 600 μm long.

Enteroplea lacustris Ehr.

Without such glands **59**

59. Head very broad and truncate anteriorly; two eyespots widely separated on protuberances; foot annulated (Fig. 38F); one carnivorous littoral species .... **Sphyrias lofuana** (Rousselet)

With other characters **60**

60. Body marked with transverse folds or annuli; elongated and spindle shaped; foot rudimentary (Fig. 38G); several littoral species; 200 μm long .... **Taphrocampa**

Body not so marked **61**

61. Toes relatively long, pointed, and usually curved; foot rudimentary and unjointed; body prismatic or spindle shaped (Figs. 34A–C); retrocerebral sac absent; lorica present, more or less sclerotized and composed of a dorsal and a ventral plate; both rami and unci of very simple construction (Figs. 10D, E); more than 50 U.S. species, some of which are very common.

Cephalodella

Toes comparatively short; foot present and distinct; retrocerebral sace present; lorica absent; rami and unci more complicated **62**

62. Foot unsegmented, though sometimes annulated (Figs. 39A–C) **63**

Foot segmented (Figs. 39D–K) **65**

63. With a single toe; retrocerebral sac absent (Fig. 39A); trophi highly specialized virgate; 1 rare species in ponds and reservoirs; 250 μm long .... **Tylotrocha monopus** (Jennings)

With two toes; retrocerebral sac present or absent; trophi not highly specialized virgate **64**

64. Retrocerebral sac absent; eyespot absent; foot short (Fig. 39B); one species ectoparasitic on oligochaetes, a second ectoparasitic on *Erpobdella,* and one other (U.S.) possibly free living; often covered by a gelatinous secretion .... **Drilophaga**

Retrocerebral sac present; eyespot present; foot long (Fig. 39C); one littoral species; 130 μm long.

Rousseletia corniculata Harring

65. With a projecting auricle on each side of the head, each auricle bearing long cilia; trophi usually asymmetrical (Figs. 10A, B; 39E) **66**

Without definite auricles but often with a tuft of long cilia in the usual location of an auricle; trophi symmetrical **67**

66. Rami roughly hemispherical (Figs. 10A, B); with a cervical eyespot only (Fig. 39D); total length 100 to 1000 μm, usually more than 300 μm; a few species covered by a gelatinous secretion; many littoral species, some very common and spectacular .... **Notommata**

Rami lyrate, broad, and strongly divergent at the base, continuing as parallel rods toward the tips (Fig. 39E); with a cervical eyespot and two frontal eyespots (Fig. 39F); about 4 uncommon littoral species .... **Itura**

67. Eyespot absent (Fig. 39G); in mountain brooks among submerged moss; one species; 200 μm long.

Dorria dalecarlica Myers

One or three eyespots present; not in mountain brooks in moss **68**

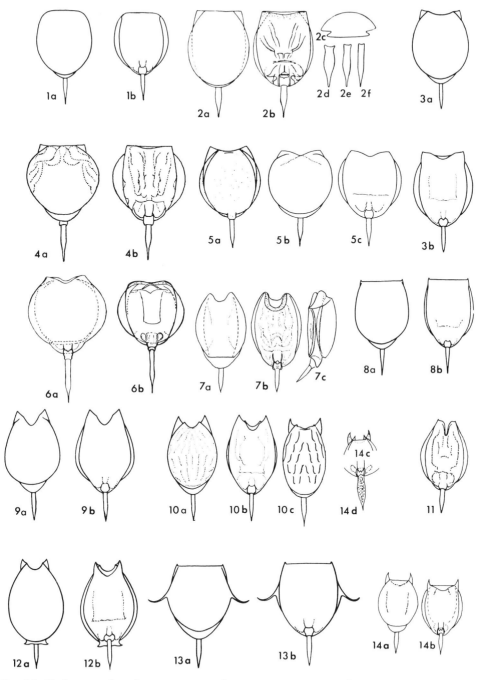

FIG. 35.–Various species of *Lecane* (=*Monostyla*). (From Koste, 1978 after various other authors.)

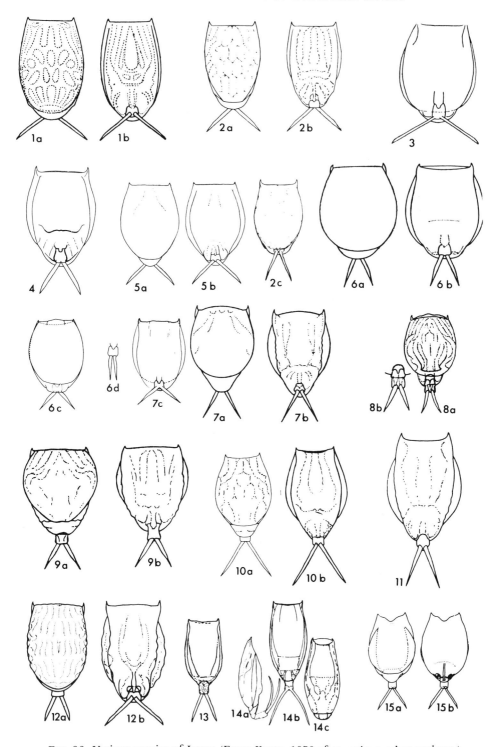

FIG. 36.–Various species of *Lecane* (From Koste, 1978 after various other authors.)

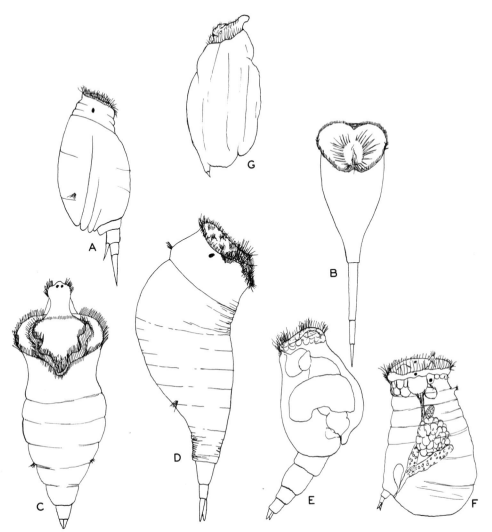

FIG. 37.–Structure of ploimate rotifers. A, *Mikrocodides chlaena*; B, *Microcodon clavus*; C, ventral view of *Rhinoglena frontalis*; D, *Cyrtonia tuba*; E, *Harringia*; F, *Asplanchnopus*; G, lateral view of *Ascomorphella volvocicola*.

68. Retrocerebral sac absent (Fig. 39H); rami without teeth (Fig. 39J); toes separate or fused; total length 110 to 250 μm; 8 littoral species .... **Pleurotrocha**
    Retrocerebral sac present (Fig. 39K); rami with one or more teeth; toes separate **69**
69. Each uncus with a single simple tooth; rami bent at right angles and with one or two teeth at midlength and many small ones more distally (Fig. 39L); total length 300 to 500 μm; about 5 littoral species .... **Eosphora**
    One or both unci with a single main tooth plus accessory teeth (Fig. 40B); rami more or less bent at right angles but with a different arrangement of teeth; total length 170 to 400 μm **70**
70. With one eyespot, or eyespot absent (Fig. 40A); each uncus with one main tooth and with accessory teeth on either the left or right uncus; probably 6 littoral species .... **Resticula**
    With three eyespots (Fig. 40C); both unci with one main tooth and accessory teeth (Fig. 40B); about 6 littoral species .... **Eothinia**
71. With a very long stylet on each side of the head (Fig. 40D); two littoral and sandy beach species; 130 μm long .... **Bryceella**

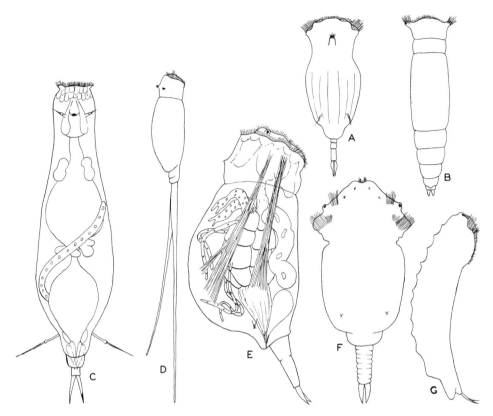

FIG. 38.–Structure of Notommatidae. A, *Birgea enantia*; B, *Lindia*; C, *Tetrasiphon hydrocora*; D, *Monommata*; E, *Enteroplea lacustris*; F, *Sphyrias lofuana*; G, *Taphrocampa*. (A–C modified from Harring and Myers, 1922; D–G modified from Harring and Myers, 1924.)

Without such stylets **72**

72. Tail region or basal foot segment with a knoblike papilla bearing a tuft of setae or a spine (Fig. 40E); 5 littoral species .... **Proalinopsis**

Without such a tuft of setae or spine (Fig. 40F); one species with a single toe; about 30 littoral species, some quite common .... **Proales**

73. Trophi asymmetrical (Fig. 40H) **74**

Trophi symmetrical (Figs. 41C, E, G, L) **75**

74. Corona ventral; retrocerebral sac present; toes long (Fig. 40G); about 12 species, usually in acid waters and sphagnum bogs .... **Aspelta**

Corona oblique; retrocerebral sac absent; toes short (Fig. 40J); 1 rare species known from sandy beaches .... **Pedipartia gracilis** (Myers)

75. Foot one fourth to one third the total length, composed of one to four segments (Fig. 41A); 3 uncommon species; most common in sandy beaches .... **Wierzejskiella**

Foot shorter **76**

76. Incus Y shaped and composed of rods; with a long naviculoid sclerite attached to the tip of each uncus (Fig. 41C); a single acid water species ..... **Streptognatha lepta** Harring and Myers

With incus and unci of different construction **77**

77. Rami slender and bent at nearly a right angle at midlength and terminating in long, slender, single teeth (Fig. 41E); about 5 littoral species .... **Erignatha**

Rami not slender and bent at nearly a right angle at midlength; retrocerebral sac present or absent; corona oblique or ventral **78**

78. With a posterior projection at the external angles of the rami (Fig. 41G); retrocerebral sac absent; corona oblique or ventral (Fig. 41F); 2 sandy beach species; 130 μm long ... **Myersinella**

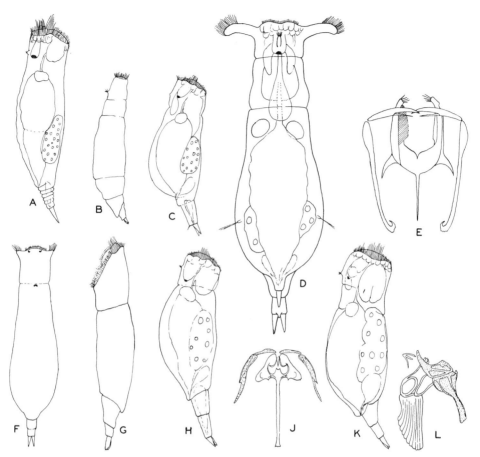

FIG. 39.—Structure of Notommatidae and Dicranophoridae. A, *Tylotrocha monopus*; B, *Drilophaga*; C, *Rousseletia corniculata*; D, dorsal view of *Notommata*; E, ventral view of *Itura* trophi; F, *Itura*; G, *Dorria dalecarlica*; H, *Pleurotrocha*; J, ventral view of *Pleurotrocha* trophi (mallei stippled); K, *Eosphora*; L, lateral view of *Eosphora* trophi (malleus stippled). (A, C, H, and J modified from Harring and Myers, 1924; B, D, K, and L modified from Harring and Myers, 1922; E and F modified from Harring and Myers, 1928; G modified from Myers, 1933.)

Without such projections on the rami 79
79. Toes minute and fused, or lacking; corona oblique; eyespots lacking; rostrum minute; retrocerebral sac absent (Fig. 41H); several species, most or all of which are ecto- or endoparasites of oligochaetes .... **Albertia**
Toes prominent, not fused; corona oblique or ventral; eyespots present or absent; rostrum conspicuous; retrocerebral sac present or absent; not parasitic 80
80. Corona essentially ventral; two frontal eyespots; retrocerebral sac present and large (Fig. 41J); rami often with shearing teeth; unci robust (Fig. 9); about 50 littoral species ... **Dicranophorus**
Corona oblique; one eyespot, or eyespots absent; retrocerebral sac present or absent (Fig. 41K); rami never with shearing teeth; unci weak and needlelike (Fig. 41L); about 15 species; chiefly littoral .... **Encentrum**
81. With a ventral gap in the ciliary wreath; mouth on the corona, near the dorsal edge; gelatinous case (Figs. 42A–C); limnetic species 82
Ciliary wreath with a dorsal gap or without a gap; mouth ventral to the corona; eyespots usually absent (Figs. 42D–L); adults usually sessile 83

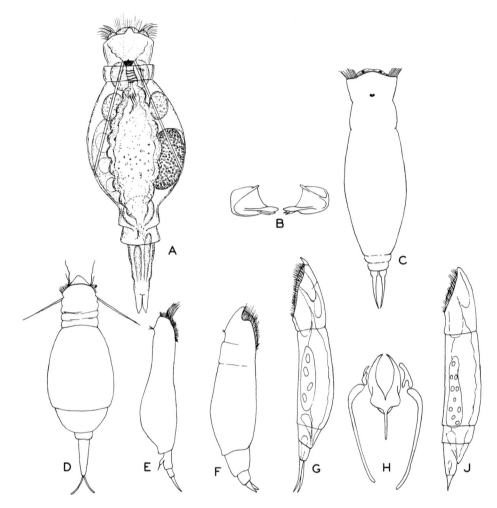

FIG. 40.–Structure of Proalidae and Notommatidae. A, *Resticula*; B, frontal view of unci of *Eothinia*; C, *Eothinia*; D, *Bryceella*; E, *Proalinopsis*; F, *Proales*; G, *Aspelta*; H, trophi of *Aspelta*; J, *Pedipartia gracilis*. (E and F modified from Harring and Myers, 1922; B and C modified from Harring and Myers, 1924; G and H modified from Harring and Myers, 1928; J modified from Myers, 1936; A from Donner, 1970.)

82. With one or two antennae on the corona; colonies composed of radiating individuals inhabiting coherent gelatinous tubes; 2 common species (Figs. 42A, B) .... **Conochilus**
    With one or two ventral antennae below the corona; solitary or colonial; probably four limnetic species, of which two are common (Fig. 42C) .... **Conochiloides**
83. Adults in spherical, sessile colonies attached to aquatic plants 84
    Adults solitary or colonial, but colonies never spherical 85
84. Colonies without tubes; trunk usually with two to four opaque protuberances; corona roughly kidney shaped (Fig. 42D); about 5 species .... **Sinantherina**
    Colonies with adhering gelatinous tubes; trunk without protuberances; corona heart shaped and with the sinus ventral (Fig. 42E); about 10 species .... **Lacinularia**
85. Corona with eight prominent lobes and one small median dorsal lobe; in a gelatinous tube; 1 poorly known species .... **Octotrocha speciosa** Thorpe
    Corona with less than eight lobes, or not lobed 86

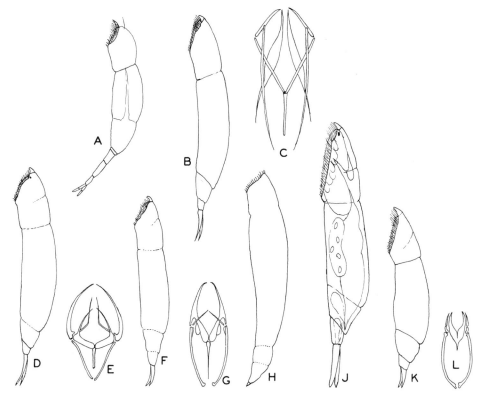

FIG. 41.–Structure of Dicranophoridae. A, *Wierzejskiella*; B, *Streptognatha lepta*; C, trophi of *S. lepta*; D, *Erignatha*; E, trophi of *Erignatha*; F, *Myersinella*; G, trophi of *Myersinella*; H, *Albertia*; J, *Dicranophorus*; K, *Encentrum*; L, trophi of *Encentrum*. (B–L modified from Harring and Myers, 1928.)

86. Corona four lobed; two large ventral antennae; with a tube composed of gelatinous material, pellets of debris, or fecal pellets (Figs. 42F, G); about 6 species .... **Floscularia**
   Corona two lobed and elliptical, circular, or kidney shaped **87**
87. Tube chitinoid, often opaque and more or less covered with debris; corona of two distinct lobes, or nearly circular; dorsal antenna short to long (Figs. 42H, J); about 6 species ..... **Limnias**
   Tube not chitinoid, or tube absent **88**
88. Tube absent (Fig. 42K); able to leave substrate and swim about; 1 rare species.
   Pseudoecistes rotifer Stenroos
   Tube present **89**
89. Dorsal antenna small or apparently absent; ventral (lateral) antennae short to long; corona a wide oval or nearly circular, indistinctly bilobed; tube made of debris, fecal pellets, or gelatinous and more or less covered with debris (Fig. 42L); one species free swimming; 20 species; 200 to 1000 μm long .... **Ptygura**
   Dorsal antenna very long; corona of two distinct lobes; tube irregular, translucent, and gelatinous; one rare species up to 500 μm long .... **Beauchampia crucigera** (Dutrochet)
90. With setae but no cilia on the corona; sessile; in a gelatinous tube; foot terminated by a long, nonretractile peduncle, ending in an adhesive disc **91**
   Without setae or cilia on the corona; gelatinous case present or absent **92**
91. Corona drawn out into long, pointed arms, which bear whorls of setae (Fig. 43A); 2 species.
   Stephanoceros
   Corona circular, lobed, or pointed (Fig. 43B); setae not arranged in whorls; some species free swimming; many species .... **Collotheca**

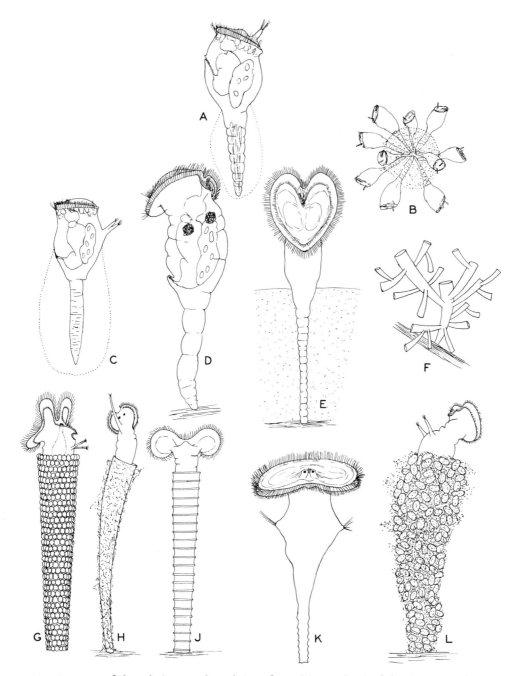

FIG. 42.–Structure of Flosculariacea. A, lateral view of *Conochilus*; B, sketch of a small colony of *Conochilus* showing common gelatinous case; C, lateral view of a solitary *Conochiloides* in gelatinous case; D, lateral view of *Sinantherina*; E, *Lacinularia* in gelatinous matrix of colony; F, cluster of empty *Floscularia* tubes; G, lateral view of *Floscularia* (note pellet-making concavity on ventral surface of head); H and J, dorsal views of *Limnias*; K, *Pseudoecistes rotifer*; L, *Ptygura*. (All greatly modified from various sources.)

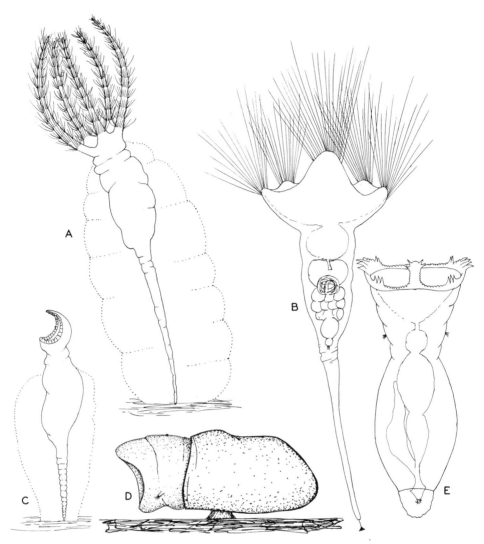

FIG. 43.–Structure of Collothecacea. A, *Stephanoceros*; B, dorsal view of *Collotheca* (gelatinous tube omitted); C, lateral view of *Acyclus inquietus*; D, lateral view of *Cupelopagis vorax*; E, dorsal view of *Atrochus tentaculatus*.

92. Body long, with a long, annulated, tapering stalk and with or without a gelatinous case; corona produced into a large dorsal lobe, and bordered by a thin, festooned membrane (Fig. 43C); lives in *Sinantherina* colonies; 1 rare species .... **Acyclus inquietus** Leidy
    Body shorter, without a gelatinous case; foot absent **93**
93. Body saclike, with a flat attachment disc on the ventral surface; corona a large concave chamber (Fig. 43D); usually on broad, flat leaves of aquatic plants; up to 800 μm long; 1 rare species.
    Cupelopagis vorax (Leidy)
    Body spindle shaped; corona with numerous small peripheral tentacles (Fig. 43E); 1 species.
    Atrochus tentaculatus Wierzejski
94. With a well-developed rostrum and corona, the latter always capable of being retracted into the mouth **95**
    With rostrum imperfect or corona absent **104**

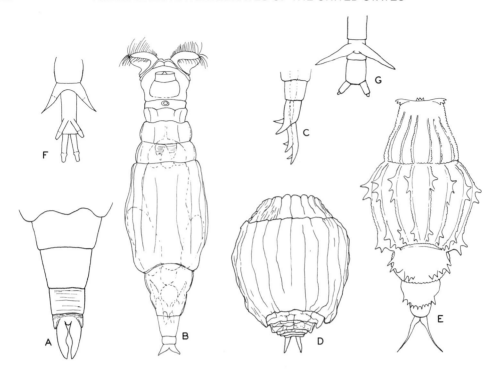

Fɪɢ. 44.–Structure of Bdelloidea. A, posterior end of *Embata* while swimming (toes retracted); B, dorsal view of *Philodina* (toes retracted); C, foot of *Philodina* (toes protruded); D, contracted *Philodina*; E, dorsal view of partly contracted *Dissotrocha*; F, dorsal view of foot of *Philodina* (toes protruded); G, toes and spurs of *Embata*. (A modified from Murray, 1906; B and E modified from Murray, 1911; C and D modified from Hickernell, 1917.)

95. Stomach with a true lumen (Fig. 13E); intestine ciliated; oviparous or viviparous **96**
    Stomach without a lumen (Fig. 13F); food formed into vacuoles in the stomach protoplasm; intestine unciliated; oviparous **102**
96. With four plain toes, of which two are dorsal and two terminal (Figs. 44C, F) **97**
    Toes otherwise **100**
97. Cuticle smooth **98**
    Cuticle coarse or leathery, and folded (Figs. 44E; 45A) **99**
98. Spurs long (Fig. 44G); ectocommensals on many aquatic invertebrates; about 3 species.
    Embata
    Spurs not particularly long (Figs. 44B, C); not usually commensals; about 20 species; common in sewage treatment processes and in old protozoan cultures .... **Philodina**
99. Cuticle coarse, with few or no transverse folds but often with few to many spines, or spines absent (Fig. 44E); viviparous; about 4 species .... **Dissotrocha**
    Cuticle leathery, with many transverse folds (Fig. 45A); oviparous; probably 4 rare species on mosses .... **Pleuretra**
100. With three plain toes, one dorsal and two terminal (Figs. 45D–J) **101**
    With toes bearing cuplike suckers or united to form a broad disc or twin discs (Figs. 45B, C); about 50 species; up to 1 mm long .... **Mniobia**
101. Eyes absent; oviparous; about 80 species (Figs. 45D–F); in mosses .... **Macrotrachela**
    Two eyes usually present on proboscis; viviparous; some species greatly elongated; about 15 species (Figs. 45G–J); up to 1.5 mm long .... **Rotaria**
102. Corona of the normal type; with three toes (Fig. 45K); many species .... **Habrotrocha**
    Corona modified (Figs. 46A, B) **103**

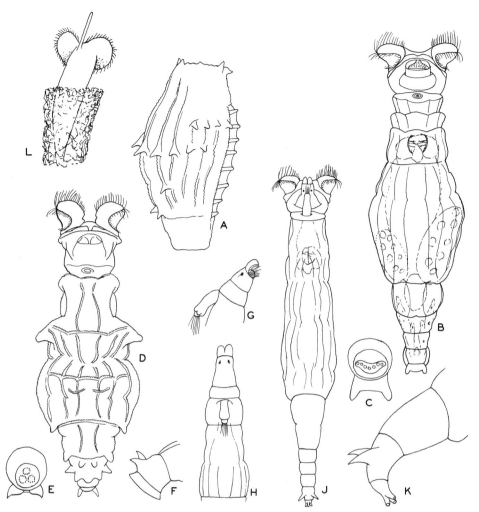

FIG. 45.–Structure of Bdelloidea. A, lateral view of contracted *Pleuretra*; B, dorsal view of *Mniobia* (toes retracted); C, ventral view of spurs and disc of *Mniobia*; D, dorsal view of *Macrotrachela* (toes retracted); E, end view of foot of *Macrotrachela* (toes retracted); F, lateral view of foot of *Macrotrachela* (toes retracted); G, lateral view of head of creeping *Rotaria*; H, dorsal view of anterior end of creeping *Rotaria*; J, dorsal view of swimming *Rotaria* (toes shown extruded); K, lateral view of foot of *Habrotrocha* (toes extruded); L, distal end of *Beauchampia crucigera*. (A–J modified from Murray, 1911.)

103. Corona with two hornlike processes (Fig. 46B); several rare species in mosses ... **Ceratotrocha**
    Corona screened by a wide dorsal membranous plate (Fig. 46A); several uncommon species, chiefly in mosses .... **Scepanotrocha**
104. Rostrum present and perfect but no corona; with four toes (Figs. 46C, D); 1 uncommon species not yet reported from the United States .... **Philodinavus paradoxus** (Murray)
    Rostrum imperfect; corona cannot be retracted within mouth; no ciliary wreaths but only with cilia distributed on corona (Fig. 46G); unable to swim **105**
105. Foot slender, with two spurs and three toes (Fig. 46E); about 10 species in mosses and shallow littoral .... **Adineta**
    Foot stout; spurs absent and replaced by a row of papilliform lobes (Figs. 46F, G); one species, not yet reported from the United States .... **Bradyscela clauda** (Bryce)

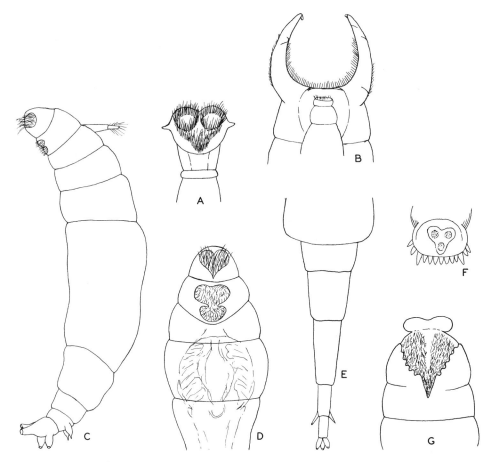

FIG. 46.–Structure of Bdelloidea. A, anterior end of *Scepanotrocha*; B, dorsal view of anterior end of *Ceratotrocha*; C, *Philodinavus paradoxus* with protruded toes; D, ventral view of anterior end of *P. paradoxus*; E, posterior end of *Adineta* with protruded toes; F, ventral view of posterior end of *Bradyscela clauda* showing papilliform lobes and region of three retracted toes; G, ventral view of anterior end of *B. clauda*. (A modified from Bryce, 1910; C and D modified from Murray, 1906.)

REFERENCES

Ahlstrom, E. H. 1940. A revision of the rotatorian genera Brachionus and Platyias with descriptions of one new species and two new varieties. *Bull. Am. Mus. Nat. Hist.* 77:143–184.

———. 1943. A revision of the rotatorian genus Keratella with descriptions of three new species and five new varieties. *Ibid.* 80:411–457.

Aloia, R. C., and R. L. Moretti. 1973. Sterile culture techniques for species of the rotifer Asplanchna. *Trans. Am. Microsc. Soc.* 92:364–371.

Bartoş, E. 1948. On the Bohemian species of the genus Pedalia Barrois. *Hydrobiologia* 1:63–77.

———. 1951. The Czechoslovak Rotatoria of the Order Bdelloidea. *Vestn. Cesk. Spol. Zool.* 15:241–500.

Beach, N. W. 1960. A study of the planktonic rotifers of the Ocqueoc River system, Presque Isle County, Michigan. *Ecol. Monogr.* 30:339–357.

Bérziņš, B. 1951. On the collothecacean Rotatoria. With special reference to the species found in the Aneboda district, Sweden. *Ark. Zool.* 1:565–592.

Brakenhoff, H. 1937. Zur Morphologie der Bdelloidea. *Zool. Jahrb. Abt. Anat. Ont. Tiere* 63:125–182.

Bryce, D. 1910. On a new classification of the bdelloid Rotifera. *J. Quek. Micros. Club.* 11:61–92.

Buchner, H. 1941. Experimentelle Untersuchungen über den Generationswechsel der Rädertiere. II. *Zool. Jahrb. Abt. Allg. Zool. Physiol. Tiere* 60:279–344.

Budde, E. 1925. Die parasitischen Rädertiere mit besonderer Berücksichtigung der in der Umgegend von Minden I. W. beobachteten Arten. *Z. Morphol. Oekol. Tiere* 3:706–784.

Calaway, W. T. 1968. The Metazoa of waste treatment processes—rotifers. *J. Water Pollut. Control Fed.* 40:R412–R422.

Campbell, R. S. 1941. Vertical distribution of the plankton Rotifera in Douglas Lake, Michigan, with special reference to depression individuality. *Ecol. Monogr.* 11:1–19.

Carlin, B. 1939. Über die Rotatorien einiger Seen bei Aneboda. *Medd. Lunds Univ. Limnol. Inst.* 2:1–68.

———. 1943. Die Planktonrotatorien des Motalaström. *Ibid.* 5:1–255.

Cori, C. 1925. Zur Morphologie und Biologie von Apsilus vorax Leidy. *Z. Wiss. Zool.* 125:557–584.

De Beauchamp, P. M. 1909. Recherches sur les rotifères: les formations tégumentaires et l'appareil digestif. *Arch. Zool. Exp. Gen.* (4) 10:1–410.

———. 1912. Instructions for collecting and fixing rotifers in bulk. *Proc. U.S. Nat. Mus.* 42:181–185.

———. 1928. Coup d'oeil sur les recherches récentes relatives aux rotifères et sur les méthodes qui leur sont applicables. *Bull. Biol. Fr. Belg.* 62:51–125.

de Paggi, S. J., and W. Koste. 1984. Checklist of the rotifers recorded from Antarctic and subantarctic areas (Rotatoria). *Senckenbergiana Biol.* 65:169–178.

Dobers, E. 1915. Über die Biologie der Bdelloidea. *Int. Rev.* 7 (Suppl. 1):1–128.

Donner, J. 1951. Erste Übersicht über die Rotatorienfauna einiger Humusboden. *Oesterreich. Zool. Z.* 3:175–240.

———. 1966. Ordnung Bdelloidea. Rotatoria, Radertiere. *Bestimmungsbucher Bodenfauna Eur.* 6:1–297.

———. 1970. Die Rädertierbestande submerser Moose der Salzach und anderer Wasser-Biotope des Flussgebietes. *Arch. Hydrobiol. Suppl.* 36:109–254.

———. 1972. Die Rädertierebestände submerser Moose und weiterer Merotope im Bereich der Stauräume der Donau an der deutsch-Österreichischen Landesgranze. *Arch. Hydrobiol. Suppl.* 44:49–114.

Dumont, H. J., and J. Green (eds.). 1980. Rotatoria: Proceedings of the 2nd International Rotifer Symposium. *Hydrobiologia* 73:1–263.

Edmondson, W. T. 1936. Fixation of sessile Rotatoria. *Science* 84:444.

———. 1939. New species of Rotatoria, with notes on heterogonic growth. *Trans. Am. Microsc. Soc.* 58:459–472.

———. 1940. The sessile Rotatoria of Wisconsin. *Ibid.* 59:433–459.

———. 1944. Ecological studies of sessile Rotatoria. Part I. Factors affecting distribution. *Ecol. Monogr.* 14:31–66.

———. 1945. Ecological studies of sessile Rotatoria. Part II. Dynamics of populations and social structures. *Ibid.* 15:141–172.

———. 1946. Factors in the dynamics of rotifer populations. *Ibid.* 16:357–372.

———. 1948. Ecological applications of Lansing's physiological work on longevity in Rotatoria. *Science* 108:123–126.

———. 1949. A formula key to the rotatorian genus Ptygura. *Trans. Am. Microsc. Soc.* 68:127–135.

———. 1960. Reproductive rates of rotifers in natural populations. *Mem. Ist. Ital. Idrobiol.* 12:21–77.

———. 1965. Reproductive rate of planktonic rotifers as related to food and temperature in nature. *Ecol. Monogr.* 35:61–111.

Finesinger, J. E. 1926. Effects of certain chemical and physical agents on fecundity and length of life, and on their inheritance in a rotifer Lecane (Distyla) inermis (Bryce). *J. Exp. Zool.* 44:63–94.

Gilbert, J. J. 1970. Monoxenic cultivation of the rotifer Brachionus calyciflorus in a defined medium. *Oecologia* 4:89–101.

———. 1973. The adaptive significance of polymorphism in the rotifer Asplanchna humps in males and females. *Ibid.* 13:135–146.

———. 1974. Dormancy in rotifers. *Trans. Am. Microsc. Soc.* 93:490–513.

———. 1975. Polymorphism in the rotifer Asplanchna sieboldi. Variability in the body-wall-outgrowth response to dietary tocopherol. *Physiol. Zool.* 48:404–419.

———. 1976. Sex-specific cannibalism in the rotifer Asplanchna sieboldi. *Science* 194:730–731.

———. 1976. Polymorphism in the rotifer Asplanchna sieboldi: biomass, growth, and reproductive rate of the saccate and campanulate morphotypes. *Ecology* 57:542–551.

———. 1985. Competition between rotifers and Daphnia. *Ecology* 66:1943–1950.

Gilbert, J. J., et al. 1979. Taxonomic relationships of Asplanchna brightwelli, A. intermedia, and A. sieboldi. *Arch. Hydrobiol.* 87:224–242.

Halbach, U. 1970. Die Ursachen der Temporal variation von Brachionus calyciflorus Pallas (Rotatoria). *Oecologia* 4:262–318.

———. 1970a. Einfluss der Temperatur auf die Populations dynamik des planktischen Rädertieres Brachionus calyciflorus Pallas. *Ibid.* 176–207.

———. 1971. Zum Adaptivwert der zyklomorphen Dornenbildung von Brachionus calyciflorus Pallas (Rotatoria). I. Räuber-Beute-Beziehung in Kurzzeit-Versuchen. *Ibid.* 6:267–288.

———. 1973. Quantitative Untersuchungen zur Assoziation von planktischen Rotatorien in Teichen. *Arch. Hydrobiol.* 71:233–254.

Hanley, J. 1949. The narcotization and mounting of Rotifera. *Microscope* 7:154–159.

Harring, H. K. 1913. Synopsis of the Rotatoria. *Bull. U.S. Nat. Mus.* 81:1–226.

———. 1916. A revision of the rotatorian genera Lepadella and Lophocharis with descriptions of five new species. *Ibid.* 51:527–568.

Harring, H. K., and F. J. Myers. 1922. The rotifer fauna of Wisconsin. *Trans. Wis. Acad. Sci. Arts Lett.* 20:553–662.

_____. 1924. The rotifer fauna of Wisconsin, II. A revision of the notommatid rotifers, exclusive of the Dicranophorinae. *Ibid.* 21:415–549.

_____. 1926. The rotifer fauna of Wisconsin, III. A revision of the genera Lecane and Monostyla. *Ibid.* 22:315–423.

_____. 1928. The rotifer fauna of Wisconsin, IV. The Dicranophorinae. *Ibid.* 23:667–808.

Hauer, J. 1935. Rotatorien aus dem Schluchseemoor und seiner Umgebung. *Verh. Naturwiss. Ver.* 31:47–130.

Hickernell, L. M. 1917. A study of desiccation in the rotifer Philodina roseola, with special reference to cytological changes accompanying desiccation. *Biol. Bull.* 32:343–407.

Hlava, S. 1904. (Monograph of the Melicertidae.) *Arch. Prir. Prozk. Cech.* 13:1–79.

Hudson, C. T., and P. H. Gosse. 1889. *The Rotifera; or wheel-animalcules, both British and foreign.* 2 vols. 272 pp. London.

Jacobs, M. H. 1909. The effects of desiccation on the rotifer Philodina roseola. *J. Exp. Zool.* 6:207–265.

Jennings, H. S. 1900. Rotatoria of the United States, with special reference to those of the Great Lakes. *Bull. U.S. Fish Comm.* 19:67–105.

_____. 1903. Rotatoria of the United States. II. A monograph of the Rattulidae. *Bull. U.S. Fish Comm.* (1902):273–352.

Jennings, H. S., and R. S. Lynch. 1928. Age, mortality, fertility, and individual diversities in the rotifer Proales sordida Gosse. I. *J. Exp. Zool.* 50:345–407.

_____. 1928a. Age, mortality, fertility, and individual diversities in the rotifer Proales sordida Gosse. II. *Ibid.* 51:339–381.

King, C. E. 1967. Food, age, and the dynamics of a laboratory population of rotifers. *Ecology* 48:111–128.

Koste, W. 1978. Rotatoria. Die Radertiere Mitteleuropas. *Monogononta.* 2 vols. 673 pp. text and 234 plates. Gebrud. Borntraeger, Berlin, Stuttgart.

Lansing, A. I. 1942. Some effects of hydrogen ion concentration, total salt concentration, calcium and citrate on longevity and fecundity of the rotifer. *J. Exp. Zool.* 91:195–212.

Maly, E. J. 1975. Interactions among the predatory rotifer Asplanchna and two prey, Paramecium and Euglena. *Ecology* 56:346–358.

Miller, H. M. 1931. Alternation of generations in the rotifer Lecane inermis Bryce. Life histories of the sexual and non-sexual generations. *Biol. Bull.* 60:345–381.

Montgomery, T. H. 1903. On the morphology of the rotatorian family Flosculariidae. *Proc. Acad. Nat. Sci. Phila.* 55:363–395.

Murray, J. 1906. Some Rotifera of the Scottish Lochs. *Trans. Roy. Soc. Edin.* 45:151–191.

_____. 1911. Rotifera Bdelloidea. *Proc. Roy. Irish Acad.* 31:1–20.

Myers, F. J. 1930. The rotifer fauna of Wisconsin, V. The genera Euchlanis and Monommata. *Trans. Wis. Acad. Sci. Arts Lett.* 25:353–411.

_____. 1931. The distribution of Rotifera on Mount Desert Island. *Am. Mus. Novit.* 494:1–12.

_____. 1933. The distribution of Rotifera on Mount Desert Island. III. New Notommatidae of the genera Pleurotrocha, Lindia, Eothinia, Proalinopsis, and Encentrum. *Am. Mus. Novit.* 660:1–18.

_____. 1934. The distribution of Rotifera on Mount Desert Island. Part V. A new species of the Synchaetidae and new species of Asplanchnidae, Trichocercidae, and Brachionidae. *Ibid.* 700:1–16.

_____. 1934a. The distribution of Rotifera on Mount Desert Island. Part VII. New Testudinellidae of the genus Testudinella and a new species of Brachionidae of the genus Trichotria. *Ibid.* 761:1–8.

_____. 1936. Psammolittoral rotifers of Lenape and Union lakes, New Jersey. *Ibid.* 830:1–22.

_____. 1937. A method of mounting rotifer jaws for study. *Trans. Am. Microsc. Soc.* 56:256–257.

_____. 1937a. Rotifera from the Adirondack region of New York. *Am. Mus. Novit.* 903:1–17.

_____. 1938. New species of Rotifera from the collection of the American Museum of Natural History. *Am. Mus. Nov.* 1011:1–17.

_____. 1941. Lecane curvicornis var. miamiensis, new variety of Rotatoria, with observations on the feeding habits of rotifers. *Not. Nat.* 75:1–8.

_____. 1942. The rotatorian fauna of the Pocono plateau and environs. *Proc. Acad. Nat. Sci. Phila.* 94:251–285.

Nachtwey, R. 1925. Untersuchungen über die Keimbahn, Organogenese und Anatomie von Asplanchna priodonta Gosse. *Z. Wiss. Zool.* 126:239–492.

Nauwerck, A. 1978. Notes on the planktonic rotifers of Lake Ontario. *Arch. Hydrobiol.* 84:269–301.

Pawlowski, L. K. 1938. Materialen zur Kenntnis der mossbewohnenden Rotatorien Polens. I. *Ann. Mus. Zool. Pol.* 13:115–159.

_____. 1970. Les Rotifères de le Rivière Grabia. *Soc. Sci. Lodz, Pr. Wydz. III - Nauk. Mat. Pryzr.* 110:1–127.

Pax, F., and K. Wulfert. 1941. Die Rotatorien deutscher Schwefelquellen und Thermen. *Arch. Hydrobiol.* 38:165–213.

Pejler, B. 1956. Introgression in plankton Rotatoria with some points of view on its causes and conceivable results. *Evolution* 10:246–261.

_____. 1957. On variation and evolution in planktonic Rotatoria. *Zool. Bidr. Uppsala* 32:1–66.

_____. 1962. Morphological studies on the genera Notholca, Kellicottia and Keratella (Rotatoria). *Ibid.* 33:295–309.

Pejler, B. et al. (eds.). 1983. Biology of rotifers. *Hydrobiologia* 104:1–396.

Pennak, R. W. 1940. Ecology of the microscopic Metazoa inhabiting the sandy beaches of some Wisconsin lakes. *Ecol. Monogr.* 10:537–615.

_____. 1949. Annual limnological cycles in some Colorado reservoir lakes. *Ibid.* 19:233–267.

Pourriot, R. 1965. Recherches sur l'écologie des rotifères. *Vie Milieu, Suppl.* 21:1–224.

Preissler, K. 1977. Do rotifers show "avoidance of the shore"? *Oekologia* **27**:253–260.

Rauh, F. 1963. Untersuchungen über die Variabilität der Radertiere. III. Die experimentelle Beeinflussung der Variation von Brachionus calyciflorus und Brachionus capsuliflorus. *Z. Morphol. Oekol. Tiere* **53**:61–106.

Remane, A. 1929. Intrazellulare Verdauung bei Rädertieren. *Z. Vgl. Physiol.* **11**:146–154.

_____. 1929a. Rotatoria. *Tierwelt Nord Ostsee.* **7**:1–156.

_____. 1929–1933. Rotatorien. *Klassen Ordnungen Tier Reichs 4 (Abt. 2, Buch 1) Lief.* **1–4**:1–576.

Ricci, C. N. 1987. Ecology of bdelloids: How to be successful. *Hydrobiologia* **147**:117–127.

Rousselet, C. F. 1902. The genus Synchaeta: a monographic study, with descriptions of five new species. *J. R. Microsc. Soc.* (1902):269–290, 393–411.

Rudescu, L. 1960. Rotatoria. *Fauna Rep. Pop. Rom.* **2**(2):1–1192.

Ruttner-Kolisko, A. 1946. Über das Auftreten unbefruchteter "Dauereier" bei Anuraea aculeata (Keratella quadrata). *Oesterreich. Zool. Z.* **1**:179–181.

_____. 1969. Kreuzungsexperimente zwischen Brachionus urceolaris und Brachionus quadridentatus, ein Beitrag zur Fortpflanzungsbiologie der heterogonen Rotatoria. *Arch. Hydrobiol.* **65**:397–412.

_____. 1972. Das Zooplankton der Binnengewässer. 1 Teil, III. Rotatoria. *Die Binnengewässer* **26**:99–234.

_____. 1983. The significance of mating processes for the genetics and for the formation of resting eggs in monogonont rotifers. *Hydrobiologia* **104**:181–190.

Salt, G. W., et al. 1978. Trophi morphology relative to food habits in six species of rotifers (Asplanchnidae). *Trans. Am. Microsc. Soc.* **97**:469–485.

Seehaus, W. 1930. Zur Morphologie der Rädertiergattung Testudinella Bory de St. Vincent. *Z. Wiss. Zool.* **137**:175–273.

Sládeček, V. 1983. Rotifers as indicators of water quality. *Hydrobiologia* **100**:169–201.

Stemberger, R. S., and J. J. Gilbert. 1985. Body size, food concentration, and population growth in planktonic rotifers. *Ecology* **66**:1151–1160.

Stossberg, K. 1932. Zur Morphologie der Rädertiergattungen Euchlanis, Brachionus und Rhinoglena. *Ecology.* **142**:313–424.

Sudzuki, M. 1955. Life history of some Japanese rotifers. I. Polyarthra trigla Ehrenberg. *Sci. Rep. Tokyo Kyoiku Daigaku, Sect. B,* **8**:41–63.

_____. 1956. (On the general structure and the seasonal occurrence of the males in some Japanese rotifers. IV.) *(Zool. Mag.)* **65**:1–6.

_____. 1959. (On the general structure and the seasonal occurrence of the males in some Japanese rotifers. VIII.) *Ibid.* **68**:1–7.

Valkanov, A. 1936. Beitrag zur Anatomie und Morphologie der Rotatoriengattung Trochosphaera Semper. *Trav. Soc. Bulg. Sci. Nat.* **17**:177–195.

Varga, L. 1966. Kerekesférgek I. Rotatoria I. *Fauna hung.* **80**:1–144.

Viaud, G. 1940. Recherches experimentales sur le phototropisme des rotifères. *Bull. Biol. Fr. Belg.* **74**:249–308; **77**:68–93, 224–242.

Wallace, R. L. 1977. Distribution of sessile rotifers in an acid bog pond. *Arch. Hydrobiol.* **79**:478–505.

Weber, E. F. 1898. Faune rotatorienne du bassin de Léman. *Rev. Suisse Zool.* **5**:263–785.

Wesenberg-Lund, C. 1923. Contributions to the biology of the Rotifera. I. The males of the Rotifera. *Det. Kgl. Vidensk. Selsk. Skr. Naturv. Math.* (8) **4**:189–345.

_____. 1930. Contributions to the biology of the Rotifera. II. The periodicity and sexual periods. *Ibid.* (9) **2**:1–230.

Williams, L. G. 1966. Dominant planktonic rotifers of major waterways of the United States. *Limnol. Oceanogr.* **11**:83–91.

Wiszniewski, J. 1934. Les rotifères psammiques. *Ann. Mus. Zool. Pol.* **10**:339–399.

Wulfert, K. 1938. Die Rädertiergattung Cephalodella Bory de St. Vincent. *Arch. Naturgesch.* **7**:137–152.

_____. 1964. Unsere gegenwärtige Kenntnis der Rotatoriengattung Macrochaetus Perty 1850. *Limnologica* **2**:281–309.

_____. 1965. Revision der Rotatorien-Gattung Platyias Harring 1913. *Ibid.* **3**:41–64.

Zacharias, O. 1902. Zum Kapital der "wurstförmigen Parasiten" bei Räderthieren. *Zool. Anz.* **25**:647–649.

9

NEMATODA
(Roundworms)

Almost any collection of sand, mud, debris, or vegetation from the bottom or wet margins of a pond, lake, brook, or river will be found to contain small roundworms, or nematodes, sometimes in great abundance. Yet as a group the fresh-water species are poorly known, and few American investigators have been concerned with them since the extensive taxonomic and ecological studies of N. A. Cobb, of the U. S. Department of Agriculture, whose many papers appeared during the first third of the present century. Nematodes have been generally avoided by aquatic biologists because of their small size and difficulties of identification. Although most fresh-water nematodes can be readily assigned to the proper genus, there are a great many undescribed species.

Most of the available information about the Phylum Nematoda is concerned with the thousands of parasitic and predatory species on plants and animals, some of which are of extreme importance to the health and economic well being of man. Less than 0.1% of all nematode literature citations are concerned with free-living aquatic species. Probably no more than 1500 free-living species have been reported from the fresh waters of the world, but this figure must represent only a small fraction of the actual number of species living in this environment. They occur in such widely differing habitats as pools of polar ice, hot springs, puddles, the bottoms of streams, and the deepest and largest lakes. As of 1967, about 500 species of fresh-water nematodes were reported from Europe in the literature.

The account of the Nematoda as given in this chapter is written specifically from the standpoint of the free-living fresh-water species of the United States, but most of the generalizations apply also to the phylum as a whole.

General characteristics. Nematodes occurring in the substrates of fresh waters are almost invariably less than a centimeter long. Their nearly constant and rapid whiplike movements in a dorsoventral plane are quite characteristic, there being no changes in body diameter and proportions. These thrashing, serpentine movements usually produce little or no locomotion in water alone, but debris and vegetation afford sufficient friction to result in forward locomotion. A few species move along the substrate like an inchworm, and the rare free-swimming forms are able to move through the water just above the substrate.

Most aquatic species have three unicellular glands in the tail. Their secretion is carried to the posterior end by one to three minute ducts, and the flow to the outside is regulated by a special pore, or spinneret, at the tip of the body (Figs. 6A, B, D). This secretion is sticky and is used for temporary attachment to objects.

Some nematodes have a pair of lateral caudal pores, or phasmids, of uncertain function (Fig. 6C), although they are probably chemoreceptors.

In cross section the nematode body is more or less circular. The mouth is terminal, but the transverse slitlike anus is subterminal. Though coloration ranges from colorless to blackish, the great majority of species are translucent and therefore fascinating objects for study under the microscope. Most of the coloration of nematodes is determined by the contents of the digestive tract and intestinal cell inclusions. The body is slightly tapered, truncate, or bluntly rounded at the anterior end, where there is sometimes a poorly defined head region, but the posterior end generally tapers to a point.

The outermost layer of the body is a dead, secreted, multilayered, proteinaceous cuticle, which is variously marked, sculptured, or scaly. Sometimes the transverse striations are composed of lines or rows of dots so fine that they may be detected only with critical illumination and the oil immersion lens. At the other extreme, the cuticle may have prominent transverse grooves, rows of dots, or even scalelike folds (Fig. 7 C). Some species have longitudinal striations or long, low, keel-like wings, or alae, which appear to aid in stiffening the animal. In the males of a few fresh-water species there is a caudal bursa composed of two more of less prominent alae (Fig. 1 B). Setae are sometimes present, especially at the anterior end; these are simply elongated projections of the general cuticular exoskeleton. Some investigators, however, speak of structures "similar to cilia" on the oral papillae and amphids. The cuticle extends inward and lines the mouth cavity, esophagus, vulva, and posterior portion of the digestive tract.

The hypodermis, or subcuticle, is usually a syncytial tissue; it secretes the cuticle and lies just below it. This layer is thin except dorsally, ventrally, and laterally, where it is greatly thickened and protruded inward to form the four chords (Fig. 2).

The remainder of the body wall consists of an innermost layer of muscle cells comprising four groups, or fields, of cells separated by the four chords. In meromyarian nematodes there are usually only two to four cells per field, but in polymyarian nematodes each field has numerous cells. A muscle cell is long and spindle shaped, and contains numerous ribbon-shaped muscle fibrils. The fibrils may all lie at the base of the cell and adjacent to the hypodermis (platymyarian arrangement), or the fibrils may extend varying distances up the sides of the cells and partially enclose the sarcoplasm (coelomyarian arrangement) (Fig. 2).

The body cavity of a nematode is properly called a pseudocoel and contains only a few connective tissue cells. There is no true mesodermal epithelium covering the digestive tract.

A nematode grows by periodically shedding the cuticle four times, including the cuticular lining of the stomodaeum and proctodaeum.

Digestive system. The mouth is characteristically surrounded by lips, each with a small papilla on its summit. Six appears to be the basic number of lips, but often there is fusion to form three, and sometimes distinct lips are absent. The lips show innumerable modifications. In some species they are muscular, mobile, and modified for sucking or gripping; in others they are modified for ripping (*Ironus*); sometimes they form toothlike odontia (*Mononchus*).

The mouth opening is minute to large, circular or triangular, and often capable of great distension. The mouth cavity is lined with cuticle and varies enormously in size and structure. Sometimes it is inconspicuous or absent; in other species it is capacious. It may be cylindrical, subglobular, cup shaped, conoid, or some other shape. Its wall is often supplied with stiffening rods or plates; rasps and special teeth, or onchia, are common. In some herbivorous and carnivorous species the pharynx is supplied with a solid or hollow stilettolike protrusile spear (Fig. 7).

The structure of the esophagus is an important taxonomic character. It is circular in cross section, is lined with cuticle, is composed mostly of radial muscle fibers, and has a small triquetrous cavity. Three unicellular salivary glands are usually imbedded in the esophageal wall; their ducts open into the lumen of the stoma or esophagus. In some species the esophagus is elongated, plain, and more or less cylindrical, but more commonly there are one or two ellipsoid or spherical bulbar swellings. If a single bulb is present, it is usually located at the posterior end of the esophagus and is termed the cardiac bulb, whereas the second (middle) bulb is located somewhere near midlength. The esophagus is a suctorial and peristaltic organ, with the bulbs serving to increase the pumping action. Sometimes esophageal valves are present that aid in sucking and prevent regurgitation.

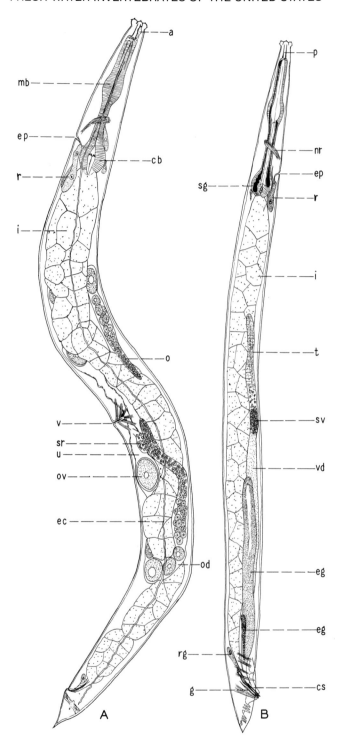

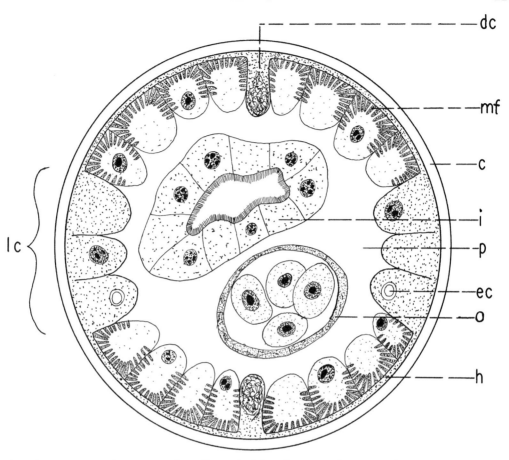

FIG. 2.–Diagrammatic cross section of a typical female nematode. c, cuticle; dc, dorsal chord of hypodermis containing dorsal nerve cord; ec, excretory canal; h, hypodermis; i, intestine, showing rodlike layer on inner surface; lc, lateral chord of hypodermis; mf, muscle fibers of muscle cell; o, oviduct with four developing eggs; p, pseudocoel.

The long intestine extends from esophagus to rectum and is devoid of caeca or loops in fresh-water species. It consists of a single layer of epithelial cells, though rarely there is a flimsy covering of connective tissue or muscle fibers. The inner surface of the intestine is lined with a layer of curious minute rodlike outgrowths, or microvilli, constituting the brush border (Fig. 2). These are sometimes mistaken for cilia, but true cilia (with basal bodies) are unknown in this phylum. Most intestinal digestion is thought to occur in the anterior half of this organ.

The junction of esophagus and intestine is supplied with an obscure valvular apparatus, and there is often a comparable structure

FIG. 1.–Structure of *Rhabditis,* a typical fresh-water nematode, ×95. A, female; B, male. A genital bursa is shown in the male, but this structure is rare in free-living species. a, amphid; cb, cardiac bulb of esophagus; cs, copulatory spicule; ec, excretory canal; eg, ejaculatory gland; ep, excretory pore; g, gubernaculum; i, intestine; mb, middle bulb of esophagus; nr, nerve ring; o, ovary; od, oviduct; ov, mature fertilized ovum; p, pharynx; r, renette; rg, rectal gland; sg, salivary gland (glands and ducts solid black; present in both sexes but shown only in figure of male); sr, seminal receptacle; sv, seminal vesicle; t, testis; u, uterus; v, vulva and genital pore; vd, vas deferens. (Modified from Chitwood, 1937, by permission.)

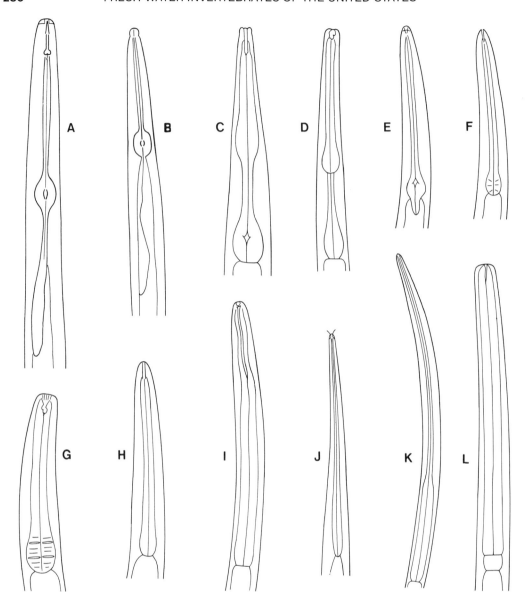

FIG. 3.–Esophagus shapes. A, *Hirschmanniella;* B, *Aphelenchoides;* C, *Rhabditis;* D, *Butlerius;* E, *Plectus;* F, *Leptolaimus;* G, *Achromadora;* H, *Cylindrolaimus;* I, *Ironus;* J, *Aphanolaimus;* K, *Alaimus;* L, *Tripyla*. (Modified from Ferris, Ferris, and Tjepkema, 1973.)

separating the intestine proper from its short, narrow, muscular posterior end, the rectum. Several unicellular rectal and anal glands are characteristic of this region of the body.

Feeding. All varieties of food habits are represented in free-living fresh-water nematodes. Some feed only on dead plant material,

others on living hydrophytes, and still others only on dead animal material. A fourth group are detritus feeders, consuming both dead plant and dead animal material. Herbivorous species are variously specialized for biting and chewing living plant tissues, and others are equipped with a rigid hollow stylet that pierces plant cells and through which the protoplasm is sucked by the action of the esophagus. Similarly, the predaceous and carnivorous forms, such as *Mononchus, Dorylaimus, Nygolaimus,* and *Actinolaimus,* show a variety of anatomical modifications for feeding. The lips may be specialized for seizing or tearing the prey, and the pharynx is often partially eversible and supplied with onchia or rasps for crushing, rasping, or macerating. Many carnivorous species have the hollow stylet type of feeding. The prey consists of Protozoa and all types of small Metazoa, including oligochaetes, rotifers, gastrotrichs, tardigrades, and other nematodes.

Material digested in the alimentary canal is absorbed into the fluid of the surrounding hemocoel and carried to all the internal organs and body wall by circulation resulting from the incessant body movements. There is no special circulatory system.

Excretion. The excretory system presents a varied picture in the phylum as a whole. In its simplest condition there is a single ventral excretory cell, or renette, which opens through an excretory pore on the midventral line in the region of the esophagus by way of a short to long duct. In other nematodes there are two lateral excretory canals imbedded in the lateral chords of the hypodermis throughout most of the body length. They are connected anteriorly and ventrally by a transverse canal, thus forming an H or U shape. A very short duct connects the transverse duct with the excretory pore. There may or may not be two special cells (the renette) associated with the transverse duct (Fig. 1).

In a few genera, including *Dorylaimus,* no excretory system has been found. Considerable excretion through the digestive tract may occur in all nematodes.

Nervous system. A ring of nerve fibers and associated indistinct ganglia encircle the esophagus. From this mass six small nerves are given off which innervate the anterior end. Four or more longitudinal nerves extend posteriorly in the hypodermis, of which the dorsal and ventral nerves are usually the most prominent. Especially at the extremities of the body, there are scattered small ganglia and numerous branch nerves that connect the longitudinal nerves with each other. In

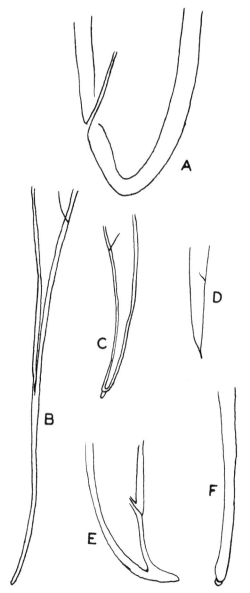

FIG. 4.—Posterior end. A, *Labronema;* B, *Mesodorylaimus;* C, *Achromadora;* D, *Aphelenchoides;* E, *Eudorylaimus;* F, *Labronema.* (Modified from Ferris, Ferris, and Tjepkema, 1973.)

general, the nervous system is considerably more complicated than most elementary textbooks indicate.

Sense organs. Paired eyespots occur in some species of *Punctodora* and *Monhystera,* as well as in a few other genera. Such eyespots are small clumps of reddish, violet, or blackish granules imbedded in the esophagus or between the esophagus and body wall. A minute lenslike device may be present in front of each clump of pigment, or it may be absent. There is some difference of opinion as to whether there are nervous connections between eyespots and nervous system.

Tactile receptors are mainly associated with the cephalic papillae and setae and with the caudal setae. The somatic setae, often present between anterior and posterior ends, are probably also tactile.

Amphids are peculiar nematode structures, and though there is still considerable controversy about their function, it appears most likely that they are chemoreceptors rather than organs of equilibration. A typical fresh-water nematode has two amphids, one on each side of the body near the anterior end. Externally, an amphid may have the shape of a narrow slit, an oval, a circle, or

spiral grooves and ridges on the cuticle. Internally, it consists of a small pouch and often a minute longitudinal posterior canal. The pouch has recently been shown to be innervated. Chitwood (1931) maintains that amphids are always present, but often they are so minute, reduced, or vestigial as to be indistinguishable to the average observer.

Reproduction, development. Both males and females are known for most species of nematodes, and reproduction is syngamic, but in many species males are rare or unknown, and reproduction is usually parthenogenetic, as in *Rhabdolaimus, Monhystera,* and *Alaimus.* Hermaphroditic species occur in *Mononchulus, Aphanolaimus, Ironus, Tobrilus, Mononchus,* and a few other genera. The female is usually the larger, but the sexes can be distinguished by the sex characters.

In general, the reproductive systems are simpler and less extensive than those of parasitic species and produce relatively few gametes. In the mature female the genital pore, or vulva, can be seen as a transverse slit on the ventral midline near the middle of the body. There is a short muscular vagina, which may connect with a single thin walled tapering reproductive tubule or, more typically, may branch to form two tubules, one anterior and one posterior. Beginning at the vagina, each tubule consists of a wide uterus, an oviduct, and a fine terminal ovary composed of very small oocytes. Sometimes the uterus has a small associated seminal receptacle (Fig. 1). A reproductive tubule may be outstretched or reflexed. Thus the reproductive system of a female nematode may have one of four configurations: double and outstretched, double and reflexed, single and outstretched (Fig. 9 D), or single and reflexed.

The male reproductive system is ordinarily a single tapering tubule, which opens ventrally into the posterior part of the digestive tract. Beginning basally, the tubule consists of a long vas deferens, short seminal vesicle, and thin testis (Fig. 1B). Rarely the testis is bifurcated or reflexed. Ejaculatory glands sometimes open into the basal part of the vas deferens. Since the rectum of the male is common to both the reproductive and digestive ducts, it is more appropriately called a

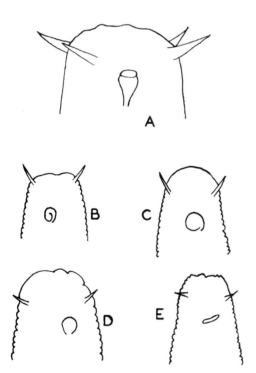

Fig. 5.–Amphid shapes. A, *Tobrilus*; B, *Achromadora*; C, *Prodesmodora*; D, *Plectus*; E, Anaplectus. (Modified from Ferris, Ferris, and Tjepkema, 1973.)

cloaca. A pair of sclerotized copulatory spicules are contained within a delicate pouch on the dorsal side of the cloaca. During copulation these spicules are extruded through the cloacal aperture, or anus, of the male and into the vulva. They are moved back and forth by muscles so that the vulva and vagina are kept open and so that sperm are more easily propelled into the vagina. The gubernaculum is a sclerotized thickening in the dorsal wall of the male cloaca that acts as a spicular guide and prevents the spicules from rupturing the spicular pouch and cloaca when exserted.

A clasping device, or copulatory bursa (Fig. 1B), is common in parasitic species but is rare in the males of fresh-water species. It consists of two thin, transparent, longitudinal flaps of cuticle near the posterior end. It is widest opposite the anus and may or may not be long enough to encompass the posterior tip of the animal.

Fertilization occurs in the upper end of the uterus and the egg shell is laid down in the lower uterus. The fertilized eggs usually leave the body before the onset of segmentation, but some species are viviparous or ovoviviparous. The embryo is formed and hatches in a few hours to a few weeks, depending on the particular species and ecological conditions. The newly hatched young are fully developed except for size, reproductive system, and certain specializations of the lips, pharynx, and esophagus. It is thought that there is almost always a series of four molts (five instars) in the complete life history. In addition to the general body covering, a molt also includes shedding of the cuticular lining of pharynx, esophagus, vulva, and rectum. The secondary sex characters are not apparent until after either the penultimate molt or the last molt. Little is known about the longevity of free-living nematodes, but males are thought to be much shorter lived than females. Little or no growth occurs below 4°C, and 20°C appears to be an optimum temperature.

Ecology. The eggs of soil and parasitic nematodes are highly resistant to desiccation (even for periods up to 10 or 20 years) and to high and low temperatures; they are easily transported by animals and blown about as dust. The larvae and adults of many soil species likewise are viable after being dried and coiled or shriveled up in a state of anabiosis or anhydrobiosis. Nevertheless, there seems to be very little information about encystment and desiccation among aquatic nematodes, and it is not known how important these mechanisms are for dissemination. Judging from the available scanty distribution records, however, most aquatic genera and many species are cosmopolitan and worldwide.

Furthermore, some species show an amazing ability to thrive in a wide variety of habitats and a wide range of ecological conditions. Among all the metazoan phyla, nematodes are perhaps the most highly adaptable from ecological and physiological standpoints. The same species, for example, may be found from the tropics to the subarctic, from warm springs to cold alpine lakes, and on many types of substrates. Only a few genera appear to be restricted to fresh waters; examples are *Cryptonchus, Chronogaster,* and *Anonchus. Oncholaimus* and *Chromadorita* are both marine and fresh-water genera. The following genera, among others, occur in both fresh waters and terrestrial soils: *Prismatolaimus, Mononchus, Tobrilus, Tripyla,* and *Ironus. Monhystera,* however, is a genus with terrestrial, marine, and fresh-water representatives. In addition to being found in fresh waters and soils, *Rhabditis* and *Cephalobus* also have parasitic representatives. A few genera show specific habitat preferences in fresh waters, and some of these are mentioned in the key that follows. Many soil nematodes often occur in fresh waters as the result of wash.

Present ideas emphasize the probable origin of nematodes in fresh waters and their subsequent migration and colonization of salt water and terrestrial environments.

As pointed out long ago by Cobb, nematode populations often reach tremendous numbers both in terrestrial soils and in the substrates of bodies of water, where they are of great mechanical importance in continuously working over and mixing soils in much the same way as oligochaetes. Even in the wet sand of lake beaches near the water's edge, which is not an especially favorable habitat for nematodes, the writer has found up to eight adult nematodes per milliliter of sand. Some of the most dense populations, especially species of *Mononchus, Ironus,* and *Tripyla,* occur in the sand and gravel filter beds of sewage disposal plants. A wide variety of species occur around the roots of aquatic plants. In natural fresh-water habitats most specimens are confined to the uppermost 5 cm of the substrate.

Unlike most fresh-water invertebrate taxa, the (micro) nematodes are ecologically tied to the substrate and do not swim freely in the water. They are chiefly members of the meio-benthos, and in addition to sewage disposal habitats, they attain their greatest densities in soft muddy substrates of lakes, ponds, and streams, often in excess of $100,000/m^2$ and to a depth of 2 cm. Such nematode assemblages commonly consist of five to 20 species, although some thorough studies report up to 150 species in one stream substrate.

Although populations of aquatic nematodes can be maintained indefinitely in low concentrations of dissolved oxygen, 2 to 10% saturation, for example, it is thought that active nematodes are capable of withstanding anaerobic conditions only for one to several weeks at a time. The eggs, however, are highly resistant and may remain viable after many months in the absence of oxygen and after repeated freezing and thawing.

Collecting, preparing, preserving. Perhaps the chief reason for the general neglect of the nematodes by investigators in aquatic zoology is the fact that they almost invariably occur in quantities of bottom mud and fine organic debris from which it is difficult to separate them. If a small quantity of the material containing nematodes is placed in a petri dish of water, they can be detected by their movements and picked out with an Irwin loop under the dissecting binoculars and transferred to clean water. This is a tedious method, however, and if a relatively large quantity of nematode-containing material is available, many worms may be concentrated by the following method: (1) wash the material through a sieve of 2- or 3-mm mesh and then (2) through a 1-mm mesh sieve to remove the larger pieces of debris; (3) wash the material through number 20 bolting silk, which will retain the nematodes and allow the fine debris to pass; (4) transfer the retained nematodes and debris to a small container of

water, agitate, and decant quickly as soon as the heavy sand particles settle out; repeat this step if desirable; (5) agitate and allow to stand two to four minutes, which is sufficient time to allow the nematodes to sink to the bottom, and then decant off the muddy water. The residue will then contain most of the nematodes with a minimum of fine debris; specimens should be picked up individually with a fine pipette, bristle, or Irwin loop for further treatment. Ferris et al. (1973) give directions for a similar method of separating nematodes from large quantities of aquatic substrate debris.

Live nematodes are so active that it is useless to study them directly with a compound microscope. They should first be narcotized with chloroform, chloral hydrate, magnesium sulfate, or other narcotic.

Hot saturated mercuric chloride or hot 70 to 80% alcohol are probably the best fixatives, but for routine fixation in the field cold 85% alcohol and 5% formalin are quite reliable. Good results are also possible by killing the worms by placing them in a water bath for 10 minutes at 57°C. Warm water prevents coiling.

Simple temporary formalin mounts ringed with petroleum jelly are excellent for routine purposes.

For permanent mounts and identification there is probably no better or more convenient mounting medium than glycerin jelly. But glycerin, like most substances, penetrates the nematode cuticle very slowly. If, however, the worms are placed in an open concave watch glass of 5% glycerin in 50% alcohol, the alcohol and water will evaporate in 3 to 10 days and the worms will be infiltrated in concentrated glycerin. They may be stored thus or mounted directly in glycerin jelly. Like all other glycerin jelly slides, the circular cover slips should be ringed with Murrayite, "Zut," or similar cement.

Stained balsam mounts are tedious to prepare and often exasperating, because unless meticulous care is taken in the reagent transfers, the final mounted nematodes collapse or become distorted and opaque. Cobb (1918) devised a glass "differentiator" for gradually transferring nematodes from one fluid to another so as to avoid distortion. Chatterji (1935) reports success with the following brief procedure: 70% alcohol, dioxan overnight, fresh dioxan 1 hour, clove oil 2 hours, and balsam. Chitwood and Chitwood (1930) give detailed directions for imbedding in paraffin.

Many aquatic nematodes may be cultured in the laboratory in 3% agar in tap water with a little added soil and debris.

Taxonomy. The taxonomic relationships of free-living nematodes, especially with respect to family designations, are uncertain and in a state of flux. The literature, furthermore, is badly scattered, often in obscure journals. Of the several modern plans that have been proposed, that of Ferris et al. (1973) is used in this manual. Following is a list of orders and genera included in the key.

Chromadorida
  Achromadora
  Chromadorita
  Ethmolaimus
  Monochromadora
  Paracyatholaimus
  Punctodora
Araeolaimida
  Anaplectus
  Anonchus
  Aphanolaimus
  Bastiania
  Chronogaster
  Cylindrolaimus
  Euteratocephalus
  Paraphanolaimus
  Plectus
  Teratocephalus
Monhysterida
  Monhystera
  Monhystrella
  Odontolaimus
Desmodorida
  Prodesmodora
Enoplida
  Cryptonchus
  Ironus
  Oncholaimus
  Prismatolaimus
  Rhabolaimus
  Tobrilus
  Tripyla
Dorylaimida
  Actinolaimus
  Alaimus
  Amphidelus
  Aulolaimoides

  Dorylaimus
  Eudorylaimus
  Labronema
  Mesodorylaimus
  Microlaimus
  Nygolaimus
  Thornia
Mononchida
  Anatonchus
  Miconchus
  Mononchulus
  Mononchus
  Mylonchulus
  Prionchulus
Tylenchida
  Apelenchoides
  Atylenchus
  Criconema
  Criconemoides
  Hemicycliophora
  Hirschmanniella
  Tylenchus
Rhabditia
  Acrobeloides
  Butlerius
  Cephalobus
  Eucephalobus
  Goffartia
  Mononchoides
  Panagrolaimus
  Rhabditis

After considerable experience, identifications to genus can sometimes be made with the 4-mm objective of the microscope, but critical illumination and the oil immersion objective are usually essential in order to study the smaller details.

The number of genera characterized as "aquatic" is often exaggerated by some investigators (e.g., Tarjan et al., 1977), because they include nematodes that are collected in damp or wet soil, lakeside situations, around the roots of plants, or in moss. The key that follows, however, is fairly restricted to genera that are found in true aquatic situations. It is greatly modified from Ferris et al. (1973) and includes the great majority of genera reported from U.S. fresh waters and should be reliable for more than 98% of all specimens collected. Some of the genera are quite rare. Family designations are not included because of the many taxonomic uncertainties among the Nematoda.

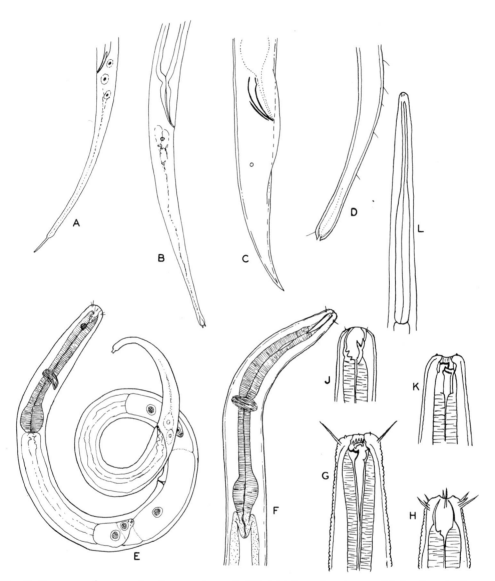

FIG. 6.–Structure of nematodes. A, posterior end of *Rhabdolaimus,* showing caudal glands and spinneret; B, posterior end of *Aphanolaimus,* showing caudal glands, spinneret, and phasmid; C, posterior end of male *Tylenchus,* showing phasmid; D, posterior end of *Tobrilus,* showing spinneret; E, female *Achromadora,* ×350; F, anterior end of *Plectus;* G, anterior end of *Punctodora;* H, anterior end of *Prismatolaimus;* J, anterior end of *Butlerius;* K, anterior end of *Mononchoides;* L, anterior end of *Nygolaimus.*

KEY TO GENERA OF NEMATODA

1. Stoma large, cup shaped; width and depth of stoma at least one half of lip region; stoma strongly cuticularized (Figs. 7A, F, K, N) 2
   Stoma not both wide and cup shaped, and either weakly or strongly cuticularized 15
2. Cephalic setae long (Fig. 6G) 3
   Cephalic setae short or absent (setose papillae may be present) 5

3. Posterior end elongate-clavate, with spinneret; four large cephalic setae (Fig. 6G) **4**

   Posterior end filiform and without spinneret; six large and four short cephalic setae (Fig. 6H); common .... **Prismatolaimus**

4. Stoma in two parts: an anterior cup-shaped region with a large tooth, and a posterior part with elongated conical cavity (Fig. 6G); rare .... **Punctodora**

   Stoma not in two parts, without a large tooth (Fig. 8B) .... **Anonchus**

5. Esophagus with two bulbs (Fig. 3D) **6**

   Esophagus cylindrical, or with only a posterior bulb **7**

6. Stoma very wide and deep; anterior edge of stoma not marked and not bearing grooves or riblike structure (Fig. 6J); uncommon .... **Butlerius**

   Stoma deep and moderately wide; anterior edge of stoma bearing riblike structures; slender tubular section of stoma extending posterior to large tooth in stoma (Fig. 6K) ..... **Mononchoides**

7. Spinneret absent **8**

   Spinneret present (Figs. 4C, F) **9**

8. Pharynx with a spear and four large teeth (Fig. 10C); common .... **Actinolaimus**

   Pharynx without a spear and four large teeth; common in decaying material ... **Panagrolaimus**

9. Spinneret subterminal (Fig. 7M) .... **Mononchulus**

   Spinneret at tip of body **10**

10. Stoma with three large teeth (Fig. 7F); rare .... **Oncholaimus**

    Stoma with another arrangement of teeth **11**

11. Dorsal tooth posteriorly directed in stoma (Fig. 7K); rare in marshy areas .... **Anatonchus**

    Dorsal tooth anterior directed in stoma **12**

12. Large dorsal tooth posteriorly placed in stoma; two large subventral teeth opposite (Fig. 7N); uncommon .... **Miconchus**

    Dorsal tooth anteriorly placed in stoma; subventral teeth absent or small **13**

13. Stoma with transverse rows of denticles opposite dorsal tooth (Fig. 12G) .... **Mylonchulus**

    Without transverse rows of denticles of opposite dorsal tooth **14**

14. Longitudinal ridge without denticles opposite dorsal tooth (Fig. 12E); common ... **Mononchus**

    Longitudinal denticulate ridge opposite dorsal tooth (Fig. 12 F) .... **Prionchulus**

15. Stoma armed with a spear (Figs. 10 E, 8 L) or spearlike tooth (Fig. 10 A) **16**

    Stoma lacking a spear or spearlike tooth **29**

16. Cuticle strongly annulated (Figs. 7C, D); body stout; swamps and acid waters; large genera **17**

    Cuticle not strongly annulated **18**

17. Annules with spines or accessory scales (Fig. 7C) .... **Criconema**

    Annules without spines or scales (Fig. 7D) .... **Criconemoides**

18. Esophagus with one bulb; amphid stirrup shaped and distinct (Fig. 5A) **19**

    Esophagus with two bulbs; amphid obscure **25**

19. Posterior end of both sexes short and blunt or elongate, never filiform (Fig. 4A) **20**

    Posterior end of female, or both female and male, filiform (Fig. 4B) **23**

20. "Tail" blunt and approximately twice as long as the anal body diameter (Fig. 7J); anterior end of body nearly square (Fig. 7H); esophagus expanding sharply at middle (Fig. 7G); uncommon.

    Thornia

    "Tail" pointed, or if blunt then distinctly shorter than twice anal body diameter; anterior end of body somewhat rounded; esophagus expanding gradually **21**

21. Toothlike spear attached to stoma wall, but stoma wall cuticularized and separated from tooth; esophagus often expanded anterior to middle (Figs. 10A, B) .... **Nygolaimus**

    With a true spear in the stoma; walls of stoma weakly developed; esophagus usually expanded at middle **22**

22. Anterior portion of esophagus much narrower than posterior portion (Fig. 8H); total length up to 2 mm; very large genus .... **Eudorylaimus**

    Anterior portion of esophagus nearly as wide as posterior portion (Fig. 8G); total length usually more than 2 mm; rare .... **Labronema**

23. Enlarged basal portion of esophagus short, barely longer than wide (Fig. 8J); rare ... **Aulolaimoides**

    Basal portion of esophagus considerably longer than wide **24**

24. Cuticle with longitudinal ridges; more than 2 mm long; abundant and common (Fig. 10E).

    Dorylaimus

    Cuticle without longitudinal ridges; less than 2 mm long; many common species.

    Mesodorylaimus

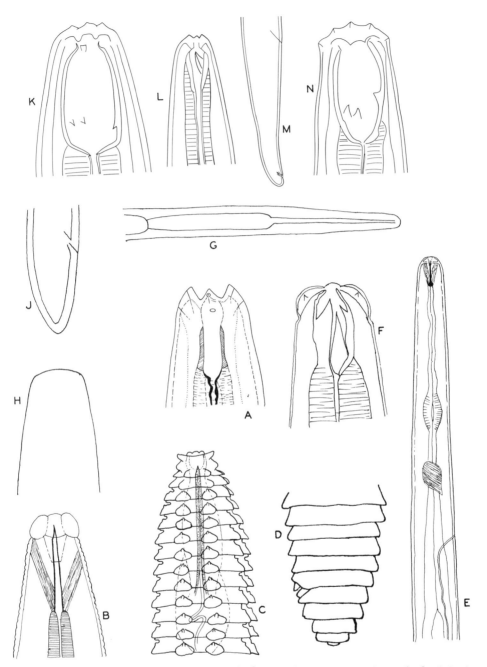

FIG. 7.–Structure of nematodes. A, anterior end of *Panagrolaimus*; B, anterior end of *Aphelenchus*; C, anterior end of *Criconema*; D, posterior end of *Criconemoides*; E, anterior end of *Tylenchus*; F, anterior end of *Oncholaimus*; G, anterior end of *Thornia*; H, anterior end of *Thornia*; J, posterior end of female *Thornia*; K, anterior end of *Anatonchus*; L, anterior end of *Mononchulus*; M, posterior end of female *Mononchulus*; N, anterior end of *Miconchus*. (A modified from Thorne, 1937; B modified from Chitwood, 1939; C modified from Cobb, 1915; G–J modified from Ferris, Ferris, and Tjepkema, 1973; K–N from Ferris, Ferris, and Tjepkema, 1973.)

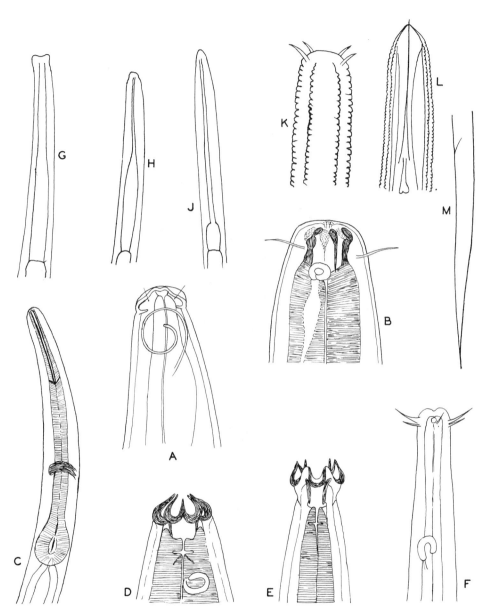

FIG. 8.–Anterior end of nematodes. A, *Aphanolaimus;* B, *Anonchus;* C, *Rhabdolaimus;* D, *Euteratocephalus;* E, *Teratocephalus* (annulations not shown); F, *Bastiana* (annulations not shown); G, *Lebronema;* H, *Eudorylaimus;* J, *Aulolaimoides;* K, *Atylenchus;* L, *Hemicycliophora;* M, posterior end of *Tylenchus.* (A, C–E modified from Cobb, 1914; B modified from Cobb, 1913; F modified from Chitwood, 1937; G–M modified from Ferris, Ferris, and Tjepkema, 1973.)

25. Cephalic setae present; cuticle with longitudinal ridges and coarse annulations (Fig. 8K); rare.
    Atylenchus
    Cephalic setae absent; cuticular ridges lacking; cuticle annulations not coarse **26**
26. Worm with an accessory loose outer cuticular layer; spear long, approximately same length as
    esophagus (Fig. 8L); uncommon .... **Hemicycliophora**
    Without an accessory cuticular layer; spear not nearly as long as esophagus (Fig. 9F) **27**
27. Posterior end filiform (Fig. 8M); esophagus not overlapping intestine; median bulb of esophagus
    small and ovate (Fig. 7E); often associated with roots .... **Tylenchus**
    Posterior end conical, not filiform (Fig. 4D); esophagus overlapping intestine, and median bulb of
    esophagus prominent and spherical (Figs. 3A, B) **28**
28. Spear slender, with or without modified basal knobs (Fig. 9F); very large genus .... **Aphelenchoides**
    Spear stout, with large basal knobs (Fig. 9G) .... **Hirschmanniella**
29. Spinneret present (Figs. 11C, G, H) **30**
    Spinneret absent or obscure **47**
30. Esophagus with basal bulb (Fig. 11D) **31**
    Esophagus cylindrical, with no basal bulb (Fig. 3H) **40**
31. Amphid multispiral (Fig. 9C), slitlike (Fig. 9H), or obscure **32**
    Amphid large, unispiral, circular, or open circle (Figs. 9A; 11E, J) **37**
32. Amphid spiral (Fig. 5B) **33**
    Amphid slitlike or obscure (Fig. 5E) **34**
33. Dorsal tooth and opposing subventral teeth large; tubular section of stoma extending posterior to
    teeth (Fig. 9C); one common species .... **Ethmolaimus**
    Dorsal tooth large, but opposing subventral teeth small or obscure; stoma ending just posterior to
    teeth; tubular section posterior to teeth lacking (Fig. 9B); common .... **Achromadora**
34. Stoma with large, sharp dorsal tooth and obscure subventral teeth (Fig. 9H); cuticle marked with
    punctations; uncommon .... **Chromadorita**
    Stoma with two or three small teeth or no teeth; cuticle smooth **35**
35. Cephalic setae present; cuticle distinctly annulated (Fig. 11A) .... **Anaplectus**
    Cephalic setae absent; cuticle weakly annulated or smooth **36**
36. Inconspicuous peglike spinneret (Fig. 11C); stoma funnel shaped (Fig. 11B); rare.
    Monochromadora
    Prominent long, conical spinneret (Fig. 6A); stoma tubular (Fig. 8C); common ..... **Rhabdolaimus**
37. Basal bulb of esophagus divided transversely in two places (Fig. 11D) .... **Prodesmodora**
    Basal bulb of esophagus not divided transversely **38**
38. Cuticle smooth (Fig. 11E); rare .... **Monhystrella**
    Cuticle annulated or striated **39**
39. Head slightly offset and rounded; stoma with a tooth and two denticles (Fig. 9A); uncommon.
    Microlaimus
    Head not offset; stoma without teeth (Fig. 11J) .... **Plectus**
40. Spinneret located ventrally at posterior end (Fig. 7M); stoma with large subventral tooth (Fig. 7L).
    Mononchulus
    Spinneret terminal; no large tooth, or large tooth dorsal in stoma **41**
41. Anterior end narrow and rounded (Fig. 3H) **42**
    Anterior end broad and square (Figs. 9E, 11M) **44**
42. Stoma distinct, elongate, and cylindrical; weak cuticular annulations (Fig. 3H); a single ovary.
    Cylindrolaimus
    Stoma a short broad cylinder or obscure; two ovaries **43**
43. Stoma greatly reduced; amphid circular and prominent (Fig. 8A) .... **Aphanolaimus**
    Stoma a short, broad cylinder; amphid unispiral and prominent (Fig. 11K) ...... **Paraphanolaimus**
44. Amphid multispiral; stoma with large dorsal tooth (Fig. 11M); rare .... **Paracyatholaimus**
    Amphid stirrup shaped (Fig. 5A), circular (Fig. 9E), or obscure; dorsal tooth lacking or inconspicuous;
    many species; common **45**
45. Amphid circular (Fig. 9E); single outstretched ovary anterior to posterior vulva ... **Monhystera**
    Amphid stirrup shaped or obscure; two ovaries, one anterior and one posterior to the median vulva.
    46
46. Stoma funnel shaped; amphid stirrup shaped (Fig. 12A) .... **Tobrilus**
    Stoma obscure, narrow; amphid reduced and obscure (Figs. 12B, C) .... **Tripyla**
47. With six sharply pointed, strongly cuticularized, incurved lips (Figs. 8D, E) **48**
    Lips not pointed, strongly cuticularized, and incurved **49**

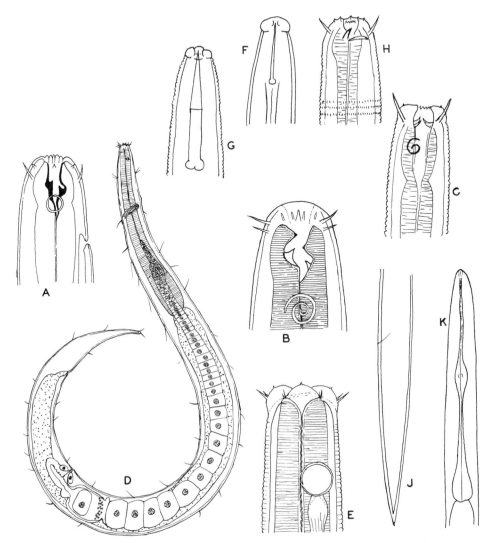

FIG. 9.–Structure of nematodes. A, anterior end of *Microlaimus;* B, anterior end of *Achromadora;* C, anterior end of *Ethmolaimus;* D, female *Monhystera,* ×110; E, anterior end of *Monhystera;* F, anterior end of *Aphelenchoides;* G, anterior end of *Hirschmanniella;* H, anterior end of *Chromadorita;* J, posterior end of female *Hirschmanniella;* K, anterior end of *Tylenchus.* (A modified from Chitwood, 1937; B and D modified from Cobb, 1914; E modified from Micoletzky, 1925; C and F–K modified from Ferris, Ferris, and Tjepkema, 1973.)

48. Cuticle strongly annulated; amphid obscure (Fig. 8E); ovary single . . . . **Teratocephalus**
    Cuticle marked only by punctations; amphid spiral (Fig. 8D); two ovaries; uncommon.

    Euteratocephalus
49. Stoma tubular, poorly cuticularized **50**
    Stoma not tubular, or if tubular then distinctly cuticularized **54**
50. Anterior margin of body almost flat; small tooth located dorsally in stoma (Figs. 12B, C); large genus;
    common . . . . **Tripyla**
    Anterior end of body rounded; no tooth in stoma **51**
51. With cephalic setae and distinct cuticular annulations (Fig. 8F); uncommon . . . . **Bastiania**
    Without cephalic setae; cuticular annulations faint or absent; uncommon species **52**

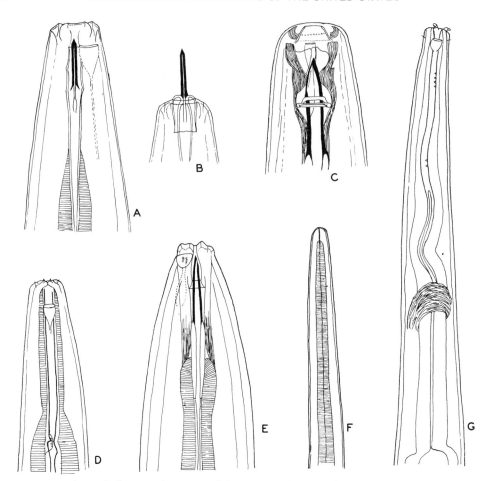

Fig. 10.–Anterior end of nematodes. A, *Nygolaimus;* B, *Nygolaimus* with protruded spear; C, *Actinolaimus;* D, *Cryptonchus;* E, *Dorylaimus;* F, *Alaimus;* G, *Ironus.* (A and B modified from Thorne, 1930; C and D modified from Cobb, 1913; E modified from Micoletzky, 1925; F and G modified from Cobb, 1914.)

52. Enlarged basal portion of esophagus sharply set off (Fig. 8J) .... **Aulolaimoides**
    Basal portion of esophagus tapered (Fig. 10F) 53
53. Amphid large, slitlike or crescent shaped (Fig. 11L) .... **Amphidelus**
    Amphid obscure .... **Alaimus**
54. Esophagus cylindrical, without a basal bulb (Fig. 10G) 55
    Esophagus with a basal bulb (Fig. 11S) 57
55. Three large, hooklike teeth at anterior end of stoma (Fig. 11N); two ovaries, one anterior and one
        posterior to vulva; common .... **Ironus**
    Without teeth at anterior end of stoma; single ovary 56
56. Posterior end bluntly rounded (Fig. 11O); amphid circular (Fig. 11F) .... **Cylindrolaimus**
    Posterior end pointed (Fig. 11P); amphid stirrup shaped (Fig. 10D); uncommon ...... **Cryptonchus**
57. Basal bulb of esophagus nonmuscular and lacking valves; median bulb present (Fig. 11X);
        uncommon 58
    Basal bulb of esophagus muscular and valvate; median bulb present or absent 59
58. Stoma moderately wide and deep; anterior edge of stoma with riblike structures; slender tubular
        section of stoma extending posterior to large tooth in stoma (Fig. 6K); amphids obscure.
                                                                          Mononchoides

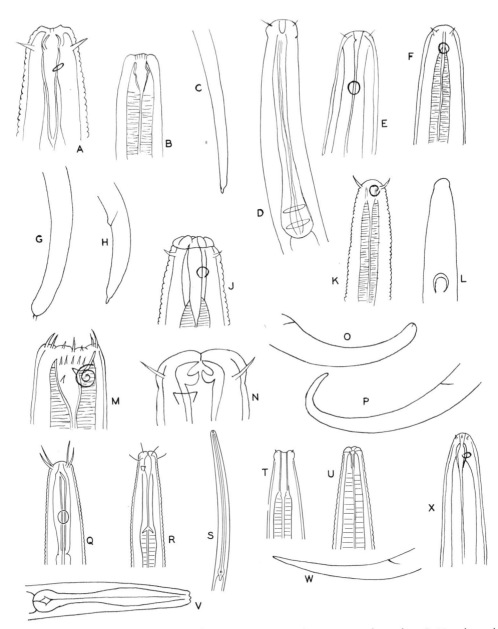

FIG. 11.–Anterior and posterior ends of nematodes. A, *Anaplectus;* B, *Monochromadora;* C, *Monochromadora;* D, *Prodesmodora;* E, *Monhystrella;* F, *Cylindrolaimus;* G, *Plectus;* H, *Plectus* female; J, *Plectus;* K, *Paraphanolaimus;* L, *Amphidelus;* M, *Paracyatholaimus;* N, *Ironus;* O, *Cylindrolaimus;* P, *Cryptonchus;* Q, *Odontolaimus;* R, *Chronogaster;* S, *Chronogaster;* T, *Rhabditis;* U, *Acrobeloides;* V, *Acrobeloides;* W, *Eucephalobus;* X, *Goffartia.* (All modified from Ferris, Ferris, and Tjepkema, 1973.)

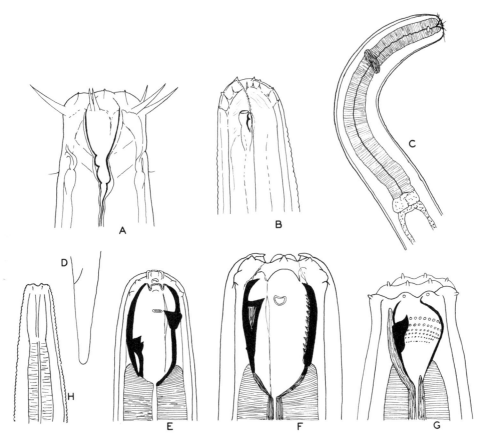

FIG. 12.–Anterior and posterior ends of nematodes. A, *Tobrilus;* B and C, *Tripyla;* D, *Cephalobus;* E, *Mononchus;* F, *Prionchulus;* G, *Mylonchulus;* H, *Cephalobus.* (A and B modified from Micoletzky, 1925; C modified from Cobb, 1914; E–G modified from Cobb, 1917; D and H modified from Ferris, Ferris, and Tjepkema, 1973.)

Stoma slender, barrel shaped or tubular; anterior edge of stoma without riblike structures; without tooth in stoma; large oval amphids (Fig. 11X) .... **Goffartia**

59. Amphid large and circular (Figs. 5C, D); uncommon **60**
 Amphid small, slitlike or porelike **61**

60. Stoma elongate, cylindrical, and strongly cuticularized (Fig. 11Q); two ovaries, one anterior and one posterior to vulva .... **Odontolaimus**
 Stoma short, funnel shaped, and moderately cuticularized (Fig. 11E); a single ovary anterior to vulva.
 Monhystrella

61. Cephalic setae present (Fig. 11R); esophagus uniformly cylindrical anterior to basal bulb (Fig. 11S); body slender and tapering very gradually anteriorly and posteriorly .... **Chronogaster**
 Cephalic setae absent; esophagus narrows between anterior portion and basal bulb (Figs. 11T, U, V); body spindle shaped, considerably thicker in middle **62**

62. Stoma elongate, open, cylindrical (Fig. 11T); one anterior and one posterior ovary, or single anterior ovary; cuticle weakly annulated; a terrestrial family (Rhabditidae) having several genera occasionally found in semiaquatic and aquatic habitats; most common genus ... **Rhabditis**
 Stoma with small, widely separated plates at anterior end followed by a slender moderately cuticularized tube (Figs. 11U, V); single ovary extending anterior to vulva, then reflexed and extending posterior to vulva; cuticle distinctly annulated **63**

63. With three cuticularized plates extending beyond lip region (Fig. 11U); esophagus expanded in middle (Fig. 11V); posterior end blunt; usually in decaying vegetation . . . . **Acrobeloides**
    Without cuticularized plates extending beyond lip region; esophagus not expanded in middle; posterior end blunt or pointed **64**
64. Posterior end pointed (Fig. 11W) . . . . **Eucephalobus**
    Posterior end blunt (Fig. 12D) . . . . **Cephalobus**

REFERENCES

Bastian, H. C. 1865. Monograph on the Anguillulidae, or free nematodes, marine, land, and fresh water; with descriptions of 100 new species. *Trans. Linn. Soc. London* **25**:74–184.

Baylis, H. A., and R. Daubney. 1926. *A synopsis of the families and genera of Nematoda.* 277 pp. British Museum of Natural History, London.

Bird, A. F. 1971. *The Structure of Nematodes.* 318 pp. Academic Press, New York.

Chatterji, R. C. 1935. Permanent mounts of nematodes. *Zool. Anz.* **109**:270.

Chitwood, B. G. 1931. A comparative histological study of certain nematodes. *Z. Morph. Oekol. Tiere* **23**:237–284.

———. 1937–1940. *An Introduction to Nematology.* 240 pp.

Chitwood, B. G., and M. B. Chitwood. 1930. A technique for the embedding of nematodes. *Trans. Am. Microsc. Soc.* **49**:186–187.

———. 1974. *Introduction to Nematology.* 315 p. University Park Press, Baltimore, Maryland.

Cobb, M. V. 1915. Some fresh-water nematodes of the Douglas Lake region of Michigan, USA. *Trans. Am. Microsc. Soc.* **34**:21–47.

Cobb, N. A. 1913. New nematode genera found inhabiting fresh water and non-brackish soils. *J. Wash. Acad. Sci.* **3**:432–444.

———. 1914. The North American free-living fresh-water nematodes. *Trans. Am. Microsc. Soc.* **33**:69–134.

———. 1914–1935. *Contributions to the science of nematology.* 490 pp. Baltimore, Md. (Issued in 26 parts, bound in one volume; some parts are reprints of previously published articles, others are original.)

———. 1917. The mononchs. A genus of freeliving predatory nematodes. *Soil Sci.* **3**:431–486.

———. 1918. Estimating the nema population of soil. *USDA Bur. Plant Ind. Agric. Tech. Circ.* **1**:1–48.

———. 1918a. Filter-bed nemas: nematodes of the slow sand filter beds of American cities (including new genera and species) with notes on hermaphroditism and parthenogenesis. *Contrib. Sci. Nematol.* **7**:189–212.

———. 1935. A key to the genera of free-living nemas. *Proc. Helminth. Soc. Wash.* **2**:1–40.

Ferris, V. R., J. M. Ferris, and J. P. Tjepkema. 1973. Genera of freshwater nematodes (Nematoda) of eastern North America. *Biota Freshwater Ecosystems Identification Manual* **10**:1–38.

Filipjev, I. N. 1934. The classification of the free-living nematodes and their relation to the parastic nematodes. *Smithson. Misc. Coll.* **89**:1–63

Goodey, T., and J. B. Goodey. 1963. *Soil and freshwater nematodes.* 544 pp. Wiley, New York.

Grundmann, A. W. 1955. Improved methods for preparing and mounting nematodes for study. *Turtox News* **33**:152–153.

Hoeppli, R. J. C. 1926. Studies of free-living nematodes from the thermal waters of Yellowstone Park. *Trans. Am. Microsc. Soc.* **45**:234–255.

Linford, N. B. 1937. The feeding of some hollow-stylet nematodes. *Proc. Helminth. Soc. Wash.* **4**:41–46.

Maggenti, A. 1981. *General Nematology.* 372 pp. Springer-Verlag, New York.

Meyl, A. 1960. Freilebenden Nematoden. *Tierwelt Mitteleuro.* **1**:(5a):1–164.

Micoletzky, H. 1925. Die freilebenden Süsswasser- und Moornematoden Dänemarks. *Det. Kgl. Danske Vidensk. Selsk. Skr. Naturv. Math. Afd.* **10**:57–310.

Prejs, K. 1970. Some problems of the ecology of benthic nematodes (Nematoda) of Mikolajskie Lake. *Ekol. Polska* **18**:225–242.

Schneider, W. 1937. Freilebende Nematoden der Deutschen Limnologischen Sundaexpedition nach Sumatra, Java, und Bali. *Arch. Hydrobiol.* 15 (Suppl):30–108.

Stefanski, W. 1938. Les nematodes libres des lacs Tatra Polanaises, leur distribution et systematique. *Arch. Hydrobiol.* **33**:585–687.

Tarjan, A. C., R. P. Esser, and S. L. Chang. 1977. An illustrated key to nematodes found in fresh water. *J. Water Poll. Control Fed.* **49**:2318–2337.

Thorne, G. 1930. Predaceous nemas of the genus Nygolaimus and a new genus Sectonema. *J. Agric. Res.* **41**:445–466.

———. 1939. A monograph of the nematodes of the superfamily Dorylaimoidea. *Capita Zool.* **8**:1–261.

10

NEMATOMORPHA
(Horsehair Worms, Gordian Worms)

Because of their fancied resemblance to animated horsehairs, members of the Phylum Nematomorpha have long been known as "horsehair worms," and because of their habit of becoming inextricably entangled in masses of two to many individuals they are also commonly called "gordian worms," after the Gordian knot episode of antiquity. Sometimes a single worm may be found tightly coiled around a small twig.

The adults move clumsily by slow undulations or writhe about in puddles and in the shallows of marshes, ponds, lakes, and streams. Sometimes they occur on the shore near the water's edge, especially if the substrate is wet. The immature stages are internal parasites of insects, usually crickets, terrestrial beetles, and grasshoppers, but occasionally aquatic insects.

General characteristics. Adult gordian worms range from about 10 to 70 cm in length with a diameter of 0.3 to 2.5 mm. The diameter is the same throughout the body length except for slight tapering at the extremities, especially anteriorly. Coloration includes opaque yellowish, gray, tan, dark brown, or blackish. There is considerable variation in size and coloration within single species. Males are consistently smaller than females and often exhibit slight coiling of the posterior end. The two sexes may be found in about equal numbers or decidedly unequal numbers.

The small mouth is terminal or subterminal, and sometimes absent. The cloacal aperture is terminal, subterminal, or ventral. Sexes of several genera can be distinguished by the structure of the posterior end. Females of *Paragordius* have three lobes. Males of *Gordius, Parachordodes, Paragordius,* and *Gordionus* have two lobes. In females of *Gordius, Gordionus,* and *Parachordodes* the posterior end is simple and rounded, and this is also the situation in both sexes for *Pantachordodes, Chordodes,* and *Pseudochordodes.* In some species the posterior end of the female is slightly bulbous.

These worms were not separated from the nematodes until 1886, and they are often confused with terrestrial mermithid nematodes. Externally, however, Mermithidae are easily differentiated by their pointed ends, tapered body, smooth cuticle, and location of the anus.

Cuticle. The gordian cuticle is unusually complex, lamellated, and fibrous (Figs. 3, 6). Areoles are irregular, rounded, or polygonal thickened surface areas of the cuticle that are present in most species. Sometimes they are poorly differentiated or flat and project only slightly from the general surface of the cuticle; in other species they may be papillate to a varying degree. Often the areoles bear small pore canals, granules, setae, or filamentous processes. The furrows or flat areas between areoles likewise may contain setae, granules, papillae, or filamentous processes. In the cloacal region of the male there are often special tracts of setae, thorns, or adhesive wartlike structures that presumably aid in copulation.

Internal anatomy, physiology. In addition to cuticle, the body wall consists of a flat to columnar hypodermis and an innermost layer of longitudinal muscle fibers. There are no circular fibers, and consequently the diameter of the body cannot be varied. Locomotion consists of localized bending and writhing movements produced by contractions of the longitudinal muscle fibers in one or two quadrants at a time. Males are able to "swim" in an undulatory manner, but such locomotion is not to be compared with the rapid wriggling of small nematodes.

Particulate food is never ingested at any time during the life history, the digestive tract being degenerate and completely functionless. A pharynx is usually differentiated but has no cavity, and the midgut is a minute narrow tubule. The posterior end of the digestive tract is a cloaca into which the two genital ducts open.

The gonads consist of a pair of long, cylindrical bodies that fill most of the pseudocoel in mature animals. Ripe eggs and sperm pass posteriorly through short paired oviducts and sperm ducts, respectively, into the cloaca, and thence to the outside. In the female there is a median ventral seminal receptacle extending forward from the anterior end of the cloaca.

A cerebral mass, or brain, encircles the digestive tract at the anterior end, and a longitudinal ventral nerve cord extends throughout the body.

One or two ocelli have been described at the anterior end of some species, but otherwise sensory receptors are limited to tactile areas near the brain and perhaps also to some of the cuticular outgrowths on the general body surface.

There are no special circulatory, respiratory, or excretory structures. The residual pseudocoel space between body wall and visceral organs is filled with loose mesenchyme.

Reproduction, life history. Probably depending on the species and ecological conditions, copulation usually occurs in spring, early summer, or autumn. Although it is not uncommon to find up to 10 or 20 worms aggregated in a single writhing mass in the springtime, only one female and one male are involved in a copulation. A male coils the posterior tip of his body around the posterior end of the female and deposits sperm at her cloacal opening. The sperm then migrate into the female cloaca, and may be stored for a short time in the seminal receptacle. Fertili-

FIG. 1.–Three female gordians, ×1.

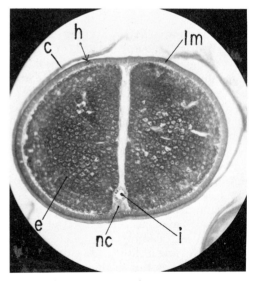

FIG. 2.–Cross section of a typical gordian, ×60. c, cuticle (partly torn away from underlying hypodermis); e, developing eggs; h, hypodermis; i, intestine; lm, longitudinal muscle; nc, nerve cord.

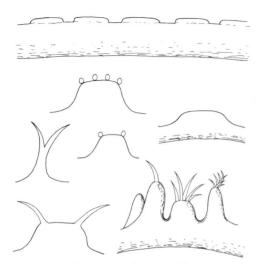

FIG. 3.–Various types of areoles, diagrammatic. (Partly modified from Müller, 1927.)

zation is internal, and presumably the eggs are fertilized just before being extruded from the cloacal opening of the female. Eggs are deposited in long gelatinous strings that swell and may exceed the size of the parent. Such masses contain up to several million eggs. Males die after copulation and females after egg deposition.

Depending on water temperature, the incubation period is usually 15 to 80 days. The resulting larva (Fig. 4) is unlike the adult and has a very brief free-swimming existence. It is cylindrical, often annulated, only about 250 μm long, and divided into a trunk and large muscular presoma. The anterior end of the presoma has an eversible proboscis bearing three long stylets and three circlets of hooks. The anterior portion of the trunk contains a large glandular mass that opens at the tip of the presoma via a long duct. Posteriorly the trunk contains a blind nonfunctional intestine which opens to the outside by means of a pore.

The immediate and ultimate fates of this larval stage have been the subject of a long and confusing controversy, and even the most recent investigations do not seem to have cleared up the problem. Within 24 hours after hatching, the larva is usually thought to encyst on vegetation or any other convenient substrate at or near the water's edge. The colder the water, the more readily encystment occurs. Such cysts may remain

viable for 2 months in water and up to 1 month in a moist habitat such as grass and other vegetation along the waterside. The latter condition may result when the water level of a stream or pond is falling.

Assuming favorable circumstances, a wide variety of crickets, grasshoppers, and terrestrial and aquatic beetles and other aquatic insects accidentally ingest these cysts along with the vegetation to which they are attached. Within the digestive tract the cyst wall disintegrates, and the released larva bores its way through the wall of the intestine and into the hemocoel of the host.

Large numbers of cysts are ingested by abnormal hosts (snails, oligochaetes, aquatic insects, fish, etc.), but in these hosts the larva dies or reencysts in the tissues instead of developing further. If such an abnormal host is eaten by an omnivorous insect that is a normal host, the contained larva then burrows through the wall of the digestive tract to the hemocoel of the latter and undergoes further development.

There is also the possibility (but one that is losing favor) that motile larvae actively penetrate through the body wall of almost any small aquatic or waterside animal but develop further only within such insects as crickets, grasshoppers, and beetles.

Within the host the worm digests and absorbs the surrounding tissues as a source of nourishment. Metamorphosis and development to the adult stage are completed in several weeks to several months, by which time the worm is a tightly coiled mass in the hemocoel. There may be one to several worms per host. Sometimes gordians overwinter in the host body.

If the host falls into water or becomes wetted at a time when the contained worm is mature or nearly mature, the latter breaks through the body wall and is henceforth free living. There is some evidence that hosts tend to seek water when their worms are ready to emerge. Newly released worms that do not have access to water soon die. If emergence from the host occurs in spring or summer, copulation usually soon follows. Sometimes emergence occurs in the autumn and then the worms remain solitary, customarily hibernating in waterside grass and roots, and become active and copulate in the water in the following spring. Length of the complete

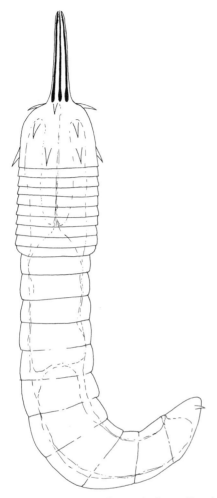

FIG. 4.–Diagram of a typical gordian larva, ×500, with protruded proboscis.

the tropics to cold latitudes, and to altitudes above timberline in mountainous areas. Some species are widely distributed throughout North and South America, and a few occur in both Europe and the United States or are Holarctic. The genus *Gordius* is essentially cosmopolitan. The most common and widely distributed species in the United States are *Gordius robustus* Leidy and *Paragordius varius* (Leidy).

There seem to be few habitat preferences, and the same species may be found in both running and standing waters. Occasionally they are on the shore just above the water's edge but more usually in water 2 to 20 cm deep. Small puddles and stock tanks are fruitful collecting areas. A few observations indicate that the worms are more active at night, but there is no evidence for a true negative phototaxis. Males are much more active than females and move about in a serpentine manner for hours in the water and on the bottom. Females are sluggish and move about but little.

Preparing. Seventy to 90% alcohol is a suitable fixative and preservative, but for more critical work hot saturated mercuric chloride containing 5 to 10% acetic acid is much better, especially if histological sections are contemplated.

For a study of the extremities, cut anterior to the cloacal aperture and posterior to the dark ring near the anterior end; mount in lactophenol, clarite, or balsam after dehydrating. Small bits of cuticle should be sliced from the middle third of the body with a sharp razor blade by stretching the worm firmly over the left forefinger. Such pieces should be placed in a drop of glycerin for about an hour. Then gently scrape off the soft tissues from the underside with needles. Rinse the cleaned cuticle in a change of glycerin and mount, outer surface up, in glycerin jelly on a slide. Lactophenol and clarite mounts are often preferable for certain cuticular structures.

Taxonomy. The Phylum Nematomorpha is divided into the Nectonematoidea and Gordioidea. The former category contains a single marine species whose larvae are parasitic

life cycle is variable, ranging from about 2 to 15 months.

Unlike most parasites, gordians generally have little host specificity, and the same species may develop in a variety of grasshoppers or beetles. A few gordians have even been reported emerging from caddis flies and dragonflies.

Adults have occasionally been recorded as human parasites, but such cases are undoubtedly fortuitous and produced by accidental ingestion of worms in food or water.

Ecology. Little is known about gordian ecology. The worms occur in suitable habitats wherever appropriate hosts are found, from

in crustaceans. All the other species are fresh water, totaling about 110 for the world, and belong in the Gordioidea. This category consists of 5 families and about 16 genera. About 12 species in the following taxa occur in the United States: Chordodidae (*Chordodes, Pantachordodes, Pseudochordodes, Euchordodes*), Parachordodidae (*Parachordodes, Gordionus, Paragordius*), and Gordiidae (*Gordius*).

Some generic characters and most species characters are based largely on the surface pattern and sculpturing of the cuticle. For studying some of these details it is often necessary to use oil immersion.

Only a negligible amount of work on American gordians has been published since the papers of Montgomery and May, and the taxonomy and distribution of our species are very poorly known. Many of the descriptions published prior to 1910 were quite brief and did not include sufficient fine details, and for this reason some such descriptions are not trustworthy. Until the gordian fauna of the United States is more thoroughly investigated, therefore, it seems fruitless to draw up a key to species that would be unreliable and soon out of date. For this reason the following key includes only generic categories. A few common European genera have been included because of the possibility of their being found in the United States eventually. Most of the delimitations and suggestions of Carvalho (1942), and Heinze (1941), have been followed. The most recent definitive North American contribution is that of Chandler (1985).

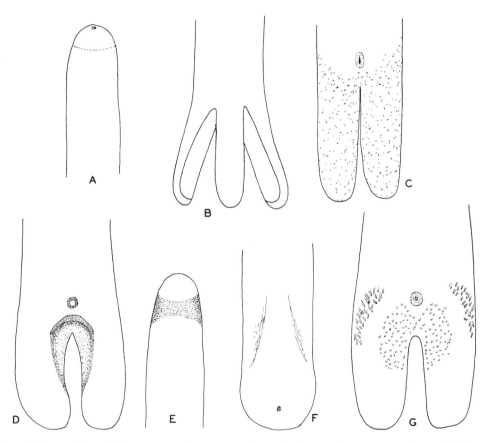

Fig. 5.–Extremities of Nematomorpha. A, anterior end of *Paragordius;* B, posterior end of female *Paragordius;* C, posterior end of male *Paragordius;* D, posterior end of male *Gordius;* E, anterior end of *Gordius;* F, posterior end of female *Gordius;* G, posterior end of male *Gordionus.*

KEY TO GENERA OF NEMATOMORPHA

1. Posterior end with three lobes (Fig. 5B); cuticle with scattered areoles of one kind (Fig. 6A); widely distributed and common; several species; females .... **Paragordius**
   Posterior end blunt or bilobed (Figs. 5C, F, G) 2
2. Posterior end clearly bilobed and often with a tendency toward coiling (Figs. 5C, D, G); males 3
   Posterior end blunt and without lobes; males and females 7
3. Postcloacal region with a crescentic fold; areoles present and indistinct (Fig. 6B); or absent; widely distributed; several species .... **Gordius**
   Postcloacal region without a crescentic fold 4

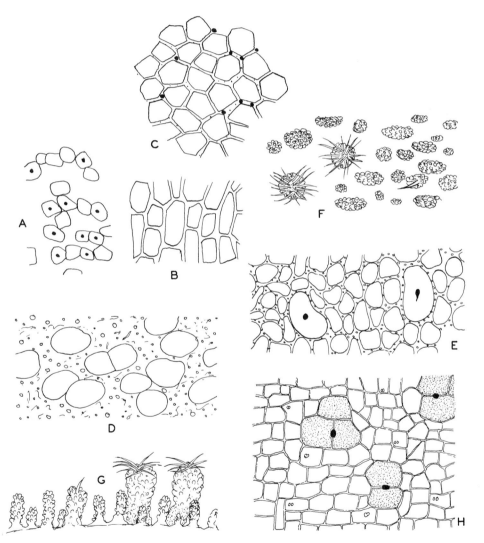

Fig. 6.–Areoles and other cuticular structures of Nematomorpha, semidiagrammatic. (Pore canals shown in solid black.) A, *Paragordius;* B, *Gordius;* C, *Gordionus;* D, *Pantachordodes;* E, *Parachordodes;* F, surface view of *Chordodes* areoles; G, side view of *Chordodes* areoles; H, *Pseudochordodes.* (E modified from Heinze, 1935; F and G modified from Camerano, 1897; H modified from Carvalho, 1942.)

4. With two kinds of areoles (Fig. 6E); poorly known in the United States **5**
   With one kind of areole (Figs. 6A, C) **6**
5. Anterior end rounded .... **Paragordius**
   Anterior end truncate .... **Parachordodes**
6. Posterior lobes about two and one half to three times as long as wide; without rows of hairs or papillae
   lateral to cloacal aperture (Fig. 5C); widely distributed and common; several species.
   Paragordius
   Posterior lobes about twice as long as wide; with rows of long papillae on either side of cloacal aperture
   (Fig. 5G); several species; should perhaps be incorporated in *Parachordodes* ..... **Gordionus**
7. Posterior end with a ventral furrow; males **8**
   Posterior end without a ventral furrow **9**
8. With two kinds of areoles; not known from the United States .... **Paragordionus**
   With one kind of areoles **10**
9. Areoles grouped; interareolar furrows with spines; not known from the United States.
   Euchordodes
   Areoles not grouped; interareolar furrows without spines; poorly known in the United States.
   Pantachordodes
10. Interareolar furrows clear and without setae, pore canals, or large granulations; areoles often absent or
    indistinct; only one kind of areole (Fig. 6B); common and widely distributed; several species;
    females .... **Gordius**
    Interareolar furrows usually with setae, pore canals, granulations, or other structures; one, two, or
    more kinds of areoles present **11**
11. With one kind of areole **12**
    With two or more kinds of areoles; pore canals in areoles or between adjacent areoles (Figs. 6E–H),
    sometimes absent **14**
12. With a few pore canals in furrows between areoles (Fig. 6C); several species; females; should perhaps
    be incorporated in *Parachordodes* .... **Gordionus**
    Without pore canals in furrows between areoles (Fig. 6D) **13**
13. Areoles not grouped; interareolar spines absent; males and females; poorly known in the United
    States .... **Pantachordodes**
    Areoles grouped; interareolar spines present; females; not known in the United States.
    Euchordodes
14. Pore canals in center of large areoles (Fig. 6E); body of mature individuals usually less than 1 mm
    thick; poorly known in the United States; females .... **Parachordodes**
    Pore canals absent, or between large areoles (Figs. 6F, H); body of mature individuals usually more
    than 1 mm thick; males and females **15**
15. Cuticle with only one type of ovoid–polygonal areole .... **Neochordodes**
    Cuticle with two or more types of areoles **16**
16. Cuticle with two or more types of prominent papillate areoles, the highest ones often bearing a tuft of
    slender filaments; interareolar furrows with scattered tubercles, bristles, and granules (Figs. 6F,
    G); several species .... **Chordodes**
    Cuticle with two types of nonpapillate areoles; interareolar furrows free of tubercles, bristles, and
    granules (Fig. 6H); uncommon in the United States .... **Pseudochordodes**

REFERENCES

Camerano, L. 1897. Monografia dei Gordii. *Mem. R. Acad. Sci. Torino* (2) **47**:339–419.

———. 1915. Revisione dei Gordii. *Ibid.* **66**:1–66.

Carvalho, J. C. M. 1942. Studies on some Gordiacea of North and South America. *J. Parasitol.* **28**:213–222.

Chandler, C. M. 1985. Horsehair worms (Nematomorpha, Gordioidea) from Tennessee, with a review of the taxonomy and distribution in the United States. *J. Tenn. Acad. Sci.* **60**:59–62.

Dorier, A. 1930. Recherches biologiques et systéma- tiques sur les Gordiacés. *Trav. Lab. Hydrobiol. Piscicult. Grenoble* **22**:1–180.

Heinze, K. 1935. Über das genus Parachordodes Camerano 1897 nebst allgemeinen Angaben über die Familie Chordodidae. *Z. Parasitenk.* **7**:657–678.

———. 1935a. Über Gordiiden. *Zool. Anz.* **111**:23–32.

———. 1937. Die Saitenwürmer (Gordioidea) Deutsch- lands. Eine systematisch-faunistische Studie über Insektenparasiten aus der Gruppe der Nematomor- pha. *Z. Parasitenk.* **9**:263–344.

———. 1941. Saitenwürmer oder Gordioidea. *Tierwelt Dtschl.* **39**:1–78.

———. 1952. Über Gordioidea, eine systematische Studie über Insektenparasiten aus der Gruppe der Nematomorpha. *Z. Parasitenk.* **15**:183–202.

Müller, G. W. 1927. Über Gordiaceen. *Z. Morphol. Oeokol. Tiere* **7**:134–219.

Montgomery, T. H., Jr. 1898. The Gordiacea of certain American collections with particular reference to the North American fauna. *Bull. Mus. Comp. Zool. Harvard Univ.* **32**:23–59.

———. 1898a. The Gordiacea of certain American collections, with particular reference to the North American fauna. II. *Proc. Calif. Acad. Sci.* (3) **1**:333–344.

Poinar, G. O., and J. J. Doelman. 1974. A reexamination of Neochordodes occidentalis (Mont.) comb. n. (Chordodidae: Gordioidea); larval penetration and defense reaction in Culex pipiens L. *J. Parasit.* **60**:327–335.

Redlich, A. 1980. Description of Gordius attoni sp. n. (Nematomorpha, Gordiidae) from northern Canada. *Can. J. Zool.* **58**:382–385.

Schuurmans Stekhoven, J. H., Jr. 1943. Contribution a l'étude des gordiides de la fauna Belge. *Bull. Mus. R. Hist. Nat. Belg.* **19**:1–28.

11

TARDIGRADA
(Water Bears)

Although tardigrades are strictly aquatic animals, they are seldom abundant in the usual types of habitats harboring fresh-water invertebrates. They are not plankton organisms; they are only occasionally collected in aquatic mosses and algae, on rooted aquatics, or in the mud and debris of puddles, ponds, and lakes. More typically, active tardigrades are found in the droplets and film of water on terrestrial wet mosses, liverworts, and certain angiosperms with a rosette growth form. They also occur in the capillary water between the sand grains of sandy beaches up to 2 or 3 m from the water's edge.

Mature individuals range from 50 to 1200 μm in length but are usually less than 500 μm long. Many species have a considerable size variation. The short, stout, cylindrical body, the four pairs of stumpy lateroventral legs, and a deliberate "pawing" sort of locomotion produce a fancied resemblance to a miniature bearlike creature, and the Tardigrada have been generally called "water bears" since 1869.

The great majority of species have been reported from lichens, liverworts, and terrestrial and fresh-water mosses. All other species have been reported from marine habitats, mostly interstitial.

Nearly all of what we know about tardigrade biology has been contributed through the efforts of British, German, Italian, and French investigators. Water bears have been badly neglected in the United States, even though we undoubtedly have a rich fauna. Within the past 20 years, however, there has been increased interest in tardigrades by American investigators. About 90 species have been reported from this country, and this figure includes marine, moss, and fresh-water forms. About 500 species have been described the world over.

General characteristics. Tardigrades are fascinating objects for electron and low-power microscopy investigations. The body consists of a distinct head and four indefinite "body segments." Each leg is armed with four claws or two pairs of double claws. Crawling and creeping are more or less sluggish, the claws being used for clinging to vegetation and debris. The mouth is anterior, subterminal, or ventral, and the cloacal aperture, or anus, lies between the bases of the last legs. The entire surface of the body is covered with a complex, multilayered cuticle secreted by the thin hypodermis, which consists of polygonal cells. Depending on the species, the cuticle may be variously thickened and marked; usually there are also various spines, papillae, setae, or hairs. Cilia do not occur in this phylum. Only liquid materials are ingested, and the anterior part of the digestive tract has a sucking pharynx and a pair of long, piercing stylets.

Sexes are separate, but the great majority of individuals are females. Usually there are four to six ecdyses in the life history.

A remarkable feature of most aquatic and semiaquatic fresh-water tardigrades is their ability to tide over dry periods in a shriveled, rounded, anabiotic state. When normal moisture conditions return, the animal resumes its normal appearance and activity.

Coloration is variously produced by the pigmentation of the cuticle and hypodermis, dissolved materials in the body fluid, reserve food bodies, and the contents of the digestive tract. Gray, bluish, yellow brown, reddish, or brown are common colors. Some species are relatively dark and opaque, especially among older individuals; others are translucent and pale.

There are no special circulatory or respiratory systems. Materials diffuse and circulate about easily in the body fluid, and respiration occurs through the general body surface.

Feeding, digestive system. Tardigrades are mostly plant feeders. The cellulose wall of moss leaflet cells and algal filaments are pierced with the stylets and the fluid contents are sucked out by pharyngeal pumping action. Occasionally, however, the body fluids of small metazoans, such as nematodes and rotifers, are used as food. *Milnesium tardigradum* is said to be chiefly carnivorous.

The periphery of the mouth is stiffened and surrounded by two or three folded and superimposed rings of cuticle. That portion of the digestive system extending from the small mouth cavity to the anterior end of the esophagus is called the buccal apparatus (Fig. 5). It consists of the tubular pharynx, the muscular or sucking pharynx, and the stylet mechanism. The tubular pharynx is narrow and sclerotized and extends into the mass of the sucking pharynx. The latter is football shaped, with thick walls composed of radial muscle fibers; its small triquetrous cavity is lined with a thin epithelium. Six longitudinal ridges or fragmented sclerotized bars give stiffness to the sucking pharynx and provide for insertions of muscle fibers. The individual sclerotized pieces in the pharynx are called macroplacoids (Figs. 5, 6). Sometimes there is a single small posterior piece, called a microplacoid or comma.

The two stylets lie at an angle on either side of the tubular pharynx and the sharp tips project into the anterior end of the pharyngeal cavity through special lateral slots, or stylet sheaths. A transverse support extends from the tubular pharynx to the basal portion of each stylet. Several pairs of small muscles extending from the base of the stylets to the tubular pharynx and the sucking pharynx are capable of withdrawing and protruding the

FIG. 1.–Scanning electron micrograph of *Parechiniscus,* ×600, a semiaquatic species. (Courtesy R. O. Schuster, University of California, Davis.)

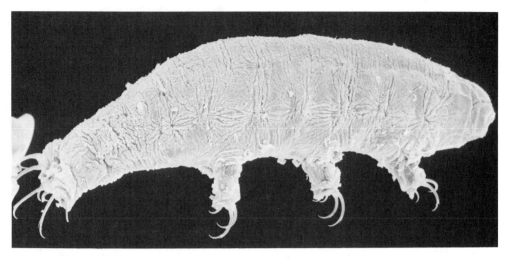

FIG. 2.–Scanning electron micrograph of *Macrobiotus,* ×150. (Courtesy of R. O. Schuster, University of California, Davis.)

stylets a considerable distance out of the mouth.

A pair of large salivary glands open into the mouth cavity. They also function in the secretion of new stylets and stylet supports at molting and for this reason are sometimes called stylet glands. Some investigators believe that they have an excretory function, especially in the Heterotardigrada. A few small unicellular glands of unknown function occur near the mouth. They are ingrowths of the hypodermis.

The esophagus ranges from long to very short. It empties into the capacious stomach, or midgut, which is usually without diverticula but may have slight lateral caeca. The hindgut, or rectum, is short and has a small cavity. In the Class Eutardigrada it is a true cloaca. The anus or cloacal aperture is a longitudinal or transverse slit.

Digestion in the anterior portion of the stomach is acid; posteriorly it is alkaline. Reserve food may be stored in hypodermal cells or in special reserve food bodies which float about in the body cavity. Feces are often released into the old cuticle during molting, especially in *Echiniscus*.

Excretion. In the Eutardigrada there are three small glands situated at the junction of midgut and hindgut, each of which consists of only three cells. One gland is dorsal and two are lateral. All supposedly have an excretory function, and perhaps they are also osmoregulatory. The paired glands are commonly called Malpighian tubules, but perhaps are more appropriately nephridia. Some excretory material is stored in the hypodermal cells, some excretory granules are left within the old exoskeleton at ecdysis, and some are given off at the anterior end when the stylet mechanism is shed just preceding an ecdysis.

Muscle system. In addition to the pharynx and the stylet muscles, a tardigrade usually has from about 40 to 140 long, thin body muscles, the number depending on the species. Each such muscle is merely a single fibrillar cell with one nucleus, or several such cells. Most origins and insertions are on the body wall (Fig. 7). Contractions of the dorsal and ventral longitudinal muscles bring about a slight shortening or curvature of the body. Each leg is moved by a set of muscles originating on the dorsal and ventral body wall and inserting near the tip of the leg. The body wall is devoid of circular muscle fibers, but the turgor of the body fluid presumably acts as an antagonist to the longitudinal and leg muscles. Muscles extending from the body wall to the

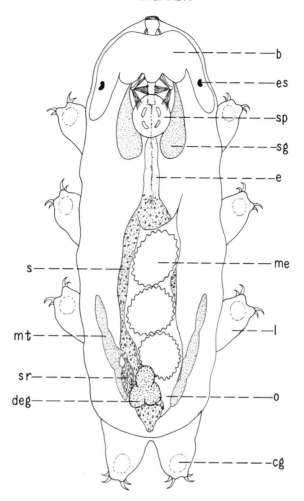

FIG. 3.–Semidiagrammatic dorsal view of *Macrobiotus*, ×220; trunk and leg muscle not shown. b, brain; cg, claw gland; deg, dorsal excretory gland; e, esophagus; es, eyespot; l, leg; me, mature egg in saclike ovary; mt, Malpighian tubule; o, oviduct; s, stomach; sr, seminal receptacle; sg, salivary gland; sp, sucking pharynx. (Greatly modified from Marcus, 1928b.)

cloaca facilitate egg deposition and defecation. A few species have longitudinal fibers associated with the stomach.

Nervous system. The large dorsal brain commonly has two long lateral lobes and two or three median lobes; it covers much of the tubular pharynx (Fig. 8). Two broad circumpharyngeal connectives pass around the pharynx and unite with a subpharyngeal ganglion. In the trunk there is a chain of four conspicuous ventral ganglia united by two longitudinal nerve strands. Frequently these ganglia are slightly bilobed. Paired nerve strands originating in the brain and ganglia innervate all parts of the body. Some of the longer strands, especially those extending into the legs, have small terminal ganglionic masses.

Sense organs. The great majority of tardigrades have a pair of eyespots in the lateral lobes of the brain. Each consists of a cup-shaped mass of black or red pigment granules.

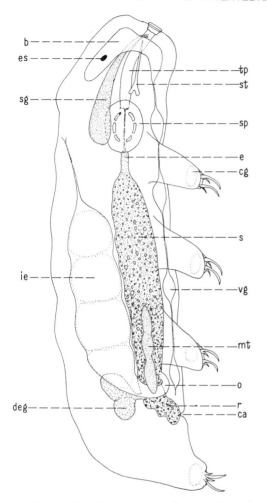

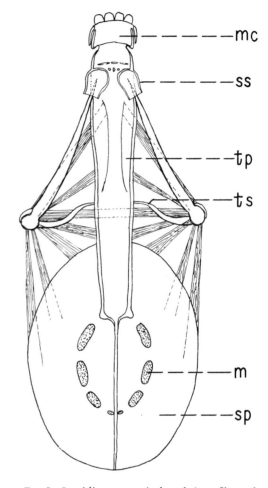

FIG. 4.–Semidiagrammatic lateral view of a typical tardigrade, ×220; all musculature omitted. b, brain; ca, cloacal aperture; cg, claw gland; deg, dorsal excretory gland; e, esophagus; es, eyespot; ie, immature egg in saclike ovary; mt, Malpighian tubule; o, oviduct; r, rectum; s, stomach; sg, salivary gland; sp, sucking pharynx; st, stylet; tp, tubular pharynx; vg, ventral ganglion.

FIG. 5.–Semidiagrammatic dorsal view of buccal apparatus and associated musculature; stylets are shown retracted. m, macroplacoids; mc, mouth cavity; sp, sucking pharynx; ss, stylet sheath; tp, tubular pharynx; ts, transverse support.

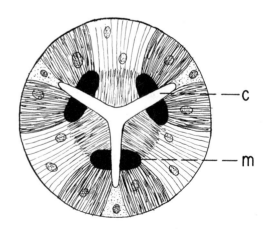

FIG. 6.–Diagrammatic cross section of a typical tardigrade sucking pharynx showing radial muscle fibers. c, triquetrous cavity; m, macroplacoid.

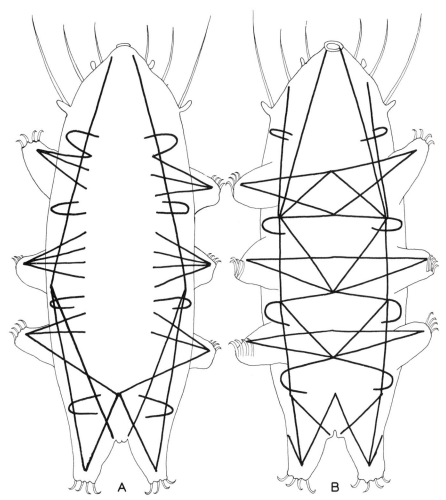

FIG. 7.–Trunk and leg musculature of *Echiniscus,* ×350, semidiagrammatic. A, dorsal view; B, ventral view. (Greatly modified from Marcus, 1928b.)

Head cirri are usually tactile, but there is some question about the presumed sensory function of body cirri and filaments.

Cuticular structures. Fresh-water species usually have a thicker cuticle than marine species, and semiaquatic species have a thicker cuticle than aquatic species. The surface may be smooth or variously sculptured, granular, or papillate. Often there are long cirri. Some species of Eutardigrada have transverse grooves or folds, but *Echiniscus* and *Pseudechiniscus* have dorsal armorlike thickened plates. Beginning at the anterior end, these are as follows (Fig. 12F): head plate, shoulder plate (first trunk plate), a small median first intercalary, second trunk plate (divided longitudinally), second intercalary, third trunk plate (usually divided longitudinally), third intercalary (often indistinct), and the anal plate (usually grooved).

Many marine species are nothing short of spectacular in the development of very long cirri, lateral trunk expansions, dorsoventral flattening, padded claw tips, great caudal processes, and other embellishments.

Reproduction, growth. The bulk of a population is always composed of females, and males reach their peak of abundance

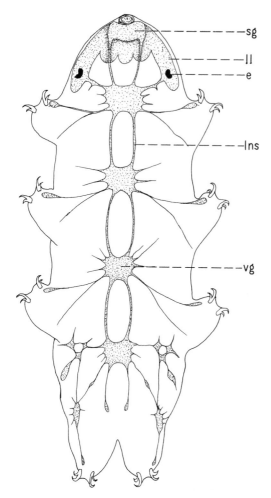

FIG. 8.–Ventral view of nervous system of *Macrobiotus,* ×220. e, eyespot; ll, lateral lobe of brain; lns, longitudinal nerve strand; sg, subpharyngeal ganglion; vg, ventral ganglion. (Modified from various sources.)

during the winter or early spring. Males are unknown in *Parechiniscus* and *Echiniscus,* and these genera probably reproduce exclusively by parthenogenesis.

Some species show sexual dimorphism in the shape and cuticular structure of the last legs, in the size and curvature of the claws, and in the smaller size of the males.

The gonad is an unpaired sac dorsal to the digestive tract and with one or two attachment fibers at the anterior end. In the male there are two vasa deferentia that curve around each side of the intestine and have a small swollen portion that acts as a seminal vesicle. The female has a single oviduct on the right side of the intestine. On the other side a blind seminal receptacle may sometimes be found in *Hypsibius* and *Macrobiotus,* especially in the late fall, winter, and early spring. In the Eutardigrada the genital ducts open into the rectum, which is therefore more appropriately a cloaca. In the Heterotardigrada there is a separate preanal genital pore.

Two distinct types of eggs have been observed in some species, thin shelled and thick shelled, and perhaps they are similar to the tachyblastic and opsiblastic eggs produced by gastrotrichs, or the summer and winter eggs produced by rotifers. At least there is some evidence to show that thick shelled eggs are produced when environmental conditions are unfavorable; there is the further possibility that thin shelled eggs are parthenogenetic and thick shelled eggs are fertilized. The peak of the reproductive period is from November to May, but females with eggs may be found at any time of the year.

In most of the true aquatic species "external" fertilization is the rule. One or more males clamber about on the body of a female before she is ready to release her eggs but after the old cuticle has become loosened in preparation for a molt. The males release sperm through the cloacal aperture or genital pore of the old cuticle, and the eggs are fertilized when they are released into the cavity between the new and old cuticle.

In typical semiaquatic species fertilization is "internal" and does not necessarily occur just before an ecdysis. Sperm enter the genital pore or cloacal aperture and fertilize the eggs before their release from the body of the female. Sometimes the sperm may be stored in the seminal receptacle.

Most female Echiniscoidea produce only 2 to 6 eggs, but in other groups a female produces anywhere from one to 30 eggs. Sometimes the whole complement may be released in 15 to 30 minutes.

In *Macrobiotus* and some species of *Hypsibius* the eggs are deposited freely and singly, or in groups. In other fresh-water forms they are contained in the newly shed cuticle. Free eggs are often sticky for attachment to the substrate and frequently faceted, spinous, tuberculate, or variously sculptured. Eggs released into the old cuticle usually have a smooth surface. Inside the true shell a thin membrane surrounds the developing embryo.

Depending on the species and ecological conditions, a thin shelled egg usually completes development in 3 to 12 days and a thick shelled egg in 10 to 14 days.

The young emerge from the egg by rupturing the shell with their stylets. Newly hatched individuals are one third to one fifth the size of mature specimens, and all growth occurs by enlargement of cells already present. Aside from their smaller size, immature tardigrades may be recognized by their underdeveloped reproductive system, slight differences in the details of the buccal apparatus, and sometimes by the fact that there are only two claws per leg.

Most investigators indicate 4 to 12 molts during the life history, and sexual maturity is usually attained sometime after the second or third molt. The old cuticle is usually shed after a period of growth, but it may also be shed as a response to hunger and unfavorable environmental conditions. Several days before a molt the sclerotized portions of the buccal apparatus are ejected and new ones (attached to the old muscles) promptly begin to be formed by the salivary glands. A day or two before a molt, the animal shrinks temporarily and slightly. Then the old cuticle ruptures at the anterior end and the animal crawls out. The lining of the rectum and the claws are lost at molting, and new claws are secreted by special claw glands in the tip of each leg.

There appears to be considerable variation in longevity, assuming that the animal is continuously active. Two months and 24 months are perhaps the extremes.

General ecology. A few species of *Macrobiotus* and *Hypsibius* are truly aquatic and some are both aquatic and semiaquatic, but the majority of species in these genera are semiaquatic only and occur chiefly on terrestrial

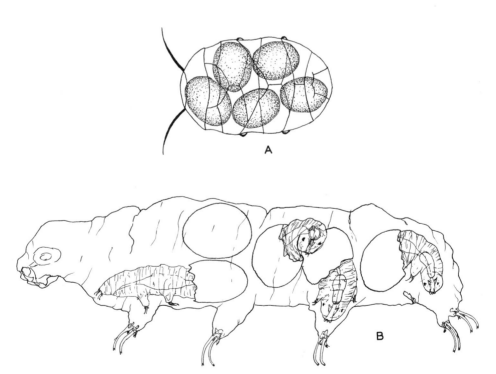

FIG. 9.–Cast tardigrade exoskeletons containing eggs. A, dorsal view of *Pseudechiniscus;* B, lateral view of *Milnesium tardigradum* (four of five eggs in process of hatching). (A modified from Ramazotti and Maucci.).

mosses and lichens. *Milnesium tardigradum* occurs in both semiaquatic and aquatic habitats. All of the species in the other fresh-water genera are semiaquatic and occur in true aquatic habitats only rarely and fortuitously.

Active tardigrades can usually be found if one searches diligently in masses of filamentous algae, washings of higher aquatic plants, or in the debris on the bottom of ponds and pools. The semiaquatic species, however, are active only when the plant has been wetted by rain or when it is splashed at streamside. Many mosses, liverworts, lichens, and rosette angiosperms contain large tardigrade populations.

Mosses on rocks or on the bark of trees contain tardigrades, even though they may be wet enough to allow them to be active for less than 5% of the time. Concentrations up to 22,000 dry and inactive anabiotic tardigrades per gram of dry moss have been reported. Mosses with especially thick cellulose cell walls, such as *Polytrichum,* usually do not harbor tardigrades. Presumably their stylets cannot penetrate such walls. Favorable portions of sandy beaches sometimes contain 300 to 400 tardigrades/10 cm^3 of wet sand. Sometimes they occur as deep as 15 cm in the sand.

Most species appear to be eurythermal and are normally active anywhere between near-freezing temperature and 25 to 30°C. Some forms have been found in warm springs where they are continuously exposed to temperatures of 40°C.

Although aquatic species are generally characteristic of the littoral, there are records of collections as deep as 100 m on lake bottoms. Semiaquatic species occur from the tropics to the Arctic and Antarctic and from sea level to mountain tops up to 6000 m high. The annual period of activity varies correspondingly, from the year round to only a week or two.

Little is known about minimal oxygen requirements, but it is usually inferred that tardigrades cannot endure low dissolved oxygen concentrations as do many other small aquatic metazoans.

Aquatic species are usually most abundant between January and May, and sometimes there is a second maximum in the autumn months. Moss dwellers do not have pronounced seasonal maxima. Individuals are active as long as the moss is damp.

Tardigrades are chiefly preyed on by amoeboid protozoans, nematodes, and perhaps by each other to a certain extent.

Anhydrobiosis. When a bit of moss dries, many of the associated Tardigrada have the unusual ability to assume an inactive anabiotic (cryptobiotic) state that may persist for as long as four to seven years. When normal ecological conditions again prevail, the tardigrades come out of their quiescent state and resume normal activity. Most marine species and true aquatic fresh-water species apparently do not have the ability to assume an anabiotic condition.

Under drying conditions the head, posterior end, and legs are retracted, and the whole body becomes more or less rounded into a "tun." Considerable water is lost from the body cavity and the anabiotic animal is therefore shriveled and wrinkled (Fig. 10). Internal

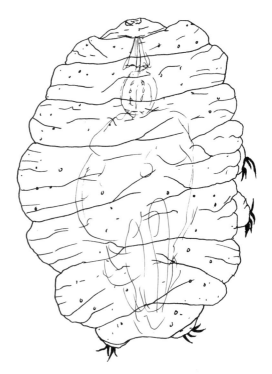

Fig. 10.–Dorsal view of a tardigrade "tun" in the anhydrobiotic state, ×300.

visceral structures are not greatly modified. Although drying is almost always responsible for bringing on anabiosis, a lack of sufficient dissolved oxygen and perhaps other unfavorable conditions may also be responsible. Tuns can be formed only when the drying process occurs slowly, and when the relative humidity is high.

Metabolic processes proceed very slowly, but the length of time a specimen can remain in continuous anabiosis depends on the amount of stored food in the body. Death occurs as soon as the supply is exhausted.

Anabiotic tardigrades are highly resistant to abnormal environmental conditions. Some remain viable even after being kept experimentally in strong brine solutions, at 100°C for 6 hours, and at −190°C for as long as 20 months.

Revival from anabiosis takes but a short time when individuals are properly wetted. Usually the interval is 4 minutes to several hours, and occasionally a day or two. The animal absorbs water, swells, and promptly becomes active. Revival is more rapid at higher temperatures, with higher dissolved oxygen, and in animals having more stored food material. Moss species usually go through numerous alternating periods of activity and anabiosis during their life cycle, depending on the nature of their habitat. One author states that the life cycle may be as long as 60 years, including anabiotic periods. Laboratory animals can easily be dried and revived 10 times or more.

Cysts. Damage, hunger, and other abnormal ecological conditions sometimes bring about the formation of resistant cysts, especially in true aquatic species of *Macrobiotus* and *Hypsibius.* The animal contracts inside the old wrinkled cuticle and forms a dark-colored, thick-walled inner cyst (Fig. 11). The internal organs undergo a variable degree of degeneration. When ecological conditions are favorable the animal becomes reconstituted, the cyst wall ruptures, and the tardigrade emerges.

A "simplex" stage has been observed in many tardigrades, especially in species of *Macrobiotus.* Such animals are inactive and have a more or less reduced buccal apparatus. The macroplacoids are abortive, the esophagus is very slender, and the stylets are nonfunctional and reduced; sometimes all of the digestive system anterior to the stomach disappears and there is no mouth. Little is known about the factors responsible for the appearance of the simplex stage, but it is thought that it may occur in animals about to encyst or in those just emerged from a cyst.

Anoxybiosis. A third type of inactive dition in tardigrades is called anoxybiosis. It usually comes about when there is insufficient oxygen. The body swells and becomes rigid and turgid. There are no movements, and it is difficult to decide whether the animal is alive. After a maximum of about 5 days in this condition the animal dies, but if food and oxygen are supplied promptly the animal again resumes normal activity. Sometimes individuals emerging from anabiosis go into the state of anoxybiosis.

Crowe (1975) recognizes two other kinds of latency among tardigrades, both of which have been only superficially investigated. One is cryobiosis, which is closely associated with tun formation and is a state of suspended animation assumed on freezing and other exceptionally low temperatures. The second

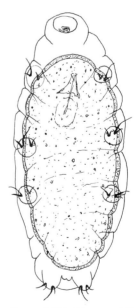

FIG. 11.–Cyst of *Hypsibius,* semidiagrammatic, ×75. (Greatly modified from Marcus, 1928.)

is osmobiosis, a cryptobiotic or tun state induced by exceptionally high osmotic pressures when the animals are placed in abnormal salt concentrations.

Everitt (1981) worked on tardigrade populations of an Antarctic pool that was frozen solid for 8 months each year. The animals (*Hypsibius*) overwintered in a "cryptobiotic state" and quickly became active in an algal mat when the pond thawed. Densities as high as 470 animals per gram of algae were found.

Geographical distribution, dispersal.
The great majority of tardigrades are undoubtedly widely distributed, and an abundance of species first found in Europe have since been found scattered over a broad range. Some are Holarctic, generally distributed in the Northern Hemisphere, or even cosmopolitan (*Macrobiotus hüfelandi* Schultze, *Hypsibius oberhäuseri* (Doyère), and *Milnesium tardigradum*). Other species, of course, have been reported only from single mainland or island localities. The tropics contain notably few species and few individuals.

Passive distribution is effected by currents, animals (usually insects), waves, and wind, especially in the anabiotic and cyst stages. There is considerable positive evidence for wind as an important distributive agency.

Collecting.
Almost any bit of suitable moss will yield tardigrades. If the moss is wet, simply rinse it out in water and examine the washings. Masses of filamentous algae and rooted aquatics may be treated similarly. Dry mosses and lichens should be soaked for 30 minutes to several days. At intervals rinse such plant material, and the tardigrades will be found as they emerge from cysts and the anabiotic state.

Culturing.
Tardigrades are not easy to culture, and little actual culture work seems to have been done. They may be kept in masses of filamentous algae or wet moss for a few days to weeks. Change a large fraction of the water every three or four days.

Preserving, preparing.
Unlike many other small fresh-water metazoans, tardigrades are easy to fix in extended form. A large percentage will be suitable when killed and fixed in 5% formalin or 85% alcohol. Especially fine specimens may be secured by inducing anoxybiosis under a cover slip sealed with petroleum jelly before killing and fixing. This procedure usually requires 10 minutes to 3 hours. Some investigators recommend the use of one of a variety of narcotics before fixation. Fixatives that are particularly useful for staining and histological work are osmic acid, saturated mercuric chloride, and 2% acetic acid. Piercing the body wall with a fine needle permits better staining and dehydration. Simple unstained permanent mounts may be made in glycerin jelly.

The sclerotized parts of the buccal apparatus may be isolated on a slide by the judicious use of 10% potassium hydroxide.

Taxonomy.
The phylogenetic affinities of tardigrades are highly debatable. During early embryology five coelomic pouches are formed from the mesoderm, but only one posterior pair persists as the gonads and their ducts; the others degenerate. The main body cavity is therefore probably a hemocoel.

Often they are placed near the mites in the Phylum Arthropoda, especially because of their indistinct segmentation, four pairs of legs, and piercing stylets. Some workers consider them crustaceans, and a further suggestion places them near the Onychophora. Other workers, however, consider the Phylum Tardigrada one of the wormlike groups of various affinities and place it after the Nematoda and either before or after the Chaetognatha and Bryozoa in the phylogenetic series. For the time being and until more definite evidence is forthcoming, this last suggestion is the one followed in the present manual.

About 50 genera are recognized in the whole phylum of which 30 contain a total of about 400 aquatic and semiaquatic fresh-water species, the others being marine. Fourteen of these genera have been reported from the United States.

The chief taxonomic structures used in generic and specific determinations include details of the buccal apparatus (especially macroplacoids), armour, claws, cirri, filaments, and eggs. *Echiniscus, Macrobiotus,* and

Hypsibius are large genera that are often subdivided into species groups or subgenera for taxonomic convenience. Most of the earlier descriptions are insufficient, owing to the false supposition that the structures of the claws alone were sufficient and trustworthy characters. The Ramazotti and Maucci monograph (1983) is the most complete reference.

The following key includes data for all valid fresh-water genera of the United States. An even greater number of genera is known from other parts of the world.

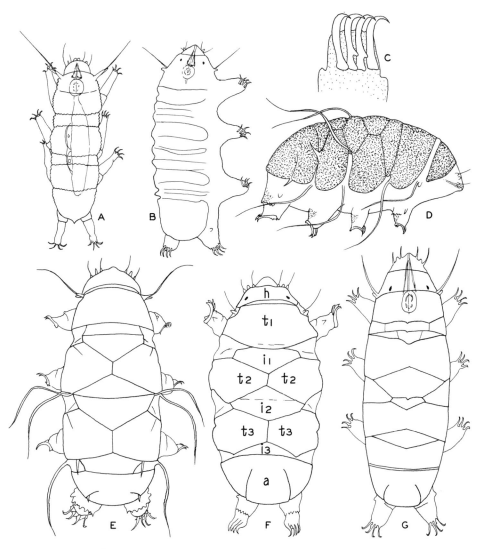

FIG. 12.–Structure of Heterotardigrada. A, dorsal view of *Oreella,* ×280, a rare genus, not yet found in the United States; B, *Parechiniscus chitonides,* ×430; C, claws of *Echiniscus;* D to F, typical species of *Echiniscus,* ×240; G, dorsal view of *Pseudechiniscus,* ×340. Armor sculpturing omitted on D, F, and G. a, anal plate; h, head plate; i1, i2, and i3, intercalaries; t1, first trunk plate; t2, second trunk plate (divided); t3, third trunk plate (divided). (Modified from various sources.)

KEY TO GENERA OF TARDIGRADA

1. Head with anterior cirri (rarely absent) and lateral filaments; with four separate but similar claws on each leg (Fig. 12); semiaquatic .... Class **HETEROTARDIGRADA,**
   Order **ECHINISCOIDEA, ECHINISCIDAE, 2**

   Head without anterior cirri and lateral filaments (Figs. 14 O–Q); each leg with two double claws or two unlike pairs of claws (Figs. 14 C–E, L–N); semiaquatic and aquatic ... Class **EUTARDIGRADA,**
   Order **MACROBIOTOIDEA, 5**

2. Body with thick dorsal plates, variously sculptured; with a variable number of cirri and filaments (Figs. 12 D–G); eggs deposited in shed exoskeleton (Fig. 9) 3

   Body with thin dorsal plates (Fig. 12); up to 200 µm long; one species reported from Utah.
   Parechiniscus

3. Armor consisting of head plate, three trunk plates, and anal plate (plus intercalaries) (Figs. 12 D–F); up to 350 µm long; about 100 species worldwide .... **Echiniscus**

   Armor consisting of head plate, four trunk plates, and anal plate (plus intercalaries) (Fig. 12) 4

4. With one very long median dorsal seta; uncommon .... **Hypechiniscus**

   Without such a dorsal seta; up to 500 µm long; about 30 species worldwide .... **Pseudechiniscus**

5. Each leg usually with two heavy double claws (Figs. 14 C–E, L–N); sucking pharynx with micro- or macroplacoids (Figs. 3, 4); eyes present or absent **6**

   Each leg with two long, slender claws and two short, heavy claws (Figs. 14 A, C); sucking pharynx elongated and without placoids (Fig. 14 B); mouth surrounded by six prominent papillae; up to 1200 µm long; one cosmopolitan species .... MILNESIIDAE, **Milnesium tardigradum** Doyère

6. Buccal tube with longitudinal anteroventral lamina (Fig. 13 A) .... MACROBIOTIDAE, 7

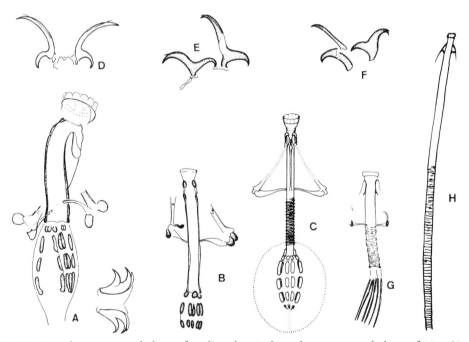

FIG. 13.–Buccal structure and claws of tardigrades. A, buccal structure and claws of *Macrobiotus;* B, buccal structure of *Hypsibius;* C, buccal structure of *Pseudodiphascon;* D, claws of *Dactylobius;* E, claws of *Isohypsibius;* F, claws of *Hypsibius;* G, buccal structure of *Diphascon;* H, buccal structure of *Itaquascon.* (C from Christenberry and Higgins, 1979; all others from Schuster et al., 1980.)

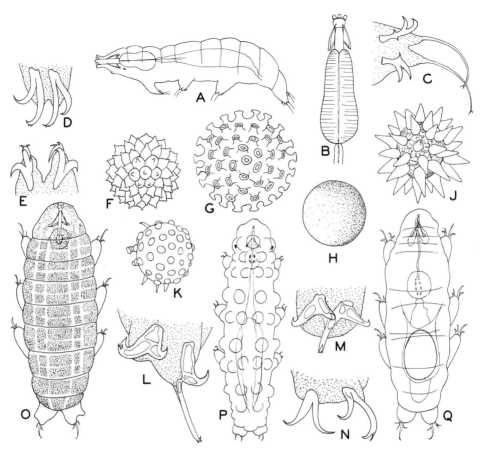

FIG. 14.–Structure of Macrobiotoidea. A, *Milnesium tardigradum,* ×75; B, buccal apparatus of *M. tardigradum;* C, claws of *M. tardigradum;* D and E, claws of *Macrobiotus;* F and G, eggs of *Macrobiotus;* H to K, eggs of *Hypsibius;* L to N, claws of *Hypsibius;* O to Q, dorsal view of typical species of *Hypsibius,* ×220. (Modified from various sources.)

Buccal tube without such a lamina (Fig. 13 B) .... HYPSIBIIDAE, 11
7. Claws single, although occasionally a basal minute spur is present also; a single species known only
     from Sonora, Mexico, and CA .... **Haplomacrobiotus hermosillensis** May
  Each claw normal and double, either similar or dissimilar in size 8
8. Buccal tube long and thin, with spiral thickenings (Fig. 13 C); reported only from AL.
                                           Pseudodiphascon
  Buccal tube of the usual construction 9
9. Fourth legs rudimentary; in wet mosses; one species reported from GA .... **Hexapodibius**
  Fourth legs normal in size; eggs mostly free and sculptured (Figs. 14 F, G) 10
10. Both branches of each claw with the tips close together (Figs. 14 D, E); up to 1100 μm long; about 100
     species .... **Macrobiotus**
  Tips of two claws of each leg remote (Fig. 13 D); uncommon .... **Dactylobiotus**
11. Claws of similar shape and size; branches solidly or broadly joined (Fig. 13 E); about 80 worldwide
     species but seldom reported from the United States .... **Isohypsibius**
  Claws dissimilar in size and shape, one branch with thin articulation (Fig. 13 F) 12
12. Buccal tube short, rigid (Fig. 13 B); about 30 species worldwide .... **Hypsibius**
  Buccal tube long, flexible, with spiral thickenings (Figs. 13 G, H) 13
13. Pharynx with normal placoids; 45 species worldwide .... **Diphascon**
  Pharyngeal placoids reduced or absent .... **Itaquascon**

REFERENCES

Baumann, H. 1921. Beiträge zur Kenntnis der Anatomie der Tardigraden (Macrobiotus hüfelandi). *Z. Wiss. Zool.* **68**:637–652.

―――. 1922. Die Anabiose der Tardigraden. *Zool. Jahrb. Abt. Syst. Geogr. Biol. Tiere* **45**:501–556.

Christenberry, D., and R. P. Higgins. 1979. A new species of Pseudodiphascon (Tardigrada) from Alabama. *Trans. Am. Microsc. Soc.* **98**:508–514.

Crowe, J. H. 1975. The physiology of cryptobiosis in tardigrades. *Mem. Ist. Ital. Idrobiol. Suppl.* **32**:37–59.

Crowe, J. H., and R. P. Higgins. 1967. The revival of Macrobiotus areolatus Murray (Tardigrada) from the cryptobiotic state. *Trans. Am. Microsc. Soc.* **86**:286–294.

Doyère, L. 1840, 1842. Mémoire sur less Tardigrades. *Ann. Sci. Nat. Zool.* (2) **14**:269–361; **17**:193–205; **18**:5–35.

Everitt, D. 1981. An ecological study of an Antarctic freshwater pool with particular reference to Tardigrada and Rotifera. *Hydrobiologia* **83**:225–239.

Higgins, R. P. 1959. Life history of Macrobiotus islandicus Richters with notes on other tardigrades from Colorado. *Trans. Am. Microsc. Soc.* **78**:137–154.

―――. (ed.). 1975. International symposium on tardigrades. *Mem. Ist. Ital. Idrobiol. Suppl.* **32**:1–469.

Marcus, E. 1928. Spinnentiere oder Arachnoidea. IV. Bärtierchen (Tardigrada). *Tierwelt Dtschl.* **12**:1–230.

―――. 1928a. Zur Ökologie und Physiologie der Tardigraden. *Zool. Jahrb. Abt. Allg. Zool.* **44**:323–370.

―――. 1928b. Zur vergleichenden Anatomie und Histologie der Tardigraden. *Ibid.* **45**:99–158.

―――. 1929. Zur Embryologie der Tardigraden. *Zool. Jahrb. Abt. Anat.* **50**:333–384.

―――. 1936. Tardigrada. *Das Tierreich* **66**:1–340.

May, R.-M. 1948. La vie des Tardigrades. *Hist. Nat.* **8**:1–133.

Morgan, C. I., and P. E. King. 1976. *British tardigrades.* 132 pp. Academic, London.

Müller, J. 1935. Zur vergleichenden Myologie der Tardigraden. *Z. Wiss. Zool.* **147**:171–204.

Murray, J. 1907. Scottish Tardigrada, collected by the Lake Survey. *Trans. R. Soc. Edin.* **45**:641–668.

―――. 1907a. Encystment of Tardigrada. *Ibid.* 837–854.

―――. 1910. Tardigrada. *Sci. Rep. Br. Antarctic Exped. 1907–1909* **1**:81–185.

Peterson, B. 1951. The tardigrade fauna of Greenland. A faunistic study with some few ecological remarks. *Medd. Grønland* **150**:1–94.

Pigon, A., and B. Weglarska. 1955. Rate of metabolism in tardigrades during active life and anabiosis. *Nature (London)* **176**:121–122.

Pilato, G. 1982. The systematics of Eutardigrada. A comment. *Z. Zool. Syst.* **20**:271–284.

Ramazzoti, G., and W. Maucci. 1983. Il Phylum Tardigrada. *Mem. Ist. Ital. Idrobiol. Dott. Marco Marchi* **41**:1–1012.

Riggin, G. T. 1962. Tardigrada of southwest Virginia: with the addition of a description of a new marine species from Florida. *Tech. Bull. Va. Agric. Exp. Sta.* **152**:1–145.

Rudescu, L. 1964. Tardigrada. *Fauna Rep. Pop. Rom.* **4**:1–401.

Schuster, T. R., and A. A. Grigarick. 1965. Tardigrada from Western North America: with emphasis on the fauna of California. *Univ. Calif. Publ. Zool.* **76**:1–67.

Schuster, R. O. et al. 1980. Systematic criteria of the Eutardigrada. *Trans. Am. Microsc. Soc.* **99**:284–303.

12

BRYOZOA
(Moss Animalcules)

Unpolluted and unsilted waters, especially ponds and the shallows of lakes and slow streams, characteristically contain colonies of Bryozoa. To the casual observer such colonies are often mistaken for a mat of moss, and certainly some of the common species do have a superficial plantlike appearance. A large, encrusting, sessile colony may consist of thousands of individuals covering an area in excess of a square meter, and a massive gelatinous species may attain the size of a grapefruit or a small watermelon. Usually bryozoans occur on the undersides of logs and stones, or on twigs and other objects where the light is dim.

Bryozoa are more generally distributed and obvious in salt-water habitats than in fresh waters, and about 4000 marine species have been described the world over, as contrasted with about 50 fresh-water species. Only about 21 species have been reported from American fresh waters, and 15 from Europe, 12 of which occur in the United States.

We use "Bryozoa" and "Ectoprocta" as equivalent names for the same phylum, but taxonomists are generally divided in their preferences. The term "Polyzoa" has been discarded except in Great Britain. This chapter also includes the Phylum Entoprocta with a single fresh-water species in the United States and a second in India.

General characteristics. The unit of organization is a microscopic, more or less cylindrical zooid (Figs. 4, 5). Superficially it resembles certain coelenterate polyps, and it wasn't until 1830 that the Bryozoa were distinguished from the Coelenterata as a separate phylum. The body wall, or cystid, consists of a thin living layer and an outer dead secreted zooecium. The Y-shaped digestive tract is suspended in the extensive coelomic cavity. The mouth is situated on a large, distal, circular, oval, or horseshoe-shaped lophophore bearing numerous ciliated tentacles, but the anus is located posterior to or below the lophophore. Sometimes the lophophore and its tentacles are collectively called the tentacular crown. The lophophore, its tentacles, the base of the lophophore, and all of the viscera constitute a polypide (collectively, the protruded living parts of an individual bryozoan). Typically the individuals of a colony are all connected in a branched twiglike manner. The coelom is more or less continuous from one polypide to another, depending on the relative development of septa.

Only the tentacular crown and distal portion of the body are exposed to the water, the rest of the body and the whole colony being invested by a zooecium, which is secreted by the body wall. This protective layer may be thin to massive, corneous, cuticular, gelatinous, delicate, or tough, depending on the particular species and variety. According to the mode of growth and type of zooecium, there are three general but intergrading growth forms: (1) branching and threadlike, (2) mat-like or crustlike, and (3) gelatinous and massive. In addition, *Plumatella casmiana* and *P. fungosa* occasionally form "honeycomb" colonies 2 to 20 mm or more thick (Fig. 3). In the older portions of colonies the zooecium is commonly opaque, brownish, and more or less covered with detritus, algae, and microzoa.

FIG. 1.–*Plumatella repens.* A, bit of colony on piece of burned wood, ×1.5; B, bit of colony, ×6.

the colony, statoblasts are distributed at random by currents and winds; they may float or remain on the bottom. A statoblast is highly resistant to drying, cold, and other adverse environmental conditions, but under appropriate conditions it germinates to produce a new colony.

The Phylum Bryozoa is divided into three classes, the Phylactolaemata, the Stenolaemata, and the Gymnolaemata. All species in the first of these classes are restricted to fresh waters, but the much larger Class Gymnolaemata is exclusively marine except for about 10 species worldwide. Four of the five orders in this class are marine, and in the fifth order, the Ctenostomata, only two American species occur in fresh waters. The Stenolaemata are all marine. Unlike many marine species, fresh-water bryozoans exhibit no polymorphism.

The Entoprocta is a phylum that is mainly marine (60 species) and possibly closely related to the Bryozoa. In addition to the true Bryozoa, the key on page 283 includes the one American fresh-water species of Entoprocta. The differentiating features of this peculiar and uncommon species are given in the key.

But the living portions of the zooid are delicate, translucent, and only slightly pigmented. Internal structures can usually be observed only in young individuals.

When undisturbed, the bryozoan tentacular crown is extended through the distal aperture of the zooecium. It is large and beautifully expanded, and a colony has the appearance of a patch of tiny and delicate flowers, but at the slightest disturbance the tentacular crowns are retracted within the cystids in a flash. After a few minutes they are again slowly extruded and expanded.

Growth and proliferation of a colony is a matter of simple asexual budding. Sexual reproduction, when evident, is restricted to a few weeks during the year. Each individual is hermaphroditic, and early development takes place within the coelom. On being released to the outside, the ciliated embryo soon becomes fixed to the substrate and gives rise to a new colony.

Statoblasts are asexual internal buds whose functional cells are enclosed by two tightly fitting convex sclerotized and sculptured valves (Figs. 9–12). On being released from

Locomotion. Although bryozoan colonies are generally sessile, a few fresh-water species are capable of sluggish creeping movements. *Cristatella mucedo* (Fig. 17 B) is the most active form, and a colony may progress as much as 1 to 10 cm/day. The small colonies of *Lophopus crystallinus* and *Lophopodella carteri,* as well as young colonies of *Pectinatella magnifica,* also move but usually less than 2 cm/day.

Pieces of the horizontal stolon of *Urnatella gracilis* may break off with attached fragments of the colony consisting of one to three zooids. Such detached portions may move about slightly on the substrate by vague peristaltic movements, or they may drift away in currents and become propagation colonies.

The mechanism of colony locomotion is obscure, but it is probable that ciliary action on the tentacles, the musculature of the body wall, and perhaps polypide retractor muscle action are of some importance.

Feeding, digestive system. When the lophophore is expanded under normal conditions, the tentacles are almost motionless, but their abundant cilia beat constantly and create miniature currents along the lophophore that bring algae, protozoans, micrometazoa, and detritus to the region of the mouth. If large and inedible particles reach the mouth, the individual tentacles may bend over in such a way as to brush the particles away and alter the ciliary currents. In all Phylactolaemata the base of the lophophore is surrounded by a membranous cup, or calyx, which is festooned or scalloped; presumably this structure increases the efficiency

of feeding. The distal end of the animal may extend upward from the substrate or downward from the underside of an object, but feeding goes on effectively regardless of position. Phylactolaemata have a flaplike epistome over the mouth. It may cover the mouth or aid in rejecting or warding off undesirable particles.

The anterior portion of the esophagus is slightly modified to form a somewhat muscular pharynx. Food passes through the mouth and pharynx and into the esophagus, and when a sufficient quantity has accumulated in

FIG. 2.—Massive colonies of *Pectinatella magnifica* photographed in natural habitat; relative size can be judged from the two tree trunks at the water's edge in the immediate background. (Photo courtesy of Percy Viosca, Jr.)

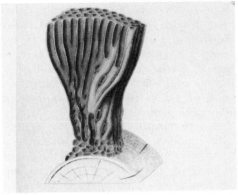

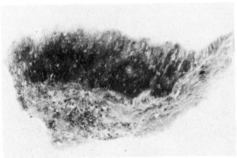

FIG. 3.–Bit of a massive colony of *Plumatella fungosa*, showing zooecium. Upper, diagram of mass of upright parallel tubes; lower, preserved material showing lower surface and upright tubes, ×1. (Lower, courtesy of John Bushnell; upper from Allman, 1856.)

the esophagus, the esophageal valve relaxes and the food passes into the stomach, where most of the digestion occurs. Below the stomach and corresponding to the basal leg of the Y is a long, baglike absorptive organ, the caecum. Connecting the lower end of the caecum and the body wall is a long strand of tissue, the funiculus. The tubular intestine passes from the upper end of the caecum to the anus; its terminal portion is slightly modified to form a rectum. The anus is located just below the lophophore. The feces are released as pellets, sometimes surrounded by a peritrophic membrane.

The Phylactolaemata alimentary canal is ciliated only around the mouth and in the pharynx but in the Gymnolaemata cilia extend as far as the upper portion of the stomach.

Much of the polypide coloration is determined by the specific contents of the alimentary canal, and by changing the diet experi-

mentally it is possible to get yellow, green, and brown coloration all in the same day.

Circulation. There is no circulatory mechanism in bryozoans, but materials are distributed quickly and efficiently via the coelomic fluid owing to the movements of the alimentary canal and retractions and extrusions of the tentacular crown.

Respiration. No special respiratory devices are present. The greater portion of the carbon dioxide and oxygen exchange undoubtedly occurs generally through the lophophore tissues.

Excretion. No excretory system occurs in the bryozoans, although some investigators claim otherwise. Coelomic amoebocytes containing engulfed excretory granules supposedly make their way through the body wall to the outside. In *Urnatella gracilis* (an entoproct) there is a simple protonephridial system.

In *Cristatella, Lophopus, Plumatella,* and possibly some other genera the epidermal cells of the lophophore apparently accumulate excretory substances and liberate them to the outside. In addition, quantities of metabolic wastes are undoubtedly retained in many of the body tissues until the animal dies and disintegrates. A true intertentacular organ is unknown in fresh-water bryozoans.

A few Phylactolaemata, including *Hyalinella, Lophopus, Cristatella,* and *Lophopodella,* sometimes undergo partial disintegration similar to true "brown-body" formation in marine forms, especially under unfavorable conditions. It is thought that this phenomenon is at least partially excretory in nature. The polypides (especially older individuals) first retract permanently and then begin to undergo degeneration. The lophophore, tentacles, digestive tract, and nervous system degenerate (Fig. 5). Such a body remains unchanged, or it may further disintegrate into small fragments that circulate about in the coelom. Sometimes, in *Plumatella casmiana,* a new polypide is then regenerated from an internal bud on the body wall; it comes to occupy the space vacated by the old polypide. Sometimes the same zooid may undergo "brown-body"

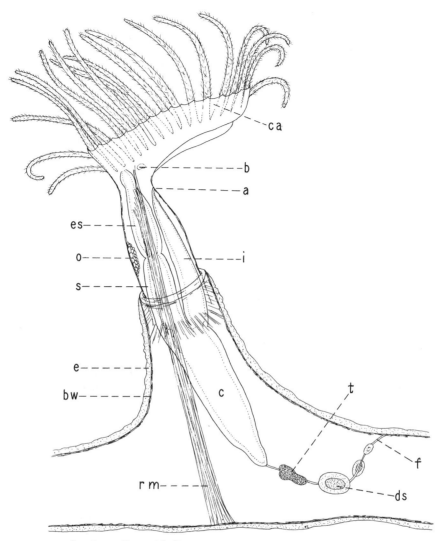

FIG. 4.–Structure of a *Plumatella* zooid, diagrammatic. Tentacles and retractor muscle shown for only one side. a, anus; b, brain; bw, body wall; c, caecum; ca, calyx; ds, developing statoblasts (3); e, cuticula; es, esophagus; f, funiculus; i, intestine; o, ovary; rm, retractor muscle; s, stomach; t, testis.

formation. In spite of certain textbook statements to the contrary, a "brown-body" has never been seen to regenerate a new cystid.

Muscle system. The most prominent muscles are the two retractors, which are somewhat stringy and not at all compact. They originate at the base of the cystid and are inserted on the base of the lophophore and uppermost portion of the digestive tract, one on each side (Fig. 4). Their contraction pulls the whole distal end of the polypide back into the coelom and cystid. Some short indistinct fibers connect the upper part of the polypide with the nearby body wall, and a few associated fibers aid in rotating the tentacular crown and depressing the lobes.

The thin body wall contains a few loosely arranged circular and longitudinal fibers, which are effective in everting the polypide. Other indistinct muscles elevate the epistome, move the alimentary canal slightly, and act as dilators and sphincters.

Nervous system. A single large ganglion, or "brain," is situated at the distal end between the mouth and anus. From it a series of flimsy fibers with a few associated nerve cell bodies radiate out to all tissues and organs. There are no proven nervous connections between adjacent zooids.

Special sense organs are absent, but bryozoans are remarkably sensitive to light and other stimuli, certain epidermal cells being specialized as receptors.

Reproduction, development. Growth of a colony is an asexual form of proliferation, often dichotomous, whereby a part of the body wall grows outward and eventually forms a new zooid. In this way, beginning with a single zooid in the spring, the resulting colony may consist of thousands of living and dead individuals by autumn. Developing *Pectinatella, Cristatella,* and *Lophopodella* colonies often undergo fission, thus giving rise to two or more new colonies (Fig. 6). A fission or fragmentation of portions of a colony has been observed in *Fredericella sultana.*

The period of sexual reproduction in a colony usually lasts about 1 to 4 weeks, most often some time between May and July. Individual zooids are hermaphroditic in all species except for *Urnatella,* which is dioecious. The single ovary develops from the peritoneum of the body wall near the distal end of the cystid, and the testis develops from the peritoneum of the funiculus. Unlike the situation in the great majority of hermaphroditic animals, there seems to be no current evidence for cross fertilization in fresh-water Bryozoa. Mature eggs and sperm are produced either simultaneously or in rapid succession. Ordinarily only a single ovum matures at a time and is fertilized by sperm in the coelomic fluid just as it leaves the ovary. The zygote then becomes implanted in a special pouch-like ooecium, which develops from the body wall in the immediate vicinity of the ovary. Here the embryo develops into a pear-shaped, oval, or spherical ciliated larva 1 to 2 mm long containing one to five immature polypides. A *Cristatella* larva, however, may have up to 25 polypides or polypide buds. This larva breaks out of the parent through a special large pore at the base of the lophophore, or it escapes when the parent degene-

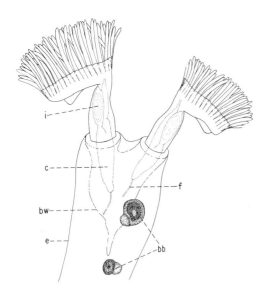

FIG. 5.–Semidiagrammatic view of a small portion of a *Lophopodella carteri* colony, ×25, showing two "brown bodies." bb, "brown bodies"; bw, body wall; c, caecum; e, zooecium; f, funiculus; i, intestine. (Greatly modified from Rogick, 1938.)

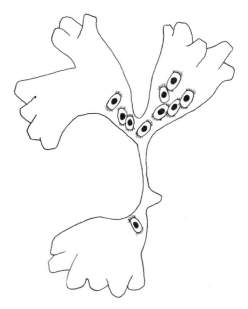

FIG. 6.–Diagram of dividing colony of *Lophopodella carteri,* ×5. Polypides are shown contracted; 10 statoblasts present. (Redrawn from Rogick, 1935.)

rates (Fig. 7). It is free swimming for only a few minutes or for as long as 20 to 24 hours and then settles on the substrate, evaginates, and eventually gives rise to a mature colony.

Little specific information is available concerning the natural longevity of individual zooids, but presumably the great majority live for 3 to 8 weeks in temperate climates. Large, old colonies contain many dead zooids in the parts of the colony that were first formed.

Although colonies customarily die out in the late summer or autumn, there are records of *Paludicella, Fredericella,* and *Lophopus* colonies being active throughout the winter, especially if it is mild and the water temperature does not drop below 2°C.

Statoblasts. The production of highly resistant statoblasts is a unique feature of fresh-water Phylactolaemata. They function mainly in tiding the species over unfavorable environmental conditions and in geographic dissemination.

Statoblasts develop by asexual budding from the funiculus, and in some zooids

several stages of development may be found simultaneously. A mature statoblast is roughly biconvex, but the central portion is considerably thickened into a capsule owing to its contained mass of undifferentiated germinative cells. The peripheral portion (annulus, or float) usually consists of air cells and is relatively thin; sometimes it bears spiny, barbed, or hooked process (Fig. 9). Except for the germinative cells, a statoblast therefore consists of two sclerotized, dead valves that are tightly attached to each other in the region of the annulus. Early stages in statoblast development are almost smooth; older ones are strongly ridged or mammillate and tan, brown, or black. Usually each zooid produces two to eight statoblasts, although more than 20 per zooid have been reported.

The literature distinguishes six general and specific types of statoblasts. Floatoblasts are free or become floating after the zooecium disintegrates. In some species the floatoblasts are released from the living zooid by means of a hyaline collarlike protrusion (Fig. 13). They have no peripheral processes; they are annulated when mature and are 0.2 to 1.0 mm long. All Phylactolaemata except

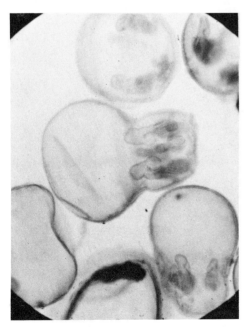

FIG. 7.–Ciliated free swimming larvae of *Plumatella,* ×27. Center embryo contains three developing zooids.

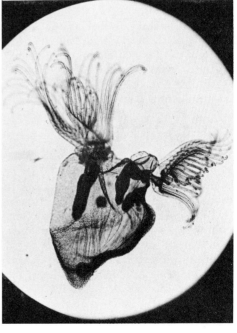

FIG. 8.–Young colony of *Pectinatella magnifica,* ×33.

Fredericella produce floatoblasts (Figs. 19 A, B; 20 B, C).

There are several specialized kinds of floatoblasts. Spinoblasts are free or floating, but are provided with spined, barbed, or hooked processes. Length ranges from 0.7 to 1.3 mm. *Cristatella, Lophopodella, Lophopus,* and *Pectinatella* produce spinoblasts (Figs. 9, 11, 14). A leptoblast is a fragile floatoblast that germinates without a diapause. A pyknoblast is thick walled (Fig. 10, top). *Plumatella casmiana* produces leptoblasts, pyknoblasts, and an "immediate form that germinates only after a diapause." All other species have a single morphotype.

Sessoblasts are sessile, fixed, or cemented to the zooecium wall; an annulus may be present or absent, well developed or reduced. The length is 0.3 to 0.6 mm. *Plumatella, Stolella,* and perhaps *Hyalinella* produce sessoblasts (Fig. 12). Some species of *Plumatella* produce both sessoblasts and floatoblasts. The two species of *Fredericella* produce piptoblasts (Fig. 10, middle), which are neither cemented to the substrate nor capable of floating. They lack an annulus and usually sink to the bottom when the zooecium disintegrates.

Formerly it was thought that statoblast production is limited to late summer and autumn just before the colonies die off, but it is now well established that statoblasts may also be regularly produced during the earlier portion of the growing season. In the tropics statoblasts usually appear with the onset of hot weather. Sexual reproduction and statoblast production may occur simultaneously in the same zooid; there is little evidence for alternation of the two methods.

On disintegration of the zooids, the floatoblasts are released to the vagaries of currents, waves, and wind. Some of them sink or remain on the bottom. Others float to the surface, and under appropriate conditions may be found in such abundance as to form windrows on the shore, especially in spring and autumn. Sessoblasts, however, remain attached to the substrate until the last bit of old zooecium has disintegrated. A few species actively release statoblasts (Fig. 13).

The length of the statoblast dormant period is highly variable, depending on the species, the individual, temperature, and other environmental conditions. In high latitudes there

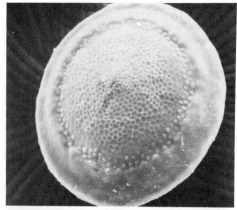
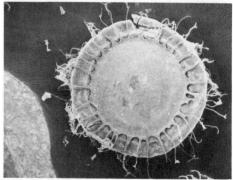
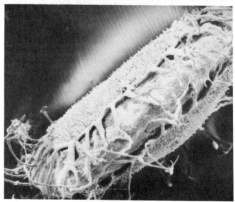

FIG. 9.–Bryozoan statoblasts. Top, *Hyalinella orbisperma,* ×160; middle, *Cristatella mucedo* spinoblast, ×40; bottom, edge view of *C. mucedo* spinoblast, ×75, showing suture zone. (SEM photos courtesy of K. S. Rao.)

is usually only one generation per year, and the majority of statoblasts winter over and germinate the following spring, but south of the 50th parallel it appears that the growing season is often sufficiently long for two or three main crops of statoblasts. Colonies

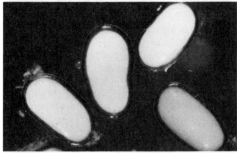

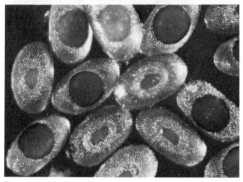

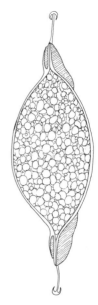

FIG. 11.–Diagrammatic section of statoblast of *Pectinatella magnifica,* showing the two valves, hooked spines, central capsule filled with germinative cells, and float.

FIG. 10.–Bryozoan statoblasts, all ×60. Top, pyknoblasts of *Plumatella casmiana*; middle, pipto-blasts of *Fredericella sultana*; bottom, floatoblasts of *Plumatella emarginata.* (Photos courtesy of T. S. Wood.)

mature in the spring, and their statoblasts hatch out and produce a second group of mature colonies by late summer or early autumn. These individuals, in turn, give rise to statoblasts that winter over. In spite of their ability to survive drying and freezing temperatures, it should be borne in mind that statoblasts have a very high natural mortality rate.

Under natural conditions the period of statoblast dormancy or desiccation extends from the end of one growing season to the beginning of the next, which is usually 4 to 9 months, and most workers are agreed that a majority of statoblasts of all species are viable for the duration of such periods.

The natural period of dormancy and germination is usually 30 to 150 days, the chief governing factors being low temperatures and desiccation. Germination begins in the spring when the water reaches 10 to 19°C or more, depending on the species. The first evidence of germination is the gradual separation of the two statoblast valves along the equator, a process that may take 1 to 8 days. A mass of undifferentiated yolky cells slowly grows and protrudes from between the valves. These cells are whitish or grayish, very contractile, and they adhere to the substrate. The primary zooid meanwhile develops between the valves, hidden from view. It utilizes the abundant stored food of the undifferentiated yolk cells, and as it enlarges it protrudes, pushes the valves farther apart, and absorbs and incorporates more and more of the yolk mass so that the latter soon becomes an indistinguishable part of the zooid. As growth proceeds, additional zooids appear by budding, and eventually a colony results.

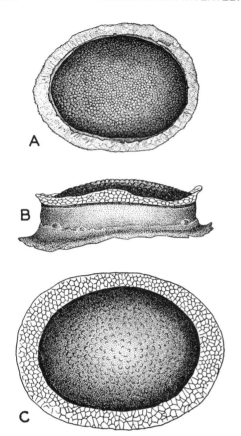

FIG. 12.—Sessoblasts of *Plumatella*, ×120. A, unattached surface view of sessoblast of *P. emarginata*; B, side view of sessoblast of *P. repens* (note sclerotized lamella that resembles the float of floatoblasts); C, unattached surface view of same. (From Rogick, 1940.)

A large amount of experimental laboratory work has been conducted on the viability of statoblasts, but the results vary so much from one species to another and from one investigator to another that it is difficult to generalize. Most workers, however, seem to agree that natural freezing and normal winter temperatures have little effect on the percentage of later germination, but little is known about the lethal effects of unusually low temperatures.

Viability of statoblasts dried and kept at room temperature differs widely from one species to another. Rogick (1940) found that all *Hyalinella*, *Pectinatella*, and *Plumatella* statoblasts died within 23 to 34 months under

such conditions. *Fredericella*, on the other hand, showed 50 to 100% germination after 24 months. Seventy percent of *Lophopodella* statoblasts were capable of germinating after being dry for 50 months, but none germinated after 74 months. In general, wet storage in the refrigerator produces a lower statoblast mortality than drying at room temperature for the same length of time. Statoblasts die within a few days when kept in a laboratory desiccator.

When statoblasts are placed in water and germinated in the laboratory the period of incubation is highly variable. Some individuals and species germinate in 4 to 10 days; others require 1 to 4 months for germination.

Hibernacula. The gymnolaemate Paludicellidae, including *Paludicella* and *Pottsiella*, do not form statoblasts, but rather hibernacula, which are similar to statoblasts in function. A hibernaculum is a specially modified external winter bud. Like statoblasts, they consist of masses of cells protected by a thick sclerotized coat (Fig. 15).

Ecology. Quiet ponds, backwaters, bays, and slows streams, especially where there are sunken logs, twigs, rocks, and aquatic vegetation, are ideal places for Bryozoa. Sandy and pebbly bottoms are poor habitats, but certain species of *Plumatella*, *Fredericella*, and *Paludicella* are often found in turbulent waters. Although they are common in stagnant waters, bryozoans are never ·found under markedly polluted conditions and only sparingly where the quantity of dissolved oxygen falls below 30% saturation. Silting also discourages establishment and growth, and strongly acid waters and bogs almost never contain bryozoans. *Lophopodella carteri* is probably the species that is most tolerant of decay and stagnation.

Light is not necessary for development and growth, as shown by the occurrence of bryozoans in water pipes and closed conduits. On the other hand, they are seldom present in bright light, although *Cristatella mucedo* occurs regularly on the upper sides of objects in the shallows. All other species found in shallow water are attached to the undersides or shaded sides of rocks, logs, vegetation, and boards.

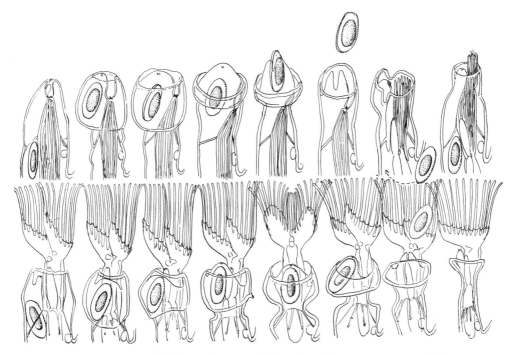

FIG. 13.–Sequence of release of two statoblasts in *Stolella.* (From Marcus, 1942.)

Most Bryozoa are collected in waters less than 1 m deep, but they commonly occur at considerably greater depths, and if appropriate collecting equipment is used, almost all of our common species may be taken in lakes at depths up to 5 or 10 m. *Paludicella* has been taken as deep as 36 m and *Fredericella* up to 214 m. *F. sultana* is sometimes found in marl lakes; it also occasionally grows up from sandy substrates in an arborescent form.

Lophopodella and *Lophopus* are generally restricted to standing waters. *Paludicella, Cristatella, Pectinatella, Fredericella,* and *Hyalinella* are more common in standing waters but are found also in slowly running waters. *Urnatella* and *Plumatella* may, in addition, be found in fairly rapid waters. *Pottsiella* is restricted to rapid waters. Sometimes *Fredericella, Plumatella, Pectinatella,* and *Cristatella* are all found in the same pond or lake at the same time. Up to eight species have been found in a single lake. *Victorella pavida* is a rare species of Gymnolaemata that is restricted to brackish waters.

Most American species attain their greatest abundance in the summer when the water temperature reaches 19 to 23°C and begin to die off in the autumn when the water cools to 10°C. *Pectinatella,* however, tolerates a much smaller range of temperature and usually dies when the water temperature goes below 16°C. *Fredericella* and *Lophopus,* on the other hand, are unusually tolerant and do well anywhere between 2 and 23°C. *Plumatella* is sometimes collected at water temperatures as high as 37°C.

Economic importance. Occasional summer plagues of large floating gelatinous colonies of *Pectinatella magnifica* clog the screens of water intakes and the grates of hydroelectric plants so effectively that it requires a full-time man to keep them clear.

In communities using unfiltered and unchlorinated water, bryozoans sometimes become established in the pipes, and dead and living masses clog the valves and meters. Occasionally fishermens' trap nets accumulate large growths. The usual offenders are *Paludicella, Fredericella,* and *Plumatella.*

Bryozoa are an element in the diet of many fresh-water invertebrates, including planarians, snails, oligochaetes, Hydracarina,

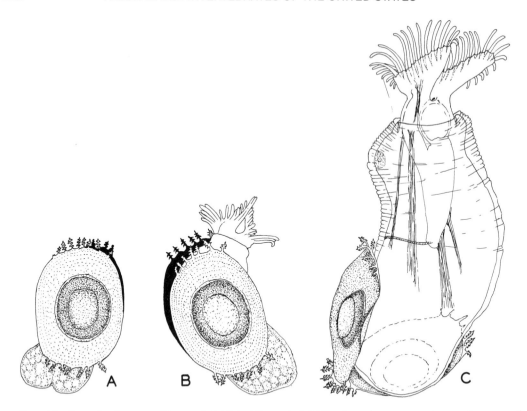

FIG. 14.–Germinating statoblasts of *Lophopodella carteri*, ×33. A, valves separating and mass of undifferentiated yolky cells extruded; B, primary polypide emerging; C, primary zooid almost mature (note small bud on left side near distal end). (Modified from Rogick, 1935.)

crustaceans, and especially Trichoptera and chironomids. Statoblasts and colonies are often eaten by small fish. The zooecium forms an excellent substrate for a variety of commensals and browsers, especially ciliates and rotifers.

Geographical distribution, dispersal.

Plumatella repens, Fredericella sultana, Plumatella emarginata, and *Hyalinella punctata* are the only known truly cosmopolitan species; they occur in suitable habitats all over the world. At the other extreme, some North American species have greatly restricted and often puzzling distributions. *Hyalinella orbisperma* is Nearctic; *Stolella evelinae* and *Pottsiella erecta* are Nearctic and Neotropical; *Cristatella mucedo* and *Plumatella fungosa* are circumpolar; *Pectinatella*

magnifica is probably native only to North America but has been recently introduced into Japan and Germany; *Hyalinella vaihiriae* is peculiar to the Nearctic and Australian regions. Europe has 13 bryozoan species and 12 of these also occur in the United States.

Unquestionably the paucity of records for some of the American species is a reflection of the fact that little systematic and intensive collecting has been done. Future collecting, especially in the South, and west of the Mississippi River, will probably show that some of our "rare" species are actually widely distributed.

Chiefly by virtue of their resistant statoblasts and hibernacula, bryozoans are one of the few fresh-water groups that are clearly capable of being transported overland from one body of water to another by animals. There are definite records of statoblast occur-

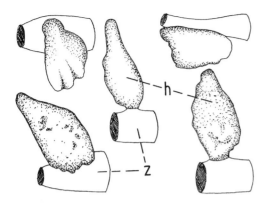

FIG. 15.–Hibernacula of *Paludicella articulata,*
×50. h, hibernaculum; z, bits of old zooecium.
(Modified from Rogick, 1935.)

rence in mud on the feet and feathers of
waterfowl, and in the fur of mammals. Some
statoblasts are capable of germinating after
passing through the alimentary canal of
waterfowl, turtles, frogs, salamanders, and
fish (see Bushnell, 1973).

Collecting. Bits of colonies, sessoblasts,
and hibernacula should be carefully removed
from their substrate with forceps, spatula,
and scalpel, but if the colonies are on small
pieces of wood, twigs, or stones, they may be
brought to the laboratory intact. A dredge or
double rake is necessary for collecting below
a depth of 1 m.

Floating statoblasts may often be obtained
in abundance at the right season by skimming
the surface of the water with a net or by
scooping them up from windrows.

Culturing. Bryozoa are among the most
difficult of all fresh-water invertebrates to
culture in the laboratory. If intact colonies are
used, they may sometimes be kept alive in
finger bowls for 5 to 15 days, provided the
water is replaced frequently with fresh pond
water, preferably from their natural habitat.
Sometimes a small quantity of fresh yeast is
beneficial to the culture.

Some investigators report partial success
using water from aquatic plant aquariums
along with a bit of the organic debris from the

bottom of such aquariums. Such debris usually
contains an abundance of suitable microorganisms for food. Both water and debris
should be changed every 2 or 3 days.

Wood (1971) presents a simple and highly
successful method of culturing in the laboratory. Statoblasts are first allowed to germinate
in petri dishes. Once the ancestrulae (first
individuals) are formed, they stick to the
glass, and the petri dish is then placed in an
aquarium where it is held upside down by
Plexiglas strips. Fecal pellets and detritus fall
away and leave the colony clean. The most
reliable food source is the suspended organic
matter normally present in a large, thriving
aquarium.

Preparing, preserving. Bryozoans are
highly sensitive to foreign substances, and if
ordinary methods of fixation are used the
tentacular crown contracts into the cystid so
strongly as to make the whole animal useless
for study. Consequently some method of
narcotization must first be used so that the
animals can be killed and fixed with the
lophophore expanded.

The colonies and a part of their substrate
should be placed in a finger bowl one third to
one half full of water from their habitat, and
the subsequent narcotization must be done
very gradually and patiently. If chloretone,
menthol, or chloral hydrate is used, drops of
saturated solution should be added slowly
over a period of 1 to 3 hours. When the
animals no longer respond appreciably to
jarring or touching, most of the fluid should
be withdrawn with a large pipette. Then
Bouin's fixative or 50% formalin should be
poured over the animals, and if the narcotization has been successful, they will be killed in
a beautifully expanded condition.

The following effective method is modified
from Hurrell (1936). Place the colonies and a
bit of their substrate in a finger bowl of water.
Sprinkle a teaspoonful of menthol crystals on
top of the water, cover the finger bowl with a
glass plate, and let stand overnight. By morning the bryozoans should be fully narcotized.
Then withdraw most of the fluid from the
finger bowl, add concentrated formalin
equivalent to the amount of water remaining
in the finger bowl, and let stand for 1 hour.

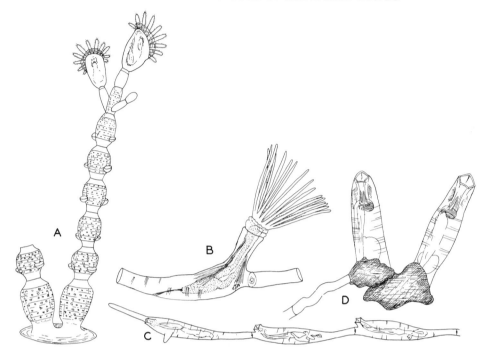

FIG. 16.–Structure of Entoprocta and Paludicellidae. A, small portion of a colony of *Urnatella gracilis,* ×32; B, zooid of *Paludicella articulata,* ×33; C, portion of a young colony of *P. articulata* showing three retracted individuals and two buds; D, young colony of *Pottsiella erecta,* ×33, showing two hibernacula. (B and C modified from Rogick, 1940; D modified from Kraepelin, 1887.)

Five percent formalin and 70% alcohol with 5% glycerin are suitable preservatives.

Bryozoa may be mounted either unstained or stained with hematoxylin or borax carmine. Before mounting or staining, however, the zooecium should be gently brushed as free of debris as possible. If glycerin jelly mounts are to be made, the animals must first be run up into glycerin. Place them in a watch glass containing a solution of 5% glycerin in 50% alcohol and cover with a loose cover to keep out dust and permit evaporation of the alcohol and water. When nearly pure glycerin remains, they are ready for mounting.

Taxonomy. Fresh-water Bryozoa are one of many groups of American fresh-water invertebrates greatly in need of much more systematic collecting and critical work. Indeed, between 1904 and 1935, this group was almost completely forgotten, but in more recent years much fundamental work has been accomplished by two outstanding American investigators, John Bushnell and Mary Rogick. Dr. Bushnell contributed much toward the anatomical and statoblast details of this chapter.

Most of our species are distinctive, especially if both the colonies and their statoblasts are available for identification. Indeed, the genera are easily distinguished with the naked eye or hand lens. The most common and widely distributed species, *Plumatella repens,* is extremely variable and puzzling. At least six different varieties and two "phases" of this species have been reported from the United States, and although all of them are perhaps best placed in the single species, there are some investigators, especially Europeans, who prefer to consider these several varieties as distinct species. Some of the more striking intergrading and variable characters are (Rogick, 1935): colony composed of two discrete masses or not, branching compact to open, zooecia furrowed or not furrowed, tubes attached to the substrate along their entire length or tubes more or less erect, keel present or absent, and statoblasts relatively long or short. Sometimes the variations differ according to the age of the animals, but often they are apparently controlled and induced by certain obscure factors of the many different types of habitats in which *P. repens* occurs. The problem awaits solution by means of controlled laboratory cultures. We may be dealing with a large cluster of incipient species.

KEY TO SPECIES OF BRYOZOA

1. Anus outside the lophophore; zooids not stalked; tentacular crown capable of being retracted within zooecium; without a protonephridial system; statoblasts present or absent.

   Phylum **BRYOZOA (ECTOPROCTA), 2**

   Anus on the lophophore; each zooid on a flexible stalk composed of 5 to 15 sclerotized urnshaped segments; 1 to 6 such stalks arising from one basal plate; stalks may be branched or unbranched and up to 5 mm long; lophophore circular, with 8 to 16 tentacles, and incapable of being retracted into zooecium (Fig. 16 A); with a simple protonephridial system; without statoblasts; stalks from the overwintering stage on objects in running waters or in large lakes; one uncommon species reported from scattered localities in eastern half of the United States, but there are also records from AZ, CA, FL, LA, NV, OK, TX, and Gulf coast rivers, including mildly polluted waters; now worldwide by introduction from United States.

   Phylum **ENTOPROCTA, Urnatella gracilis** Leidy
2. Lophophore circular and without an epistome (Fig. 16 B); colony covered with a corneous or membranous cuticle; statoblasts absent; hibernacula produced (Fig. 16 D).

   Class **GYMNOLAEMATA, PALUDICELLIDAE, 3**

   Lophophore horseshoe shaped, oval, or circular, and with an epistome (Figs. 17, 19); statoblasts produced .... Class **PHYLACTOLAEMATA, 4**
3. Colony threadlike and consisting of a series of club-shaped zooecia arranged end to end and separated from each other by partitions; 16 to 18 tentacles; cystid openings on upper surface; zooids about 2 mm long; colony recumbent or partially erect (Figs. 16 B, C); standing and slowly flowing waters; reported from scattered localities chiefly east of the Mississippi River but probably cosmopolitan in the United States; collected as far north as Alaska.

   Paludicella articulata (Ehr.)

   Colony consisting of a stolon from which single individuals arise; cystid openings terminal; 19 to 21 tentacles; zooids about 1.5 mm long (Fig. 16 D); on upper surface of stones in rapid water; reported from streams in CT, OH, LA, PA, VA, TN, and TX ...... **Pottsiella erecta** (Potts)
4. Statoblasts all free, annulated, and with processes (Figs. 11, 14, 17 C, D; 18 D, E); colony soft, gelatinous, transparent, without septa, and not dendritic (Figs. 17 B, E; 18 H, J) 5

   Often with two types of statoblasts; one attached and without float cells (sessoblasts); the other unattached, funicular, and annulated (floatoblasts) (except *Fredericella*); statoblasts without processes; colony most often dendritic **8**
5. With a transient secretion elaborated by flat base of colony; colony elongated and slowly motile (Fig. 17 B); statoblasts with two circlets of hooked spines; dorsal circlet with 10 to 34 hooks and ventral circlet with 20 to 50 hooks (Figs. 17 C, D); zooids on upper surface and arranged marginally in three rows; all polypides retracted into a common cavity; usually 80 to 90 tentacles; long dimension of colony sometimes as much as 10 to 30 cm but more typically 2 to 6 cm; in standing waters or slow currents, on upper side of submerged objects and vegetation; widely distributed but not common .... CRISTATELLIDAE, **Cristatella mucedo** Cuvier

   Case present throughout life of colony and sometimes very well developed; colony not usually motile; statoblasts never with two circlets of hooked spines 6
6. Colonies in the form of rosettes on gelatinous base or surrounding a gelatinous core; very young colonies motile; colonies commonly investing twigs and other submerged objects, especially in shaded places in quiet waters (Fig. 17 E); 60 to 84 tentacles; free statoblasts with single row of marginal anchorlike spines (Fig. 11); widely distributed but uncommon west of the Mississippi Valley .... PECTINATELLIDAE, **Pectinatella magnifica** Leidy

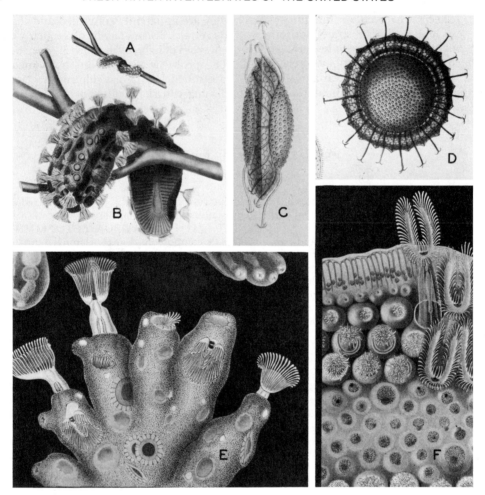

FIG. 17.–Structure of *Cristatella* and *Pectinatella.* A, colony of *Cristatella mucedo,* ×5; B, colony of *C. mucedo,* ×40; C, edge view of *C. mucedo* statoblast, ×50; D, surface view of *C. mucedo* statoblast; E, rosette of *Pectinatella magnifica*; F, edge of colony of *Cristatella mucedo* showing three rows of marginal polypides. (A–D from Allman; E and F from Kraepelin, 1887.)

Colonies in the form of small gelatinous masses with radiate branching; a few to 70 zooids per colony; free statoblasts pointed or with hooked spines at each end (Figs. 14; 18 D, E).
                                                             LOPHOPODIDAE, 7
7. Free statoblasts broadly oval, with one or several hooked polar spines (Fig. 14); cystid translucent; colonies small, circular to oval, and mound shaped when only a few polypides are present, but radiating lobate when many polypides are present; average colony 5 mm wide, 8 mm long, and with 20 to 45 polypides; 52 to 82 tentacles; rare, reported from northeastern quarter of the United States .... **Lophopodella carteri** (Hyatt)
    Free statoblasts elongated, attenuated to a point at each pole (Figs. 18 D, E); colony sac shaped, erect, often lobed, and usually less than 1 cm across; cystid gelatinous (Figs. 18 H, J); with about 60 tentacles; usually attached to vegetation in standing water; rare in eastern states.
                                                        Lophopus crystallinus Pallas
8. Lophophore circular or elliptical; colony tubular, branched, and antlerlike (Fig. 18 B); zooecium usually opaque and brown; all statoblasts (piptoblasts) free, without annulus (Fig. 18 A); no floatoblasts .... FREDERICELLIDAE, **Fredericella, 9**
    Lophophore horseshoe shaped; floatoblasts present; sessoblasts also present in most species.
                                                           PLUMATELLIDAE, 10

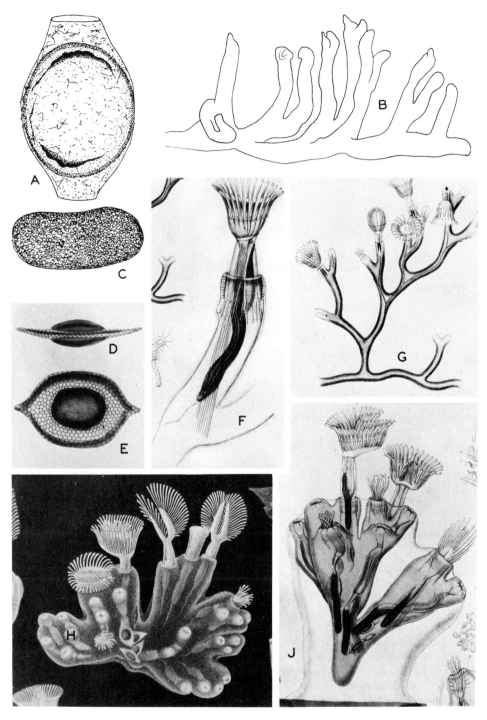

FIG. 18.–Structure of Plumatellidae. A, cuticula enclosing sessoblast of *Fredericella australiensis,* ×100; B, branching in colony of *F. australiensis,* ×12 (polypides shown retracted); C, sessoblast of *F. sultana,* ×85; D, edge view of statoblast of *Lophopus crystallinus,* ×40; E, surface view of same statoblast; F, zooid of *Fredericella sultana,* ×15; G, small colony of *F. sultana,* ×3; H and J, small colonies of *Lophopus crystallinus,* ×7. (A and B modified from Rogick, 1945; C modified from Rogick, 1937; D–G, and J from Allman; H from Kraepelin, 1887.)

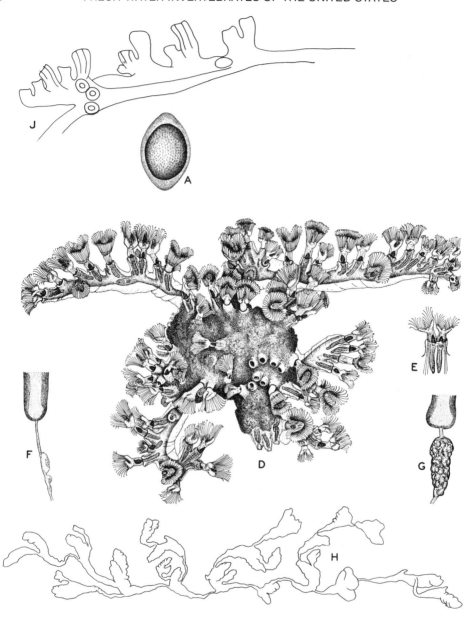

FIG. 19.–Structure of Plumatellidae. A, floatoblast of *Hyalinella vaihiriae,* ×60; B and C, surface view of floatoblast of *H. punctata,* ×65; D, entire colony of *H. punctata,* ×6; E, twin polypides of *H. punctata*; F, three developing statoblasts on funiculus of *H. punctata*; G, immature sperm mass on funiculus of *H. punctata*; H, outline of colony of *Stolella indica,* ×6 (all polypides shown retracted); J, portion of a colony of *S. evelinae* (all polypides retracted). (A from Rogick and Brown, 1942; B and C from Rogick, 1940; D–G from Rogick, 1935; H modified from Rogick, 1943; J from Bushnell, 1965.)

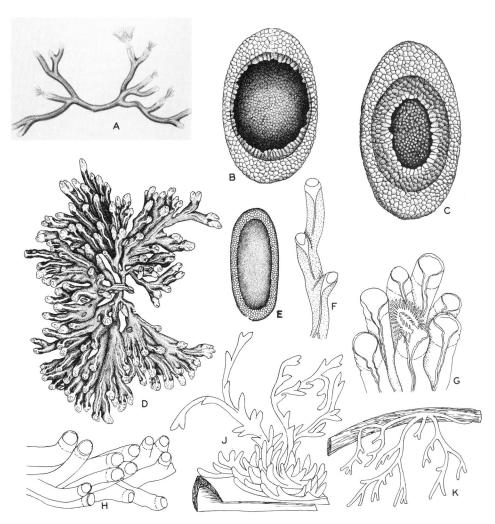

Fig. 20.–Structure of *Plumatella*. A, small colony of *Plumatella repens*, ×4; B and C, opposite surfaces of floatoblasts of *P. repens*, ×100; D, colony of *P. casmiana*, ×5; E, floatoblast of *P. casmiana*, ×65; F, diagram of small portion of colony of *P. repens* (polypides all retracted); G, small portion of colony of *P. emarginata* (only one polypide partially expanded); H, small portion of colony of *P. repens* (all polypides retracted); J, diagram of small portion of colony of *P. repens* (all polypides retracted); K, diagram of small portion of colony of *P. fruticosa* (all polypides retracted). (A from Allman; B to D from Rogick, 1941; E from Rogick, 1943; F modified from Newcomer; G and H modified from Rogick, 1935; J modified from Rogick, 1940.)

9. With 24 to 30 tentacles; polypides short; zooecium not especially slender; cuticula thin, chitinoid, and opaque (Fig. 18B); most piptoblasts circular or broadly elliptical (Fig. 18A); reported from a pond in Uinta County, WY, and otherwise known only from Mexico in North America.

Fredericella australiensis Goddard

With 17 to 27 tentacles; polypides longer; cystid slender, with antlerlike branching (Fig. 18G); most piptoblasts reniform or elongated (Fig. 18C); on submerged objects and vegetation in standing and slowly flowing waters; common and widely distributed; possibly should be called *F. indica*.

Fredericella sultana (Blumenbach)

10. Zooecium in the form of elongated pseudostolons (Figs. 19H, J) . . . . **Stolella, 11**
    Zooecium not in the form of elongated pseudostolons **12**
11. Basal portion of the zooecium of each zooid slender and strongly tapered (Fig. 19H); reported only from PA and MI . . . . **Stolella indica** Annandale
    Basal portion of the zooecium of each zooid not slender and strongly tapered (Fig. 19J); reported only from MI . . . . **Stolella evelinae** (Marcus)
12. With soft, hyaline, gelatinous cuticula; thickly branched, often completely covering the substrate (Fig. 19D) . . . . **Hyalinella, 13**
    With thin, firm, and sometimes horny cuticula; zooecium long and tubular, pale and transparent to dark brown; in masses up to 5 cm high or closely adherent to substrate (Fig. 20); in a variety of habitats . . . . **Plumatella, 15**
13. Floatoblasts circular, diameter about 0.34 mm; colony flat, massive, or irregularly clustered; reported only from MI . . . . **Hyalinella, orbisperma** (Kellicott)
    Floatoblasts not circular **14**
14. Floatoblasts averaging 0.37 mm wide and 0.54 mm long (Fig. 19B); cuticula soft, swollen, colorless, or yellowish (Fig. 19D); widely distributed and common . . . . . **Hyalinella punctata** (Hancock)
    Floatoblasts averaging 0.25 mm wide and 0.36 mm long (Fig. 19A); reported only from Bear River, UT.

Hyalinella vaihiriae Hastings

15. With sessoblasts and three types of floatoblasts; one floatoblast type, the leptoblast, is somewhat narrow, very thin walled, and has the narrow annulus the same width all around the capsule (Fig. 20E); another type, the pyknoblast (Fig. 10) with greatest float coverage, formed in early or mid-summer; a third type, strong walled, and intermediate in float coverage, formed in late summer and autumn; reported only from Lake Erie, IN, MI, and CO . . . **Plumatella casmiana** Oka
    With sessoblasts plus only thick-walled floatoblasts; annulus not of the same width all around capsule (Figs. 20B, C); on underside of vegetation and objects **16**
16. Floatoblasts twice or more as long as wide; sessoblasts with wide, lacy annulus; cystids keeled; zooids slender and adherent; Rocky Mountains, IL, MI, and PA . . . . **Plumatella fruticosa** Allman
    Length of floatoblasts less than twice the width; sessoblasts not as above **17**
17. Cystid strongly keeled, furrowed, and with an emarginate orifice (Fig. 20G); cuticula dark; dorsal surface of floatoblasts largely covered by float cells; common and widely distributed.

Plumatella emarginata Allman

    Cystid not strongly keeled; floatoblasts not as above **18**
18. Colony translucent to light colored; zooecium adherent for only part of the length of each zooid; colony loosely branched (Figs. 20A, H, J); the most common U.S. bryozoan.

Plumatella repens (L.)

    Colony dark colored; zooecia extending upright and cemented together into masses of parallel tubes (Fig. 3); IL, ME, CO, MA; some investigators report overlapping of morphological features between this form and *P. repens*, and accordingly consider this form a variety of *P. repens*.

Plumatella fungosa (Pallas)

REFERENCES

Agrawal, V., and K. S. Rao. 1981. Studies on the body wall of some phylactolaematous Ectoprocta. 1. Plumatella casmiana Oka. *Bioresearch* 5:41–46.

Allman, J. 1856. *A monograph of the fresh-water Polyzoa.* 119 pp. Ray Society, London.

Brown, C. J. D. 1933. A limnological study of certain fresh-water Polyzoa with special reference to their statoblasts. *Trans. Am. Microsc. Soc.* 52:271–316.

Bushnell, J. H. 1965. On the taxonomy and distribution of fresh-water Ectoprocta in Michigan. Part I. *Trans. Am. Microsc. Soc.* 84:231–244.

———. 1965a. On the taxonomy and distribution of fresh-water Ectoprocta in Michigan. Part II. *Ibid.* 339–358.

_____. 1965b. On the taxonomy and distribution of freshwater Ectoprocta in Michigan. III. *Ibid.* 529–548.

_____. 1966. Environmental relations of Michigan Ectoprocta, and dynamics of natural populations of Plumatella repens. *Ecol. Monogr.* **36**:95–123.

_____. 1968. Aspects of architecture, ecology, and zoogeography of freshwater Ectoprocta. *Atti Soc. Ital. Sci. Nat. Mus. Civ. St. Nat. Milano* **108**:129–151.

_____. 1973. The freshwater Ectoprocta: a zoogeographical discussion. *In: Living and fossil bryozoa: Recent advances in research,* pp. 503–521. E. P. Larwood (ed.), Academic, London.

Bushnell, J. H., and K. S. Rao. 1974. Dormant or quiescent stages and structures among the Ectoprocta: Physical and chemical factors affecting viability and germination of statoblasts. *Trans. Am. Microsc. Soc.* **93**:524–543.

_____. 1979. Freshwater Bryozoa: micro-architecture of statoblasts and some Aufwuchs animal associations. *Syst. Assoc. Spec. Vol.* "Advances in Bryozoology": **13**:75–91.

Dendy, J. W. 1963. Observations on bryozoan ecology in farm ponds. *Limnol. Oceanogr.* **8**:478–482.

Emschermann, P. 1987. Creeping propagation stolons— an effective propagation system of the freshwater entoproct Urnatella gracilis Leidy (Barentsiidae). *Arch. Hydrobiol.* **108**:439–448.

Everitt, B. 1975. Fresh-water Ectoprocta: distribution and ecology of five species in southeastern Louisiana. *Trans. Am. Microsc. Soc.* **94**:130–134.

Geiser, S. W. 1937. Pectinatella magnifica Leidy an occasional river-pest in Iowa. *Field & Laboratory* **5**:65–76.

Hurrell, H. E. 1936. Freshwater Polyzoa in English lakes and rivers. *Turtox News* **14**:1–2, 20–21.

Ingold, J. L. et al. 1984. Ecology and population genetics of the freshwater bryozoan, Pectinatella magnifica Leidy. *J. Freshwater Ecol.* **2**:499–508.

Kraepelin, K. 1887. Die deutschen Süsswasserbryozoen. Eine Monographie. I. Anatomisch-systematischer Teil. *Abh. Natw. Vereins. Hamburg* **10**:1–168.

_____. 1892. Die deutschen Süsswasserbryozoen. Eine Monographie. II. Entwicklungsgeschichtlicher Teil. *Ibid.* **12**:1–68.

Lacourt, A. W. 1968. A monograph of the fresh-water Bryozoa–Phylactolaemata. *Uitg. Rijksmus. Nat. Hist. Leiden* **93**:1–159.

Marcus, E. 1925. Bryozoa. *Biol. Tiere Deutschl.* **47**:1–46.

_____. 1926. Beobachtungen und Versuche an lebenden Süsswasserbryozoen. *Zool. Jahrb. Abt. Syst. Oekol. Geogr. Tiere* **52**:279–350.

_____. 1942. Sôbre Bryozoa do Brasil II. *Univ. Sao Paulo. Bol. Zool.* **6**:57–96.

Mayr, E. 1968. Bryozoa versus Ectoprocta. *Syst. Zool.* **17**:213–216.

Mukai, H. 1974. Germination of the statoblasts of a freshwater Bryozoan, Pectinatella gelatinosa. *J. Exp. Zool.* **187**:27–40.

Mundy, S. F. 1980. A key to the British and European freshwater bryozoans. *Sci. Publ. Freshwater Biol. Assoc. U. K.* **41**:1–31.

Newcomer, W. S. 1950. Additions to the description of Plumatella repens furcifer Jullien. *Am. Midl. Nat.* **44**:205–210.

Oda, S. 1959. Germination of the statoblasts in freshwater Bryozoa. *Sci. Rep. Tokyo Kyoiku Daigaku, B,* **135**:90–135.

Raddum, G. G., and T. M. Johnsen. 1983. Growth and feeding of Fredericella sultana (Bryozoa) in the outlet of a humic acid lake. *J. Freshwater Ecol.* **2**:499–508.

Rogick, M. D. 1935. Studies on fresh-water Bryozoa. II. The Bryozoa of Lake Erie. *Trans. Am. Microsc. Soc.* **54**:245–263.

_____. 1937. Studies on fresh-water Bryozoa. VI. The finer anatomy of Lophopodella carteri var. typica. *Ibid.* **56**:367–396.

_____. 1938. Studies on fresh-water Bryozoa. VII. On the viability of dried statoblasts of Lophopodella carteri var. typica. *Ibid.* **57**:178–199.

_____. 1940. Studies on fresh-water Bryozoa. IX. Additions to New York Bryozoa. *Ibid.* **59**:187–204.

_____. 1940a. Studies on fresh-water Bryozoa. XI. The viability of dried statoblasts of several species. *Growth* **4**:315–322.

_____. 1941. Studies on fresh-water Bryozoa. X. The occurrence of Plumatella casmiana in North America. *Trans. Am. Microsc. Soc.* **60**:211–220.

_____. 1943. Studies on fresh-water Bryozoa. XIV. The occurrence of Stolella indica in North America. *Ann. N.Y. Acad. Sci.* **45**:163–178.

_____. 1945. Studies on fresh-water Bryozoa. XV. Hyalinella punctata growth data. *Ohio J. Sci.* **45**:55–79.

Rogick, M. D., and C. J. D. Brown. 1942. Studies on fresh-water Bryozoa. XII. A collection from various sources. *Ann. N.Y. Acad. Sci.* **43**:123–144.

Rogick, M. D., and H. van der Schalie. 1950. Studies on fresh-water Bryozoa. XVII. Michigan Bryozoa. *Ohio J. Sci.* **50**:136–146.

Toriumi, M. 1956. Taxonomical study on fresh-water Bryozoa. XVII. General consideration: interspecific relation of described species and phylogenetic consideration. *Sci. Rep. Tohoku Univ. Ser. 4, Biol.* **22**:57–88.

_____. 1971. Additional observations on Plumatella repens (L.) (a fresh-water bryozoan), IV. Reexamination on the field materials of P. repens and P. fungosa. *Bull. Mar. Biol. Sta. Asamushi* **14**:117–126.

Twitchell, G. B. 1934. Urnatella gracilis Leidy, a living Trepostomatous bryozoan. *Am. Midl. Nat.* **15**:629–655.

Williams, S. R. 1921. Concerning "larval" colonies of Pectinatella. *Ohio J. Sci.* **21**:123–127.

Wood, T. S. 1971. Laboratory culture of fresh-water Ectoprocta. *Trans. Am. Microsc. Soc.* **90**:92–94.

13

ANNELIDA
(Aquatic Earthworms, Branchiobdellida, Polychaetes, Archiannelida, Leeches)

Typical segmented worms, constituting the Phylum Annelida, are constructed on a tube-within-tube plan. The body wall is soft, muscular, and covered with a very thin cuticle. The specialized digestive tract has a terminal mouth and anus and is supported in the coelom by thin transverse septa that mark the internal segmental divisions. Unlike the condition in arthropods, segmentation is homonomous. There is an extensive closed circulatory system, and a few to most segments have a pair of long, tubular nephridia, or excretory organs. The two ventral nerve cords and their segmental ganglia are fused along the median line; the brain typically consists of a pair of cerebral ganglia, and these connect with the nerve cord by two circumpharyngeal connectives. Anterior eyespots are present in some genera. Except for certain aquatic species that reproduce by budding, reproduction is syngamic, and individuals are hermaphroditic. Cross fertilization is the rule.

Five classes of annelids are represented in fresh waters. Oligochaeta, or earthworms, have many fresh-water representatives; the Hirudinea, or leeches, are dominantly a fresh-water group; the Polychaeta, however, which

are so varied and richly represented in salt water, occur in fresh water only as a few rare and sporadic species. The Branchiobdellida are highly modified commensals or parasites of crayfishes. Fresh-water Archiannelida are rare. These five classes are easily distinguished from each other, and because they are so different they are considered in separate sections of this chapter.

Polychaeta have muscular, paired, lateral projections from the body wall, called parapodia; these are absent in the other four classes occurring in fresh water. Groups of setae accompany the parapodia of polychaetes, and setae are also present in the Oligochaeta and archiannelids, though not in the Hirudinea and branchiobdellids. Polychaetes usually have accessory tentacles or other appendages at the anterior end but these are absent in the other two groups. Leeches are dorsoventrally flattened, have an oral and a caudal sucker, and the body segments are subdivided into superficial annuli. None of these features occur in the oligochaetes, polychaetes, archiannelids, and branchiobdellids, although the branchiobdellids have a caudal sucker.

OLIGOCHAETA (Aquatic Earthworms)

Aquatic oligochaetes have the same fundamental structure as the common terrestrial earthworms, and the anatomy, physiology, and behavior of the latter are so familiar to everyone who has been exposed to an ele-

mentary course in college zoology that it would be superfluous to repeat these details here. Instead, the various specializations and peculiarities of aquatic oligochaetes will be emphasized.

General characteristics. Representatives of 10 families of oligochaetes occur in the fresh waters of the United States. The Aeolosomatidae, Naididae, and Tubificidae are strictly aquatic. The Lumbriculidae are semiaquatic. Some species of Haplotaxidae are semiaquatic or amphibious and occur especially in debris at the edges of streams, ponds, or in marshes and soggy ground; other species, however, are essentially terrestrial and never occur in water. One genus in the family Glossoscolecidae is regularly found in fresh waters, but other genera, as well as the Megascolecidae and Lumbricidae, are terrestrial and may be accidental in aquatic habitats. The Opistocystidae are extremely rare.

The great majority of true aquatic species are common in the mud and debris substrate of stagnant pools and ponds and in streams and lakes everywhere. They are sometimes especially abundant in masses of filamentous algae. Tubificidae occur even in the deepest parts of large lakes, but the shallows down to a depth of a meter are the usual habitat for the great majority of species.

Compared with amphibious and terrestrial forms, the true aquatic oligochaetes are much more delicately constructed and small, the usual length range being 1 to 30 mm. The body wall is thin, and the internal organs can be easily seen in living specimens.

The prostomium is a half-segment at the extreme anterior end of the body; it projects in a rooflike fashion above the mouth and is sometimes elongated into a proboscis. In locating various external and internal structures of oligochaetes it is customary to designate the body segments serially by Roman numerals, beginning at the anterior end. Thus the first complete body segment immediately behind the prostomium is designated as I.

The number of segments is quite variable, even in mature individuals, and commonly there is a plus or minus 25% variation from the average for various specimens of a particular species. The threadlike Haplotaxidae have the largest number of segments, sometimes up to 500. At the opposite extreme the Aeolosomatidae and Naididae usually have between 7 and 40 segments. The other families have an intermediate number of segments,

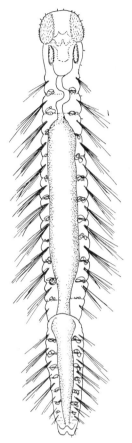

Fig. 1.–Dorsal view of *Aeolosoma,* a typical aquatic oligochaete showing separating bud, ×50. (Modified from Marcus, 1944.)

tubificids, for example, usually consisting of 40 to 200 segments.

The chitinoid setae are arranged in four bundles, two dorsolateral and two ventrolateral. Setae are never found on segment I, but usually begin on II, although in some genera they are absent from several successive segments, either dorsally, ventrally, or both. Detailed structure of the setae and their arrangement are important as taxonomic characters. A bundle may consist of from 1 to as many as 20 setae, and the number in a particular bundle may be constant from one specimen to another within a species, or it may vary. Setae may be long, short, straight, curved, sigmoid, hair (capilliform), aciculate, pectinate, bifurcate, hooked bifurcate, or uncinate (Fig. 3). Two thirds or more of the total length of a capilliform seta usually projects from the body wall, but only about a quarter or less of an aciculate seta is exposed. Sigmoid or crotchet setae are usually called needle setae when the nodule is small or absent and when they are more or less straight.

Locomotion. Usually locomotion is a crawling movement on or in the superficial layers of the substrate; it is similar to that of the earthworm and involves contractions of the muscular body wall and obtaining a purchase with the setae. A few naids, such as *Pristina* and *Stylaria,* are effective swimmers that move along just above the substrate in a serpentine manner. Some species of *Aeolosoma* swim about with the aid of the beating of the cilia of the anterior end.

Feeding. The great majority of aquatic oligochaetes obtain nutriment by ingesting quantities of the substrate after the manner of the earthworm, the organic component being digested as it passes through the alimentary canal. Sometimes the food is ingested at a depth of 2 or 3 cm below the water-substrate interface. Under some circumstances the food may consist largely of filamentous algae, diatoms, or simply miscellaneous plant and animal detritus. *Chaetogaster* is one of the few carnivorous forms. It feeds on entomostraca, insect larvae, protozoans, and other oligochaetes. *Aeolosoma* feeds on particulate debris and microorganisms, which are swept toward the mouth by ciliary currents set up at the anterior end and by a type of suction feeding.

Respiration. Because the body wall is thin and often well supplied with capillaries, most of the carbon dioxide and oxygen exchange occurs through the general body surface. In most naids and in some tubificids water is taken into the anus and passed forward for a variable distance by antiperistalsis and ciliary action, and these processes are thought to constitute an accessory respiratory mechanism.

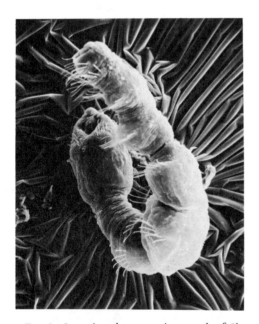

FIG. 2.–Scanning electron micrograph of *Chaetogaster diastrophus,* ×240. Anterior end in ventral view; posterior end in lateroventral view. The fission zone separates the two zooids. Background ripples are the gold-plated surface of the stud. (Photo courtesy of Donald L. Elrick.)

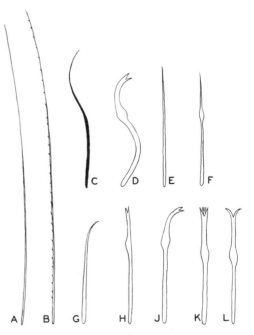

FIG. 3.–Typical setae of aquatic Oligochaeta. A and B, capilliform; C through L, aciculate; B, serrate; C and D, sigmoid; D, H, and J, bifurcate; G, uncinate; J, hooked; K, pectinate; L, biuncinate.

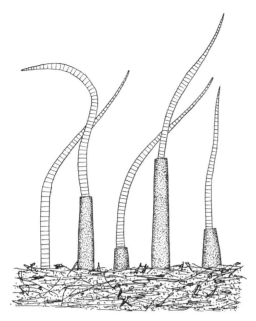

FIG. 4.–*Tubifex,* a typical tube builder, ×3.

Dero has handsome ciliated gills surrounding the anal region. The posterior extremity has the form of a shallow cup, and the gills originate within the cup or from its margin. *Branchiura* has a fingerlike dorsal and ventral gill on each of the posterior segments. All oligochaete gills are well supplied with blood vessels. Gilled species remain quietly in position with the anterior portion of the body hidden in the substrate and in a tube constructed of fine debris and projecting up from the substrate for a variable distance. The posterior end projects vertically out of the tube and into the water. When disturbed, the worm withdraws into its tube and the substrate in a fraction of a second.

Most tubificids also build tubes, the projecting posterior end of the animal being waved about vigorously in the water, presumably to circulate the water and make more oxygen available to the body surface. Tubificids are commonly red, owing to dissolved erythrocruorin in the blood.

Epidermal sensory structures. Many species of Naididae have a pair of eyespots. They are entirely epidermal, the overlying cuticle being unmodified. Each eye consists of a transverse row of five or six small visual cells and a large number of pigment granules that cover the sensory cells medially and posteriorly. Other epidermal sense organs, consisting of either single cells or aggregates and often arranged symmetrically, occur over the general body surface of all oligochaetes but are especially concentrated at the anterior end. Sensory hairs are common in the Naididae, and in *Slavina appendiculata* they are borne on prominent sensory papillae. Epidermal sensory cells are generally assumed to be sensitive to tactile, thermal, and chemical stimuli.

Reproduction. According to the published literature, asexual reproduction by budding is the rule in the Aeolosomatidae and Naididae, only a few instances of syngamic reproduction having been reported. It is our experience, however, that sexual reproduction may sometimes be very common, especially in *Chaetogaster*. In asexual reproduction, the budding zone is localized in one of the segments toward the posterior end (Figs. 2 and 3), and the number of segments anterior to the budding segment is often designated by n. The location of the budding segment is usually constant within a single individual, but under experimental conditions it is possible to alter its position slightly; favorable temperature, food, and oxygen conditions will move the zone anteriorly a segment or two; unfavorable conditions may move it posteriorly a segment or two. On the other hand, the specific location of the fission zone varies greatly from one individual to another within a species. Thus, in *Nais elinguis, n* varies from 12 to 20, and in *Stylaria fossularis, n* varies from 8 to 23 but is usually 18.

The budding segment grows and undergoes repeated transverse divisions that result in four or more anterior segments of a new posterior worm and several new posterior segments of the anterior parent worm. These regions rapidly become di..erentiated (Fig. 5). The two animals remain attached for a variable length of time, and the more posterior individual may in turn produce a third individual posterior to its budding zone. This process may be repeated so that in some species there may be chains of as many as four to eight zooids in various stages of develop-

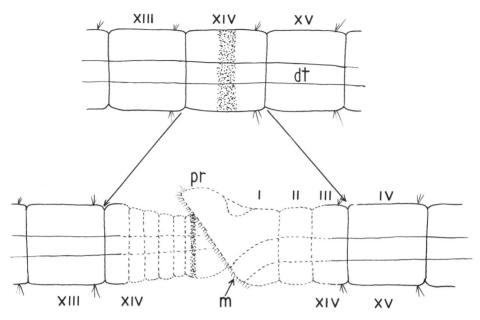

FIG. 5.–Budding in *Aeolosoma,* showing segmental derivatives in daughter zooids. dt, digestive tract; m, mouth; pr, prostomium. (Modified from Marcus, 1944.)

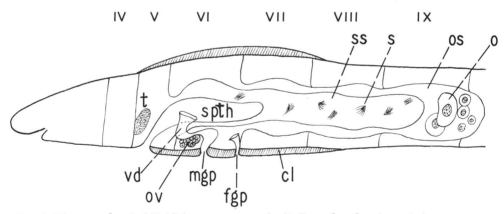

FIG. 6.–Diagram of typical Naididae sex organs. cl, clitellum; fgp, female genital pore; mgp, male genital pore; o, ovum; os, ovisac; ov, ovary; s, sperm; spth, spermatheca; ss, sperm sac; t, testis; vd, vas deferens.

ment. After an interval, however, two adjacent and mature zooids separate and become completely independent individuals. Under favorable conditions most naids reproduce a new individual every 2 or 3 days. In *Dero* it is unusual to find a chain of more than two zooids. In general, zooid chains are longest in spring and autumn.

Syngamic reproduction is essentially simi-

lar to the familiar story in the earthworm. In the true aquatic species, however, the clitellum is often thin and inconspicuous. Cocoons containing embryos are deposited on rocks, vegetation, and debris, especially in the late summer and early autumn when a large percentage is consumed by small fish. Syngamic reproduction has often been reported as early as July. Under rare circumstances reproduction in isolated tubificids indicates self-fertilization. Life cycles of macroscopic species usually take 1 or 2 years.

The capacity to regenerate lost or damaged anterior or posterior segments varies within the aquatic oligochaetes, but it is best developed in the Naididae and Tubificidae.

Ecology. Aquatic oligochaetes occupy a niche equivalent to that occupied by terrestrial species, and they feed on bottom mud and mix it much as earthworms effectively mix the surface layers of garden and meadow soils.

Temperature is not usually a limiting factor, but it often determines the relative abundance of oligochaetes. Some amphibious species, especially enchytraeids and lumbriculids, are notable because they are found on the shores of cold mountain streams and brooks originating from snowbanks and glaciers. Many microannelids occur in hyporheic, phreatic, and psammolittoral situations as a part of the meiobenthos. Some species of *Aeolosoma* are known to encyst when the water temperature is 6°C or lower. A translucent, hardened coat of mucus is formed around the tightly coiled worm. Emergence from the cyst occurs at about 20°C. Several Naididae are known to produce cocoons in times of environmental stress.

Tubificidae are the dominant forms at depths exceeding 1 m, and in the deep waters of lakes there are sometimes more than 8000 individuals/m². By far the most concentrated populations are found in streams and rivers that are polluted with sewage. Here tubificids, especially the common and cosmopolitan *Tubifex tubifex,* occur in dense waving masses. In fact, this species is usually considered an "indicator" of organic pollution, especially where the water is between 10 and 60% saturated with oxygen. *Tubifex,* some enchytraeids, and a few other macroscopic forms are common inhabitants of sewage settling tanks and trickling filters. *Tubifex* often exhibits negative aerotaxis; that is, it actively migrates from regions of high dissolved oxygen to low oxygen.

Two to four species of *Chaetogaster* may commonly be found in the same collection, but, in general, little is known about interspecies competition.

Most of the true aquatic species are able to thrive in low concentrations of dissolved oxygen, and many are also able to withstand the complete absence of oxygen for extended periods. Tubificids, for example, can endure the summer and winter periods of stagnation in lakes. Nevertheless, populations cannot be maintained indefinitely in the absence of oxygen. Dausend (1931) found that only one third of the specimens of *Tubifex* that he used were able to survive anaerobic conditions for 48 days at 0 to 2°C, and at higher temperatures the fraction was progressively smaller. Other workers report the survival of very small percentages of *Tubifex* populations after 120 days under anaerobic conditions. In general, the lower the oxygen concentration the more the animal projects from its tube and the more vigorous are the waving aeration movements of the posterior part of the body, but when the oxygen concentration nears complete exhaustion most tubificids become relatively quiescent.

Of the three known species of luminescent fresh-water invertebrates, one, an enchytraeid annelid, has been reported from the shores of a river in the USSR in concentrations of 100 to 250/m².

Dispersal, geographic distribution. Oligochaetes undoubtedly migrate easily throughout whole drainage systems, but passive and accidental overland transport presumably is most effective during the cocoon stage in those species that have syngamic reproduction. *Aeolosoma* and *Tubifex,* however, form a resistant mucus cyst stage that may be easily transported from place to place.

Juveniles of *Tubifex* are known to produce temporary protective tests composed of mucus, sand particles, and organic detritus.

The aquatic oligochaete fauna of the United States is poorly known. Collecting has been scanty and sporadic, and there have been less than 50 major American papers published since 1900. Nevertheless, as additional records slowly accumulate, it is becoming increasingly apparent that many of our species are cosmopolitan and occur the world over or are widely distributed in the Northern Hemisphere. *Aeolosoma variegatum,* several species of *Chaetogaster, Nais obtusa, Slavina appendiculata, Dero limosa,* and *Tubifex tubifex* are all examples

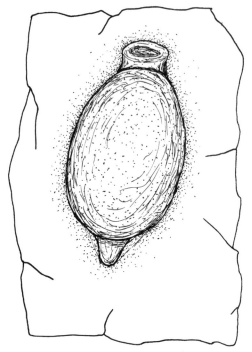

F<small>IG.</small> 7.–Cocoon of *Stylaria* deposited on a bit of debris, ×60.

of common and cosmpolitan species. Of the species of Aeolosomatidae and Naididae occurring in the United States, less than 20% have not yet been reported from other continents.

The Lumbricidae are restricted to temperate and cold-temperate zones, and most of the Megascolecidae and Glossoscolecidae are tropical to subtropical. Some few species have an interesting sporadic distribution. The "rare" *Branchiura sowerbyi,* for example, has been collected only one to several times in most states. A few "fresh-water" genera also have salt-water representatives, especially *Peloscolex, Tubifex,* and *Clittelio.*

Collecting. If bottom material is finely divided, a sieve or strainer may be used for field collecting. If the straining process is too vigorous, the more delicate species will be damaged and fragmented. Otherwise, the bottom debris should be brought into the laboratory where it may be examined immediately or allowed to stand until the worms can be seen at the surface of the debris or

crawling up the sides of the container. Masses of filamentous algae can be rinsed or washed out with a stream of water. The small species can best be found with the aid of dissecting binoculars. Amphibious and large species can usually be picked up with tweezers by sorting stream and lakeside trash.

Culturing. The success of a laboratory culture often depends on the amount and kind of water supply and the available oxygen, but many species can be grown in their natural substrate when it is kept in shallow pans into which water drips or slowly flows. Water should never be changed rapidly. Needham et al. (1937) give directions for raising Naididae in rice–agar cultures. Naids often become abundant in old protozoan cultures.

Preparation. For gross structure and even for many fine details of small oligochaetes there is no substitute for the examination of living specimens. Nevertheless, preserved specimens are unavoidable for lengthy examinations and comparisons. Perhaps the greatest difficulty encountered in working with Aeolosomatidae, Naididae, Tubicifidae, and Lumbriculidae is the usual strongly contracted and distorted condition, which makes careful anatomical work impossible. Chloretone, chloroform vapors, and magnesium sulfate have all been suggested as narcotic agents, but almost invariably the posterior end of the worm being narcotized disintegrates long before the anterior end has become quiet. Nevertheless, the writer has had some sucess with 2% hydroxylamine hydrochloride. Regardless of the method used, it is almost impossible to kill *Dero* with expanded gills.

Small annelids are best fixed and preserved in the field with 5% formalin. If necessary, Amman's lactophenol is a good clearing agent. It consists of 40 g of carbolic acid, 40 ml of lactic acid, 80 ml of glycerol, and 40 ml of water. Rinse preserved worms in water, place on a slide, add a few drops of Amman's fluid, and then a circular coverslip. In 30 minutes to a day or two the worms will be sufficiently cleared to distinguish setae and other details. For permanent mounts, replace Amman's

FIG. 8.–Mixed colony of *Tubifex* (long, flexible individuals) and *Lumbriculus* (shorter, less flexible individuals), ×1. (From Von Wagner.)

fluid with polyvinyl lactophenol or warm glycerine jelly and ring the coverslip with Murrayite.

Most investigators advocate killing and fixing small oligochaetes without the use of a narcotic. Some tubificids will die in an extended condition if diluted fixatives are used. If many specimens are being fixed, they may be placed in a shallow dish, and then the water is poured off. The worms thereupon elongate and crawl about, and if they are flooded with hot Schaudinn's fixative, a good percentage of the worms will remain elongated and in good condition. Permanent glycerin jelly mounts are generally satisfactory, either with or without staining.

Large amphibious or terrestrial species are best killed and preserved in an extended condition in accordance with standard methods. If the specimens are to be sectioned, they should be first kept in water or on wet filter paper until the digestive tract is free of grit.

The number, arrangement, and structural details of the setae are essential for the identification of Aeolosomatidae, Naididae, Tubificidae, and some Lumbriculidae, but whole mounts do not often allow an examination of the finer details. If living specimens are available, it is best to place one in water on a slide with a circular cover slip. Then remove most of the water at the edge of the cover slip with blotting paper and allow the mount to dry. The worm will be compressed, and then it will burst and flatten, leaving the body wall and setae nicely in place. If a drop of glycerin is placed at the edge of the cover slip, capillarity will draw it under, and such a slide can

be made semipermanent by ringing with Murrayite.

Similar preparations of larger dead or preserved specimens of Tubificidae and some Lumbriculidae may be made if they are left in water for a day or two until they begin to decompose and are soft enough for setae preparations.

Taxonomy. In addition to setae, other characters of taxonomic significance in small species are the location of the budding zone, gill structure, shape and size of the prostomium, and the arrangement of certain blood vessels. Body length is of little value since preserved specimens exhibit varying degrees of contraction. In addition, the published literature is often confusing because it is not made clear whether lengths refer to a single anterior zooid or a whole chain of zooids.

It is unfortunate that the identification of Lumbricidae, Megascolecidae, Glossoscolecidae, and some Lumbriculidae depends on internal details of the reproductive system, and though careful dissections are often adequate, it is frequently necessary to make stained serial sections of the segments containing the reproductive structures. Usually cross sections are sufficient, but some workers advocate longitudinal sections in addition. It is chiefly this tedious procedure that has discouraged taxonomic investigations of these annelids.

Because oligochaetes of the United States are relatively poorly known, the following key has certain limitations. Nevertheless, with the exception of a few rare and poorly described and unrecognizable species, this key is effective for essentially all genera of Aeolosomatidae, Naididae, and Tubificidae. The Lumbricidae, Megascolecidae, Enchytraeidae, and Glossoscolecidae are almost exclusively terrestrial and are not considered beyond the family characters.

It should also be emphasized that aquatic oligochaetes are constantly undergoing taxonomic manipulation; changes in genus and species names are frequent, and, to the nonspecialist, confusing.

KEY TO GENERA AND SPECIES OF OLIGOCHAETA

1. Reproduction normally by fission, producing chains of individuals, but sexual reproduction may occur occasionally; living animals more or less translucent and delicate; with 1 to 6 setae per bundle; less than 25 mm long; cosmopolitan and common on the substrate of pools, ponds, and lakes. (part) .... Order **HAPLOTAXIDA, 2**
   Reproduction always sexual; no chains of individuals; living animal more or less opaque; more than 25 mm long **4**
2. Single zooids usually less than 5 mm long, but a few species sometimes reaching 10 mm; usually with minute red, yellow, or greenish pigment globules in the epithelium of living animals; with cilia on ventral and sometimes lateral surfaces of prostomium (Figs. 1, 9D); cerebral ganglia permanently connected with epidermis; this family has uncertain fundamental taxonomic affinities, and may not even be annelids .... AEOLOSOMATIDAE, **Aeolosoma, 10**
   Usually 5 to 25 mm long; without colored oil globules in epithelium of living animals; without cilia anteriorly; with septa; cerebral ganglia free in coelom .... Order **HAPLOTAXIDA, 3**
3. Posterior end with two lateral and one median process; extremely rare.
   OPISTOCYSTIDAE, **Opistocysta, Crustipellis**
   Posterior end not with two lateral and one median process; common and abundant everywhere.
   NAIDIDAE, **14**
4. Body threadlike, 0.5 to 1.0 mm in diameter and 100 to 300 mm long; with two large ventral and two small dorsal sigmoid setae per segment; up to 500 segments; semiaquatic; in wet earth, marshes, ditches, mud, and under rocks at streamside; several rare and 1 common species.
   Order **HAPLOTAXIDA,** HAPLOTAXIDAE, **Haplotaxis gordioides** (Hartmann)
   Not threadlike; with a different pattern of setae **5**
5. Large, thick-bodied worms that are essentially terrestrial but occasionally occur in mud or debris of stream and lake margins or in trickling filters of sewage plants; setae sigmoid; true earthworms; usually 5 to 30 cm long .... Order **HAPLOTAXIDA, 8**
   Smaller, thin worms, mostly aquatic; setae sigmoid or some other shape; usually less than 8 cm long but a few up to 15 cm **6**
6. Whitish to pinkish; spermathecae opening on IV or V; without a tube; usually two to six setae per bundle; setae all similar; usually 10 to 30 mm long; terrestrial; many species; common in wet debris along margins of streams and lakes ...... Order **HAPLOTAXIDA,** ENCHYTRAEIDAE
   Red to brown; spermathecae opening behind VI; anterior end of worm sometimes hidden in a tube that projects vertically from the substrate and is composed of mud and debris; two setae per bundle, or an indeterminate number per bundle; dorsal and ventral setae similar or dissimilar; typically aquatic **7**
7. Setae all of one form, single pointed or sometimes more or less distinctly bifurcate; four bundles of two setae each per segment; usually reddish or brownish worms; 8 to 100 mm long; prostomium present or absent; setae single pointed or bifid; in mud or debris of pools, ponds, and lakes.
   Order **LUMBRICULIDA,** LUMBRICULIDAE, **74**
   Dorsal and ventral setae different; ventral setae bifurcate, dorsal setae bifurcate or a mixture of two or three types; four bundles of an indeterminate number each per segment; worm usually reddish.
   Order **HAPLOTAXIDA,** TUBIFICIDAE, **45**
8. With a well-formed gizzard at the beginning of the intestine; four pairs of setae per segment; clitellum beginning at or behind XVIII .... LUMBRICIDAE
   Without a gizzard at the beginning of the intestine; clitellum beginning before XVIII; mostly tropical and subtropical **9**

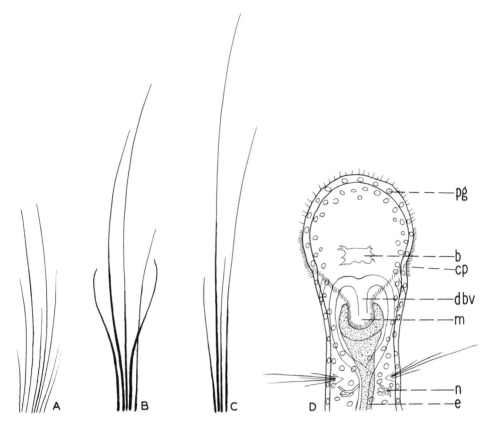

FIG. 9.–Structure of *Aeolosoma*. A, setal bundle of *Aeolosoma hemprichi*; B, setal bundle of *A. leidyi*; C, setal bundle of *A. headleyi*; D, ventral view of anterior end of *A. leidyi*. b, brain; cp, ciliated pit; dbv, dorsal blood vessel; e, esophagus; m, mouth; n, nephridium; pg, pigment globules.

9. Male genital pores on XVII, XVIII, or XIX; 8, 12, or more setae per segment, in the latter case forming rings that may be either closed or broken dorsally and ventrally; clitellum beginning with or in front of XV .... MEGASCOLECIDAE
   Male genital pores in anterior portion of clitellum or in front of it; eight setae per segment; clitellum beginning with or in front of XXV .... GLOSSOSCOLECIDAE
10. Setae bundles composed of long, flexible hair setae and short, stiff sigmoid setae (Fig. 9B) 11
    Sigmoid setae absent (Fig. 9C) 12
11. Sigmoid setae may occur in all bundles but mostly absent from II; epidermal glands green.
                                                          Aeolosoma leidyi Cragin
    Sigmoid setae only in V and posteriorly; epidermal glands greenish yellow to olive green.
                                                          Aeolosoma tenebrarum Vejd.
12. Longest hair setae more than 175 μm; body 150 to 300 μm wide ..... **Aeolosoma headleyi** Beddard
    Longest hair setae less than 160 μm; body 40 to 175 μm wide 13
13. Epidermal glands orange red .... **Aeolosoma hemprichi** Ehr.
    Epidermal glands green or yellow .... **Aeolosoma variegatum** Vejd.
14. Dorsal setae completely lacking; carnivorous; mouth and pharynx large (Figs. 2; 10E, F) 15
    With dorsal setae in a few to all segments 20
15. Ventral setae in all segments; mouth ventral with a prominent prostomium; four dark anterior pigment bands (Fig. 10A); body more or less covered with detritus; usually in mud substrates; up to 30 mm long; widely distributed and common .... **Ophidonais serpentina** (Müller)

Ventral setae in all except segment III; mouth terminal or subterminal; prostomium slightly developed .... **Chaetogaster,** 16

16. Tip of setae with a pair of distinct teeth (Figs. 10C, G) 17
    Tip of setae simple or with a very fine bifurcation (Fig. 12L); northeastern.
                                               Chaetogaster setosus Svetlov

17. Teeth of setae strongly reflexed (Fig. 10); epizoic on surface or in mantle cavity of snails, fingernail clams, and unionid clams; chain of zooids 2 to 6 mm long; usually eight segments in first animal of a chain; common and widespread .... **Chaetogaster limnaei** von Baer
    Teeth of setae not reflexed **18**

18. Setae of segment II 120 or more μm long; mouth terminal, prostomium not developed 19
    Setae of segment II less than 120 μm long; mouth subterminal, prostomium conspicuous (Fig. 10D); widespread and common .... **Chaetogaster diastrophus** (Gruit.)

19. Setae of segment II less than 200 μm long; northeastern ...... **Chaetogaster crystallinus** Vejd.
    Setae of segment II more than 200 μm long; widely distributed.
                                     Chaetogaster diaphanus (Gruit.)

20. Some or all dorsal fascicles with hairlike setae (Fig. 12K) 21
    All dorsal fascicles without hairlike setae 23

21. Segment II with dorsal setae 22
    Segment II without dorsal setae 28

22. Prostomium a long thin proboscis (Fig. 12K); about 6 species; common and widely distributed.
                                               Pristina
    Prostomium not in the form of a proboscis 42

23. Dorsal setae straight, with a slight swelling or minutely cleft (Fig. 10B); common and widely distributed (see couplet 2) .... **Ophidonais serpentina**
    Dorsal setae at least slightly sigmoid, apex bifurcate 24

24. Segments III and IV with dorsal setae; widely distributed .... **Amphichaeta**
    Segments III and IV without dorsal setae 25

25. Segment V with dorsal setae 26
    Segment V without dorsal setae 41

26. One tooth of anterior ventral setae about twice as long as the other; uncommon in southern rivers and streams and estuaries; several species .... **Paranais**
    One tooth of anterior ventral setae only slightly longer than the other 27

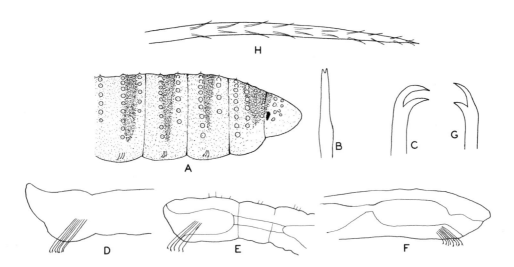

FIG. 10.–Structure of Naididae. A, anterior end of *Ophidonais serpentina*; B, dorsal seta of *O. serpentina*; C, tip of seta of *Chaetogaster limnaei*; D and E, anterior end of *C. diastrophus*; F, anterior end of *C. limnaei*; G, terminal bifurcation of seta of *C. diaphanus*.

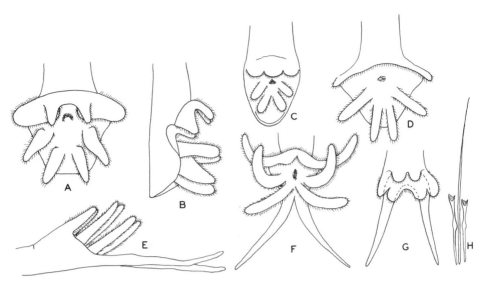

FIG. 11.–Structure of Naididae. A, dorsal view of posterior end of *Dero digitata*; B, lateral view of posterior end of *D. digitata*; C, dorsal view of posterior end of *D. obtusa*; D, dorsal view of *Dero* sp.; E, lateral view of posterior end of *D. furcatus*; F, dorsal view of posterior end of *D. furcatus*; G, dorsal view of posterior end of *D. vagus*; H, dorsal bundle of setae of *D. vagus*. (B and E modified from Bousfield; D modified from Chen, 1940; F modified from Marcus, 1943.)

27. Ventral setae of segments II to IV about 60 μm long; eyes absent; widely distributed but uncommon.
    Paranais litoralis (Müller)
    Ventral setae of segments II to IV more than 60 μm long; widespread.
    Uncinais uncinata (Ørsted)
28. Dorsal setae not present in segments I to IX; eastern states .... **Haemonais waldvogeli** Bretscher
    Dorsal setae present in some or all of segments I to IX 29
29. Lobes or palps present on anal segment (Fig. 11); eyes absent; about 10 species; a common and widely
    distributed genus .... **Dero**
    No lobes or palps on anal segment 30
30. Prostomium with a proboscis (Figs. 13E, F) 31
    Prostomium without a proboscis 33
31. Capilliform setae of segments VI to VIII more than three times as long as those on segment IX; widely
    distributed .... **Arcteonais lomondi** (Martin)
    Capilliform setae of segments VI to VIII less than three times as long as those on segment IX; widely
    distributed .... **Stylaria, 32**
32. Prostomium appearing as a prostomial invagination (Fig. 13F) .... **Stylaria lacustris** (L.)
    Prostomium not invaginated (Fig. 13E) .... **Stylaria fossularis** Leidy
33. Segment III with dorsal setae (Fig. 12B); rare; NJ, FL .... **Bratislavia bilongata** (Chen)
    Segment III with no dorsal setae 34
34. Capilliform setae of segment VI three times as long as those of other segments; body covered with
    detritus; widely distributed .... **Slavina appendiculata** (d'Udekem)
    Capilliform setae of segment VI not conspicuously longer than those of other segments 35
35. Acicular setae bifurcate or pectinate 36
    Acicular setae with simple tips 39
36. Acicular setae pectinate (Figs. 12O–Q) 37
    Acicular setae bifurcate 38
37. Shape of ventral setae of segment II different from those in segment VI and posteriorly (Fig. 13H);
    rare; TX, LA .... **Dero**
    Shape of ventral setae of segment II similar to those in segment VI and posteriorly; eastern rivers.
    Allonais pectinata (Stephenson)

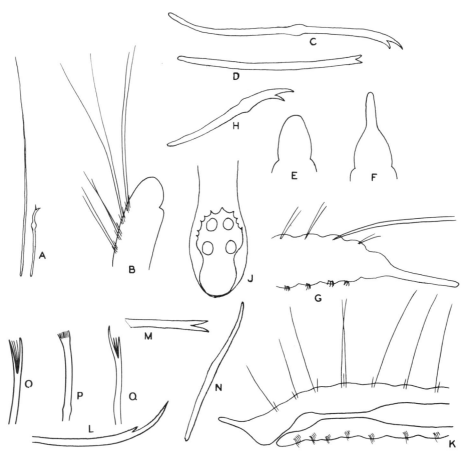

FIG. 12.–Structure of Naididae. A, dorsal bundle of *Pristina osborni*; B, anterior end of *Bratislava bilongata*; C, ventral seta of *Pristina aequiseta*; D, dorsal seta of *P. aequiseta*; E and F, extreme variations in anterior end of *P. breviseta*; G, anterior end of *P. longiseta*; H, thick seta of VI of *Nais pardalis*; J, branchial apparatus of *Dero nivea*; K, anterior end of *Pristina plumaseta*; L, tip of seta of *Chaetogaster setosus*; M, dorsal seta of *Nais elinguis*; N, dorsal seta of *N. simplex*; P, and Q, examples of pectinate setae (tips). (B modified from Chen, 1944; G modified from Marcus, 1943; K modified from Leidy, 1883; C, D, H, J modified from Brinkhurst et al., 1980; O–Q modified from Hiltunen and Klemm, 1980.)

38. Acicular setae slightly sigmoid; eyes absent; widely distributed .... **Specaria josinae** (Vejdovsky)
    Acicular setae straight or only slightly bent (Figs. 12M, N); eyes usually present (Fig. 13A); several common and widely distributed species .... **Nais**
39. More than three capilliform setae per fascicle on segments II to VI; northeastern.
                                               Vejdovskyella, 40
    No more than three capilliform setae per fascicle on segments II to VI; several common and widely distributed species .... **Nais**
40. Segments beyond V with ventral fascicles composed of three or more setae; eyes present.
                                 Vejdovskyella comata (Vejdovsky)
    Segments beyond V with a single seta in ventral fascicles; eyes absent.
                                 Vejdovskyella intermedia Bretscher
41. Distal tooth of setae of segment II longer than proximal tooth (Fig. 13J); widely distributed.
                                 Uncinais uncinata (Ørsted)
    Distal tooth of setae of segment II equal to or shorter than proximal tooth (Fig. 13K); northeastern.
                               Piguetiella michiganensis Hiltunen

FIG. 13.–Structure of Naididae. A, dorsal view of anterior end of *Nais communis*; B, dorsal bundle of setae of *N. communis*; C, ventral bundle of setae of *N. communis*; D, dorsal bundle of setae of *N. elinguis*; E and F, anterior end of *Stylaria lacustris*; G, dorsal bundle of setae of *S. lacustris*; H, ventral setae of *Dero* II, VI, and X; J, seta of *Uncinais uncinata* II; K, seta of *Piguetiella michiganensis* II; L, ventral setae of *Bratislavia unidentata* II and VI. (H–L from Hiltunen and Klemm, 1980.)

42. Tip of acicular setae simple; prostomium variable (Figs. 12E–G) **43**
    Tip of acicular setae bifurcate (Figs. 12C, D); widely distributed and common ...... **Pristinella**
43. Anterior ventral setae unlike those in segment VI (Fig. 13L) .... **Bratislavia unidentata** (Harman)
    All ventral setae similar **44**
44. Distal tooth of ventral setae twice as long as proximal tooth; widely distributed but uncommon.
                                      Stephensoniana trivandrana (Aiyer)
    Distal tooth of ventral setae less than twice as long as proximal tooth; rare in southeast.
                                                     Pristina

45. Capilliform setae present in II to V in dorsal fascicles (Figs. 3A, B) 46
    Capilliform setae not present in anterior dorsal fascicles 64
46. With a dorsal and a ventral gill on each segment in the posterior 25 to 40% of the body; 40 to 140 pairs of such gills (Fig. 14F); 20 to 185 mm long; fragments easily; pinkish gray; widely distributed in rivers .... **Branchiura sowerbyi** Bedd.
    Without such gills 47
47. Dermis opaque, densely papillate and encrusted with detritus, or dermis slightly opaque, with sparsely scattered papillae 48
    Dermis clear or slightly opaque; without papillae or encrusted detritus 49
48. Dermis coarse but not opaque, and not encrusted; scattered short fingerlike papillae; common in organic substrates in eastern states .... **Quistadrilus**
    Dermis opaque, encrusted, and densely papillate; several species; widely distributed and common.
                                              Spirosperma
49. Anterior dorsal fascicles composed of capilliform and bifurcate crotchet setae without intermediate teeth 50
    Anterior dorsal fascicles composed of capilliform and pectinate setae (Figs. 14G–J) 53
50. Dorsal crotchet setae on VII and posteriorly spatulate (Figs. 14K, L); silty substrates; common and widespread .... **Aulodrilus pigueti** Kowalewski
    Dorsal crotchet setae not spatulate 51
51. Teeth of anterior dorsal crotchet setae of equal length and not strongly divergent; northeastern states.
                                              Potamothrix
    Teeth of anterior dorsal crotchet setae of unequal length (Fig. 14M), but if teeth are similar in length, then they are strongly divergent 52
52. Distal tooth of ventral crotchet setae of VIII distinctly shorter and/or thinner than proximal tooth (Fig. 14M); with a mucoid tube; northeastern states .... **Aulodrilus**
    Distal tooth of ventral crotchet setae of VIII neither shorter nor thinner than proximal tooth; Great Lakes drainages .... **Rhyacodrilus**
53. Teeth of anterior dorsal pectinate setae long, distal tooth often much longer than proximal tooth (Fig. 14H); Great Lakes drainages; cold profundal waters .... **Rhyacodrilus**
    Distal tooth of anterior dorsal pectinate setae not distinctly longer than proximal tooth (Figs. 14B, C).
                                                    54
54. Posterior dorsal and ventral fascicles with some strongly sigmoid crotchet setae (Fig. 14N); scattered in eastern states .... **Tubifex harmani** Loden
    Posterior dorsal and ventral fascicles having no strongly sigmoid crotchet setae; anterior and posterior ventral setae not pectinate; setae generally capilliform; identification difficult unless worm is sexually mature 55
55. With no modified genital setae; 30 to 100 mm long; anterior end imbedded in mud bottoms; sometimes in a special tube, and waves the posterior end about in the water for aeration; common in polluted waters everywhere, sometimes in enormous numbers; this genus contains a few other rare species which are not included here .... **Tubifex tubifex** (O.F.M.)
    With modified genital setae in at least one segment between VII and XI 56
56. Specialized genital setae in one or more of segments VII to XI 57
    No genital setae in segment X; ventral setae in segment XI similar to somatic setae 61
57. Scalpel-shaped setae in some or all of segments VII to XI (Fig. 14O); Great Lakes drainages.
                                              Potamothrix
    Genital setae not scalpel shaped, present only in segment X or XI 58
58. With penial setae in segment XI; without spermathecal setae in segment X 59
    Without penial setae in segment XI; with spermathecal setae in segment X 60
59. Penial setae with enlarged bifurcate tip; no coelomocytes; Lake Tahoe.
                               Psammoryctides minutus Brinkhurst
    Penial setae blunt and simple pointed; with coelomocytes; several rare species scattered over northern states in cold waters .... **Rhyacodrilus**
60. Spermathecal setae in segment X parallel sided, with teeth fused to form a groove (Fig. 14 P); rare, scattered distribution .... **Psammoryctides**
    Spermathecal setae not parallel sided but elongated and spade shaped (Figs. 15A, B); rare, scattered distribution .... **Potamothrix**
61. Penis sheath well developed, elongated (Figs. 15C, D) 62
    Penis sheath indistinct or short and stout; known only from Lake Tahoe and the Mississippi River.
                                         Varechaetadrilus

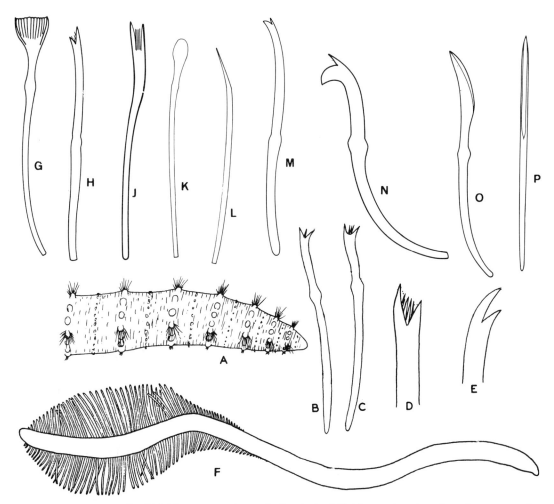

FIG. 14.–Structure of Tubificidae. A, anterior end of typical tubificid; B and C, dorsal pectinate setae of *Tubifex tubifex*; D, distal end of dorsal pectinate seta of *Ilyodrilus*; E, tip of seta of *Limnodrilus*; F, lateral view (diagrammatic) of *Branchiura sowerbyi*, ×1.1, showing dorsal and ventral gills; G, anterior dorsal pectinate seta of *Psammoryctides*; H and J, anterior dorsal pectinate seta of *Rhyacodrilus*; K, Facial view of posterior dorsal crotchet seta of *Aulodrilus pigueti*; L, lateral view of same; M, ventral crotchet seta of *Aulodrilus*; N, posterior dorsal crotchet seta of *Tubifex harmani*; O, seta of *Potamothrix*; P, *Psammoryctides* spermathecal setae of X. (G–P from Stimpson et al., 1975.)

62. Penis sheath a cylindrical reflexed tube (Fig. 15E); rare in Great Lakes drainages ...... **Tubifex**
     Penis sheath with different structure 63
63. Penis sheath conical and uniformly tapered (Fig. 15C); extremely common in rivers and lakes; easily
     confused with *Tubifex tubifex* .... **Ilyodrilus templetoni** (Southern)
     Penis sheath conical and abruptly tapered (Fig. 15D); uncommon in cold deep waters of Great Lakes
     drainages .... **Tubifex**
64. All setae of first seven segments simple to minutely bifurcate 65
     Anterior setae never simple, teeth distinct 66
65. Posterior to segment VII the dorsal setae are expanded to form several minute teeth (Fig. 15F); tube
     dweller; northeastern states .... **Aulodrilus americanus** Brinkhurst and Cook
     Posterior to segment VII the dorsal setae are bifurcate; CA .... **Telmatodrilus vejdovskyi** Eisen

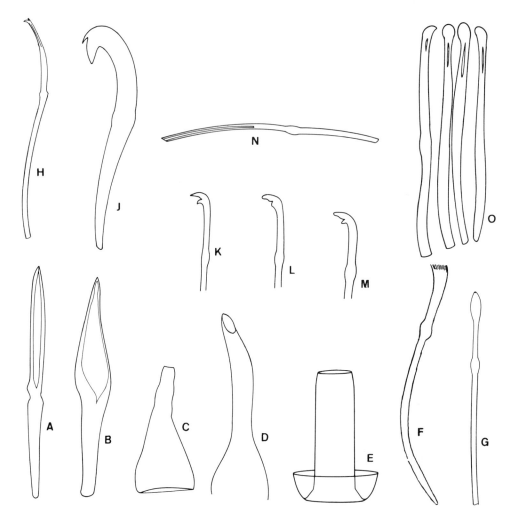

FIG. 15.–Structure of Tubificidae. A and B, spermathecal setae of *Potamothrix*; C, penis sheath of *Ilyodrilus templetoni*; D and E, same for *Tubifex*; F, posterior dorsal crotchet seta of *Aulodrilus americanus*; G and H, aspects of posterior dorsal crotchet seta of *Aulodrilus limnobius*; J, posterior dorsal seta of *Isochaetides curvisetosus*; K to M, variations in tip of ventral setae of *Limnodrilus udekemianus*; N, spermathecal seta of *Isochaetides freyi*; O, knobbed setae in XI of *Rhizodrilus lacteus*. (Chiefly from Stimpson et al., 1975.)

66. Posterior to segment VII the dorsal crotchet setae are spatulate or compressed, with tiny teeth, depending on the angle (Figs. 15G, H); often in a tube; widely distributed.
    Aulodrilus limnobius Bretscher
    Posterior to segment VII the dorsal setae have a different structure 67
67. Setae of segments II to IX generally 3 to 4/fascicle; sandy substrates in large rivers.
    Bothrioneurum vejdovskyanum Stolc
    Setae otherwise 68
68. Dorsal crotchet setae beyond segment II large and sharply arcuate (Fig. 15J); occasional in oligotrophic waters .... **Isochaetides curvisetosus** (Brinkhurst and Cook)
    Dorsal crotchet setae beyond segment XX small and not sharply arcuate 69
69. Distal tooth of ventral setae in segments II through IV longer than proximal tooth and bent at 90° (Figs. 15 K–M); widely distributed .... **Limnodrilus udekemianus** Claparède
    Distal tooth not bent at 90° 70

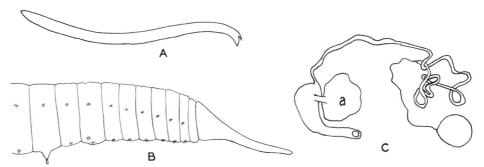

FIG. 16.–Structure of Tubificidae. A, ventral seta from posterior part of body of *Lumbriculus*; B, anterior end of *Rhynchelmis*; C, male duct of *Tubifex*. (B modified from Altman; C modified from Brinkhurst et al., 1980.)

70. With penis sheath in segment XI; spermathecal setae long and thin (Fig. 15N), in segment X; widely distributed but not common . . . . **Isochaetides freyi** (Brinkhurst)
    With spermathecal setae in segment X and/or penial setae in XI or penis sheath, but no combination of genital setae and penis sheath **71**
71. Four or five knobbed setae in segment XI (Fig. 15O); rare . . . . **Rhizodrilus lacteus** (Smith)
    Knobbed setae absent **72**
72. With penis sheaths; about 12 species; widely distributed and very common . . . . . . **Limnodrilus**
    Without penis sheaths **73**
73. No spermathecal setae in segment X; rare and poorly known, in lakes Huron and Superior.
    Phallodrilus hallae Cook and Hiltunen
    Spermathecal setae present in segment X; uncommon in Great Lakes drainages.
    Potamothrix moldaviensis Vejdovsky and Mrazek
74. One pair of male gonopores on segment X **75**
    One pair of male gonopores on VII to XII **78**
75. Fifty to 100 mm long; several rare species . . . . **Rhynchelmis**
    Much smaller worms **76**
76. Twenty to 50 mm long **77**
    Less than 20 mm long; very rare in caves . . . . **Trichodrilus**
77. Proboscis absent, setae single pointed, bifid, or both; rare . . . . **Stylodrilus**
    Prostomium rounded or with a proboscis; setae single pointed; widely distributed but uncommon.
    Eclipidrilus
78. Setae single pointed or bifid; rare . . . . **Kincaidiana**
    Setae with an upper reduced tooth (Fig. 16A); several species but only one common.
    Lumbriculus variegatus (Müller)

BRANCHIOBDELLIDA

In these commensals of crayfishes there is always a distinct head region. They are so similar to leeches that they were considered a leech family until 1884. From a morphological standpoint, however, the Branchiobdellida form a connecting link between typical oligochaetes and true leeches. Most of the 130 species occur only in the United States.

The body ranges from 1 to 12 mm in length and consists of head, trunk, and muscular caudal sucker. The cylindrical head consists of four fused segments, which are sometimes superficially annulated. A prostomium is absent. The mouth is surrounded by a dorsal and a ventral lip, and just inside the mouth is a pair of sclerotized jaws.

The trunk consists of 11 segments, although the last three are often indistinct and somewhat fused into the base of the sucker. Trunk segments are usually subdivided externally by transverse sulci. Setae are absent, and there are only two pairs of nephridia. The anus is on the dorsal side of segment X. In most species a spermatheca opens on the ventral midline of V, and there is a male genital pore on the ventral midline of VI. There is a pair of testes in each of V and VI. At maturity, the testes disassociate into clumps of spermatids, which, with developing and mature spermatozoa, fill the coelomic cavities of V and VI in American members of the group. A pair of ovaries is present in VII, and there are two lateroventral female genital pores. Large eggs may be seen in sexually mature specimens. The glandular clitellum forms around V, VI, and VII. Cocoons are formed, but almost nothing is known about the embryological development.

Branchiobdellids in the United States are always associated with crayfish, except for some species of *Cambarincola* that are found on cavernicolous *Asellus* and on *Callinectes sapidus* acclimated to fresh water. They occur on the gills, on the inner surface of the gill chambers, and on the general anterior external surface of the body and appendages. Sometimes a single crayfish may harbor as many as 200 or 300 worms. It is thought that the worms are essentially commensals since their digestive tracts usually contain only diatoms, other algae, and organic debris. Nevertheless, some species almost surely feed on blood of the host taken from gills or intersegmental membranes, and the species found on *Asellus* apparently feed on the eggs of the host. Branchiobdellids have survived up to 240 days in the laboratory, separated from their hosts, but have never been observed to produce cocoons under such conditions.

Except for *Xironogiton*, species of branchiobdellids reported from western crayfish (*Pacifastacus*) in Pacific drainages have never been found on eastern crayfish (Cambaridae), and vice versa; otherwise there seems to be no host specificity. One species of *Cambarincola*, for example, has been reported from at least 11 species of Cambaridae. Nevertheless, most species are known only from small geographic areas, often only one state, or one cave. Sometimes, when a crayfish carries two or more species of branchiobdellids, the worms are distributed in a specific and localized manner. In a study by Gelder and Smith (1987), for example, the crayfish *Cambarus bartoni* had *Bdellodrilus* only in the branchial chambers, *Cambarincola* only on the oral and ventral surfaces of the body, and *Xironogiton* only on the chelipeds.

A convenient place to find branchiobdellids is in the debris at the bottom of jars of preserved crayfish. If living specimens are collected, they are best killed and fixed in a solution consisting of 4 parts formalin and 96 parts of 70% alcohol. Some workers relax living worms with magnesium chloride. Either

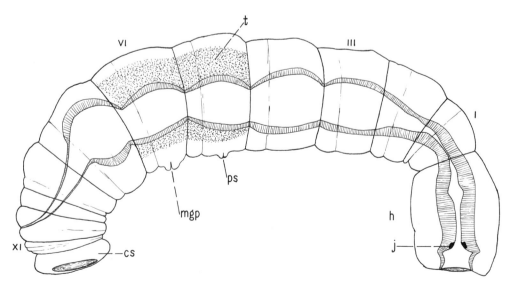

FIG. 17.–Diagram of *Cambarincola*, lateral view, showing digestive tract, ×70. cs, caudal sucker; h, head; j, jaw; mgp, male genital pore; ps, pore of spermatheca; t, testes (shaded areas).

stained or unstained specimens may be dehydrated and mounted permanently on slides in a suitable medium.

Because of the rate at which undescribed species are being found by specialists, a key to the U.S. species now known would be out of date in a short time. We are therefore including a key that goes only to the generic level. Unfortunately, the determination of genera in this key is based chiefly on differences in the anatomy of the male reproductive system. In many cases this requires careful observation of the internal reproductive organs, which can be done with whole mounts or by dissection of large worms. Even so, considerable experience and effort is required. See Holt (1960a) for methods and definitions of anatomical terms.

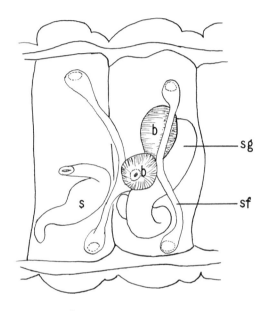

FIG. 18.–Ventral view of reproductive organs of V and VI of *Ankyrodrilus* (testes not shown). b, bursa; sf, sperm funnel (4); sg, spermiducal gland; s, spermatheca. (Modified from Holt, 1965.)

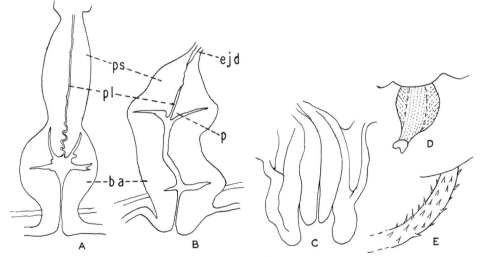

FIG. 19.–Male copulatory structures in Branchiobdellida. A, male copulatory apparatus in *Triannulata*; B, male copulatory apparatus in *Magmatodrilus*; C, protrusible penis in shape of a muscular cone; D, eversible penis with internal strands; E, cuticular tube type of penis. ba, bursal atrium; ejd, ejaculatory duct; p, penis; pl, lumen of penis; ps, penial sheath. (A and B modified from Holt, 1974; C to E courtesy of Perry C. Holt.)

KEY TO GENERA OF BRANCHIOBDELLIDA*

1. Body dorsoventrally flattened (depressed); paired nephridiopores on dorsal surface of segment III.

                                                                       2

      Body cylindrical (terete); single nephridiopore opening on dorsal surface of segment III 4
2. Spermiducal gland with ental bifurcation; vasa deferentia enter spermiducal gland near ectal end; spermatheca prominent (Fig. 18); body 2 to 3 mm[†] long; two species in VA and TN.

                                                                   Ankyrodrilus

      Spermiducal gland not bifurcated entally 3

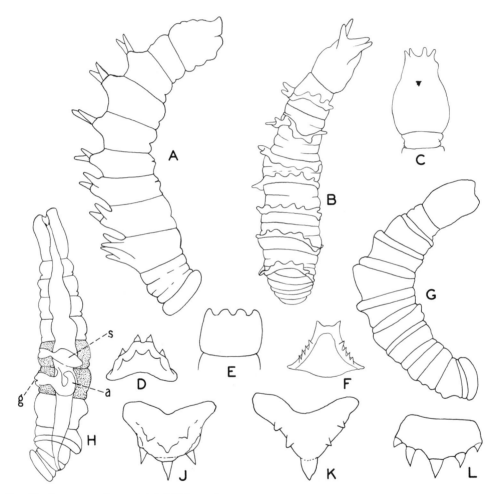

FIG. 20.–Structure of Branchiobdellida. A, lateral view of *Pterodrilus,* ×85; B, dorsal view of *Ceratodrilus,* ×35; C, dorsal view of head of *Cambarincola*; D to L, structure of *Cambarincola*; D, lower jaw; E, upper lip; F, lower jaw; G, lateral view, ×24; H, lateral view, ×30; J, upper jaw; K, upper jaw; L, upper jaw. a, accessory sperm duct; g, male genital pore; s, spermatheca; testes shown as shaded area in H. (B modified from Hall, 1915; D, J, and L modified from Ellis, 1919; F redrawn from Moore, 1898; G modified from Goodnight, 1941; H modified from Goodnight, 1940.)

---

*Prepared by Perry C. Holt.

[†]Lengths given are of preserved specimens and are approximate; extended living worms may be two or three times as long.

3. Vasa deferentia enter tubular spermiducal gland near ectal end; spermatheca disproportionately small; 2 to 8 mm long (Fig. 21B); five species on crayfishes of both eastern and western mountains.
Xironogiton

Vasa deferentia enter tubular or subspherical spermiducal gland basally; spermatheca prominent; up to 6 mm long; 5 species on eastern crayfish .... **Xironodrilus**

4. Penis eversible (Figs. 19D, E) 5
Penis a protrusible muscular cone (Fig. 19C) 12

5. Eversible penis a cuticular tube, often very long, with or without spines (Fig. 19E) 6
Eversible penis with internal strands (Fig. 19D), or an epithelial or muscular wall or both 7

6. Vasa deferentia enter spermiducal gland ectad to ental end of latter; prostate absent; about 2 mm long; one species in GA .... **Cronodrilus ogygius** Holt
Vasa deferentia enter ental end of spermiducal gland; prostate present, often rudimentary; 1 to 3 mm long; three species in TN and KY .... **Oedipodrilus**

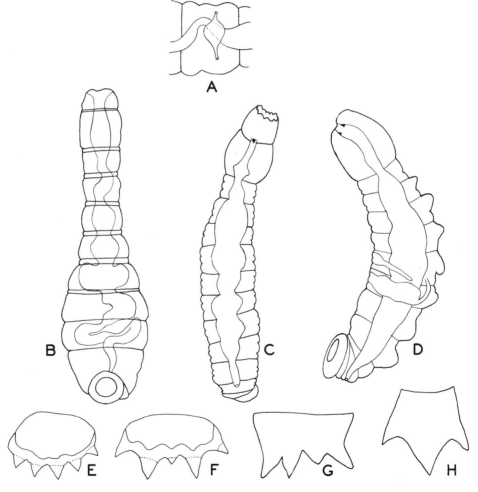

FIG. 21.–Structure of Branchiobdellida. A, lateral view of fourth trunk segment of *Bdellodrilus* showing lateral gland; B, ventral view of *Xironogiton,* ×18; C, dorsal view of *Triannulata,* ×21; D, lateral view of *Magmatodrilus,* ×30; E to H, upper jaws of *Xironodrilus.* (B redrawn from Moore, 1898; C and D modified from Goodnight, 1940; E and F modified from Ellis, 1919; G and H modified from Goodnight, 1943.)

7. With cylindrical projections borne dorsally on transverse folds of segments II to VIII (Fig. 20B); upper lip with four tentacles; 3 mm long; 2 species in UT, ID, WA, OR, WY . . . . **Ceratodrilus**
   Without such dorsal projections **8**
8. With pairs of large, clear glands in lateral body wall of segments I to IX (Fig. 21A); upper jaw a longitudinal ridge fitting concavity of lower jaw; bifid spermatheca; gill-inhabiting delicate worms; 2 to 3 mm long; eastern states . . . . **Bdellodrilus illuminata** Moore
   Without lateral glands; spermatheca not bifid **9**
9. Penis and ejaculatory duct a continuous heavily muscular eversible tube (Fig. 19A); 4 to 5 mm long (Fig. 21C); 4 species in CA, OR, WA . . . . **Triannulata magna** Goodnight
   Penis and ejaculatory duct distinct regions; ejaculatory duct not eversible **10**
10. Bursal atrium long and heavily muscular (Fig. 19B); penial sheath and eversible penis short; spermiducal gland very long with prominent anterior deferent lobe; prostate absent (Fig. 21D); about 3 mm long; 1 species in CA . . . . **Magmatodrilus obscurus** (Goodnight)
    Bursal atrium shorter than penial sheath; noneversible penial sheath **11**
11. Spermiducal gland small and slender, subequal to ejaculatory duct in length and diameter; prostate short; penis cuticular, enclosed when retracted by muscular extension of ejaculatory duct, which projects into bursal atrium; 2 to 3 mm long; 1 species in TN . . . . . **Tettodrilus friaufi** Holt
    Spermiducal gland various but always larger than ejaculatory duct; prostate present or absent; penis eversible, cuticular with strands connecting its outer everted surface to wall of penial sheath; 1 to 5 mm long; 15 species; FL, WA . . . . **Sathodrilus**
12. Spermatheca absent; bursa asymmetrically subglobular; about 1.5 mm long; 2 species; KY, IN, MI.
                                                                                           Ellisodrilus
    Spermatheca present **13**
13. Large worms; without dorsal appendages, though peristomial lobes or tentacles may be present; dorsal ridges, if present, not confined to segment VIII; 2.5 to 8 mm long; 50 species; generally distributed . . . . **Cambarincola**
    Small worms; usually with finger- or fanlike projections on some dorsal ridges (Fig. 20A), but sometimes a single prominent dorsal ridge on segment VIII; 1 to 2 mm long; 8 species; uplands of eastern United States and Ozark Mountains . . . . **Pterodrilus**

POLYCHAETA

Polychaetes are almost exclusively marine annelids, and an amazing variety of species and enormous numbers of individuals are found in many types of marine habitats. Some species (especially Nereidae) are euryhaline and occur also in estuaries and brackish waters, but less than 50 species of polychaetes are known to be restricted to the fresh waters of the world.

The colonization of inland waters by these few polychaetes is by no means restricted to one family or a few genera. Eight families are represented, and all species have close marine relatives. In addition to being found in the United States, fresh-water polychaetes have been reported sporadically from Yugoslavia, Lake Baikal in Asia, Viet Nam, Gabon, Java, Sumatra, China, Great Britain, Trinidad, and Brazil. With few exceptions, they have been collected from fresh-water rivers, lakes, and springs within 20 miles of the seacoast, and this seems to be an indication that

polychaetes, in spite of their phylogenetic antiquity, are relative newcomers to fresh waters. Presumably the physiological and morphological adjustments necessary for the marine–brackish–fresh-water transition constitute almost insurmountable barriers. Perhaps the most striking incidence of polychaete distribution is the occurrence of two fresh-water species in Lake Baikal, a body of water that is geologically ancient and more than 1600 km from the nearest salt water.

Most fresh-water polychaetes look like the familiar *Nereis* of invertebrate zoology courses. About 13 species have been reported from the United States. *Nereis limnicola* Johnson, one of the Nereidae, was first collected in 1895 from Lake Merced near San Francisco, a body of fresh water less than 0.4 km from the ocean and formed in very recent geological times. It may be taken as far north as Oregon. *N. succinea* Frey and Leuckart is widely distributed in brackish to fresh waters on both east and west coasts. Another nereid, *Lycastoides alticola* Johnson, was described from a single specimen collected in 1892 in a mountain stream at an altitude of 2100 m in "Sierra Laguna, Lower California." It has apparently not been collected since. *Laeonereis culveri* (Webster) occurs in brackish to fresh waters along the east and gulf coasts, and *Lycastopsis hummelincki* Augener is in similar habitats on the west coast of Florida. *Namanereis hawaiiensis* (Johnson), however, has been taken from only a single pond in southern California.

A single serpulid, *Ficopomatus enigmaticus,* is

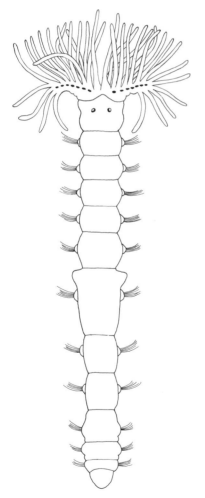

FIG. 22.–*Manayunkia speciosa, ×*37, a freshwater polychaete. (Modified from Leidy, 1883.)

known from marine to brackish and fresh waters near Oakland, California, and from a few streams entering the Gulf of Mexico. It builds a calcareous tube.

In 1859 and in several later papers the great microscopist Leidy described an interesting sabellid polychaete, *Manayunkia speciosa* Leidy, from the Schuylkill River near Philadelphia; it was later found in other areas in southeastern Pennsylvania and New Jersey. In 1939 Krecker described another species in the same genus, *M. eriensis* Krecker, from mud samples taken at a depth of 17 m in Lake Erie, near Put-in-Bay, Ohio. More recently this species has been considered a synonym of *M. speciosa.* It has since been found to be widely distributed in the Great Lakes, Georgia, New York, Oregon, California, Alaska, Vermont, North Carolina, South Carolina, and the Gulf Coast. This is undoubtedly the polychaete most likely to be found by the fresh-water biologist, and undoubtedly it is generally distributed in the United States.

Manayunkia speciosa is 3 to 5 mm long and inhabits a tube built of mud or sand and mucus. The head bears two large lateral lophophorelike structures, each of which has about 18 long, ciliated tentacles used in feeding. There is a pair of eyes near the median line of the head, and also a linear group of pigmented spots on each half of the lophophore, which are presumed to be light sensitive. Each of the first 11 of the 12 trunk segments has a pair of reduced, knoblike parapodia bearing a bundle of 4 to 10 setae. The circulatory system is well developed, and the bright green chlorocruorin of the blood colors the whole body. Leidy showed that development is direct and that there is no free-swimming trochophore larva. The large size and lateral swellings of the sixth trunk segment may be an indication that these polychaetes also reproduce by budding. Unlike marine relatives, there is no larval stage.

Most recently, a small (1.3 mm) polychaete (*Hesionides riegerorum*) has been described from the sandy beaches of a North Carolina river; it has very close marine relatives.

In addition to the species just mentioned, 5 or 10 additional polychaetes have occasionally been found in brackish waters in estuaries. It is doubtful whether any of these can reproduce in fresh waters.

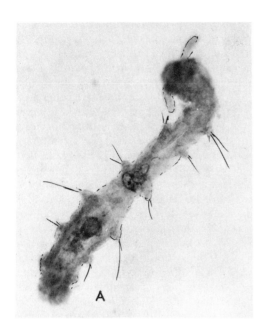
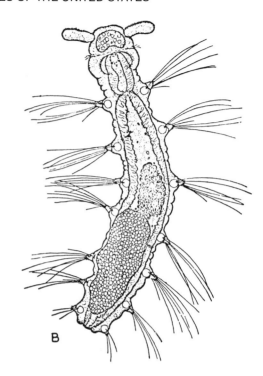

FIG. 23.–*Troglochaetus beranecki,* ×100. A, retouched photograph; B, showing general anatomy. (B from Andrassy.)

ARCHIANNELIDA

One of the most remarkable recent phylogenetic discoveries among fresh-water invertebrates is the collection of what are presumably two species of archiannelids.

Troglochaetus beranecki Delachaux (Fig. 23) was first described from underground waters of caverns in Switzerland (Delachaux, 1921), and more recently this worm has been taken in Hungary, Romania, Austria, and Germany. In 1969 the writer found what is thought to be the same species in the interstitial water 20 cm below the surface of gravels of mountain streams in Colorado at altitudes of 2640 and 3050 m. More recently, hundreds of specimens have been found in similar Colorado habitats. Undoubtedly this species will be found elsewhere in the United States. *Troglochaetus* averages only 630 μm in length. There are two blunt prostomial tentacles and seven pairs of parapodia. Each parapodium has a dorsal and a ventral fascicle of long, thin setae that are easily broken off during field collecting. There are no segmental septa. The species is closely related to several marine genera, especially the littoral genus *Thalassochaetus,* from the North Sea.

A problematical microannelid, *Rheomorpha neiswestnovae* Lastochkin, has been found in the interstitial zone of fresh-water sandy beaches in Russia, Poland, and Italy. It may be an archiannelid, but the worm also has certain oligochaete affinities.

HIRUDINEA (Leeches)

The leeches are predominantly a fresh-water group, although there are many marine as well as numerous terrestrial species, the latter mostly in the tropics. About 60 fresh-water species are known from the United States, and they are common inhabitants of ponds, marshes, lakes, and slow streams, especially in the northern half of the country. To the layman, leeches are collectively known as "bloodsuckers," but many forms are predators and scavengers, and only a small minority of our species will take blood from warm-blooded animals. Unlike the great majority of fresh-water invertebrates, leeches are often brightly colored and patterned, and in some species the coloration is highly variable. The length of mature leeches ranges from minute species only 5 mm long to some of the giant *Haemopis* that have been reported to reach a length of 45 cm when extended and swimming.

During recent years, more than 90% of leech literature citations have been of a physiological or biochemical nature.

General characteristics. Unlike oligochaetes, the Hirudinea are dorsoventrally flattened. The mouth is surrounded by an oral sucker that may be large, or small, fused with the anterior end, and liplike. The caudal sucker usually faces ventrally and is much larger, discoid, powerful, and expanded over a short central attachment pedestal. The anus is dorsal and in front of the sucker. The surface of the body may be smooth, wrinkled, tuberculate, or papillated. Although most genera have characteristic shapes, it should be borne in mind that leeches are highly muscular and contractile, and the body outline may vary greatly with locomotion, reaction to stimuli, and method of killing and fixation.

From the standpoint of basic body segmentation, external appearances are deceiving. All leeches are composed of only 34 true segments, but each segment is subdivided into a definite and constant number of superficial annuli. In the middle portion the body a segment typically has three, five, or six annuli, but in some of the Piscicolidae (fish leeches) there are 7, 12, or 14 annuli. Toward the extremities of the body annulation is incomplete so that there are fewer annuli per segment than in the midbody segments. In *Haemopis marmorata,* for example, midbody segments all have five annuli, but segments I through VII are represented by 1, 1, 1, 2, 2, 3, and 4 or 5 annuli, respectively; and XXIII

through XXVII are represented by 5, 4, 3, 2, and 1 annuli. Furthermore, in all leeches there is considerable fusion of the true segments of the body at the extremities, and especially at the posterior end. In *H. marmorata* segments XXVIII through XXXIV are all fused to form the caudal sucker.

The true body segmentation has been precisely established by a study of the nervous system. Each of the 34 segments is represented by a ganglion and paired nerves, but these structures are crowded together and partially fused at the anterior and posterior ends.

Some investigators distinguish the following major body regions in leeches: head (I–VI), preclitellar (VII–IX), clitellar (X–XIII), postclitellar (XIV–XXIV), anal (XXV–XXVII), and caudal sucker (XXVIII–XXXIV). The degree of differentiation of these regions is highly variable from one species to another.

The Class Hirudinea is divided into three orders. Each member of the Order Rhynchobdellida has a small, porelike mouth in the oral sucker through which a muscular proboscis can be protruded; jaws are absent; the blood is colorless; and there are three or six or more annuli per segment in the middle of the body. Members of the orders Gnathobdellida and Pharyngobdellida have a medium to large mouth, no proboscis, jaws present or absent, red blood, and usually five annuli per segment in the middle of the body.

The Order Rhynchobdellida is subdivided into two American families. The Glossiphoniidae generally have a flattened body, the anterior sucker more or less fused with the body, one to four pairs of eyes, and typically three annuli in each of the midbody segments. The American Piscicolidae, on the other hand, are somewhat cylindrical, with a distinct anterior sucker, one or two pairs of eyes, and usually six or more annuli per midbody segment; these leeches are almost invariably found as parasites on the body of fish or crustaceans.

The Order Gnathobdellida contains only the Family Hirudinidae in North America. The Family Erpobdellidae in the Order Pharyngobdellida have three or four pairs of eyes (never arranged in an arch), no jaws, and no gastric caeca. The Hirudinidae, however, have five pairs of eyes (forming a regular arch), toothed jaws, and one to several pairs of gastric caeca.

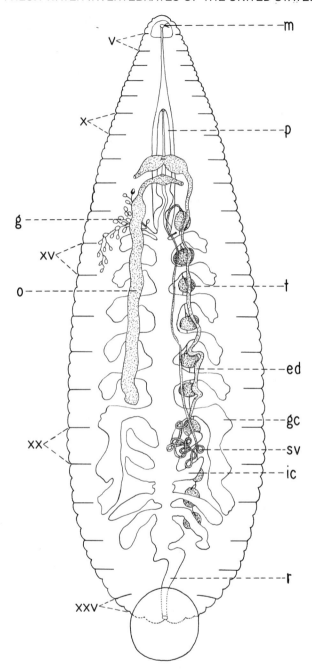

FIG. 24.–Ventral diagram of *Glossiphonia complanata,* ×10, showing annulations, reproductive system, and digestive system. Reproductive system stippled; female organs shown only on the left; male organs shown only on the right. ed, ejaculatory duct; ic, intestinal caecum (second); g, salivary glands; gc, gastric caecum (sixth); m, mouth; o, ovary; p, proboscis; r, rectum; t, testis; sv, seminal vesicle. (Modified from Harding and Moore, 1927.)

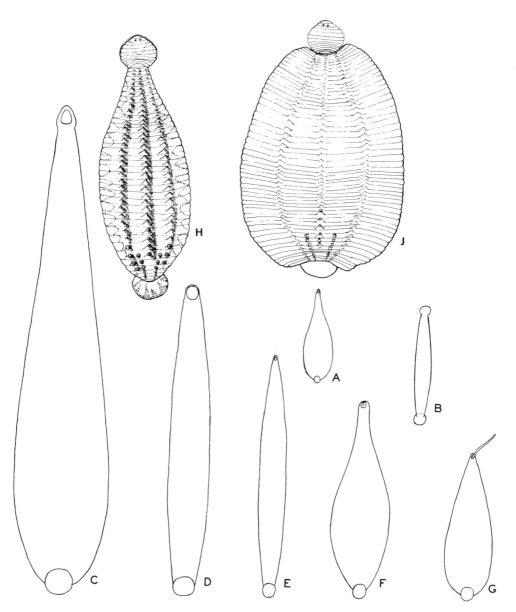

FIG. 25.–Relative shape and size of typical American leeches, ×1. A, *Helobdella*; B, *Piscicola*; C, *Haemopis*; D, *Macrobdella*; E, *Erpobdella*; F, *Placobdella*; G, *Glossiphonia,* with extruded proboscis; H, *Placobdella montifera,* normal shape; J, same, contracted. (H and J from Klemm, 1972.)

Color. When collecting specimens in the field, notes on coloration should always be taken, since leeches quickly fade in preservatives, with only red, brown, and black pigments persisting. Uniform coloration is a rare condition, most leeches being brightly mottled, spotted, or segmentally striped; sometimes the color pattern is arranged transversely, sometimes longitudinally. The ventral surface is almost always paler than the dorsal. The Glossiphoniidae range from translucent and pinkish to an opaque and brightly colored condition. *Glossiphonia complanata,* for example, is dull green to brownish, with six longitudinal rows of yellowish spots and two dark longitudinal paramedian lines. *Macrobdella decora,* the American medicinal leech, is very showy; the general coloration is green with obscure longitudinal stripes or short median lines; in the midline are bright orange or reddish spots, while the marginal spots are black. The ventral surface is rich orange, either plain or spotted with black. *Placobdella parasitica* is richly and variably colored, and it is difficult to find two specimens with exactly the same coloration and intensity. The general ground color is some shade of greenish brown, but the body is variously spotted, striped, or blotched with yellow and orange. A histological study of this species has revealed three types of pigment granules: (1) a pale yellowish granular reflecting substance, (2) chromatophores that range from black (contracted) to green (expanded), and (3) reddish-brown chromatophores that form a fine network in the expanded state. Leeches show no background color adaptation, but usually become paler in darkness and darker under illumination. Color changes are thought to be governed through the central nervous system.

Locomotion. All leeches can move about on the substrate with characteristic creeping, looping, or inchworm movements, by getting a firm grip alternately with the oral and caudal suckers. The body shape changes greatly during such locomotion. The Hirudinidae and Erpobdellidae are excellent swimmers, moving along rapidly with graceful up-and-down or side-to-side undulations. The Glossiphoniidae are poor swimmers, however, and when disturbed they usually roll up like a pillbug and fall to the substrate.

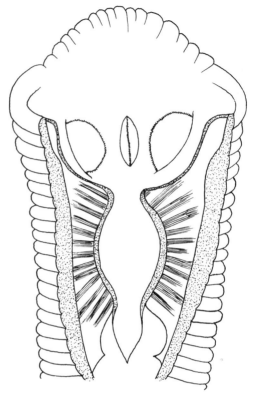

FIG. 26.–Diagrammatic ventral dissection of the anterior end of *Hirudo.* Stippling denotes cut edges of oral chamber, pharynx, and body wall. (Greatly modified from Pfurtscheller.)

Feeding, digestive system. The structure of the digestive system is partially correlated with food habits. In the Rynchobdellida there are no jaws, and the mouth is normally a small pore in the oral sucker. The anterior portion of the digestive tract contains an elongated, muscular, protrusible proboscis that lies in a proboscis sheath and is thrust forward and out of the mouth during feeding. There is a very short esophagus at the base of the proboscis, but the unicellular salivary glands in this region have long ducts that carry their secretion through the proboscis to its tip. The large stomach extends approximately through the middle third of the body and has from 1 to 10 pairs of lateral gastric caeca (often branched), in which quantities of ingested food may be stored for long periods prior to digestion. The gastric caeca are capable of great distension, and after a large meal a leech may weigh two to four times as much as it did before the meal. Digestion occurs in the intestine, and this part of the digestive tract is also supplied with caeca; usually, however, there are only four small pairs. The rectum is an unspecialized part of the digestive tract just before the anus. The Glossiphoniidae have a wide variety of food habits; many species are scavengers, feeding on dead animal matter, but more commonly they are carnivorous and feed on snails, insects, oligochaetes, and other small invertebrates. Some species have a notable preference for aquatic insects. *Glossiphonia complanata* and *Helobdella stagnalis* are almost always associated with snails on which they feed and are consequently commonly called "snail leeches." *Placobdella* and *Helobdella* are often temporary parasites on fish, frogs, and turtles, and they have occasionally been reported as taking a blood meal from man, especially from an open abrasion on the skin. Except during the breeding season, the Piscicolidae are almost invariably found as parasites on fish or crustaceans; they feed on the surface mucus as well as on blood and tissue fluids. The fact that certain Rynchobdellida are capable of taking blood from vertebrates does not seem to have been satisfactorily explained, especially since these leeches have no teeth with which to make an incision in the host epidermis. It is probable, however, that digestive enzymes are released through the pore at the tip of the proboscis where it is in contact with the skin of the host and that these enzymes digest the superficial layers and permit blood to flow.

The general plan of the Gnathobdellida and Pharyngobdellida digestive system is similar to that of the Rynchobdellida, but there are several striking differences. For one thing, there is no proboscis, and in its place is a strong, muscular, sucking pharynx. Some of the Hirudinidae, or true bloodsuckers, have three sharp-toothed jaws behind the large mouth. These are used in making incisions in the skin of the host. Other Hirudinidae, especially *Haemopis,* have no teeth or only vestigial teeth and normally feed on small invertebrates or dead animal matter, though *H. marmorata* will occasionally attack man. The latter species is called the "horse leech" because it commonly lives in ponds and troughs, where it attacks drinking cattle and horses. The Pharyngobdellida never have

Fig. 27.–*Hirudo medicinalis* L., X1. (From Klemm, 1972.)

true jaws. They are scavengers or feed mostly on small invertebrates, which they sometimes swallow whole; occasionally they suck blood from fish, frogs, and man, especially at bleeding epidermal abrasions. None of the American Gnathobdellida and Pharyngobdellida have intestinal caeca. The Hirudinidae have one to several pairs of gastric caeca, but the Erpobdellidae have none.

Macrobdella and *Philobdella* are the only common American leeches that regularly take human blood. Like all other bloodsuckers, they attach to the host with the caudal sucker and explore with the anterior end until a suitable spot is located, especially where the skin is thin. The oral sucker is then attached tightly and three fine painless incisions are made by back-and-forth rotary motions of the jaws. Sufficient blood is taken to distend the stomach and its caeca greatly so that the leech may be five times as heavy as it was when it began feeding. Much of the fluid of the ingested blood is excreted by the nephridia during feeding, the stored organic components, both cellular and noncelluar, being considerably concentrated. When the leech has filled its digestive tract, it leaves the host voluntarily, but the incisions keep on bleeding for a variable time because of the persistence of small quantities of the salivary anticoagulant, hirudin, in the adjacent tissues.

True bloodsucking leeches require only an occasional full meal, the stored blood being digested and utilized very slowly. Specimens have been kept for more than 2 years without feeding.

Free-living species may feed actively at temperatures as low as 2°C, but food intake rate usually stabilizes between 7 and 12°C. A "typical" small carnivorous leech requires only 1 to 4 chironomid larvae per day, for example. Predaceous species are usually nocturnal feeders.

Circulatory system. Unlike the condition in oligochaetes, where there is a spacious coelom regularly divided by transverse septa, the coelom of leeches is largely obliterated and reduced by greatly developed muscular, botryoidal, and parenchymatous tissues. The remaining sinuslike cavities consist of several longitudinal canals interconnected by smaller branch canals. The nerve cord, larger blood vessels, and reproductive organs are located in these sinuses.

The main portion of the circulatory system of typical Rynchobdellida consists of a large dorsal longitudinal blood vessel and a large ventral longitudinal blood vessel, which are connected near the two extremities by a series of convoluted branch vessels. The muscular wall of the anterior portion of the dorsal vessel acts as a heart. Near the posterior end the dorsal vessel is greatly expanded into an intestinal blood sinus that envelops the intestine and its caeca, and muscular contractions of the intestine are thought to aid in forcing the blood anteriorly out of this sinus and through the dorsal vessel proper. All the body tissues are well supplied with capillaries. In addition to the blood circulatory system, the coelomic sinuses contain a lymphlike fluid that circulates slowly. The two systems are unconnected, blood flowing in one and

lymph in the other, both fluids being colorless. In some of the Piscicolidae the lymphatic system has a series of paired lateral pulsating vesicles that force the lymph around the body.

In the other two orders the true blood vascular system has disappeared, and the system of coelomic sinuses alone remains in a modified form. It contains blood that is red because of dissolved erythrocruorin.

Respiration. Some marine leeches have special lateral gill-like structures, but none of the American fresh-water species have any special respiratory structures. The oxygen and carbon dioxide exchange readily occurs through the general body surface since there is a rich network of capillaries just below the epidermis. Many species attach to the substrate with one or both suckers and undulate the body, probably to facilitate respiration. Little is known about the ability of active leeches to tolerate low concentrations of dissolved oxygen, but many species will live through several days without free oxygen. They do not appear to be able to tolerate highly acid conditions or decomposition gases at low oxygen concentrations.

Excretion. There are never more than 17 pairs of nephridia and these are all segmentally arranged in the middle body segments. The internal openings, located in the sinus system, are simple ciliated funnels in some species and simple to complex, variable "ciliated organs" in others. The paired external openings, or nephridiopores, are in the ventral furrows between annuli.

Nervous system. The brain is in the region of segments V and VI and consists of the fused ganglia and basal portions of the nerves of the first six body segments. There is a small suprapharyngeal mass, a small connective passing around each side of the pharynx, and a very large subpharyngeal mass. The posterior ganglionic mass, in segments XXV and XXVI, represents the more or less fused ganglia of XXV through XXXIV. The ventral nerve cord connects brain and posterior ganglionic mass, and each of VII through

XXIV has a ganglion with three pairs of nerves.

Sense organs. Typical leeches have a variable number of small papillae arranged in transverse rows or extending completely around the body. These are of two general types. Ordinary cutaneous papillae are tactile and possibly chemical receptors. Sensillae, however, are more specialized, light colored or whitish, and sensitive to light. Although there may be numerous transverse rows or rings of cutaneous papillae in a single segment, there is never more than one row or ring of sensillae per segment. Usually a ring of sensillae consists of a dorsal transverse row of six or eight sensillae and a ventral row of six sensillae. Such a ring of sensillae is often called the neural ring, or sensory ring, and is always situated on the neural or sensory annulus of every somite. The term is usually applied to the annulus and not to the sense organs themselves.

True paired eyes are found at the anterior end of the body. Each eye consists of a clump of dark-colored pigment granules, often cup shaped, and an adjacent sensory area. Glossiphoniidae have one to four pairs of eyes, Piscicolidae one or two or more, Erpobdellidae three or four, and Hirudinidae five. A few leeches have eyespots on the caudal sucker or body somites.

Reproduction. In a very few leeches there is a common genital pore for both sets of reproductive organs, but usually there are two genital pores located on the median ventral line in XI and XII and separated by one to five or more annuli, the male genital pore always being more anterior than the female (see Fig. 35).

The female reproductive system is much simpler than that of the male. It consists of two saclike ovaries of variable size and two very short, narrow oviducts that unite to form a vagina. The vagina, in turn, opens to the outside through the female genital pore. In the Hirudinidae the female reproductive system also has an albumen gland, vaginal caecum, and muscular vagina.

Usually there are 5 to 12 pairs of segmentally arranged spherical testes, but in

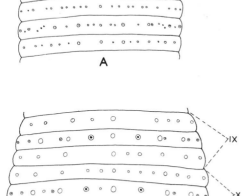

FIG. 28.–Arrangement of sensillae and papillae in typical segments in two species of *Placobdella.* Sensillae are shown with central dots; all other areas are cutaneous papillae. Note that sensillae occur only on the middle annulus of each segment and that they are symmetrically arranged.

FIG. 29.–Spermatophore of *Glossiphonia,* showing two oval bundles of sperm. (Modified from various sources.)

some of the Erpobdellidae they may be minutely subdivided like linear bunches of grapes. On each side of the body the testes are connected to a long vas deferens by means of short vasa efferentia. Each vas deferens has a modified seminal vesicle region that, in turn, becomes an ejaculatory duct. The two ejaculatory ducts unite to form a median atrium that is often complex and glandular.

The relative development and size of the seminal vesicles and ejaculatory ducts vary greatly from one species to another. They may be short to long, straight or much convoluted, and small to massive. In the seminal vesicles sperm become cemented together and stored for future use.

An eversible muscular bursa is associated with the male atrium. In the Glossiphoniidae it is rudimentary, but in the other families it is generally larger and functional. In the Hirudinidae, however, the bursa region contains a coiled filamentous tubular penis, often having a complicated structure.

As the sperm leave the seminal vesicle and pass down the ejaculatory ducts into the region of the atrium, they are formed into spermatophores (except in the Hirudinidae) for conveyance to another individual during copulation. These cylindrical or double club-shaped structures, which have a chitinoid covering, consist of two more or less adherent tubes containing sperm bundles (Fig. 29). Spermatophores are formed by glandular areas of the atrium and adjacent portions of the ejaculatory ducts.

Like most hermaphroditic animals, an individual leech does not fertilize itself, but copulates with another individual and exchanges sperm. In the Rhynchobdellida and

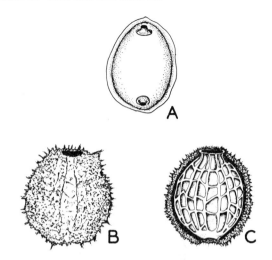

FIG. 31.–Leech cocoons. A, *Erpobdella*; B, complete cocoon of *Cystobranchus*; C, cocoon of *Cystobranchus* with spongy layer removed from upper surface. (B and C modified from Hoffman, 1955.)

Erpobdellidae copulating pairs implant spermatophores on the ventral or dorsal surface of each other, usually in the clitellar region, but sometimes only one member of a copulating pair releases spermatophores. The sperm issue out of the spermatophores and penetrate the superficial tissues to the sinuses of the recipient leech with the aid of local histolysis. Soon the sperm reach the ovaries in the ventral sinuses and fertilize the eggs.

In the Hirudinidae the penis of each member of a copulating pair is extruded so that sperm are deposited at the genital pore or in the vagina of the recipient.

The clitellum of a leech is relatively inconspicuous, but in the breeding season it secretes ringlike structures that slip off the anterior end of the body to form cocoons, just as in the earthworm. As the cocoon passes the female genital pore it receives and encloses from one to several fertilized eggs. The Piscicolidae and most Gnathobdellida and Pharyngobdellida fasten their cocoons to the substrate, but some of the Hirudinidae superficially bury their cocoons in soft substrates, often in or on exposed mud along the shores of bodies of water. Cocoons are usually 2 to 15 mm long, more or less oval, and flattened on the attachment surface. The outer surface may be gelatinous, spongy, tough and membranous, horny, or chitinoid (Figs. 30, 31). After a variable period of growth, the young

FIG. 30.–Cocoons of *Haemopis,* ×1.5.

leeches emerge from the cocoons and become independent.

The Glossiphoniidae do not form true cocoons but carry their fertilized eggs in membranous capsules on the ventral surface of the body. After hatching, the young remain on the body of the parent in the same area. They are attached by means of mucous threads and probably feed on mucus until they attain quite a large size, at which time they leave the parent.

Copulation usually begins in the spring when temperatures are changing and after a good meal has been obtained; most cocoons are deposited between May and August. Some species produce batches of eggs for periods of 5 to 6 months, and empty cocoons may be found as late as October or November.

Little is known about the length of life of leeches. *Macrobdella decora* takes 2 to 3 years to mature, certain other Hirudinidae may require 4 to 5 years, and it is claimed that some specimens may live for 10 to 15 years. *Helobdella stagnalis,* on the other hand, may have 2 generations/year.

Ecology. Leeches abound in warm, protected shallows where there is little wave action and where plants, stones, and debris afford concealment. In such habitats population densities of 700 or more leeches per square meter of bottom have been reported. Exceptional densities exceed 50,000/m^2. Superficial daylight examination of a likely body of water may be disappointing, but it must be borne in mind that leeches are chiefly nocturnal, and that they are hidden under stones, vegetation, and debris during the hours of daylight. The great majority of specimens are collected between the water's edge and depths of 2 m, but a few species have been collected as deep as 50 m. Leeches require substrates to which they can adhere and consequently are uncommon on pure mud, clay, or sand bottoms.

Few species and few individuals occur in strongly acid lakes, and they are generally absent from sphagnum bogs. There is some evidence indicating that low calcium content of waters may be a limiting factor in leech distribution. They are notably absent from soft-water high mountain lakes, but this may be due to food shortages in such waters.

Unlike most groups of fresh-water invertebrates, there is very little habitat preference among leeches. The same species (*Helobdella stagnalis,* for example) may occur in lakes, ponds, springs, slow streams, marshes, and on all kinds of suitable substrates. A few species of Erpobdellidae and Glossiphoniidae may be collected in swift streams, and some of them even tolerate some degree of pollution.

The leech fauna of intermittent ponds persists because some species are able to tide over dry periods by burrowing into the mud bottom, losing considerable weight, and constructing a small mucus-lined cell. Leeches have been known to aestivate in this manner for more than 4 weeks at a time. Winters are spent in a lethargic dormant state buried in the upper parts of the substrate but below the frost line. During summer stagnation, leeches often migrate to shallow waters near shore where there is more dissolved oxygen.

Leeches serve as food for dytiscid larvae, caddis larvae, and odonate nymphs, and fish, but the predation rate is generally low.

Although American leeches are all considered aquatic organisms, there are some that are actually amphibious. *Haemopis,* for example, commonly crawls out of the water and wanders about on the moist shore and ground, sometimes as much as a quarter of a mile from the water, where it feeds on living and dead invertebrates, especially earthworms.

Semipermanent and permanent parasites, such as the Piscicolidae and *Placobdella parasitica,* remain on the body of the host except during the breeding season when they become free-living long enough to deposit their eggs on the substrate. But some of the fish leeches are completely dependent and deposit their cocoons on the host fish.

Economic importance. The common medical practice of bloodletting as treatment for a wide variety of human ills supported a considerable leech industry in eighteenth and nineteenth century Europe. *Hirudo medicinalis* (see Fig. 27) was raised commercially in so-called leech ponds, especially in France, Hungary, and Russia, and it is estimated that 25 million leeches were used in France in 1846 and 7 million in London hospitals in 1863.

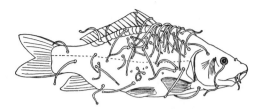

FIG. 32.–A small carp heavily infected with *Piscicola,* ×0.5.

Leeching never attained such great importance in the United States, and between 1840 and 1890 it is doubtful if more than 1.5 million leeches were used in any one year, including both the imported and native species (usually *Macrobdella decora*). The maximum wholesale price was about 100 dollars per thousand. By 1890 the use of leeches had dwindled greatly and by the turn of the century it was difficult to find a drugstore that still carried leeches in stock. Bloodletting by the use of leeches is chiefly confined to a few older European elements of our population. The practice is mostly resorted to as a home remedy for boils, contusions, and abscessed teeth. Very recently, however, several large American centers have gone to the use of *Hirudo* for removing blood from areas where tissue has been transplanted or reattached in microsurgery. In England, there is a large leech-producing firm which keeps a stock of 30,000 leeches, which are worth about five dollars each. In addition to the anticoagulant hirudin, other organic substances are derived from salivary secretions of the animals and are used for a variety of biochemical research.

Leeches sometimes become so abundant in northern lake resort areas as to constitute a nuisance for bathers. Disturbances in the water, such as splashing and currents, are known to attract bloodsuckers quickly. Unfortunately there are usually no effective remedial measures that can be taken to reduce leech populations, but if the water level of a lake can be lowered at least a meter in the late fall, dormant leeches in the exposed muddy shallows will be killed by temperatures of −6°C or less. Leeches may be temporarily controlled in localized bathing beach areas by applying 100 lb of powdered lime per acre per day in the shallows.

Hook and line fishermen commonly use a wide variety of leeches as bait, the most common species probably being *Nephelopsis obscura.*

With reference to fish populations, the Glossiphoniidae, Hirudinidae, and Erpobdellidae can be dismissed as being of no real importance as parasites. Heavy infestations of Piscicolidae, however, have sometimes caused considerable mortality in both hatcheries and the natural habitat. Meyer (1969) summarizes the generally unsuccessful methods for treating fish infestations.

Dispersal, geographic distribution. Since many leeches temporarily attach to or definitely parasitize ducks and shore birds, they are easily transported overland from one body of water or drainage system to another. In fact, the leeches are one of the new freshwater groups in which such transport is definitely known to occur. Some investigators report the recovery of viable cocoons from duck feces. Within a single drainage system, passive dispersal by attachment to fish is undoubtedly of great importance.

In addition, leeches actively and readily spread from place to place and colonize suitable habitats within individual drainage systems. Peculiar seasonal upstream mass migrations of great numbers of leeches have been reported several times. As shown in the key, some American leeches are known from only one or two states, whereas others are generally distributed in suitable habitats. Two species of *Glossiphonia* occur throughout Europe and much of Asia; *Helobdella stagnalis* and *Theromyzon tessulatum* are widely distributed in the north and south temperate zones. The common European bloodsucker, *Hirudo medicinalis,* has been recorded a few times from the northeastern states, but is probably not well established in the wild.

Collecting, preparing. Usually specimens are collected individually with forceps by turning over rocks, logs, and debris in the shallows. A small dipnet should be carried to capture the rapid swimmers and to retrieve sluggish glossiphoniids as they roll up and fall to the bottom. Some collectors have had success by placing the body of a dead and bleeding small animal as bait in the water where there is a slight current.

If leeches are kept in aquariums, they must

be provided with food, especially the Hirudinidae, which otherwise have a habit of crawling out and wandering about on the floor.

For close examination in the living state leeches should be narcotized. Carbonated water, chloretone, chloroform vapors, chloral hydrate, nicotine sulfate, and magnesium sulfate are all variously successful for different species. Specimens to be killed and preserved should receive similar treatment until they no longer respond to pinching. If carefully done, exceptionally good specimens may require 1 to 5 hours or more for narcotization. They are then placed in a flat dish and flooded with the killing agent, which may be any one of a variety of standard fixatives. Schaudinn's fluid, 0.5% chromic acid, and 2% formalin in 50% alcohol are all suitable. If an acid-containing fixative is used, care must be exercised to remove all acid after fixation; otherwise the connective tissues swell greatly. Permanent preservation should be in 80% alcohol or 5% formalin. Preserved leeches should be straight, moderately extended, undistorted, and not too hard or too soft.

Bennike (1943) advises narcotizing leeches in 15% alcohol for 15 minutes. Then they should be removed, stretched to normal size, and placed in 10% formalin or 70% alcohol. If they bend or contract, they have not been kept in the weak alcohol long enough. Zinn and Kneeland (1964) obtained best results with 1% eucaine hydrochloride.

Laboratory cultures. With the exception of the Piscicolidae, laboratory cultures can be easily maintained, especially since leeches need to be fed only infrequently. It is most important that the culture water be kept cool and clean. Although leeches are tolerant of a wide range of temperatures, they do not usually survive in the laboratory in direct sunlight or water that is too warm. A wide variety of common substances are toxic, even in minute quantities; copper and chlorine are especially harmful. Dead leeches, feces, and unused food material should always be removed promptly. Water should be changed as soon as it shows signs of becoming foul.

Small Glossiphoniidae may be kept in finger bowls or similar dishes of pond water containing a few shoots of *Anacharis* or other aquatic plants. Pieces of shell or pebbles afford additional concealment and protection. Small balanced aquariums also make excellent containers for leeches in the laboratory. Living snails should be added occasionally for food. Large bloodsucking glossiphoniids should be kept in larger containers and require a feeding of turtle, frog, or toad blood every month or two. Glossiphoniids thrive and readily reproduce under suitable conditions, but care should be taken to prevent overcrowding.

Erpobdellidae are very active, especially at night, and require large containers with tightly fitting covers or screens so that they will not crawl out. Stones and pieces of shell are a suitable substrate for the deposition of cocoons. Earthworms, immature insects, and ground fresh meat are appropriate foods for maintaining satisfactory cultures.

Hirudinidae may be kept in large, low aquariums or earthenware jars, but similar precautions must be taken against wandering. They do not require vegetation, but since they are somewhat amphibious there should be room for them to crawl up on the side of the container above the water level, and it is well to have a sloping bank of sandy soil at one end with a partial cover of moss, stones, or sticks. A vertebrate blood meal should be supplied at intervals of about six months, but the diet may be varied with frog eggs, immature insects, and earthworms. Cocoons are mostly deposited in wet soil just above the water's edge.

Taxonomy. Most of the United States has had only sporadic collecting, and much remains to be done. Extensive work has been accomplished only in the eastern half of the country. Consequently, aside from those leeches that are undoubtedly generally distributed, geographic distributions are necessarily uncertain.

External key characters are used wherever possible, but in a few couplets it has been necessary to use internal anatomical features that require dissection. Lengths are for average specimens when at rest. The 63 species listed in this key are believed to be essentially a complete list for the United States. Only a few poorly described, unrecognizable, and rare species are not included. The key is modified from Sawyer (1972) and Klemm (1972).

KEY TO SPECIES OF HIRUDINEA

1. Mouth a small pore in the oral sucker through which a muscular proboscis can be protruded; jaws absent (Figs. 24, 25G); blood colorless .... Order **RHYNCHOBDELLIDA, 2**
   Mouth medium to large, without a proboscis (Figs. 25C–E); blood red **3**
2. Body flat and much wider than head, never cylindrical (Figs. 24, 34); oral sucker ventral and more or less fused to body; eggs in membranous sacs on ventral surface of body; young clinging to ventral side of adult; poor swimmers .... **GLOSSIPHONIIDAE, 4**
   Body cylindrical, often divided into a narrow anterior and a wide posterior region; oral sucker distinctly separated from body (Fig. 36); 6 to 30 mm long; almost invariably found as parasites on fishes; fish leeches .... **PISCICOLIDAE, 31**
3. Five pairs of eyes forming a regular arch (Fig. 33B); jaws typically present and toothed, but may be toothless or absent; one to several pairs of large gastric caeca; medium to very large size.
   Order **GNATHOBDELLIDA, HIRUDINIDAE, 38**
   Three or four pairs of eyes (rarely absent) never arranged in a regular arch (Fig. 33C); jaws absent; gastric caeca absent; body linear .... Order **PHARYNGOBDELLIDA, ERPOBDELLIDAE, 51**
4. With three or four pairs of eyes (Figs. 34G–J) **13**
   With one or two pairs of eyes (except for *Placobdella hollensis,* which has accessory eyes) (Figs. 34B, L). **5**
5. Mouth within anterior sucker cavity, not on rim **12**
   Mouth on rim of anterior sucker (Fig. 34 M) **6**
6. Caudal sucker with marginal papillae and glands, their positions marked by faint radiating ridges (Fig. 34F); uncommon in northeastern states; usually found on fish .... **Actinobdella, 14**
   Caudal sucker without marginal papillae and glands **7**
7. With one pair of separated eyes; 15 to 22 mm long; reported only from MI.
   Marvinmeyeria lucida (Moore)
   With one pair of fused eyes (Fig. 34 K) or almost fused (Fig. 34 L) **8**
8. Anterior segments forming a discoidal "head" (Fig. 34B); free living or parasitic on fish.
   Placobdella, 9
   Anterior segments not forming a discoidal "head" **10**
9. Body with three dorsal ridges (Fig. 25H); eastern half of the United States.
   Placobdella montifera Moore
   Body without three dorsal ridges; 25 mm long; coastal plain of Carolinas.
   Placobdella nuchalis Sawyer and Shelley
10. Body smooth, without conspicuous tubercles; parasitic on amphibians; rare **11**
    Body papillate or tuberculate .... **Placobdella, 15**
11. Translucent pale green; 2 to 7 mm long .... **Oligobdella biannulata** (Moore)
    Base color brownish; 10 to 20 mm long; NC, SC, LA.
    Placobdella translucens Sawyer and Shelley
12. One or two pairs of eyes; if only one pair, they are fused or close together; parasitic on amphibians, occasionally on fish; seldom free living; widely distributed but rare ..... **Batracobdella, 20**
    One pair of eyes, well separated .... **Helobdella, 23**
13. Three pairs of eyes (Figs. 34G, H) **28**
    Four pairs of eyes (Fig. 34J) widely distributed but not common .... **Theromyzon, 29**
14. Five rows of dorsal tubercles; three annuli per segment; parasitic on cold-blooded vertebrates.
    Actinobdella triannulata Moore
    One middorsal row of tubercles; six unequal annuli per segment; host unknown.
    Actinobdella inequiannulata Moore
15. Caudal sucker set off from the body by a peduncle (Fig. 34C); parasitic on fish; rare midwestern species .... **Placobdella pediculata** Hemingway
    Caudal sucker not set off from the body **16**
16. Accessory eyes present (Fig. 34E); main coloration olive green with small brown, yellow, and colorless areas; usually free living; rare in eastern half of the United States.
    Placobdella hollensis (Whitman)
    Without accessory eyes **17**
17. With ventral stripes of blue, brown, or green; dorsal coloration variable, usually greenish brown; dorsal tubercles inconspicuous or absent; 35 to 65 mm long; commonly found clinging to the legs of the common snapping turtle; widely distributed .... **Placobdella parasitica** (Say)

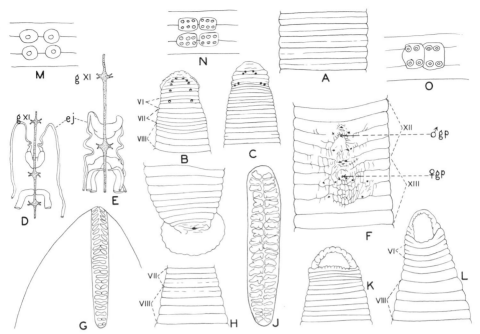

FIG. 33.–Structure of Gnathobdellida and Pharyngobdellida. A, typical midbody segment of *Nephelopsis obscura,* showing subdivided annuli; B, anterior and posterior ends of *Macrobdella decora*; C, anterior end of *Dina parva*; D, dorsal view of part of reproductive system of *Mooreobdella parva*; E, dorsal view of part of reproductive system of *D. fervida*; F, ventral view of XII and XIII of *Philobdella gracilis,* showing copulatory gland region and genital pores; G, edge view of median jaw of *P. gracilis*; H, segments VII and VIII of *Haemopis marmorata*; J, edge view of teeth of median jaw of *H. terrestris*; K, ventral view of anterior end of *H. plumbea*; L, ventral view of anterior end of *H. grandis*; M, N, and O, arrangement of copulatory gland pores in *Macrobdella decora, M. sestertia,* and *M. ditetra,* respectively. *gXI,* eleventh ganglion; *ej,* ejaculatory duct; nerve tissue stippled. (B–D, K, and L modified from Nachtreib et al., 1912; E–G, and J modified from Moore, 1901; M–O modified from Klemm, 1972.)

Without ventral stripes; dorsal tubercles prominent; smaller **18**

18. Dorsal surface warty; papillae large, numerous, and randomly arranged; dorsal surface a mixture of brown, green, and yellow, with or without a median interrupted stripe; ventral surface unstriped; free living or parasitic; widely distributed and common .... **Placobdella ornata** (Verrill)

Dorsal surface with more uniform tubercles and papillae **19**

19. Dorsal surface uniformly brown with a darker brown median longitudinal stripe; ventral surface with about 30 dark lines; 40 to 55 mm long; widely distributed and rare in eastern states.

Placobdella multilineata Moore

Dorsal surface brownish but with five longitudinal rows of tubercles; ventral surface with two bluish longitudinal stripes; 15 to 45 mm long .... **Placobdella papillifera** (Verrill)

20. With white patches in eye, neck, genital, and anal areas; body translucent, with conspicuous middorsal and lateral segmental dots or papillae **21**

Without white patches in genital and anal areas; slight and variable patches sometimes in eye and neck areas; body smooth, opaque; parasitic on amphibians, especially in small ponds in northern states .... **Batracobdella picta** Verrill

21. Body convex, thick; three longitudinal rows of papillae; uncommon, widely distributed, eastern.

Batracobdella phalera (Graf)

Body very thin and flat **22**

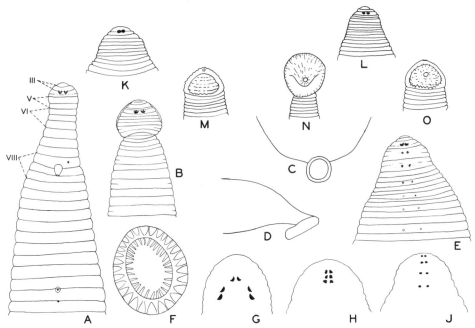

FIG. 34.–Structure of Rhynchobdellida. A, dorsal view of anterior end of *Helobdella stagnalis* also showing position of genital pores on ventral surface; B, dorsal view of anterior end of *Placobdella montifera*; C, ventral view of posterior end of *P. pediculata* (annulations not shown); D, lateral view of same; E, dorsal view of anterior end of *P. hollensis* (note transitions between well-developed pair of eyes and sensillae); F, caudal sucker of *Actinobdella inequiannulata*; G, eyes of *Glossiphonia heteroclita*; H, eyes of *G. complanata*; J, eyes of *Theromyzon*; K and L, anterior end of *Placobdella,* showing eyes fused and adjacent; M to O, ventral views of anterior suckers of Rhynchobdellida showing various locations of the porelike mouth. (A, C–E modified from Nachtreib et al., 1912; B and F modified from Moore, 1901; K–O modified from Klemm, 1972.)

22. Body length about twice the width; with longitudinal rows of prominences; MI.
    Batracobdella michiganensis Sawyer
    Body length about four times the width; with another type of color pattern; body excessively thin; southeastern .... **Placobdella translucens** Sawyer and Shelley
23. With a brown chitinoid plate on the dorsal surface of VIII (Fig. 34A); brown, pink, or green color; 9 to 14 mm long; the most cosmopolitan of all American leeches .... **Helobdella stagnalis** (L.)
    Without a chitinoid plate **24**
24. Dorsal surface smooth **25**
    Dorsal surface with three to seven longitudinal series of papillae **27**
25. Body translucent, elongate, and cylindrical; lateral margins almost parallel; feeds on aquatic insects and parasitic on fish; common east of the Continental Divide ..... **Helobdella elongata** (Castle)
    Body pigmented, flattened **26**
26. Dorsal surface with transverse brown bands alternating with white bands; Michigan.
    Helobdella transversa Sawyer
    Dorsal surface with six major longitudinal white stripes alternating with brown stripes; widely distributed and common .... **Helobdella fusca** Castle
27. Dorsal surface with three (incomplete) series of small, black-tipped papillae; a common, variable species .... **Helobdella triserialis** (Blanchard)
    Dorsal surface with five to seven longitudinal series of papillae; lightly pigmented; uncommon in Great Lakes area .... **Helobdella papillata** (Moore)

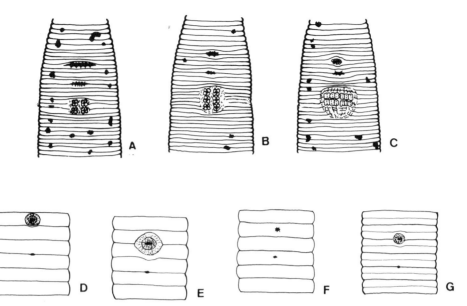

FIG. 35.–External reproductive structures. A–C show the male genital pore, female genital pore, and copulatory glands along the midline for *Macrobdella decora, M. diplotertia,* and *M. sestertia,* left to right; D–G show relative positions of male genital pore (anterior) and female genital pore (posterior) for *Erpobdella punctata, E. punctata, Mooreobdella bucera,* and *Nephelopsis obscura,* left to right. (Modified from Klemm, 1972.)

28. Eyes arranged in a roughly triangular pattern of three groups of two each (Fig. 34G); body translucent, with little pigmentation; body 6 to 9 mm long; rare in northeast.

                Alboglossiphonia heteroclita (L.)

  Eyes in two nearly parallel rows near the median line (Fig. 34H); body opaque, deeply pigmented; 14 to 25 mm long; very common . . . . **Glossiphonia complanata** (L.)

29. With two annuli between genital pores; olive-green to brown, translucent, flecked with black spots; 20 to 26 mm long; free living or parasitic on waterfowl; central and eastern states.

                Theromyzon biannulatum Klemm

  With three or four annuli between genital pores 30

30. With three annuli between gonopores; color variable, with spots of yellow, orange, or brown; 20 to 30 mm long; central and western . . . . **Theromyzon rude** (Baird)

  With four annuli between gonopores; transparent amber or greenish color; 15 to 30 mm long; CO.

                Theromyzon tessulatum (O.F.M.)

31. Caudal sucker flattened, as wide or wider than the widest part of body; vesicles on margin of trunk; suckers distinctly set off from body (Figs. 36A–C, G, H) 32

  Caudal sucker concave, weakly developed, and narrower than widest part of body; vesicles absent.

                       36

32. Vesicles small, difficult to see on preserved leeches; body not divided into anterior and posterior parts (Fig. 36C) . . . . **Piscicola,** 33

  Vesicles large, obvious after preservation; body divided into anterior and posterior parts (Figs. 36A, B); uncommon . . . . **Cystobranchus,** 35

33. With eyespots on caudal sucker (Figs. 36G, H) 34

  Without eyespots on caudal sucker (Fig. 36F); translucent greenish color; 14 to 16 mm long.

                  Piscicola punctata (Verrill)

34. Crescent-shaped spots on caudal sucker (Fig. 36G); 10 to 31 mm long; western.

                Piscicola salmositica Meyer

  Oculiform spots on caudal sucker (Fig. 36H); 16 to 24 mm long; Great Lakes area.

                Piscicola milneri (Verrill)

35. Gonopores separated by two annuli; 10 to 30 mm long; northeastern states.
                                              Cystobranchus verrilli Meyer
    Gonopores separated by one or two annuli; 9 to 15 mm long; VA.
                                              Cystobranchus virginicus Hoffman
36. Body divided into narrow anterior region and wider posterior region (Fig. 36J); one pair of eyes;
    colorless or with scattered pigment cells; 24 to 26 mm long; widely distributed.
                                              Myzobdella lugubris Leidy
    Body not divided into two regions **37**
37. Body thin, without pigment; 9 to 30 mm long; scattered distribution; one variable species, may be
    identical with *Myzobdella lugubris* .... **Illinobdella moorei** Meyer
    Body thicker; six longitudinal rows of brown-black pigments; 6 to 8 mm long; scattered distribution
    in eastern half of United States .... **Piscicolaria reducta** Meyer
38. Copulatory glands behind gonopores on ventral surface (Figs. 33M–O, 35A–C).
                                              Macrobdella, 39
    Copulatory glands absent **42**
39. With about 21 red or orange dots in middorsal line **40**
    Without red spots in middorsal line; 10 to 15 cm long; rare in southeastern states.
                                              Macrobdella ditetra Moore
40. With four or six copulatory gland pores on ventral surface (Figs. 35A, B) **41**
    With 24 copulatory gland pores (Fig. 35C); ventral surface reddish brown with a few black flecks; 50 to
    100 mm long; rare in MA .... **Macrobdella sestertia** Whitman
41. With four copulatory gland pores; ventral surface red or orange; dorsal surface green; 11 to 15 cm
    long; the "American medicinal leech"; feeds almost exclusively on vertebrate blood; widely
    distributed and common .... **Macrobdella decora** (Say)
    With six copulatory gland pores; light gray background on dorsal surface with reddish and black
    spots; ventral surface light yellow or gray; 10 to 15 cm long; KS, MO.
                                              Macrobdella diplotertia Meyer
42. Gonopores separated by five to five and one-half annuli; female gonopore small **43**
    Gonopores separated by six and one-half to seven annuli; female gonopore large and nipplelike; up to
    20 cm long; NC, SC, VA .... **Haemopis septagon** Sawyer and Shelley
43. Dorsal surface with one or more colored stripes **44**
    Dorsal surface without colored stripes .... **Haemopis, 48**
44. Dorsal surface with a single median stripe **45**
    Dorsal surface with longitudinal reddish-yellow stripes and color pattern; ventral surface black with
    white and gray markings; each jaw with about 50 small teeth (Fig. 26); up to 100 mm long; the
    introduced European medicinal leech; rare, probably not established ... **Hirudo medicinalis** L.
45. Dorsal surface with a median black stripe .... **Haemopis, 46**
    Dorsal surface with a median stripe of another color .... **Philobdella, 47**
46. Each jaw with 9 to 12 pairs of teeth; 60 to 110 mm long; uncommon; IA, CO.
                                              Haemopis kingi (Mathers)
    Each jaw with 20 to 25 pairs of teeth (Fig. 33J); 150 to 200 mm long; one variety has been found as
    much as 1 km from water; feeds on aquatic invertebrates and earthworms; widely distributed and
    common .... **Haemopis terrestris** (Forbes)
47. Each jaw with 20 teeth; 40 to 85 mm long; known only from southern FL.
                                              Philobdella floridana (Verrill)
    Each jaw with 35 to 50 teeth (Fig. 33G); 40 to 85 mm long; widely distributed but uncommon in
    southern states .... **Philobdella gracilis** Moore
48. Jaws and teeth absent or vestigial; incapable of taking vertebrate blood; often found away from water.
                                              49
    Jaws small, with few teeth **50**
49. Lip broad and flat (Fig. 33K); dorsal surface grayish, with few or no blotches; margins with reddish or
    orange band; 140 to 200 mm long; Great Lakes area .... **Haemopis plumbea** (Moore)
    Lip narrow and arched (Fig. 33L); dorsal surface usually green, gray, or plain, but always more or less
    blotched with black; 150 to 300 mm long; widely distributed ...... **Haemopis grandis** (Verrill)
50. Each jaw with 10 or 14 pairs of teeth; caudal sucker about three-quarters of the body width; 50 to 85
    mm long; known only from IA, MN .... **Haemopis lateromaculata** Mathers
    Each jaw with 10 to 16 pairs of teeth; caudal sucker about one half of body width; 75 to 100 mm long;
    often in mud at edge of water; the common "horse leech"; occasionally attacks man.
                                              Haemopis marmorata (Say)

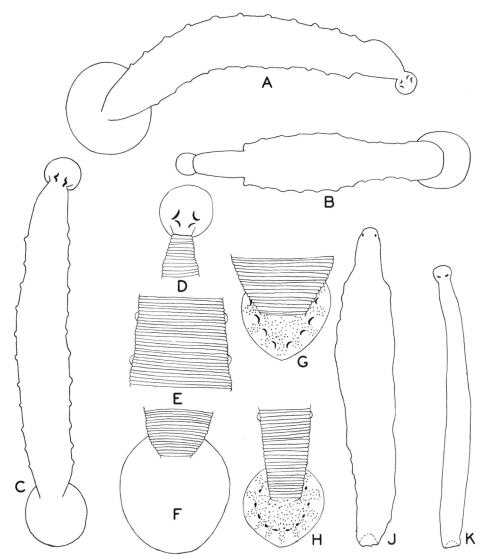

FIG. 36.–Structure of Piscicolidae. A, *Cystobranchus verrilli,* ×10; B, *Cystobranchus vivida,* ×4; C, *Piscicola punctata,* ×6.5; D to F, anterior, middle, and posterior portions of *P. punctata*; G, posterior end of *P. salmositica*; H, posterior end of *P. milneri*; J, *Myzobdella lugubris,* ×6.5; K, *Illinobdella moorei,* ×8. (A, C, J, and K modified from Meyer, 1940; D–F, and H modified from Meyer, 1946; B modified from Moore, 1898; G modified from Meyer, 1946a.)

51. Segments mostly composed of five annuli; dorsal surface with two or four longitudinal stripes, but some forms are barred with black or are lightly pigmented; three pairs of eyes; 6 to 10 cm long.
    Erpobdella, 52
    Segments mostly composed of six or seven annuli; three or four pairs of eyes 53
52. Eyes all of similar size; feeds on fish, frogs, invertebrates, and occasionally man; sometimes a scavenger; the most common and widely distributed leech ... **Erpobdella punctata** (Leidy)
    Second and third pairs of eyes smaller than the first pair; known only from Montezume Well, AZ.
    Erpobdella montezuma Davis, Singhal, and Blinn

53. With four eyes arranged in an anterior horizontal fashion, and the posterior four eyes similarly arranged in a horizontal line; feeds on aquatic invertebrates, also a scavenger; up to 10 cm long; northern states .... **Nephelopsis obscura** Verrill

    With six or eight or eight eyes arranged in a different manner **54**

54. With three pairs of eyes (rarely less) **55**

    With four pairs of eyes **62**

55. Uniformly gray, black, barred, or striped; never with scattered pigment areas **56**

    Whitish, with scattered pigmented areas; up to 55 mm long; East Coast states.

                           Mooreobdella melanostoma Sawyer and Shelley

56. With two or two and one-half annuli between gonopores, in furrows or on the rings (Figs. 35D–F).
                                                 57

    With three to four and one-half annuli between gonopores, in furrows or on rings **61**

57. Gonopores separated by two annuli **58**

    Gonopores separated by two and one-half annuli (see couplet 52) .... **Erpobdella punctata**

58. Dorsal surface densely marked with black **59**

    Dorsal surface variable, lacking pigment, gray, or with two longitudinal bands **60**

59. With two or four black pigment rows on dorsal surface forming longitudinal stripes; up to 10 cm long (see couplet 52) .... **Erpobdella punctata**

    With four longitudinal dorsal stripes of gray or black; 10 to 15 mm long; rare in Pacific Coast states.

                                           Dina anoculata Moore

60. Color uniform gray, without black pigment; 10 to 30 mm long; known only from MI.

                                        Mooreobdella bucera Moore

    Color gray, often with black pigment in various patterns; 20 to 50 mm long; common and widely distributed in northern half of the United States .... **Mooreobdella fervida** Verrill

61. Gonopores separated by three annuli, usually in furrows; reddish color; 30 to 50 mm long; common and widely distributed .... **Mooreobdella microstoma** (Moore)

    Gonopores separated by four to four and one-half annuli, usually on rings; grayish color; up to 40 mm long; Atlantic and Gulf Coast states ...... **Mooreobdella tetragon** Sawyer and Shelley

62. Gonopores separated by two annuli (see couplet 60) .... **Mooreobdella fervida**

    Gonopores separated by three and one-half (seldom two and one-half or four) annuli **63**

63. Dorsal surface with a dark brown or black stripe on a greenish background; northern states; 20 to 60 mm long .... **Dina dubia** Moore and Meyer

    Dorsal surface without a middorsal stripe; gray, with dark and light pigment spots; 25 to 30 mm long; widely distributed .... **Dina parva** Moore

REFERENCES

Alsterberg, G. 1922. Die respiratorischen Mechanismen der Tubificiden. *Lunds Univ. Årsskr. Avd. 2* 18:1–176.

Andrassy, I. 1956. Troglochaetus beranecki Delachaux, ein Repräsentant der für die Fauna Ungarns neuen Tierklasse Archiannelida. *Ann. Hist. Nat. Mus. Nat. Hung.* 7:371–375.

Appleby, A. G., and R. O. Brinkhurst. 1970. Defecation rates of three tubificid oligochaetes found in the sediment of Toronto Harbour, Ontario. *J. Fish. Res. Board Can.* 27:1971–1982.

Aston, R. J. 1968. The effect of temperature on the life cycle, growth and fecundity of Branchiura sowerbyi (Oligochaeta: Tubificidae). *J. Zool. London* 154:29–40.

———. 1973. Tubificids and water quality; a review. *Environ. Pollut.* 5:1–10.

Becker, C. D., and M. Katz. 1965. Distribution, ecology, and biology of the salmonid leech, Piscicola salmositica (Rhynchobdellae: Piscicolidae). *J. Fish. Res. Board Can.* 22:1175–1195.

Beddard, E. F. 1895. *A Monograph of the Order Oligochaeta.* 769 pp. Oxford, England.

Bennike, S. A. B. 1943. Contributions to the ecology and biology of the Danish freshwater leeches (Hirudinea). *Fol. Limn. Scandinavica* 2:1–109.

Bogatov, V. V. et al. 1980. The luminescence of freshwater Oligochaeta. *Hydrobiol. J.* 16:50–51. (Translated from Russian.)

Bousfield, E. C. 1887. The natural history of the genus Dero. *J. Linn. Soc. (Zool.)* 20:91–107.

Brandt, E. 1978. Anpassungen von Tubifex tubifex Müller (Annelida, Oligochaeta) an die Temperatur, den Sauerstoffgehalt und den Ernährungszustand. *Arch. Hydrobiol.* 84:302–338.

Brinkhurst, R. O. 1963. A guide for the identification of British aquatic Oligochaeta. *Sci. Publ. Freshwater Biol. Assoc.* 22:1–52.

———. 1964. Studies on the North American aquatic Oligochaeta. I. Naididae and Opistocystidae. *Proc. Acad. Nat. Sci. Philadelphia* 116:195–230.

———. 1966. The Tubificidae (Oligochaeta) of polluted waters. *Proc. Int. Assoc. Theor. Appl. Limnol.* 16:854–859.

_____. 1970. Distribution and abundance of tubificid (Oligochaeta) species in Toronto harbour, Lake Ontario. *J. Fish. Res. Board Can.* **27**:1961–1969.

_____. 1985. The generic and subfamilial classification of the Naididae (Annelida: Oligochaeta). *Proc. Biol. Soc. Wash.* **98**:470–475.

_____. 1986. Guide to the freshwater aquatic microdrile oligochaetes of North America. *Can. Spec. Publ. Fish. Aquatic Sci.* **84**:1–259.

Brinkhurst, R. O., and K. A. Coates. 1985. The genus Paranais (Oligochaeta: Naididae) in North America. *Proc. Biol. Soc. Wash.* **98**:303–313.

Brinkhurst, R. O., and D. G. Cook (eds.). 1980. *Aquatic Oligochaete Biology.* 530 pp. Plenum Press, New York.

Brinkhurst, R. O., and B. G. M. Jamieson. 1971. *Aquatic Oligochaeta of the World.* 860 pp. University of Toronto Press, Toronto.

Brinkhurst, R. O., and M. J. Wetzel. 1984. Aquatic Oligochaeta of the World: supplement. A catalogue of new freshwater species, descriptions, and revisions. *Can. Tech. Rep. Hydrogr. Ocean Sci.* **44**:1–101.

Bunke, D. 1967. Zur Morphologie und Systematik der Aeolosomatidae Beddard 1895 und Potamodrilidae nov. fam. (Oligochaeta). *Zool. Jahrb. Abt. Syst. Oekol. Geogr. Tiere* **94**:187–368.

Chen, Y. 1940. Taxonomy and faunal relations of the limnetic Oligochaeta of China. *Contr. Biol. Lab. Sci. Soc. China, Zool. Ser.* **14**:1–132.

Cook, D. G. 1975. Cave-dwelling aquatic Oligochaeta (Annelida) from the Eastern United States. *Trans. Am. Microsc. Soc.* **94**:24–37.

Dausend, K. 1931. Über die Atmung der Tubifiziden. *Z. vergl. Physiol.* **14**:557–608.

Davies, R. W. 1971. A key to the freshwater Hirudinoidea of Canada. *J. Fish. Res. Board Can.* **28**:543–552.

Davies, R. W., and R. P. Everett. 1975. The feeding of four species of freshwater Hirudinoidea in southern Alberta. *Proc. Int. Assoc. Theor. Appl. Limnol.* **19**:2816–2827.

Davies, R. W. et al. 1982. Passive dispersal of four species of freshwater leeches (Hirudinoidea) by ducks. *Freshwater Invert. Biol.* **1**:40–44.

_____. 1985. Erpobdella montezuma (Hirudinoidea: Erpobdellidae), a new species of freshwater leech from North America. *Can. J. Zool.* **63**:965–969.

Delachaux, T. 1921. Un Polychète d'eau douce cavernicole. Troglochaetus beranecki nov. gen. nov. spec. (Note Prél.). *Bull. Soc. Sci. Nat. Neuchatel* **45**:3–11.

Elliott, J. M., and K. H. Mann. 1979. A key to the British freshwater leeches. *Freshwater Biol. Assoc. Sci. Publ.* **40**:1–72.

Elliott, J. M., and P. A. Tullett. 1984. The status of the medicinal leech Hiro medicinalis in Europe and especially in the British Isles. *Biol. Conserv.* **29**:15–26.

Ellis, M. M. 1919. The branchiobdellid worms in the collections of the United States National Museum, with descriptions of new genera and new species. *Proc. U.S. Nat. Mus.* **55**:241–265.

Famme, P., and J. Knudsen. 1985. Aerotaxis by the freshwater oligochaete Tubifex sp. *Oecologia* **65**:599–601.

Foster, N. 1972. Freshwater polychaetes (Annelida) of North America. *Biota Freshwater Ecosystems Identification Manual* **4**:1–15.

Gelder, S. R. 1980. A review of the symbiotic Oligochaeta (Annelida). *Zool. Anz.* **204**:69–81.

Gelder, S. R., and R. C. Smith. 1987. Distribution of branchiobdellids (Annelida: Clitellata) in northern Maine, USA. *Trans. Am. Microsc. Soc.* **106**:85–88.

Goodnight, C. J. 1940. The Branchiobdellidae (Oligochaeta) of North American crayfishes. *Ill. Biol. Monogr.* **17**:1–75.

_____. 1941. The Branchiobdellidae (Oligochaeta) of Florida. *Trans. Am. Microsc. Soc.* **60**:69–74.

_____. 1943. Report on a collection of Branchiobdellidae. *J. Parasitol.* **29**:100–102.

Greene, K. L. 1974. Experiments and observations on the feeding behavior of the freshwater leech Erpobdella octoculata (L.) (Hirudinea: Erpobdellidae). *Arch. Hydrobiol.* **74**:87–99.

Gruffydd, Ll. D. 1965. Notes on a population of the leech, Glossiphonia heteroclita, infesting Lymnaea periger. *Ann. Mag. Nat. Hist.* **8**:144–150.

Harding, W. A., and J. P. Moore. 1927. Hirudinea. *Fauna Br. India.* 302 pp.

Hartman, O. 1936. New species of polychaetous annelids of the family Nereidae from California. *Proc. U.S. Nat. Mus.* **83**:467–480.

_____. 1938. Brackish and freshwater Nereidae from the northeast Pacific with the description of a new species from central California. *Univ. Calif. Publ. Zool.* **43**:79–82.

_____. 1959. Capitellidae and Nereidae (marine annelids) from the Gulf side of Florida, with a review of freshwater Nereidae. *Bull. Mar. Sci. Gulf Carib.* **9**:153–168.

Herlant-Meewis, H. 1950. Cyst-formation in Aeolosoma hemprichi (Ehr.). *Biol. Bull.* **99**:172–180.

Herrmann, S. J. 1970. Systematics, distribution, and ecology of Colorado Hirudinea. *Am. Midl. Nat.* **83**:1–37.

Herter, K. 1932. Hirudinea. *Biol. Tiere Deutschl.* **12b**:1–158.

_____. 1968. *Der Medizinische Blutegel und seine Verwandten.* 199 pp. Ziemsen Verlag, Wittenberg, East Germany.

Hiltunen, J. K., and D. J. Klemm. 1980. A guide to the Naididae (Annelida: Clitellata: Oligochaeta) of North America. *EPA Publ.* 600/4-80-031, 47 pp.

Hobbs, H. H., et al. 1967. The crayfishes and their epizootic ostracod and branchiobdellid associates of the Mountain Lake, Virginia, Region. *Proc. U.S. Nat. Mus.* **123**:1–84.

Hoffman, R. L. 1963. A revision of the North American annelid worms of the genus Cambarincola (Oligochaeta: Branchiobdellidae). *Proc. U.S. Nat. Mus.* **114** (3470):271–371.

Holmquist, C. 1973. Fresh-water polychaete worms of Alaska with notes on the anatomy of Manayunkia speciosa Leidy. *Zool. Jahrb. Syst.* **100**:497–516.

Holt, P. C. 1960. On a new genus of the Family Branchiobdellidae (Oligochaeta). *Am. Midl. Nat.* **64**:169–176.

_____. 1960a. The genus Ceratodrilus Hall (Branchiobdellidae, Oligochaeta) with the description of a new species. *Va. J. Sci.* **11**:53–77.

_____. 1965. On Ankyrodrilus, a new genus of branchiobdellid worms (Annelida). *Ibid.* **16**:9–21.

_____. 1967. Oedipodrilus oedipus, n. g., n. sp. (Annelida, Clitellata: Branchiobdellida). *Trans. Am. Microsc. Soc.* **86**:58–60.

_____. 1967a. Status of genera Branchiobdella and Stephanodrilus in North America with description of a new genus (Clitellata: Branchiobdellida). *Proc. U.S. Nat. Mus.* **124**:1–10.

_____. 1968. The Branchiobdellida: epizootic annelids. *The Biologist* **50**:79–94.

_____. 1968a. New genera and species of branchiobdellid worms (Annelida: Clitellata). *Proc. Biol. Soc. Wash.* **81**:291–318.

_____. 1968b. The genus Pterodrilus (Annelida: Branchiobdellida). *Proc. U.S. Nat. Mus.* **125**:1–44.

_____. 1969. The relationships of the branchiobdellid fauna of the Southern Appalachians. *Va. Polytech. Inst. Res. Div. Monogr.* **1**:191–219.

_____. 1973. Branchiobdellids (Annelida: Clitellata) from some eastern North American caves, with descriptions of new species of the genus Cambarincola. *Int. J. Speleol.* **5**:219–256.

_____. 1973a. Epigean branchiobdellids (Annelida: Clitellata) from Florida. *Proc. Biol. Soc. Wash.* **86**:79–104.

_____. 1974. The genus Xironogiton Ellis, 1919 (Clitellata: Branchiobdellida). *Va. J. Sci.* **25**:5–19.

_____. 1981. A resume of the members of the genus Cambarincola (Annelida: Branchiobdellida) from the Pacific drainage of the United States. *Proc. Biol. Soc. Wash.* **94**:675–695.

_____. 1986. Newly established families of the Order Branchiobdellida (Annelida: Clitellata) with a synopsis of the genera. *Proc. Biol. Soc. Wash.* **99**:676–702.

Hotz, H. 1938. Protoclepsis tesselata (O. F. Müller); ein Beitrag zur Kenntnis von Bau und Lebensweise der Hirudineen. *Rev. Suisse Zool. (Suppl.)* **45**:1–380.

Howmiller, R., and M. S. Loden. 1976. Identification of Wisconsin Tubificidae and Naididae. *Trans. Wis. Acad. Sci. Arts Lett.* **64**:185–197.

Jennings, J. B., and V. M. Van Der Lande. 1967. Histochemical and bacteriological studies on digestion in nine species of leeches (Annelida: Hirudinea). *Biol. Bull.* **133**:166–183.

Kaster, J. L. 1979. Morphological development and adaptive significance of autotomy and regeneration in Tubifex tubifex Müller. *Trans. Am. Microsc. Soc.* **98**:473–477.

_____. 1980. The reproductive biology of Tubifex tubifex Müller (Annelida: Tubificidae). *Am. Midl. Nat.* **104**:355–364.

Kaster, J. L., and J. H. Bushnell. 1981. Cyst formation by Tubifex tubifex (Tubificidae). *Trans. Am. Microsc. Soc.* **100**:34–41.

_____. 1981a. Occurrence of tests and their possible significance in the worm, Tubifex tubifex (Oligochaeta). *Southwest Nat.* **26**:307–310.

Keyl, F. 1913. Beiträge zur Kenntnis von Branchiura sowerbyi Bedd. *Z. Wiss. Zool.* **107**:199–304.

Klemm, D. J. 1972. Freshwater leeches (Annelida: Hirudinea) of North America. *Biota Freshwater Ecosystems Identification Manual* **5**:1–53.

_____. 1975. Studies on the feeding relationships of leeches (Annelida Hirudinea) as natural associates of mollusks. *Sterkiana* **58**:1–51.

_____. 1977. A review of the leeches (Annelida: Hirudinea) in the Great Lakes Region. *Mich. Academician* **9**:397–418.

_____. 1982. Leeches (Annelida: Hirudinea) of North America. *EPA Publ.* 600/3-82-025, 177 pp.

_____ (ed.). 1985. *A Guide to the Freshwater Annelida (Polychaeta, Naidid and Tubificid Oligochaeta, and Hirudinea of North America).* 198 pp. Kendall/Hunt, Dubuque, Iowa.

Learner, M. A. et al. 1978. A review of the biology of British Naididae (Oligochaeta) with emphasis on the lotic environment. *Freshwater Biol.* **8**:357–375.

Leidy, J. 1883. Manayunkia speciosa. *Proc. Philadelphia Acad. Nat. Sci.* **1883**:204–212.

Loden, M. S. 1981. Reproductive ecology of Naididae (Oligochaeta). *Hydrobiologia* **93**:115–123.

Maloney, S. D., and C. M. Chandler. 1976. Leeches (Hirudinea) in the Upper Stones River drainage of Middle Tennessee. *Am. Midl. Nat.* **95**:42–48.

Mann, K. H. 1955. The ecology of the British freshwater leeches. *J. Anim. Ecol.* **24**:98–119.

_____. 1962. *Leeches (Hirudinea). Their structure, physiology, ecology and embryology.* 201 pp. Pergamon, New York.

Marcus, E. 1942. Sôbre algumas Tubificidae do Brasil. *Univ. Sao Paulo, Bol. Fac. Filos, Cien. Lett. Zool.* **6**:153–228.

_____. 1943. Sôbre Naididae do Brasil. *Ibid.* **7**:1–247.

_____. 1944. Sôbre Oligochaeta limnicos do Brasil. *Ibid.* **8**:5–135.

Meyer, F. P. 1969. A potential control of leeches. *Prog. Fish Cult.* **31**:160–163.

Meyer, M. C. 1940. A revision of the leeches (Piscicolidae) living on fresh-water fishes of North America. *Trans. Am. Microsc. Soc.* **59**:354–376.

_____. 1946. Further notes on the leeches (Piscicolidae) living on fresh-water fishes of North America. *Ibid.* **65**:237–249.

_____. 1946a. A new leech, Piscicola salmositica n. sp. (Piscicolidae), from steelhead trout (Salmo gairdneri gairdneri Richardson, 1838). *J. Parasitol.* **32**:467–473.

Miller, J. A. 1929. The leeches of Ohio. *Contrib. Franz Theodore Stone Lab.* **2**:1–38.

Moore, J. P. 1895. The anatomy of Bdellodrilus illuminatus, an American discodrilid. *J. Morphol.* **10**:497–540.

_____. 1898. The leeches of the U.S. National Museum. *Proc. U.S. Nat. Mus.* **21**:543–563.

_____. 1901. The Hirudinea of Illinois. *Bull. Ill. State Lab. Nat. Hist.* **5**:479–547.

_____. 1912. Leeches of Minnesota. *Geol. Nat. Hist. Surv. Minn. Zool.* **5**, III:64–150.

_____. 1923. The control of blood-sucking leeches, with

an account of the leeches of Palisades Interstate Park. *Roosevelt Wild Life Bull.* 2:1–53.

Nachtreib, H. F., E. E. Hemingway, and J. P. Moore. 1912. The leeches of Minnesota. *Geol. Nat. Hist. Surv. Minn. Zool. Ser.* 5:1–150.

Pennak, R. W. 1971. A fresh-water archiannelid from the Colorado Rocky Mountains. *Trans. Am. Microsc. Soc.* 90:372–375.

Peterson, D. L. 1983. Life cycle and reproduction of Nephelopsis obscura Verrill (Hirudinea: Erpobdellidae) in permanent ponds of northwestern Minnesota. *Freshwater Invert. Biol.* 2:165–172.

Poe, T. P., and D. C. Stefan. 1974. Several environmental factors influencing the distribution of the fresh-water polychaeta, Manayunkia speciosa Leidy. *Chesapeake Sci.* 12:235–237.

Reynoldson, T. B. 1947. An ecological study of the enchytraeid worm population of sewage bacteria beds. *J. Anim. Ecol.* 16:26–37.

Richardson, L. R. 1969. A contribution to the systematics of the hirudinid leeches, with description of new families, genera and species. *Acta Zool. Acad. Sci. Hung.* 15:97–149.

Rupp, R. W., and M. C. Meyer. 1954. Mortality among brook trout, Salvelinus fontinalis, resulting from attacks of freshwater leeches. *Copeia* 1954:294–295.

Sapkarev, J. A. 1967–1968. The taxonomy and ecology of leeches (Hirudinea) of Lake Mendota, Wisconsin. *Trans. Wis. Acad.* 56:225–253.

Sawyer, R. T. 1970. Observations on the natural history and behavior of Herpobdella punctata (Leidy) (Annelida: Hirudinea). *Am. Midl. Nat.* 83:65–80.

———. 1972. North American freshwater leeches, exclusive of the Piscicolidae, with a key to all species. *Ill. Biol. Monogr.* 46:1–154.

———. 1986. *Leech Biology and Behaviour.* 3 vols. 1100 pp. Oxford Univ. Press, Oxford, England.

Sawyer, R. T., and R. M. Shelley. 1976. New records and species of leeches (Annelida: Hirudinea) from North and South Carolina. *J. Nat. Hist.* 10:65–97.

Singer, R. 1978. Suction-feeding in Aeolosoma (Annelida). *Trans. Am. Microsc. Soc.* 97:105–110.

Smith, D. G. 1977. The rediscovery of Macrobdella sestertia Whitman (Hirudinea: Hirudinidae). *J. Parasit.* 63:759–760.

Smith, R. I. 1942. Nervous control of chromatophores in the leech Placobdella parasitica. *Physiol. Zool.* 15:410–417.

———. 1959. The synonymy of the viviparous polychaete Neanthes lighti Hartman (1938) with Nereis limnicola Johnson (1903). *Pac. Sci.* 13:349–350.

Soos, A. 1963. Identification key to the species of the genus Dina R. Blanchard, 1892 (emend. Mann, 1962) (Hirudinea: Erpobdellidae). *Acta Univ. Szeged. Acta Biol.* 9:253–261.

———. 1965. Identification key to the leech (Hirudinoidea) genera of the world, with a catalogue of the species. I. Family: Piscicolidae. *Acta Zool. Acad. Sci. Hung.* 11:417–463.

———. 1966. On the genus Glossiphonia Johnson, 1816, with a key and catalogue to the species. *Ann. Hist. Nat. Mus. Nat. Hung. Zool.* 58:271–279.

———. 1966a. Identification key to the leech (Hirudinoidea) genera of the world, with a catalogue of the species. III. Family: Erpobdellidae. *Acta Zool. Acad. Sci. Hung.* 12:371–407.

———. 1967. On the genus Batracobdella Viguier, 1879, with a key and catalogue to the species. *Ann. Hist. Nat. Mus. Nat. Hung. Zool.* 59:243–257.

———. 1969. Identification key to the leech (Hirudinoidea) genera of the world, with a catalog of the species: V. Family: Hirudinidae. *Acta Zool. Acad. Sci. Hung.* 15:151–201.

Spencer, D. R. 1976. Occurrence of Manayunkia speciosa (Polychaeta: Sabellidae) in Cayuga Lake, New York, with additional notes on its North American distribution. *Trans. Am. Microsc. Soc.* 95:127–128.

———. 1980. The aquatic Oligochaeta of the St. Lawrence Great Lakes region. *Aquatic Oligochaete Biol.* 1:115–164.

Spencer, W. P. 1932. A gilled oligochaete Branchiura sowerbyi new to America. *Trans. Am. Microsc. Soc.* 51:267–272.

Sperber, C. 1948. A taxonomical study of the Naididae. *Zool. Bidrag f. Uppsala* 28:1–296.

Stimpson, K. S., et al. 1975. Distribution and abundance of inshore oligochaetes in Lake Michigan. *Trans. Am. Microsc. Soc.* 94:384–394.

———. 1982. A guide to the freshwater Tubificidae of North America. *EPA Publ.* 600/3-82-033.

Taylor, W. D. 1980. Observations on the feeding and growth of the predaceous oligochaete Chaetogaster langi on ciliated Protozoa. *Trans. Am. Microsc. Soc.* 99:360–367.

Von Wagner, P. 1916. Zur Oecologie des Tubifex und Lumbriculus. *Zool. Jahrb. Abt. Syst.* 23:295–318.

Walker, J. G. 1970. Oxygen poisoning in the annelid Tubifex tubifex. I. Response to oxygen exposure. *Biol. Bull.* 138:235–244.

Westheide, W. 1979. Hesionides riegerorum n. sp., a new freshwater polychaete from the United States. *Int. Rev. gesamten Hydrobiol.* 64:273–280.

Young, J. O. 1983. A stimulus for egg production in Glossiphonia complanata (L.) (Hirudinoidea: Glossiphoniidae). *Freshwater Invert. Biol.* 8:112–115.

Young, J. O., and S. M. Spelling. 1986. The incidence of predation on lake-dwelling leeches. *Freshwater Biol.* 16:465–477.

Young, W. 1966. Ecological studies of the Branchiobdellida (Oligochaeta). *Ecology* 47:571–577.

Zinn, D. J., and I. R. Kneeland. 1964. Narcotization and fixation of leeches (Hirudinea). *Trans. Am. Microsc. Soc.* 83:275–276.

14

INTRODUCTION TO THE CRUSTACEA

Four obvious features clearly distinguish the members of the Class Crustacea: With few exceptions, they are all aquatic arthropods; respiration occurs through gills or through the general body surface; they all have two pairs of antennae; most of the body segments, or at least the more anterior ones, bear paired, jointed appendages that are fundamentally biramous.

The great majority of the 36,000 known species of the world are marine, and of the many orders occurring in the United States and surrounding waters, 13 are represented in fresh waters, but only 5 are restricted to our fresh waters. The number of species represented for each of the 13 orders in the United States ranges from one (Thermosbaenacea) to about 350 (Decapoda). Eleven orders are treated in detail in Chapters 15 through 22, so the present chapter is merely a summary and tabulation of their chief distinguishing characteristics and affinities.

Table VI shows the environmental affinities of major groups of Crustacea in the area of the continental United States. On a worldwide basis, however, some of these entries are not strictly true because of exceptional rarities. For example, several species of Cumacea are known from fresh-water habitats in Kerala; and several fresh-water species of Tanaidacea have been found in South America and the Kurile Islands.

Some carcinologists distinguish fine gradations in meaning in the usage of "somite," "metamere," and "body segment" to denote the primary body divisions of adult Crustacea. Nevertheless, these expressions are more generally used interchangeably and without confusion; in this manual "body segment," "trunk segment," or simply "segment" are used consistently.

In the primitive condition the trunk segments are distinct, but the segments forming the head are completely fused. The trunk is usually subdivided into thorax and abdomen so that there are three body regions, though fusion of some thoracic and abdominal segments is the common condition. One or more thoracic segments, for example, are often fused with the head to form a cephalothorax, which in many crustaceans is completely or partly covered by a shieldlike carapace. In groups such as the Cladocera and Ostracoda body segmentation has disappeared. The general body plan of the various orders is outlined in Table VII.

In the strict sense, the most posterior division of the body, which bears the anus, is not considered a true body segment and is usually referred to as the telson. This term is universally used in the orders of Malacostraca, where this structure is comparatively small and flattened and does not bear any processes. In the other orders, however, the telson typically resembles the preceding body segment; furthermore, it bears a pair of terminal caudal rami, cercopods, or heavy claws. In common usage it is often referred to as a "segment" in the nonmalacostracan orders. The writer can see no real objection to this practice. However, see details of these terminology arguments in Bowman (1971) and Schminke (1976).

All crustaceans go through a free nauplius stage or at least show some evidence of an egg-nauplius phase in early development, but among fresh-water forms only eubranchiopods, ostracods, and copepods have free-swimming nauplii.

The paired appendages are serially homologous; each is fundamentally biramous and consists of a basal protopod, a segmented

TABLE VI. Environmental Affinities of Groups of Living Crustacea Found in the United States and Adjacent Waters

Restricted to marine and brackish waters	Chiefly marine; few fresh-water species	Well represented in both marine and fresh waters	Chiefly in fresh waters; few marine species	Exclusively fresh water	Approximate number of species living in U.S. fresh waters
Amphionidacea					
Caridea	Mysidacea				3
Cephalocarida		Amphipoda			150
Cirripedia		Copepoda			230
Cumacea		Decapoda			350
Dendrobranchiata		Isopoda			130
Euphausiacea		Ostracoda			300
Leptostraca			Cladocera		150
Mystacocarida				Anostraca	30
Nebaliacea				Bathynellacea	5(?)
Nectiopoda				Conchostraca	28
Stomatopoda				Notostraca	6
Tanaidacea				Thermosbaenacea (fresh to brackish)	1

endopod, and a segmented exopod. They are usually more or less specialized for a wide variety of functions, many such modifications being figured and considered rather completely in Chapters 15 through 22.

Because of variations in early embryology from one order to another and because of the variety of interpretations that may be drawn from such embryological material, the "head problem" of arthropods is still unsettled. Some investigators maintain that the adult crustacean head represents the fusion of five segments or four segments plus a large anterior cephalic lobe (Snodgrass, 1951); others hold to an older theory that the head represents six fused segments. Henry (1948) maintains that the older theory is the more acceptable and that none of the six segments forming the head are "preoral." Aside from this controversy, the typical adult crustacean head bears five pairs of serially homologous appendages. Beginning at the anterior end, these are first antennae, second antennae, mandibles, first maxillae, and second maxillae. Compound eyes are not considered true segmental appendages. In the Cladocera and Ostracoda the second maxillae are lacking. More posteriorly the situation is variable, with some groups having an additional complement of thoracic or body appendages modified as mouthparts, and others not. Further, the number of thoracic and abdominal append-

ages and their modifications vary widely from one group to another, even within single genera.

The Crustacea may be conveniently divided into two groups, Entomostraca and Malacostraca. The latter group is a clearly defined and natural section of the Crustacea, all of its members having abdominal appendages, a gastric mill, an eight-segmented thorax, and an abdomen of six (rarely seven or eight) segments. The Entomostraca, however, is a heterogeneous assemblage of orders that differ from one another as widely as each of them does from the Malacostraca, and for this reason the use of "Entomostraca" as a specific taxonomic category should be discouraged. Nevertheless, "entomostraca" is a convenient general descriptive term that might be legitimately used to designate collectively all small fresh-water Crustacea. It is interesting, however, to note that hemoglobin has been reported from all entomostracan orders, but is unknown in the Malacostraca.

Carcinologists differ in their ideas about systematic arrangement and hierarchical terminology for the many major taxa of Crustacea. From the student's standpoint it is often a disturbing melange of groups, divisions, superorders, and suborders. We are avoiding a discussion of this whole problem and are restricting our systematic array to the plan shown in Table IX, which involves only those

TABLE VII. Gross Morphological Features of the Orders of Fresh-Water Crustacea of the United States

Order	Main body features	Length range (mm)	Number of thoracic segments fused with head to form cephalothorax	Number of free thoracic segments	Number of abdominal segments	Number of pairs of thoracic appendages		Number of pairs of abdominal appendages	Common terminology for posterior end
						Modified for use as mouth parts	Not used as mouth parts		
Anostraca	Head and trunk; body cylindrical	7–100	0	19 Trunk segments		0	11 Pairs of trunk appendages	Pairs of trunk appendages	Telson and caudal rami
Notostraca	Head and trunk; shieldlike carapace covering most of body; depressed dorsoventrally	10–58	0	25–44 Trunk segments		0	35–71 Pairs of trunk appendages	Pairs of trunk appendages	Telson and caudal rami
Conchostraca	Head and trunk; body enclosed in a bivalve carapace; laterally compressed	2–16	0	10–32 Trunk segments		0	10–32 Pairs of trunk appendages	Pairs of trunk appendages	Telson and caudal rami
Cladocera	Head, trunk, and postabdomen; segmentation obscure; body laterally compressed and covered with a folded carapace	0.2–18.0				0	5 or 6	0	Postabdomen and claws
Ostracoda	Head indistinctly set off from trunk; segmentation obscure; body laterally compressed and covered with a bivalve shell	0.3–4.2					3	0	Caudal rami
Copepoda	Cephalothorax, thorax, and abdomen; body cylindrical	0.3–3.2	1 or 2	Usually 5	3–5	1	5	0	Caudal rami

338

Taxon	Body regions	Size (mm)							Caudal rami
Branchiura	Cephalothorax, thorax, and abdomen; large, expanded cephalothoracic carapace; body strongly depressed; parasitic	5–25	1 or 2	3 (4th and 5th fused with abdomen)	Fused	1	4	0	
Mysidacea	Cephalothorax, thorax, and abdomen; carapace covering cephalothorax and most of thorax; body slightly compressed laterally	10–30	1	7	6	2	6	6	Telson
Isopoda	Cephalothorax, thorax, and abdomen; body depressed	5–20	1	7	6 (usually fused)	1	7	6	Telson (obscure)
Amphipoda	Cephalothorax, thorax, and abdomen; body slightly compressed laterally	5–25	1	7	6	1	7	6	Telson
Decapoda	Cephalothorax and abdomen; carapace covering all of cephalothorax; body subcylindrical	15–150	8	0	6	3	5	6	Telson
Thermosbaenacea	Cephalothorax, thorax, and abdomen; body cylindrical	1–3.5	1–2	6–7	6	1	7	3	Telson may or may not be fused with last abdominal segment
Bathynellacea	Head, thorax, and abdomen	0.5–2.0	0	8	6	0	8	1–3	Telson and caudal rami (uropods)

TABLE VIII. Arrangement and Terminology of Segmental Appendages of Typical Fresh-Water Crustacea

Body region	Eubranchiopoda	Cladocera	Copepoda	Ostracoda	Mysidacea
Head	1st antennae	1st antennae	1st antennae	1st antennae	1st antennae
	2d antennae	2d antennae	2d antennae	2d antennae	2d antennae
	mandibles	mandibles	mandibles	mandibles	mandibles
	1st maxillae	maxillae	1st maxillae	maxillae	1st maxillae
	2d maxillae	—	2d maxillae	—	2d maxillae
Trunk segments 1		1st legs	maxillipeds[b]	1st legs	1st maxillipeds
2		2d legs	1st legs	2d legs	2d maxillipeds
3		3d legs	2d legs	3d legs	1st legs
4		4th legs	3d legs		2d legs
5		5th legs[a]	4th legs		3d legs
6		6th legs	5th legs		4th legs
7	10–71 pairs of				5th legs
8	appendages on 10–				6th legs
9	44 trunk segments				1st pleopods
10					2d pleopods
11					3d pleopods
12					4th pleopods
13					5th pleopods
14					uropods

[a]These appendages are present only in three small families of Cladocera.
[b]Some copepod specialists consider the maxillipeds as head appendages.

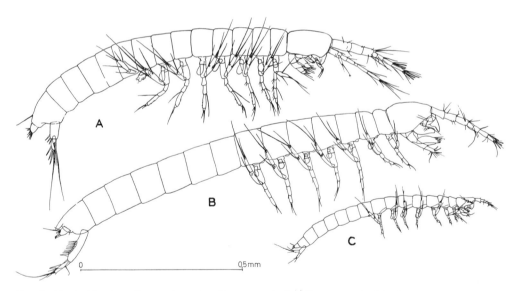

FIG. 1.–Typical Bathynellacea. A, *Bathynella;* B, *Parabathynella;* C, *Leptobathynella.* (From Noodt, 1964.)

Isopoda	Amphipoda	Decapoda	Thermosbaenacea	Bathynellacea
1st antennae	1st antennae	1st antennae	1st antennae	1st antennae
2d antennae	2d antennae	2d antennae	2nd antennae	2nd antennae
mandibles	mandibles	mandibles	mandibles	mandibles
1st maxillae	1st maxillae	1st maxillae	1st maxillae	1st maxillae
2d maxillae	2d maxillae	2d maxillae	2nd maxillae	2nd maxillae
Maxillipeds	Maxillipeds	1st maxillipeds	Maxillipeds	1st legs (pereiopods)
1st legs (gnathopods)	1st legs (gnathopods)	2d maxillipeds	1st legs (pereiopods)	2nd legs (pereiopods)
2d legs (pereiopods)	2d legs (gnathopods)	3d maxillipeds	2nd legs (pereiopods)	3rd legs (pereiopods)
3d legs (pereiopods)	3d legs (pereiopods)	1st legs (pereiopods)	3rd legs (pereiopods)	4th legs (pereiopods)
4th legs (pereiopods)	4th legs (pereiopods)	2d legs (pereiopods)	4th legs (pereiopods)	5th legs (pereiopods)
5th legs (pereiopods)	5th legs (pereiopods)	3d legs (pereiopods)	5th legs (pereiopods)	6th legs (pereiopods)
6th legs (pereiopods)	6th legs (pereiopods)	4th legs (pereiopods)	6th legs (pereiopods)	7th legs (pereiopods)
7th legs (pereiopods)	7th legs (pereiopods)	5th legs (pereiopods)	7th legs (pereiopods)	8th legs (pereiopods)
1st pleopods	1st pleopods	1st pleopods	1st pleopods	1st pleopods (often absent)
2d pleopods	2d pleopods	2d pleopods	2nd pleopods	2nd pleopods (often absent)
3d pleopods	3d pleopods	3d pleopods		
4th pleopods	1st uropods	4th pleopods		
5th pleopods	2d uropods	5th pleopods		
uropods	3d uropods	uropods	uropods	uropods

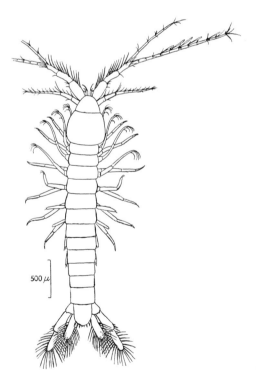

FIG. 2.–Dorsal view of *Monodella argentarii* Stella, a typical member of the Thermosbaenacea. (From Fryer, 1964.)

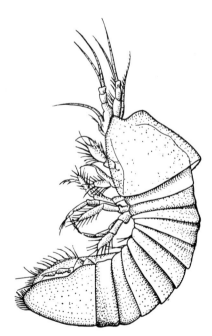

FIG. 3.–Lateral view of *Thermosbaena mirabilis* Monod. (From Absolon and Bruun.)

TABLE IX. Classification of Crustacea Occurring in the Fresh Waters of the United States

Class Crustacea
  Subclass Branchiopoda
    Division Eubranchiopoda
      Order Anostraca (fairy shrimps)
      Order Notostraca (tadpole shrimps)
      Order Conchostraca (clam shrimps)
    Division Oligobranchiopoda
      Order Cladocera (water fleas)
  Subclass Ostracoda
    Order Podocopa (seed shrimps)
  Subclass Copepoda
    Order Eucopepoda (copepods)
    Order Branchiura (fish lice)
  Subclass Malacostraca
    Division Peracarida
      Order Mysidacea (opossum shrimps)
      Order Isopoda (aquatic sow bugs)
      Order Amphipoda (scuds, sideswimmers)
    Division Eucarida
      Order Decapoda (fresh-water shrimps, crayfish)
    Division Pancarida
      Order Thermosbaenacea
    Division Syncarida
      Order Bathynellacea

taxa found in the fresh waters of the United States.

Crustacean rarities. Among the Syncarida, the fresh-water Order Stygocaridacea is known only from sand and gravel interstitial stream habitats in Argentina, Chile, and New Zealand. The Order Anaspidacea occurs only in ponds and streams of Tasmania and Australia. The Order Bathynellacea, however, is almost worldwide in temperate and tropical latitudes, chiefly in wells, subterranean waters, artesian systems, and interstitial waters of streams. The bathynellid literature is large and chiefly European. Only since 1974 have bathynellids been reported from the United States, and they are now known from Wyoming, California, Texas, Colorado, Montana, Kansas, Oklahoma, Kansas, and Georgia. The author has collected thousands from Colorado streamside sandy interstitial waters. Undoubtedly they are even more widely distributed in this country. Bathynellids are microscopic, blind, vermiform, and well adapted to interstitial situations. Some typical species are shown in Fig. 1.

Another crustacean rarity in fresh waters is the Order Thermosbaenacea. These are likewise small crustaceans, normally found in interstitial and subterranean waters and in springs. The earliest representative, *Thermosbaena mirabilis* (Fig. 3) was first found in saline hot springs near Gabes, Tunisia, in 1924. Other species have more recently been described from Italy, Yugoslavia, Canary Islands, West Indies, Somali, Mexico, and Israel. All these species occur in salty or brackish waters relatively close to the sea, but Fryer (1964) reports keeping thermosbaenaceans in fresh water for long periods. In 1965, however, Maguire described *Monodella texana* from fresh, cool water in a cave near San Marcos, Texas, about 200 km from the sea. Undoubtedly, careful searching should reveal Thermosbaenacea in other subterranean habitats. Superficially, they are similar to bathynellids, but they have a shieldlike carapace that roofs the anterior part of the thorax (Fig. 2). The female uses this device as a marsupium.

REFERENCES

Absolon, K. 1935. O zené fossilii Thermosbaena mirabilis z korkych vod Sahary. *Priorda* **38**:1–11.

Barker, D. 1959. The distribution and systematic position of the Thermosbaenacea. *Hydrobiologia* **13**:209–237.

———. 1962. A study of Thermosbaena mirabilis (Malacostraca, Peracarida) and its reproduction. *Q. J. Microsc. Sci.* **103**:261–286.

Bowman, T. E. 1971. The case of the nonubiquitous telson and the fraudulent furca. *Crustaceana* **21**:165–175.

Bruun, A. F. 1939. Observations on Thermosbaena mirabilis Monod from hot springs of El-Hamma, Tunisia. *Vid. Medd. Dansk Nat. For.* **103**:493–501.

Crampton, G. C. 1928. The evolution of the head region in lower arthropods and its bearing upon the origin and relationships of the arthropodan groups. *Can. Ent.* **40**:284–302.

Delamare Debouteville, C., et al. 1975. Découverte de la famille des Parabathynellidae (Bathynellacea) en Amérique du Nord. Texanobathynella bowmani n.

g. n. sp. *C. R. Hebd. Seances Acad. Sci. Paris, Ser. D,* **280**:2223–2226.

Fitzpatrick, J. F. 1983. *How to Know the Freshwater Crustacea.* 227 pp. W. C. Brown, Dubuque, Iowa.

Fryer, G. 1964. Studies on the functional morphology and feeding mechanism of Monodella argentarii Stella (Crustacea, Thermosbaenacea). *Trans. R. Soc. Edinburgh* **66**:51–90.

Gordon, I. 1957. On Spelaeogriphus, a new cavernicolous crustacean from South Africa. *Bull. Br. Mus. Nat. Hist. Zool.* **5**:29–47.

Henry, L. M. 1948. The nervous system and the segmentation of the head in the Annulata. *Microentomology* **13**:1–26.

Husmann, S. 1967. Ökologie, Systematik und Verbreitung zweier in Norddeutschland sympatrisch lebender Bathynella-Arten (Crustacea, Syncarida). *Int. J. Speleol.* **3**:111–114.

Kurian, C. V. 1962. Three species of Cumacea from the lakes of Kerala. *Bull. Cent. Res. Inst. Trivandrum 8ʹC,* **1961**:55–61.

Maguire, B. 1965. Monodella texana n. sp., an extension of the range of the crustacean order Thermosbaenacea to the Western Hemisphere. *Crustaceana* **9**:149–154.

Mañé-Garzón, F. 1943. Tres especies de Tanais de las aquas dulces de Sud América. *Com. Zool. Mus. Hist. Nat. Montevideo* **1**:(4):1–15.

Noodt, W. 1964. Natürliches System und Biogeographie der Syncarida (Crustacea Malacostraca). *Gewässer Abwässer* **37**:77–186.

_____. 1967. Biogeographie der Bathynellacea (Syncarida). *Proc. Symp. Crustacea* **1**:411–417.

_____. 1968. Deuten die Verbreitungsbilder reliktärer Grundwasser-Crustaceen alte Kontinentzusammenhänge an? *Naturwiss. Rundschau* **21**:470–476.

Pennak, R. W., and J. V. Ward. 1985. Bathynellacea (Crustacea: Syncarida) in the United States, and a new species from the phreatic zone of a Colorado mountain stream. *Trans. Am. Microsc. Soc.* **104**:209–215.

Schminke, H. K. 1976. The ubiquitous telson and the deceptive furca. *Crustaceana* **30**:292–299.

Schram, F. R. 1986. *Crustacea.* 606 pp. Oxford University Press, New York.

Siewing, R. 1957. Anatomie und Histologie von Thermosbaena mirabilis. *Akad. Wiss. Lit. Abh. Math.-Naturwiss. Kl.* **7**:195–270.

Snodgrass, R. E. 1938. Evolution of the Annelida, Onychophora, and Arthropoda. *Smithson. Misc. Coll.* **97**:1–159.

_____. 1951. *Comparative studies on the head of mandibulate arthropods.* 118 pp. Comstock, Ithaca, New York.

_____. 1956. Crustacean metamorphoses. *Smithson. Misc. Coll.* **131**:1–78.

Zilch, R. 1972. Beitrag zur Verbreitung und Entwicklungsbiologie der Thermosbaenacea. *Int. Rev. ges. Hydrobiol.* **57**:75–107.

15

EUBRANCHIOPODA
(Fairy, Tadpole, and Clam Shrimps)

The Eubranchiopoda are among the most characteristic inhabitants of temporary ponds and pools, especially during spring and early summer. They move along the bottom or swim and glide about gracefully, often with the ventral side uppermost, and frequently in dense concentrations. They are absent from running waters, and there are no true marine representatives, although one species, *Artemia salina*, is restricted to highly saline lakes and evaporation basins used for the commercial production of salt.

Except for the immature stages in some families of insects, this is the only major taxonomic category found exclusively in fresh waters in the United States. All others have marine representatives, terrestrial representatives, or both.

General characteristics. All members of this primitive group are distinctly segmented and have 10 to 71 pairs of delicate, flat, lobate swimming and respiratory appendages. Because of the leaflike appearance of these appendages, the members of this division are collectively known as phylopods. The last body segment bears a pair of short to long cercopods. Body length, exclusive of cercopods, ranges between 2 and 100 mm. Females are usually more abundant than males in natural populations, and sometimes males are rare.

The Division Eubranchiopoda consists of three sharply defined orders (Figs. 1, 3). The Anostraca (fairly shrimps) have stalked compound eyes, 11 pairs of swimming legs (in American species), and no carapace. The Notostraca (tadpole shrimps) have sessile

compound eyes, a large, shieldlike carapace covering most of the body, and 35 to 71 pairs of legs. Neither the number of legs nor the number of segments is constant within a species. The Conchostraca (clam shrimps) are laterally compressed, have sessile compound eyes, 10 to 32 pairs of legs, and are enclosed in a carapace consisting of two lateral valves. This carapace is flexible along the middorsal line, but rarely is there a well-defined hinge line. Many species have concentric growth lines that are important in taxonomy. Opening and closing of the valves are governed by a short transverse adductor muscle.

Coloration ranges from translucent or whitish through gray, blue, green, orange, and reddish. Coloration of some species is highly variable and probably governed to a large extent by the type of food being ingested. All the individuals in a particular pond usually have the same general coloration at any one time, although some species exhibit progressive color changes during the adult instars.

The first antennae are small, uniramous, and often unsegmented (Figs. 3, 4). In female Anostraca the second antennae are cylindrical and elongated, but in the males they are greatly enlarged and specialized for clasping the females during copulation (Figs. 4, 13). The males of some species of Anostraca have a median frontal appendage on the head (Fig. 14B), or an antennal appendage arising near the base of each second antenna (Figs. 4, 14E). In the Notostraca the second antennae are minute or absent in both sexes, but in the Conchostraca they are long, biramous, and setose (Fig. 3C).

In ventral view the mouthparts are largely

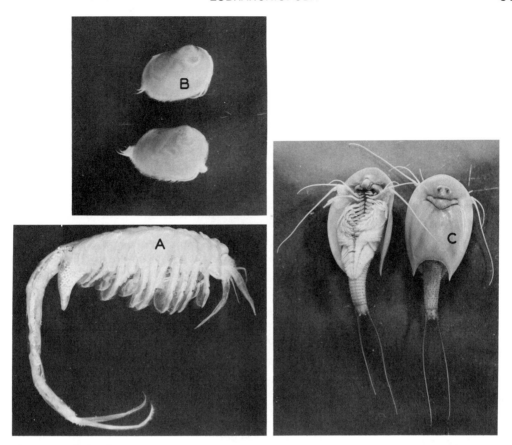

Fig. 1.–Typical Eubranchiopoda. A, female *Branchinecta gigas,* ×1; B, large clam shrimps, ×2.5; C, *Triops longicaudatus,* ×1.5.

concealed by the labrum. The mandibles have undivided molar surfaces and no palp, the first maxillae are triangular plates with setose biting edges, and the second maxillae are small simple plates that are of little importance in feeding (Fig. 2).

According to prevailing usage, the boundary between head and trunk of phyllopods lies immediately behind the segment that bears the second maxillae, the first trunk segment being the one that bears the first pair of legs. Although the head forms a well-defined body region, the trunk is not clearly divisible into thorax and abdomen as it is in most entomostraca. The trunk segments are all similar morphologically and vary widely in number from one genus to another. Furthermore, the number of pairs of legs and the proportion of trunk segments bearing them also vary widely.

The location of the genital pores, however, is sometimes used as a basis for loosely designating a "thorax" (pregenital segments) and an "abdomen" (postgenital segments) in some families. In the Anostraca the first two abdominal segments (collectively the genital segment) are partially fused and bear the external uterine chamber and base of the ovisac in the female, and penes in the male.

All Anostraca of the United States have 20 trunk segments, including 11 that bear legs, the two genital segments, and seven other posterior segments, the last of which is the telson. The Conchostraca have 10 to 32 trunk segments and 10 to 32 pairs of legs.

The body divisions, segmentation, and leg arrangement of the Notostraca represent a peculiar situation among fresh-water crustaceans. Posterior to the second maxillae there

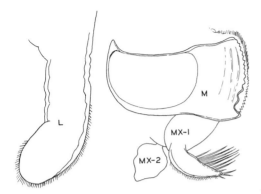

FIG. 2.–Mouthparts of *Lynceus brachyurus.* L, lateral view of labrum; M, MX-1, and MX-2, right mandible, first maxilla, and second maxilla as seen from below. (Modified from Sars, 1896.)

are 25 to 44 so-called segments, not counting the terminal telson. Unlike typical segments, however, many of them have two or more pairs of legs, and especially posteriorly some of them bear 5, 10, or more pairs of legs per segment (Figs. 1C, 5). The total number of pairs of legs ranges from 35 to 71 among the American species of Notostraca. Following the suggestions of Linder (1952), it therefore seems preferable to refer to "body rings" in the Notostraca rather than "segments." The first 11 rings are sometimes considered the thorax, the remainder being the abdomen. In general, all body rings are of comparable length. A variable number of posterior abdominal rings are without legs. A single species usually shows considerable variation with respect to: number of body rings, total number of legs, number of legs on a particular body ring, number of legs bearing abdominal rings, and number of legless abdominal rings. In *Triops longicaudatus,* for example, the number of legless abdominal rings varies from 4 to 16. The number of body rings is usually more constant in a parthenogenetic population than in one containing males.

Tadpole shrimps sometimes have an incomplete body ring just anterior to the telson. Only the dorsal or ventral part of the ring may be present. About 1% of the specimens in many collections show "spiral rings" (Fig. 6), usually near the telson. These consist of one to four spirals, running either to the left or to the right.

The telson of the fairy shrimps bears two terminal platelike plumose cercopods (caudal rami, furcal rami). In the tadpole shrimps the cercopods are very long, filamentous, segmented, and spiny. In the clam shrimps the telson is laterally compressed and bears two flattened upward curved processes and two stout anal spines (Fig. 17F).

Most phyllopod trunk appendages, or swimming legs, are basically similar (Fig. 7). They are biramous, flat, translucent, lobed, and setose. The median margin bears a series of small lobes, or endites, the basal endite of the Notostraca and Conchostraca being a more robust gnathobase. Near the base of the lateral margin are one or two large lobes, the proepipodites, and more distally is the thin, oval or elongated, nonsetose branchia (epipodite). The two large distal lobes are the exopodite (flabellum) and endopodite.

A few of the legs of all phyllopods are modified from this basic plan. In the Anostraca the more posterior pairs are definitely smaller and less lobate than preceding pairs. In the Notostraca the first one or two pairs of legs have elongated filamentous rami that probably serve as tactile structures (Fig. 8A). The flabellum of the eleventh pair of legs of the female is modified into a cuplike brood pouch that carries the eggs after they leave the reproductive tract (Fig. 8C). The more posterior legs of the Notostraca taper progressively to a very small size. The first one or two pairs of legs of male Conchostraca are hooked and prehensile (Fig. 8D), and in the female elongated flabella of two or three pairs of posterior legs aid in keeping the developing eggs in place between body and carapace (Fig. 3C).

Locomotion. Fairy shrimps and tadpole shrimps glide or swim gracefully by means of complex beating movements of the legs that pass in a wavelike anterior-posterior direction. Sometimes they drift along slowly; other times they dart rapidly or come to rest on the bottom. Lowndes (1933) presents evidence to show that locomotion in the Anostraca is not produced by beating of the legs as a whole, but rather by propellerlike movements of the exopodites, the speed of the animal being determined by the angle at which the exopodites are held. In the clam shrimps locomotion

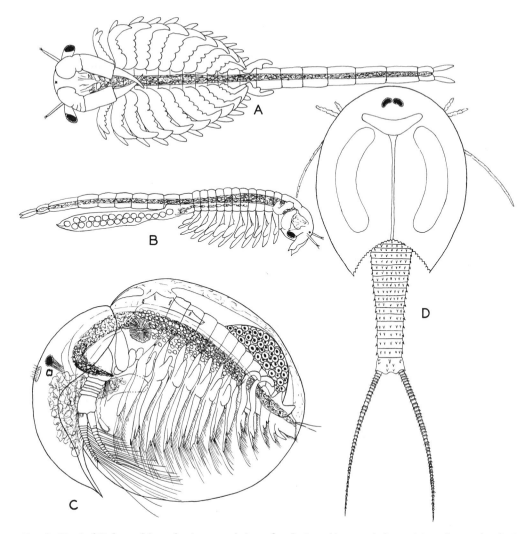

FIG. 3.–Typical Eubranchiopoda. A, ventral view of male *Branchinecta paludosa,* ×8 (setation omitted); B, lateral view of female *B. paludosa,* ×6 (setation omitted) showing brood sac; C, female *Lynceus brachyurus,* ×25, with left valve removed (diagrammatic); D, dorsal view of *Triops,* ×3. (A–C modified from Sars, 1896.)

is accomplished mainly by "rowing" movements of the large biramous second antennae, the legs being of little importance. Fairy shrimps habitually swim with the ventral side upward, but members of the other two orders are usually seen swimming ventral side downward. Tadpole shrimps creep or burrow superficially in soft substrates much of the time. Clam shrimps also burrow or move along the surface of the substrate in a clumsy manner. When the antennae stop beating they fall on their side.

Food, feeding. Food consists mostly of algae, bacteria, Protozoa, rotifers, and bits of detritus. Movements of the legs serve as a food-getting mechanism, plankton and materials from the substrate being strained nonselectively from the water or scraped from the substrate by the setose appendages (especially by the basal endites) and concentrated and agglutinated in a median ventral groove running most of the length of the body. This stream of food moves forward, activated chiefly by the gnathobases, and is further

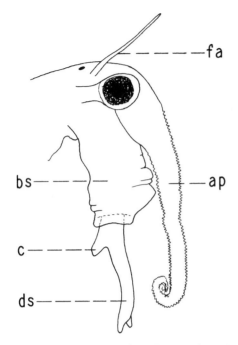

FIG. 4.–Diagrammatic lateral view of head of male *Eubranchipus bundyi*. ap, antennal appendage; bs, basal segment of second antenna; c, calcar of distal segment of second antenna; ds, distal segment of second antenna; fa, first antenna.

agglutinated at the anterior end by a sticky secretion produced by labral glands. Mastication occurs outside the digestive tract in an atrium formed by the overhanging labrum. It is probable that phyllopods feed continuously, but all the food reaching the mouth is not necessarily ingested, the excess being sloughed off. Although microscopic organisms and detritus constitute the sole food of Anostraca and Conchostraca, the Notostraca also utilize larger particles and have even been observed gnawing on dead tadpoles, earthworms, mollusks, and frog eggs. They are best characterized as omnivores. *Branchinecta gigas* is unique among American Anostraca. It is not a filter-feeder and subsists on large food masses as a carnivore, chiefly other crustaceans.

Internal anatomy, physiology. The digestive tract consists of a vertical esophagus, a small lobate stomach lying in the head, a long intestine, a short rectum, and an anus. Two digestive glands in the head empty into the stomach.

A dorsal tubular heart extends throughout most or all of the body segments in the Anostraca, but in the Notostraca and Con-

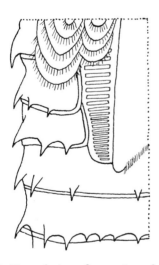

FIG. 5.–Ventral view of a portion of the right side of a female *Lepidurus lemmoni* from which the last 25 legs have been removed. The small rectangles represent the attachment areas of the removed legs. (Redrawn from Linder, 1952.)

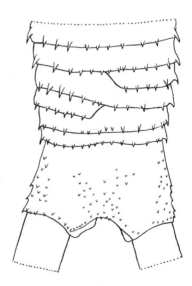

FIG. 6.–Ventral view of telson and a portion of the abdomen of *Triops longicaudatus* showing a spiral of two rounds located between normal body rings and beginning and ending in the midventral line. (Modified from Linder, 1952.)

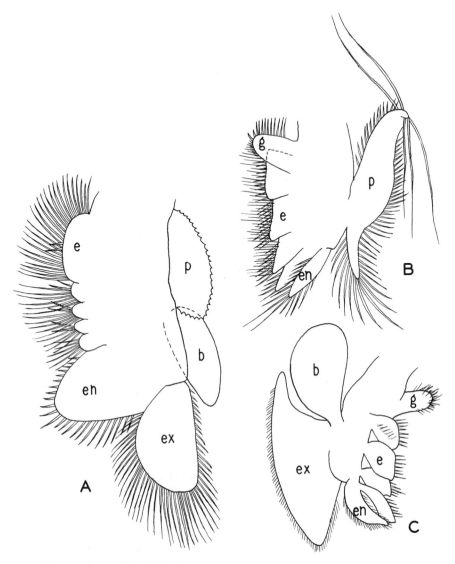

Fɪɢ. 7.–Trunk appendages of typical Eubranchiopoda. A, *Branchinecta paludosa;* B, *Lynceus brachyurus;* C, *Lepidurus.* b, branchia; e, endite; en, endopodite; ex, exopodite; g, gnathobase; p, proepipodite. (A and B modified from Sars, 1896; C modified from Packard.)

chostraca it is much shorter, sometimes extending through only four segments. A variable number of paired, lateral, slitlike ostia bring blood from the hemocoel into the heart, and blood is forced anteriorly and out of the opening at the anterior end of the heart by peristaltic contractions. Colorless amoebocytes occur in phyllopod blood, and an erythrocruorin is known to be dissolved in the plasma of some species.

The exchange of oxygen and carbon dioxide probably takes place through all exposed surfaces of the body but especially through the surfaces of the legs and their branchiae.

The chief excretory organs are the two coiled shell glands that empty their contents at the base of the second maxillae. It is possible that the paired antennal glands and mandibular glands also function in excretion during the early instars of some species.

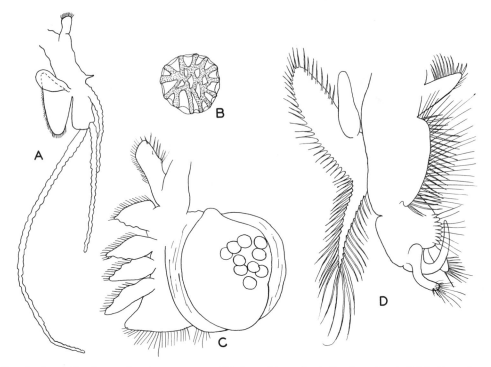

FIG. 8.–Specialized reproductive structures of Eubranchiopoda. A, first leg of male *Triops;* B, resting egg of *Eulimnadia;* C, eleventh leg of female *Triops;* D, first leg of male *Lynceus brachyurus.* (A and C modified from Packard; B redrawn from Mattox, 1939; D modified from Sars, 1896.)

Osmotic regulation presumably occurs chiefly through the wall of the digestive tract and through the branchiae of the first 10 pairs of legs. In some species there is an active uptake of sodium by the gills. *Artemia* continually swallows water, regardless of its salt concentration.

The ladderlike nervous system is diagrammatic and primitive (Fig. 9). The two widely separated nerve cords run the full length of the body, and except for the more posterior segments, there are two ganglia and two transverse commissures per segment.

In the Anostraca the compound eyes are stalked and far apart, but in the other two orders they are quite close together, sometimes touching or partially fused. A small median ocellus of the nauplius persists in the adult. Experiments have shown that the upside-down swimming position of some species is a phototropic response governed by stimuli received by both the ocellus and compound eyes, and if the upper side of the

container is covered and weak illumination is supplied from below, many of the animals will turn over and swim with the ventral side downward. Swimming on the back is normal in darkness. Immature individuals usually react more strongly to a light source than do the mature animals. Taste areas appear to be concentrated in the region of the mouthparts, and touch is distributed over the whole body. The "dorsal organ" is a small patch of modified cephalic epithelium of unknown function, but it is probably sensory.

Reproduction. None of the Eubranchiopoda are known to be hermaphroditic, although there is some evidence that in some of the Notostraca the gonads contain a mixture of both ovarian and testicular tissues. Testes and ovaries are two posterior tubules, one on each side of the digestive tract; in the Notostraca the gonads are ramified. Males have two short vasa deferentia and two small seminal vesicles

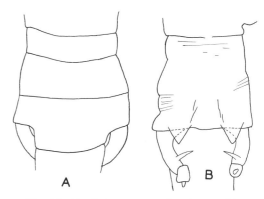

Fig. 10.–Genital segments of male *Branchinecta paludosa*. A, dorsal view; B, ventral view showing penes. (Redrawn from Linder, 1941.)

Fig. 9.–Nervous system of *Branchinecta*. (Modified from Sars, 1896.)

in the genital segment; sperm leave the body through two simple genital pores in the Notostraca and Conchostraca, but through penes in the Anostraca (Fig. 10).

Many of the reproductive phenomena in the phyllopods are poorly understood. The occurrence of males, for example, cannot usually be correlated with any seasonal, ecological, or reproductive events. In many species males are not abundant in natural populations. Parthenogenesis is therefore common, and, in a few species of Notostraca, sometimes the rule. Even in a mixed population, however, both parthenogenetic and syngamic reproduction may occur at the same time.

Clasping, paired males and females may swim about firmly united together for several days at a time. In the Anostraca the male assumes a dorsal position with the second antennae clasped around the female in the region of the genital segment. The actual processes of copulation and transfer of sperm are completed in a few seconds or minutes. In the fairy shrimps the male turns the body at an angle to that of the female and recurves the posterior end so that the genital segment is brought into contact with the external uterine chamber of the female. The process is somewhat similar in the tadpole shrimps, and in the clam shrimps the body of the male is placed at a right angle and vertical to that of the female before sperm can be transferred. Males usually die a few hours after copulation.

Both fertilized and parthenogenetic eggs are retained externally for one to several days on the body of the female before being released. The eggs of Anostraca are carried in an oval or elongated ventral brood sac. In the Notostraca they are carried in a curious receptable formed by modified and overlapping flabellum and branchia on each eleventh trunk appendage. In the Conchostraca the eggs are held together in a mucous mass between body and shell (Fig. 3C).

The number of eggs produced by a single female varies widely. They are released in a series of fertilized or unfertilized clutches at intervals of 2 to 6 or more days. Usually there are one to six clutches per female and 10 to 250 eggs/clutch, depending on the species. In syngamic reproduction, each clutch of eggs is fertilized by a separate copulation. The female carries the eggs for one to several days, during which time they undergo early development. Thereafter they may be dropped to the bottom or remain attached until the female dies and sinks to the bottom. Hatching appears to be governed chiefly by salinity and the level of dissolved oxygen.

It appears that some phyllopods produce two distinct types of eggs: (1) thin-shelled "summer" eggs, which hatch almost immediately, and (2) thick shelled, brown "resting" or "winter" eggs, which are capable of withstanding unusual heat, cold, and prolonged desiccation (Fig. 8 B); the thick shell affords protection from mechanical injury and sunlight. Both types are produced in either the presence or absence of males in the population. There is some evidence to indicate that the type of egg depends on the nature and amount of secretion coming out of the shell glands at the time of egg formation. A copious brown secretion results in thick shelled resting eggs; slightly less secretion produces thinner

shells, and the eggs hatch within several days after deposition; if the secretion is scanty and colorless, the shell is thin and transparent, and development is so rapid that the larva may hatch before the egg is released from the body of the female. It is thought that both thin and thick shelled eggs may produce either males or females; the genetic and physiological mechanisms involved are not clear.

Resting eggs. When it is recalled that phyllopods are chiefly inhabitants of vernal pools and ponds, which dry up completely in the dry, warm months, it is obvious that resting eggs constitute the sole device for tiding a population over from one season to another. It is probable that resting eggs of all species are capable of withstanding desiccation. During the summer the resting eggs become dried in the bottom mud, and during the winter they are frozen for varying periods. Until rather recently it was commonly believed that *both* drying and freezing were necessary

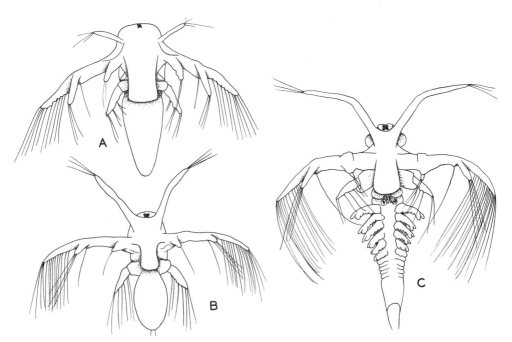

FIG. 11.–Immature Eubranchiopoda. A, first instar of *Artemia salina,* ×110; B, first instar (nauplius) of *Linderiella occidentalis,* ×80; C, third instar of *L. occidentalis,* ×30. (Modified from Heath.)

prerequisites for further development and hatching of these eggs, but experimental evidence has been accumulated to show that this is sometimes not the case. Resting eggs have been hatched in the laboratory after drying without freezing and freezing without drying as well as after both freezing and drying. Groups of eggs subjected to rapid or slow freezing and thawing all showed hatching. There is also some evidence to show that resting eggs of a few species will hatch without either drying or freezing.

The observation is sometimes made that phyllopods occur in ponds that never dry up completely during the summer months. This is unquestionably true, but it must be borne in mind that the active phyllopod population disappears with the onset of warm or cool weather and that the animals appearing in the following spring have undoubtedly hatched from resting eggs that overwintered on the bottom or in the dried mud along the pond margin above the lowered water level.

Although freezing and drying usually occur naturally, neither is indispensable. The resting period usually lasts from 6 to 10 months in temperate latitudes, but such a long interval is neither necessary nor inhibitory; dry eggs of certain species may be placed in water and hatched at any time of the year in the laboratory. The viability of resting eggs is very striking. Some species have been described from adults reared from eggs in dried mud sent from distant parts of the world. Viable eggs have been kept in dried pond mud on the laboratory shelf for as long as 15 years. Viable eggs of *Artemia* have withstood a temperature of 81°C for 1 hour, −190°C for 24 hours, and an air pressure of 0.000001 mm of mercury for 6 months. Eggs are more viable when stored in the refrigerator, as compared with room temperature. Eggs of other species can withstand experimental temperatures as high as 99°C.

Development, life cycle. Depending on the species, the eggs hatch into the typical nauplius or the more advanced metanauplius larvae with three pairs of appendages representing the first antennae, second antennae, and mandibles of the adults (Fig. 11). A metanauplius usually shows the faint beginning of two or three trunk segments. Although summer eggs hatch while they are still held on the body of the female or on the bottom shortly after their release, development of resting eggs is much slower. It is thought that early segmentation begins soon after the eggs reach the bottom of the drying pond and before the water temperature gets too low. Development ceases or is greatly retarded by desiccation and low temperatures, however, and is not resumed until early the following spring, when water is available and temperatures become favorable. Hatching of resting eggs in temperate latitudes therefore normally occurs sometime between late January and early May. Development proceeds comparatively rapidly in the early spring, and phyllopods may often be collected immediately after the ice melts. They have even been seen swimming about under ice, especially after an early thaw and subsequent freeze, or in the winter months in deep pools.

Beginning with the newly hatched nauplius, there is a long series of instars, each following a complete ecdysis, or shedding of the exoskeleton. Changes in size from one instar to the next are gradual, and there is progressive appearance of more segments, more appendages, and increasing complexity of appendages. *Linderiella occidentalis,* for example, has 17 instars in the life history. By the third instar the length averages 1.1 mm and all appendages through the fifth or sixth trunk segments are represented to some degree. By the sixth instar the average length is 2.1 mm, and all appendages are present, though most of the most posterior ones are completely developed. Complete development, sexual maturity, and copulation are attained in the 16th instar. *Artemia salina* probably has 14 instars, sexual maturity being attained in the twelfth instar. *Streptocephalus seali* has 18 preadult instars. Some investigators have found that the number of instars is variable for each species, depending on temperatures and food conditions.

The active portion of the life cycle may be completed in a surprisingly short time. Mattox (1937), for example, found that the life history of *Eulimnadia diversa* from hatching until death was completed during the 15 days that a small pond had water in it. He also found (1939) that *Lynceus brachyurus* lived about 26 days and

Cyzicus mexicanus about 2 months. *Artemia salina,* on the other hand, reaches sexual maturity in 18 to 21 days after hatching and has a normal life span of four months, although specimens have been kept alive for 9 months in the laboratory. *Eubranchipus oregonus* has an exceptionally long natural life span, sometimes up to 25 weeks.

In small ponds that contain water for only a few weeks during the spring and early summer, phyllopods usually have one generation per year. The resting eggs hatch early in the spring, and the animals mature rapidly and produce resting eggs that fall to the bottom and do not hatch until the following spring. Some summer eggs are usually produced during the short period of activity, and although they may hatch, there is not sufficient time for the second generation to mature, and such immature individuals and any remaining summer eggs are presumably destroyed by the adverse conditions accompanying the disappearance of water from the pond.

In Great Salt Lake, in permanent ponds, and in ponds that persist for several months, however, there are usually two or more complete generations per year, the summer generations being completed in a short time because the summer eggs hatch almost immediately.

Ecology. It seems quite clear that the development of phyllopod populations in the spring and their sudden disappearance in summer or early autumn are governed largely by temperature conditions. Most species do not appear until water temperatures exceed 4°C, and 13 to 30°C, depending on the species, is the upper limit beyond which individuals die quickly. In Great Salt Lake, *Artemia salina* disappears in the autumn when the water temperature drops below 6°C; *Eubranchipus bundyi* and *Eubranchipus serratus* are seldom found in water warmer than 15°C. At the other extreme, phyllopod populations have been recorded from desert ponds where the mid-day temperatures exceeded 42°C.

Most species exhibit little in the way of specific habitat preferences. The widely distributed *Eubranchipus vernalis,* for example, occurs in roadside ditches, grassy vernal ponds, cattail marshes, and woodland pools. Some species, however, appear to be generally restricted to clear or muddy waters. *Eubranchipus bundyi, Eubranchipus oregonus,* and *Branchinecta coloradensis* occur only in clear ponds and pools, while many clam shrimps, *Triops, Branchinecta gigas,* and *Thamnocephalus platyurus* are almost invariably found in muddy, alkaline waters.

I have often collected phyllopods in mountain locations in puddles in pothole depressions of a boulder, sometimes where the entire water volume was less than 20 liters.

A curious feature about phyllopod ecology is their absence from lakes. Seldom are they collected from bodies of water having areas exceeding one hectare. Phyllopods are almost defenseless, and it is also notable that they are not often abundant in ponds containing carnivorous insects, and are rarely present along with carnivorous fishes. In fact they attain greatest densities in vernal ponds and prairie pools that are generally lacking in other macrometazoans.

In keeping with widely fluctuating water levels and the consequent rapid changes in the dissolved salt content of pond waters, phyllopods have highly efficient but poorly known means of making physiological adjustments to varying osmotic pressures. In prairie pools, for example, the total salt or alkali content may vary between 0.05 and 1.00% within several weeks, yet their phyllopod populations maintain themselves successfully. Similarly, prairie phyllopods are usually capable of tolerating wide ranges in hydrogen ion concentration. *Eubranchipus serratus* has been found in a pH range of 6.6 to 9.5, *E. vernalis* from 5.2 to 7.4, and *Streptocephalus seali* in 5.2 to 10.0. As a whole, however, phyllopods have alkaline affinities; pH 6.8 seems to be about the usual lower limit for most species.

One part per million of dissolved oxygen seems to be a typical limiting factor for many species.

Excellent cultures of *Artemia salina* can be maintained in the laboratory, using natural or synthetic sea water, and much experimental work has been done on this species. Mature specimens will tolerate salinities ranging from the saturation point of sodium chloride down to about 3% sea water, though they will not reproduce in salinities below 3% sodium chloride. Adults reared in high salt concentrations reach a larger size than those reared in lower salt concentrations, and sexual maturity is attained more rapidly in populations reared at low salt concentrations. Europeans have found that the pregenital: postgenital body-length ratio and relative development of the cercopods and their setation are correlated with salinity, but these results have not been duplicated with American strains. *Artemia* populations have variously been found to be parthenogenetic, bisexual, diploid, triploid, tetraploid, pentaploid, or octoploid.

Year-to-year field notes on the occurrence of phyllopods in particular ponds present a puzzling problem. Sometimes a species is abundant for several successive years, and then one year it is unaccountably absent. Other ponds may have populations present or absent from year to year in a completely sporadic manner. Furthermore, where there is a group of small ponds and pools in a restricted area, phyllopods are usually present only in certain scattered pools and absent from most or all of the bodies of water immediately surrounding each inhabited one. With few exceptions, a pond never contains more than one species of a particular genus at a time, but Huggins (1976) reports finding *Streptocephalus, Thamnocephalus,* and *Triops* at the same time in a flooded soybean field in Kansas.

The bodies and appendages of phyllopods sometimes become so completely covered with sessile ciliates and green algae that they swim sluggishly and with difficulty.

Geographical distribution. A few genera and species of Eubranchiopoda are truly cosmopolitan. *Artemia salina* is found on all continents, and in the United States is best known from Great Salt Lake and man-made California salterns. *Lepidurus* occurs everywhere except in South America. The species of *Streptocephalus* are quite restricted; there are several species in Asia, 21 in Africa and Madagascar, 1 in Europe, 2 in the West Indies, and 6 in continental North America. The species of *Eubranchipus* are all restricted to North America, and, as extreme examples of

endemism, *Eulimnadia diversa* is known only from a single Illinois pond and *E. stoningtonensis* was found year after year in a single small pool in Connecticut. Similar examples are evident from the key that follows. Temporary ponds and pools of the Great Plains constitute a rich collecting ground, the majority of the species reported from the United States occurring in this area.

Phyllopods are found as far as extreme northern Alaska, Canada, and Greenland, but they have never been taken from the transient fresh-water and brackish ponds on the periphery of parts of Antarctica.

Most species are probably effectively distributed through their resting eggs, which may be blown about as dust or transported by birds and insects. The presence of populations of *Branchinecta coloradensis* above timberline in rock pools having a volume of only a few liters may be easily attributed to both of these agencies.

Enemies. Amphibia, dytiscid larvae, caddis larvae, and perhaps a few other insects are the chief predators of phyllopods. They are of no importance in the diet of fish since they occur in bodies of water too alkaline, too small, or too temporary to support fish.

Economic importance. Many years ago, the Indians in the vicinity of Great Salt Lake are said to have used dried *Artemia* extensively as food, and *Triops* is at present said to be used as food by a few natives of the Federal District in Mexico. Occasionally *Triops* becomes a pest in rice fields, notably in California. There are several reports of clam shrimps becoming a nuisance in fish rearing ponds, especially where the fish do not feed on the shrimp (goldfish and pike). Shrimp becomes so numerous as to clog outlet screens.

Collection, preparation. A coarse dipnet is the only necessary implement for collecting adults, but a plankton net is required for the early immature stages. Either 4% formalin or 70 to 80% alcohol is a satisfactory killing agent and preservative.

With few exceptions, phyllopods are usually difficult to maintain and rear in the laboratory. *Artemia,* however, is easy to culture; dry brine shrimp eggs are simply added to sea water or a 0.1 to 6.0% brine solution. The salt should not be iodized. Best results are obtained at 20 to 30°C. See Galtsoff et al. (1937), Cole and Brown (1967), and Dempster (1953) for more details about salt tolerances. *Artemia salina* is one of the very few species of metazoans that can be raised in axenic cultures (Provasoli and Shiraishi, 1959).

Dissection is not often necessary for identification, but specimens being studied for any length of time should be manipulated in glycerin. Conchostraca and isolated appendages may be permanently mounted in glycerin jelly.

Taxonomy. The Anostraca, Notostraca, Conchostraca, and Cladocera constitute the Subclass Branchiopoda and have the following characters in common: leafy trunk appendages, compound eyes, reduced mouthparts, and usually parthenogenesis, fertilized eggs, and a carapace. The Anostraca, Notostraca, and Conchostraca, however, form a convenient taxonomic group, the Division Eubranchiopoda, having several fundamental characters in common and differing from the Cladocera (Division Oligobranchiopoda) in some important features. The bodies of Eubranchiopoda are elongated, distinctly segmented, and have 10 or more pairs of trunk appendages, whereas the Cladocera are short and compact, with no sign of segmentation, and have only five or six pairs of trunk appendages. The abdomen of Cladocera is greatly reduced, but in Eubranchiopoda the abdomen (postgenital region) is prominent. It is also significant that the phyllopods have nauplius stages, whereas the Cladocera have direct development. Other less important differences are the occurrence of ephippia in the Cladocera and a single compound eye in Cladocera as compared with two in phyllopods.

Instead of Branchiopoda, the name "Phyllopoda" is occasionally used for the whole subclass; other references use "Phyllopoda" to include all Branchiopoda exclusive of Cladocera. Preuss uses the term "Phyllopoda" to include all Branchiopoda except Anostraca. Because of this ambiguity "Phyllopoda" is best not used in systematic nomenclature to designate a definite taxonomic category. On the other hand, there is no such serious objection to using "phyllopod" as a common, descriptive, inclusive term for Anostraca, Notostraca, and Conchostraca.

Because of brief descriptions, inadequate figures, and a reluctance to recognize wide variations within a species as to size, color, and the number and arrangement of spines and spinules, many phyllopods have an uncertain and controversial taxonomy. Formerly most investigators maintained that there were four to six species of *Triops* in the United States. These presumed species are represented by surprisingly few collections and specimens, and furthermore their differentiation was based on size, the number of exposed trunk segments behind the carapace, and the number and disposition of spinules on the telson. The revision of Linder (1952), however, has shown that none of these characters are reliable taxonomically, and that it is quite likely that there is only a single highly variable species of *Triops* in the United States. *T. longicaudatus. Lepidurus,* the only other American notostracan genus, is represented by five recognizable species.

Morphological variations within single species of Notostraca are so pronounced that Linder has been led to make the following statement: "A new species should never be described from less than 100 specimens from the same district, preferably taken at varying times of the year."

Intraspecific variations are further complicated by the occurrence of hybrids. Wiman (1979), for example, cites evidence for "natural" hybrids in the genus *Streptocephalus,* notably *S. mackini* × *S. dorothae* and *S. mackini* × *S. texanus.*

Unfortunately the general outline of the conchostracan shell is somewhat variable and does not have the taxonomic importance formerly attributed to it. Furthermore, the original descriptions of many American conchostracans are too generalized to be used for critical work. Hence this section of the following key is subject to considerable revision.

The taxonomic status of *Artemia* has long been controversial, especially in Europe, where *Artemia* shows much greater variability than it does in the United States. The reproductive habits of different populations vary markedly. In parts of Europe parthenogenesis is the rule, males being rare or absent; some such populations are diploid, but others are tetraploid and octoploid. *Artemia* in Great Salt Lake is diploid and males are common.

Thus, *Artemia salina* is thought to contain at least five sibling species, including *A. franciscana* from North and Central America. Most recently, however, Bowen et al. (1985) have studied *Artemia* strains from 15 North American waters of differing ionic compositions. Although these strains differ by only subtle morphological details, they are more clearly reproductively isolated by the specific ionic concentrations of the habitat. They may be designated as cross-fertile population clusters, which are either incipient species or sibling species. Currently, perhaps it is best to use *Artemia franciscana* for North American *Artemia*,

but if "*Artemia salina*" is used alternatively, there will be no misunderstanding. On a more detailed local and worldwide basis the *Artemia* problem must be enormously complicated. The fact that more than 300 "*Artemia salina*" papers are now appearing each year is not simplifying the situation.

To use the following keys effectively, mature males and females are required. The scanty and spotty distribution records are usually the result of inadequate collecting; undoubtedly such ranges will eventually be greatly extended.

Fryer (1987) has suggested that the Eubranchiopoda, as used in this volume, should be divided into four orders: Anostraca, Notostraca, Laevicaudata (including only the lynceid conchostracans), and the Spinicaudata (all other conchostracans). It remains to be seen whether this suggestion is generally accepted.

KEY TO EUBRANCHIOPODA

1. Eyes stalked; no carapace; body elongated; 11 pairs of swimming legs; second antennae uniramous, greatly enlarged and used as clasping organs in male; some genera with one frontal appendage or two antennal appendages in male (Figs. 1A, 3A, B) (most of the Anostraca couplets are based on detailed morphology of the male second antennae).
      Order **ANOSTRACA** (fairy shrimps); **page 357**
   Eyes sessile; with a large carapace 2
2. With a low, arched, univalve carapace covering head and much of trunk; 35 to 78 pairs of trunk appendages; second antennae minute or absent; two long cercopods (Figs. 1C, 3D); living animals often brightly colored, blue-green, red, brown, or marbled; total body length 11 to 58 mm, not including supraanal plate; not known east of the Mississippi River.
      Order **NOTOSTRACA** (tadpole shrimps); **page 365**
   Body compressed laterally and completely enclosed in a carapace composed of two lateral, shell-like valves; 10 to 32 pairs of legs; second antennae large and biramous (Figs. 1B, 3C).
      Order **CONCHOSTRACA** (clam shrimps); **page 362**

■ ■ ■

KEY TO SPECIES OF ANOSTRACA

1. Second antennae of male long and with a long, folded cheliform outgrowth on basal segment (Fig. 12C); with a small median frontal appendage between the bases of the second antennae.
      STREPTOCEPHALIDAE, Streptocephalus, 4
   Second antennae of male shorter and more simply constructed 2
2. Frontal and antennal appendages small or absent (Fig. 13A) 9
   Male with a long, branched, median frontal appendage (Fig. 14B); females with a long, unbranched frontal appendage (Fig. 14C); up to 50 mm long; central and southwestern states.
      THAMNOCEPHALIDAE, Thamnocephalus, 3

3. With a small cylindrical process near the base of the primary branch of the frontal appendage, sometimes tipped with a spine (Fig. 14B); animal up to 50 mm long; common in central and southwestern states .... **Thamnocephalus platyurus** Packard
Process broad and leaflike, not with a spine at the tip; a small species; rare in southern AZ and TX.
Thamnocephalus mexicanus Linder
4. Male cercopods bowed and with setae along proximal half and with short, curved spines distally (Fig. 12A) **5**
Male cercopods straight and setose along entire margins (Fig. 12B) **6**
5. Inner branch of male second antenna with two basal teeth on anterior margin; up to 45 mm long; widely distributed in pools and ponds .... **Streptocephalus seali** Ryder
Inner branch of male second antenna with three basal teeth; known only from TX.
Streptocephalus similis Baird
6. Distal tooth at base of finger shorter than proximal tooth; tip of finger serrate (Fig. 12F); southwestern TX .... **Streptocephalus linderi** Moore
Distal tooth at base of finger longer than proximal tooth; tip of finger variously formed **7**
7. Inner branch of male second antenna without a process near distal end (Fig. 12D); 10 to 18 mm long; OK, TX, NM, WY .... **Streptocephalus dorothae** Mackin
Inner branch of male second antenna notched, or with a posterior process or spine **8**
8. Distal end of finger as in Fig. 12C; up to 23 mm long; in pools and ponds of the plains in the southwest quarter of the United States .... **Streptocephalus texanus** Packard
Distal end of finger as in Fig. 12E; rare in TX, AZ, NM ...... **Streptocephalus mackini** Moore
9. Male second antennae somewhat fused basally and with a laminate terminal segment (Fig. 13A); average length 10 mm; in saline lakes and salterns of CT and western states.
ARTEMIIDAE, **Artemia salina** Leach, or **A. franciscana** Kellogg
Male second antennae not fused basally and terminal segment not laminate (Fig. 15); common and widely distributed west of the Mississippi River **10**

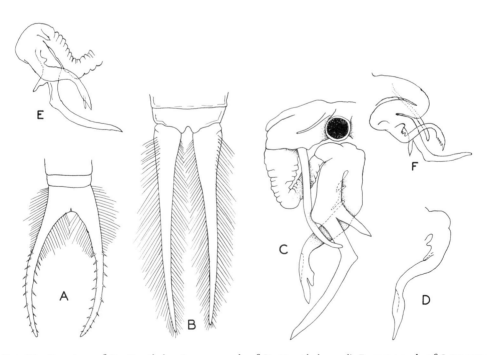

FIG. 12.–Structure of *Streptocephalus.* A, cercopods of *Streptocephalus seali;* B, cercopods of *S. texanus;* C, lateral view of anterior part of head and appendages of male *S. texanus;* D, inner branch of male second antenna of *S. dorothae;* E, second antenna of male *S. mackini;* F, second antenna of male *S. linderi.* (B–D modified from Mackin, 1942; E and F modified from Moore, 1966.)

10. Frontal appendage complex and longer than second antennae (Figs. 15A, B).
THAMNOCEPHALIDAE, Branchinella, 11
    Frontal appendage shorter than second antennae or absent 12
11. Frontal appendage with two terminal branches (Fig. 15B); 8 to 11 mm long; known only from rock pools in De Kalb County, GA .... **Branchinella lithaca** (Creaser)
    Frontal appendage with three terminal branches (Fig. 15A); 6 to 7 mm long; FL.
Branchinella alachua Dexter
12. Each abdominal segment with one spine on each side; 6 to 10 mm long; known only from west TX.
THAMNOCEPHALIDAE, Branchinella sublettei Sissom
    Without such abdominal spines 13
13. With antennal appendages (Fig. 16) 14
    Without antennal appendages .... BRANCHINECTIDAE, 24
14. Basal antennal appendages conical, with short median spines (Fig. 13D); CA.
LINDERIELLIDAE, Linderiella occidentalis (Dodds)
    Basal antennal appendages lamelliform, with conical projections along at least part of one or both edges (Figs. 16A–E) .... CHIROCEPHALIDAE, 15
15. Basal antennal appendage not extending beyond distal end of basal segment of second antenna (Fig. 16A) 16
    Basal antennal appendage longer than basal segment of second antenna (Fig. 16B) 18
16. Teeth on either side of male antennal appendage not greatly different in size (Figs. 16A, E) 17
    Teeth on one side of male antennal appendage much longer than those on the other side (Fig. 16C); average length 23 mm; known only from KY, OH (this may not be a good species).
Eubranchipus neglectus Garman

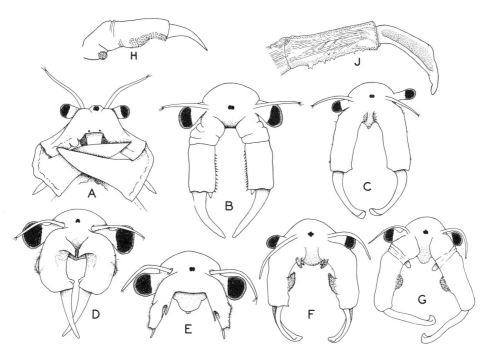

FIG. 13.–Structure of Artemiidae, Branchinectidae, and Linderiellidae. A, head of male *Artemia salina;* B, head of male *Branchinecta paludosa;* C, head of male *B. lindahli;* D, head of male *Linderiella occidentalis;* E, head of female *L. occidentalis;* F, head of male *Branchinecta packardi;* G, head of male *B. coloradensis;* H, second antenna of male *B. cornigera;* J, second antenna of male *B. dissimilis.* (A modified from Heath; B modified from Sars, 1896; C modified from Shantz; D and E modified from Dodds; H modified from Lynch, 1958; J from Lynch, 1972.)

17. Basal antennal appendage of male with large, separate teeth (Fig. 16E); 12 to 18 mm long; WA, OR, OK .... **Eubranchipus oregonus** Creaser
    Basal antennal appendage of male with small teeth (Fig. 16A); 20 to 26 mm long; widely distributed and common in northern states .... **Eubranchipus vernalis** (Verrill)
18. Basal antennal appendage without conical processes along lateral margin (Fig. 14F) 19
    Basal antennal appendage with conical processes along at least part of lateral and median margins (Figs. 16B–E) 20
19. Basal antennal appendage robust (Fig. 15C); a rare species known only from LA.
    Eubranchipus moorei Brtek
    Basal antennal appendage slim (Fig. 14F); 8 to 17 mm long; CT, NJ, VA, NC, GA, TN, OH, MD.
    Eubranchipus holmani Ryder
20. Basal antennal appendage long, tapering, and with small serrations on medial and lateral margins (Fig. 14D) 21
    Basal antennal appendage with a different shape 23
21. Labrum with a knoblike protuberance at anterior end (Fig. 15D); 10 to 18 mm long; widely distributed in northern states .... **Eubranchipus bundyi** Forbes
    Labrum without such a protuberance 22
22. Basal segment of second antenna with a fingerlike process on the median surface near midlength; FL.
    Dexteria floridana (Dexter)
    Basal segment of second antenna without such a process; uncommon but widely distributed in temporary pools of low salinity .... **Eubranchipus intricatus** Hartland-Rowe
23. Terminal segment of male second antenna with a calcar half as long as the segment (Fig. 16B); 10 to 35 mm long; IL, IN, NE, KS, OK, MT, WA, OR .... **Eubranchipus serratus** Forbes

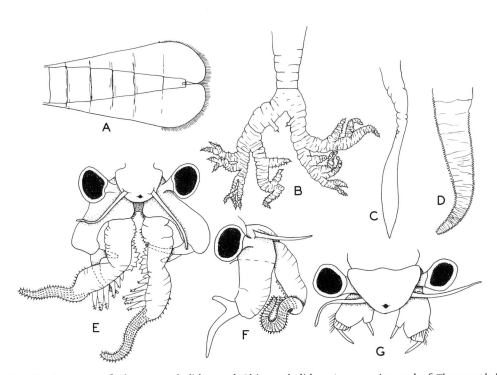

FIG. 14.–Structure of Thamnocephalidae and Chirocephalidae. A, posterior end of *Thamnocephalus platyurus;* B, frontal appendage of male *T. platyurus;* C, frontal appendage of female *T. platyurus;* D, antennal appendage of *Eubranchipus bundyi;* E, anterior view of head of male *Eubranchipus holmani;* F, lateral view of head of male *E. holmani;* G, anterior view of head of female *E. holmani.* (C redrawn from Packard; E–G modified from Mattox, 1936.)

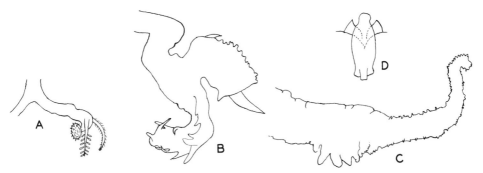

FIG. 15.–Structure of *Branchinella* and *Eubranchipus*. A, half of frontal appendage of *Branchinella alachua;* B, half of frontal appendage of *B. lithaca;* C, antennal appendage of *Eubranchipus moorei;* D, ventral view of labrum of *E. bundyi*. (A modified from Dexter, 1953; C modified from Brtek, 1967; D modified from Brtek, 1966.)

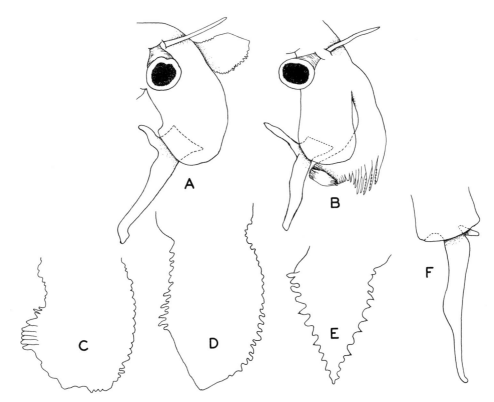

FIG. 16.–Structure of *Eubranchipus*. A, lateral view of head of *Eubranchipus vernalis;* B, lateral view of head of *E. serratus* (antennal appendage partly coiled); C, antennal appendage of *E. neglectus;* D, antennal appendage of *E. vernalis;* E, antennal appendage of *E. oregonus;* F, male second antenna of *E. ornatus*. (A and B modified from Mattox, 1939; C–F redrawn from Creaser, 1930.)

Terminal segment of male second antenna with a calcar about one eighth or less as long as the segment (Fig. 16F); length 12 mm; WI, MN, ND, MT, NE .... **Eubranchipus ornatus** Holmes

24. Length 50 to 100 mm (Fig. 1A); known only from a few temporary alkali ponds in WA, UT, MT, NV.
    Branchinecta gigas Lynch
    Usually 8 to 25 mm long 25

25. Basal segment of second antenna with a prominent dorsally directed peglike process just below midlength of the median surface (Fig. 13F); CO, TX, OK, WY, KS, NE, NM, AZ.
    Branchinecta packardi Pearse
    Basal segment of second antenna otherwise 26

26. Distal segment of second antenna tapering to a point (Figs. 13B, H) 27
    Distal segment of second antenna otherwise 28

27. Basal segment of male second antenna with a row of spines along much of the median margin (Fig. 13B); a subarctic species, reported only from WY, UT, AZ .... **Branchinecta paludosa** (O.F.M.)
    Basal segment of male second antenna with a spinous pad, a median protuberance, and many small distal spines (Fig. 13H); 21 to 35 mm long; known only from WA.
    Branchinecta cornigera Lynch

28. Basal segment of second antenna with a pad covered with minute spines near proximal end of segment (Fig. 13C) 29
    Basal segment of second antenna with no pad or a different type of pad 30

29. Basal pad prominent and anteromedial; a eurythermal and euryhaline species reported from IA, KS, NE, and most of the western states .... **Branchinecta lindahli** Packard
    Basal pad small and median; known only from potassium-rich waters in NE.
    Branchinecta potassa Belk

30. Distal segment of second antenna curved medially, laterally, flattened, and slightly concave on the lateral (anterior) surface (Fig. 13J); 11 to 24 mm long; known only from ponds in OR.
    Branchinecta dissimilis Lynch
    Distal segment of second antenna otherwise 31

31. Basal segment of second antenna with a spinous tubercle at about midlength, and usually with a small cylindrical process at the base of the segment (Fig. 13G); usually in ponds and pools at elevations of 1500 to 4000 m; CA, TX, UT, CO, WY, OR .... **Branchinecta coloradensis** Packard
    Basal segment of second antenna otherwise 32

32. Female first antenna equal to or shorter than second antenna; 15 to 30 mm long; known only from alkali ponds in WA, WY .... **Branchinecta campestris** Lynch
    Female first antenna longer than second antenna; NV, WA, CA ...... **Branchinecta mackini** Dexter

■   ■   ■

KEY TO SPECIES OF CONCHOSTRACA

1. Shell without growth lines 2
   Shell with growth lines 3

2. Head not entirely covered by carapace .... LYNCEIDAE, 5
   Head entirely covered by carapace; known only from Arkansas rice fields.
   Eulimnadia alineata Mattox

3. Head with a frontal appendage (Fig. 17E) .... LIMNADIIDAE, 9
   Head without a frontal appendage 4

4. Apex of rostrum armed with a heavy spine (Fig. 17C); about 11 mm long; KS, TX, CO, UT, OK, CA.
   LEPTESTHERIIDAE, **Leptestheria compleximanus** (Packard)
   Apex of rostrum unarmed .... CAENESTHERIIDAE, 20

5. Second male right or left thoracopod obviously modified (Fig. 18B); rostral carina bifurcate near top of head region, as seen in frontal view; known only from two ponds near Llano, TX.
   Paralimnetus texana Martin and Belk
   Thoracopods not obviously modified (Fig. 18A); rostral carina not bifurcate (Figs. 17A, B).
   Lynceus, 6

6. Second antenna 29 segmented; 4 to 7 mm long; KS, CO, NM, TX .... **Lynceus brevifrons** (Packard)
   Second antenna with 22 segments or less 7
7. Second antenna 16 to 22 segmented; 6 mm long; rare, reported only from TX.
   Lynceus gracilicornis (Packard)
   Second antenna with less than 18 segments 8
8. Front of head of male narrow (Fig. 17 B); second antenna 16 segmented; 2 to 6 mm long; the only common species in this genus; often in permanent ponds; widely distributed except for southern states .... **Lynceus brachyurus** Müller
   Front of head of male broad and square; rami of second antenna 14 to 17 segmented; 4 mm long; MT.
   Lynceus mucronatus (Packard)
9. Shell with about 7 to 18 growth lines, broadly oval, and 10 to 17 mm long; this genus is currently being revised, and it appears that there are really only five valid North American species; MA.
   Limnadia lenticularis (L.)
   Shell with 2 to 11 growth lines (Fig. 17 D); narrow, oval, and 3 to 11 mm long .... **Eulimnadia, 10**
10. Shell with one to four growth lines 11
    Shell with 5 to 12 growth lines 15
11. Telson with 9 or 10 dorsal spines 12
    Telson with 12 to 16 dorsal spines 13

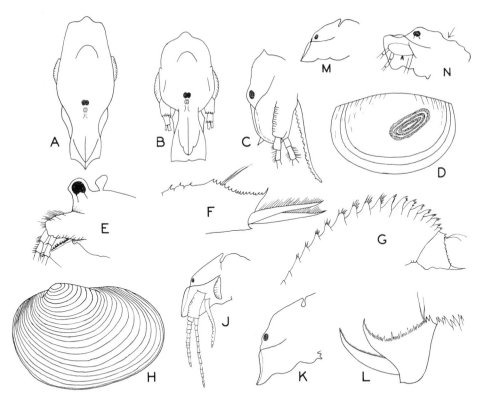

FIG. 17.–Structure of Conchostraca. A, frontal view of female *Lynceus brachyurus;* B, frontal view of male *L. brachyurus;* C, lateral view of head of *Leptestheria compleximanus;* D, lateral view of male shell of *Eulimnadia diversa,* ×9; E, head of female *E. diversa;* F, telson of *E. texana;* G, telson and dorsal portion of abdomen of *Cyzicus mexicanus;* H, lateral view of *Caenestheriella belfragei,* ×3.5; J, head of male *Cyzicus mexicanus;* K, head of female *C. mexicanus;* L, telson of *C. mexicanus;* M, anterior end of *Caenestheriella* showing acute rostrum; N, anterior end of *Eocyzicus concavus* showing shallow occipital notch. (A, B, D, G, H, J, and K redrawn from Mattox, 1939; C, F, and L modified from Packard.)

12. Shell 5 to 6 mm long and 3 to 4 mm wide; rostrum rounded; known only from LA.
    Eulimnadia antillarum (Baird)
    Shell 4.3 mm long and 2.5 mm wide; rostrum pointed; known only from MD.
    Eulimnadia francescae Mattox
13. Shell with two growth lines (Fig. 17 D); male 4.2 mm long and 2.5 mm wide; forked filament between third and fourth spinules; known only from IL .... **Eulimnadia diversa** Mattox
    Shell with three or four growth lines; 6 to 7.5 mm long **14**
14. Telson with 12 dorsal spines; forked filament between first and second spines; shell averages 6.2 by 3.8 mm; Woods Hole, MA, area .... **Eulimnadia agassizi** Packard
    Telson with 16 dorsal spines; forked filament between sixth and seventh spines; shell averages 7.3 by 4.3 mm; IL, OH .... **Eulimnadia inflecta** Mattox
15. Female with five growth lines; less than 8 mm long when mature **16**
    Female with 7 to 12 growth lines; more than 8 mm long when mature **17**
16. Seven to nine telson spines; shell averages 5 by 3 mm; OK .... **Eulimnadia antlei** Mackin
    Sixteen to 20 telson spines (Fig. 17 F); shell averages 7 by 4 mm; widely distributed in FL, TX, KS, AZ, LA, NE, OK .... **Eulimnadia texana** Packard
17. Forked filament of telson between third and fourth spines; male first antenna exceeds scape of second antenna **18**
    Forked filament of telson between fifth and sixth spine; male first antenna reaches only to end of scape of second antenna; CT .... **Eulimnadia stoningtonensis** Berry
18. Male first antennae reach to second segment of second antennae; average of seven growth lines; IL.
    Eulimnadia thompsoni Mattox
    Male first antennae reach to third segment of second antennae; 10 to 12 growth rings **19**
19. Shell length–height ratio 1.3:1.0; transparent amber color; MD, VA, GA.
    Eulimnadia ventricosa Mattox
    Shell length–height ratio 1.5:1.0; transparent colorless; Arkansas rice fields.
    Eulimnadia oryzae Mattox
20. Occipital notch shallow and inconspicuous (Fig. 17 N); rare .... **Eocyzicus, 21**
    Occipital notch acute and pronounced (Fig. 17 M) **22**
21. Shell with 18 to 22 growth lines; telson with 16 or 17 spines; OK, TX .... **Eocyzicus concavus** (Mackin)
    Shell with 14 to 16 growth lines; telson with 12 to 15 spines; reported only from KS, NV, CA.
    Eocyzicus digueti (Richard)
22. Rostrum of both male and female acute (Fig. 17 M); second antennae 14 or 15 segmented.
    Caenestheriella, 23
    Rostrum of male spatulate (Fig. 17 J); rostrum of female acute (Fig. 17 K); second antennae 16 to 22 segmented .... **Cyzicus, 25**
23. Shell compressed **24**
    Shell thick and globose, with 21 to 25 growth lines (Fig. 17 H); telson with 17 to 25 spines; total length up to 11 mm; KS, OK, TX, SD, MT, NE .... **Caenestheriella belfragei** (Packard)
24. Umbones one-third length from anterior end; 13 to 21 growth lines; up to 8 mm long; NE, OK, OR, SD, TX, MO .... **Caenestheriella setosa** (Pearse)
    Umbones one-fifth length from anterior end; 15 to 26 growth lines; up to 11 mm long; reported only from OH .... **Caenestheriella gynecia** Mattox
25. With 35 or more crowded, indistinct growth lines; shell globose, 5 to 6 mm thick and 9 to 13 mm long; SD, NE, IA, CO, OK, ND .... **Cyzicus morsei** (Packard)
    Growth lines distinct; shell not globose; umbo near anterior end **26**
26. With 25 to 35 growth lines; shell swollen; hinge line short and straight; 9 to 12 mm long; telson with 40 to 50 spines (Fig. 17 L); common from east coast to the Rockies and Canada to Mexico.
    Cyzicus mexicanus (Claus)
    With less than 25 growth lines; shell not swollen; rare, known only from CA **27**
27. Hinge line in an arc; 15 to 25 growth lines; up to 16 mm long; telson with 30 to 50 dorsal spines.
    Cyzicus californicus (Packard)
    Hinge line straight; with about 18 growth lines; average length 7 mm; telson with about 31 dorsal spines .... **Cyzicus elongatus** Mattox

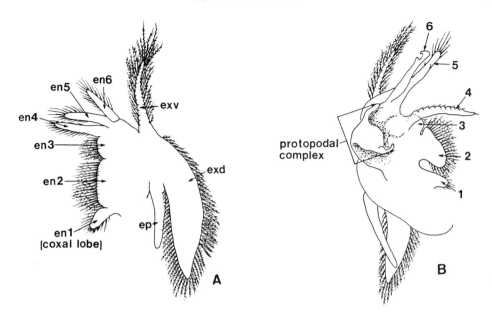

FIG. 18.–Structure of Lynceidae. A, thoracopod of *Lynceus;* B, modified male thoracopod of *Paralimnetus.* (From Martin and Belk, 1980.)

■ ■ ■

KEY TO SPECIES OF NOTOSTRACA

1. Telson extended as a flat, paddle-shaped protuberance (supra-anal plate) (Figs. 19A–D); 25 to 34 body rings, of which two to eight are legless; second maxillae well developed ...... **Lepidurus, 2**
   Telson not extended as a supra-anal plate; 35 to 44 body rings, of which 5 to 15 are legless; second maxillae reduced in adult; a single highly variable species generally distributed west of meridian 99 or 100; temporary waters and rice fields .... **Triops longicaudatus** (LeConte)
2. Anterior part of nuchal organ between posterior part of eye tubercles, not far from hind margin of eyes (Figs. 19F–H) 3
   Nuchal organ considerably behind eye tubercles (Fig. 19E); 30 to 34 body rings; 60 to 78 pairs of legs; supra-anal plate simple or bilobed; alkali ponds in CA, WA, NV, MT, CA, OR, WY; a highly variable species .... **Lepidurus lemmoni** Holmes
3. With 9.5 to 13 leg-bearing abdominal rings; usually 26 to 29 body rings 4
   With 15 to 18 leg-bearing abdominal rings; 30 to 44 body rings; supra-anal plate slightly bilobed (Fig. 19C); 66 to 71 pairs of legs; males unknown; a poorly known species reported from CO, AZ, UT.
   Lepidurus bilobatus Packard
4. Third to fifth endites of first leg rather similar in size, projecting very little or not at all beyond edge of carapace; 41 to 46 pairs of legs; reported from temporary pools in subarctic America and not yet reported from the United States .... **Lepidurus arcticus** (Pallas)
   Third to fifth endites of first leg dissimilar in size, the fifth endite projecting well beyond the carapace margin (Fig. 19J) 5
5. Median spines of supra-anal plate similar in size, few in number, and not placed on a keel (Fig. 19D); about 35 pairs of legs; a poorly known species reported from only two CA localities.
   Lepidurus packardi Simon
   Median spines of supra-anal plate dissimilar in size, numerous, and placed on a keel (Fig. 19A); about 35 to 40 pairs of legs; reported from alkali ponds in western states and MN.
   Lepidurus couesi Packard

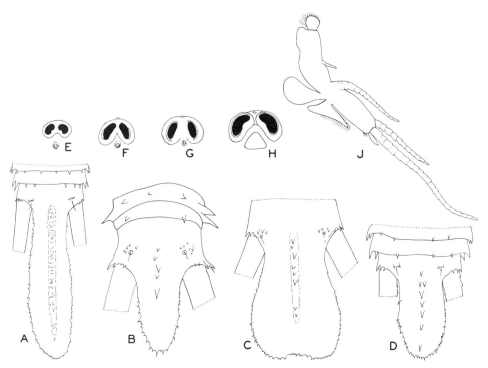

FIG. 19.–Structure of Notostraca. A, dorsal view of telson and last two body rings of *Lepidurus couesi;* B, dorsal view of telson and last two body rings of *L. lemmoni;* C, dorsal view of telson of *L. bilobatus;* D, dorsal view of telson and last two body rings of *L. packardi;* E, eyes and nuchal organ of *L. lemmoni;* F, eyes and nuchal organ of *L. packardi;* G, eyes and nuchal organ of *L. couesi;* H, eyes and nuchal organ of *Triops longicaudatus;* J, first leg of *Lepidurus couesi* (note long fifth endite). (A–H modified from Linder, 1952.)

REFERENCES

Baqui, I. U. 1963. Studies on the postembryonic development of the fairy shrimp Streptocephalus seali Ryder. *Tulane Stud. Zool.* 10:91–120.

Belk, D. 1975. Key to the Anostraca (fairy shrimps) of North America. *Southwest Nat.* 20:91–103.

_____. 1977. Zoogeography of the Arizona fairy shrimps (Crustacea: Anostraca). *J. Ariz. Acad. Sci.* 12:70–78.

Bowen, S. T. et al. 1985. Ecological isolation in Artemia: population differences in tolerance of anion concentrations. *J. Crust. Biol.* 5:106–129.

Brtek, J. 1967. Eubranchipus (Creaseria) moorei n. sp. Annot. *Zool. Botan.* 36:1–7.

Cannon, H. G., and F. M. C. Leak. 1933. On the feeding mechanism of the Branchiopoda. *Philos. Trans. R. Soc. London Ser. B,* 222:267–352.

Cole, G. A., and R. J. Brown. 1967. The chemistry of Artemia habitats. *Ecology* 48:858–861.

Creaser, E. P. 1930. Revision of the phyllopod genus Eubranchipus with description of a new species. *Occas. Pap. Mus. Zool. Univ. Mich.* 208:1–13.

_____. 1930a. The North American phyllopods of the genus Streptocephalus. *Ibid.* 217:1–10.

Daborn, G. R. 1975. Life history and energy relations of the giant fairy shrimp Branchinecta gigas Lynch 1937 (Crustacea: Anostraca). *Ecology* 56:1025–1039.

Daday de Deés, E. 1910. Monographie systématique des phyllopodes anostracés. *Ann. Sci. Nat. Zool. (9)* 11:91–492.

_____. 1915. Monographie systématique des phyllopodes conchostracés. *Ibid.* 20:39–330.

_____. 1923. Monographie systématique des phyllopodes conchostracés (2e partie). *Ibid.* (10) 6:255–390.

_____. 1925. Monographie systématique des phyllopodes conchostracés. Troisième partie. *Ibid.* 8:143–184.

_____. 1926. Monographie systématique phyllopodes conchostracés. Troisième partie (suite). *Ibid.* 9:1–81.

_____. 1927. Monographie systématique des phyllopodes conchostracés. Troisième partie (fin.). *Ibid.* 10:1–112.

Dempster, R. P. 1953. The use of larval and adult brine shrimp in aquarium fish culture. *Calif. Fish and Game* 39:355–364.

Dexter, R. W. 1946. Further studies on the life history and distribution of Eubranchipus vernalis (Verrill). *Ohio J. Sci.* 46:31–44.

_____. 1953. Studies on North American fairy shrimps with the description of two new species. *Am. Midl. Nat.* **49**:751–771.

_____. 1956. A new fairy shrimp from Western United States, with notes on other North American species. *J. Wash. Acad. Sci.* **46**:159–165.

Dexter, R. W., and D. B. McCarraher. 1967. Clam shrimps as pests in fish rearing ponds. *Prog. Fish Cult.* **29**:105–107.

Eriksson, S. 1936. Studien über die Fangapparate der Branchiopoden nebst einigen phylogenetischen Bemerkungen. *Zool. Bidr. Uppsala* **15**:23–287.

Eriksen, C. H., and R. J. Brown. 1980. Comparative respiratory physiology and ecology of phyllopod Crustacea. III. Notostraca. *Crustaceana* **39**:22–32.

Fryer, G. 1966. Branchinecta gigas Lynch, a nonfilter-feeding raptatory anostracan, with notes on the feeding habits of certain other anostracans. *Proc. Linn. Soc. London* **177**:19–34.

_____. 1987. A new classification of the branchiopod Crustacea. *Zool. J. Linn. Soc.* **91**:357–383.

Heath, H. 1924. The external development of certain phyllopods. *J. Morph.* **38**:453–483.

Huggins, D. G. 1976. The sympatric occurrence of three species of Eubranchiopoda in Douglas County, Kansas. *Southwest Nat.* **20**:577–578.

Jennings, R. H., and D. M. Whitaker. 1941. The effect of salinity upon the rate of excystment of Artemia. *Biol. Bull.* **80**:194–201.

Jensen, A. C. 1918. Some observations on Artemia gracilis, the brine shrimp of Great Salt Lake. *Biol. Bull.* **34**:18–32.

Linder, F. 1941. Contributions to the morphology and the taxonomy of the Branchiopoda Anostraca. *Zool. Bidr. Uppsala* **20**:101–302.

_____. 1952. Contributions to the morphology and taxonomy of the Branchiopoda Notostraca, with special reference to the North American species. *Proc. U.S. Nat. Mus.* **102**:1–69.

Littlepage, J. L., and M. N. McGinley. 1965. A bibliography of the genus Artemia (Artemia salina) 1812–1962. *Spec. Publ. San Francisco Aquarium Soc., Calif. Acad. Sci.* **1**:1–73 (mimeographed).

Longhurst, A. R. 1955. A review of the Notostraca. *Bull. Br. Mus. Nat. Hist. Zool.* **3**:1–57.

Lowndes, A. G. 1933. The feeding mechanism of Chirocephalus diaphanus Prévost, the fairy shrimp. *Proc. Zool. Soc. London* (B) (1933): 1093–1118.

Lynch, J. E. 1937. A giant new species of fairy shrimp of the genus Branchinecta from the State of Washington. *Proc. U.S. Nat. Mus.* **84**:555–562.

_____. 1966. Lepidurus lemmoni Holmes: a redescription with notes on variation and distribution. *Trans. Am. Microsc. Soc.* **85**:181–192.

_____. 1972. Lepidurus couesii Packard (Notostraca) redescribed with a discussion of specific characters in the genus. *Crustaceana* **23**:43–49.

_____. 1972a. Branchinecta dissimilis n. sp. a new species of fairy shrimp, with a discussion of specific characters in the genus. *Trans. Am. Microsc. Soc.* **91**:240–243.

Mackin, J. G. 1940. A new species of conchostracan phyllopod, Eulimnadia antlei, from Oklahoma. *Am. Midl. Nat.* **23**:219–221.

_____. 1942. A new species of phyllopod crustacean from the southwestern short-grass prairies. *Proc. U.S. Nat. Mus.* **92**:33–39.

_____. 1952. On the correct specific names of several North American species of Branchinecta Verrill. *Am. Midl. Nat.* **47**:61–65.

Martin, J. W., and D. Belk. 1988. A review of the clam shrimp family Lynceidae Stebbing, 1902 (Branchiopoda, Conchostraca) in the Americas. *J. Crust. Biol.* **8**:451–482.

Martin, J. W. et al. 1986. Redescription of the clam shrimp Lynceus gracilicornis (Packard) (Branchiopoda, Conchostraca, Lynceidae) from Florida, with notes on its biology. *Zool. Scripta* **15**:221–232.

Mattox, N. T. 1937. Studies on the life history of a new species of fairy shrimp, Eulimnadia diversa. *Trans. Am. Microsc. Soc.* **56**:249–255.

_____. 1939. Descriptions of two new species of the genus Eulimnadia and notes on the other Phyllopoda of Illinois. *Am. Midl. Nat.* **22**:642–653.

_____. 1950. Notes on the life history and description of a new species of conchostracan phyllopod, Caenestheriella gynecia. *Trans. Am. Microsc. Soc.* **69**:50–53.

_____. 1953. A new conchostracan phyllopod, Eulimnadia alineata, from Arkansas. *Am. Midl. Nat.* **49**:210–213.

_____. 1953a. Two new species of Eulimnadia from Maryland and Virginia (Crustacea: Conchostraca) *J. Wash. Acad. Sci.* **43**:57–60.

_____. 1954. A new Eulimnadia from the rice fields of Arkansas with a key to the American species of the genus. *Tulane Stud. Zool.* **2**:1–10.

_____. 1954a. Description of Eocyzicus concavus (Mackin) with a review of other North American species of the genus (Crustacea: Conchostraca). *J. Wash. Acad. Sci.* **44**:46–49.

_____. 1957. A new estheriid conchostracan with a review of the other North American forms. *Am. Midl. Nat.* **58**:367–377.

McCarraher, D. B. 1970. Some ecological relations of fairy shrimps in alkaline habitats of Nebraska. *Am. Midl. Nat.* **84**:59–68.

Moore, W. G. 1951. Observations on the biology of Streptocephalus seali. *Proc. La. Acad. Sci.* **14**:57–65.

_____. 1957. Studies on the laboratory culture of Anostraca. *Trans. Am. Microsc. Soc.* **76**:159–173.

_____. 1966. New World fairy shrimps of the genus Streptocephalus (Branchiopoda, Anostraca). *Southwest Nat.* **1**:24–48.

Moore, W. G., and J. B. Young. 1964. Fairy shrimps of the genus Thamnocephalus (Branchiopoda, Anostraca) in the United States and Mexico. *Ibid.* **9**:68–77.

Packard, A. 1883. A monograph of the phyllopod Crustacea of North America, with remarks on the Order Phyllocarida. *Ann. Rep. U.S. Geog. Surv. Terr.* **12**:295–592.

Provasoli, L., and K. Shiraishi. 1959. Axenic cultivation of the brine shrimp Artemia salina. *Biol. Bull.* 117:347–355.

Relyea, G. M. 1937. The brine shrimp of Great Salt Lake. *Am. Nat.* 71:612–616.

Rosenberg, L. E. 1946. Fairy shrimps in California rice fields. *Science* 104:111–112.

Sars, G. O. 1896. Phyllocarida og Phyllopoda. *Fauna Norv.* 1:1–140.

Saunders, J. F. 1980. A redescription of Lepidurus bilobatus Packard (Crustacea: Notostraca). *Trans. Am. Microsc. Soc.* 99:179–186.

Sissom, S. L. 1976. Studies on a new fairy shrimp from the playa lakes of West Texas (Branchiopoda, Anostraca, Thamnocephalidae). *Crustaceana* 30:39–42.

Warren, H. S. 1938. The segmental excretory glands of Artemia salina Linn. var. principalis Simon. (The brine shrimp.) *J. Morphol.* 62:263–298.

Weaver, C. R. 1943. Observations on the life cycle of the fairy shrimp. Eubranchipus vernalis. *Ecology* 24:500–502.

Whitaker, D. M. 1940. The tolerance of Artemia for cold and high vacuum. *J. Exp. Zool.* 83:391–399.

Wiman, F. H. 1979. Hybridization and the detection of hybrids in the fairy shrimp genus Streptocephalus. *Am. Midl. Nat.* 102:149–156.

16

CLADOCERA
(Water Fleas)

Because of their interesting habits, their availability in nearly all types of fresh-water habitats, and their complex but easily studied anatomy, the water fleas have been favorite subjects of observation by both amateur and professional biologists ever since the invention of the microscope.

General characteristics. Most members of the Order Cladocera are between 0.2 and 3.0 mm long. The body is not clearly segmented, and in the great majority of species the thoracic and abdominal regions are covered by a secreted shell or carapace that has a general bivalved appearance but is actually a single folded piece that gapes ventrally. In lateral view the shell is variously shaped; it may be oval, circular, elongated, or angular. There are often surface reticulations, striations, or other types of markings. In many species the posterior end has a spinule or spine, and the ventral edges of the valves usually bear setae. The inner surface of the valves is lined with the delicate body wall.

The head is a compact structure and does not open ventrally as the valves do. It is bent downward, and dorsally it is sometimes indistinctly set off from the body by a cervical sinus or notch. The most conspicuous internal structure of the head is the large compound eye. It consists of a few to many small hyaline lenses surrounding a mass of pigment granules. It is rotated constantly and jerkily by three pairs of small muscles. A single small ocellus often lies posterior or ventral to the compound eye. The first antennae (antennules) are inserted on the ventral side of the head near its posterior margin. They are usually inconspicuous, unsegmented, and bear olfactory setae. The second antennae, however, are very large and are inserted laterally near the posterior margin of the head. Each consists of a stout basal segment, a segmented dorsal ramus, and a segmented ventral ramus. The two rami bear a variable number of plumose setae. Setation formulas are often used as a means of identifying genera and species. Thus the setation formula for *Daphnia* is 0–0–1–3/1–1–3; it indicates that the dorsal ramus of the antenna is four segmented and that the basal, second, third, and fourth segments bear 0, 0, 1, and 3 setae, respectively; it also indicates a three-segmented ventral ramus of which the basal, second, and third segments bear 1, 1, and 3 setae, respectively. The antennae are activated by a set of powerful muscles that originate dorsally in the "neck" region. Externally, there is a strengthening ridge, or fornix, above the base of each antenna. The beak, or rostrum, is a more or less well-defined projection of the head between or in front of the antennules. The vertex of the head is that part that lies anterior to the compound eye.

The small mouthparts are situated near the junction of head and body. Beginning at the anterior end, they consist of (1) a median labrum; (2) a pair of stout, sclerotized, toothed, or ridged grinding mandibles; (3) a pair of small, pointed maxillae, used in pushing food between the mandibles; and (4) a single median labium. In some of the Macrothricidae and Chydoridae the labium bears a median keel that is of systematic value.

There are five or six pairs of lobed, leaflike thoracic legs, bearing numerous hairs and

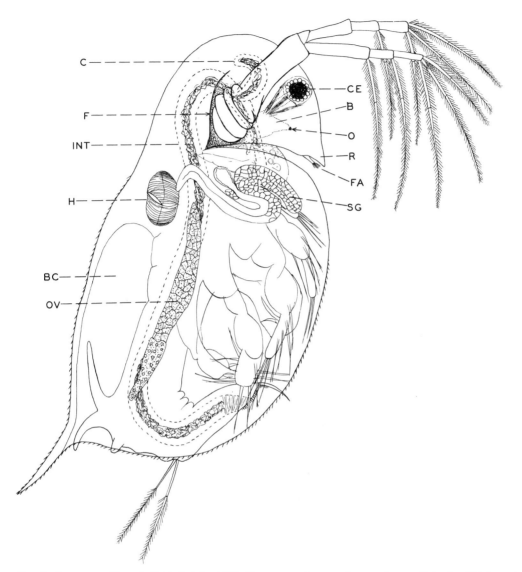

FIG. 1.–Anatomy of female *Daphnia pulex,* ×70, diagrammatic, muscles not shown. B, brain; BC, brood chamber; C, digestive caecum; CE, compound eye; F, fornix; FA, first antenna (antennule); H, heart; INT, intestine; O, ocellus; OV, ovary; R, rostrum or beak; SG, shell gland.

setae (Fig. 5). Although the legs are fundamentally biramous, this condition is not clearly evident because of their considerable modification. In the Sididae and Holopedidae all legs are similar, but in the other families the first two pairs are more or less prehensile and may aid in clinging to a substrate.

The true abdomen is suppressed, but there is a large postabdomen at the posterior end of the body. It is usually bent forward so that the dorsal side is downward. It bears two long abdominal setae, two terminal claws, and usually a series of marginal or lateral teeth or denticles. The postabdomen seems to be used chiefly for cleaning debris from the thoracic legs, although it may aid in locomotion.

During the greater part of the year Cladocera populations consist almost exclusively of females, the males being abundant only in

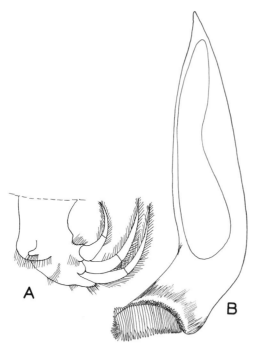

FIG. 2.–Mouthparts of *Daphnia pulex*. A, maxilla; B, mandible. (Redrawn from Lilljeborg, 1900.)

the autumn or spring. In many species, however, males are rare or unknown. Anatomically, males are distinguished from females by their smaller size, larger antennules, modified postabdomen, and first legs armed with a stout hook used in clasping (Fig. 6E).

Limnetic species are usually light-colored and translucent. Pond, littoral, and bottom species are darker in color, ranging from light yellowish brown to reddish brown, grayish, or almost black. Pigmentation occurs in both the carapace and general body tissues. *Holopedium* has chromatophores in the carapace.

Locomotion. For the most part, Cladocera keep intermittently in motion, the antennae being the chief organs of locomotion. In some genera, such as *Latona*, single vigorous strokes of the antennae produce powerful leaps. *Daphnia* and its relatives move by a series of "hops," produced by more rapid and less vigorous strokes. In some other genera the swimming movements have been characterized as "uncertain" and "tottering." A few genera, including *Holopedium* and

Scapholeberis, habitually swim upside down. Species occurring on a substrate have less frequent periods of active swimming and often use the antennae incidentally for obtaining a purchase on vegetation and bottom debris. The postabdomen is also useful for locomotion in bottom-inhabiting species. A cladoceran may literally kick itself along, sometimes by quite respectable hops.

A few species have become remarkably adapted for living in damp moss and debris of tropical forest floors; they are unable to swim and only creep about (Frey, 1980). A few species (chiefly Chydoridae) inhabit the interstices of stream gravel deposits; they show no obvious adaptations to this habitat.

Feeding. Complex movements of the highly setose thoracic legs (Figs. 3, 4) produce a constant current of water between the valves. These movements further serve to filter food particles from the water and collect them in a median ventral groove at the base of the legs. This stream of food is fed forward to the mouth parts where the particles may be ground between the surfaces of the mandibles before being taken into the mouth.

Algae and Protozoa have often been assumed to be the chief foods, to the exclusion of other materials, but it is now well known that organic detritus of all kinds, as well as bacteria, are very important and commonly form the great bulk of material ingested. It has also been successfully demonstrated that the feeding movements of the legs of plankton species are so efficient that they separate some colloidal organic particles from the water. Although there is a little evidence that certain types of food, such as particular groups of algae, Protozoa, or bacteria may be selected by some species, it is generally believed that all organic particles of suitable size are ingested without any selective mechanism. When undesirable material or large tangled masses are introduced between the mandibles, they may be removed by spines at the base of the first legs and then kicked out of the shell by the postabdomen. In *Daphnia*, the higher the concentration of algal food, the more narrow is the gap between the valves. This may be a mechanism for surviving algal blooms. Although the food stream in the median groove is continuous, ingestion may cease for varying periods. Minute amounts of dissolved organic materials may be absorbed by the general body surface, but they are of no metabolic significance.

A few genera, such as *Polyphemus* and *Leptodora,* are predaceous and have the legs modified for seizing. Their prey consists mostly of other entomostraca and rotifers.

The finely divided food is passed from the mouth parts to the esophagus in small masses of definite size. The time required to fill the digestive tract varies considerably, depending on the species, temperature, concentration of food, and other factors. Laboratory observations show a range of 10 to 240 minutes. Hasler (1937) demonstrated the presence of fat, carbohydrate, and protein digesting enzymes in *Daphnia* and *Polyphemus.* He found the pH of the digestive tract of *Daphnia* to range between 6.8 and 7.2. The indigestible residue leaves the body through the anus, which is usually situated on the lower (dorsal) border of the postabdomen.

Internal anatomy. Although the complex muscular system obscures some of the smaller anatomical features, the essential parts of most of the organ systems can be easily distinguished. The digestive system is relatively unspecialized. In the head region the narrow esophagus opens into a stomach that is often indistinguishable from the remainder of the tubular digestive tract, the intestine. In some genera there are one or two digestive caeca that open into the anterior part of the intestine. The course of the intestine may be relatively straight or convoluted. It can always be distinguished in both living and preserved specimens because of its contained dark food mass. In some of the Chydoridae there is a posterior ventral caecum.

The posterior part of the digestive tract is somewhat specialized as a rectum. It is now known that many algae cells are still viable after traversing the cladoceran digestive tract. Many chydorids and macrothricid cladocerans exhibit "anal drinking," which would appear to prevent loss of digestive products and to increase the efficiency of food utilization.

The simple oval or football-shaped heart lies behind the head on the dorsal side. Blood enters the heart through two lateral ostia and leaves through an anterior opening. At room temperature the heart beats about as rapidly as one can open and close his fist. There are no blood vessels, the blood being roughly guided about in the hemocoel by a series of thin mesenteries. The blood plasma is usually colorless or has a faint yellowish cast; it contains numerous colorless corpuscles.

Some pond species, however, are pinkish, owing to dissolved hemoglobin that may be formed when dissolved oxygen concentrations are low. This appears to be associated with an increased life span, more energetic swimming, and faster embryonic development. Presumably the hemoglobin functions in the respiration of embryos that are retained in a more or less closed brood pouch with little access to the outside water.

The exchange of oxygen and carbon dioxide occurs through the general body surface but especially along the inner surfaces of the valves and through the surfaces of the legs.

Irregular or looped shell glands, situated near the anterior end of the valves, are thought to function as excretory organs. In some of the Chydoridae there is a long tubular gland that arises from the rectum and lies freely in the hemocoel. Its function is unknown.

The nervous system consists of the usual ventral double nerve cord, relatively few ganglia, paired nerves, and a brain just anterior (dorsal) to the esophagus. Both the compound eye and the ocellus appear to be important in orientation to light sources and light intensity. Olfactory stimuli are thought to be received by the setae at the edges of the valves, by the antennules, and by areas around the mouth. The sense of touch is centered particularly in the abdominal setae and sensory hairs on the basal segment of the antennae.

All Cladocera have small structures on or near the middorsal line known as "head pores." These are borne on the head shield, an unpaired plate covering the frontal and lateral surfaces of the head. In adult Chydoridae there are one to five minute head pores on the midline of the posterior portion of the head shield (Frey, 1959). The number

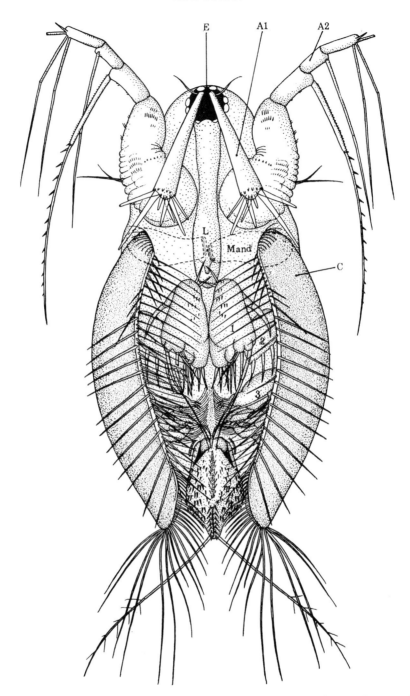

FIG. 3.–Ventral view of female *Acantholeberis curvirostris,* showing the complex setation. (Modified from Fryer, 1974.)

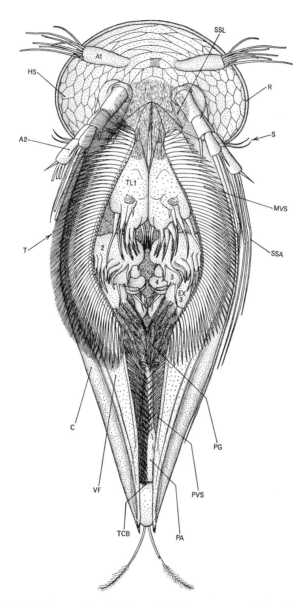

Fig. 4.–*Graptoleberis testudinaria* seen from below as it glides over a surface. The median ventral setae (MVS) form a tight chamber between the body and the substrate surface. The first and second trunk limbs scrape food particles from the surface, and there are no significant currents within the carapace chamber. (From Fryer, 1968.)

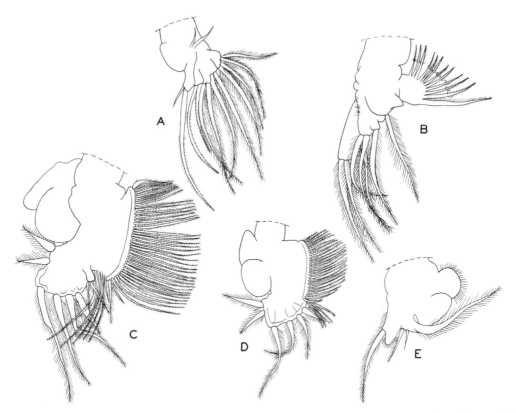

FIG. 5.–Legs of adult female *Daphnia pulex*. A, first leg; B, second leg; C, third leg; D, fourth leg; E, fifth leg. (Modified from Uéno.)

and arrangement of these structures may be used as taxonomic characters in the Chydoridae. All the other families of Cladocera have (larger) head pores, at least in some stage of their ontogeny, but here they are associated with certain underlying tissues. Such structures have variously been called, for example, neck organs, dorsal organs, and cervical glands. It is possible that they have a special respiratory function in early instars.

Gonads are easily distinguished only in sexually mature specimens. The two elongated ovaries lie lateral or somewhat ventral to the intestine in the thoracic region of the body (Fig. 7B). Depending on the stage in reproduction, an ovary is densely granular, with large oogonial nuclei, or it contains a mass of large, pigmented, granular, yolky eggs. Each oviduct is located dorsally at the posterior end; it is so delicate that it cannot be

distinguished except when an egg is passing through it. The testes are smaller and contain small, semitransparent, granulelike sperm (Fig. 7A). The sperm ducts are posterior continuations of the testes, which pass laterally to the intestine and rectum and open on the postabdomen near the anus or claws. A part of the postabdomen is sometimes specialized to form a copulatory organ.

Reproduction in general. Reproduction is parthenogenetic during the greater part of the year in most habitats, and only female young are produced. The eggs undergo a single maturation division in the ovary and a number are released at a time via the oviducts into the brood chamber. The latter is a cavity dorsal to the body proper and between the valves; it is effectively closed posteriorly by

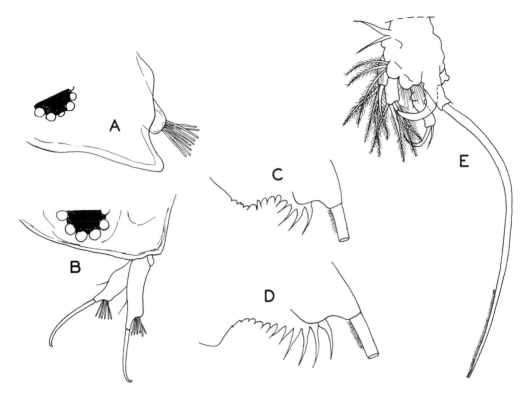

FIG. 6.–Sex characters of Cladocera. A, rostrum region of female *Daphnia pulex*; B, rostrum region of male *D. pulex*; C, postabdomen of male *Simocephalus exspinosus*; D, postabdomen of female *S. exspinosus*; E, first leg of male *D. pulex*. (C and D redrawn from Banta, 1930; E modified from Uéno.)

the abdominal processes. Depending on the species and environmental conditions, the number of eggs per clutch varies considerably; usually there are between 2 and 40, and most frequently between 10 and 20. The parthenogenetic eggs undergo further development in the brood chamber and hatch into young similar in form to the adult. The parent then liberates them to the outside by moving the postabdomen downward. Normally, one clutch of eggs is released into the brood chamber during each adult instar.

Seasonal abundance. In early spring relatively few cladocerans are to be found in lakes and ponds. Such populations consist of females that survived over the winter or recently hatched from winter (resting) eggs. As the water reaches a temperature of 6 to 12°C, active reproduction begins and subsequently speeds up tremendously so that large populations result (in exceptional cases as high as 200 to 500 individuals per liter of water). Populations in ponds then soon begin to wane, so that few individuals can be found during the summer months. In the autumn there may or may not be a second population pulse, but during the winter the population is invariably low, with little or no reproduction. Thus in a pond where the common species is *Daphnia pulex* or *D. magna* the population may be monocyclic or dicyclic, that is, having one or two population maxima during the year.

In larger bodies of water, however, seasonal variations in the abundance of Cladocera are not usually so pronounced. There is commonly a spring maximum and sometimes a less well-defined fall maximum, but summer and winter populations are large compared with those of ponds. In some lakes *D. pulex* is monocyclic, in others dicyclic. *Daphnia rosea* may be monocyclic, dicyclic, and in some cold lakes, acyclic, that is,

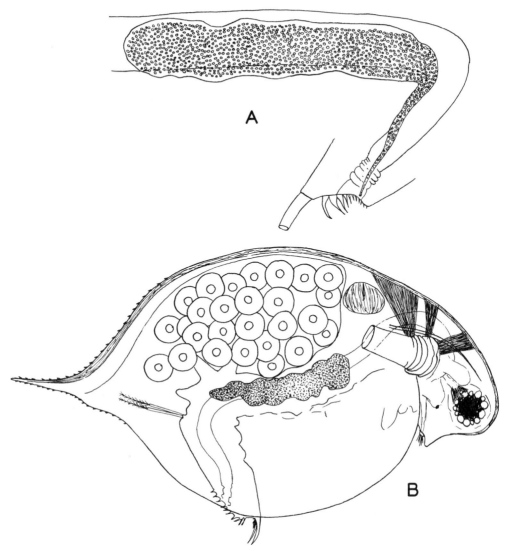

FIG. 7.–Cladoceran structures. A, postabdomen region of male *Simocephalus exspinosus* showing testis, sperm duct, intestine, and rectum; B, mature female *Daphnia pulex* (×70) showing ovary, parthenogenetic eggs in brood chamber, heart, antennal muscles, and compound eye. (A modified from Banta, 1930; B modified from Uéno.)

without any pronounced population maxima during the entire year. Other common forms exhibiting monocycly or dicycly are *Simocephalus* and *Ceriodaphnia*. Occasionally, as in *Sida crystallina* in some lakes, there may be a single autumn pulse. There are also many Cladocera that exhibit single long population pulses during the warmer months in certain lakes. Examples are *Diaphanosoma brachyurum,*

Chydorus sphaericus, Bosmina longirostris and *Moina.*

It is useless to predict or formulate any preconceived notions concerning the seasonal abundance of Cladocera in a particular lake or pond. Species differ greatly from one another in their seasonal abundance; a single species may have quite different population curves in two adjacent bodies of water; furthermore, relative abundance and the specific time of maximum and minimum populations may vary considerably within a single species in the same lake from one year to the next. Predation pressure is being given increasing importance in the seasonal abundance problem.

Male production, resting eggs. Numbers of males begin to appear in ponds in the spring, late in the intensive reproductive cycle. Sometimes only 5% of the population may consist of males; sometimes more than half are males. The factors responsible for the appearance of males in cladoceran populations have been the object of long and intensive studies by many investigators. We do not as yet have a detailed explanation of the phenomenon. By way of partial explanation, however, it is probable that a complex interplay of factors is responsible. Production of male eggs seems to be induced mostly by (1) crowding of the females and the subsequent accumulation of excretory products, (2) a decrease in available food, (3) a water temperature of 14 to 17°C, and (4) light intensity. These conditions (and probably others also), by altering the metabolism, appear to affect the chromosome mechanism in such a way that parthenogenetic male eggs rather than parthenogenetic female eggs are released into the brood chamber.

The same conditions responsible for male production, if continued for a longer time, appear to induce the appearance of sexual eggs. Females producing such eggs are morphologically similar to parthenogenetic females, but they produce only one, two, or sometimes several large opaque "resting" eggs, and, unlike the parthenogenetic females, they are capable of copulation with males. The fertilized eggs pass into the brood chamber, the walls of which then become thickened and darkened to form an ephippium (Fig. 8), which usually contains only one egg, sometimes two. In the Daphnidae the ephippium separates from the rest of the shell at the subsequent molt. In the Chydoridae it remains attached to the shell at the molt. Sometimes as many as a third or half of the females in a population may bear ephippia.

On being released, the ephippia sink to the bottom, become attached to the substrate (certain Macrothricidae), or float on the surface. In the last event, they may be blown ashore and accumulated in diminutive windrows. Ephippia and their contained eggs are capable of withstanding drying and freezing, and their production is clearly an adaptation to adverse environmental conditions. It has been shown that ephippia are sometimes viable even after passing through the digestive tract of fish. Ephippia are of special value in small ponds that dry up during the summer months. When such basins become filled with water in the early autumn, a large percentage of the ephippial eggs hatch into parthenogenetic females. Then there may be an autumn pulse followed by the appearance of males, sexual females, and ephippia that winter over and give rise to at least a part of the small seed population in the following spring. There is considerable evidence to show that ephippial females are exceptionally susceptible to predation.

Thus it is seen that during most of the year the population consists entirely of parthenogenetic females, but for short periods the population may be complex, consisting of parthenogenetic females, males, unfertilized sexual females, and fertilized sexual females.

In bodies of water larger than ponds, ephippia are produced by smaller percentages of the populations, and in limnetic populations in large lakes it is thought that reproduction may be entirely parthenogenetic the year round, especially in *Daphnia rosea*.

The expressions "monocyclic" and "dicyclic" are really used in two different ways in the literature. They may refer to an annual

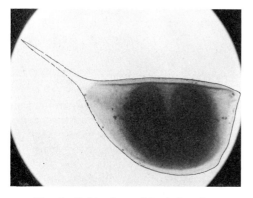

FIG. 8.–Ephippium of *Daphnia pulex.*

population curve having one or two pronounced maxima, without reference to types of reproduction, or they may be used to designate populations having one or two periods of syngamic reproduction during the year.

The older idea of internal genetic or physiological "rhythms" being responsible for the periodic occurrence of syngamic reproduction has been discarded. The experimental work of Banta and others has demonstrated conclusively that the mode of reproduction is influenced by environmental factors. By changing the culture solutions frequently, Banta reared several common Cladocera for 800 to 1600 successive parthenogenetic generations (more than 27 years). Males, sexual females, and subsequent ephippia may be easily produced by crowding the mothers in a small amount of culture solution or by chilling the culture.

Laboratory observations have shown that under appropriate conditions a mature parthenogenetic female may begin to produce sexual eggs, or a sexual female may change over and produce parthenogenetic eggs. Males usually appear in both natural and controlled populations before females begin the formation of sexual eggs in the ovary. When a female passes an unfertilized sexual egg into the developing ephippium it disintegrates. There are, however, a few records in the literature of such eggs developing into parthenogenetic females.

Wood (1938) has contributed much to our understanding of the induction of hatching of ephippial eggs. Working with *Daphnia,* she found that (1) the considerable majority of eggs died when stored either wet or dry for one to many weeks; (2) the longer the period of storage, the greater the mortality; (3) a few eggs stored wet for seven years were viable; (4) exposure to low temperatures decreased the percentage of hatch; (5) placing eggs in fresh culture medium or allowing them to dry out only a day or two gave the highest percentage of hatch, usually 35 to 45%; and (6) aeration and changing the culture solution frequently induced hatching of a comparable percentage of eggs that had not been allowed to dry out.

Development, life cycle. Length of life, from release of the egg into the brood chamber until the death of the adult, is highly variable, depending on the species and environmental conditions. *Daphnia rosea* usually lives for 28 to 33 days in laboratory cultures. MacArthur and Baillie (1929) found that *D. magna* lived an average of 26, 42, and 108 days at 28, 18, and 8°C, respectively. Limnetic individuals undoubtedly live through the entire winter at low temperatures, although high-latitude and high-altitude populations overwinter by means of ephippial eggs. Poor food supply also increases length of life.

Four distinct periods may be recognized in the life history of a cladoceran: egg, juvenile, adolescent, and adult. When a clutch of eggs is released into the brood chamber, segmentation begins promptly; the young, in the first juvenile instar and similar in form to the adult, are released from the brood chamber in about two days. There are but few juvenile instars, although greatest growth occurs during these stages. *Moina macrocopa* has two juvenile instars, *Daphnia rosea* has three, *D. pulex* has three or four, and *D. magna* three to five. The adolescent period is a single instar between the last juvenile instar and the first adult instar; during this instar the first clutch of eggs reaches full development in the ovary. As soon as the animal molts at the end of the adolescent instar and enters the first adult instar, the first clutch of parthenogenetic eggs is released into the brood chamber. During the first adult instar the second clutch of eggs is developing in the ovary. Successive adult instars and new clutches of young are produced in a similar manner. However, there is often a sterile period during the last few instars of life.

As in all other Crustacea, growth, in terms of increase in size, becomes apparent only immediately after each molt. During juvenile instars there may be almost a doubling of size after each molt, the increase in volume occurring within a few seconds or minutes and before the new exoskeleton hardens and loses its elasticity.

The number of adult instars is much more variable than the number of juvenile instars. *Daphnia pulex* usually has 18 to 25 adult instars; *D. rosea,* 10 to 19; and *D. magna,* 6 to 22. As already inferred, the duration of a single adult instar is highly variable, from a day to several weeks, though about two days under favorable conditions. At the close of each adult instar four events follow one another in rapid succession, usually in a few minutes to a few hours. These are the release of young from the brood chamber to the outside, molting, increase in size, and the release of a new clutch of eggs to the brood chamber.

Recent laboratory work has shown that growth and reproduction of *Daphnia* may be biochemically slowed or inhibited by dense populations of blue-green algae (e.g., Infante and Abella, 1985).

Cyclomorphosis. One of the most intriguing and puzzling of all cladoceran problems is the matter of cyclomorphosis, or seasonal changes in morphology, especially in females of limnetic species such as *Daphnia pulex* and *D. rosea*. A population of a cyclomorphic species has a homogeneous "normal," or round headed, form during the late fall, winter, and early spring. As the water becomes warmer and the population develops, however, there is commonly a progressive increase in the longitudinal axis produced by a general elongation of the head and the appearance of a "helmet" (Figs. 9E, P, Q). Characteristically, the helmets become fully developed by midsummer, when they may be quite bizarre (Figs. 9G, H, Q). Beginning in the late summer or early autumn, the morphology of the head progressively reverts so that the "normal" head condition prevails by late autumn. In addition to changes in the shape of the head, cyclomorphosis may also involve changes in size of eye and length of the posterior spine.

The problem is greatly complicated by the fact that the degree of summer helmet development differs widely in the same species, even in two neighboring lakes. Cyclomorphosis is less pronounced in ponds and shallow lakes, and the degree of helmet development is relatively consistent from one individual to another at any one time, but in larger and deeper lakes cyclomorphosis and the degree of helmet development in a population is much more variable, and both strongly developed and poorly developed helmets, as well as intergrades, may be found in the same townet sample.

Explanations of these seasonal changes in form have naturally long been sought for in the changing seasonal ecological conditions in bodies of water. Between 1900 and 1910 Wesenberg-Lund and Ostwald elaborated their "Buoyance Theory," which attempted to explain the occurrence of helmets as a flotation adaptation to decreased viscosities of water at summer temperatures. Subsequent observations and experimental work have discredited this viewpoint. More recent proposals have centered around nutrition, internal cycles, and the accumulation of waste products, but none of these now receives favorable consideration by investigators in this field.

One of the more logical explanations of cyclomorphosis appears to center around temperature, and particularly the temperature during the early stages of development. The controlled experiments of Coker and Addlestone (1938) are illuminating. These investigators found that the last one third of the period of development of the first juvenile instar in the brood chamber was the critical period. *Daphnia rosea* raised at a temperature at 10°C or below during this period all had round heads; 15% of those raised at 12°C had pointed heads; 33% of those raised at 13°C had pointed heads; 87% at 14°C had pointed heads; and all of those raised at 16°C or more had pointed heads. In general, the higher temperatures produced more prominent helmets. Brooks (1946) found comparable temperature relationships for helmet formation in *Daphnia retrocurva* and head spine formation in *D. galeata*. Kiang (1942) demonstrated a relation between mean temperature of a lake and relative head length in *D. rosea*.

Nevertheless, the early work of Banta (1939) indicates that temperature is by no means the whole story and that genetic factors probably play an important role. He collected 10 distinct morphological types of *Daphnia pulex* in the field and cultured them in the laboratory. Under such circumstances the pronounced helmets were lost or minimized, but each of the 10 types remained recognizable and peculiar to itself. Other workers have shown that turbulence and high light intensities may increase the degree of helmet formation. By selective predation on larger individuals, plankton-feeding fish may also influence helmet development in a popu-

FIG. 9.–Some variations in head form in *Daphnia*. A to C, J, and K, *D. pulex*; D to H, *D. retrocurva;* L, M, *D. galeata mendotae*; N, O, *D. ambigua*; P, Q, *D. dubia*.

lation. Dodson (1988) has shown that predators "induced" longer helmets and sometimes tail spines in two species of *Daphnia*. There are literally hundreds of papers giving more or less attention to cladoceran cyclomorphosis. Seldom do any two investigators agree on the details of this complex phenomenon.

However, in temperate lakes, we may broadly summarize cyclomorphosis phenomena as environmental—genetic—predation pressure interrelationships. In the anticipatory words of Coker (1939),

"The changes in form are not simple functions of external conditions or of any inherent cycle, but rather of a combination of internal and external conditions in a way that becomes exceedingly baffling the more we know about it."

Vertical migrations. Most limnetic cladocerans undergo vertical migratory movements during each 24-hour period. These "drifting" movements of the general population are commonly upward with the onset of darkness and downward with the coming of the light of dawn. In some species, however, and in some habitats, there is a rapid upward and downward cycle near dusk and a similar upward and downward cycle around dawn. In rare instances some species show "reverse" migratory movements. The average amplitude ranges from 2 to 10 m, depending on the species and many environmental factors. This whole phenomenon defies a simple explanation, since the vertical movements of the cladocerans appear to be variously governed by species, age, size, temperature, food supply, water chemistry, water color, season of year, light intensity, and turbulence.

Much of the experimental work on vertical migrations has been done with *Daphnia magna,* and the results have been transposed to other species. This is an unwarranted generalization because *D. magna* is a species of shallow ponds, and it is likely that the responses of this species differ from those of species inhabiting deep, limnetic habitats.

Most of the conclusions about vertical migrations have been based on light sensitivity plus a cluster of environmental variables. The recent discovery that experimentally blinded *Daphnia* still exhibit vertical movements suggests that some part of the body other than the eyes may be light sensitive. Furthermore, some investigators have revived the "endogenous rhythm" explanation because they have found that *Daphnia* migrates even when kept in continuous darkness.

General ecology. Cladocera are primarily fresh-water organisms, the several American marine species belonging to *Evadne* and *Podon* (Polyphemidae). Aside from rapid streams, brooks, and grossly polluted waters, they are abundant everywhere. Common open-water and limnetic forms are *Daphnia rosea, D. pulex, Bosmina, Diaphanosoma, Chydorus sphaericus, Ceriodaphnia,* and *Holopedium.* Such forms are seldom found in vegetated areas, and it has been demonstrated experimentally that rooted aquatics have a repellent effect on limnetic species, especially *Daphnia* (Pennak, 1973). The greatest abundance of species may be collected in the vegetation at margins of lakes and rivers, including *D. pulex, Sida crystallina, Ophryoxus gracilis,* most Chydoridae, and most Macrothricidae. A careful collector may find

as many as 25 species in a single day's collecting. The two most common inhabitants of ponds, permanent pools, and temporary pools are *D. pulex* and *D. magna. Alona quadrangularis, Drepanothrix, Ilyocryptus,* and *Monospilus* are among the most common forms to be found on or near the bottom in weedy littoral areas. Specific affinities are indicated for most species in the key at the end of this chapter.

It is unusual for a natural population of *Daphnia* and other genera to consist of more than one numerically dominant species. Pennak (1957), for example, in reviewing the literature, found that in limnetic situations it is unusual to find more than one species in a single genus at the same time, but when two species in the same genus do occur together, one is usually 20 or more times as abundant as the other. At the other extreme, Ranta (1979), working on very small rock pools, found three species of *Daphnia: D. magna* (common), *D. longispina,* and *D. pulex* (rare). No pool contained all three species, and two of these species were found together in only 10% of the pools.

The great majority of species and nearly all the common ones are eurythermal. Only a few appear to have distributions limited by temperature. *Latona, Holopedium gibberum,* and *Daphnia longiremis* are cold stenotherms, whereas *Pseudosida bidentata, Ceriodaphnia rigaudi,* and *Euryalona occidentalis* are restricted to warm waters of southern states.

Little information is available concerning chemical limiting factors. Dissolved oxygen is seldom of any significance except in the hypolimnion of lakes during summer and winter periods of complete oxygen exhaustion. Many species can withstand oxygen concentrations of less than 1 ppm. Most Cladocera occur in waters containing a wide range of concentrations of calcium. *Holopedium,* however, is widely distributed and confined to calcium-poor waters. Hutchinson (1932) found that magnesium may act as a limiting factor by inhibiting reproduction; critical concentrations for *Daphnia magna, D. pulex,* and *D. rosea* were found to be 240, 120, and 30 to 60 mg/l. Although a few species are restricted to acid and bog waters, most Cladocera occur over a wide range in pH, nearly all species being found at 6.5 to 8.5.

In littoral and substrate environments two

or more species in the same genus usually coexist in the same local habitat. In the open-water limnetic environment, however, it is unusual to find two species of *Daphnia* co-existing. But occasionally when two (or more) species are found in the same limnetic sample, then one of them is always much more abundant than the other(s).

There is increasing evidence to show that fishes are selectively predaceous on populations by ingesting larger and more obvious individuals in preference to smaller individuals. In this way the species makeup in lakes and ponds may be significantly altered.

Many laboratories have used *Daphnia,* especially *D. magna,* as an assay organism for the detection of toxic substances in water. Unfortunately, however, most such laboratories have evolved their own techniques, and we do not as yet have a "standard method" for such assays.

Probably no other major fresh-water taxon is currently the object of such investigative activity as the Cladocera. The literature is vast and controversial, with variable and debatable information. Many aspects of cladoceran biology are thus deserving of critical published reviews, but few investigators (e.g., Hebert, 1978) have had the courage to undertake such chores. Unfortunately, experimental studies based on laboratory cultures are often inconsistent and give results that cannot be transposed to natural, open-water situations.

Most recent interests are centered around cyclomorphosis, clone genetics, species flocks, competition, selective food habits, microanatomy as shown by ultramicroscopy, effects of trace elements, and population fluctuations. Any detailed discussion of these topics is beyond the scope of this volume.

Geographical distribution. Because of their resistant ephippial eggs, Cladocera are easily transported overland to new habitats. As a result, the majority of species are very widely distributed, some being truly cosmopolitan. The general occurrence of a few common species is indicated in Table X. Less than 20 of these species are known only from the United States, and most of these have been collected in only one or two localities. Intensive collecting has been done in relatively few areas, however, and our knowledge of the

TABLE X. GENERAL GEOGRAPHIC DISTRIBUTION OF SOME COMMON SPECIES OF CLADOCERA

Cosmopolitan
  Alona guttata, A. rectangula
  Ceriodaphnia laticaudata
  Chydorus
  Daphnia magna, D. pulex
  Eurycercus
  many bdelloid species
North America, Europe, Asia, and South America
  Alona affinis
  Bosmina longirostris
  Ceriodaphnia quadrangula
  Macrothrix laticornis
  Simocephalus vetulus
North America, Europe, and Asia
  Acroperus harpae
  Bosmina coregoni
  Camptocercus rectirostris
  Daphnia rosea
  Diaphanosoma brachyurum
  Eurycerus lamellatus
  Holopedium gibberum
  Leptodora kindti
  Moina macrocopa
  Pleuroxus strictus
  Polyphemus pediculus
  Scapholeberis mucronata
  Sida crystallina

geographical limits of many species in the United States is decidedly incomplete.

Bythotrephes (or *Cercopagis*) is a large bizarre European carnivorous cladoceran that has recently been found in fish stomachs in lakes Erie and Huron (Fig. 10). Undoubtedly it is a common plankton species in this area.

Frey (1982) has clearly pointed out that many species formerly considered true "cosmopolitans" actually are clusters of two or more species or morphs, each of which has a more restricted distribution.

Culturing. Several species of *Daphnia* are among the most easily cultured of all fresh-water invertebrates. Fish hatcheries and tropical fish fanciers raise them in tanks or in concrete or plastic outdoor wading pools. Cottonseed meal, manure, some kinds of commercial agricultural fertilizers, dried yeast, dried milk, dried and chopped hay are all suitable foods, especially for the suspended organic particles and the bacteria that become abundant. It is important to avoid adding too much food to the cultures; otherwise the medium becomes foul. The writer has kept

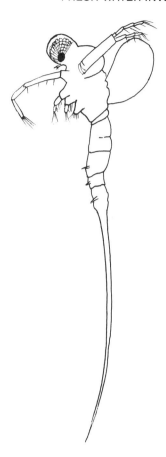

FIG. 10.–*Bythotrephes* (or *Cercopagis*), a bizarre cladoceran that has recently made its way to the United States from Europe; ×18.

20 cultures of *Daphnia rosea* going continuously in 1- to 10-gal aquariums for 8 years. Once per week each aquarium receives dried yeast suspended in 20 to 200 ml of water. Each aquarium is then thoroughly stirred for a few seconds.

Economic significance. The great importance of Cladocera in the aquatic food chain as food for both young and adult fish was emphasized first by Forbes in 1883, and since then by innumerable investigators. Various studies of the stomach contents of young fish show from 1 to 95% Cladocera by volume, and very few studies show less than 10%. Other groups of less importance that utilize Cladocera in the diet are *Hydra* and immature and mature insects. Goulden (1971), for example, found that midsummer populations of littoral Chydoridae were greatly reduced by Diptera larvae predation.

Collecting, preserving. An ordinary plankton townet is suitable for limnetic species but is of little use for collecting in littoral vegetation. For such habitats the Birge cone net does an excellent job. Bottom forms may be taken by bringing the top centimeter of mud and debris into the laboratory and allowing it to settle. By looking horizontally along the surface toward a light source, the Cladocera may be located and picked up with a long pipette. In small ponds and pools, and to some extent in vegetation, a small dipnet is convenient.

The best general killing agent is 95% ethyl alcohol. Because of their delicate body and strong muscles, however, *Moina* and most of the Sididae are badly distorted when placed in alcohol. These forms should first be narcotized with chloral hydrate or a similar deadening agent. Seventy percent alcohol is the standard preservative. The writer always adds 5% glycerin to his storage vials so that specimens will not be destroyed through evaporation if the corks dry out or the vials break.

Fortunately, little or no dissection is necessary for identification purposes, but permanent slide mounts in glycerin or glycerin jelly are highly desirable. Cover slips should be supported by bits of cardboard of appropriate thickness so that specimens will not be crushed. Sometimes it is necessary to disarticulate and clear the exoskeletons of cladocerans in order to make out minute anatomical details. Judicious use of cold dilute hydrochloric acid for 24 hours or hot acid for a few minutes usually will give the desired results.

Taxonomy. Brooks (1957) has given the genus *Daphnia* its most thorough treatment, and most of his suggestions have been followed in this chapter. Nevertheless, much work still remains to be done on this troublesome taxon. Most of the species are featured by irksome seasonal and geographic variations in carapace and head morphology, and details of setation and the postabdomen. The retrocurva, carinata, rosea-longispina, and *pulex-pulicaria* problems, as well as several other questions, deserve further consideration. The situation is confused by what seems to be "hybridization" and species (or subspecies) "flocks" in some geographical areas. Our major difficulty is the fact that definitive species descriptions are often based on specimens collected from only a limited number of field sites. We need widely distributed massive collecting and careful analyses of large numbers of specimens.

Wolf (1987) has demonstrated naturally coexisting hybrids between *Daphnia hyalina, D. galeata,* and *D. cucullata* by means of gel electrophoresis. Even more puzzling is the report of Hebert and Crease (1980) who found, by electrophoresis, 22 clones of *D.*

pulex in 11 separate populations, with as many as 7 clones being present in a single population. Dodson (1981) concludes that *D. pulex* is either one highly variable species or a cluster of many species differing from each other only slightly.

Some species found in the United States have long been considered Holarctic and identical with those found in Europe. More careful scrutiny of our American forms is needed, however, because they are sometimes sufficiently different from their European counterparts as to require different species designations.

Aside from the *Daphnia* problem, certain other genera will be clarified as more specimens are collected from many additional field sites, notably *Ceriodaphnia, Diaphanosoma,*

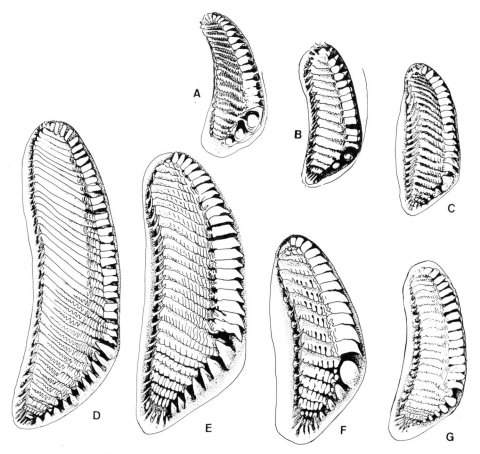

FIG. 11.–Right mandibles of species of *Daphnia*. A, *rosea*; B, *ambigua*; C, *dubia*; D, *magna*; E, *middendorfiana*; F, *retrocurva*; G, *pulex*. (From Edwards, 1980).

Moina, Alonella, Alona, Pleuroxus, and *Chydorus.* The papers of Frey, especially, have pointed out the minute anatomical and taxonomic distinctions of species in the chydorids. Edwards (1980) has developed a new means of studying problems of *Daphnia* systematics. It involves examining the complicated sculpturing of the mandibular surfaces of newly molted daphnids with the scanning electron microscope, aided by canonical analyses. Figure 11 shows right mandibular details of some common species.

In the key that follows, lengths refer only to females and do not include spines or antennae. Some species appear in two different places in the key because of variable anatomical characters.

Fryer (1987) has suggested that "Cladocera" as an Order designation should be abandoned, and that this taxon should be split into four orders: Ctenopoda, Anomopoda, Onychopoda, and Haplopoda. It remains to be seen whether this suggestion is generally accepted.

KEY TO SPECIES OF CLADOCERA

1. Body and legs covered with a bivalve carapace; legs foliaceous, not clearly segmented 2
   Body and legs not covered with a carapace; legs subcylindrical or somewhat flattened, clearly segmented, and prehensile; northern states 144
2. With six pairs of foliaceous legs; first and second pairs not prehensile (Figs. 12, 13) 3
   With five or six pairs of legs; first and second pairs more or less prehensile (Figs. 15 ff.) . . . 12
3. Carapace of the usual type; antennae of female biramous and flattened (Figs. 12F; 13A, C).
   SIDIDAE, 5
   Animal enclosed in a large gelatinous mantle, open ventrally and forming two valves (Fig. 12A); the mantle is shed during ecdysis but regenerated within 2 hours; antennae of female simple and cylindrical; swims ventral side up; found in low-calcium waters; 1 to 2 mm long.
   HOLOPEDIDAE, Holopedium, 4
4. Ventral margins of valves with fine spines; up to 25 anal spines and spinules; anal claws with a basal spine (Fig. 12B); spotty distribution in open waters of northern states and southward in mountains . . . . **Holopedium gibberum** Zaddach
   Ventral margins of valves smooth; 7 to 13 anal spines and spinules; anal claws without basal spine (Fig. 12D); east of LA and south of northern tier of states.
   Holopedium amazonicum Stingelin
5. Dorsal ramus of antenna three segmented (Fig. 12E); 3 to 4 mm long; common among aquatic plants in lakes and ponds . . . . **Sida crystallina** (O.F.M.)
   Dorsal ramus of antenna two segmented 6
6. With a lateral expansion on basal segment of dorsal ramus of antenna; with a tonguelike projection on ventral surface of head (Figs. 12F, G); 2 to 3 mm long; in littoral vegetation; chiefly in northern states and mountainous areas . . . . **Latona, 7**
   Without lateral expansion on antenna 8
7. Antennal expansion very large (Fig. 12F) . . . . **Latona setifera** (O.F.M.)
   Antennal expansion small (Fig. 12G) . . . . **Latona parviremis** Birge
8. Without spines on postabdomen; ocellus absent; 0.8 to 1.2 mm long; common and widely distributed (Fig. 13A) . . . . **Diaphanosoma, 9**
   With spines on postabdomen (Fig. 13B); ocellus present; about 2 mm long 10
9. Head about two thirds the valve length; eye in the middle of the head near the ventral margin; limnetic . . . . **Diaphanosoma birgei** Kořínek
   Head not more than half the valve length; eye anterior (Fig. 13A); littoral.
   Diaphanosoma brachyurum (Liéven)
10. Rostrum present (Fig. 13B); southern states . . . . **Pseudosida bidentata** Herrick
    Rostrum absent (Fig. 13C); usually in vegetation . . . . **Latonopsis, 11**
11. Postabdomen with about nine small spines; common in Gulf states, sporadic farther north (Fig. 13C).
    Latonopsis occidentalis Birge
    Postabdomen with 12 to 14 small clusters of lancet-shaped anal spines (Fig. 13D); Gulf states.
    Latonopsis fasciculata Daday

FIG. 12.–Structure of Holopedidae and Sididae. A, *Holopedium gibberum,* ×40; B, postabdomen of *H. gibberum*; C, first leg of *H. gibberum*; D, postabdomen of *H. amazonicum*; E, *Sida crystallina,* ×14; F, antenna of *Latona setifera*; G, anterior end of *L. parviremis*. (B and D modified from Ward and Whipple, 1918; C modified from Sars; E and F modified from Lilljeborg, 1900.)

12. Antennules attached to ventral side of head, not covered by fornices (Figs. 15–23) 13
    Fornices extended so as to cover antennules more or less, and united with rostrum into a beak that projects ventrally in front of antennules (Figs. 24 ff.) .... CHYDORIDAE, 76
13. Antennules of female small, often rudimentary; if large, then never inserted at anterior end of ventral edge of head (Figs. 15–19) 15
    Antennules of female large, inserted at anterior end of ventral edge of head (Figs. 20–23) 14
14. Antennules of female fixed (Fig. 20); intestine simple; 0.3 to 0.5 mm long ...... BOSMINIDAE, 53
    Antennules of female freely movable (Fig. 21); intestine simple or convoluted.
                                                      MACROTHRICIDAE, 58
15. Rostrum present .... DAPHNIDAE, 16
    Rostrum absent 39
16. Dorsal and ventral margins lacking spines; pubescence on the middle part of the ventral carapace margin; no dorsal carina; no posterior spine; rare, known only from small ponds in CT, IL.
                                     Daphniopsis ephemeralis Schwartz and Hebert
    Dorsal and ventral margins more or less spinous; pubescence absent; dorsal carina present; posterior spine usually present 17
17. Without a cervical sinus .... **Daphnia,** 18
    With a cervical sinus (Figs. 18B–D) 34
18. Carapace extends anteriorly along middorsal line as a broad strip between halves of head shield (Fig. 16H); thick bodied; up to 5 mm long; scattered distribution in western half of the United States.
                                                                             19
    Median line of head shield continues along middorsal line onto carapace; marked lateral compression; up to 3 mm long, usually less than 2 mm long 20

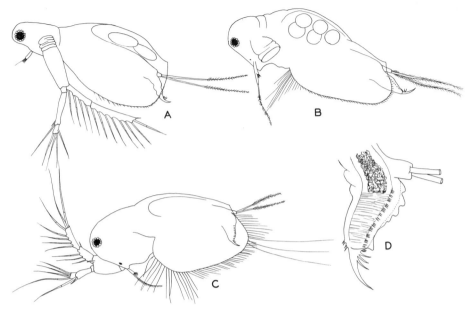

FIG. 13.–Structure of Sididae. A, *Diaphanosoma brachyurum,* ×45; B, *Pseudosida bidentata,* ×27; C, *Latonopsis occidentalis,* ×22; D, *L. fasciculata.* (A modified from Lilljeborg, 1900; B modified from Birge, 1910; C and D modified from Ward and Whipple, 1918.)

19. Margin of postabdomen sinuate (Fig. 16E); up to 5 mm long; usually in ponds containing much suspended organic matter .... **Daphnia magna** Straus
    Margin of postabdomen not sinuate (Fig. 14A); up to 3 mm long; usually in temporary, saline, or alkaline ponds .... **Daphnia similis** Claus
20. Teeth of proximal and middle pectens of postabdominal claw larger than teeth of distal pecten (Figs. 17G, H, K) 21
    Teeth of all three pectens of claw about the same size (Fig. 17J) 28
21. Teeth of middle pecten of postabdominal claw stout, the largest at least three times as long as teeth of distal pecten (Figs. 17G, K); ocellus present 22
    Teeth of middle pecten of postabdominal claw not more than twice as long as teeth of distal pecten (Fig. 17H); ocellus absent 27
22. Ventral margin of head concave; optic vesicle touching margin of head (Figs. 17B–E) 23
    Ventral margin of head sinuate or more or less straight; optic vesicle not touching margin of head. 26
23. Head longest over optic vesicle; ventral margin of head separated from anterior margin of valves by a prominent gap (Fig. 15E); dorsal part of head usually dark brownish; up to 3 mm long; ponds and lakes; TX to MT and westward; easily confused and sometimes hybridized(?) with *D. pulex;* perhaps a variety of *D. pulex;* not common .... **Daphnia middendorffiana** Fischer
    Head longest near midline; ventral part of head close to anterior margin of valves (Figs. 15B–D); entire body occasionally brownish, but head alone never dark brown; up to 2.5 mm long; extremely common and generally distributed in lakes and ponds 24
24. With an antennular mound (Fig. 14B); spinules of ventral margin widely spaced (Fig. 14D); rare in central states .... **Daphnia obtusa** Kurz
    Without an antennular mound (Fig. 14C); spinules of ventral margin closely spaced (Fig. 14E); extremely common 25
25. Tip area of rostrum with roughly symmetrical polygonal markings (Fig. 14G); marginal denticles of postabdomen extending slightly less than half of the length of the postabdomen (Fig. 16C); variable species; easily confused and sometimes hybridized with *D. middendorffiana.*

Daphnia pulex (Leydig)

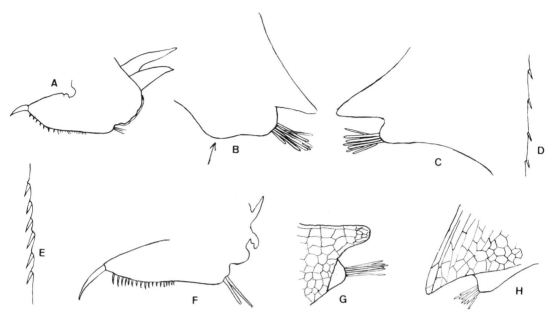

FIG. 14.–Structure of *Daphnia*. A, postabdomen of *D. similis*; B, tip of rostrum of *D. obtusa*, showing antennular mound (arrow); C, tip of rostrum of *D. pulex*; D, spinules of ventral margin of *D. obtusa*; E, spinules of ventral margin of *D. pulex*; F, postabdomen of *D. pulicaria*; G, tip of rostrum of *D. pulex*; H, tip of rostrum of *D. pulicaria*.

Tip area of rostrum with more elongated polygonal markings (Fig. 14H); marginal denticles of postabdomen extending slightly more than half of the length of the postabdomen (Fig. 14F); rare .... **Daphnia pulicaria** Forbes

26. Spinules on dorsal margin of body long, interspinule distance less than 1.5 times spinule length (Fig. 17B); middle pecten of claw with three to seven teeth (Fig. 17K); up to 2 mm long; uncommon; found west of the Mississippi River but occurs also in the Great Lakes; may be a variant of *D. pulex* or *D. pulicaria* .... **Daphnia schødleri** Sars

    Spinules on dorsal margin of body short, interspinule distance two to three times spinule length (Fig. 17A); middle pecten of claw with two to five teeth (Fig. 17G); up to 1.5 mm long; uncommon; in eastern states .... **Daphnia catawba** Coker

27. Anterior margin of head broadly rounded; if a small crest is present, it is longest in the midline; posterior spine less than one quarter of valve length (Fig. 15D); up to 1.2 mm long; lakes and ponds; widely distributed except for Rocky Mountain ranges (often confused with *D. retrocurva*).
    Daphnia parvula Fordyce

    Anterior margin of head with small to large helmet, longest dorsal to midline (Figs. 9D–H); posterior spine at least one-third valve length; northeastern quarter of the United States, and WA and OR (often confused with *D. parvula*) .... **Daphnia retrocurva** Forbes

28. Swimming hair at base of second segment of three-segmented ramus does not reach end of ramus (Fig. 17C); ocellus absent; swimming hairs of antennae extend beyond posterior margin of valves; limnetic, confined to hypolimnion during stratification; up to 1.2 mm long; northeastern states, but as far west as MN .... **Daphnia longiremis** Sars

    Swimming hair at base of second segment of three-segmented ramus extends beyond tip of ramus (Fig. 17D); ocellus present; swimming hairs of antennae not extending beyond posterior margins of valves 29

29. Small, head and valves up to 1.3 mm long; almost circular in outline; head often with a small anterior point (Figs. 9N, O); highly variable morphological details; may not be a good species; ponds and lakes east of the Continental Divide, but also along the Pacific coast.
    Daphnia ambigua Scourfield

    Medium to large; head and valves more than 1.3 mm long 30

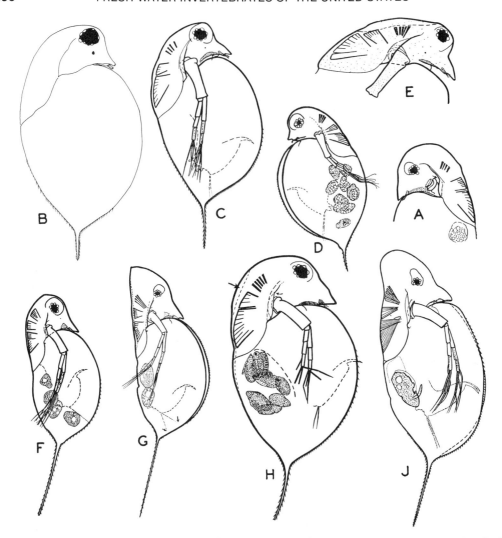

FIG. 15.–Structure of *Daphnia*. A, head of *D. schødleri*; B and C, *D. pulex*; D, *D. parvula*; E, head of *D. middendorffiana*; F, *D. laevis*; G, *D. dubia*; H, *D. rosea*; J, *D. thorata*. (A, C–J from Brooks, 1957; B from Brandlova, et al.)

30. Second abdominal process about one quarter as long as first abdominal process (Fig. 17E) 31
    Second abdominal process about one half as long as first abdominal process (Fig. 17F) 32
31. With spines over slightly more than half of ventral margin of valve; anterior margin of head rounded
    or with a low, blunt crest (Fig. 15F); up to 1.7 mm long; ponds in southern states but extending
    up along both coasts .... **Daphnia laevis** Birge
    With spines over posterior three quarters of ventral margin of valve; head usually helmeted (Fig.
    15G); up to 1.8 mm long; extreme northeastern states, west to WI.
                                                                  Daphnia dubia Herrick
32. Anterior margin with a low, rounded crest; head twice as deep as long (Fig. 15H); posterior spine
    usually less than one-third valve length; morphological details variable; up to 2 mm long;
    common in ponds and lakes of western states (usually referred to in the past as *D. longispina*).
                                                                  Daphnia rosea Sars
    Anterior margin usually helmeted; head length more than half its greatest depth (Fig. 15J); posterior
    spine more than one third valve length 33

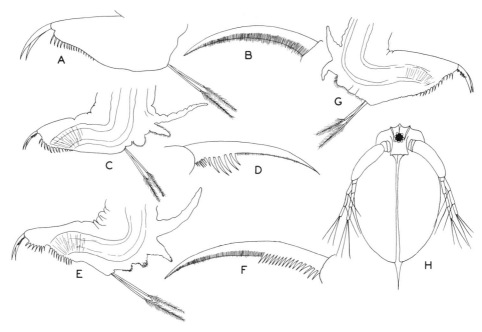

Fig. 16.–Structure of *Daphnia*. A, postabdomen of *D. rosea*; B, claw of *D. laevis*; C, postabdomen of *D. pulex*; D, claw of *D. pulex*; E, postabdomen of *D. magna*; F, claw of *D. magna*; G, postabdomen of *D. middendorffiana*; H, ventral view of *D. magna*, diagrammatic, ×11.

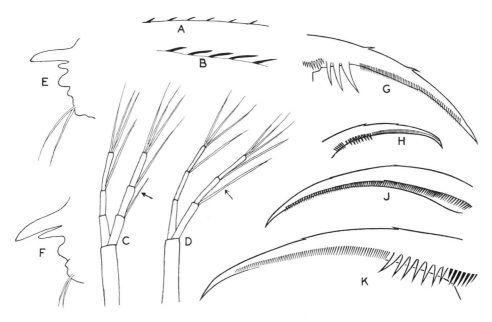

Fig. 17.–Structure of *Daphnia*. A, spinules on dorsal margin of carapace of *D. catawba*; B, same for *D. schødleri*; C, second antenna of *D. longiremis*; D, same for *D. ambigua*; E, abdominal processes of *D. dubia*; F, same for *D. rosea*; G, postabdominal claw of *D. catawba*; H, same for *D. retrocurva*; J, same for *D. rosea*; K, same for *D. schødleri*. (B from Brandlova, Brandl, and Fernando; G–K from Brooks, 1957.)

33. Dorsal margin of head with a concavity; anterior margin of head not pointed (Fig. 15J); ventral spines of valves extend over at least the posterior two thirds; up to 1.8 mm long; large lakes of WA, OR, ID, MT .... **Daphnia thorata** Forbes
    Dorsal margin of head without a concavity; anterior margin of head usually pointed; head more or less helmeted except during early spring (Figs. 9L, M); ventral spines of valves extending over less than the posterior two thirds; up to 3 mm long; northeastern states and Pacific coast states.
    .... **Daphnia galeata mendotae** Birge
34. Valves transversely striated (Fig. 18B); up to 3 mm long; the species in this genus are known to hybridize .... **Simocephalus, 35**
    Valves not transversely striated; up to 1 mm long; usually in vegetation .... **Scapholeberis, 37**
35. Claws with a pecten (Fig. 18A); uncommon; reported from scattered localities in the eastern half of the United States .... **Simocephalus exspinosus** (Koch)
    Claws without a pecten **36**
36. Vertex evenly rounded about the eye and without serrations or spinules (Fig. 18B); widely distributed but not common .... **Simocephalus vetulus** (O.F.M.)
    With serrations or spinules in front of or below the eye; vertex more or less angulate (Fig. 18C); common everywhere .... **Simocephalus serrulatus** (Koch)
37. Rostrum of female pointed; color whitish or greenish, translucent or opaque; uncommon in northern states .... **Megafenestra aurita** (Fischer)
    Rostrum more pointed **38**

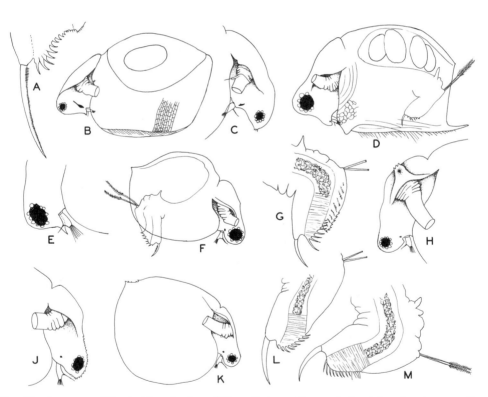

Fig. 18.–Structure of Daphnidae. A, claw of *Simocephalus exspinosus*; B, *S. vetulus*, showing a portion of transverse striations, ×15; C, head of *S. serrulatus*; D, *Scapholeberis mucronata*, ×50; E, head of *Ceriodaphnia rigaudi*; F, *C. reticulata,* ×25; G, postabdomen of *C. megalops*; H, head of *C. lacustris*; J, head of *C. rotunda*; K, *C. laticaudata,* ×50; L, postabdomen of *C. quadrangula*; M, postabdomen of *C. pulchella*. (B–H, and L modified from Ward and Whipple, 1918; J and M modified from Lilljeborg, 1900.)

38. Front of rostrum rectilinear (Fig. 18D); common and widely distributed.

                     Scapholeberis mucronata (O.F.M.)

 Front of rostrum trilobate or triangularly produced; uncommon in eastern states.

                       Scapholeberis armata (Herrick)

39. Antennules small; head small and depressed (Figs. 19F, H–K); up to 1.4 mm long.

                   DAPHNIDAE, **Ceriodaphnia**, 40

 Antennules large; head large and extended (Figs. 19A, B, D) . . . . MOINIDAE, 47

40. Head with a short spine (Fig. 18E); a variable species, 0.4 to 0.5 mm long; pools in southern states.

                     Ceriodaphnia rigaudi Richard

 Head without a spine **41**

41. Claws with pecten (Fig. 18F); up to 1.4 mm long; common and widely distributed.

                     Ceriodaphnia reticulata (Jurine)

 Claws without pecten **42**

42. Postabdomen abruptly incised (Fig. 18G); widely distributed but uncommon.

                     Ceriodaphnia megalops Sars

 Postabdomen not incised **43**

43. Fornices projecting to form spinous processes (Fig. 18H); 0.8 to 0.9 mm long; limnetic; scattered east of the Mississippi River . . . . **Ceriodaphnia lacustris** Birge

 Fornices of the usual form; up to 1 mm long **44**

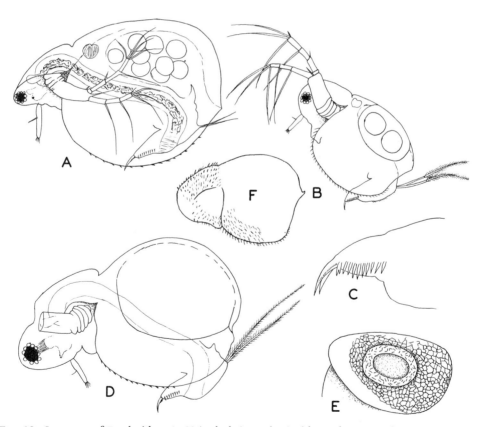

FIG. 19.–Structure of Daphnidae. A, *Moinodaphnia macleayi* with parthenogenetic eggs, ×60; B, *Moina macrocopa* with ephippium, ×24; C, postabdomen of same; D, *M. micrura*, ×40; E, ephippium of *M. micrura*; F, head and carapace of *Moina affinis* showing distribution of hairs. (A, D, and E modified from Ward and Whipple, 1918; B and C modified from Uéno.)

44. Vertex with spines (Fig. 18J); in aquatic vegetation; reported only from WI.
Ceriodaphnia rotunda Sars
Vertex without spines; widely distributed **45**
45. General form round (Fig. 18K); up to 1 mm long; in vegetation.
Ceriodaphnia laticaudata P. E. Müller
General form not especially round; littoral and limnetic **46**
46. With seven to nine anal spines (Fig. 18L); up to 1 mm long.
Ceriodaphnia quadrangula (O.F.M.)
With 7 to 10 anal spines plus 3 to 5 anal setae (Fig. 18M); up to 0.7 mm long.
Ceriodaphnia pulchella Sars
47. Body laterally compressed; valves completely covering body (Fig. 19A); up to 1 mm long; on bottom and in vegetation; reported only from LA . . . . **Moinodaphnia macleayi** (King)
Body thick and heavy; valves not completely covering body (Figs. 19B, D); usually in pools and ponds . . . . **Moina, 48**
48. Postabdomen without a distal bident tooth; up to 1.6 mm long; highly saline basins; WA, CA, NE, NV . . . . **Moina hutchinsoni** Brehm
Postabdomen with a distal bident tooth (Fig. 19C) **49**
49. Head almost invariably with a supraocular depression (Fig. 19D); first leg of female with anterior setae on penultimate and ultimate segments feathered or with fine hairs **50**
Head without a supraocular depression (Fig. 19B); first leg of female with anterior setae of penultimate and ultimate segments toothed; widely distributed but not common.
Moina macrocopa (Straus)
50. Female with long hairs on head and carapace (Fig. 19F) **51**
Female with long hairs only on ventral surface of head, or hairs absent **52**
51. Ephippium reticulated and with a single egg; up to 1.2 mm long; widely distributed.
Moina affinis Birge
Ephippium knobby and with two eggs; up to 1.5 mm long; reported only from AZ, CA, CO, KS, TX.
Moina wierzejskii Richard
52. Head narrow and with a distinct supraocular depression (Fig. 19D); up to 1.2 mm long; one ephippial egg (Fig. 19E); widely distributed and common . . . . **Moina micrura** Kurz
Head large and broad, without a supraocular depression; up to 1.4 mm long; two ephippial eggs; one record from the southern CA desert . . . . **Moina brachycephala** Goulden
53. Antennules united at base and diverging at apex (Fig. 20B); widely distributed in southern states; uncommon . . . . **Bosminopsis dietersi** Richard
Antennules not united at base, parallel; highly variable; littoral and limnetic; widely distributed and common **54**
54. Female postabdominal claw with a very fine distal pecten (Fig. 20H); frontal sensory bristle near midpoint between eye and tip of rostrum (Fig. 20C); widely distributed and common.
Bosmina longirostris (O.F.M.)
Female postabdominal claw with proximal pecten only, emarginate near tip (Fig. 20J); frontal sensory bristle near tip of rostrum, near or at base of antennule (Fig. 20D); sometimes considered a subgenus of *Bosmina* . . . . **Eubosmina, 55**
55. Mucro absent (Fig. 20K); Great Lakes and northeastern states . . . **Eubosmina coregoni** (Baird)
Mucro present (Figs. 20E–G) **56**
56. Mucro with minute ventral notches, at least on immature forms (Fig. 20G); reported only from extreme northeastern states . . . . **Eubosmina longispina** (Leydig)
Mucro with minute dorsal notches, at least on immature forms (Fig. 20F) **57**
57. Rostrum becomes relatively shorter with age; mucro grows longer with age; frontal bulge present at base of antenna (Fig. 20L); reported from northeastern states . . . . **Eubosmina tubicen** (Brehm)
Rostrum becomes longer with age; mucro grows shorter with age; frontal bulge absent (Fig. 20M); western states . . . . **Eubosmina hagmanni** (Stingelin)
58. Intestine convoluted (Figs. 21A, B) **59**
Intestine simple **64**
59. Valves with spine at posterior dorsal angle (Fig. 21A); up to 2 mm long; in vegetation; widely distributed . . . . **Ophryoxus gracilis** Sars
Valves without such a spine **60**
60. Antennal setae 0–0–1–3/1–1–3; 0.2 to 0.5 mm long . . . . **Streblocerus, 61**
With another antennal setae formula; more than 0.5 mm long **62**

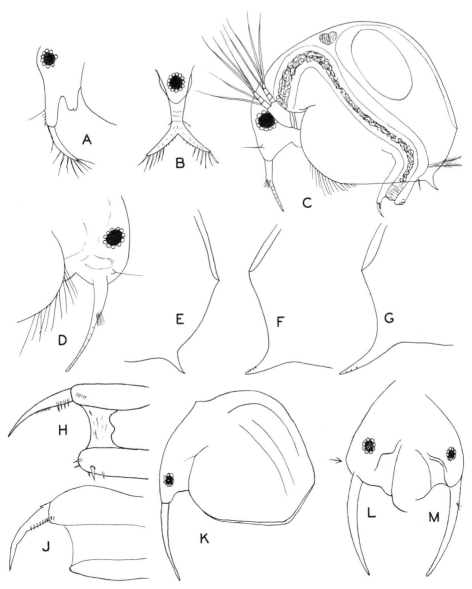

FIG. 20.–Structure of Bosminidae. A, lateral view of *Bosminopsis dietersi*; B, dorsal view of same; C, *Bosmina longirostris,* ×140; D, anterior end of *Eubosmina coregoni*; E, mucro of *B. longirostris*; F, mucro of *E. hagmanni*; G, mucro of *E. longispina*; H, postabdomen of *B. longirostris*; J. postabdomen of *E. longispina*; K, *E. coregoni*; L, anterior end of *E. tubicen*; M, anterior end of *E. hagmanni*. (J–M modified from Deevey and Deevey.)

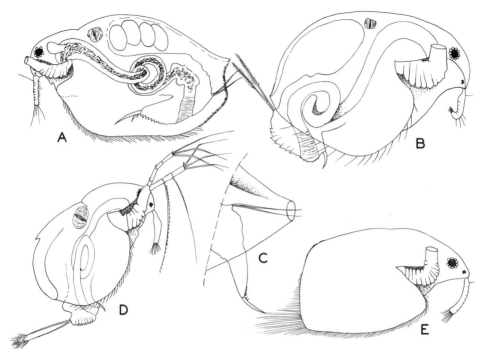

FIG. 21.–Structure of Macrothricidae. A, *Ophryoxus gracilis,* ×30; B, *Streblocerus serricaudatus,* ×100; C, posterior end of *Parophryoxus tubulatus*; D, *Drepanothrix dentata,* ×60; E, *Acantholeberis curvirostris,* ×26. (Modified from Ward and Whipple, 1918.)

61. Dorsal margin of valves smooth (Figs. 21B, 22A); uncommon but widely distributed in littoral zones.
　　　　　　　　　　　　　　　　　　　Streblocerus serricaudatus (Fischer)
　　　Dorsal margin of valves serrate; in vegetation in pools; LA ...... **Streblocerus pygmaeus** Sars
62. Valves narrowed posteriorly and prolonged into a short tube (Fig. 21C); up to 1.2 mm long; in vegetation; New England states and rare farther west ...... **Parophyroxus tubulatus** Doolittle
　　　Valves not narrowed posteriorly; widely distributed **63**
63. Dorsal and posterior margin arched; with a conspicuous dorsal tooth (Fig. 21D); up to 0.7 mm long; in littoral zones; uncommon in northern states .... **Drepanothrix dentata** (Eurén)
　　　Without a dorsal tooth (Fig. 21E); up to 1.8 mm long; in littoral, especially in *Sphagnum* bogs.
　　　　　　　　　　　　　　　　　　Acantholeberis curvirostris (O.F.M.)
64. With a wide crest on dorsal margin of valves (Fig. 23B); up to 1 mm long; reported from northern states .... **Bunops serricaudata** (Daday)
　　　Without such a crest **65**
65. Postabdomen with numerous long spines (Figs. 23C–E); antennal setae 0–0–0–3/1–1–3; often more or less covered with detritus; usually found creeping about on the substrate, especially in vegetated areas .... **Ilyocryptus, 67**
　　　Postabdomen without numerous long spines; with another setae formula **66**
66. Hepatic caeca present: 0.9 mm long; in vegetation in shallow water; LA.
　　　　　　　　　　　　　　　　　　　　Grimaldina brazzai Richard
　　　Hepatic caeca absent **71**
67. Anus on dorsal margin of postabdomen (Fig. 23D) **68**
　　　Anus terminal (Fig. 23E); widely distributed but uncommon ..... **Ilyocryptus acutifrons** Sars
68. Molting complete; a European species now found along the Atlantic Coast states.
　　　　　　　　　　　　　　　　　　　　Ilyocryptus agilis Kurz
　　　Molting incomplete **69**

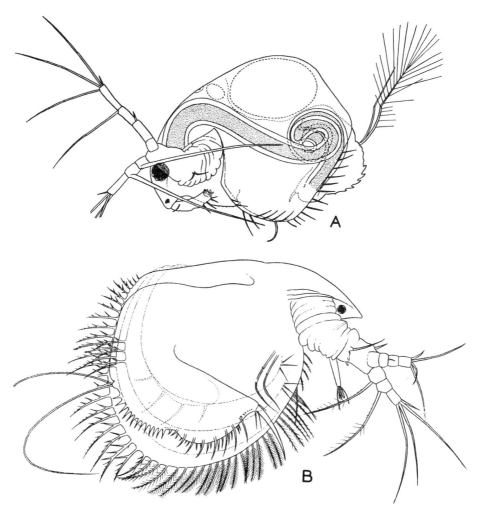

FIG. 22.–Structure of Macrothricidae. A, *Streblocerus serricaudatus*; B, *Ilyocryptus sordidus*. (From Fryer, 1974.)

69. With five to seven preanal spines (Fig. 23C); common and widely distributed.

                                            Ilyocryptus spinifer Herrick

    With eight or more preanal spines (Figs. 22B, 23D); uncommon 70

70. Postabdominal claw with a flexure; VA, TX . . . . **Ilyocryptus gouldeni** Williams

    Postabdominal claw smoothly curved; uncommon . . . . **Ilyocryptus sordidus** (Liéven)

71. Antennal setae 0–0–1–3/1–1–3; setae of basal segment of lower ramus stout (Fig. 23G) 72

    Antennal setae 0–1–1–3/1–1–3; all setae slender; 1 mm long (Fig. 23F); widely distributed but uncommon in vegetation of littoral . . . . **Lathonura rectirostris** (O.F.M.)

72. Dorsal margin of head evenly rounded . . . . **Macrothrix, 73**

    Dorsal margin of head not evenly rounded (Fig. 23G); 0.7 mm long; in vegetation; widely distributed . . . . **Echinisco rosea** Liéven

73. Head depressed, rostrum close to margin of valves (Fig. 23H); up to 1.1 mm long; southwestern.

                                      Macrothrix borysthenica Matile

    Head extended, rostrum far from margin of valves (Fig. 23K) 74

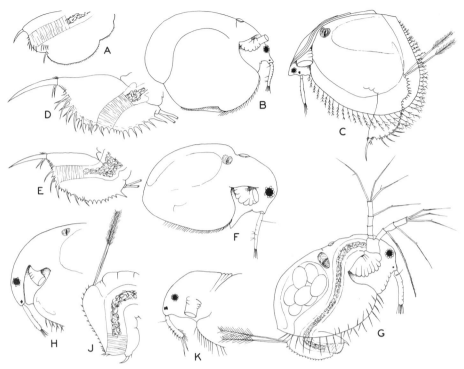

FIG. 23.–Structure of Macrothricidae. A, postabdomen of *Grimaldia brazzai*; B, *Bunops serricaudata*, ×35; C, *Ilyocryptus spinifer,* ×40; D, postabdomen of *I. sordidus*; E, postabdomen of *I. acutifrons*; F, *Lathonura rectirostris*, ×38; G, *Echinisca rosea*, ×55; H, anterior end of *M. borysthenica*; J, postabdomen of *M. laticornis*; K, anterior end of *M. montana*. (B–H and K modified from Ward and Whipple, 1918; J modified from Lilljeborg, 1900.)

74. Postabdomen not bilobed (Fig. 23J); up to 0.7 mm long; widely distributed.
    Macrothrix laticornis (Jurine)
    Postabdomen bilobed; 0.55 mm long 75
75. With conspicuous folds at cervical sinus (Fig. 23K); Rocky Mountains.
    Macrothrix montana Birge
    Without such folds; New England to CO ...... **Macrothrix hirsuticornis** Norman and Brady
76. Anus terminal; up to 5 mm long (Fig. 24A); often with many ephippial eggs; common everywhere in
    weed beds; a complex of closely related species, designations uncertain ...... **Eurycercus**
    Anus on dorsal side of abdomen (Figs. 24B, C) 77
77. Compound eye present 78
    Compound eye absent but ocellus large; 0.5 mm long; rare; northern states.
    Monospilus dispar Sars
78. Eye and ocellus of usual size 79
    Compound eye and ocellus extremely large and of similar size; rare in southern states.
    Dadaya macrops (Daday)
79. Posterior margin of valves not greatly less than maximum height (Figs. 24, 25) 80
    Posterior margin of valves considerably less than maximum height (Figs. 27, 28) 108
80. Claws with secondary tooth in the middle (Figs. 24C, G); secondary tooth sometimes very small.
    81
    Claws without secondary tooth in the middle 88
81. Postabdomen with both marginal and lateral denticles (Figs. 24C, G) 83
    Postabdomen without both marginal and lateral denticles 82

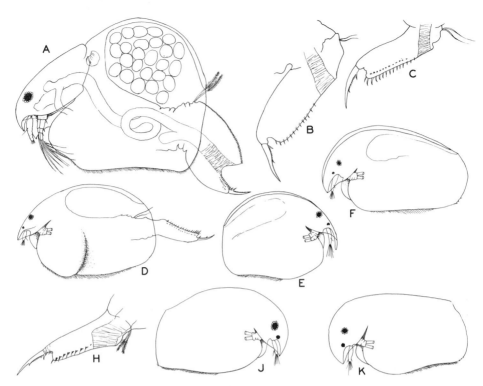

FIG. 24.–Structure of Eurycercinae and Chydorinae. A, *Eurycercus lamellatus,* ×17; B, postabdomen of *Alonopsis aureola*; C, postabdomen of *A. americana*; D, *Euryalona occidentalis,* ×38; E, *Kurzia latissima,* ×48; F, *Camptocercus rectirostris,* ×40; H, postabdomen of *Acroperus harpae*; J, *A. harpae,* ×42; K, *A. angustatus,* ×46. (A modified from Lilljeborg, 1900; B and C modified from Doolittle; D–F, J, and K modified from Ward and Whipple, 1918; H modified from Uéno.)

82. Postabdomen with lateral denticles only, sometimes inconspicuous (Fig. 24H); 0.8 to 0.9 mm long (Fig. 24J); common everywhere in vegetation . . . . **Acroperus harpae** Baird
   Postabdomen with marginal denticles only (Fig. 24B); up to 1.9 mm long; in littoral vegetation; known only from an old ME record . . . . **Alonopsis aureola** Doolittle
83. Postabdomen relatively broad (Fig. 24C) **84**
   Postabdomen relatively narrow (Fig. 24G) **85**
84. Terminal claw with two medium-sized teeth (Fig. 24C); 0.8 mm long; ME, NH.
   Alonopsis americana Kubersky
   Terminal claw with one large tooth; 0.4 mm long; reported only from MS.
   Alonella fitzpatricki Chien
85. Anterior portion of valves swollen (Fig. 24D); 1 mm long; in vegetation; southern states.
   Euryalona occidentalis Sars
   Anterior portion of valves not swollen; usually in vegetation in shallows **86**
86. Crest on head and valves (Fig. 24F); about 1 mm long . . . . **Camptocercus, 87**
   Crest on valves only (Fig. 24E); 0.6 mm long; common in temperate America.
   Kurzia latissima (Kurz)
87. Postabdomen with 20 to 30 marginal denticles; widely distributed but rare.
   Camptocercus macrurus (O.F.M.)
   Postabdomen with 45 to 65 minute marginal denticles; in temporary ponds in OK, KS.
   Camptocercus oklahomensis Mackin
88. Rostrum not greatly exceeding antennules (Figs. 25A, G, H) **89**
   Rostrum much longer than antennules (Fig. 27B) **107**

89. Rostrum abruptly narrowed and pointed near tip (Figs. 25G, H) **91**
    Rostrum not abruptly narrowed near tip (Fig. 25A) **90**
90. Ventroposterior angle with teeth (Fig. 25A); 0.5 to 0.7 mm long; usually on bottom in shallows;
    common and widely distributed . . . . **Graptoleberis testudinaria** (Fischer)
    Ventroposterior angle without teeth; up to 0.4 mm long; reported from WI, IN, FL, GA, MN.
                                                                                      Alona setulosa Megard
91. Ventroposterior angle without teeth **92**
    Ventroposterior angle with one to four small teeth **99**
92. Postabdomen with clusters of large spines (Fig. 25B); about 1 mm long; uncommon.
                                                                                      Leydigia, 93
    Postabdomen without clusters of large spines **94**
93. Valves without markings (Fig. 25C); widely distributed but uncommon among vegetation.
                                                                               Leydigia leydigi (Schödler)
    Valves with longitudinal striations; rare, in southern states.
                                                                        Leydigia acanthocercoides (Fischer)
94. Postabdomen with marginal and lateral denticles (Figs. 25D, L) **95**
    Postabdomen with marginal denticles only (Figs. 25E, F) **97**
95. Marginal denticles markedly longer distally (Fig. 25D); about 0.5 mm long . . . . . **Oxyurella, 96**
    Marginal denticles not markedly longer distally . . . . **Alona, 101**
96. Penultimate marginal denticle largest (Fig. 25D); widely distributed but not common east of the
    Rockies . . . . **Oxyurella brevicaudis** Frey
    Ultimate marginal denticle largest, serrate; rare, in southern states.
                                                                               Oxyurella longicaudis (Birge)

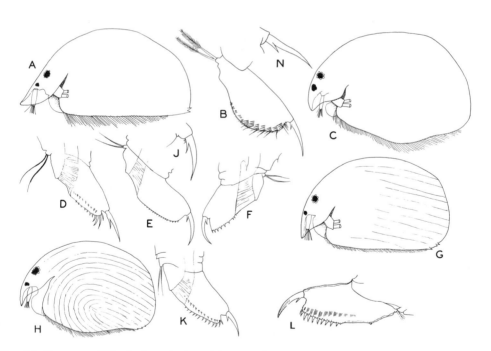

FIG. 25.–Structure of Chydorinae. A, *Graptoleberis testudinaria,* ×80; B, postabdomen of *Leydigia leydigi;* C, *L. leydigi,* ×45; D, postabdomen of *Oxyurella brevicaudis;* E, postabdomen of *Alonella diaphana;* F, postabdomen of *Alona guttata;* G, *A. monacantha,* ×100; H, *Biapertura karua,* ×75; J, claw of *Biapertura affinis;* K and L, postabdomen of *A. costata;* N, claw of *A. quadrangularis* (A, E, G, and H modified from Ward and Whipple, 1918; B modified from D'Ancona; C, F, and K modified from Lilljeborg, 1900; D modified from Birge, 1910.)

97. Denticles minute (Fig. 25E); 0.5 mm long; rare; southern states .... **Alonella diaphana** (King)
Denticles of usual size (Fig. 25F); 0.4 mm long 98
98. Postabdomen with 8 to 10 marginal denticles (Fig. 25F); common everywhere.
Alona guttata Sars
Postabdomen with more than 10 marginal denticles; widely distributed.
Disparalona leei (Chien)
99. With a fringe of setae along most of the ventral margin (Figs. 25G, H); southern states 100
With a fringe of setae confined to the posterior ventral margin; up to 0.4 mm long; rare; reported
from MN, IN, SC, FL, NM .... **Alona circumfimbriata** Megard
100. Valves with longitudinal striae (Fig. 25G); 0.4 mm long; in vegetation in pools.
Alona monacantha Sars
Valves with oblique striae (Fig. 25H); about 0.5 mm long .... **Biapertura karua** (King)
101. Postabdomen with marginal denticles only (Fig. 25F); 0.4 mm long; common everywhere.
Alona guttata Sars
Postabdomen with both marginal and lateral denticles (Figs. 25L; 26B) 102
102. With 14 or more marginal denticles; up to 1 mm long 103
With less than 14 marginal denticles; up to 0.5 mm long; cryptic species 104
103. With a cluster of fine spinules at base of claw (Fig. 25J); 1 mm long; abundant everywhere in
vegetation of littoral .... **Biapertura affinis** (Leydig)
Without spinules at base of claw (Fig. 25N); up to 0.9 mm long; in vegetation of littoral and on
bottom in deeper water; common everywhere .... **Alona quadrangularis** (O.F.M.)
104. Anal groove of postabdomen unarmed (Fig. 26B) 105
Anal groove of postabdomen completely edged by groups of spinules (Fig. 26C) 106
105. Female postabdomen with about 12 subequal marginal denticles (Fig. 25L); common and widely
distributed .... **Alona costata** Sars
Female postabdomen with about seven subequal marginal denticles; WY, MN.
Alona barbulata Megard
106. Rostrum not projecting beyond setae of antennae (Fig. 26D); carapace unicolored; widely
distributed .... **Alona rustica** Scott
Rostrum projecting beyond tips of setae of antennae (Fig. 26E); carapace strongly amber or straw
colored, brood chamber area colorless; reported only from New England.
Alona bicolor Frey
107. Postabdomen with marginal denticles only (Fig. 25E) .... **Alonella,** 116
Postabdomen with numerous lateral denticles, only two to four marginal denticles (Fig. 27A); 0.5
mm long; northern states .... **Rhynchotalona falcata** (Sars)
108. Body elongated 109
Body spherical or broadly ellipsoidal 124
109. Posterior margin with teeth along entire length (Fig. 27D); 0.5 mm long; rostrum bent sharply into a
hook (Figs. 27B, H); common in northern states .... **Pleuroxus procurvatus** Birge
Posterior margin without teeth along entire length 110

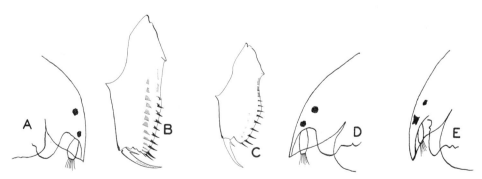

FIG. 26.–Structure of *Alona*. A, anterior end of *A. setulosa*; B, postabdomen of *A. costata*; C, postabdomen of *A. rustica*; D, anterior end of *A. rustica*; E, anterior end of *A. bicolor*. (B and C from Frey, 1965; D and E modified from Frey, 1965.)

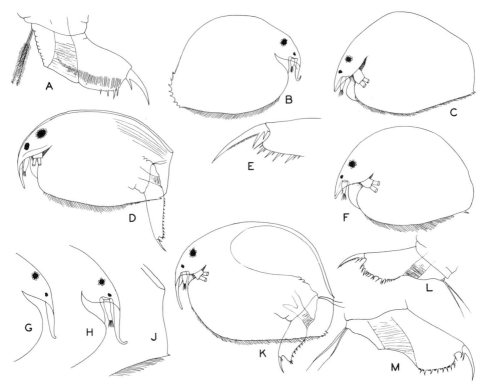

FIG. 27.–Structure of Chydorinae. A, postabdomen of *Rhynchotalona falcata*; B, *Pleuroxus procurvatus,* ×70; C, *Alonella dentifera,* ×95; D, *Pleuroxus striatus,* ×55; E, claw of *P. striatus*; F, *Dunhevedia crassa,* ×75; G and H, anterior end of *Pleuroxus*; J, posterior end of *P. laevis*; K, *P. denticulatus,* ×85; L, postabdomen of *P. trigonellus*; M, postabdomen of *P. aduncus.* (A modified from Lilljeborg, 1900; B, C, and F–M modified from Ward and Whipple, 1918; D and E redrawn from Uéno.)

110. Without teeth at the ventroposterior angle 111

With one or more teeth at the ventroposterior angle 117

111. Claws with two basal spines (Figs. 27L, M) .... **Pleuroxus,** 112

Claws with one basal spine (Fig. 25E) .... **Alonella,** 115

112. Postabdomen long and slender; ventroposterior angle rounded (Fig. 27D); 0.8 mm long; common in vegetation .... **Pleuroxus striatus** Schödler

Postabdomen of moderate length (Figs. 27L, M); 0.5 to 0.6 mm long 113

113. Angle of postabdomen sharp; in pools and vegetation; probably generally distributed.

Pleuroxus hamulatus Birge

Angle of postabdomen rounded (Figs. 27L, M) 114

114. With the row of marginal denticles longer than anal emargination (Fig. 27L); widely distributed but uncommon .... **Pleuroxus trigonellus** (O.F.M.)

With the row of marginal denticles about equal to anal emargination (Fig. 27M); in vegetation; reported from several northern and western states .... **Pleuroxus aduncus** (Jurine)

115. Postabdomen with marginal and lateral denticles; in vegetation; LA.

Alonella globulosa Daday

Postabdomen with marginal denticles only (Fig. 25E) 116

116. Marginal denticles minute (Fig. 25E); rostrum reaching not more than two thirds the distance to the ventral margin; rare; southern states .... **Alonella diaphana** (King)

Marginal denticles of usual size; rostrum long, recurved, and reaching the ventral margin (Fig. 29D).

Disparalona rostrata (Koch)

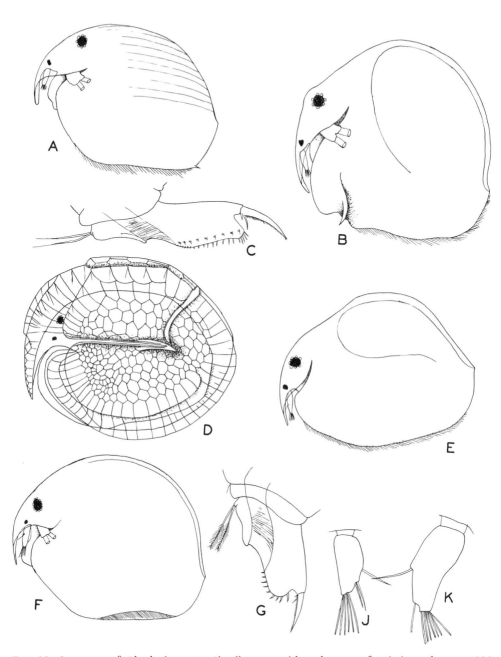

FIG. 28.–Structure of Chydorinae. A, *Alonella nana* with only part of striations shown, ×200; B, *Anchistropus minor*, ×160; C, postabdomen of *Pseudochydorus globosus*; D, *Chydorus bicornutus*, ×120; E, *C. gibbus*, ×110; F, *C. sphaericus*, ×130; G, postabdomen of *C. sphaericus*; J, antennule of *C. piger*; K, antennule of *C. ovalis*. (A–E modified from Ward and Whipple, 1918; G–K modified from Lilljeborg, 1900.)

117. Claws with two major basal spines 118
     Claws with one major basal spine 135
118. Rostrum long, extending well beyond the tips of the terminal olfactory setae (Figs. 27G, H); teeth on
     ventroposterior angle variable .... **Pleuroxus,** 120
     Rostrum shorter, extending only slightly, if at all, beyond the tips of the olfactory setae; lowermost
     part of posterior margin excised or notched, or ventroposterior angle with one or two
     prominent teeth (Fig. 27C) .... **Alonella,** 119
119. Postabdomen angled at apex (Fig. 29F); common everywhere in vegetation.
                                                           Alonella excisa (Fischer)
     Postabdomen rounded at apex (Fig. 29G); rare; northern states .... **Alonella exigua** (Lillj.)
120. Postabdomen long and slender (Fig. 27D) 121
     Postabdomen of moderate length (Figs. 27K–M); 0.5 to 0.6 mm long 122
121. Ventroposterior angle a sharp point, with a very small tooth (Fig. 27J); 0.6 mm long; rare but widely
     distributed .... **Pleuroxus laevis** (Sars)
     Ventroposterior angle rounded, with a small tooth anterior to it (Fig. 27D); 0.8 mm long; common
     everywhere in vegetation .... **Pleuroxus striatus** Schödler
122. Angle of postabdomen sharp; ventroposterior angle with teeth (Fig. 27K); a highly variable species,
     common everywhere in vegetation .... **Pleuroxus denticulatus** Birge
     Angle of postabdomen rounded (Figs. 27L, M) 123
123. With a row of marginal denticles longer than anal emargination (Fig. 27L); two to three small teeth on
     inferoposterior angle; widely distributed but uncommon .... **Pleuroxus trigonellus** (O.F.M.)
     With a row of marginal denticles about equal to anal emargination (Fig. 27M); occasionally two or
     three small teeth on inferoposterior angle; in vegetation; reported from several northern and
     western states .... **Pleuroxus aduncus** (Jurine)
124. With a small to large spine at ventroposterior angle (Fig. 28A) 125
     Without a spine at ventroposterior angle 126
125. Valves conspicuously striated (Fig. 28A); 0.2 to 0.3 mm long; rare; northern states.
                                                            Alonella nana (Baird)
     Valves not striated 132
126. Valves with conspicuous projection at anteroventral margin (Fig. 28B); 0.35 mm long; widely
     distributed .... **Anchistropus minor** Birge
     Valves without such a projection 127
127. Animal spherical or ovate; postabdomen with prominent preanal angle (Fig. 28); 0.3 to 0.8 mm long.
                                                                                 128
     Animal more elongated; postabdomen without prominent preanal angle; 0.2 to 0.5 mm long.
                                                                                 138
128. Postabdomen long, narrow (Fig. 28C); widely distributed; in vegetation.
                                                    Pseudochydorus globosus (Baird)
     Postabdomen short, broad (Fig. 28G) .... **Chydorus,** 129
129. Carapace deeply sculptured 130
     Carapace not deeply sculptured 138
130. Carapace covered with deep polygonal cells; northeastern states ... **Chydorus faviformis** Birge
     Carapace covered with deep polygonal cells plus pronounced ridges (Fig. 28D) 131
131. Two lowermost rows of polygonal cells elongated (Fig. 28D); northeastern states.
                                                       Chydorus bicornutus Doolittle
     Two lowermost rows of polygonal cells not elongated; FL to NJ .... **Chydorus bicellaris** Frey
132. Ventral edge of keel of labrum smooth 133
     Ventral edge of keel of labrum toothed (Figs. 29A–C) 135
133. Dorsoanterior surface of head and valves somewhat flattened (Fig. 28A); rare in northern states.
                                                            Chydorus gibbus Lillj.
     Dorsoanterior surface of head and valves not flattened 134
134. Rostrum pointed (Fig. 28F); total length 0.3 to 0.5 mm; the most common of all Cladocera; widely
     distributed; undoubtedly this is a group of closely related species and morphs, only slightly
     differing from each other .... **Chydorus sphaericus** (O.F.M.)
     Rostrum blunt (Fig. 29A); uncommon in TX, LA, OK, CA ...... **Ephemeroporus poppei** Richard
135. Labrum with a single tooth; rare in TX, LA, OK (Fig. 29C) ...... **Ephemeroporus acanthodes** Frey
     Labrum with three to four teeth 136

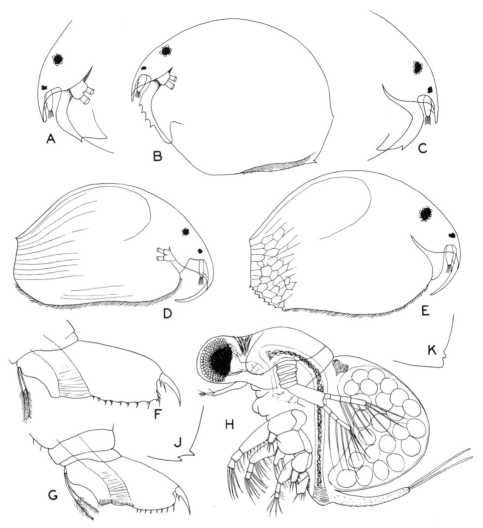

FIG. 29.–Structure of Chydorinae and Polyphemidae. A, anterior end of *Ephemeroporus poppei*; B, *E. barroisi*, ×140; C, anterior end of *E. hybridus*; D, *Disparalona rostrata* with only a portion of valve reticulations shown, ×110; E, *Alonella dadayi* with only a portion of valve reticulations shown, ×200; F, postabdomen of *Alonella excisa*; G, postabdomen of *A. exigua*; H, *Polyphemus pediculus*, ×50; J and K, ventroposterior angle of *Alonella excisa*. (A–E modified from Ward and Whipple, 1918; F–H modified from Lilljeborg, 1900.)

136. Usually with three to five teeth at posterior ventral corner of carapace; rare in FL.

Ephemeroporus archboldi Frey

Carapace with one such tooth 137

137. Labrum with four teeth (Fig. 29B); rare in LA, FL; undoubtedly a group of species.

Ephemeroporus barroisi Richard

Labrum with a single tooth (Fig. 29C); a group of species; rare along Gulf Coast.

Ephemeroporus hybridus Daday

138. Rostrum short, extending only slightly, if at all, beyond the tips of the terminal olfactory setae (Figs. 27C, F); lateral denticles present on postabdomen 139

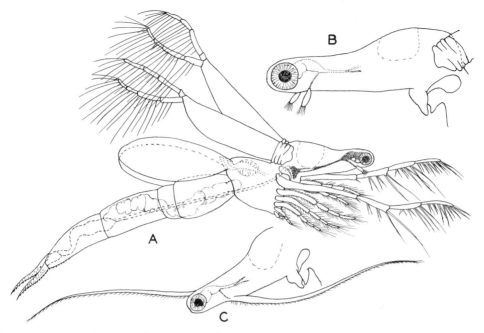

Fig. 30.–Structure of *Leptodora kindti*. A, adult female with winter eggs, ×85; B, anterior end of female; C, anterior end of male. (Modified from Sebestyén.)

Rostrum long, extending well beyond the tips of the terminal olfactory setae (Figs. 28A; 29E) lateral denticles absent . . . . **Alonella, 143**

139. Postabdomen bent abruptly behind anus; one or two teeth at inferoposterior angle.

Dunhevedia, 140

Postabdomen not bent abruptly behind anus **141**

140. Ventral margin of keel of labrum smooth; dorsal margin of body arched (Fig. 28F); up to 0.5 mm long; widely distributed . . . . **Dunhevedia crassa** King

Ventral margin of keel of labrum toothed; dorsal margin of body only slightly arched; up to 0.7 mm long; LA, TX . . . . **Dunhevedia serrata** Daday

141. Valves reticulated; one to three larger teeth at inferoposterior angle (Fig. 28C); LA, TX.

Alonella dentifera Sars

Valves striated; usually one to four small teeth at inferoposterior angle (Fig. 25H); up to 0.5 mm long.

142

142. Postabdomen expanded beyond anus; one to four (usually three) small teeth at inferoposterior angle (Fig. 25H); LA, TX, AR . . . . **Alonella karua** (King)

Postabdomen not especially expanded beyond anus; usually one minute tooth at inferoposterior angle; rostrum variable; rare but probably widely distributed.

Disparalona rostrata (Koch)

143. Rostrum very long and strongly recurved; valves reticulated; inferoposterior angle rounded, with several minute teeth (Fig. 29E); up to 0.3 mm long; in pools in vegetation; LA, TX.

Alonella dadayi Birge

Rostrum shorter and only slightly recurved; one small tooth at inferoposterior angle (Fig. 28A); up to 0.3 mm long; rare; northern states . . . . **Alonella nana** Baird

144. Body short; with four pairs of stout legs bearing branchial appendages (Fig. 29H); up to 1.5 mm long; common in northern lakes, ponds, and marshes.

POLYPHEMIDAE, Polyphemus pediculus (L.)

Body long; with six pairs of legs; branchial appendages absent; up to 17 mm long (Fig. 30); limnetic; common in northern states but found as far south as TX.

LEPTODORIDAE, Leptodora kindti (Focke)

REFERENCES

Anderson, B. G., and J. C. Jenkins. 1942. A time study of events in the life span of Daphnia magna. *Biol. Bull.* 83:260–272.

Angino, E. E., K. E. Armitage, and B. Saxena. 1973. Population dynamics of pond zooplankton. II. Daphnia ambigua Scourfield. *Hydrobiologia* 42:491–507.

Applegate, E. R., and J. W. Mullan. 1970. Ecology of Daphnia in Bull Shoals Reservoir. *Res. Rept. U.S. Bur. Sport Fish. Wildlife* 74:1–23.

Banta, A. M. 1939. Studies on the physiology, genetics, and evolution of some Cladocera. *Publ. Carnegie Inst. Wash.* 513:1–285.

Berg, K. 1931. Studies on the genus Daphnia O. F. Müller with especial reference to the mode of reproduction. *Vidensk. Medd. Dansk. Naturhist. Foren.* 92:1–222.

_____. 1934. Cyclic reproduction, sex determination and depression in the Cladocera. *Biol. Rev.* 9:139–174.

Birge, E. A. 1910. Notes on Cladocera. IV. *Trans. Wis. Acad. Sci. Arts and Lett.* 16:1017–1066.

Brandlova, J., et al. 1972. The Cladocera of Ontario with remarks on some species and distribution. *Can. J. Zool.* 50:1373–1403.

Brooks, J. L. 1946. Cyclomorphosis in Daphnia. I. An analysis of D. retrocurva and D. galeata. *Ecol. Monogr.* 16:409–447.

_____. 1953. A re-description of typical Daphnia clathrata Forbes and Daphnia arcuata Forbes. *Am. Midl. Nat.* 49:193–209.

_____. 1953a. Re-description of Daphnia pulex var. pulicaria Forbes, D. thorata F. and D. dentifera F. *Ibid.* 772–800.

_____. 1957. The systematics of North American Daphnia. *Mem. Conn. Acad. Arts and Sci.* 13:1–180.

Cannon, H. G., and F. M. C. Leak. 1933. On the feeding mechanism of the Branchiopoda. *Philos. Trans. R. Soc. London Ser. B,* 222:267–352.

Coker, R. E. 1939. The problem of cyclomorphosis in Daphnia. *Q. Rev. Biol.* 14:137–148.

Coker, R. E., and H. H. Addlestone. 1938. Influence of temperature on cyclomorphosis of D. longispina. *J. Elisha Mitchell Sci. Soc.* 54:45–75.

Davison, J. 1969. Activation of the ephippial egg of Daphnia pulex. *J. Gen. Physiol.* 53:562–575.

Deevey, E. S., and G. B. Deevey. 1971. The American species of Eubosmina Seligo (Crustacea, Cladocera). *Limnol. Oceanogr.* 16:201–218.

Dodson, S. J. 1981. Morphological variation of Daphnia pulex Leydig (Crustacea: Cladocera) and related species from North America. *Hydrobiologia* 83:101–114.

_____. 1988. Cyclomorphosis in Daphnia galeata mendotae Birge and D. retrocurva Forbes as a predator-induced response. *Freshwater Biol.* 19:109–114.

Duigan, C. A., and D. A. Murray. 1987. A contribution to the taxonomy of C. sphaericus sen lat. (Cladocera, Chydoridae). *Hydrobiologia* 145:113–124.

Dumont, H. J., and J. Pensaert. 1983. A revision of the Scapholeberinae (Crustacea: Cladocera). *Hydrobiologia* 100:3–45.

Edmondson, W. T. 1955. The seasonal life history of Daphnia in an arctic lake. *Ecology* 36:439–451.

Edmondson, W. T., and A. H. Litt. 1982. Daphnia in Lake Washington. *Limnol. Oceanogr.* 27:292–293.

Edwards, C. 1980. The anatomy of Daphnia mandibles. *Trans. Am. Microsc. Soc.* 99:2–24.

Eriksson, S. 1936. Studien über die Fangapparate der Branchiopoden nebst einigen phylogenetischen Bemerkungen. *Zool. Bid. Upps.* 15:23–287.

Evans, M. S. 1985. The morphology of Daphnia pulicaria, a species newly dominating the offshore southeastern Lake Michigan summer daphnia community. *Trans. Am. Microsc. Soc.* 104:223–231.

Frey, D. G. 1959. The taxonomic and phylogenetic significance of the head pores of the Chydoridae (Cladocera). *Int. Rev. gesamten Hydrobiol.* 44:27–50.

_____. 1962. Supplement to: The taxonomic and phylogenetic significance of the head pores of the Chydoridae (Cladocera). *Int. Rev. gesamten Hydrobiol.* 47:603–609.

_____. 1965. Differentiation of Alona costata Sars from two related species (Cladocera, Chydoridae). *Crustaceana* 8:159–173.

_____. 1973. Comparative morphology and biology of three species of Eurycercus (Chydoridae, Cladocera) with a description of Eurycercus macrocanthis sp. nov. *Int. Rev. gesamten Hydrobiol.* 58:221–267.

_____. 1974. Reassignment of Alonella fitzpatricki and A. leei Chien, 1970 (Cladocera, Chydoridae). *Trans. Am. Microsc. Soc.* 93:162–170.

_____. 1980. The non-swimming chydorid Cladocera of wet forests, with descriptions of a new genus and two new species. *Int. Rev. gesamten Hydrobiol.* 65:613–641.

_____. 1980a. On the plurality of Chydorus sphaericus (O. F. Müller) (Cladocera, Chydoridae) and designation of a neotype from Sjaelsø, Denmark. *Hydrobiologia* 69:83–123.

_____. 1982. Relocation of Chydorus barroisi and related species (Cladocera, Chydoridae) to a new genus and description of two new species. *Hydrobiologia* 86:231–269.

_____. 1982a. The honeycombed species of Chydorus (Cladocera, Chydoridae): comparison of C. bicornutus and C. bicollaris n. sp. with some preliminary comments on faviformis. *Can. J. Zool.* 60:1892–1916.

_____. 1982b. Questions concerning cosmopolitanism in Cladocera. *Arch. Hydrobiol.* 93:484–502.

_____. 1987. The taxonomy and biogeography of the Cladocera. *Hydrobiologia* 145:5–17.

_____. 1987a. The North American Chydorus faviformis (Cladocera, Chydoridae) and the honeycombed taxa of other continents. *Philos. Trans. R. Soc. London* 315:353–402.

Fryer, G. 1968. Evolution and adaptive radiation in the

Chydoridae (Crustacea: Cladocera): a study in comparative functional morphology and ecology. *Philos. Trans. R. Soc. London, Ser. B, Biol. Sci.* 254:221–385.

_____. 1968a. Tubular and glandular organs in the Cladocera, Chydoridae. *Zool. J. Linn. Soc.* 48:1–8.

_____. 1970. Defaecation in some macrothricid and chydorid cladocerans, and some problems of water intake and digestion in the Anomopoda. *Ibid.* 49:225–269.

_____. 1972. Observations on the ephippia of certain macrothricid cladocerans. *Zool. J. Linn. Soc.* 51:79–96.

_____. 1974. Evolution and adaptive radiation in the Macrothricidae (Crustacea: Cladocera): a study in comparative functional morphology and ecology. *Philos. Trans. R. Soc. London, Ser. B, Biol. Sci.* 269:137–274.

_____. 1987. A new classification of the branchiopod Crustacea. *Zool. J. Linn. Soc.* 91:357–383.

Gellis, S. S., and G. L. Clarke. 1935. Organic matter in dissolved and in colloidal form as food for Daphnia magna. *Physiol. Zool.* 8:127–137.

Goulden, C. E. 1968. The systematics and evolution of the Moinidae. *Trans. Am. Philos. Soc.* 58:(6):1–101.

_____. 1971. Environmental control of the abundance and distribution of the chydorid Cladocera. *Limnol. Oceanogr.* 16:320–331.

Goulden, C. E., and D. G. Frey. 1963. The occurrence and significance of lateral head pores in the genus Bosmina (Cladocera). *Int. Rev. gesamten Hydrobiol.* 48:513–522.

Harris, J. E., and P. Mason. 1956. Vertical migration in eyeless Daphnia. *Proc. R. Soc. London.* 145B:280–290.

Hasler, A. D. 1937. The physiology of digestion in plankton Crustacea. II. Further studies on the digestive enzymes of (A) Daphnia and Polyphemus; (B) Diaptomus and Calanus. *Biol. Bull.* 72:290–298.

Hebert, P. D. N. 1978. The population biology of Daphnia (Crustacea, Daphnidae). *Biol. Rev.* 53:387–426.

Hebert, P. D. N., and T. J. Crease. 1980. Clonal coexistence in Daphnia pulex (Leydig): another planktonic paradox. *Science* 207:1363–1365.

Hellors, W. K. 1975. Selective predation of ephippial Daphnia and the resistance of ephippial eggs to digestion. *Ecology* 56:974–980.

Hrbacek, J. 1959. Circulation of water as a main factor influencing the development of helmets in Daphnia cucullata Sars. *Hydrobiologia* 13:170–185.

Hutchinson, G. E. 1932. Experimental studies in ecology. I. The Mg tolerance of Daphniidae and its ecological significance. *Int. Rev.* 28:90–108.

Idris, B. A. G. 1983. *Freshwater zooplankton of Malaysia (Crustacea: Cladocera).* 153 pp. Penerbit Universiti Pertanian Malaysia.

Infante, A., and S. E. B. Abella. 1985. Inhibition of Daphnia by Oscillatoria in Lake Washington. *Limnol. Oceanogr.* 30:1046–1052.

Ingle, L., T. R. Wood, and A. M. Banta. 1937. A study of longevity, growth, reproduction, and heart rate in

Daphnia longispina as influenced by limitations in quantity of food. *J. Exp. Zool.* 76:325–352.

Jacobs, J. 1961. Cyclomorphosis in Daphnia galeata mendotae Birge, a case of environmentally controlled allometry. *Arch. Hydrobiol.* 58:7–71.

_____. 1962. Light and turbulence as co-determinants of relative growth rates in cyclomorphic Daphnia. *Int. Rev. gesamten Hydrobiol.* 47:146–156.

Kerfoot, W. C. 1975. The divergence of adjacent populations. *Ecology* 56:1298–1313.

_____. 1975a. Seasonal changes of Bosmina (Crustacea, Cladocera) in Frains Lake, Michigan; laboratory observations of phenotypic changes induced by inorganic factors. *Freshwater Biol.* 5:227–243.

Kiang, H. M. 1942. Über die Cyclomorphose der Daphnien einiger Voralpenseen. *Int. Rev.* 41:345–408.

Kubersky, E. S. 1977. Worldwide distribution and ecology of Alonopsis (Cladocera: Chydoridae) with a description of Alonopsis americana sp. nov. *Int. Rev. gesamten Hydrobiol.* 52:649–685.

Lilljeborg, W. 1900. Cladocera sueciae. *Nova Acta Regiae Soc. Sci. Ups. Ser. 3,* 19:1–701.

MacArthur, J. W., and W. H. T. Baillie. 1929. Metabolic activity and duration of life. I: Influence of temperature on longevity in Daphnia magna. *J. Exp. Zool.* 53:221–242.

McMahon, J. W., and F. H. Rigler. 1963. Mechanisms regulating the feeding rate of Daphnia magna Straus. *Can. J. Zool.* 41:321–332.

McNaught, D. C., and A. D. Hasler. 1963. Rate of movement of populations of Daphnia in relation to changes in light intensity. *J. Fish. Res. Board Can.* 21:291–318.

Megard, R. O. 1965. A chemical technique for disarticulating the exoskeletons of chydorid Cladocera. *Crustaceana* 9:208–210.

_____. 1967. Three new species of Alona (Cladocera, Chydoridae) from the United States. *Int. Rev. gesamten Hydrobiol.* 52:37–50.

Meijering, M. P. D. 1975. Notes on the systematics and ecology of Daphnia pulex Leydig in northern Canada. *Int. Rev. gesamten Hydrobiol.* 60:691–703.

Negra, S. 1983. Cladocera. *Fauna Rep. Soc. Romania, Crustacea* IV (12):1–399.

Pacaud, A. 1939. Contribution à l'écologie des Cladocères. *Bull. Biol. Fr. Belg., Suppl.* 25:1–260.

Pejler, B. 1973. On the taxonomy of limnoplanktic Daphnia species in northern Sweden. *Zoon* 1:23–27.

Pennak, R. W. 1957. Species composition of limnetic zooplankton communities. *Limnol. Oceanogr.* 2:222–232.

_____. 1973. Some evidence for aquatic macrophytes as repellants for a limnetic species of Daphnia. *Int. Rev. gesamten Hydrobiol.* 58:569–576.

Pratt, D. M. 1943. Analysis of population development in Daphnia at different temperatures. *Biol. Bull.* 85:116–140.

Ranta, E. 1979. Niche of Daphnia species in rock pools. *Arch. Hydrobiol.* **87**:205–223.

Ringelberg, J. 1964. The positively phototactic reaction of Daphnia magna Straus. A contribution to the understanding of diurnal vertical migration. *Neth. J. Sea Res.* **2**:319–406.

Schindler, D. W. 1968. Feeding, assimilation and respiration rates of Daphnia magna under various environmental conditions and their relation to production estimates. *J. Anim. Ecol.* **37**:369–385.

Schwartz, S. S. et al. 1985. Morphological separation of Daphnia pulex and Daphnia obtusa in North America. *Limnol. Oceanogr.* **30**:189–197.

Scourfield, D. J. 1942. The "pulex" forms of Daphnia and their separation into two distinct series represented by D. pulex (De Geer) and D. obtusa Kurz. *Ann. Mag. Nat. Hist.* **9**:202–219.

Scourfield, D. J., and J. P. Harding. 1966. A key to the British species of freshwater Cladocera. *Sci. Publ. Freshwater Biol. Assoc.* **5**:1–55.

Sebestyen, O. 1931. Contribution to the biology and morphology of Leptodora kindtii (Focke) (Crustacea, Cladocera). *Arb. Ung. Biol. Inst.* **4**:151–170.

Smirnov, N. N. 1966. Alonopsis (Chydoridae, Cladocera): morphology and taxonomic position. *Hydrobiologia* **27**:113–136.

_____. 1966a. Pleuroxus (Chydoridae): morphology and taxonomy. *Ibid.* **28**:161–194.

_____. 1968. On comparative functional morphology of limbs of Chydoridae (Cladocera). *Crustaceana* **14**:76–96.

_____. 1972. Detailed morphology of trunk limbs of some Aloninae. *Hydrobiologia* **40**:393–422.

_____. 1974. Chydoridae of the World's fauna. *Fauna of the U.S.S.R. Crustacea* **1**(2):1–644.

Smirnov, N. N., and B. V. Timms. 1983. A revision of the Australian Cladocera (Crustacea). *Rec. Aust. Mus. Suppl.* **1**:1–132.

Tappa, D. W. 1965. The dynamics of the association of six limnetic species of Daphnia in Aziscoos Lake, Maine. *Ecol. Monogr.* **35**:395–423.

Thomas, I. F. 1961. Review of the genera Pseudosida Herrick, 1884 and Latonopsis Sars, 1888 (Cladocera). *Crustaceana* **3**:1–8.

Ueno, M. 1927. The freshwater Branchiopoda of Japan. I. *Mem. Coll. Sci. Kyoto Imp. Univ. Ser. B*, **2**:259–311.

Ward, H. B., and G. C. Whipple. 1918. *Fresh-water Biology.* 1111 pp. Wiley, New York.

Wesenberg-Lund, C. 1926. Contributions to the biology and morphology of the genus Daphnia with some remarks on heredity. *Mem. Acad. R. Sci. Lett. Dan., Sect. Sci., Ser. 8,* **11**:89–251.

Wolf, H. G. 1987. Interspecific hybridization between Daphnia hyalina, D. galeata, and D. cucullata and seasonal abundance of these species and their hybrids. *Hydrobiologia* **145**:213–217.

Woltereck, R. 1932. Races, associations and stratification of pelagic daphnids in some lakes of Wisconsin and other regions of the United States and Canada. *Trans. Wis. Acad. Sci. Arts and Lett.* **27**:487–522.

Wood, T. R. 1938. *Activation of the dormant form of freshwater animals, with special reference to Cladocera.* 87 pp. Unpublished thesis, Brown University, Providence, Rhode Island.

Wood, T. R., and A. M. Banta. 1937. Hatchability of Daphnia and Moina sexual eggs without drying. *Int. Rev. gesamten Hydrobiol.* **35**:229–242.

Zaret, T. M., and W. C. Kerfoot. 1975. Fish predation on Bosmina longirostris: body-size selection versus visibility selection. *Ecology* **56**:232–237.

17

COPEPODA

Like the Cladocera, the Copepoda are almost universally distributed in the plankton, benthic, and littoral regions of fresh waters. They are either absent or present in small numbers in rapid headwater brooks and streams. Including marine and fresh-water species, there are probably 8000 to 10,000 described species world-wide.

General characteristics. The body length of the American species ranges from 0.3 to 3.2 mm, but the great majority are less than 2.0 mm long. Most species are drab grayish or brownish, but others, especially littoral species in the spring of the year and at high altitudes, are brilliant orange, purple, or red. Such colorations are usually produced by a wide variety of carotenoids, and are thought to be a means of photoprotection in some species. Conversely, pigmented individuals are more prone to fish predation. As a group, the members of the Subclass Copepoda are much more homogeneous in their general structure than the Cladocera, and accurate identification is based largely on anatomical details of the appendages.

The American fresh-water copepods comprise two orders, Eucopepoda and Branchiura, and six suborders: Caligoida, Lernaeopodoida, Arguloida, Calanoida, Cyclopoida, and Harpacticoida. The first three of these suborders are exclusively parasitic, and the morphology is greatly modified and specialized. In the three free-living suborders, however, the body is clearly segmented, more or less elongated, cylindrical, and is divided into a head, thorax, and abdomen. The thoracic region is here considered to be composed of seven segments, but the first one or two are fused with the head to form a cephalothorax, or cephalosome, which is covered with a carapace. In addition, the fourth and fifth or fifth and sixth thoracic segments are also occasionally fused. The abdomen consists of three to five segments, typically four. Commonly the last thoracic (genital) segment and the first abdominal segment are fused. Conversely, there is subdivision of some segments in a few species.

Each segment is a rigid, sclerotized cylinder that is attached to adjacent segments by short, flexible, annular portions. One of the articulations is particularly movable, and, for convenience, the copepod body is often divided into metasome and urosome, the former being that part anterior to this articulation and the latter posterior (Fig. 1). The urosome includes all the abdomen, the seventh thoracic (genital) segment, and sometimes the sixth thoracic segment.

Appendages. The true head has five pairs of appendages: first antennae (antennules), second antennae, mandibles, first maxillae, and second maxillae. The first thoracic segment (fused with the head) bears a pair of maxillipeds, and each of the five subsequent thoracic segments bears one pair of swimming legs. In a few species the seventh thoracic (genital) segment bears a pair of vestigial swimming legs. There are no abdominal appendages.

In Figs. 1A and C the first two thoracic segments are fused with the head; the metasome therefore consists of cephalothorax and free thoracic segments III to V; the urosome consists of thoracic segment VI, thoracic segment VII (fused with abdominal segment I), and free abdominal segments II to IV. In Fig. 1B only the first thoracic segment is fused with the head; the metasome therefore consists of the cephalothorax and free thoracic

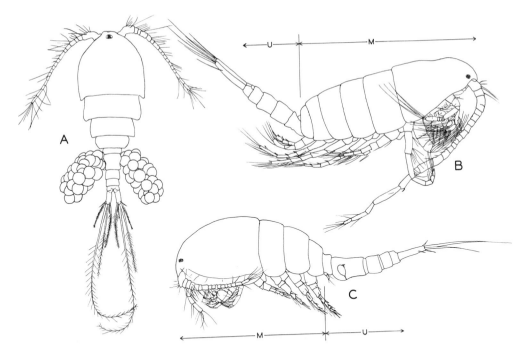

FIG. 1.–Typical copepods. A, female *Macrocyclops ater* with egg masses, ×18; B, male *Limnocalanus macrurus,* ×33; C, female *Cyclops,* ×42. M, metasome; U, urosome. (A modified from Coker, 1938; B modified from Gurney, 1931–1933.)

segments II to VI; the urosome consists of thoracic segment VII (genital segment) and abdominal segments I to IV.

Some investigators consider the maxillipeds to be a part of the true head, so that there are only six thoracic segments. Others consider the genital segment to be a part of the abdomen.

In studying species with confusing fusion or subdivision of the thoracic segments, it is easy to orient oneself by locating the first and subsequent swimming legs. They correspond to thoracic segments II to VI. Fundamentally, all head and thoracic appendages are biramous, and a primitive copepod swimming leg consists of a three-segmented basipod having attached at its distal end a median three-segmented endopod and a lateral three-segmented exopod. A typical and relatively unmodified swimming leg of a cyclopoid is shown in Fig. 4.

The appendages of the head are highly modified for various special functions. The first antennae are uniramous and long, con-

sisting of as many as 25 segments; they are sensory appendages but are used also for locomotion. In male cyclopoids and harpacticoids both first antennae are geniculate and modified for copulation (Figs. 3, 5A). In male calanoids only the right is geniculate (Fig. 1).

The second antennae (Fig. 5C) are much shorter, either biramous or uniramous, and are probably most important as sensory structures, although they are prehensile in male harpacticoids. The mandibles, first maxillae, second maxillae, and maxillipeds (occurring in that order) are highly modified for feeding (Figs. 5D–G).

The segmentation of the legs varies considerably from the primitive condition, and the modifications are most pronounced in the last two pairs. In the free-living species the sixth legs are always lacking in the female and are either rudimentary or lacking in the male. The fifth legs are reduced or vestigial in both sexes in the Cyclopoida and Harpacticoida, but in the Calanoida they are well developed and symmetrical in the female and asymmetri-

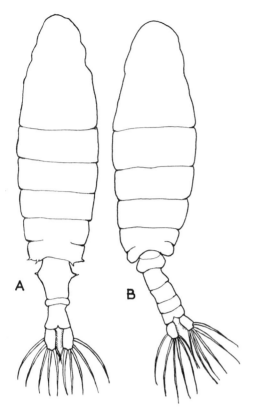

FIG. 2.–Dorsal view of *Leptodiaptomus siciloides* (appendages and most setation omitted). A, female; B, male. (From Comita and Tommerdahl, 1960.)

cal and modified for clasping in the male (Figs. 11–16).

The last abdominal (anal) segment bears two posteriorly directed caudal rami. These are simple, more or less cylindrical structures that are neither biramous nor serially homologous with the head and thoracic appendages.

In general, there are five types of slender outgrowths from the exoskeleton. One type, the aesthetasks, presumably function as sensory receptors. They are delicate, elongated, blunt tipped, and occur on the first and second antennae. Aesthetasks are shown in Fig. 5A. The other five types of outgrowths occur on all copepod appendages and less commonly on the body segments; they form a morphological series: pegs–spines–spinules–setae–hairs. Spines and spinules are generally short, stout, strong, and inflexible. Setae, on the other hand, are relatively long, flexible, and thin. Hairs are the very finest outgrowths and they usually occur as fringes or rows on setae and spines. None of the categories spines, spinules, setae, and hairs can be precisely defined to the exclusion of the others, and there are wide variations in size and all possible and imperceptible intergrades. All of them, however, may have one or several functions, depending on their location. Generally they aid in swimming, crawling, and feeding, but some are sensory, and it is possible that the long setae of the caudal rami may act as balancers or stabilizers during locomotion. Pegs are a type of sensory organelle that can be distinguished only with the electron microsccope. The same is true for many fine hairs. Sensillae, another type of organelle, are very short triangular sensory spines found especially on the trunk and legs.

Feeding. Harpacticoid mouth parts are adapted from raking, seizing, and scraping food from the bottom. Calanoids feed chiefly by filtration of plankton. The first antennae function as screws and the second antennae beat continuously to produce a current, from which the particles are filtered by the mouth parts, chiefly the maxillae. There is considerable evidence to show that calanoids exhibit some selectivity in the size and kind of algae ingested, and species differ markedly in the sizes of particles that they remove from the water. Differential feeding has been called the "leaky sieve" concept. There is also some evidence to show that calanoids combine raptorial feeding with filter feeding. Cyclopoids have the mouth parts modified for seizing and biting. The food consists mostly of unicellular plants and animals, small metazoans, and especially other crustaceans, as well as organic debris; and it has now been well established that debris may, under some circumstances, form the majority of material ingested. Cannibalism on immature stages is common. The course of the food through the digestive tract can easily be seen in the living animal under the microscope. The anus is on the last abdominal segment between the bases of the caudal rami. The accumulation of reserve food in the form of oil globules is usually chiefly responsible for the brilliant red coloration of some diaptomids.

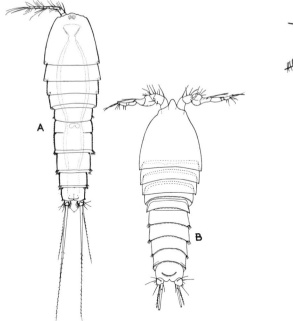

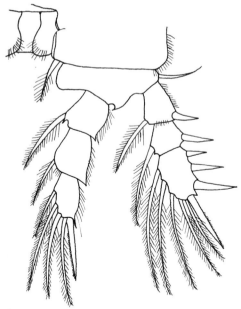

FIG. 3.–Harpacticoid copepods. A, female *Attheyella*; B, male *Attheyella*. (From Coker, 1934.)

FIG. 4.–Typical swimming leg (right, second) of a cyclopoid, showing two-segmented basipod, three-segmented exopod (right), and three-segmented endopod (left). (Redrawn from Gurney, 1931–1933.)

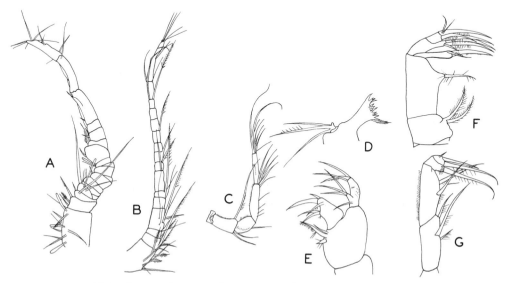

FIG. 5.–Copepod head appendages. A, first antenna of male *Mesocyclops leuckarti*; B, first antenna of female *Macrocyclops fuscus*; C, second antenna of same; D, mandible of same; E, first maxilla of same; F, second maxilla of same; G, maxilliped of same. (Redrawn from Gurney, 1931–1933.)

Locomotion. Harpacticoids crawl or run on, and to some extent in, the substrate, and although many cyclopoids and a few calanoids also occur in close association with the substrate, they are characteristically swimmers. The swimming mechanisms of *Diaptomus* have been carefully analyzed by Storch (1929) and Kerfoot et al. (1980). A constant, smooth, and relatively slow type of locomotion is produced by the feeding movements of the mouth parts and second antennae. This is punctuated at frequent intervals by a jerky type of locomotion caused by rapid backward movements of the legs. Between beats of the legs the first antennae are kept thrust out stiffly and laterally; they are thus thought to act chiefly as parachutes that greatly retard sinking. A fraction of a second before each beat of the legs, however, the first antennae are bent medially into the general longitudinal plane of the body, the beat of the legs thereby being much more effective. Immediately following each beat of the legs, the antennae are again extended laterally. According to Storch, a single beat of the legs and the accompanying antennal movements are completed in less than one twelfth of a second. Cyclopoid mouth parts are of little importance, locomotion being brought about by relatively frequent beats of the legs. The accompanying antennal movements are similar to those of diaptomids. In both cyclopoids and calanoids the abdomen is thought to function as a rudder.

Internal anatomy. Internal details are generally difficult to study because they are largely obscured by the very complicated system of muscles. A saccular heart occurs only in the Calanoida, and in the other suborders the blood simply circulates about in the hemocoel as the result of the movements of muscles, appendages, and the digestive tract. The oxygen and carbon dioxide exchange occurs through the general body surface and probably to some extent through the posterior part of the digestive tract where water is drawn in and out by extrinsic muscles. Excretion is thought to occur through the maxillary glands near the anterior end, but it is possible that the general body surface and the posterior part of the digestive tract may function in excretion also. The nervous system

exhibits considerable fusion and cephalization. Aside from tactile structures, the only recognizable sensory area is the eyespot. In free-living species this is a small, median dorsal, clear or pigmented area at the anterior end of the cephalothorax. With the aid of a scanning electron microscope, however, Strickler (1975) reports an abundance of "sensory pegs" and "sensory hairs" scattered over the copepod body.

Reproduction. Although reproductive habits are similar throughout the free-living copepods, the various species differ in their detailed behavior and breeding periods. Calanoids, especially, have complicated mating behaviors. Some species, such as *Acanthocyclops vernalis* and *Eucyclops agilis,* reproduce throughout the year; *Tropocyclops prasinus* breeds between July and October. Other species are monocyclic and breed for short periods: *Macrocyclops fuscus* and *M. ater,* for example, reproduce only during the summer months, and some species of diaptomids and *Limnocalanus macrurus* have only one generation per year. Little is known about the breeding habits of harpacticoids. Some have only one generation per year, but others undoubtedly have three or more. Three species of harpacticoids are known to reproduce by parthenogenesis.

The antennae and, in some genera, the modified fifth legs of the male, are used in clasping the female. The period of clasping may last only a few minutes or it may persist for several days. Sometimes the male clasps the female before she has had her final molt, and such clasping pairs have been observed for as long as 10 days.

Male calanoids have a single pore located asymmetrically on the genital segment, but female calanoids and both sexes in the other free-living suborders have paired genital pores. During clasping, the male transfers the sperm to the female in small, packetlike spermatophores, usually with the aid of the legs. The sperm are stored in a special ventral area of the female genital segment which serves as a seminal receptacle.

Actual fertilization occurs sometime after the two sexes have separated and as the eggs leave the female reproductive tract. This process may be completed within a few minutes or as long as two months after copulation. Fertilized eggs are carried by the female in one or two ovisacs (Fig. 1A). Ovisacs usually contain from 5 to 40 eggs each and are attached to the genital segment ventrally, laterally, or subdorsally. In some species the clutch size varies seasonally, with the largest numbers of eggs being produced in the spring months. Possibly this variation is correlated with temperature or food conditions.

In cyclopoids the approximate incubation period ranges from 12 hours to 5 days, and when the newly hatched larvae leave the disintegrating ovisacs the female releases another group of eggs from the reproductive tract; these, in turn, are fertilized and retained in a new pair of ovisacs. In this way from 7 to 13 pairs of ovisacs full of eggs may be produced successively at intervals of 1 to 6 days, all resulting from the sperm retained from a single copulation. Sometimes unfertilized females release sterile, nonhatching eggs, and a single European species of *Attheyella* is known to reproduce by parthenogenesis.

Although it is believed that ordinary fertilized eggs may occasionally overwinter in an extended incubation period, adverse environmental conditions are usually withstood by special thick-walled "resting" eggs. Such eggs do not occur in the cyclopoids, but they are known to be produced by several species of diaptomids and some harpacticoids, and it is

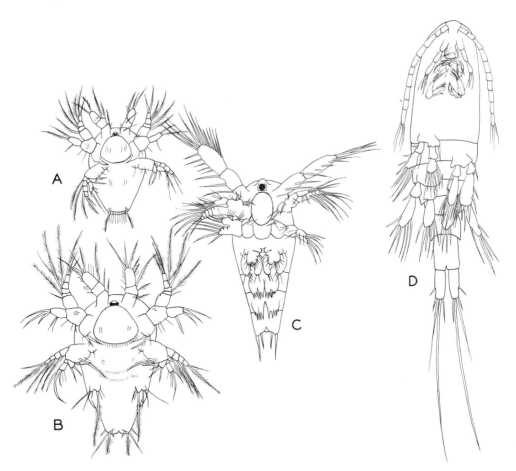

Fig. 6.–Immature copepods. A, cyclopoid nauplius I, ×100; B, cyclopoid nauplius IV, ×100; C, diaptomid nauplius VI, ×80; D, diagram of cyclopoid copepodid, ×90 (only alternate legs are shown on each side). (A and B redrawn from Gurney, 1931–1933; C modified from Gurney, 1931–1933.)

the only type of egg produced by *Limnocalanus macrurus.*

Metamorphosis. The copepod egg hatches into a small, compact, active larva called a nauplius, which has three pairs of abbreviated appendages representing the first antennae, second antennae, and mandibles (Fig. 6A). After a period of feeding, it molts and becomes the nauplius II stage, which usually has the first maxillae in addition. In a similar manner there are four additional nauplius stages and five to seven copepodid stages before the last molt, which results in the sexually mature adult. As metamorphosis proceeds, the larvae become progressively larger and more elongated, and acquire additional appendages (Figs. 6B–D). Nauplius VI, for example, usually has all appendages through the second pair of legs, and copepodid I has four distinct thoracic segments and all appendages through the fourth legs.

In summary, the typical life cycle consists of the egg, six nauplius stages, five to seven copepodid stages, and the adult. However, in all parasitic species and some free-living species this series is abbreviated. Thus, there are five nauplius stages in some Cyclopoida, and only four or five in certain Harpacticoida.

The time necessary for the complete life cycle from egg to egg is highly variable, depending on the species and environmental conditions. In cyclopoids it lasts from 7 to about 180 days, whereas some species of diaptomids and *Limnocalanus macrurus* have a 1-year cycle.

Some cyclopoids are known to reproduce year round, even in the winter months, although maturation and growth proceed very slowly in cold water. Thus a copepod may have four or more generations during the eight warmest months of the year but only one generation durng the four coldest months. Some cyclopoid copepods regularly go into an inactive diapause on lake and pond bot-

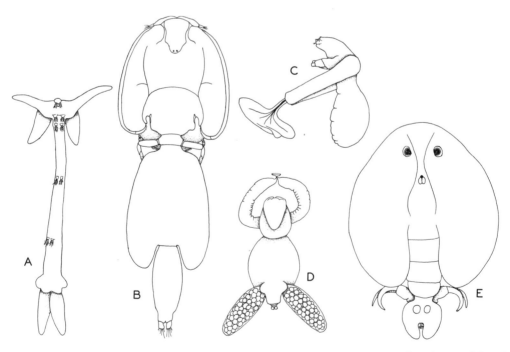

FIG. 7.–Parasitic copepods. A, ventral view of female *Lernaea cruciata,* ×6.5; B, dorsal view of female *Lepeophtheirus salmonis,* ×6; C, lateral view of female *Salmincola inermis,* ×6; D, dorsal view of female *Achtheres microptera,* ×8; E, dorsal view of female *Argulus flavescens,* ×10. (A and E modified from Wilson, 1944; B modified from Gurney, 1931–1933; C and D modified from Wilson, 1915.)

toms during the cold months. This is simply a nonencysted dormant period found in the copepodid II to the copepodid V stages, especially after the fall overturn. Most commonly diapause occurs in copepodid IV. In the spring these stages resume their active existence when the water becomes warmer. Concentrations of more than a million diapausing copepods per square meter of bottom have been reported.

Evidence is accumulating to show that some cyclopoids may have a 2- or 3-year life cycle.

Considerably less is known about the life cycles of harpacticoids. Littoral populations of *Canthocamptus staphylinus,* however, have the greatest reproductive interval in January. The new generation becomes adult in May or June, and the adults then aestivate as cysts, to emerge again in the autumn. Apparently encystment is induced by high temperatures and long days.

Parasitic copepods. Copepod parasites occur on most common species of American fish and they are among the most highly modified and bizarre of all fresh-water animals. Adult stages of some genera do not even resemble copepods, and, indeed, some investigators insist that they are not copepods. They are found on the general body surface, fins, and gills, and obtain nourishment from the tissues of the host.

Only a single genus, *Ergasilus,* of the Suborder Cyclopoida is parasitic and, incidentally, it is the least specialized for its parasitic existence. Only the adult female is parasitic, the immature females and all stages of males being free living. Superficially, the female looks like a *Cyclops* but may be easily distinguished by the greatly enlarged three-segmented clawlike second antennae (Fig. 8) used for clinging to the gills of fish where it feeds on blood and gill tissues.

The Caligoida is a relatively large suborder composed entirely of parasites. Only one species in one genus and about a dozen species in a second genus occur in American fresh waters, all others being parasitic on marine fishes. The former, *Lepeophtheirus salmonis* (the sea louse), is scarcely recognizable as a copepod in the adult stage (Fig. 7B); there is a large, broad, flat suckerlike cephalo-thorax, and most of the segments are fused, and the legs reduced. This species is only an incidental one in fresh waters; it is found attached to the surface of the body near the anus of salmon which have recently migrated from the sea, and may live only a week or two in fresh water. Coloration is blackish and they are thus easily seen. Males are very rare. *Lepeophtheirus* feeds by rasping the surface tissues.

The other genus of the Suborder Caligoida occurring in the United States is *Lernaea.* The nauplii are free-livng plankters, but the copepodids seek out a temporary host fish, cling to the gills, and copulate in this immature stage. Soon thereafter the male copepodid dies, but

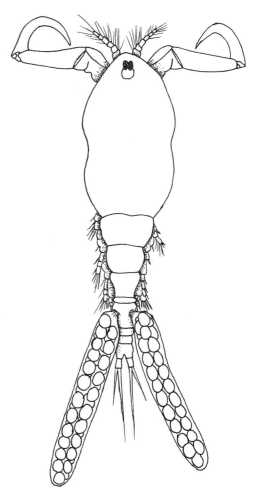

FIG. 8.–Dorsal view of female *Ergasilus versicolor,* ×70. (Modified from Wilson, 1911.)

the female leaves the host and is free living for a short time. She then attaches to the general body surface of any one of a great variety of fresh-water fishes and becomes completely altered morphologically into a long wormlike creature. The anterior half of the body is buried in the superficial host tissues and produces several large anchoring processes (Fig. 7A), which cause inflammation and suppuration.

The genus *Argulus* of the parasitic Suborder Arguloida is the only one occurring in the United States. There are about 15 species parasitizing many fresh-water fishes and occasionally amphibians. The adults, familiarly known as fish lice, range from 5 to 25 mm in length and are found attached in the branchial chamber or on the general body surface. Fish lice (Figs. 7E; 9) are greatly flattened dorsoventrally, and the head and thorax are covered with an expanded carapace whose lateral areas have a respiratory function. The abdominal segments are fused. The mouth parts are greatly modified or reduced; the second maxillae are completely transformed into curious suction cups; and the region of

the mouth has a piercing organ used in obtaining the host blood. The first four pairs of legs are well developed and used for swimming, but the fifth and sixth legs are absent. Dorsally, there are two large movable compound eyes. *Argulus* often leaves its host and swims about freely, especially during the breeding season when the eggs are attached to stones and sticks of the substrate. There are seven to nine immature stages.

Of the parasitic Suborder Lernaeopodoida, only *Achtheres* and *Salmincola* occur in American fresh waters. These two genera are found in the gills and fins of many fishes and are perhaps the ultimate among Copepoda in modification for the parasitic mode of life. The soft-bodied adult females (Figs. 7C, D) have no legs, little evidence of segmentation, and greatly reduced mouth parts. The second maxillae, however, are large cylindrical structures that have a single saucer- or umbrella-shaped organ at their distal end. The latter is imbedded in the host tissues and functions in absorbing nutriment. Little is known about the males; they are dwarfed, attached to the host fish during the immature stages, but

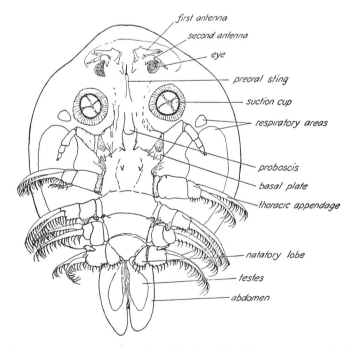

FIG. 9.–Ventral view of *Argulus,* ×23. (Modified from Meehean, 1940.)

cling to the body of the female before and after copulation. The nauplii stages are passed within the egg, but the first copepodid is a free-swimming plankter for a brief period before becoming attached to a fish.

Economic significance.

Under natural conditions parasitic copepods are rarely present in sufficient numbers to cause serious injury to the host. In hatchery ponds, however, where fish are crowded in a limited area, there is much greater opportunity for the free-swimming immature stages to find a host, and, as a result, heavy and serious infections may break out. When firmly established, copepod parasites are difficult to control, chiefly because the sclerotized exoskeleton of the adult is resistant to chemical solutions. Newly attached larval parasites are relatively delicate, however, and sometimes it is feasible to combat infections by giving the fish frequent salt or acid baths. In extreme cases there is no alternative but to get rid of the parasitized fish, clean the hatchery ponds, and start anew.

Free-living copepods constitute an essential link in the aquatic food chain. They are on an intermediate trophic level between bacteria, algae, and protozoans, on the one hand, and small and large plankton predators (chiefly fish), on the other. In general, however, fresh-water copepods are not as important an element in the fish diet as Cladocera.

Copepods are of further importance as intermediate hosts of parasites of higher animals, particularly some tapeworms of fish, waterfowl, and carnivores; some flukes of fish, amphibians, and birds; and a few nematodes of fish and birds. In Africa, southern Asia, the West Indies, and parts of the Guianas and Brazil, *Cyclops* is of great importance in the transmission of the nematode parasite of man, *Dracunculus medinensis* (the Guinea worm), and some species of both cyclopoids and diaptomids are intermediate hosts of the broad or fish tapeworm of man and numerous carnivores.

Recently it has been shown that non-parasitic cyclopoid copepods may occasionally be a nuisance in hatcheries. They may merely nibble at the fins of fingerling fish, or they may be so abundant as to feed on and kill the fingerlings.

Ecology.

Of the free-living copepods, the limnetic cyclopoid and calanoid species are by far the best known and most frequently collected. Sometimes they have been found in almost unbelievable numbers, up to 1000 individuals per liter having been recorded. In most lakes the limnetic copepod plankton is monotonous; usually it is composed of one dominant cyclopoid, such as *Diacyclops bicuspidatus, Acanthocyclops vernalis,* or *Mesocyclops leuckarti.* Often there is a dominant calanoid present also. In exceptional instances a limnetic population may consist of two or more cyclopoid species and two or more calanoid species all at the same time. Under such circumstances one cyclopoid is the numerical dominant and one calanoid is the numerical dominant. Coexistence of two species in the same genus is usually explained by vertical, seasonal, and food particle size preferences.

Perhaps because of the greater variety of ecological niches, the shallow littoral, especially in aquatic vegetation, is much richer in cyclopoid and calanoid species, but the population density is not usually high.

In rivers and lakes the Harpacticoida are restricted to bottom debris of both littoral and benthic regions. In addition, however, there are many Cyclopoida and Calanoida that have become adapted to this habitat. Common bottom forms are *Paracyclops fimbriatus, Eucyclops agilis, Macrocyclops fusus, M. albidus,* and species of *Canthocamptus, Bryocamptus,* and *Attheyella.* Very few copepods are found in rapid streams.

As ecological groups, the deep-water and limnetic species are somewhat distinct from those occurring in the shallow littoral. The former are generally smaller, slender, and more translucent, whereas the latter are larger, stout, more deeply pigmented, and have more robust appendages.

Harpacticoid copepods have been reported from some unusual habitats. Large numbers (up to $277/10$ cm^3 of sand) have been found in the interstitial waters of sandy lake beaches, with greatest concentrations at depths of 1 to 6 cm, and at distances of 1.0 to 2.5 m from the water's edge (Pennak, 1940). European workers have described unusual endemic species from springs, underground streams, and cave systems, but little collecting has been done in such habitats in this country. Harpacticoids have also been reported from damp moss in forests far from any body of water.

Small cyclopoids and harpacticoids form important elements of the meiobenthic communities of streams and rivers, especially in hyporheic and shoreline gravels (Pennak and Ward, 1985). This copepod fauna is so poorly known, however, that it is meaningless to include the few American species in keys in this chapter. Many new species await collection and description. European investigators have often reported collecting common substrate cyclopoids and cyclopoids in hypogeic habitats, including some species that commonly occur in bottom fauna collections in both Europe and United States (e.g. Lescher-Moutoué, 1973).

It is thought that temperature plays an important role in copepod distribution and activity. Some species, such as *Limnocalanus macrurus, Senecella calanoides, Epischura lacustris, Leptodiaptomus minutus, L. sicilis,* and *Skistodiaptomus oregonensis,* are definitely cold-water forms occurring only in the northern states and in deep lakes. Others, such as *S. mississippiensis* and *Arctodiaptomus dorsalis,* are restricted to the southern states. The rare *Acanthocyclops crassicaudis* occurs only in temporary, warm pools.

In some interesting experiments, Coker (1933) and Aycock (1942) reared *Acanthocyclops vernalis* and *Eucyclops agilis* under controlled conditions and found that at low temperatures (7.7 to 8.0°C) adults were as much as 50% longer than adults that had been reared at high temperatures (28.0 to 30.0°C). These workers believe that such size variations with temperature are internally controlled and are not dependent on the available food supply.

In general, copepods are much more tolerant of a deficiency of oxygen than are Cladocera. Cyclopoids and *Canthocamptus* have been collected many times from the surface of bottom muds in stratified lakes during summer and winter periods of stagnation and oxygen depletion. Although it is possible for cyclopoids to swim up into the upper oxygenated layers intermittently, *Canthocamptus,* not being a swimmer, is necessarily restricted to the substrate. The whole problem of the metabolism of copepods and other organisms in anaerobic benthic regions deserves much more attention.

Some adult or advanced copepodids may form cysts or cocoons during unfavorable environmental conditions, and the fact that cysts have been found in abundance on lake bottoms in midsummer is an indication that they may be formed as a response to anaerobic conditions. In cyclopoids, cysts also constitute an aestivation mechanism and are resistant to drying. The cyst wall is composed of a secreted organic membrane and an enveloping mass of detritus.

Like the Cladocera, some common species of plankton copepods show a daily rhythmic cycle of vertical migrations in lakes, with a greater concentration of individuals in the upper waters during the hours of darkness and correspondingly large numbers in the bottom waters during the hours of daylight. Individuals of some species may move upward at dusk and downward at dawn a vertical distance of as much as 10 or 20 m. It is believed that the primary stimulus for vertical migrations is the daily cycle of subsurface illumination.

Geographical distribution. The Copepoda undoubtedly originated in the ocean, and in this environment they have undergone much greater evolutionary specialization and speciation than in fresh water. The accompanying physiological divergence has been quite complete, and only a very few American species found in fresh water are known also to inhabit brackish or salt water. Some of these are *Limnocalanus macrurus, Eurytemora affinis, Harpacticus gracilis, Nitocra spinipes,* and *Cletocamptus.*

Limnocalanus macrurus is generally considered a relict species, that is, a saltwater form that has gradually become acclimated to fresh water. It is found only in the deep waters of larger lakes such as Green Lake (Wisconsin), the Finger Lakes, and the Great Lakes. It is also widely distributed in similar European lakes and in salt water. *Eurytemora* has close relatives in salt water and is probably a recent migrant. *Senecella* is thought to be of postglacial origin; it has been reported from several localities in Siberia, including some saline lakes. *Epischura* is confined to North America except for two Asiatic species.

Cyclops and its relatives have a worldwide distribution, some species, such as *Diacyclops bicuspidatus,* being essentially cosmopolitan. The closest marine and brackish relative is probably *Halicyclops.* Europe has about 110 species of cyclopoids, of which about half are known only from subterranean and interstitial waters.

The Diaptomidae are confined to fresh waters, and there are no close marine relatives. Unlike the cyclopoids, they show much evidence of recent speciation, and many of them seem to be confined to small geographic areas; numerous examples are given in the keys. Europe has about 70 species of calanoid copepods. Only 10 of these occur also in the United States.

Most fresh-water harpacticoids belong to the family Canthocamptidae, a group that has a worldwide distribution in salt, brackish, and fresh waters. Many species occur in both North American and European fresh waters, but so little systematic collecting has been done on the other continents that there are insufficient records to warrant the use of the term "cosmopolitan" for them. The great majority of species in the other harpacticoid families covered by this manual are typical of coastal, brackish, and saline waters.

Unquestionably, active migration is responsible for the geographic dispersal of species throughout a single drainage system, but overland transport to adjacent systems presents some difficult questions. Calanoida and Harpacticoida are probably dispersed as resting eggs, and although cysts are produced by some of the latter, they are not resistant to drying. The Cyclopoida, on the other hand, are not known to produce resting eggs, but resistant copepodid cysts are presumably of common occurrence, and it is likely that passive dispersal is brought about in this stage. It is significant that the first cyclopoids to appear in dried mud cultures are in the copepodid stages. Many copepods are found year after year in vernal ponds which are completely dried up during the greater part of the year, and we may safely assume that resting eggs and cysts in the dry mud and debris tide the species over.

Although passive transfer by waterfowl, aquatic insects, and in wind-blown dust explains the extensive distribution of some species, it is difficult to account for the fact that others are restricted to very small geographic areas, sometimes a single pond.

Collecting. Littoral and plankton copepods are easily collected with a townet or dipnet. An abundance of species may often be obtained by drawing nets through rooted aquatic vegetation and by lightly skimming the bottom. Collecting harpacticoids is a more tedious procedure. The top centimeter or two of mud and debris should be scooped up, brought into the laboratory, and allowed to settle. If the material is examined by looking toward a light source along the surface of the settled mud, the harpacticoids can be seen moving about at the mud–water interface and can be removed with a long pipette. Sometimes washings of aquatic mosses produce surprising numbers of specimens.

Phreatic and other interstitial species are easily collected from lakeside or streamside situations by digging a hole up to a meter deep and straining the accumulating water through a fine net. Hyporheic species may be taken with a Bou-Rouch sampler in shallow water (see figure on page 610). The Bou-Rouch sampler is also useful as a semiquantitative device for lakeside and streamside sampling.

Preservation. The best specimens are obtained by killing in 95% alcohol. Seventy percent alcohol is a suitable preservative and much superior to formalin solutions.

Culturing. Copepods are much more difficult than Cladocera to maintain as laboratory cultures, and they apparently never do well in dense numbers. Cyclopoids should be fed large protozoans or small metazoans, preferably crustaceans. Calanoids will sometimes thrive if a bit of algal culture suspension is added to the culture twice a week. Most copepods do best in cultures kept below 15°C.

Identification. It is unfortunate that the identification of species is usually based on minute anatomical details. Some characters, such as the structure of the antennae, first legs, and caudal rami, may be studied from the whole animal, but it is often necessary to dissect the anmal in order to study the mouth parts and legs, and it is an imperative procedure in formulating descriptions of new species. Harpacticoids must be dissected more frequently because of their general opacity and smaller appendages. Specimens to be dissected should be placed in a solution of 20% glycerin in 70% alcohol in a partially covered watch glass. After one to several days, most of the alcohol and water will have evaporated, leaving the undistorted specimens in glycerin. Then they may be easily picked up and transferred with an Irwin loop. The animals should be dissected, one at a time, in a drop of glycerin on a slide on the stage of a high power binocular microscope. The necessary appendages are removed with a pair of sharp, mounted "minuten nadeln." As a fair warning, it should be said that requisite skill and patience in dissection come only with considerable practice! Appendages and whole animals are conveniently and permanently mounted for examination in glycerin jelly under a cover slip ringed with Murrayite. Mounting in fluid glycerin is more

of a nuisance and less permanent. Immature forms of the great majority of species cannot be identified.

If an abundance of specimens is available, the writer has found that all of the necessary anatomical details can usually be seen, even in difficult species, if 5 to 10 whole specimens are mounted, ventral side up, on the same slide. The appendages will become spread out sufficiently in a variety of ways so that all structures can be observed by studying several specimens under the high powers of the compound microscope.

Sometimes it is helpful if the animals are stained before dissection and mounting. From 70% alcohol they should be transferred to alum cochineal overnight, washed in 70% alcohol for 5 minutes, and then transferred to the glycerin–alcohol mixture.

Body length extends from the anterior margin of the cephalothorax to the posterior end of the caudal rami. Accurate measurements of body length are of limited use as a taxonomic character because of the fact that the trunk segments may be more or less retracted into each other.

KEY TO MAJOR TAXA OF COPEPODA

1. With no constriction, a slight constriction, or a marked constriction between the segments bearing the fourth and fifth legs (Figs. 1A, C; 3; 9); first antennae of female with 5 to 18 segments; depressed or cylindrical; parasitic, commensal, free swimming, or benthic 6
   With a marked constriction between segments bearing fifth legs and genital segment (Fig. 1B); first antennae of female with 22 to 25 segments; free living or fixed parasites.
   Order **EUCOPEPODA**, 2
2. With a movable articulation between sixth and seventh thoracic segments (Fig. 1B); segmentation well defined; male left antenna geniculate in *Senecella,* male right antenna geniculate in all other genera; free living .... Suborder **CALANOIDA**, p. 428
   Body rigidly fused, without movable articulations, often without segmentation; fixed parasites of fishes 3
3. Adults long and wormlike (Fig. 7A); on gills and general external surface of many common fresh-water fishes; copepodid clings to gills of temporary host fish, then leaves and is free swimming for a short time before becoming attached to final host; adult females 7 to 15 mm long; about a dozen species.
   Suborder **CALIGOIDA**, LERNAEIDAE, **Lernaea**
   Adults not long and wormlike; males rare 4
4. Cephalothorax forming a broad flat disc (Fig. 7B); first to fourth legs well developed; males not reduced in size, rare; attached to surface of body near anus of salmon that have recently migrated from the sea; may live for a week or more in fresh water; 6 to 16 mm long; sea lice.
   Suborder **CALIGOIDA**, CALIGIDAE, **Lepeophtheirus salmonis** (Kröyer)
   Cephalothorax not forming a broad flat disc; legs absent; males small to minute pygmies clinging to females; early immature stages passed in egg; first copepodid free living for a brief period before

seeking host; attached to gills and fins of host by greatly enlarged second maxillae; adult females 3 to 8 mm long without egg sacs ..... Suborder **LERNAEOPODOIDA,** LERNAEOPODIDAE, 5

5. Posterior end of trunk rounded (Fig. 7C); about eight American species parasitic on Salmonidae and Coregonidae .... **Salmincola**
Posterior end of trunk conical or pointed; segmentation present or absent (Fig. 7D); about 15 American species parasitic on a few species of Coregonidae, Centrarchidae, and Ameiuridae.
Achtheres

6. Body greatly flattened dorsoventrally; abdominal segments fused; no fifth or sixth legs (Fig. 9); fixed parasites in branchial chamber or on general body surface of many species of fish; from 5 to 25 mm long but usually less than 10 mm; fish lice; a single American genus and about 15 species.
Order **BRANCHIURA,** Suborder **ARGULOIDA,** ARGULIDAE, **Argulus**
Body not greatly flattened dorsoventrally; abdominal segments not fused; fifth and often sixth legs present; less than 3 mm long; all genera free living except *Ergasilus* .... Order **EUCOPEPODA, 7**

7. Metasome much wider than urosome; basal segment of fifth legs without inner expansion; first antennae with 6 to 18 segments (Figs. 1A, C) .... Suborder **CYCLOPOIDA, 8**
Urosome about as wide as metasome and both more or less cylindrical; basal segment of fifth legs with inner expansion; first antennae with five to nine segments (Fig. 3); 0.3 to 1.0 mm long.
Suborder **HARPACTICOIDA,** p. 437

8. Second antennae large, three segmented, and with a large apical claw used for prehension (Fig. 8); adult females parasitic on gills of many common species of fresh-water fishes; 0.6 to 1.0 mm long; males and immature females free living; about 20 species in the United States.
ERGASILIDAE, **Ergasilus**
Second antennae small, without a large apical claw; 0.6 to 3.0 mm long .... CYCLOPIDAE, p. 424

■   ■   ■

KEY TO SPECIES OF CYCLOPOIDA

This key is drawn up specifically for identifying mature females, rather than males, although males can sometimes be successfully run through the couplets because the fifth legs and caudal rami are similar in both sexes. Male cyclopoids often have been misidentified, confused, or not identified at all because the investigator looks at the *sixth* pair of legs and mistakes them for the *fifth* legs. The sixth legs are not present in females. Both of the first antennae are geniculate in the male, neither in the female. Mature females often carry egg sacs.

Immature (copepodid) cyclopoids are difficult to identify, even to the genus level. In plankton populations, however, they may often be identified by association with the adults. In copepodid stages the last body segment is about twice as long as the preceding segment; in adults the last body segment is shorter than the preceding segment.

Except for several very rare species, usually reported only once or twice, our key includes essentially all cyclopoids reported from the 48 contiguous states. According to the number of segments in the first antenna, a copepod will fall under one of the A to K categories. The remainder of the key consists of ordinary numbered alternative couplets.

Some species that "typically" have 17 segments in the first antenna (J in the key) often will appear to have only 14, 15, or 16, owing to partial fusion, complete fusion, or unusual turgidity of adjacent segments. Such variations, however, can be easily checked against the other characteristics given in items G and H in the key. Those species having other variable characters are included in two or more corresponding places in the key. Figure 10 is semidiagrammatic and does not show minute details such as plumose ornamentation on setae and spinules on certain of the spines.

A. First antenna six segmented; last segment of fifth leg with four or five setae and/or spines; several species in brackish coastal waters .... **Halicyclops**

B. First antenna eight segmented; body 0.7 to 0.9 mm long; caudal ramus with a dorsal longitudinal row of spinules; widely distributed in bottom debris of littoral areas.
    Paracyclops fimbriatus poppei (Rehberg)

C. First antenna occasionally 9 segmented (usually 11 segmented, occasionally 10 segmented); fifth leg rudimentary and not distinct from body; in the female in the form of a plate with three stout setae (Fig. 10B); in the male a small plate with two terminal setae plus a stout seta on the body at the base of the plate; 0.7 to 1.3 mm long; a widely distributed but not abundant species of the littoral.
    Ectocyclops phaleratus (Koch)

D. First antenna 10 segmented; littoral species 1

E. First antenna 11 segmented 2

F. First antenna 12 segmented 11

G. First antenna 14 segmented; inner margin of caudal ramus hairy; length of caudal ramus more than five times the width; 2.5 to 5.0 mm long; rare and poorly known .... **Cyclops insignis** Claus

H. First antenna 16 segmented; fifth leg composed of three segments (the only U.S. species with this characteristic); 0.7 to 1.3 mm long; common in aquatic vegetation.
    Orthocyclops modestus (Herrick)

J. First antenna 17 segmented 18

K. First antenna 18 segmented (rare); 0.8 mm to 1.8 mm long; a variable and common species; an alkaline water species; may be a group of closely related species, including *A. robustus,* an acid water form.
    Acanthocyclops vernalis (Fischer)

1. Caudal ramus less than twice as long as wide C
    Caudal ramus about four times as long as wide; 0.5 to 1.8 mm long; scattered and rare.
    Cryptocyclops bicolor (Sars)

2. Caudal ramus less than twice as long as wide C
    Caudal ramus more than twice as long as wide 3

3. Exopod and endopod of fourth leg two segmented 4
    Exopod and endopod of fourth leg three segmented 8

4. Inner of the two large terminal setae of each caudal ramus at least as long as all abdominal segments and caudal rami combined; 0.5 to 1.0 mm long; widely distributed in weedy ponds and in hypogean waters .... **Microcyclops varicans rubellus** (Lillj.)
    Inner of the two large terminal setae of each caudal ramus much shorter than all abdominal segments and caudal rami combined; rare 5

5. Outer apical spine of endopod of fourth leg small and rudimentary (Fig. 10A); 0.5 to 0.8 mm long; a littoral species; second antenna four segmented .... **Cryptocyclops bicolor** (Sars)
    Outer apical spine of endopod of fourth leg more than one quarter as long as inner apical spine; second antenna three segmented 6

6. Body 0.6 to 0.7 mm long; a slender species in coastal brackish waters of the southern tier of states.
    Cyclops panamensis Marsh
    Body more than 0.7 mm long 7

7. Body 0.7 to 0.9 mm long; brackish waters of NJ, DE .... **Cyclops spartinus** Ruber
    Body 0.9 to 1.2 mm long; a robust species known only from the Salton Sea, CA.
    Cyclops dimorphus Kiefer

8. Lateral seta of caudal ramus inserted at about midlength of ramus (Fig. 10C); 0.4 to 0.9 mm long; a rare species of bottom debris .... **Diacyclops nanus** (Sars)
    Lateral seta of caudal ramus inserted about two thirds or more from base of ramus 9

9. Fifth leg consisting of one broad segment armed with an inner spine and two outer setae (Fig. 10F); 0.5 to 0.9 mm long; a rare creeping species of the weedy shallows ... **Paracyclops affinis** (Sars)
    Fifth leg two segmented; 0.7 to 0.9 mm long 10

10. Terminal segment of fifth leg with a seta and a small spur; rare, NC, NY.
    Cyclops exilis Coker
    Terminal segment of fifth leg with a seta and a spine that is longer than the segment; rare, known only from NC well .... **Cyclops jeanneli putei** Yeatman

11. Caudal ramus with spinules along the outer margin (Fig. 10D); in shallows and littoral 12
    Caudal ramus without such spinules 14

12. First antenna not reaching the posterior margin of the cephalothorax; 0.7 to 1.0 mm long; rare, ponds and creeks in a few eastern states .... **Eucyclops prionophorus** Kiefer
    First antenna reaching beyond the posterior margin of the cephalothorax; 0.7 to 1.6 mm long ... 13

13. Length of caudal ramus more than five times the width; widespread but uncommon.
     Eucyclops speratus (Lillj.)
     Length of caudal ramus less than five times the width (Fig. 10D); common everywhere (=*serrulatus*).
     Eucyclops agilis (Koch)
14. Fifth leg one segmented (Figs. 10E, G); 0.5 to 1.0 mm long **15**
     Fifth leg two segmented **16**
15. Fifth leg small and cylindrical, with a single apical seta and with or without a tiny spine on the inner margin (Fig. 10E); widely distributed in shore debris and vegetation.
     Microcyclops varicans (Sars)
     Fifth leg broad, with two long setae and one long spine (Fig. 10G); a common and widespread limnetic species .... **Tropocyclops prasinus** (Fischer)
16. Inner margins of caudal rami with small hairs; 1.5 to 1.9 mm long; rare.
     Cyclops venustoides Coker
     Inner margins of caudal rami without hairs **17**
17. Innermost terminal seta of caudal ramus longer than outermost; 0.7 to 0.9 mm long; rare, NC, NY.
     Cyclops exilis Coker
     Innermost terminal seta of caudal ramus shorter than outermost; 0.7 to 1.1 mm long; uncommon, in stagnant waters and puddles .... **Acanthocyclops crassicaudis** (Sars)
18. With a hyaline lamella on each of the last two segments of the first antennae (lamellae sometimes difficult to see; use critical lighting and tap the coverslip to turn the antennae if necessary) (Figs. 10M–O) **19**
     Without such lamellae **25**
19. Fifth leg composed of one segment bearing one spine and two setae (Fig. 10H); 1.8 to 2.9 mm long; uncommon but widely distributed .... **Macrocyclops ater** (Herrick)
     Fifth leg composed of two segments (Figs. 10J, K) **20**
20. Terminal segment of fifth leg with a terminal seta and a long spine (or stout seta) on each side of the seta (Fig. 10K) **21**
     Terminal segment of fifth leg with one seta and one spine, or two setae **23**
21. Inner margin of caudal ramus hairy **22**
     Inner margin of caudal ramus not hairy; 1.0 to 2.5 mm long; common and widely distributed, especially in vegetation and on bottom .... **Macrocyclops albidus** (Jurine)
22. Hyaline lamella on last segment of first antenna toothed (Fig. 10M); 1.2 to 4.0 mm long; common and widely distributed .... **Macrocyclops fuscus** (Jurine)
     Hyaline lamella smooth or minutely toothed; 1.2 to 2.2 mm long; widely distributed but not yet reported from North America .... **Macrocyclops distinctus** (Richard)
23. Hyaline lamella of last segment of first antenna easily seen **23a**
     Hyaline lamella of last segment of first antenna extremely narrow and very difficult to see; small lakes, ponds, and pools; uncommon .... **Diacyclops navus** (Herrick)
23a. Hyaline lamella of last segment of first antenna distinctly notched (Figs. 10N, O); 0.8 to 1.5 mm long.
     24
     Hyaline lamella of last segment of first antenna not notched; 0.6 to 1.1 mm long; a rare limnetic species .... **Thermocyclops dybowskii** (Lande)
24. Inner margins of caudal rami hairy (Fig. 10L); hyaline lamella of antenna with several sharp notches (Fig. 10N); a common limnetic species .... **Mesocyclops edax** Forbes
     Inner margins of caudal rami not hairy; hyaline lamella of antenna with one deep rounded notch and sometimes many indistinct small notches (Fig. 10O); an uncommon but widely distributed limnetic species; probably a group of closely related species, each with a definite geographic range .... **Mesocyclops leuckarti** (Claus)
25. Terminal segment of fifth leg with both a long seta and a long spine (Fig. 10R); 1.1 mm long; rare, reported only from AZ .... **Mesocyclops tenuis** (Marsh)
     Terminal segment of fifth leg with a seta and a short spine or spur, the spine never being greater than one and one-half times the length of the segment (Fig. 10P) **26**
26. Three distal segments of first antenna with a dense marginal row of fine hyaline spines; caudal ramus with a longitudinal dorsal ridge and with a hairy inner margin; 1.0 to 1.9 mm long; locally abundant in the hypolimnion of limnetic habitats .... **Cyclops scutifer** Sars
     Antenna without such spines; caudal ramus without a longitudinal ridge, inner margin with or without hairs **27**
27. Inner margins of caudal rami hairy **28**
     Inner margins of caudal rami not hairy **29**

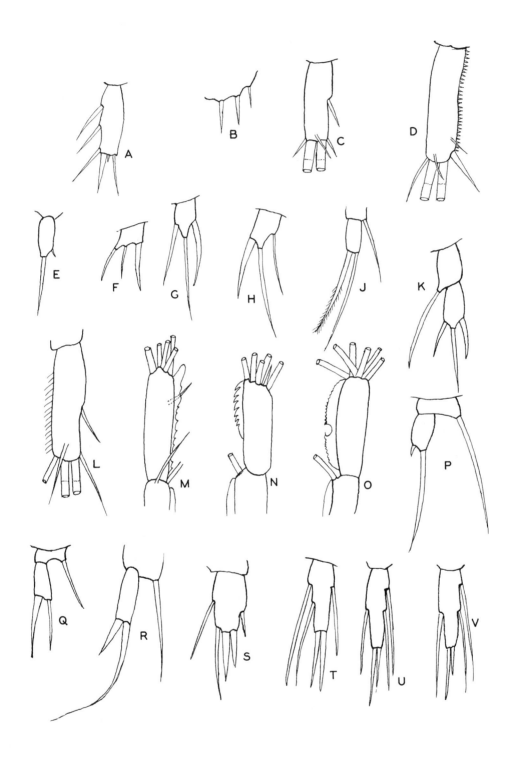

28. Innermost of the four terminal setae of caudal ramus nearly twice the length of the outermost; 1.5 to 2.5 mm long; sporadic and poorly known .... **Cyclops latipes** Lowndes

Innermost of the four terminal setae of caudal ramus about three times the length of the outermost; 1.45 mm long; known only from an Indiana cave ...... **Cyclops donnaldsoni** Chappuis

29. Terminal segment of fifth leg with a long seta and a very short spur that is always less than half the length of the segment (Fig. 10P); 0.8 to 1.8 mm long; a highly variable and common species; in alkaline waters; may be a group of related species, including *A. robustus,* an acid water form.
Acanthocyclops vernalis (Fischer)

Terminal segment of fifth leg with a long seta and a spine that is always half or more than the length of the segment **30**

30. Length of the caudal ramus more than six times the width; 1.1 to 1.4 mm long; a rare species in temporary pools .... **Cyclops haueri** Kiefer

Length of caudal ramus less than six times the width **31**

31. Lateral seta of caudal ramus attached at a point three qauarters to four fifths the distance from base to end of ramus **32**

Lateral seta of caudal ramus attached at a point two thirds or less the distance from the base to the end of ramus **35**

32. Terminal endopod segment of fourth leg with a short spine on the lateral margin (Fig. 10S); this spine is shorter than the segment bearing it; 0.5 to 0.8 mm long; in small streams and wells.
Cyclops nearcticus Kiefer

Terminal endopod segment of fourth leg with a seta on the lateral margin; this seta is about as long or longer than the segment bearing it **33**

33. Terminal spine of fifth leg only one half or less as long as the terminal seta **34**

Terminal spine of fifth leg about as long as the terminal seta (Fig. 10Q); 0.8 to 1.2 mm long; in ponds, small lakes, and wells; this species has an extremely narrow lamella on the last segment of the first antenna, almost impossible to see with the ordinary light microscope.
Cyclops navus Herrick

34. Last endopod segment of fourth leg with a terminal spine and a longer terminal seta; 0.9 mm long; known only from Marengo Cave, IN .... **Cyclops jeanneli** Chappuis

Last endopod segment of fourth leg with two terminal spines and one terminal seta; 0.9 to 1.0 mm long; known only from subterranean waters in TN, IL ... **Cyclops clandestinus** Yeatman

35. Inner terminal spine of endopod of fourth leg about two thirds the length of the outer terminal spine (Fig. 10U); penultimate thoracic segment with rounded posterolateral angles; 0.9 to 1.6 mm long; rare in the United States .... **Diacyclops bicuspidatus** (Claus)

Inner terminal spine of endopod of fourth leg about one half the length of the outer terminal spine (Fig. 10V); penultimate thoracic segment with papillate posterolateral processes; 0.9 to 1.2 mm long; very common in limnetic habitats; a highly variable species.
Diacyclops bicuspidatus thomasi (Forbes)

■ ■ ■

FIG. 10.–Structure of cyclopoid copepods (fine setation and ornamentation usually omitted). A, terminal endopod segment of fourth leg of *Cryptocyclops bicolor*; B, fifth leg of female *Ectocyclops phaleratus*; C, caudal ramus of *Diacyclops nanus*; D, caudal ramus of *Eucyclops agilis*; E, fifth leg of *Microcyclops varicans*; F, fifth leg of *Paracyclops affinis*; G, fifth leg of *Tropocyclops prasinus*; H, fifth leg of *Macrocyclops ater*; J, fifth leg of *Mesocyclops tenuis*; K, fifth leg of *Macrocyclops albidus*; L, caudal ramus of *Mesocyclops edax*; M, last segment of first antenna of *Macrocyclops fuscus*; N, terminal segment of first antenna of *Mesocyclops edax*; O, terminal segment of first antenna of *Mesocyclops leuckarti*; P, fifth leg of *Acanthocyclops vernalis*; Q, fifth leg of *Cyclops navus*; R, fifth leg of *Mesocyclops tenuis*; S, terminal endopod segment of fourth leg of *Cyclops nearcticus*; T, terminal endopod segment of fourth leg of *Cyclops navus*; U, terminal endopod segment of fourth leg of *Diacyclops bicuspidatus*; V, terminal endopod segment of fourth leg of *Diacyclops bicuspidatus thomasi*. (R modified from Yeatmen, 1964.)

KEY TO SPECIES OF CALANOID COPEPODA

The critical taxonomic characters of calanoids are usually fine details of the first antennae and the fifth legs of males. Taxonomic characters of the females are less definitive, and in keys they too often amount to unclear judgment decisions. Fortunately, however, when field collections are made there are almost invariably adequate numbers of both males and females in the populations, so that the general facies of associated males and females clearly show that they are in the same species.

Several rare species are not included in this key. Most of them have been reported only once and more than 40 years ago.

Most of the species in this key were formerly considered simply *Diaptomus* species, with several subgenera. More recently, however, the subgenera have been elevated to full generic status (a practice which I have followed, even though I am not enthusiastic about this splitting.)

1. Endopod of first leg two segmented; 0.9 to 4.5 mm long but usually 1.2 to 2.5 mm long.
   DIAPTOMIDAE, 9
   Endopod of first leg with one or three segments; limnetic 2
2. Endopod of first leg one segmented 3
   Endopod of first leg three segmented .... CENTROPAGIDAE, 5
3. Endopods of second to fourth legs one segmented; abdomen of male asymmetrical (Figs. 11D, F); 1.1 to 3.6 mm long .... TEMORIDAE, Epischura, 6
   Endopods of second to fourth legs two or three segmented; abdomen of male symmetrical ... 4
4. Endopods of second to fourth legs two segmented; male right antenna geniculate; 1.0 to 1.5 mm long; in salt and brackish waters, as well as fresh-water ponds and lakes along all coasts; also reported from Lake Erie .... TEMORIDAE, **Eurytemora affinis** (Poppe)
   Endopod of second leg two segmented, those of third and fourth legs three segmented; male left antenna not geniculate; 2.4 to 2.9 mm long (Fig. 11A); deep, cold lakes of Great Lakes region and MT .... SENECELLIDAE, **Senecella calanoides** Juday
5. Caudal rami short (Fig. 11B); 1.5 to 2.5 mm long; uncommon and erratically distributed over most of the United States .... **Osphranticum labronectum** Forbes
   Caudal rami long (Fig. 11C); 2.0 to 3.2 mm long; occurs in deep, cold lakes in northeast, such as Great Lakes, Finger Lakes, and Green Lake (WI), but also in Cascade Range lakes in WA; usually considered a relict marine species .... **Limnocalanus macrurus** Sars
6. Female abdomen bent to the right (Fig. 11G); deep, cold waters of Great Lakes region.
   Epischura lacustris Forbes
   Female abdomen straight 7
7. Last abdominal segment of male without special processes; terminal segment of female fifth leg armed with seven spines; reported only from Massachusetts ponds.
   Epischura massachusettsensis Pearse
   Last abdominal segment of male with special processes 8
8. Last abdominal segment of male with two processes, one dorsal and one ventral; terminal segment of female fifth leg armed with six spines; MA and western mountainous areas, especially WA to CA.
   Epischura nevadensis Lillj.
   Last abdominal segment of male with one dorsal process (Fig. 11D); terminal segment of female fifth leg with five spines; Atlantic coast from ME to NC .... **Epischura nordenskioldi** Lillj.
9. Antepenultimate segment of male right first antenna without distinct appendage 10
   Antepenultimate segment of male right first antenna with lateral or terminal process (Figs. 12J, K; 13K) 19
10. Male right and left fifth legs nearly equal in length; terminal hook of right leg symmetrical (Fig. 12A); common and widely distributed, especially in northern states; 1.2 to 1.4 mm long.
    Skistodiaptomus oregonensis (Lillj.)
    Male left fifth leg shorter than right 11

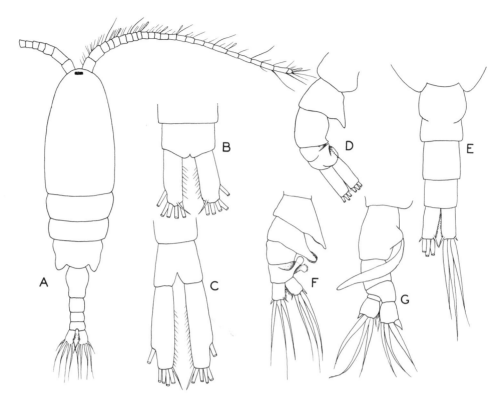

FIG. 11.–Structure of calanoid copepods. A, female *Senecella calanoides,* ×26; B, caudal rami of female *Osphranticum labronectum*; C, caudal rami of male *Limnocalanus*; D, abdomen of male *Epischura nordenskioldi*; E, female abdomen of same; F, abdomen of male *Epischura lacustris*; G, female abdomen of same. (A modified from Juday, 1925; D–G modified from Marsh, 1933.)

11. Male left fifth leg reaching beyond first segment of right exopod (Fig. 12B) 12
    Male left fifth leg reaching end of first segment of right exopod or only slightly exceeding it (Fig. 12H).
    14
12. Endopod of male right fifth leg equal in length to first exopod segment (Fig. 12B); 1 mm long; Great
    Lakes drainages .... **Skistodiaptomus reighardi** (Marsh)
    Endopod of male right fifth leg longer than first exopod segment (Fig. 12C) 13
13. Terminal claw of right fifth exopod angled (Fig. 12C); 1.2 to 1.3 mm long; GA and FL to TX.
                                              Skistodiaptomus mississippiensis (Marsh)
    Terminal claw of right fifth exopod not angled (Fig. 14A); 1 mm long; lakes and ponds of
    northeastern states .... **Skistodiaptomus pygmaeus** Pearse
14. Antepenultimate segment of male right first antenna produced into a blunt point at its distal end (Fig.
    12D) 15
    Antepenultimate segment of male right first antenna not produced into a blunt point at its distal end.
    16
15. First segment of right exopod of male fifth leg with marked quadrangular hyaline appendage (Fig.
    12E); erratic distribution east of Mississippi River ...... **Onychodiaptomus birgei** (Marsh)
    First segment of right exopod of male fifth leg with triangular projection (Fig. 12M); rare; VA, MS,
    LA .... **Onychodiaptomus virginiensis** (Marsh)
16. Inner process of terminal exopod segment of male left fifth leg falciform (Fig. 12F); Mississippi Valley
    generally; 1 mm long .... **Skistodiaptomus pallidus** (Herrick)
    Inner process of terminal exopod segment of male left fifth leg digitiform (Fig. 12H) 17

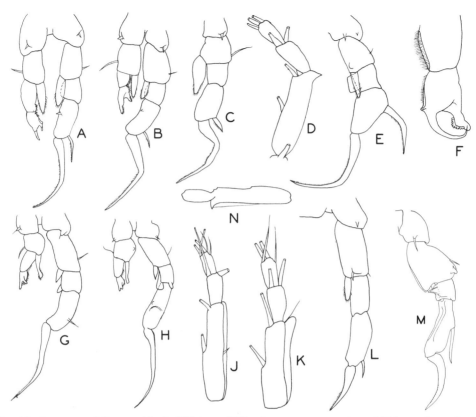

FIG. 12.–Structure of diaptomids. A, fifth legs of *Skistodiaptomus oregonensis*; B, fifth legs of *S. reighardi*; C, right fifth leg of *S. mississippiensis*; D, terminal right antennal segments of *Onychodiaptomus birgei*; E, right fifth leg of same; F, left fifth exopod of *Skistodiaptomus pallidus*; G, fifth legs of *Leptodiaptomus tyrrelli*; H, fifth legs of *Onychodiaptomus coloradensis*; J, terminal right antennal segments of *Aglaodiaptomus clavipes*; K, terminal right antennal segments of *A. leptopus*; L, right fifth leg of same; M, right fifth leg of *Onychodiaptomus virginiensis*; N, antepenultimate and penultimate segments of right first antenna of *O. hesperus*. (A–E, G, and J–L modified from Marsh, 1907; F modified from Light 1939; H modified from Dodds; M from Wilson and Moore, 1953.)

17. Lateral spine of second segment of male right fifth exopod strongly curved (Fig. 12H); 1.1 to 1.3 mm long; scattered in high-altitude lakes and ponds of western states.
                              Leptodiaptomus coloradensis (Marsh)
    Lateral spine of second segment of male right fifth exopod essentially straight (Fig. 12G) **18**
18. Median margin of second basal segment of male left fifth leg with a three-lobed hyaline expansion (Fig. 14B); 1.0 to 1.6 mm long; rare in ponds and lakes of mountains and northern states.
                              Leptodiaptomus pribilofensis (Juday and Muttkowski)
    Without such an expansion (Fig. 12G); 1.1 to 1.8 mm long; occasional in western mountains.
                              Leptodiaptomus tyrrelli (Poppe)
19. Antepenultimate segment of male right antenna with a hyaline lamella (Figs. 12J, K) **20**
    Antepenultimate segment of male right antenna with a slender process **26**
20. Hyaline lamella broad, extending beyond end of segment (Fig. 12K); 1.2 to 2.4 mm long; widely distributed .... **Aglaodiaptomus leptopus** (Forbes)
    Hyaline lamella narrow, extending beyond end of segment slightly if at all (Fig. 12J) **21**
21. Antepenultimate segment of male right first antenna with a short, hook-shaped process (Fig. 12N); 1.1 mm long; OR .... **Onychodiaptomus hesperus** Wilson and Light
    Antepenultimate segment of male right first antenna without a hook-shaped process (Figs. 12J, K).
                                                                                              22

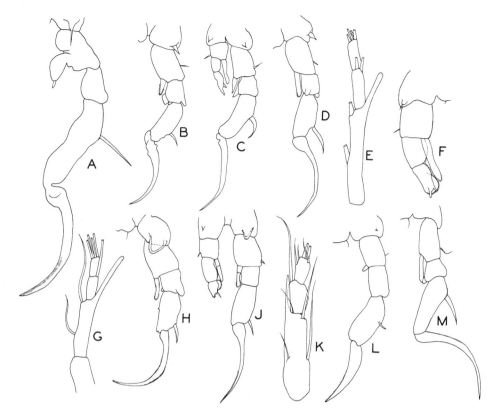

FIG. 13.–Structure of diaptomids. A, right fifth leg of *Leptodiaptomus trybomi*; B, right fifth leg of *L. connexus*; C, fifth legs of *L. judayi*; D, right fifth leg of *L. sicilis*; E, terminal antennal segments of same; F, left fifth leg of same; G, terminal antennal segments of *Hesperodiaptomus wardi*; H, right fifth leg of *H. caducus*; J, fifth legs of *H. shoshone*; K, terminal antennal segments of same; L, right fifth leg of *Leptodiaptomus minutus*; M, right fifth leg of *L. ashlandi*. (A, D–F, K–M modified from Marsh, 1907; B and H modified from Light, 1938; C modified from Ward and Whipple, 1918.)

22. Basal segment of right fifth leg armed with a sclerotized hook that is as long as the first exopod segment (Fig. 14C) 23
    Basal segment of right fifth leg without such a hook 24
23. Setae on male left antenna segments 17, 19, 20, and 22 hooked; 1.2 to 2.2 mm long; erratically distributed between the Mississippi River and the Continental Divide.
    Aglaodiaptomus clavipes (Schacht)
    Setae on male left antenna segments 17, 19, 20, and 22 not hooked at end; 1.8 to 2.0 mm long; temporary ponds in OK, TX .... **Aglaodiaptomus kingsburyae** Robertson
24. Both terminal processes of left fifth exopod blunt (Fig. 14D); 1.5 mm long; a Eurasian species found in northern Rocky Mountain lakes and ponds .... **Acanthodiaptomus denticornis** (Wierzejski)
    Both terminal processes not blunt; 1.3 mm long 25
25. One terminal process of left fifth exopod pointed, the other blunt (Fig. 14E); rare in ponds of FL, LA.
    Onychodiaptomus lousianensis (Wilson and Moore)
    Both terminal processes pointed (Fig. 14F); LA ponds.
    Skistodiaptomus bogalusensis (Wilson and Moore)
26. Slender process generally straight and blunt (Figs. 13E, K) 27
    Process curved and usually pointed 42

27. Process shorter than the penultimate segment 28
  Process as long or longer than the penultimate segment 31
28. Male right fifth endopod rudimentary (Fig. 13A); OR; rare ... **Leptodiaptomus trybomi** (Lillj.)
  Male right fifth endopod not rudimentary 29
29. Posterior face of the terminal right exopod segment of the fifth leg with a large, cuticular, spinelike
  structure (Fig. 14G); 1.3 to 1.7 mm long; Rocky Mountain lakes.
  Arctodiaptomus arapahoensis (Dodds)
  Posterior face of the terminal right exopod segment of the fifth leg without such a structure 30
30. First segment of male right fifth exopod produced as a rounded lobe on its lateral distal margin, its
  inner margin bearing a rounded hyaline lamella (Fig. 13B); 0.9 to 1.5 mm long; western states.
  Leptodiaptomus connexus (Light)
  First segment of male right fifth exopod not produced on its lateral distal margin, its inner margin
  bearing a triangular hyaline lamella (Fig. 13C); 0.9 mm long; Rocky Mountain lakes.
  Leptodiaptomus judayi (Marsh)
31. Process about as long as the penultimate segment (Fig. 13E); 1.1 to 1.5 mm long; common in northern
  half of the United States, usually in larger lakes .... **Leptodiaptomus sicilis** (Forbes)
  Process longer than the penultimate segment (Figs. 13G, K) 32
32. Antennae at most reaching the proximal end of the caudal rami 33
  Antennae reaching beyond the distal end of the caudal rami 40
33. Process of antepenultimate segment of male right antenna not attaining the distal end of the terminal
  segment (Fig. 13K) 36
  Process of antepenultimate segment of male right antenna extending to or beyond the distal end of
  the terminal segment (Fig. 13G) 34
34. Second basal segment of male right fifth leg with one small median lamella (Fig. 14H); 2.5 mm long;
  rare in northern Rocky Mountain states .... **Hesperodiaptomus schefferi** Wilson
  Second basal segment of male right fifth leg otherwise 35
35. Second basal segment of male right fifth leg with two small median lamellae (Fig. 14J); 2.4 mm long; a
  poorly known species from ponds in eastern WA ..... **Hesperodiaptomus kiseri** (Kincaid)
  Second basal segment of male right fifth leg with no lamellae at all; 1.2 to 1.6 mm long; ponds in WA,
  MT .... **Hesperodiaptomus wardi** (Pearse)
36. Dorsal surface of caudal rami of *female* thickly set with hairs (Fig. 14K); 1.8 mm long; California
  mountain lakes and ponds .... **Hesperodiaptomus hirsutus** Wilson
  Dorsal surface of caudal rami of *female* with sparse hairs or no hairs 37
37. Male 1.0 to 1.1 mm long; NV, CA, WA .... **Leptodiaptomus spinicornis** (Light)
  Male 1.5 to 3.3 mm long 38
38. Male 1.5 to 2.3 mm long; LA, GA, NC .... **Hesperodiaptomus augustaensis** (Turner)
  Male longer than 2.3 mm; western distribution 39
39. Terminal hook of right fifth exopod sharply angled (Fig. 13H); 2.4 to 2.6 mm long; Pacific coast ponds.
  Hesperodiaptomus caducus (Light)
  Terminal hook of right fifth exopod not sharply angled (Fig. 13J); 2.6 to 3.3 mm long; above 1500 m
  in western states .... **Hesperodiaptomus shoshone** (Forbes)
40. Right endopod rudimentary (Fig. 13L); 0.9 to 1.0 mm long; common in cold waters in northeastern
  states, south to NC and west to WY .... **Leptodiaptomus minutus** (Lillj.)
  Right endopod not rudimentary (Fig. 13M) 41
41. Lateral spine of second segment of male right fifth exopod long; lateral claw of terminal right exopod
  segment near base (Fig. 13M); 0.9 to 1.2 mm long; scattered in northern states.
  Leptodiaptomus ashlandi (Marsh)
  Lateral spine of second segment of male right fifth exopod short; lateral claw of terminal right
  exopod segment subterminal (Fig. 14L); 1.9 to 2.0 mm long; common in western mountain
  lakes .... **Hesperodiaptomus kenai** Wilson
42. Process equal to or exceeding the length of the penultimate segment 43
  Process shorter than the penultimate segment (Fig. 15K) 47
43. Process about as long as the last two segments 45
  Process not attaining the tip of the last antennal segment 44
44. Process only slightly longer than the penultimate segment (Fig. 15A); 1.1 to 2.0 mm long; west coast
  states .... **Hesperodiaptomus franciscanus** (Lillj.)
  Process reaching to the middle of the apical segment or beyond; 3.2 to 4.0 mm long; a poorly known
  species reported from the prairies of Nebraska and Saskatchewan.
  Hesperodiaptomus breweri Wilson

FIG. 14.–Structure of male diaptomids (K is female). A, right fifth leg of *Skistodiaptomus pygmaeus*; B, left fifth leg of *Leptodiaptomus pribilofensis*; C, right fifth leg of *Aglaodiaptomus clavipes*; D, left fifth leg of *Acanthodiaptomus denticornis*; E, left fifth leg of *Onychodiaptomus louisianensis*; F, fifth legs of *Skistodiaptomus bogalusensis*; G, exopod and endopod of right fifth leg of *Arctodiaptomus arapahoensis*; H, right fifth leg of *Hesperodiaptomus schefferi*; J, right fifth leg of *H. kiseri*; K, left caudal ramus of female *H. hirsutus*; L, right fifth leg of *H. kenai*; M, right fifth leg of *Leptodiaptomus moorei*; N, tip of right fifth leg of *Aglaodiaptomus dilobatus*; O, tip of right fifth leg of *A. clavipoides*; P, left fifth leg of *A. marshianus*. (C, D, G, and J modified from Edmondson, Ward, and Whipple, 1959; E from Wilson and Moore, 1953; F from Wilson and Moore, 1953a; H, J, K, and P modified from Wilson, 1953; M modified from Wilson, 1954; N modified from Wilson, 1958; O modified from Wilson, 1955.)

45. Male left fifth exopod with a long, stiff pinnate terminal seta (Fig. 15B); 1.1 to 1.4 mm long; Pacific coast states .... **Aglaodiaptomus forbesi** (Light)
    Male left fifth exopod without such a seta **46**
46. Terminal hook of male right fifth leg abruptly angled (Fig. 15C); rare; northern states and west coast; 2.6 to 3.5 mm long .... **Hesperodiaptomus eiseni** (Lillj.)
    Terminal hook of male right fifth leg curved, not angled; 3.5 mm long; rare; CA, NV, ND, WA; saline lakes .... **Hesperodiaptomus nevadensis** (Light)
47. Both terminal processes of male left fifth exopod digitiform and blunt **48**
    Both terminal processes of male left fifth exopod not digitiform and blunt **53**
48. Male right fifth endopod as long as or longer than the first segment of the exopod (Fig. 15D) **49**
    Male right fifth endopod shorter than the first segment of the exopod (Fig. 15H) **50**
49. Median margin of first exopod segment of right fifth leg with a hyaline lamella (Fig. 14M); 1.2 mm long; ponds in FL to TX .... **Leptodiaptomus moorei** (Wilson)
    Without such a lamella (Fig. 15D); 1.0 to 1.7 mm long; southwestern Rocky Mountain area.
    Leptodiaptomus novamexicanus (Herrick)

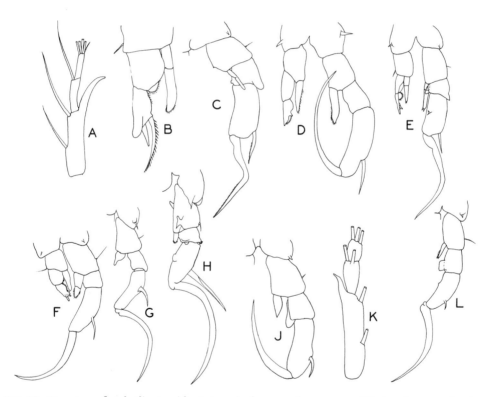

FIG. 15.–Structure of male diaptomids. A, terminal antennal segments of *Hesperodiaptomus franciscanus*; B, left fifth leg of *Aglaodiaptomus forbesi*; C, right fifth leg of *Hesperodiaptomus eiseni*; D, fifth legs of *Leptodiaptomus novamexicanus*; E, fifth legs of *Hesperodiaptomus franciscanus*; F, fifth legs of *Leptodiaptomus novamexicanus*; G, right fifth leg of *L. nudus*; H, right fifth leg of *Mastigodiaptomus albuquerquensis*; J, right fifth leg of *Leptodiaptomus signicauda*; K, terminal antennal segments of *L. siciloides*; L, right fifth leg of same. (A modified from De Guerne and Richard, 1889; B modified from Light, 1938; C, E–H, J, and L modified from Marsh, 1907; D modified from Herrick and Turner, 1895; K modified from Dodds.)

50. With hyaline appendages on either the second basal segment of male right fifth leg or first segment of male right fifth exopod 51

Without such appendages (Fig. 15G); 1.1 mm long; Rocky Mountain states, rare elsewhere.

Leptodiaptomus nudus (Lillj.)

51. With hyaline appendages on second basal segment of male right fifth leg (Fig. 15H); 0.9 to 1.5 mm long; Rocky Mountain states but found as far east as OK.

Mastigodiaptomus albuquerquensis (Herrick)

With a hyaline appendage on first segment of male right fifth exopod 52

52. Hyaline lamella on median surface of male right fifth exopod roughly triangular and projecting posteriorly (Fig. 15J); 1.0 to 1.3 mm long; occasional in western states, rare elsewhere.

Leptodiaptomus signicauda (Lillj.)

Hyaline lamella on median surface of male right fifth exopod not triangular and not projecting posteriorly (Fig. 15L); 1.0 to 1.1 mm long; common and widely distributed.

Leptodiaptomus siciloides (Lillj.)

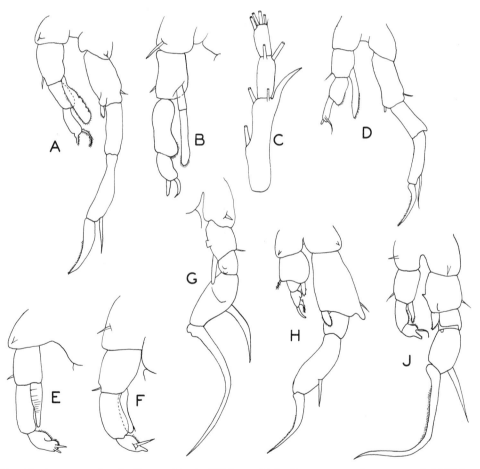

FIG. 16.–Structure of male diaptomids. A, fifth legs of *Aglaodiaptomus spatulocrenatus*; B, left fifth leg of *A. conipedatus*; C, terminal antennal segments of *A. lintoni*; D, fifth legs of same; E, left fifth leg of *A. stagnalis*; F, left fifth leg of *Arctodiaptomus floridanus*; G, right fifth leg of *A. dorsalis*; H, fifth legs of *Onychodiaptomus sanguineus*; J, fifth legs of *Arctodiaptomus saltillinus*. (A and E modified from Marsh, 1929; B, G, H, and J modified from Marsh, 1907; C and D modified from Dodds; F modified from Marsh, 1926.)

53. Male left fifth endopod marked with transverse striae (Fig. 16E); 3.0 to 4.0 mm long; eastern states.
    Aglaodiaptomus stagnalis (Forbes)
    Male left fifth endopod not marked with transverse striae **54**
54. Male right fifth endopod rudimentary (Fig. 16D) **55**
    Male right fifth endopod not rudimentary **59**
55. Both terminal processes of male left fifth exopod about the same length (Fig. 16B); 1.3 mm long; LA.
    Aglaodiaptomus conipedatus (Marsh)
    Terminal processes of male left fifth exopod of unequal length (Figs. 14N, O) **56**
56. Male left fifth leg reaching much beyond end of second basal segment of right leg **57**
    Male left fifth leg not reaching much beyond end of second basal segment of right leg (Fig. 16D).
    58
57. Terminal spine of right fifth exopod stout (Fig. 14N); 1.6 to 1.7 mm long; rare; FL.
    Aglaodiaptomus dilobatus Wilson
    Terminal spine of right fifth exopod slim (Fig. 14O); 2.0 to 2.1 mm long; rare; LA, FL.
    Aglaodiaptomus clavipoides Wilson
58. Chiefly in Rocky Mountain lakes; 1.5 to 2.0 mm long .... **Aglaodiaptomus lintoni** (Forbes)
    Reported only from Louisiana and Saskatchewan; 1.2 mm long.
    Aglaodiaptomus saskatchewanensis Wilson
59. One of the terminal processes of the male left fifth exopod in the form of a straight or slightly curved
    sharp spine (Fig. 16F) **60**
    One of the terminal processes of the male left fifth exopod distinctly falciform **61**
60. Male left fifth endopod not extending beyond end of first exopod segment (Fig. 16F); 0.9 mm long;
    FL, GA; sometimes placed in *Arctodiaptomus* .... **Diaptomus floridanus** Marsh
    Male left fifth endopod longer (Fig. 14P); 1.3 to 1.6 mm long; reported only from Florida.
    Aglaodiaptomus marshianus Wilson
61. Male right fifth endopod shorter than first segment of exopod **62**
    Male right fifth endopod distinctly longer than first segment of exopod (Fig. 16G); 1 mm long;
    uncommon; southern states; sometimes placed in *Arctodiaptomus*.
    Diaptomus dorsalis Marsh
62. Terminal segment of male right fifth exopod slender and elongated (Fig. 16A); 1.1 to 1.3 mm long;
    ME to MD .... **Aglaodiaptomus spatulocrenatus** (Pearse)
    Terminal segment of male right fifth exopod of the usual proportions **63**
63. Male left fifth leg not attaining the end of the first exopod segment of the right leg (Fig. 16H); 1.4 to 2.1
    mm long; widely distributed in ponds .... **Onychodiaptomus sanguineus** (Forbes)
    Male left fifth leg extending beyond the end of the first exopod segment of the right leg (Fig. 16J); NE
    to TX .... **Arctodiaptomus saltillinus** Brewer

■ ■ ■

KEY TO GENERA OF HARPACTICOIDA

The following key is to genus only. We feel that a key to the species of harpacticoids of the United States is a fruitless and misleading effort, chiefly because the group is so poorly known and because there are so many undescribed species in this country. The writer, for example, has seen more than 30 undescribed species during his field work since the publication of the first edition of this book.

Most of the key characters apply to both sexes. When they apply to only one sex, the exception is indicated in the couplet. Body length of harpacticoids ranges from 0.3 to 1.4 mm, but the great majority of species are less than 0.8 mm long. Their small size may be the chief reason they have been neglected. An exceptionally large number of new species are awaiting discovery in the interstitial habitat of sandy beaches and in the phreatic zone of streams.

1. Male body composed of 11 segments; female, 10 segments (Fig. 3) 2
   Male body composed of 10 segments; female, 9 segments (Fig. 17G) 3
2. First antenna eight segmented; a small genus with one cosmopolitan species found in many kinds of habitats .... **Phyllognathopus viguieri** (Maupas)
   First antenna seven segmented; several uncommon and widely distributed species.
   　　　　　　　　　　　　　　　　　　　　　　　　　　　　　　　　　　Chappuisius
3. Distal exopod segment of first leg with three to five short clawlike spines; fresh and brackish water.
   　　　　　　　　　　　　　　　　　　　　　　　　　　　　　　　　　　　　　11
   Distal exopod segment of first leg with the usual long spines and setae 4
4. Exopod of first leg three segmented 5
   Exopod of first leg two segmented 12
5. Terminal segment of exopod of first leg with five or six spines and setae (Fig. 18M) 20
   Terminal segment of exopod of first leg with four spines and setae 6
6. Second exopod segment of first leg with an inner seta (Fig. 18C) 10
   Second exopod segment of first leg without an inner seta 7
7. Second exopod segment of first leg with an outer spine 8
   Second exopod segment of first leg without an outer spine; vermiform (Fig. 17A); a large worldwide genus; usually found in interstitial and subterranean waters .... **Parastenocaris**
8. Endopod of first leg one segmented 13
   Endopod of first leg with more than one segment 9
9. Endopod of first leg two segmented 14
   Endopod of first leg three segmented 24
10. Endopod of first leg two segmented (Fig. 18C) 25
    Endopod of first leg three segmented 26
11. Basal expansion of female fifth leg with three or four setae (Fig. 18A); New England coast.
    　　　　　　　　　　　　　　　　　　　　　　　　　　　　　　　　　　Harpacticus
    Basal expansion of female fifth leg with five or six setae; Pacific coast .... **Tigriopus**

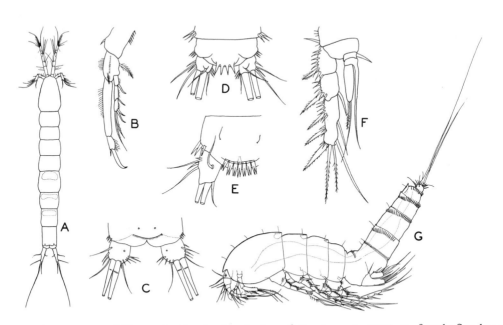

Fig. 17.–Structure of Harpacticoida. A, dorsal view of *Parastenocaris,* ×100; B, female first leg of *Pseudonychocamptus*; C, D, and E, posterior end of *Bryocamptus* spp.; F, male third leg of *Bryocamptus*; G, female *Bryocamptus,* ×85. (B modified from Wilson, 1932; C, D, F, and G modified from Coker, 1934; E redrawn from Gurney, 1931–1933.)

12. Endopod of first leg about twice or more as long as exopod (Fig. 18D); several species reported from fresh and inland saline waters .... **Onychocamptus**
    Endopod of first leg no longer than exopod 13
13. Endopod of first leg two segmented (Fig. 18E); in vegetation at margins of lotic and lentic waters; uncommon but widely distributed .... **Maraenobiotus**
    Endopod of first leg one segmented and small; a marine genus reported from Bear Lake, Utah; undoubtedly occurs elsewhere .... **Huntemannia**
14. Endopod of first leg not extending beyond second exopod segment (Fig. 17B) 15
    Endopod of first leg longer 18
15. Anterior end of body expanded, tapered posteriorly (Fig. 18G); brackish and fresh waters of Atlantic coast .... **Metis**

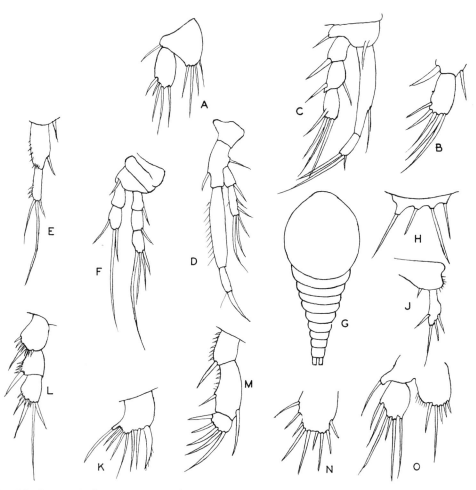

Fig. 18.–Structure of Harpacticoida (fine hairs and other ornamentation not shown). A, fifth leg of *Harpacticus*; B, terminal segment of exopod of first leg of *Nitocra*; C, first leg of *Mesochra*; D, fifth leg of *Onychocamptus*; E, endopod of first leg of *Maraenobiotus*; F, first leg of *Epactophanes*; G, dorsal view of body of *Metis*; H, fifth leg of *Heterolaophonte*; J, fifth leg of male *Pseudonychocamptus*; K, fifth leg of female *Stenocaris*; L, exopod of second leg of *Moraria*; M, exopod of first leg of *Paradactylopodia*; N, fifth leg of *Tachidius*; O, fifth leg of *Nitocra*. (O redrawn and modified from Edmondson, Ward and Whipple, 1959.)

Body slender and vermiform 16
16. First segment of first exopod more than three times as long as wide (Fig. 17B) 17
   First segment of first exopod no more than twice as long as wide (Fig. 18F); one cosmopolitan species
      in aquatic mosses and other vegetation; 0.3 to 0.5 mm long.
                                                             Epactophanes richardi Mrazek
17. Fifth leg of male unsegmented (Fig. 18H); brackish ponds on northeast coast ..... **Heterolaophonte**
   Fifth leg of male segmented (Fig. 18J); fresh and brackish ponds on northeast coast.
                                                             Pseudonychocamptus
18. Fifth leg unsegmented (Fig. 18K); vermiform; brackish and fresh waters of northeastern coast.
                                                             Stenocaris
   Fifth leg segmented 19

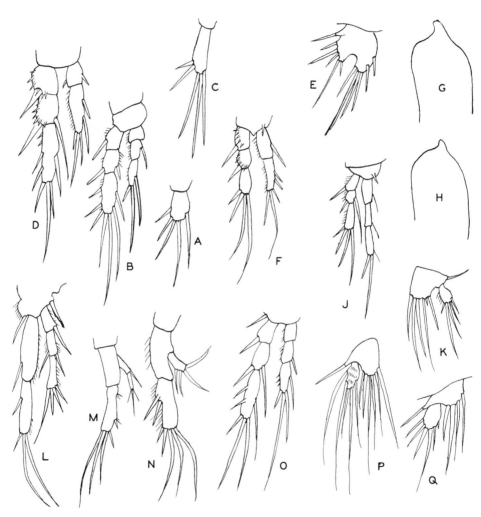

Fig. 19.–Structure of Harpacticoida (fine hairs and other ornamentation not shown). A, third exopod segment of first leg of *Tachidius*; B, second leg of *Nitocra*; C, third exopod segment of third leg of *Nitocrella*; D, second leg of *Bryocamptus*; E, fifth leg of *Cletocamptus*; F, first leg of *Bryocamptus*; G, lateral view of *Mesochra* cephalothorax outline; H, lateral view of *Bryocamptus* cephalothorax outline; J, first leg of *Bryocamptus*; K, fifth leg of *Canthocamptus*; L, first leg of *Attheyella*; M, second antenna of *Bryocamptus*; N, second antenna of *Attheyella*; O, second leg of *Bryocamptus*; P, fifth leg of *Attheyella*; Q, fifth leg of *Elaphoidella*. (K, M, N, and P modified from Coker, 1934.)

19. Second exopod segment of second to fourth legs without a median seta (Fig. 18L); many species, widely distributed .... **Moraria**
    Second exopod segment of second to fourth legs with a median seta (Fig. 17F); many species, widely distributed, very common on all kinds of substrates .... **Bryocamptus**
20. Third exopod segment of first leg about one third as long as the second exopod segment (Fig. 18M); uncommon, euryhaline species of Atlantic coast .... **Paradactylopodia**
    Third exopod segment of first leg about as long as the second exopod segment 21
21. Fifth leg unsegmented (Fig. 19A) 22
    Fifth leg segmented (Fig. 18O) 23
22. Third exopod segment of first leg with six spines and setae; several euryhaline species of Atlantic coast. **Microarthridion**
    Third exopod segment of first leg with five spines and setae (Fig. 19A); several fresh-water and brackish coastal species .... **Tachidius**
23. Third exopod segment of second and third legs with three lateral marginal spines (Fig. 19B); mostly coastal brackish and fresh-water species .... **Nitocra**
    Third exopod segment of second and third legs with two lateral marginal spines (Fig. 19C); expect to find this genus in phreatic and subterranean waters .... **Nitocrella**
24. Endopod of second leg two segmented (Fig. 19D); many species, widely distributed, and very common on all kinds of substrates .... **Bryocamptus**
    Endopod of second leg three segmented; one common cosmopolitan species in many habitats. **Phyllognathopus viguieri** (Maupas)
25. Exopod of fifth leg separated from base by only a notch or gap (Fig. 19E); several species in brackish and saline lakes and ponds .... **Cletocamptus**
    Exopod of fifth leg clearly separated from base (Fig. 19K) 26
26. Rostrum greatly enlarged (Fig. 19G); endopod of first leg with two or three segments (Fig. 18C); coastal brackish and fresh-water ponds, and inland ponds; several species ...... **Mesochra**
    Rostrum small to moderate size (Figs. 17G; 19H) 27
27. Rostrum small (Fig. 17G); endopod of first leg two segmented (Fig. 19F); many widely distributed species in a variety of fresh-water habitats .... **Bryocamptus**
    Rostrum small to moderate size (Fig. 19H); endopod of first leg three segmented (Fig. 19J) 28
28. Second seta of basal expansion of female fifth leg greatly reduced (Fig. 19K); several species in a wide variety of fresh-water habitats .... **Canthocamptus**
    Second seta of basal expansion of female fifth leg not reduced 29
29. First endopod segment of first leg reaching or exceeding middle of third exopod segment (Fig. 19L); many species in a wide variety of habitats .... **Attheyella**
    First endopod segment of first leg not extending beyond second exopod segment 30
30. Palp of second antenna one segmented (Fig. 19N) 31
    Palp of second antenna two segmented (Fig. 19M); endopods of female second and third legs three segmented (Fig. 19O); many widely distributed species in a variety of fresh-water habitats .... **Bryocamptus**
31. Fifth leg of female with rows of fine setae on the ventral surface, and basal expansion with six setae; several species .... **Attheyella**
    Fifth leg of female without rows of fine setae on the ventral surface, and basal expansion with four setae (Fig. 19Q); several uncommon and widely distributed species .... **Elaphoidella**

REFERENCES

Aycock, D. 1942. Influence of temperature on size and form of Cyclops vernalis Fischer. *J. Elisha Mitchell Sci. Soc.* **58**:84–93.

Brandl, Z. 1973. Laboratory culture of cyclopoid copepods on a definite food. *Vestn. Cesk. Spol. Zool.* **37**:81–88.

Brandl, Z., and C. H. Fernando. 1975. Food consumption and utilization in two freshwater cyclopoid copepods (Mesocyclops edax and Cyclops vicinus). *Int. Rev. gesamten Hydrobiol.* **60**:471–494.

Byron, E. R. The adaptive significance of calanoid copepod pigmentation: a comparative and experimental analysis. *Ecology* **63**:1871–1886.

Chappuis, P. A. 1927. Freilebende Süsswasser-Copepoden aus Nordamerika. 2. Harpacticiden. *Zool. Anz.* **74**:302–313.

———. 1929. Die Unterfamilie der Canthocamptinae. *Arch. Hydrobiol.* **20**:471–516.

———. 1929a. Copépodes cavernicoles de l'Amérique du Nord. *Bull. Soc. Sci. Cluj* **4**:51–58.

_____. 1957. Le genre Parastenocaris Kessler. *Vie et Milieu* 8:423–432.

Coker, R. E. 1933. Influence of temperature on size of freshwater copepods (Cyclops). *Int. Rev.* 29:406–436.

_____. 1934. Contribution to knowledge of North American freshwater harpacticoid copepod Crustacea. *J. Elisha Mitchell Sci. Soc.* 50:75–141.

_____. 1938. Anomalies of crustacean distribution of the general region of Chapel Hill, N.C. *J. Elisha Mitchell Sci. Soc.* 54:76–87.

Comita, G. W., and S. J. McNett. 1976. The postembryonic developmental instars of Diaptomus oregonensis Lilljeborg, 1889 (Copepoda). *Crustaceana* 30:123–163.

Comita, G. W., and D. M. Tommerdahl. 1960. The postembryonic developmental instars of Diaptomus siciloides Lilljeborg. *J. Morphol.* 107:297–355.

Cressey, R. F. 1972. The genus Argulus (Crustacea: Branchiura) of the United States. *Biota Freshwater Ecosystems, Identification Manual* 2:1–14.

Davis, C. C. 1959. Osmotic hatching in the eggs of some fresh-water copepods. *Biol. Bull.* 116:15–29.

Deevey, E. S. 1941. Notes on the encystment of the harpacticoid copepod Canthocamptus staphylinoides Pearse. *Ecology* 22:197–200.

De Guerne, J., and J. Richard. 1889. Révision des Calanides d'eau douce. *Mém. Soc. Zool. Fr.* 2:53–144.

Dodds, G. S. 1915. Descriptions of two new species of Entomostraca from Colorado, with notes on other species. *Proc. U.S. Nat. Mus.* 49:97–102.

Dussart, B., and D. Defaye. 1983. *Repértoire mondial des crustacés copépodes des eaux intérieures.* Editions du CNRS, Paris, 224 pp.

Edmondson, W. T., H. B. Ward, and G. C. Whipple. 1959. *Fresh-water Biology,* 2nd ed. 1248 pp. Wiley, New York.

Elgmork, K. 1967. Ecological aspects of diapause in copepods. *Proc. Symp. Crustacea* 3:947–954.

_____. 1967a. On the distribution and ecology of Cyclops scutifer Sars in New England (Copepoda, Crustacea). *Ecology* 48:967–971.

Ewers, L. A. 1930. The larval development of freshwater Copepoda. *Ohio State Univ., Contrib. Franz Theodore Stone Lab.* 3:1–43.

_____. 1936. Propagation and rate of reproduction of some freshwater Copepoda. *Trans. Am. Microsc. Soc.* 55:230–238.

Forbes, S. A. 1897. A contribution to a knowledge of North American fresh-water Cyclopidae. *Bull. Ill. State Lab.* 5:27–83.

Frenzel, P. 1980. Die Populationsdynamik von Canthocamptus staphylinus (Jurine) (Copepoda, Harpacticoida) im Littoral des Bodensees. *Crustaceana* 39:282–286.

Fryer, G. 1982. The parasitic Copepoda and Branchiura of British freshwater fishes. *Freshwater Biol. Assoc. Publ.* 46:1–87.

_____. 1985. An ecological validation of a taxonomic distinction: The ecology of Acanthocyclops vernalis and A. robustus (Crustacea: Copepoda). *Zool. J. Linn. Soc.* 84:165–180.

Gurney, R. 1931–1933. *British fresh-water Copepoda.* 3 vols. 958 pp. Ray Society, London.

Harding, J. P., and W. A. Smith. 1974. A key to the British fresh-water cyclopoid and calanoid copepods. *Sci. Publ. Freshwater Biol. Assoc.* 18:1–56.

Herrick, C. L., and C. H. Turner. 1895. A synopsis of the Entomostraca of Minnesota. *2d Rept. State Zool., Geol. Nat. Hist. Surv. Minn.* 1–337.

Herzig, A. 1983. The ecological significance of the relationship between temperature and duration of embryonic development in planktonic freshwater copepods. *Hydrobiologia* 100:65–91.

Hill, L. I., and R. E. Coker. 1930. Observations on mating habits of Cyclops. *J. Elisha Mitchell Sci. Soc.* 45:206–220.

Humes, A. G. 1954. The specific validity of Epischura massachusettsensis Pearse (Copepoda: Calanoida). *Am. Midl. Nat.* 52:154–158.

Juday, C. 1925. Senecella calanoides, a recently described fresh-water copepod. *Proc. U.S. Nat. Mus.* 66:1–6.

Kabata, Z. 1969. Revision of the genus Salmincola Wilson 1915 (Copepoda: Lernaeopodidae). *J. Fish. Res. Board Can.* 26:2987–3041.

Kerfoot, W. C. et al. 1980. Visual observations of live zooplankters: evasion, escape, and chemical defenses. In W. C. Kerfoot (ed.), *Evolution and Ecology of Zooplankton Communities.* pp 10–27. University Press of New England, Hanover, New Hampshire.

Kiefer, F. 1967. Copepoda. *Limnofauna Europaea* 173–185.

Lang, K. 1948. *Monographie der Harpacticiden.* Vols. 1 and 2, 1683 pp. Lund, Sweden.

Lescher-Moutoué, F. 1973. Sur le biologie et l'écologie des copépodes cyclopides hypogés (Crustaces). *Ann. Speleol.* 28:429–502, 581–674.

Light, S. F. 1938. New subgenera and species of diaptomid copepods from the inland waters of California and Nevada. *Univ. Calif. Publ. Zool.* 43:67–78.

_____. 1939. New American subgenera of Diaptomus Westwood (Copepoda, Calanoida). *Trans. Am. Microsc. Soc.* 58:473–484.

Marsh, C. D. 1907. A revision of the North American species of Diaptomus. *Trans. Wis. Acad. Sci. Arts and Lett.* 15:381–516.

_____. 1926. On a collection of copepods from Florida with a description of Diaptomus floridanus, new species. *Proc. U.S. Nat. Mus.* 70:1–4.

_____. 1929. Distribution and key to the North American copepods of the genus Diaptomus, with the description of a new species. *Ibid.* 75:1–27.

_____. 1933. Synopsis of the calanoid crustaceans, exclusive of the Diaptomidae, found in fresh and brackish waters, chiefly of North America. *Ibid.* 82:1–58.

Meehean, O. L. 1940. A review of the parasitic Crustacea of the genus Argulus in the collections of the United States National Museum. *Proc. U.S. Nat. Mus.* 88:459–522.

Papinska, K. 1985. Carnivores and detritivores feeding of Mesocyclops leuckarti Claus (Cyclopoida, Copepoda). *Hydrobiologia* 120:249–257.

Pennak, R. W. 1940. Ecology of the microscopic metazoa

inhabiting the sandy beaches of some Wisconsin lakes. *Ecol. Monogr.* 10:537–615.

Pennak, R. W., and J. V. Ward. 1985. New cyclopoid copepods from interstitial habitats of a Colorado mountain stream. *Trans. Am. Microsc. Soc.* 104:216–222.

Reed, E. B. 1986. Esteval phenology of an Acanthocyclops (Crustacea, Copepoda) in a Colorado tarn with remarks on the vernalis-robustus complex. *Hydrobiologia* 139:127–144.

Roberts, L. S. 1970. Ergasilus (Copepoda: Cyclopoida): revision and key to species in North America. *Trans. Am. Microsc. Soc.* 89:134–161.

Ruber, E. 1968. Descriptions of a salt marsh copepod Cyclops (Apocyclops) spartinus n. sp. and a comparison with closely related species. *Trans. Am. Microsc. Soc.* 87:368–375.

Rylov, V. M. 1948. Freshwater Cyclopoida. Crustacea. *Fauna USSR* 3:(3):1–314.

Sandercock, G. A. 1967. A study of selected mechanisms for the coexistence of Diaptomus spp. in Clarke Lake, Ontario. *Limnol. Oceanogr.* 12:97–112.

Sarvala, J. 1979. Effect of temperature on the duration of egg, nauplius and copepodite development of some freshwater benthic Copepoda. *Freshwater Biol.* 9:515–534.

———. 1979a. A parthenogenetic life cycle in a population of Canthocamptus staphylinus (Copepoda, Harpacticoida). *Hydrobiologia* 62:113–129.

Siefken, M., and K. B. Armitage. 1968. Seasonal variations in metabolism and organic nutrients in three Diaptomus (Crustacea: Copepoda). *Comp. Biochem. Physiol.* 24:591–609.

Smyly, W. J. P. 1957. Observations on the life-history of the harpacticoid copepod Canthocamptus staphylinus (Jurine). *Ann. Mag. Nat. Hist.* 10(115):509–512.

Storch, O. 1929. Die Schwimmbewegung der Copepoden, auf Grund von Mikro-Zeitlupenaufnahmen analysiert. *Zool. Anz., Suppl. Bd.* 4:118–129.

Strickler, J. R. 1975. Intra- and interspecific information flow among planktonic copepods: receptors. *Proc. Int. Assoc. Theor. Appl. Limnol.* 19:2951–2958.

Walter, E. 1922. Über die Lebensdauer der freilebenden Süsswasser-Cyclopiden und andere Fragen ihrer Biologie. *Zool. Jahrb. Abt. Syst.* 44:375–420.

Ward, H. B., and G. C. Whipple. 1918. *Fresh-water Biology.* 1111 pp. Wiley, New York.

Wilson, C. B. 1911. North American parasitic copepods.–Part. 9. The Lernaeopodidae. *Proc. U.S. Nat. Mus.* 39:189–226.

———. 1911a. North American parasitic copepods belonging to the family Ergasilidae. *Ibid.* 263–400.

———. 1915. North American parasitic copepods belonging to the Lernaeopodidae, with a revision of the entire family. *Ibid.* 47:565–729.

———. 1916. Copepod parasites of fresh-water fishes and their economic relations to mussel glochidia. *Bull. U.S. Bur. Fish.* 34:333–374.

———. 1917. North American parasitic copepods belonging to the Lernaeidae with a revision of the entire family. *Proc. U.S. Nat. Mus.* 53:1–150.

———. 1918. The economic relations, anatomy, and life history of the genus Lernaea. *Bull. U.S. Bur. Fish.* 35:163–198.

———. 1932. The copepods of the Woods Hole region, Massachusetts. *Bull. U.S. Nat. Mus.* 158:1–635.

———. 1944. Parasitic copepods in the United States National Museum. *Proc. U.S. Nat. Mus.* 94:529–582.

Wilson, M. S. 1941. New species and distribution records of diaptomid copepods from the Marsh collection in the United States National Museum. *J. Wash. Acad. Sci.* 31:509–515.

———. 1953. New and inadequately known North American species of the copepod genus Diaptomus. *Smithson. Misc. Coll.* 122:1–30.

———. 1954. A new species of Diaptomus from Louisiana and Texas with notes on the subgenus Leptodiaptomus (Copepoda, Calanoida). *Tulane Stud. Zool.* 2:51–60.

———. 1955. A new Louisiana copepod related to Diaptomus (Aglaodiaptomus) clavipes Schacht. (Copepoda, Calanoida). *Ibid.* 3:37–47.

———. 1956. North American harpacticoid copepods. I. Comments on the known fresh water species of the Canthocamptidae. 2. Canthocamptus oregonensis, n. sp. from Oregon and California. *Trans. Am. Microsc. Soc.* 75:290–306.

———. 1958. New records and species of calanoid copepods from Saskatchewan and Louisiana. *Can. J. Zool.* 36:489–497.

———. 1958a. The copepod genus Halicyclops in North America, with description of new species from Lake Pontchartrain, Louisiana, and the Texas coast. *Tulane Stud. Zool.* 6:176–189.

———. 1958b. North American harpacticoid copepods. 4. Diagnoses of new species of fresh-water Canthocamptidae and Cletodidae (genus Huntemannia). *Proc. Biol. Soc. Wash.* 71:43–48.

———. 1972. Copepods of marine affinities from mountain lakes of western North America. *Limnol. Oceanogr.* 17:762–763.

———. 1975. North American harpacticoid copepods II. New records and species of Elaphoidella (Canthocamptidae) from the United States and Canada. *Crustaceana* 28:125–138.

Wilson, M. S., and S. F. Light. 1951. Description of a new species of diaptomid copepod from Oregon. *Trans. Am. Microsc. Soc.* 70:25–30.

Wilson, M. S., and W. G. Moore. 1953. Diagnosis of a new species of diaptomid copepod from Louisiana. *Ibid.* 72:292–295.

———. 1953a. New records of Diaptomus sanguineus and allied species from Louisiana, with the description of a new species (Crustacea: Copepoda). *J. Wash. Acad. Sci.* 43:121–127.

Yeatman, H. C. 1944. American cyclopoid copepods of the viridis-vernalis group (including a description of Cyclops carolinianus n. sp.). *Am. Midl. Nat.* 32:1–90.

———. 1959. Some effects of temperature and turbulence on the external morphology of Cyclops carolinianus. *J. Elisha Mitchell Sci. Soc.* 75:154–167.

———. 1964. A new cavernicolous cyclopoid copepod from Tennessee and Illinois. *J. Tenn. Acad. Sci.* 39:95–98.

18

OSTRACODA
(Seed Shrimps)

Although ostracods are abundant and widely distributed, they have received much less attention than the Cladocera and Copepoda. They inhabit all types of substrates in both standing and running waters, including rooted vegetation, algal mats, debris, mud, sand, and rubble. In Europe, many species have been found in hyporheic and hypogean waters (e.g., Danielopol, 1978). Similar assemblages undoubtedly occur in the United States. A few species swim about actively above the substrate, especially *Cypridopsis, Physocypria,* and *Cyprinotus.* Representatives of one family occur only as commensals (or parasites) on the gills of crayfish. Ostracods are comparatively difficult to study, chiefly because of their somewhat opaque bivalve shell (more accurately, a carapace). Identification usually involves dissection. Ostracods are richly represented as living marine groups and as fossils; indeed the fossil species record is enormous.

Superficially, the members of the Subclass Ostracoda resemble miniature mussels, and "mussel shrimps" is an old European vernacular name. However, this name is so easily confused with "clam shrimps," which is used for the Conchostraca, that it does not seem advisable to adopt it. "Seed shrimps" is suggested as an appropriate alternative, for indeed, without a lens ostracods do look much like small seeds.

General characteristics. The American fresh-water species are seldom more than 3 mm long and usually less than 1 mm long. A South African fresh-water species reaches a length of nearly 8 mm and the largest marine species is 21 mm long.

Coloration ranges from chalky white through yellow, green, gray, red, brown, and blackish. Light-colored valves are often blotched with darker colors, and species occurring among algae and rooted aquatics are usually gray, green, or brown.

Each sclerotized lime-impregnated valve consists of an inner and an outer plate that are fused along the anterior, posterior, and ventral margins. The space between the two plates is occupied by a thin skin fold that secretes the valve material. Dorsally, the skin folds are continuous with the main body of the animal. The valves are connected on the dorsal margin by an elastic band and may be tightly closed by a group of adductor muscle fibers passing transversely through the body of the ostracod and attached to the inner surfaces of the valves (Figs. 9, 11, 12). When the animal is active, however, the valves gape and the locomotor appendages protrude. The outer surface of the valves may be smooth, pitted, papillate, or setose, and the margins are often tuberculate, crenulate, or lipped. There are no concentric growth lines as in some Conchostraca.

All traces of body segmentation are lost (Fig. 2), but the region corresponding to the head bears four paired appendages: first antennae (antennules), second antennae (antennae), mandibles, and maxillae. The five- to seven-segmented first antennae are uniramous, the exopod being lost; they bear short, stiff, clawlike bristles for digging and climbing or long setae for swimming. The four- to six-segmented second antennae are also uniramous, the exopod being reduced to a scale or seta; these appendages are used in locomotion and feeding, and in the male they are modified for clasping the female during copulation. The mouth is surrounded by an upper and a lower lip and two lateral mandibles. Each mandible consists of a strongly toothed and sclerotized base, a branchial plate, and a three-segmented palp. Each maxilla has a large branchial plate and four basal processes, the outermost of which is largest and palplike.

The thoracic region bears three pairs of legs. In the Limnocytheridae and Entocytheridae all pairs are morphologically similar; in the Darwinulidae the second and third are similar; and in the Cypridae all legs are different. The first legs may be modified for mastication, prehension, copulation, or respiration (respiratory plate). The second legs usually have a long terminal claw. The third legs of the Cypridae are bent dorsally and used in keeping the body and inner surface of

The abdomen is represented only by two long caudal rami (furcal rami, or furca), which are articulated to the body. Each ramus usually has two terminal claws and two terminal setae. In some species the rami are reduced or lacking.

Locomotion. The great majority of individuals are found moving about on the substrate by means of beating movements of the first and second antennae, and to some extent by kicking of the caudal rami. Such locomotion ranges from creeping and uncertain, weak, tottering movements to rapid bouncing or scurrying. Some of the Candonidae burrow superficially in soft substrates. Those species that leave the bottom and swim about actively are characterized by long, plumose antennules and antennae.

Food, feeding. Food consists mostly of bacteria, molds, algae, and fine detritus, but some of the larger Cypridae have been observed feeding on living and dead animals. Ecologically, ostracods are omnivorous scavengers. Beating movements of the setose mandibular palps, maxillary processes, and the branchial plates of the first legs create a current of water between the valves. The current picks up fine particles from the substrate that are strained out by the setae and brought to the mouth. Inedible particles are mostly rejected and removed by the respiratory plates of the maxillae. Large edible bits of food are pulled or pushed toward the mouth by the antennae, mandibles, and first legs. The mandibles are used to rasp and break up

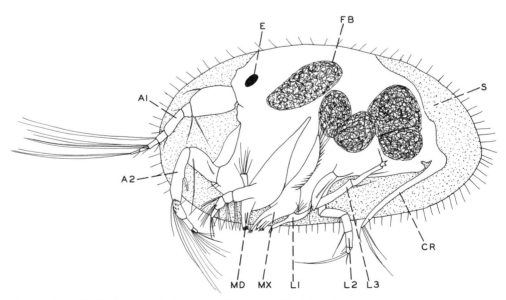

FIG. 1.–Female *Cypricercus reticulatus* (Zadd.), ×80, with left valve removed to show appendages, diagrammatic. A1, first antenna; A2, second antenna; CR, caudal ramus; E, eye; FB, food ball in digestive tract; L1, first leg; L2, second leg; L3, third leg; MD, mandible; MX, maxilla; S, right valve of shell. (Modified from Hoff, 1942.)

the larger pieces. Small balls of finely divided food are swept into the mouth by the mandibular palps.

Internal anatomy. The digestive tract consists of a short esophagus, a secretory and absorptive midgut, and a hindgut (Fig. 4). One digestive gland on each side between the lamellae of the valves empties into the midgut. The anus lies at the base of the caudal rami.

Respiration occurs through the general body surface, and a constant supply of oxygenated water is ensured by the movements of the branchial plates.

The hemocoel is filled with a circulating fluid, but no heart has ever been described for fresh-water species.

Three distinct, paired glands are found in ostracods, but for no one of them has a definite function been assigned. The large convoluted shell glands lie between the shell lamellae and in the hemocoel; they open near the base of the second antennae and have an

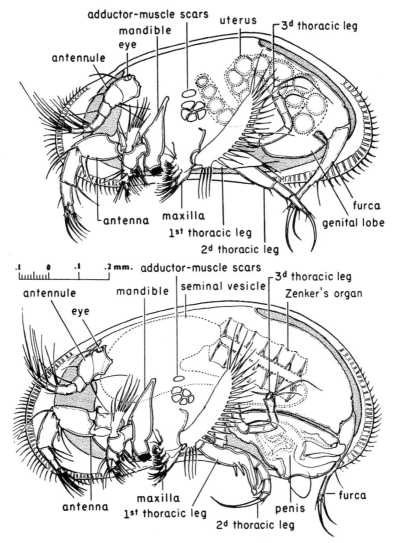

FIG. 2.–General anatomy of *Candona suburbana* Hoff. Upper, female with left valve removed; lower, male with left valve removed. (From Kesling (1956), courtesy Turtox/Cambosco, Macmillan Science Co., Chicago, Ill.)

unknown function. The small antennal glands open near the base of the first antennae; presumably they have an excretory function, especially during immature stages. The small maxillary glands are imbedded in the tissues just posterior to the lower lip; some investigators believe that they have an excretory function.

The nervous system consists of a large supraesophageal ganglion (brain), two esophageal connectives, a subesophageal ganglion, and a ventral chain of two paired ganglia.

Two or three prominent eyespots, each with a small lens, and fused in varying degrees, are mounted on a protuberance near the bases of the first antennae. In a few species the eyespots are absent. Many setae of the appendages undoubtedly have a tactile function. In the Cypridae the second antenna bears a small clublike sensory structure similar to the aesthetasks of copepods (Figs. 1, 8M).

Reproduction. A superficial examination of any field collection of ostracods usually reveals a great majority of females. The various species differ widely in their reproductive habits, depending on the relative occurrence of males, and Hoff (1942) has accordingly recognized four general groups of ostracods. One group, including some species of *Candona* and *Cypricercus,* always reproduces by parthenogenesis, the males being unknown. In some species of *Candona, Cyprinotus, Darwinula,* and *Ilyocypris* males are rare and reproduction is usually parthenogenetic. In some *Cypricercus, Potamocypris, Cypridopsis,* and *Darwinula* males occur in small numbers, and reproduction is both syngamic and parthenogenetic. A fourth group includes species in which males are always present and reproduction is usually syngamic: *Candona, Cyclocypris, Cypria, Cypricercus, Cyprois, Limnocythere, Notodromas,* and *Physocypria.*

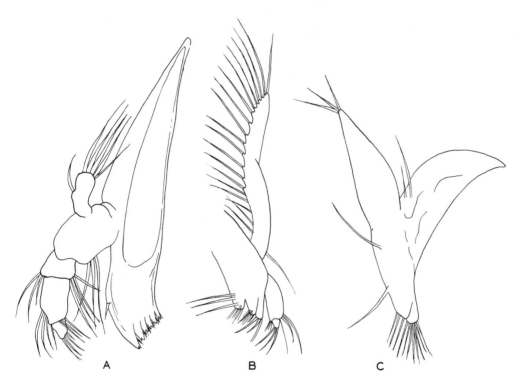

FIG. 3.–Some ostracod appendages. A, mandible of *Cypridopsis vidua,* showing the long, sclerotized, toothed basal portion, the single-segmented branchial plate, and the larger segmented palp; B, maxilla of *Candona acuta* Hoff, showing the four basal processes and the long branchial plate; C, first leg of *C. fluviatilis* Hoff. (A modified from Sars, 1928; B and C redrawn from Hoff, 1942.)

The female reproductive system (Fig. 5A) is simple, consisting of paired tubular ovaries, oviducts, seminal receptacles, and genital openings. In the Cypridae the ovaries lie between the valve lamellae, but in the Cytheridae they are lateral to the midgut. The genital openings are located between the bases of the third legs and the caudal rami.

The paired male reproductive system is more complicated, especially in the Cypridae

(Figs. 5B; 6). Each testis consists of four long, tubular branches lying between the valve lamellae. In the body proper these unite to form a vas deferens that enters the dorsal and posterior ejaculatory duct, a spiny, sclerotized cylinder (Zenker's organ) that forces sperm into the penis during copulation. The two penes are large, irregular, lobed, complex, sclerotized structures located near the posterior end of the animal. In the Cytheridae

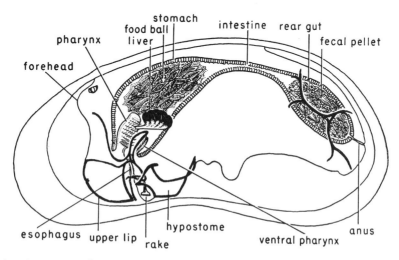

FIG. 4.–Digestive system of *Candona suburbana,* left valve removed. Sclerotized body framework of left side is shown in heavy lines. (From Kesling (1956), courtesy Turtox/Cambosco, Macmillan Science Co., Chicago, Ill.)

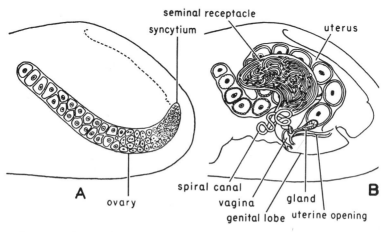

FIG. 5.–Female ostracod reproductive system. A, ovary in position in hypodermis; B, right half of female reproductive system. (From Kesling (1956), courtesy Turtox/Cambosco, Macmillan Science Co., Chicago, Ill.)

there is no ejaculatory duct, and the testes are lateral to the midgut in the body proper. The detailed anatomy of the penis is of importance in separating species taxonomically. The sperm are the largest in the animal kingdom.

Reproduction is usually restricted to the summer months, but a few species produce more than one generation per year.

Copulation involves the attachment of the male to the dorsal and posterior margin of the female carapace. The second antennae of the male aid in this attachment, and the penes are inserted between the valves and into the female genital pores. Fertilization occurs in the seminal receptacles.

The careful investigations of Lowndes (1935) have led to some interesting conclusions. He maintains that fertilization of the egg does not actually occur in spite of copulation and highly specialized copulatory apparatus, that all ostracod reproduction is parthenogenetic, and that copulation is merely an instinctive behavior pattern that is a relic of past times when the union of sperm and egg was probably the sole means of reproduction.

Darwinula stevensoni retains the fertilized

eggs in the dorsal part of the shell cavity until embryonic development is completed, but all other American fresh-water species are thought to be oviparous. The spherical eggs have a delicate, double-walled, limy shell and are usually white, yellow, orange, red, or green. They are deposited singly or in clumps or rows on rocks, twigs, other bits of debris, or aquatic vegetation.

Development, life history. Egg development is usually suspended during the cold months and unfavorable moisture conditions, and there are numerous records of ostracods hatching from eggs kept in dried pond mud in the laboratory for many weeks and months. One report indicates that viable eggs may be kept for more than 20 years in the desiccated condition.

In normal habitats, however, the eggs usually hatch in a few days to several months. The liberated larva (Fig. 7) is a shelled nauplius with three pairs of appendages, representing the first antennae, second antennae, and mandibles of the adult at intervals of 1 to 3 days. Eight molts and eight additional instars

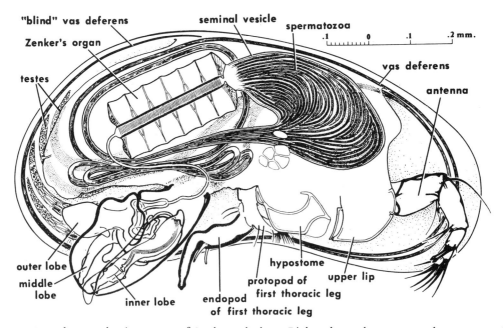

FIG. 6.–Male reproductive system of *Candona suburbana*. Right valve and most appendages removed. Hemipenis slightly lowered from normal noncopulatory position. (From McGregor and Kesling, 1969.)

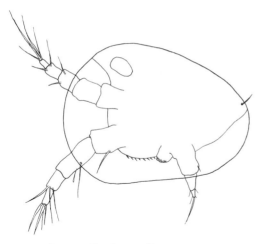

FIG. 7.–Nauplius larva of *Cyprinotus incongruens* Ramdohr, ×260. (Redrawn from Schreiber, 1922.)

occur during the life history, the adult being the ninth instar. Other appendages and increased structural complexity develop with each successive molt. The caudal rami, for example, first appear as two small bristles in the fourth instar. The copulatory organs appear in the eighth instar, but sexual maturity and copulation do not occur until the ninth and last instar.

Many species exhibit a definite seasonal periodicity. Ferguson (1944), working in Missouri, found that *Cypridopsis vidua* was active between February and December, *Potamocypris smaragdina* between March and October, and *Physocypria pustulosa* between March and September. McGregor found that *Darwinula stevensoni* had its reproductive period between May and October. Many species of *Candona* may be collected only in the spring. A few species occur only in the autumn. Aside from eggs and some adults that winter over, the usual life span is completed in several weeks to 7 or 8 months, depending on the species and environmental conditions. Sometimes the adult instar may last as long as 6 months. A single generation per year is the common condition in vernal ponds; a few species have two or three generations per year.

Ecology. The nature of the substrate and the general type of environment seem to have little influence on the distribution of most seed shrimps. In many cases the same species may be found on algae, decaying vegetation, rooted aquatics, mud, and gravel. *Cyprinotus incongruens* is exceptional, however, and may be found in swift streams, ponds, and puddles. Only a minority of species are restricted to such places as vernal ponds and temporary streams. Out of 31 species collected in Illinois by Hoff (1942), 4 were restricted to temporary still waters, 6 were restricted to permanent still waters, 2 were found only in temporary running waters, and 1 in permanent running waters. The other 18 species were collected in two or more of these habitats. The largest numbers of species may be collected on mud bottoms where there is little or no current, often in aquatic vegetation; few species occur on bare rocks. Habitats having wave action are usually without ostracods. Core samples taken in soft substrates show few ostracods deeper than 5 cm. Many forms, especially clamberers and active swimmers, such as *Cyprinotus* and *Cypricercus,* are more active on lighted than on shaded bottoms. Most species occur in water less than 1 m deep, but some are found regularly as deep as 15 m. *Cypridopsis vidua* has been found in depths ranging from 1 cm to 300 m. *Notodromas monacha* is unusual in that it clings to the surface film with the dorsal surface downward.

On several occasions, I have found dense clusters of ostracods, amounting to several thousand individuals, covering 100 to 200 cm^2 on the substrates of slow streams. These did not appear to be feeding aggregations, but may have been a temporary reproductive phenomenon. As a group, ostracods occur in waters having pH readings of 4.0 to 8.0, but some species are definitely restricted to acid conditions and others to alkaline conditions.

In general, ostracods tolerate wide ranges of ecological factors usually regarded as limiting for entomostracans. The normal ranges of temperature and water chemistry are of little significance. Though they do not occur in grossly polluted waters, such genera as *Candona* and *Cypris* survive long periods of stagnation and oxygen exhaustion on lake bottoms. It is now known that advanced instars may tide over dry and cold periods by being buried in a torpid state in the mud of dried ponds. Aquaria and old protozoan cultures sometimes become thick with them.

A few hypogean species occur in the waters of underground caves and in interstitial gravels.

The Entocytheridae are common commensals on the gills and articular membranes of crayfishes, sometimes in great densities. They may be removed from their hosts and will live under such artificial conditions for at least several months. However, they will not reproduce unless they are attached to the crayfish gills.

Geographical distribution. Active and passive distribution are brought about much as in the Cladocera and Copepoda. Most genera and many species are cosmopolitan or Holarctic, *Chylamydotheca* and the Entocytheridae being the only important major North American taxa not found in the Palaearctic region. Even in New Zealand most of the ostracod fauna consists of cosmopolitan species.

In South Africa and New Zealand two species have become specialized for living in the damp humus of forest floors.

Because little intensive collecting has been done in the United States, about half of the reported species are known from only a single state or collecting locality. The known distribution of some species is hopelessly incomplete, for example, "Alaska and Florida"!

Economic significance. With occasional exceptions, ostracods are only a minor element in the diet of young and adult fish. A few are the intermediate hosts of Acanthocephala parasitizing fish. Hoff (1942) estimates that 1.0 to 1.5% of the ostracods in permanent waters are infected. Ostracods are also the intermediate hosts of some tapeworms of water fowl.

Collection, preparation. Perhaps the chief reason for the general neglect of the American Ostracoda lies in the fact that they are nearly always necessarily collected in masses of debris from which they must be laboriously separated. Surface debris of the bottom and its contained ostracods may be scooped up with a dipnet, but the Birge cone net towed along the bottom, or better, the

Birge cone net mounted on a tobogganlike piece of sheet metal is more useful. Ekman dredge samples are equally useful. By washing the debris thoroughly through a fine net or sieve, the smaller particles may be removed, and if the remaining mass is allowed to stand quietly in jars of water, the ostracods soon become active at the debris–water interface, where with a little patience they may be picked up under the dissecting binoculars with a long pipette or Irwin loop. Also, by stirring the contents well, some of the ostracods get a small bubble of air caught between the valves and rise to the surface, where they may be easily removed.

Specimens should be killed in 50% alcohol and preserved in 85% alcohol. Levinson (1951) gives a good method for anesthetizing and killing with the shell gaping. A few drops of 10% chloral hydrate or 0.1% chloretone are good relaxers. Formaldehyde makes the animals brittle and may decalcify the valves.

Stained or unstained whole animals may be mounted in balsam or glycerin jelly, but cover-glass supports should be used to avoid crushing. Dissections of adult specimens are usually necessary for accurate identification, and the operation is best carried out in glycerin with fine needles. The valves should be removed first, leaving the body intact, and then the appendages should be carefully dissected off. Valves and appendages are conveniently mounted on separate slides.

Culturing. Little is known about culturing ostracods in the laboratory, but some investigators report success by feeding dried yeast and bits of lettuce, celery, or vegetable debris.

Taxonomy. The Subclass Ostracoda is divided into four major living taxa, variously called orders or suborders: Myodocopa, Cladocopa, Platycopa, and Podocopa. The first three of these are confined to marine waters and are not considered here. The Podocopa are found in both marine and fresh-water environments; the family Bairdiidae is exclusively marine, the Cytheridae contains only a few fresh-water genera. Cyprididae are well represented in both environments, and the Darwinulidae are confined to fresh waters.

In the United States, fresh-water seed shrimps have received more than casual consideration only in Ohio, Massachusetts, Illinois, Florida, and Washington, and each new comprehensive study reveals many undescribed species. Ward and Whipple (1918) listed 54 species from the entire United States, but in 1933 Furtos listed 57 species from Ohio alone. Dobbin (1941) collected 33 species in Washington, of which 14 were new, and Hoff (1942) collected 39 in Illinois, of which 10 were new. There are now more than 300 species known from the United States. Delorme (1967) lists 74 Canadian species.

Candona is the only large U.S. genus, with about 40 species. Most of the other genera have less than 15 U.S. species. Obviously, it is fruitless to include here a key to the known American species, since such a key would soon be obsolete. The key given below therefore goes only as far as genera. At the species level the details of the male reproductive apparatus are important. It should be borne in mind that it can be used successfully only for specimens in the late instars, preferably the last. Almost all ostracod research during the past 20 years in the United States has been restricted to the Entocytheridae.

KEY TO GENERA OF OSTRACODA

1. Exopod of second antenna in the form of a long, hollow seta carrying the secretion from a gland located at the base of the antenna (Fig. 12E); all legs morphologically similar 30
   Exopod otherwise; all legs not morphologically similar 2
2. Second and third legs similar (Figs. 8A, B); caudal rami lacking, body terminating in a single process (Fig. 8C); 0.70 to 0.80 mm long; on mud bottoms of larger bodies of water; widely distributed but rare .... DARWINULIDAE, **Darwinula stevensoni** (B. and R.)
   Second and third legs not similar (Figs. 8D, E); caudal rami present but sometimes reduced 3
3. Terminal segment of third leg short, usually armed with beaklike claw and reflexed seta (Fig. 11H).
   　　　　4
   Terminal segment of third leg more or less cylindrical (rarely spherical), armed with three setae and no claw (Fig. 8E) 20
4. Caudal rami well developed, usually with two claws and two setae (Figs. 9C, D) ..... CYPRIDIDAE, 6
   Caudal rami rudimentary, terminating in a simple flagellum (Fig. 8G); 0.50 to 0.80 mm long.
   　　　　CYPRIDOPSIDAE, 5
5. Tumid (Fig. 8H); valves nearly equal in size; widely distributed and very common in many types of habitats .... **Cypridopsis**
   Compressed; right valve higher than left (Fig. 8J); common in standing and running waters.
   　　　　Potamocypris
6. Third leg with a cylindrical terminal segment without a claw and armed with three setae, one of which is reflexed (Fig. 12D) .... CANDOCYPRINAE, 7
   Third leg with a short terminal segment, a beaklike claw, and a reflexed seta (Fig. 10D).
   　　　　CYPRINAE, 8
7. Natatory setae of second antenna rudimentary (Fig. 11F); 0.80 to 1.00 mm long; a burrower and creeper; reported from brooks in OH .... **Candocypria osburni** Furtos
   Natatory setae of second antenna well developed; with two large setae on the penultimate segment of third leg (Fig. 12D); 0.45 mm long; WA .... **Cyclocypria kincaidia** Dobbin
8. Shell with reticulate patterns anteriorly and posteriorly; caudal ramus with three claws and one seta (Fig. 9C); 1.10 to 1.20 mm long; rare in ponds and slow streams.
   　　　　Ilyodromus pectinatus Sharpe
   Shell smooth (margins tuberculate in some species) 9
9. Caudal rami slightly different in shape (Fig. 10A); one rare FL species .... **Stenocypris**
   Caudal rami similar 10
10. Natatory setae of second antenna well developed, extending beyond midlength of claws (Fig. 9B).
    　　　　11
    Natatory setae of second antenna greatly reduced, not attaining midlength of claws (Fig. 9H, J) 17
11. Second segment of second leg with two large setae; 2.60 to 3.30 mm long; several uncommon species in southern half of the United States .... **Chlamydotheca**

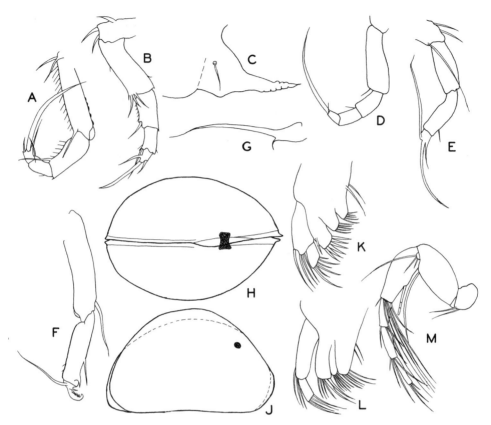

FIG. 8.–Structure of Ostracoda. A, second leg of female *Darwinula stevensoni*; B, third leg of same; C, end of thorax of same; D, second leg of female *Candona punctata* Furtos; E, third leg of same; F, third leg of female *Cypridopsis vidua*; G, caudal ramus of female *Potamocypris smaragdina* Vavra; H, dorsal view of female *Cypridopsis vidua,* ×70; J, lateral view of female *Potamocypris hyboforma* Dobbin, ×40; K, maxilla of *Cyprois marginata* (Straus); L, maxilla of female *Cyprinotus incongruens*; M, second antenna of female *Notodromus monacha.* (A, B, F, and G redrawn from Furtos, 1933; C–E redrawn from Hoff, 1942; H modified from Furtos, 1933; J and L modified from Dobbin, 1941.)

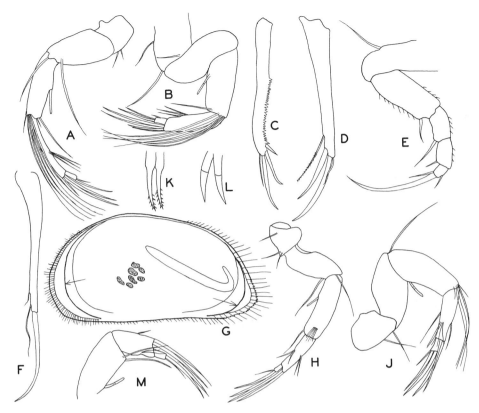

FIG. 9.–Structure of Ostracoda. A, second antenna of female *Cyprois marginata*; B, second antenna of female *Cyprinotus incongruens*; C, caudal ramus of *Ilyodromus pectinatus*; D, caudal ramus of female *Cyprinotus incongruens*; E, second leg of female *Eucypris virens* (Jurine); F, caudal ramus of *Cypretta turgida* (Sars); G, lateral view of female *Candonocypris pugionis* Furtos, ×15 (anterior and posterior ridges indicated by arrows); H, second antenna of female *Prionocypris longiforma*; J, second antenna of female *Herpetocypris repetans* (Baird); K, spines of third maxillary process of *H. repetans*; L, spines of third maxillary process of *Megalocypris gigantea* (Dobbin); M, end of female second antenna of *Candona caudata* Kaufmann. (B, H, J, and M modified from Dobbin, 1941; C redrawn from Ward and Whipple, 1918; D redrawn from Furtos, 1933; E modified from Sars, 1928; F redrawn from Furtos, 1933, 1936; G modified from Furtos, 1936; K and L redrawn from Dobbin, 1941.)

17. Natatory setae of second antenna very small (Figs. 9H, K) **19**
     Natatory setae of second antenna longer, reaching nearly halfway to tip of terminal claws (Fig. 9J); 2.0 to 4.0 mm long; several widely distributed but uncommon species **18**
18. Second antenna robust (Fig. 10E); valves 0.60 to 0.80 mm long; common in Europe but reported only from a TN spring .... **Scottia**
     Second antennae not robust (Fig. 9H); valves 1.25 mm long; known only from WA.
                                                                       Prionocypris longiforma Dobbin
19. Spines of third maxillary process distinctly denticulate (Fig. 9K); less than 2.5 mm long.
                                                                                      Herpetocypris
     Spines of third maxillary process smooth (Fig. 9L); 2.5 to 5.0 mm long .... **Megalocypris**
20. Natatory setae of second antenna lacking (Fig. 9M); shell white when dry; creeping and burrowing forms .... **CANDONIIDAE, 21**
     Natatory setae of second antenna present; shell not white when dry **23**
21. Shell ornamented with polygonal areas and tubercles (Fig. 11D); 0.50 to 0.60 mm long; rare, scattered distribution .... **Paracandona euplectella** (B. and N.)
     Shell not ornamented (Fig. 11E) **22**

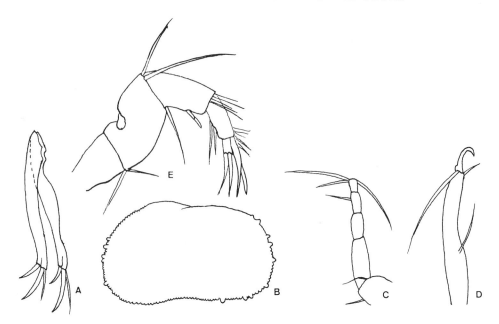

FIG. 10.–Structure of Ostracoda. A, caudal rami of *Stenocypris*; B, outline of *Pelocypris* shell; C, third leg of *Candona*; D, tip of third leg of *Herpetocypris*; E, second antenna of female *Scottia*.

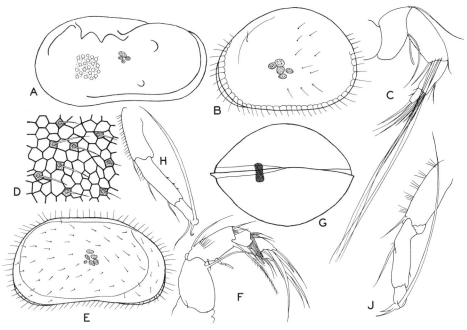

FIG. 11.–Structure of Ostracoda. A, left valve of *Ilyocypris gibba* (Ramdohr), ×47, showing a small portion of sculpturing; B, left valve of *Cyclocypris sharpei* Furtos, ×75; C, second antenna of female *Cypria elegantula* (Lillj.); D, valve sculpturing of *Paracandona euplectella*; E, right valve of female *Candona parallela* Müller, ×55; F, second antenna of male *Candocypria osburni*; G, dorsal view of female *Cyclocypris sharpei*, ×75; H, third leg of female *C. sharpei*; J, third leg of female *Cypria elegantula*. (A, B, D, E, and G modified from Furtos, 1933; F, H, and J redrawn from Furtos, 1933; C modified from Dobbin, 1941.)

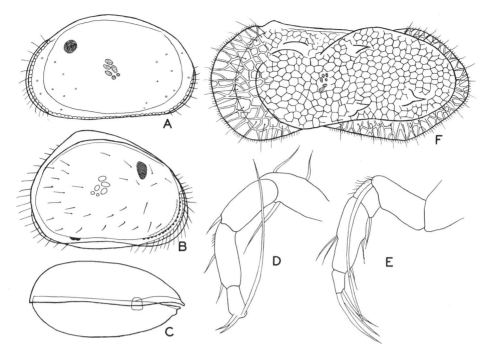

FIG. 12.–Structure of Ostracoda. A, left view of male *Cypria obesa* Sharpe, ×60; B, right view of female *Physocypria pustulosa* Sharpe, ×70; C, dorsal view of same, ×70; D, third leg of female *Cyclocypria kincaidia*; E, second antenna of female *Limnocythere verrucosa* Hoff; F, left valve of male *Limnocythere ornata* Furtos, ×80. (A, C, and F redrawn from Furtos, 1933; B modified from Furtos, 1933; D redrawn from Dobbin, 1941; E redrawn from Hoff, 1942.)

22. Terminal segment of third thoracic leg with three unequal setae (Fig. 10C); 0.60 to 2.00 mm long; unable to swim; widely distributed and very common; about 40 species .... **Candona**
    Terminal segment of third thoracic leg with two setae; rare, FL .... **Candocyprinotus**
23. Shell pitted or reticulate 24
    Shell smooth 26
24. Natatory setae of second antenna used for swimming but extending only slightly beyond tips of terminal claws .... ILYOCYPRIDIDAE, 25
    Antenna stout and used for walking; one species in deep, cold lakes.
                          CYTHERIDEIDAE, *Cytherissa lacustris* (Sars)
25. Margins of shell without mammillary pustules (Fig. 11A); several rare species in streams.
                          Ilyocypris
    Margins of shell with numerous mammillary pustules (Fig. 10B); rare .... **Pelocypris**
26. Third maxillary process with six prominent spines (Fig. 8K) .... NOTODROMADIDAE, 27
    Third maxillary process with only two or three prominent spines plus some setae.
                          CYCLOCYPRIDIDAE, 28
27. Second antenna six segmented (Fig. 8M); 1.10 to 1.20 mm long; active swimmers; rare; usually in aquatic vegetation .... **Notodromus monacha** (O.F.M.)
    Second antenna five segmented (Fig. 9A); 1.10 to 1.70 mm long; widely distributed in creeks and vernal ponds .... **Cyprois**
28. Shell tumid (Fig. 11B); last segment of third leg at least twice as long as wide (Fig. 11H); 0.80 to 1.40 mm long; widely distributed .... **Cyclocypris**
    Shell compressed; last segment of third leg less than twice as long as wide (Fig. 11J) 29
29. Valves nearly equal in size; margins of both valves smooth (Fig. 12A); 0.40 to 1.00 mm long; many widely distributed species .... **Cypria**

Valves unequal in height or length or both; margin of either valve more or less tuberculate (Figs. 12B, C); 0.40 to 0.70 mm long; widely distributed .... **Physocypria**

30. Free margins of valves flattened, with many long pore canals (Fig. 12F); free living in a variety of habitats; 0.50 to 0.90 mm long .... LIMNOCYTHERIDAE, **Limnocythere**

Free margins of valves not flattened, with many long pore canals; commensals on gills and articular membranes of crayfish; 0.30 to 0.70 mm long; about 170 U.S. species in 20 genera, with the great majority of species being restricted to the southeastern states (see Hart and Hart, 1974 and more recent studies) .... ENTOCYTHERIDAE

REFERENCES

Danielopol, D. L. 1978. Über Herkunft und Morphologie der Süsswasser-hypogäischen Candoninae (Crustacea, Ostracoda). *Sitzungsber. Öster. Akad. Wiss. Math. Naturw. Kl. Abt.* 1:1–160.

Delorme, L. D. 1967. Field key and methods of collecting freshwater ostracodes in Canada. *Can. J. Zool.* 45:1275–1281.

———. 1969. On the identity of the ostracode genera Cypriconcha and Megalocypris. *Ibid.* 47:271–281.

———. 1970. Freshwater ostracodes of Canada. Part I. Subfamily Cypridinae. *Ibid.* 48:153–168.

———. 1970a. Freshwater ostracodes of Canada. Part II. Subfamily Cypridopsinae and Herpetocypridinae, and Family Cyclocyprididae. *Ibid.* 253–266.

———. 1970b. Freshwater ostracodes of Canada. Part III. Family Candonidae. *Ibid.* 1099–1127.

———. 1970c. Freshwater ostracodes of Canada. Part IV. Families Ilyocyprididae, Notodromadidae, Darwinulidae, Cytherideidae, and Entocytheridae. *Ibid.* 1251–1259.

———. 1971. Freshwater ostracodes of Canada. Part V. Families Limnocytheridae, Ooxochonchidae. *Ibid.* 49:49–64.

Delorme, L. D., and D. Donald. 1969. Torpidity of freshwater ostracodes. *Ibid.* 47:997–999.

Dobbin, C. N. 1941. Fresh-water Ostracoda from Washington and other western localities. *Univ. Wash. Publ. Biol.* 4:175–245.

Ferguson, E. 1944. Studies of the seasonal life history of three species of freshwater Ostracoda. *Am. Midl. Nat.* 32:713–727.

———. 1957. Ostracoda (Crustacea) from the northern lower Peninsula of Michigan. *Trans. Am. Microsc. Soc.* 76:212–218.

———. 1958. A supplementary list of species and records of distribution for North American freshwater Ostracoda. *Proc. Biol. Soc. Wash.* 71:197–202..

———. 1959. A synopsis of the ostracod (Crustacea) genus Cypridopsis with the description of a new species. *Ibid.* 72:59–68.

———. 1959a. The ostracod genus Potamocypris with the description of a new species. *Ibid.* 133–138.

———. 1964. Stenocyprinae, a new subfamily of freshwater cyprid ostracods (Crustacea) with description of a new species from California. *Ibid.* 77:17–24.

———. 1964a. The ostracod (Crustacea) genus Cypridopsis in North America and a description of

Cypridopsis howei, sp. nov. *Trans. Am. Microsc. Soc.* 83:380–384.

———. 1967. Cyprinotus newmexicoensis, a new cyprid ostracod. *Am. Midl. Nat.* 78:248–250.

———. 1967a. New ostracods from the playa lakes of eastern New Mexico and western Texas. *Trans. Am. Microsc. Soc.* 86:244–249.

———. 1968. Recently described species and distributional records for North American freshwater Ostracoda. *Am. Midl. Nat.* 79:499–506.

Furtos, N. C. 1933. The Ostracoda of Ohio. *Bull. Ohio Biol. Surv.* 29:413–524.

———. 1935. Fresh-water Ostracoda from Massachusetts. *J. Wash. Acad. Sci.* 25:530–544.

———. 1936. Freshwater Ostracoda from Florida and North Carolina. *Am. Midl. Nat.* 17:491–522.

Hart, C. W. 1962. A revision of the ostracods of the Family Entocytheridae. *Proc. Acad. Nat. Sci.* 114:121–147.

Hart, C. W., and D. G. Hart. 1969. Evolutionary trends in the ostracod Family Entocytheridae, with notes on the distributional patterns in the southern Appalachians. *Va. Polytech. Inst. Res. Div. Monogr.* 1:179–190.

Hart, D. G., and C. W. Hart. 1974. The ostracod family Entocytheridae. *Monogr. Acad. Nat. Sci.* 18:1–239.

Hoff, C. C. 1942. The ostracods of Illinois. *Univ. Ill. Biol. Monogr.* 19:1–196.

———. 1942a. The subfamily Entocytherinae, a new subfamily of fresh-water cytherid Ostracoda, with descriptions of two new species of the genus Entocythere. *Am. Midl. Nat.* 27:63–73.

———. 1943. Seasonal changes in the ostracod fauna of temporary ponds. *Ecology* 24:116–118.

———. 1944. The origin of Nearctic fresh-water ostracods. *Ibid.* 25:369–372.

Howe, H. V. 1962. *Handbook of ostracod taxonomy.* 386 pp. Louisiana State University Press, Baton Rouge.

Kesling, R. V. 1951. The morphology of ostracod molt stages. *Ill. Biol. Monogr.* 21:1–324.

———. 1956. The ostracod. A neglected little crustacean. *Turtox News* 34:82–86, 90–94, 114–115.

———. 1957. Notes on Zenker's organs in the ostracod Candona. *Am. Midl. Nat.* 57:175–182.

Kosmal, A. 1968. On the distribution of Ostracoda in the littoral of Lake Kisajno (Masurian Lake District). *Pol. Arch. Hydrobiol.* 15:87–102.

Levinson, S. A. 1951. A technique for anesthetizing fresh water ostracodes. *Am. Midl. Nat.* **46**:254–255.

Lowndes, A. G. 1935. The sperms of freshwater ostracods. *Proc. Zool. Soc. London* (1935) 1:35–48.

McGregor, D. L. 1969. The reproductive potential, life history and parasitism of the freshwater ostracod Darwinula stevensoni (Brady and Robertson). *In:* Taxonomy, morphology and ecology of recent Ostracoda (ed.) J. W. Neale, pp. 194–221, Oliver and Boyd, Edinburgh.

McGregor, D. L., and R. V. Kesling. 1969. Copulatory adaptations in ostracods. Part I. Hemipenes of Candona. *Univ. Mich. Contrib. Paleontol.* 22:169–191.

———. 1969a. Copulatory adaptations in ostracods. Part II. Adaptations in living ostracods. *Ibid.* 221–239.

Sars, G. O. 1928. Ostracoda. *An Account of the Crustacea of Norway* 9:1–277. Bergen Museum, Bergen, Norway.

Scheerer-Ostermeyer, E. 1940. Beitrag zur Entwick-

lungsgeschichte der Süsswasserostrakoden. *Zool. Jahrb. Abt. Anat. Ontog. Tiere* 66:349–370.

Schreiber, E. 1922. Beiträge zur Kenntnis der Morphologie, Entwicklung und Lebensweise der Süsswasser-Ostrakoden. *Zool. Jahrb. Abt. Anat. Ontog. Tiere.* 43:485–538.

Sohn, I. G., and L. S. Kornicker. 1973. Morphology of Cypretta kawarai Sohn and Kornicker, 1972. (Crustacea, Ostracoda), with a discussion of the genus. *Smithson. Contrib. Zool.* 141:1–28.

Tressler, W. L. 1947. A check list of the known species of North American freshwater Ostracoda. *Am. Midl. Nat.* 38:698–707.

Turpen, J. B., and R. W. Angell. 1971. Aspects of molting and calcification in the ostracod Heterocypris. *Biol. Bull.* 140:331–338.

Ward, H. B., and G. G. Whipple. 1918. *Fresh-water Biology.* 1111 pp. Wiley, New York.

Young, W. 1971. Ecological studies of the Entocytheridae (Ostracoda). *Am. Midl. Nat.* 85:399–409.

19

MYSIDACEA
(Opossum Shrimps)

The opossum shrimps form a large order that is almost exclusively marine, only three species having become adapted to fresh and mildly brackish waters in the United States. *Mysis relicta* Lovén (formerly *Mysis oculata relicta*) is a native plankter in a few deep, cold, oligotrophic lakes of the northern states east of the Great Plains, in the Great Lakes, and in many Canadian lakes. *Neomysis mercedis* Holmes occurs in lakes, rivers, and brackish estuaries of the Washington, Oregon, and California coasts. *Taphromysis louisianae* Banner is known from roadside ditches in Louisiana and Texas. *T. bowmani* Băcescu is known only from mildly brackish waters in Florida and Alabama. In contrast, Europe has about 20 fresh-water, troglobitic, and mildly brackish species.

Superficially, mysids resemble miniature crayfish, but a closer examination reveals some striking differences. The carapace for example, is thin and does not completely cover the thorax, the last two segments being exposed dorsally and, to some degree, laterally. In place of the five pairs of walking legs of decapods, mysids have long, thin, setose, biramous, many-segmented appendages. The first thoracic segment bears a pair of maxillipeds, and is fused with the head, but the other seven thoracic segments are distinct. The second thoracic segment bears the second maxillipeds, and each of the remaining six segments has a pair of swimming legs. The stalked compound eyes are extremely large, the scale of the second antenna is large, but the pleopods are proportionately much more reduced than those of decapods. *Mysis relicta* reaches a length of 30 mm, *Neomysis mercedis* reaches 15 mm, and *Taphromysis* only 8 mm.

Mature male mysids may be distinguished from females by their long and specialized fourth pleopods, which may extend as far as the posterior end of the telson, whereas females are characterized by a marsupium consisting of four ventral oostegites originating at the bases of the last two pairs of legs (hence the name "opossum shrimps"). Females are generally larger and more abundant than males.

The exopods of the six pairs of legs project laterally somewhat and their beating produces smooth, rapid swimming. The endopods of these appendages extend more medially, and their movements create a current of water that passes anteriorly over the mouthparts. The two pairs of maxillipeds are particularly efficient in straining zooplankton, phytoplankton, and particulate debris from this current. Such food is then passed on toward the second maxillae, first maxillae, mandibles, and mouth. *Mysis* has been characterized as an "opportunistic" feeder. If zooplankton is abundant, *Mysis* uses it as a primary food source. If zooplankton is sparse, then *Mysis* appears to obtain food from suspended organic detritus or from the surface of bottom deposits. Thus, the feeding habits are both raptorial and filter feeding.

Internal anatomy and physiology are generally comparable to conditions found in the crayfish. There are no gills, however, most respiration occurring through the thin lining of the carapace. A current of water is drawn under the carapace by the action of the epipodites of the first maxillipeds. A unique feature is the occurrence of a statocyst in the basal portion of the endopod of each uropod.

A female may produce up to 40 eggs per clutch, depending on her size. Developing

FIG. 1.–Male *Mysis relicta*, ×2.

eggs and young are carried in the marsupium for 1 to 3 months, the young leaving the mother when 3 or 4 mm long. Development is direct and the life cycle is thought to extend through 2–4 years in northern latitudes or high altitudes but only 1 year farther south. Undoubtedly food availability is also involved in this range of life cycles. Little is known about the number of clutches of eggs that may be produced by a female during her life cycle.

Fresh-water mysids, and especially *Mysis relicta*, are cold-water forms. During the warm months this species is almost restricted to the hypolimnion, where temperatures range as low as 4°C. It is thought that 14°C is the maximum temperature that can be tolerated for any length of time, and certainly *Mysis* has never been reported from lakes where the hypolimnial maximum exceeds this figure. Newly released juveniles migrate in spring to warm shallow waters. Later they move back to deep waters.

Of all the fresh-water Crustacea known to exhibit daily vertical migrations, *Mysis* has the most extensive and rapid movements. Populations are confined to the meter of water just above the bottom during the middle of the day and no amount of sampling above that depth will take specimens. At dusk, however, most of the individuals, especially the older ones, migrate into the surface waters where temperatures as high as 20°C may be tolerated for a few hours. Reverse downward movements begin at dawn.

M. relicta has been recorded a few times from water that was only about 15% saturated with oxygen or 1 ppm, but the great majority of investigations show that 30 or 40% saturation is the usual lower limit. In exceptional years the oxygen content of the water just above the bottom of oligotrophic lakes falls below this limit, and populations are then confined to strata above the deepest waters during the daytime.

Original native populations of *M. relicta* have been reported from the Great Lakes (as deep as 270 m in Lake Superior); Geneva, Trout, and Green lakes in Wisconsin; some of the Finger Lakes in New York; a few other lakes in the northern states; and many deep Canadian lakes. It is especially abundant in Waterton Lake on the border between Montana and Canada. Undoubtedly it is much more common and widely distributed in the United States than the published literature indicates. This form is circumpolar and occurs also in the British Isles, the Scandinavian countries, northern Germany, and northern Russia. A few zoogeographers contend that it is a marine relict from glacial times when many northern lake basins were presumably connected with the sea and filled with salt water, but a more acceptable theory, postulating postglacial migration from the sea, is now favored. In addition to being found in fresh waters, *M. relicta* has been reported from northern brackish waters and estuaries.

M. relicta is closely related to *Mysis oculata* (Fabr.), a circumpolar, arctic and subarctic, sublittoral, marine species, from which it is undoubtedly derived. Adults of the two forms differ slightly in the detailed structure of the telson and antennal scale. The immature stages are almost indistinguishable, however.

Neomysis mercedis is a species that has apparently evolved from marine ancestry very recently. It occurs in brackish bays and estuaries on the west coast where the salinity is usually less than 20.0°/oo and in strictly fresh waters of rivers and lakes near the coast.

Fresh-water mysids are an extremely important food source for a group of commercially important fishes inhabiting deep, cold waters—especially lake trout and coregonids. Food studies show that often 80 to 100% of the stomach contents of these fishes may be *Mysis*. Because of its value in the trout diet, *Mysis* is a logical artificial transplant, especially because the "*Mysis*-niche" is unoccupied in most lakes. The genus has already been successfully established in several western Canadian lakes, a few Colorado mountain lakes and some in California. Experimental plants have also been made in Wisconsin, Minnesota, and New York, Nevada, and Idaho; and undoubtedly elsewhere in the western states. Similar transplants have been tried in Canada, Sweden, and Russia. Some transplants are known to be successful or provisionally successful; others are still uncertain or unsuccessful.

Often the introduction of *Mysis* has resulted in drastic decreases in densities of *Daphnia* and *Bosmina*, as well as shifts in density ratios of other cladoceran zooplankters, much to the detriment of trout and salmon populations. Such interactions seem to be more common in shallow (rather than very deep) oligotrophic lakes. Less is known about inter-

actions between *Mysis* and copepod zooplankters. One group of investigators established the following order of prey preferences: cladocerans > copepod copepodites > copepod nauplii > adult copepods.

The American species belong to the Subfamily Mysinae in the Family Mysidae. Aside from their distinctive geographic ranges and size differences, they are easily distinguished morphologically. The telson of *Mysis relicta* and *Taphromysis louisianae* has a wide, bifurcated tip, but that of *Neomysis* has a narrow truncated tip. The exopod of the fourth pleopod of the mature male *Neomysis* is two segmented; that of *M. relicta* is six segmented; that of *Taphromysis* is indistinctly seven segmented. Further, in the male *Mysis* the exopod of the third pleopod is well developed and consists of six segments; in *Taphromysis* this ramus is lacking.

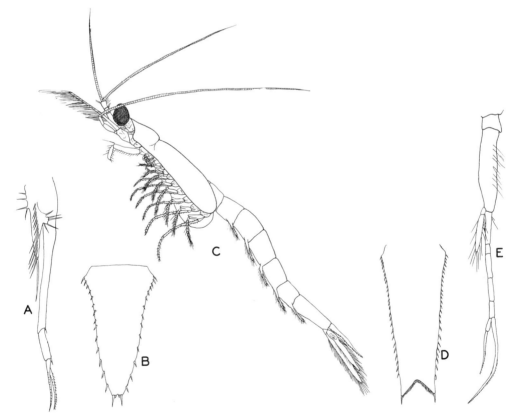

FIG. 2.–Mysidacea. A, male fourth pleopod of *Neomysis mercedis*; B, telson of *N. mercedis*; C, female *Mysis relicta*, ×8; D, telson of same; E, male fourth pleopod of same. (A and B modified from Tattersall.

REFERENCES

Banner, A. H. 1953. On a new genus and species of mysid from southern Louisiana. *Tulane Stud. Zool.* 1:1–8.

Juday, C., and E. A. Birge. 1927. Pontoporeia and Mysis in Wisconsin lakes. *Ecology* 7:445–452.

Kinsten, B., and P. Olsen. 1981. Impact of Mysis relicta Lovén introduction on the plankton of two mountain lakes, Sweden. *Rep. Inst. Freshwater Res. Drottningholm* 59:64–74.

Kost, A. L. B., and A. W. Knight. 1975. The food of Neomysis mercedis Holmes in the Sacramento–San Joaquin estuary. *Calif. Fish and Game* 61:35–46.

Morgan, M. D. 1981. Impact of introduced populations of Mysis relicta on zooplankton in oligotrophic subalpine lakes. *Verh. Int. Ver. Theor. Angew. Limnol.* 21:339–345.

———. (ed.) 1982. Ecology of Mysidacea. *Hydrobiologia* 93:1–222.

Murtagh, P. A. 1981. Selective predation by Neomysis mercedis in Lake Washington. *Limnol. Oceanogr.* 26:445–453.

Pesce, G. L. 1976. Stato attuale delle conoscenze sui misidacei cavernicoli e freatici (Crustacea). *Notiz. Circolo Speleol. Romano* 1:47–57.

Rieman, B. E., and C. M. Falter. 1981. Effects of the establishment of Mysis relicta on the macrozooplankton of a large lake. *Trans. Am. Fish. Soc.* 110:613–620.

Schulze, P. 1926. Schizopoda. *Biol. Tiere Deutschl.* 17:1–18.

Tattersall, W. M. 1951. A review of the Mysidacea of the United States National Museum. *Bull. U.S. Nat. Mus.* 201:1–292.

Thienemann, A. 1925. Mysis relicta. *Z. Morphol. Ökol. Tiere* 3:389–440.

20

ISOPODA
(Aquatic Sow Bugs)

Pill bugs and sow bugs are chiefly terrestrial and marine, only about 5% of the total known North American species being found in fresh waters. Approximately 130 fresh-water species occur in the United States, and for the most part these are restricted to springs, spring brooks, streams, and hyporheic, interstitial, and subterranean waters. A decided minority are found in ponds and the shallows of lakes. The biology of *Asellus aquaticus* L. and of several other European species has been thoroughly investigated, but relatively little is known about our American species.

General characteristics. Unlike amphipods, the members of the Order Isopoda are strongly flattened dorsoventrally. The "head" of an isopod is actually a cephalothorax, since it represents the fusion of the true head and the first thoracic segment. The seven remaining thoracic segments are all similar and are expanded laterally in the form of eavelike lamellae over the basal portions of the thoracic appendages. In all American fresh-water species the last four abdominal segments and true telson are completely fused into a large shieldlike region, often incorrectly referred to as a "telson." The first two abdominal segments are greatly reduced and scarcely visible at the anterior end of the abdomen. In the Sphaeromidae the abdomen consists of two segments, and in *Cirolanides* there are six distinct segments.

Total length, from the anterior margin of the cephalothorax to the posterior margin of the abdomen, usually ranges between 5 and 20 mm.

Coloration may be blackish, brown, gray, dusky, reddish, or yellowish. Some species are variously marked or mottled. Subterranean species are generally whitish or creamy.

The first antennae arise near the median line and anterior margin of the cephalothorax. Each consists of a three-segmented peduncle and a many-segmented flagellum. The second antennae are much longer than the first; they arise ventral and lateral to the first antennae and have five-segmented peduncles.

The eyes are dorsal, unstalked, compound, and usually reduced or absent in subterranean species.

In typical, common species the mouth parts form a compact buccal mass that is covered anteriorly by an upper lip. The sclerotized mandibles are strong, toothed, and may or may not have a palp. Each first maxilla has a basal portion and two elongated, toothed and setose palps. The second maxillae are smaller, weaker, more platelike and have three short, terminal divisions. The maxillipeds, representing the appendages of the fused thoracic segment, are flattened and palped; they fit closely together at the median line, thus effectively forming the lower and posterior surfaces of the buccal mass.

Each of the seven free thoracic segments bears a pair of long walking legs or peraeopods (also spelled pereiopods and pereopods), all of which are similar except the first pair (gnathopods), which are usually subchelate and used for grasping. In general, the more posterior legs are the largest. The coxae are completely fused with the body, but the basis, ischium, merus, carpus, propodus, and dactylus are distinct in all the legs. Mature females have large, platelike oostegites attached at the inner bases of several of the anterior

FIG. 1.–*Caecidotea,* ×6.5. (Setation omitted.)

pairs of legs. These oostegites extend inward and collectively form a shallow chamber, or marsupium, on the ventral side of the thorax in which incubating eggs and young are carried. A similar device occurs in the amphipods.

The first five pairs of abdominal appendages (pleopods) are considerably modified and hidden beneath the abdomen. The sixth pair of appendages (uropods), however, are unmodified and project well beyond the posterior end of the body.

In the female the first pleopods are absent, and the second pleopods are reduced and one segmented. The third, fourth, and fifth pleopods, however, are large, flattened, and biramous. The exopod of the third pleopod forms a sclerotized protective operculum, but the third endopod and both rami of the fourth and fifth pleopods are delicate and respiratory.

The male first pleopods are small, elongated, uniramous, and two segmented. The second are biramous and greatly specialized for copulation and the transfer of sperm to the female. The third to sixth abdominal appendages are similar to those of the female.

Males may be distinguished from females by their larger body size, larger and more specialized gnathopods, absence of oostegites, and the structure of the first and second pleopods.

Locomotion is restricted to slow crawling in fresh-water species. Only the Sphaeromidae are capable of rolling up into a ball as do some of the land isopods.

Species inhabiting subterranean waters are notable for their more elongated body,

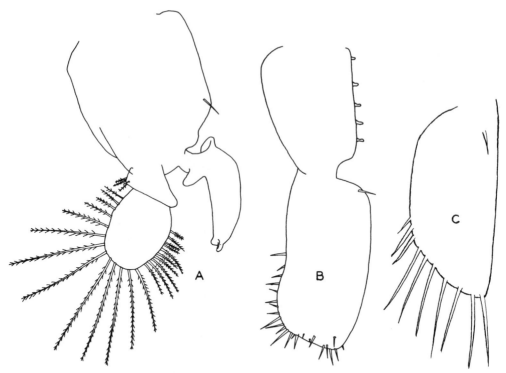

FIG. 2.–*Caecidotea* appendages. A, second pleopod of male *C. communis*; B, first pleopod of male *C. communis*; C, "first" pleopod of female *Caecidotea* (actually this is the second pleopod, the true first pleopod being absent in all females.) (A and B from Williams, 1972.)

long weak legs, and long antennae, as well as for their abundance of tactile hairs.

Food, feeding. Isopods are perhaps best characterized as scavengers, since they have been observed eating dead and injured aquatic animals of all kinds and both green and decaying leaves, grass, and aquatic vegetation.

Internal anatomy, physiology. The digestive system consists of a short esophagus, a stomach containing a gastric mill, a long intestine, and a rectum. Four long caeca arise at the junction of the stomach and intestine.

The large heart is located in the posterior part of the thorax. It receives blood from the pericardial chamber via one to three pairs of ostia and sends it out to the hemocoel through a surprisingly large number of arteries (up to 11 in some species). In at least one species, however, the heart has no arteries and no ostia. Most of the carbon dioxide–oxygen exchange occurs through the thin-walled

third to fifth pleopods, although it is also possible that the general body surface may also be of some respiratory significance.

Two coiled maxillary glands, probably excretory, open at the base of the second maxillae. In addition, other small paired glands in the cephalothorax have sometimes been interpreted as excretory devices.

The nervous system is quite simple in structure; it consists of a brain, circumesophageal connectives, subesophageal ganglia, a double ventral nerve cord with double segmental ganglia in the thorax, and a group of fused ganglia in the anterior portion of the abdomen.

Reproduction. Males have three pairs of ovoid testes in the middle of the thorax and dorsal to the intestine. The vasa deferentia open at the tips of two fingerlike projections on the ventral surface of the seventh thoracic segment. (These structures should not be mistaken for the first pleopods, which imme-

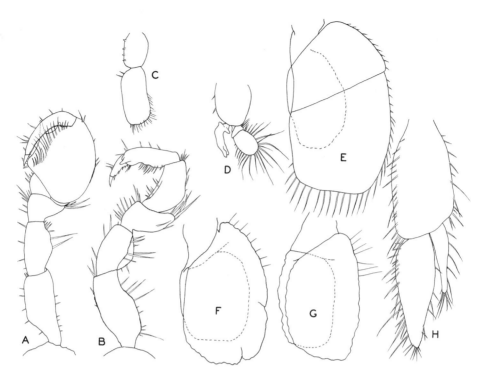

FIG. 3.–Appendages of a male *Caecidotea*; all in ventral view, from left side, and all to same magnification. A, first leg or pereiopod (gnathopod); B, fourth leg or pereiopod; C, first pleopod; D, second pleopod; E, third pleopod; F, fourth pleopod; G, fifth pleopod; H, uropod.

diately follow.) The two ovaries are dorsal, long, and baglike, and the oviducts open through two slitlike genital pores on the ventral surface of the fifth thoracic segment. Just beneath the surface of the body each oviduct is swollen to form a seminal receptacle.

Reproductive habits have been observed for only a few species. In general, breeding may occur throughout the year and does not seem to be governed by temperature, but very few gravid females can be found between October and February. There are one or two broods of young per year.

When ready for copulation, a mature male seizes a female and carries her, ventral side down, beneath his body with his second to fourth legs. Such females either may be in the preadult instar or may already have shed the posterior half of their exoskeleton in the preadult molt. If a male seizes an immature female or one that has already been impregnated, he releases her almost immediately.

Little is known about the duration of the clasping and copulatory period, but in *Lirceus*, the American genus whose reproductive habits have been most carefully studied, it lasts up to 24 hours.

The female genital pores are not exposed until the preadult exoskeleton has been shed, and if the female has not already lost the posterior half of her exoskeleton, the clasping male usually aids in pulling it off. The male then assumes a copulatory position by sliding sideways so that his ventral surface is pressed against the side of the body of the female. This places the male genital pores and the first and second pleopods in the vicinity of the female genital pore on that side. As sperm are released, the male first and second pleopods undergo rapid vibratory movements that probably aid in transmitting the sperm into the female genital pore and into a seminal receptacle. The process is then repeated on the other side of the female's body, the entire bilateral copulation being completed in less than 2 hours.

The two individuals then separate, and the female promptly sheds the anterior half of her preadult exoskeleton, the oostegites thereby being transformed from small buds to large functional plates.

After an interval of as long as several days, the mature eggs pass down the oviducts, through the seminal receptacles, where they are fertilized, out the genital pores, and into the marsupium. Depending on the species and size of the female, the number of eggs per brood ranges between 20 and 250. Incubating eggs and newly hatched young are retained in the marsupium for 20 to 30 days. They are aerated by slow up-and-down movements of the oostegites and maxilliped movements that produce an anterior-posterior current of water through the marsupium. The young eventually find their way out of the posterior end of the marsupium by trial and error.

Development, life cycle. First instar young have the general characteristics of adults. Little is known about the total number of instars during the isopod life cycle. Presumably there may be at least 15, of which the first five to eight are probably preadult. Casting of the exoskeleton is completed in a few minutes to as much as three days. The process is initiated by a transverse split on the fourth thoracic segment. The posterior half of the exoskeleton is lost first, then, after a variable interval, the anterior half. The life span is thought to be about one year or less. Little specific information is available concerning the number of broods that may be produced during the female life cycle.

Ecology. Fresh-water isopods seldom come into open waters but remain secreted under rocks, vegetation, and debris. They are primarily inhabitants of the unpolluted shallows, rarely being found in water more than a meter deep. *Lirceus lineatus* is a notable exception in having been collected at depths up to 55 m, and *Caecidotea racovitzai* has been found in the Great Lakes as deep as 42 m.

Some species are indicators of organic pollution, inasmuch as they may be abundant in the "recovery" zone of streams polluted by domestic sewage.

A single brook or pond almost invariably contains only one species, and it is very unusual to find two species in the same habitat.

A few species, such as *Caecidotea forbesi, C. obtusus,* and *C. scrupulosus,* may occur in temporary ponds, where they burrow into the substrate during periods of drought, but adaptations for withstanding adverse environmental conditions are not generally developed in this order.

Reynoldson (1961), working in Great Britain, has found a positive correlation between the occurrence of *Asellus* and the concentration of calcium and dissolved solids.

Striking masses, or aggregations, of *Caecidotea*, consisting of scores to thousands of individuals have been observed in small streams. Such aggregations are thought to be formed chiefly as the result of reactions to current velocity. The migration of some individuals upstream as far as possible, or if the current is too swift, being washed downstream until they are able to maintain a footing has the net result of concentrating the isopods in a segment of the stream where the current is neither too fast nor too slow.

Except for *Sphaeroma terebrans,* all of our species of fresh-water isopods are restricted to North America. Furthermore, aside from about 10 species that are widely distributed in this country, they are known mostly from single localities (especially caves and springs) or single states. The common European species *Asellus aquaticus* (L.) does not occur in the United States.

Though the loss of sight is generally characteristic of species confined to subterranean waters, and functional eyes are characteristic of surface water species, there are some interesting exceptions and variations. *Caecidotea oculata* is found in springs and streams, yet the eyes are reduced. *Lirceus hoppinae*, with functional eyes, occurs in both caves and surface waters. *Caecidotea hobbsi* (Maloney) is subterranean and has eyes present or absent. *C. spatulata* (Mackin and Hubricht) is a secretive surface form, but the eyes may be present or absent.

Economic significance. With few exceptions, isopods are restricted to small lakes and streams, and consequently they are of little importance in the diet of fishes. The writer, however, knows of several small trout lakes where there are large populations of *Caecidotea communis* and where the rainbow trout populations exhibit extremely rapid growth rates. Economically, *Sphaeroma terebrans* is perhaps the most important species. It is a borer in salt and brackish waters, as well as in fresh-water estuaries of the Gulf coast, and causes extensive damage to wharves and piling. This species has numerous important relatives restricted to salt water. *Lirceus brachyurus* sometimes becomes a pest in commercial beds of water cress. Some isopods serve as intermediate hosts for parasitic nematodes and Acanthocephala of birds and fishes and Acanthocephala of amphibians.

Collection, preparation. Isopods are easily collected by hand picking, washing out aquatic vegetation, or with a small net. Seventy percent alcohol is a suitable killing agent and preservative. Appendages may be dissected off under the binoculars and permanently mounted in glycerin jelly.

Taxonomy. Many of the older descriptions of isopods are useless because they stress highly variable characters such as the number of segments in the antennae and the body length–width ratio. In recent years greater emphasis has been placed on the detailed structure of the mature male second pleopods and gnathopods. These features are constant within each species and there are considerable differences from one genus and species to another.

Most studies in the United States have been scattered and uncorrelated. In 1918 only 21 species were known; now there are about 130, and the several active American investigators in this group have specimens of many more undescribed species in their collections. There are still differences of opinion as to which characters are of generic and specific rank. The status of *Caecidotea* and *Asellus* has been the object of a long controversy; some workers contend that the two should be united under *Caecidotea*, and this policy has been followed in this manual.

The key to epigean species of *Caecidotea* is modified from Williams (1970). The hypogean species are poorly known and are not included here; most are known only from single localities. It also appears fruitless to attempt a complete key for *Lirceus* until this genus has been more carefully studied; our key includes the six most common species. Unfortunately, identification of most species of *Caecidotea* is dependent on minute anatomical differences in the sclerotized tip of the endopod of the second pleopod of the male. Furthermore, this structure usually shows morphological variability.

KEY TO ISOPODA

1. Free living; morphologically unspecialized 2
   Parasitic in branchial chambers of Palaemonidae (*Macrobrachium* and *Palaemonetes*); greatly modified for parasitic existence, appendages reduced, female asymmetrical (Fig. 4); 1 species.
   Suborder **EPICARIDEA, BOPYRIDAE, Probopyrus**
2. Uropods terminal (Fig. 1); epigean and hypogean .... Suborder ASELLOTA, ASELLIDAE, 6
   Uropods broad and inserted laterally on abdomen, forming with it a large, horizontally expanded, fanlike structure (Fig. 5B) .... Suborder **FLABELLIFERA, 3**
3. Abdomen with two distinct segments (Fig. 5B); eyes present; up to 10 mm long.
   SPHAEROMATIDAE, 5
   Abdomen with six distinct segments (Fig. 5A); eyes absent .... CIROLANIDAE, 4
4. Exopod of second pleopod and endopods of third to fifth pleopods two segmented; up to 17 mm long; cave systems of south central TX .... **Cirolanides texensis** Benedict
   Exopod of second pleopod and endopods of third to fifth pleopods one segmented; up to 12 mm long; found only in Madison Cave, August County, VA .... **Antrolana lira** Bowman
5. Second, third, and fourth segments of palp of maxilliped not produced into lobes (Fig. 5D); outer margins of uropods serrate; a borer in timbers, roots, and mangroves in fresh, brackish, and salt waters; FL to TX .... **Sphaeroma terebrans** Bate

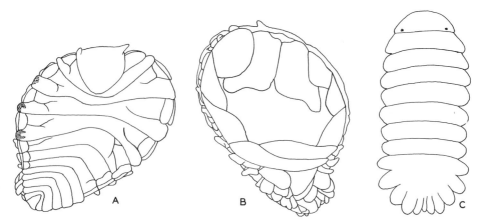

FIG. 4.–*Probopyrus bithynis*, parasitic on *Palaemonetes ohionis*. A, dorsal view of female, ×16; B, ventral view of female, ×16; C, dorsal view of male, ×40. (Modified from Richardson, 1904.)

Second, third, and fourth segments of palp of maxilliped produced into lobes (Fig. 5C); outer margins of uropods smooth; one species in hot springs near Socorro, NM, and another in hot springs in western TX . . . . **Thermosphaeroma**

6. First pereiopods normal and subchelate, used for grasping (Figs. 6G, J, L) **8**
    First pereiopods simple, not subchelate (Fig. 6P); blind, unpigmented **7**

7. Pereiopods broad and fringed with long setae; able to swim weakly; FL cave systems.
    Remasellus parvus (Steeves)
    Pereiopods not broad and fringed; small, blind, unpigmented; two hypogean TX species . . . **Lirceolus**

8. Suture between exopod segments of third pleopod transverse; troglobitic, blind, unpigmented; body long and thin; known only from Shaver Lake area in Sierra National Forest, CA.
    Calasellus longus Bowman
    Suture between exopod segments of third pleopod more or less oblique (Fig. 6H, 3E) **9**

9. Anterior margin of cephalothorax with a low but distinct carina (Figs. 6A–C); suture between exopod segments of third pleopod running from the median posterior angle very obliquely toward the lateral margin (Fig. 6H); up to 25 mm long; eyes present; only one hypogean species; eastern states as far west as the Great Plains; at least 15 epigean species; usually common in highly restricted ranges; six of these species are known only from single localities in single states; *Lirceus lineatus* (Say) is most widespread and has been taken in at least 10 eastern states; the 6 most common and widely distributed species are keyed below . . . . **Lirceus, 10**
    Without a carina on the anterior margin of cephalothorax (Fig. 7F); suture between exopod segments of third pleopod running from the median margin less obliquely toward the lateral margin (Fig. 3E); up to 20 mm long; eyes present or absent; about 80 species in the United States; only males may be identified with this key; only a few typical species are keyed here . . . . **Caecidotea, 17**

10. Mandible with a one- to three-segmented palp, often quite small and stumplike **11**
    Mandible without a palp **12**

11. Abdomen distinctly broader than long (Fig. 6D); reported from streams and springs in MO, AR, OK.
    Lirceus hoppinae (Faxon)
    Abdomen only slightly wider than long (Fig. 6F); ponds, springs, and spring brooks in eastern states; common . . . . **Lirceus fontinalis** Raf.

12. Middle and distal propodal processes absent (Fig. 6D); usually in water cress; PA, VA.
    Lirceus brachyurus (Harger)
    Middle process and usually also the distal process of the propodus present (Fig. 6J) **13**

13. Endopod of each uropod as long or longer than the basal segment (Fig. 6E) **14**
    Endopod of each uropod shorter than the basal segment (Fig. 6M) **16**

14. Proximal process of propodus as large or larger than the middle process (Fig. 6J); wells and permanent or temporary streams in MO, AR, KS, LA . . . **Lirceus garmani** Hubricht and Mackin

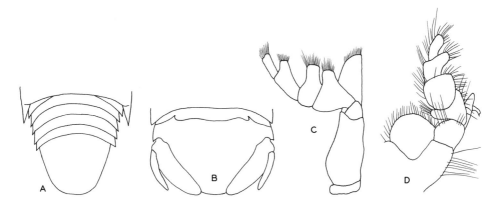

FIG. 5.–Structure of Flabellifera. A, abdomen of *Cirolanides texensis*; B, posterior end of *Thermosphaeroma*; C. palp of maxilliped of same; D, palp of maxilliped of *Sphaeroma terebrans*. (A–C modified from Richardson, 1905; D modified from Van Name, 1936.)

Proximal process of propodus smaller than middle process (Fig. 6L) **15**

15. Uropods flattened (Fig. 6E) **11**
    Uropods cylindrical (Fig. 6M); the most common and widely distributed species in this genus; Great Lakes region and probably in all states east of the Great Plains ...... **Lirceus lineatus** (Say)

16. Abdomen longer than broad (Fig. 6O); streams, sloughs, ditches, and marshes in IL, MO, AR, LA.
    Lirceus louisianae (Mackin and Hubricht)
    Abdomen as broad as long or distinctly broader (Figs. 6D, N) **15**

17. Eyes absent; translucent, whitish, or creamy coloration; abdomen longer than broad (Fig. 7A); subterranean springs, wells, and caves; rarely in surface waters; about 50 poorly known species, not included in this key .... **Caecidotea**
    Eyes present; usually gray or brownish coloration; abdomen broader than long (Fig. 1); up to 16 mm long; not subterranean .... **Caecidotea, 18**

18. Palm of propodus of first pereiopod without a triangular process near the midpoint (Fig. 8N); running and standing waters in northwestern states .... **Caecidotea occidentalis** (Williams)
    Palm of propodus of first pereiopod with a small to large triangular process near the midpoint (Figs. 7C, D) **19**

19. First pleopod distinctly longer than second **20**
    First pleopod equal to or shorter than the second pleopod **25**

20. Endopod of second pleopod twisted so that the ventral groove is not visible in ventral aspect (Figs. 8D, L) **21**
    Endopod of second pleopod not twisted; ventral groove clearly visible in ventral aspect **22**

21. Endopodial armature of second pleopod forming a terminal spiral structure (Fig. 8L); AR, OK.
    Caecidotea montanus (Mackin and Hubricht)
    Endopodial armature of second pleopod consisting of two large, heavily sclerotized structures showing little torsion (Fig. 8D); known only from MD ...... **Caecidota nodulus** (Williams)

22. Lateral process at tip of endopod of second pleopod not developed, but medial process large and bifid, and caudal process wide and dentate (Fig. 8B) small creeks in AR, LA.
    Caecidotea dentadactylus (Mackin and Hubricht)
    Lateral process well developed, and caudal process either absent or broadly rounded **23**

23. Uropod never more than 0.7 as long as telson; springs and streams in AR, IL, KY, MO.
    Caecidotea brevicaudus (Forbes)
    Uropod about as long as telson **24**

24. Medial process at tip of endopod of second pleopod dentate, lateral process sclerotized and pointed, caudal process not developed (Fig. 8C); WV pools ..... **Caecidotea scrupulosus** (Williams)

FIG. 6.–Structure of asellids. A, cephalothorax of *Lirceus fontinalis*; B, cephalothorax of *L. brachyurus*; C, cephalothorax of *L. lineatus*; D, abdomen of *L. hoppinae*; E, uropod of *L. hoppinae*; F, abdomen of *L. fontinalis*; G, gnathopod of *L. brachyurus*; H, third pleopod of *L. brachyurus*; J, gnathopod of *L. garmani*; K, uropod of *L. garmani*; L, gnathopod of *L. lineatus*; M, uropod of *L. lineatus*; N, abdomen of *L. lineatus*; O, abdomen of *L. louisianae*; P, pereiopod of *Remasellus*. (A and C modified from Racovitza, 1920a; D, E, G, and J–O modified from Hubricht and Mackin, 1949; O from Bowman and Sket, 1985).

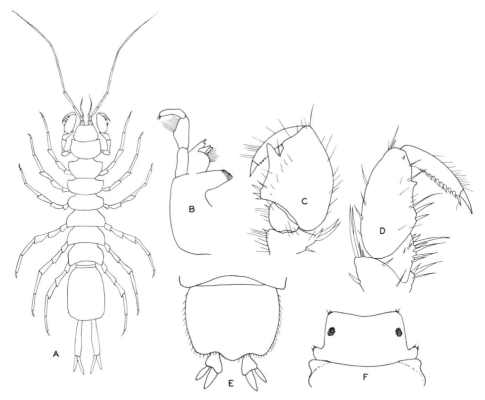

FIG. 7.–Structure of *Caecidotea*. A, *C. stygius* (Packard), ×7 (setation omitted), a subterranean species; B, mandible of *Caecidotea*; C, gnathopod of male *C. communis*; D, gnathopod of female *C. communis*; E, posterior end of *C. brevicaudus*; F, anterior end of *C. intermedius*. (A modified from Hay, 1903; B modified from Huntsman, 1915; C and D modified from Racovitza, 1920.)

      Medial process not dentate, lateral process rounded, caudal process developed and broadly rounded with a few irregularities (Fig. 8A); small streams and spring-fed creeks in D.C., MD, PA, VA.

                    Caecidotea kenki (Bowman)

25. Medial process absent **26**
    Medial process present **28**
26. Caudal process absent (Fig. 8J); creeks and ponds in KY, LA .... **Caecidotea laticaudatus** (Williams)
    Caudal process present **27**
27. Caudal process usually with acutely pointed apex; cannula (tube) short and wide (Fig. 8H); common in many kinds of waters in east central states .... **Caecidotea intermedius** (Forbes)
    Caudal process usually broadly rounded; cannula long and narrow (Fig. 8K); very common in northeastern states but known as far west as CO .... **Caecidotea communis** (Say)
28. Cannula long and narrow (Fig. 8G) **29**
    Cannula short and wide (Figs. 8E, F) **30**
29. First and second pleopods about the same length; common in northeastern states but reported also from WA .... **Caecidotea racovitzai** (Williams)
    First pleopod shorter than second; reported only from Dismal Swamp, VA.

                Caecidotea attenuatus (Richardson)

30. Second antennae about as long as body; medial process short and wide, cannula very wide (Fig. 8E); southeastern states .... **Caecidotea obtusus** (Williams)
    Second antennae shorter than body; medial process long and narrow (Fig. 8F); common and widely distributed in east central states .... **Caecidotea forbesi** (Williams)

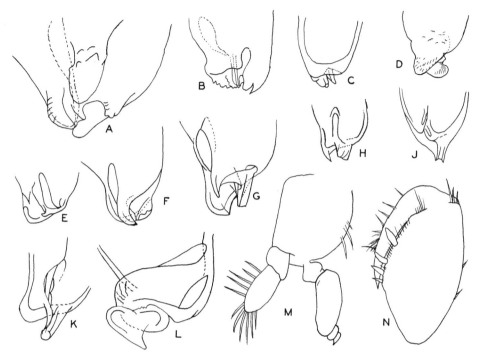

FIG. 8.–Appendage structure of male *Caecidotea*. A to L, tip of endopod of second pleopod. A, *C. kenki*; B, *C. dentadactylus*; C, *C. scrupulosus*; D, *C. nodulus*; E, *C. obtusus*; F, *C. forbesi*; G, *C. racovitzai*; H, *C. intermedius*; J, *C. laticaudatus*; K, *C. communis*; L. *C. montanus*; M, second pleopod of *C. montanus*; N, gnathopod of *C. occidentalis*. (All redrawn from Williams, 1972.)

REFERENCES

Allee, W. C. 1929. Studies in animal aggregations: natural aggregations of the isopod, Asellus communis. *Ecology* **10**:14–36.

Bowman, T. E. 1964. Antrolana lira, a new genus and species of troglobitic cirolanid isopod from Madison cave, Virginia. *Int. J. Speleol.* **1**:229–236.

_____. 1967. Asellus kenki, a new isopod crustacean from springs in the eastern United States. *Proc. Biol. Soc. Wash.* **80**:131–140.

_____. 1974. The California freshwater isopod, Asellus tomalensis, rediscovered and compared with Asellus occidentalis. *Hydrobiologia* **44**:431–441.

_____. 1981. Calasellus longus, a new genus and species of troglobitic asellid from Shaver Lake, California (Crustacea: Isopoda: Asellidae). *Proc. Biol. Soc. Wash.* **94**:866–872.

_____. 1981a. Thermosphaeroma milleri and T. smithi, new sphaeromatid isopod crustaceans from hot springs in Chihuahua, Mexico, with a review of the genus. *J. Crust. Biol.* **1**:105–122.

Bowman, T. E., and G. Longley. 1976. Redescription and assignment to the new genus Lirceolus of the Texas troglobitic water slater, Asellus smithii (Ulrich)

(Crustacea: Isopoda: Asellidae). *Proc. Biol. Soc. Wash.* **88**:489–496.

Bowman, T. E., and B. Sket. 1985. Remasellus, a new genus for the troglobitic asellid isopod, Asellus parvus Steeves. *Proc. Biol. Soc. Wash.* **98**:554–560.

Cole, G. A., and C. A. Bane. 1978. Thermosphaeroma subequalum, n. gen., n. sp. (Crustacea: Isopoda) from Big Bend National Park, Texas. *Hydrobiologia* **59**:223–228.

Ellis, R. J. 1961. A life history study of Asellus intermedius Forbes. *Trans. Am. Microsc. Soc.* **80**:80–102.

_____. 1971. Notes on the biology of the isopod Asellus tomalensis Harford in an intermittent pond. *Ibid.* **90**:51–61.

Fleming, L. E. 1973. The evolution of North American isopods of the genus Asellus (Crustacea: Asellidae). *Int. J. Speleol.* **5**:283–310.

Hatchett, S. P. 1947. Biology of the Isopoda of Michigan. *Ecol. Monogr.* **17**:47–79.

Hay, W. P. 1903. Observations on the crustacean fauna of Nickajack Cave, Tennessee, and vicinity. *Proc. U.S. Nat. Mus.* **25**:417–439.

Henry, J. P., and G. Magniez. 1970. Contribution a la systématique des asellides (Crustacea Isopoda). *Ann. Speleol.* **25**:335–367.

Holsinger, J. R., and T. E. Bowman. 1973. A new troglobitic isopod of the genus Lirceus (Asellidae) from southwestern Virginia, with notes on its ecology and additional cave records for the genus in the Appalachians. *Int. J. Speleol.* **5**:261–271.

Holsinger, J. R., and H. R. Steeves. 1971. A new species of subterranean isopod crustacean (Asellidae) from the central Appalachians, with remarks on the distribution of other isopods of the region. *Proc. Biol. Soc. Wash.* **84**:189–199.

Hubricht, L., and J.G. Mackin. 1949. The fresh-water isopods of the genus Lirceus (Asellota, Asellidae). *Am. Midl. Nat.* **42**:334–349.

Huntsman, A. G. 1915. The fresh-water Malacostraca of Ontario. *Contrib. Canad. Biol. ii (Suppl.)* **47**:145–163.

Lewis, J. J. 1983. The assignment of the Texas troglobitic water slater Caecidotea pilus to the genus Lirceolus, with an amended diagnosis of the genus (Crustacea: Isopoda: Asellidae). *Proc. Biol. Soc. Wash.* **96**:145–148.

Lewis, J. J., and T. E. Bowman. 1981. The subterranean asellids (Caecidotea) of Illinois (Crustacea: Isopoda: Asellidae). *Smithson. Contrib. Zool.* **335**:1–66.

Mackin, J. G., and L. Hubricht. 1938. Records of the distribution of species of isopods in Central and Southern United States, with descriptions of four new species of Mancasellus and Asellus (Asellota, Asellidae). *Am. Midl. Nat.* **19**:628–637.

_____. 1940. Descriptions of seven new species of Caecidotea (Isopoda, Asellidae) from central United States. *Trans. Am. Microsc. Soc.* **59**:383–397.

Markus, H. C. 1930. Studies on the morphology and life history of the isopod, Mancasellus macrourus. *Trans. Am. Microsc. Soc.* **49**:220–237.

Menzies, R. J. 1954. A review of the systematics and ecology of the genus "Exosphaeroma," with the description of a new genus, a new species, and a new subspecies (Crustacea, Isopoda, Sphaeromidae). *Am. Mus. Novit.* **1683**:1–24.

Miller, M. A. 1933. A new blind isopod, Asellus californicus, and a revision of the subterranean asellids. *Univ. Calif. Publ. Zool.* **39**:97–110.

Minckley, W. L. 1961. Occurrence of subterranean isopods in epigean waters. *Am. Midl. Nat.* **66**:452–455.

Needham, A. E. 1941. Abdominal appendages of Asellus, II. *Q. J. Microsc. Sci.* **83**:61–89.

Racovitza, E. G. 1920. Notes sur les isopodes. 6. Asellus communis Say. 7. Les pléopodes I and II dés asellides; morphologie et développement. *Arch. Zool. Exp. Gen.* **58**:79–115.

_____. 1920a. Notes sur les isopodes. 8. Mancasellus tenax (Smith). 9. Mancasellus macrourus (Garman). *Ibid.* **59**:28–66.

Reynoldson, T. B. 1961. Observations on the occurrence of Asellus (Isopoda, Crustacea) in some lakes of northern Britain. *Proc. Int. Assoc. Theor. Appl. Limn.* **14**:988–994.

Richardson, H. 1904. Contributions to the natural history of the Isopoda. *Proc. U.S. Nat. Mus.* **27**:1–89.

_____. 1905. A monograph of the isopods of North America. *Bull. U.S. Nat. Mus.* **54**:1–727.

Schultz, T. W. 1973. Digestive anatomy of Lirceus fontinalis (Crustacea: Isopoda). *Trans. Am. Microsc. Soc.* **92**:13–25.

_____. 1973a. Functional morphology of the oral appendages and foregut of Lirceus garmani (Crustacea: Isopoda). *Ibid.* 349–364.

Shuster, S. M. 1981. Life history characteristics of Thermosphaeroma thermophilum, the Socorro isopod (Crustacea: Peracarida). *Biol. Bull.* **161**:291–302.

Steeves, H. R., III. 1963. The troglobitic asellids of the United States: the Stygius group. *Am. Midl. Nat.* **69**:470–481.

_____. 1964. The troglobitic asellids of the United States; the Hobbsi group. *Ibid.* **71**:445–451.

_____. 1966. Evolutionary aspects of the troglobitic asellids of the United States: the hobbsi, stygius and cannulus groups. *Ibid.* **75**:392–403.

_____. 1968. Three new species of troglobitic asellids from Texas. *Ibid.* **79**:183–188.

_____. 1969. The origin and affinities of the troglobitic asellids of the southern Appalachians. *Va. Polytech. Inst. Res. Div. Monogr.* **1**:51–65.

Steeves, H. R., III, and J. R. Holsinger. 1968. Biology of three new species of troglobitic asellids from Tennessee. *Va. Polytech. Inst. Res. Div. Monogr.* **80**:75–83.

Styron, C. E. 1969. Taxonomy of two populations of an aquatic isopod, Lirceus fontinalis Raf. *Am. Midl. Nat.* **82**:402–416.

Unwin, E. E. 1921. Note upon the reproduction of Asellus aquaticus. *J. Linn. Soc.* **34**:335–343.

Van Name, W. G. 1936. The American land and fresh-water isopod Crustacea. *Bull. Am. Mus. Nat. Hist.* **71**:1–535.

_____. 1940. A supplement to the American land and fresh-water Isopoda. *Ibid.* **77**:109–142.

_____. 1942. A second supplement to the American land and fresh-water isopod Crustacea. *Ibid.* **80**:299–329.

Williams, W. D. 1970. A revision of North American epigean species of Asellus (Crustacea: Isopoda). *Smithson. Contrib. Zool.* **49**:1–80.

_____. 1972. Freshwater isopods (Asellidae) of North America. *Biota of Freshwater Ecosystems, Identification Manual* **7**:1–45.

21

AMPHIPODA
(Scuds, Sideswimmers)

Like the Decapoda, the Amphipoda are chiefly marine, only about 150 described American species being confined to fresh waters. World-wide there are about 900 fresh-water species. They occur in a wide variety of unpolluted lakes, ponds, streams, brooks, springs, and subterranean waters. With the exception of one species that may swim about as a plankter in lakes, they are all more or less confined to the substrate.

The layman often calls amphipods "fresh-water shrimp" but of course they are only distantly related to (marine) shrimp.

General characteristics. The great majority of species are 5 to 20 mm long. The body is laterally compressed and consists of a cephalothorax (first thoracic segment fused with the head), seven free thoracic segments, a six-segmented abdomen, and a small, terminal telson. The pleura of the first four abdominal segments are prolonged ventrally so that these segments are deeper than any of the other body segments. In some species the last three abdominal segments are more or less fused.

Eyes are usually well developed in species living in surface waters, but subterranean species show varying degrees of degeneration or absence of eyes. They are unstalked, compound, and circular, oval, reniform, or elongated.

Both pairs of antennae range from short to long. Each first antenna consists of a three-segmented peduncle and a long flagellum; sometimes there is a very small accessory flagellum (Figs. 3A, B). Each second antenna has a flagellum and a five-segmented peduncle, but the basal segment of the peduncle is fused with the head, and the second and third segments are very short; only the last two segments are well developed.

The mouth parts are relatively small, compactly arranged, and hidden by the basal segments of the appendages of the first thoracic segment. They include an upper lip; a pair of tearing and cutting mandibles; a pair of laminar and spinous first maxillae; a pair of small, flexible second maxillae; a lower lip; and a pair of maxillipeds. The maxillipeds are homologs of the first maxillipeds of decapods; each consists of a small inner plate, a small outer plate, and a long palp.

The seven pairs of thoracic legs all have seven segments corresponding to the segments of crayfish legs. Beginning proximally, these are as follows: coxa, basis, ischium,

FIG. 1.–*Gammarus lacustris,* ×4.

474

merus, carpus, propodus, and dactylus. The coxae are usually called coxal plates; they are greatly enlarged extensions of the coxae, flattened, relatively immovable, and project ventrally at the edges of the body.

The first two pairs of legs, known as gnathopods, are subchelate and adapted for grasping (Fig. 4D), but the third to seventh legs are relatively unspecialized pereiopods (also spelled *peraeopods* and *pereopods*). The first two pereiopods are usually flexed forward, the last three backward.

The coxal gills are flattened oval sacs extending downward from the inner surface of the upper posterior corner of the coxal plates. They occur on the second to seventh or second to sixth legs. In addition, there are sometimes very small lateral sternal gills on some of the thoracic segments (Fig. 6B).

The first three abdominal segments bear paired pleopods. Each consists of a basal peduncle and two flexible, many-segmented rami. The last three abdominal segments bear paired uropods. These appendages are never very movable, always directed posteriorly, and closely approximated. The peduncles are unsegmented, but the rami are sometimes composed of two segments, the terminal one being small.

Coloration in living amphipods is occasionally brilliant, but preserved specimens bleach out so that they are whitish, gray, or cream colored. Some common species show a wide range in color from one place or time to another. *Hyalella azteca*, for example, is usually light brown to greenish, but bluish, purple, dark brown, or reddish populations are often found. *Synurella dentata* is usually gray blue but sometimes lavender, brown, or greenish. *Gammarus fasciatus* is whitish, with the body and proximal segments of the appendages banded with green or brown. Very little is known about the causes of color variations within single species, but presumably diet, temperature, and age may be important factors. Subterranean amphipods are almost invariably whitish, cream colored, gray, or translucent.

The sexes are most readily separated on the basis of large marsupial plates, or oostegites, on the inner surface of the coxal plates of some of the legs (usually second to fifth) of the female. Males often have the larger gnathopods.

Locomotion. In general, amphipods are much more active at night than during daylight. The pereiopods are used in crawling and walking but are aided by pushing with the

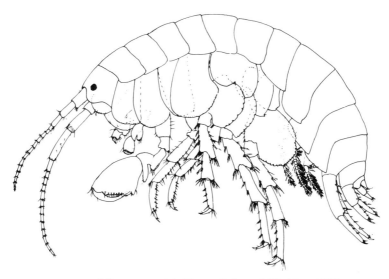

FIG. 2.–*Hyalella azteca*, ×14. (From Cole and Watkins, 1977.)

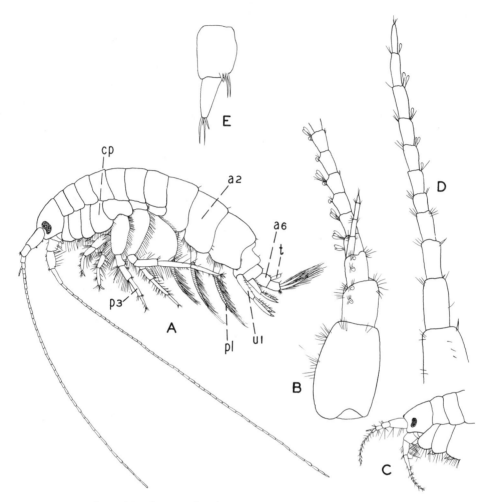

FIG. 3.–Structure of Amphipoda. A, male *Pontoporeia hoyi,* ×9; B, basal portion of first antenna of male *P. hoyi*; c, anterior end of female *P. hoyi*; D, first antenna of female *Hyalella azteca*; E, third uropod of *H. azteca*; a2, second abdominal segment; a6, sixth abdominal segment; cp, coxal plate; p3, third pereiopod; pl, third pleopod; t, telson; ul, first uropod. (A–C modified from Segerstråle, 1937; D redrawn from Geisler, 1944; E redrawn from Weckel, 1907.)

uropods and grasping and pulling with the gnathopods. The last two pereiopods are flexed outward somewhat and are important in a clumsy, skittering sort of locomotion when the animal is on its side. Flexing and extending movements of the entire body aid in crawling and walking. The pleopods are the chief locomotor appendages for rapid undulatory swimming just above the substrate, the body being kept well straightened and the pereiopods directed backward. As an amphipod swims it often rolls over on its side or back (hence the name "sideswimmer").

Especially when disturbed, many amphipods will wriggle into the superficial layers of soft substrates.

Food, feeding. Amphipods are voracious feeders, all kinds of animal and plant matter being consumed. Only rarely do they attack and feed on living animals, but freshly killed animals are consumed readily. Species occurring in aquatic vegetation may often be seen browsing on the film of microscopic plants, animals, and organic debris covering leaves, stems, and other substrates (aufwuchs). Like the decapods, then, scuds are omnivorous, general scavengers or detritus feeders. *Ponto-poreia* is a notable detritus feeder.

The food mass is held by the gnathopods and anterior pereiopods and chewed directly without being first torn into smaller pieces. A few species, however, are filter feeders, at least in part.

Internal anatomy, physiology. Food passes through a short esophagus and into a small stomach containing a gastric mill. The long, undifferentiated intestine opens via the anus just below the telson. Six or nine caeca are associated with the digestive tract.

The tubular heart is in the dorsal portion of the thorax. It has three pairs of ostia and anterior and posterior aortas.

Although it is possible that some respiration occurs through the general body surface, the thin-walled coxal gills are undoubtedly the chief organs for the oxygen–carbon dioxide exchange. The beating of the pleopods creates a current of water over the gills.

Antennal glands, probably osmoregulatory and excretory, open at the base of the second antennae.

The typical nervous system consists of two ventral nerve cords united at intervals by large segmental ganglia, two circumesophageal connectives, and a supraesophageal mass, or "brain."

Both pairs of antennae are olfactory and highly sensitive to touch. They sometimes bear minute, stalked, clublike, or wine-glass shaped sensory appendages (Figs. 3B, D), presumably comparable to the aesthetasks of copepods.

Reproduction. Testes and ovaries are paired, elongated strands ventral to the heart. The oviducts open at the base of the fifth coxal plates, and the male genital pores open on papillae on the ventral side of the last thoracic segment.

With the exception of *Hyalella azteca* and two or three other species, our knowledge of the details of reproductive behavior is rather sketchy. Syngamic reproduction is the rule, but males are sometimes rare. Most common species breed some time between February and October, depending largely on water temperatures. Mature males pair for the first time when they are in their ninth instar, females in their eighth (nuptial) instar. When two males meet both are active and fail to pair and soon separate, but a female is passive when seized by a male.

In *Gammarus, Hyalella azteca*, and one species of *Synurella*, pairing consists of the males carrying the females on their backs and keeping them in place with their gnathopods. Paired individuals feed and swim about for one to as much as 7 days, or until the female is ready to molt to the first adult instar. The two animals then separate for a few minutes to several hours while the female loses her old exoskeleton. A male then returns, the two individuals pair again, and copulation usually occurs within the subsequent 24 hours.

Copulation and transfer of sperm are completed in less than a minute. The male extends the posterior part of his body around and to the ventral side of the female until the tips of his uropods touch the marsupium which is formed by the marsupial plates, or oostegites. Ejected sperm are then swept into the marsupium by movements of the female pleopods. This process of sperm transfer may be repeated several times at intervals of a few minutes, the male holding on to the female all the time.

The animals finally separate, and the female promptly releases her eggs from the oviducts into the marsupium where they are fertilized. The number of eggs released by the female depends on her size and age as well as the particular species. In general, the larger and older females release the larger number of eggs. One species of *Stygobromus* releases only one or two eggs at a time, but the more common species usually average from 15 to 50 eggs per brood. *Hyalella azteca* averages about 18 eggs per brood.

This same species has been shown to have "strains" in different lakes; these do not interbreed.

Development, life cycle. The oostegites are bordered with stiff hairs that are hooked distally and entangled somewhat, and the developing eggs are therefore retained securely within the marsupium. In the few species that have been carefully studied, the incubation period ranges from one to three weeks. Newly hatched young are retained in the marsupium an additional one to eight days and are released to the outside when the mother has her first molt following copulation.

Judging from the limited observations that have been made, the females of most species produce only a single brood during the life cycle, but in a few species each female customarily produces a series of broods during the breeding months. *Hyalella azteca*, for example averages 15 broods in 152 days (Embody, 1912). Copulation is necessary for the production of each brood, so that females pair for the second or subsequent time while they are still carrying their previous brood of eggs or young in the marsupium. These events therefore closely follow each other within a few hours: (1) male leaves female, (2) female molts and releases young, (3) male and female pair again, (4) release of eggs to marsupium, (5) repeated copulation and fertilization, and (6) male leaves female.

Hyalella azteca has a minimum of nine instars in its life history (Geisler, 1944). As in all other species, development is direct, and the newly hatched young have all the adult appendages. The first five instars form the juvenile period, during which the sexes are indistinguishable. In the sixth and seventh instars, or adolescent period, the sexes can be differentiated; the female has small oostegites and a few small eggs in the ovaries, and the male has slightly enlarged gnathopods. Pairing occurs for the first time in the eighth, or nuptial, instar, and the ninth and subsequent instars form the adult period. The number of molts that may occur during the adult period is variable but may be as high as 15 or 20.

When the old exoskeleton is ready to be cast, a dorsal, transverse split appears halfway around the body between the first and second thoracic segments or occasionally between the cephalothorax and first thoracic segment. Sometimes there are accessory longitudinal splits along the upper margins of the coxal plates. The cephalothorax is withdrawn first and then the rest of the body. Molting is completed in less than an hour.

The interval between molts ranges from a minimum of 3 to a maximum of 40 days, depending on food conditions, temperature, and the species. Immature amphipods molt at much shorter intervals than adults.

Most species complete the life cycle in a year or less. *Pontoporeia hoyi* is unusual in that it is thought to have a life cycle of 30 months or more. This species breeds between December and April, and the young are released from the marsupium in the spring.

Ecology. *Hyalella azteca, Gammarus pseudolimnaeus, G. fasciatus,* and *Crangonyx gracilis* constitute the great majority of specimens taken by casual collectors. These species are widely distributed and common in unpolluted clear waters, including springs, spring brooks, streams, pools, ponds, and lakes. Less common species are more restricted to certain types of environments. *Pontoporeia hoyi,* for example, occurs only in deep, cold, oligotrophic, northern lakes. *Gammarus minus* occurs only in caves, springs, and small streams from Pennsylvania, Maryland, Virginia, and West Virginia west–southwest to Illinois, Missouri, and Arkansas. Many species are restricted to seeps, springs, and subterranean waters in one or a few states, or, indeed, to a single cave or cave system.

As a group, amphipods are cold stenotherms, strongly thigmotactic, and react negatively to light. Consequently, during the daytime they are in vegetation or hidden under and between debris and stones. They usually congregate in the corners of culture jars in the laboratory. The true burrowing habit has not evolved in fresh-water species.

Amphipods are sometimes unbelievably abundant. The writer has collected *Gammarus* from spring brooks rich in rooted vegetation where populations have exceeded $10,000/m^2$. Juday and Birge (1927) reported an average population of 4553 *Pontoporeia hoyi* per square meter at a depth of 50 to 60 m on the bottom of Green Lake, Wisconsin.

Except for *P. hoyi*, which is both a benthic and plankton organism and which has been found as deep as 300 m in Lake Superior, amphipods are restricted to shallow waters. *Hyalella azteca* is about the only other species ever found at depths exceeding 1 m. Large rivers contain few amphipods, except in the shallow backwater and overflow pond areas.

Although it is generally true that the majority of species appear to be restricted to waters of low or medium carbonate content, there are a few notable exceptions. *Gammarus lacustris* is common in hard waters, and *Hyalella azteca* is sometimes found in alkaline and brackish waters.

An abundance of dissolved oxygen appears to be an environmental necessity. The only reported notable exception is *Pontoporeia hoyi,* which has been collected from the bottom waters of lakes where the concentration of dissolved oxygen was less than 7% saturation. *Gammarus* might prove to be a useful bioassay organism for studies on pesticides.

The restriction of many species to subterranean waters and springs strongly suggests that they are cold stenotherms.

About 100 species are terrestrial or semiterrestrial on a worldwide basis. Some occur in the debris of the high-tide line on marine shores. Others may be found in commercial greenhouses or in terrestrial habitats in the tropics.

A few species burrow in the substrate during times of drought and high temperatures; with the onset of normal conditions they resume activities in the water.

Geographic distribution, dispersal. Most of the scuds of the United States are restricted to this continent. *Hyalella azteca* occurs also in South America. *Gammarus fasciatus* and *Crangonyx gracilis* have recently been reported from England; undoubtedly they have been introduced from the United States. *Niphargus,* the very large Old World genus, does not occur in the United States.

Very little is known about the passive transport of amphipods from one drainage system to another. Unlike the entomostraca, amphipods are not generally adapted for withstanding drought and other adverse environmental conditions. *Crangonyx gracilis, C. shoemakeri,* and *Synurella bifurca,* however, are inhabitants of temporary as well as permanent ponds and streams, and it has been suggested that these species tide over unfavorable conditions by burrowing into the substrate.

C. forbesi undergoes interesting seasonal migrations. In the spring, populations move upstream into spring brooks and springs. There the populations often become so dense that cannibalism results. In the autumn the remnants of the population migrate back downstream. Breeding occurs during the winter months.

Pontoporeia hoyi is probably derived from *P. femorata,* an arctic circumpolar and marine species. *P. affinis* is a European and Asiatic fresh-water species also thought to have been derived from *P. femorata.*

Subterranean amphipods. The central and south central states, from eastern Kansas and Oklahoma to Indiana, Kentucky, and Tennessee is a region abundant in large caves and cave systems, most of which contain brooks, streams, and pools. It is in such bodies of water that interesting subterranean amphipods are found, but usually as sparse populations. Texas has a rich fauna of subterranean species, and recently many species have been reported from interstitial habitats in the Rocky Mountain region. Undoubtedly interstitial and subterranean species are generally distributed in the United States. Although the cave habit has evolved in many genera, the species, as a group, are similarly modified. They are generally whitish, creamy, or straw colored, the antennae and tactile hairs are well developed, the body is often relatively fragile, and the eyes are usually reduced, vestigial, or absent. Their food consists of bits of dead vegetation washed into the caves and the thin bacterial scum covering submerged surfaces. Some subterranean species are generally distributed throughout large areas, but others are rare and have been reported only from single or a few localities.

Subterranean amphipods are often collected just outside of caves and underground streams where the water forms surface seeps, wells, and springs. Some species are represented by two distinct varieties, one subterranean and the other a surface form. The cave variety of *Crangonyx gracilis,* for example, has degenerate eyes and no pigmentation, whereas its pond and stream counterpart has functional eyes and good pigmentation. Comparable differences occur in the varieties of *C. forbesi* and *Gammarus troglophilus* Hubricht and Mackin. *G. minus* is unusual in having three varieties. The one occurring in springs has large eyes, short antennae, and brown coloration. A cave variety has slightly reduced eyes, long antennae, and bluish coloration. Another rare cave variety has a fragile body, greatly reduced eyes, long antennae, and bluish coloration.

Enemies, commensals, parasites. Fishes are the chief predators, although only a few species of amphipods occur in streams and ponds large enough to support natural fish populations. The planting of amphipods in small western lakes has, in a few instances, been at least temporarily successful in augmenting the natural food supply of trout. Birds, predaceous aquatic insects, and amphibians probably take an appreciable toll.

Like crayfish, amphipods support an amazing population of algae and sessile Protozoa on the external body surfaces.

Scuds serve as intermediate hosts for a wide variety of parasites, including tapeworms of waterfowl and fishes, and a few nematodes, trematodes, and Acanthocephala of birds, fishes, and amphibians.

Collection, preparation. Where specimens are abundant they may be easily taken with a dipnet and by rinsing out masses of aquatic vegetation and bottom debris. Stony bottoms require hand picking with forceps or a small aquarium net. Seventy percent alcohol is a satisfactory killing agent and preservative.

Live specimens may be maintained in aquaria, especially if they are well supplied with aquatic vegetation.

Taxonomy. Of the three suborders of the Order Amphipoda, only one, the Gammaroidea, is represented in American fresh waters. The list of species described has grown from 16 in 1907 (Weckel) to about 150 at the present time. Our American specialists have representatives of more than 100 undescribed fresh-water species in their possession.

For accurate identification it is important that mature males and females be used, since the keys below are based largely on structures which are best differentiated in the later instars.

Pontoporeia has been a particularly puzzling genus, and several species have been described from the United States. The work of Segerstråle, however, has shown that there is only one valid species, *P. hoyi,* which is closely related to *P. affinis* Lindstrom occurring in Europe. The previous confusion resulted from the fact that this species shows marked changes in the later instars, especially with respect to length of the antennae and structure of the uropods.

We use reflected light and watch glasses containing black wax for our dissections under binoculars. Watchmaker's forceps are mandatory. The following key is greatly modified from that of Holsinger (1972) and Holsinger and Longley (1980).

KEY TO GENERA AND SPECIES OF AMPHIPODA

1. First antenna with accessory flagellum (Figs. 3A, B) and either longer or shorter than second antenna; third uropod with or without rami; telson cleft or entire 2

   First antenna without accessory flagellum (Fig. 3D) and shorter than second antenna; third uropod uniramous (Fig. 3E); telson entire; 4 to 8 mm long; springs, brooks, pools, and lakes; widely distributed and common (Fig. 2) .... TALITRIDAE, **Hyalella azteca** (Saussure)

2. First antenna slightly shorter than second antenna, very short in female (Fig. 3C) and very long in mature male (Fig. 3A); accessory flagellum of first antenna consisting of three or four segments (Fig. 3B); fifth pereiopod much shorter than fourth and with second segment greatly expanded; up to 9 mm long; on the bottom and in the plankton of deep, cold lakes, including Great Lakes; Lake Nipigon; Green Lake, WI, Lake Washington, Seattle; some Finger Lakes, NY; and others. Mature males of variety *typica* have 31 to 57 segments in the flagellum of the first antenna and 42 to 72 segments in the flagellum of the second antenna; mature males of variety *brevicornis* have 14 to 19, and 19 to 24 segments, respectively .... HAUSTORIIDAE, **Pontoporeia hoyi** Smith

   First antenna either longer or shorter than second, but both antennae long and slender; accessory flagellum very short, consisting of one or two short segments, or well developed and consisting of from three to seven segments; fifth pereiopod longer or slightly shorter than fourth, and with the second segment only moderately expanded (Fig. 4A) 3

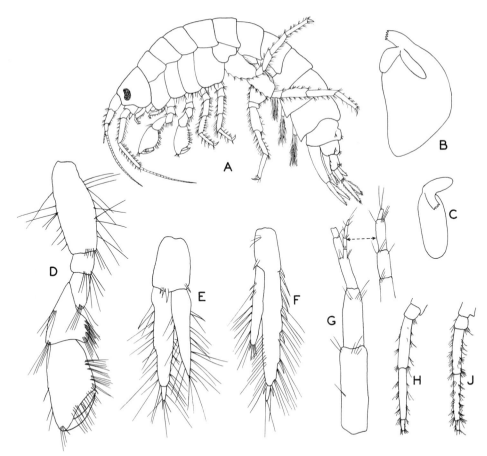

FIG. 4.–Structure of Gammaridae. A, male *Gammarus fasciatus,* ×5; B and C, coxal gills of *Anisogammarus oregonensis* Shoemaker; D, second gnathopod of male *Gammarus lacustris*; E, third uropod of *G. lacustris*; F, third uropod of *G. minus*; G, basal portion of first antenna of *G. lacustris*; H, basal segments of second antenna of *G. minus*; J, basal segments of second antenna of *G. fasciatus*. (A modified from Kunkel; B and C modified from Shoemaker, 1944; D–G modified from Weckel, 1907.)

3. Accessory flagellum of first antenna three- to seven segmented (Fig. 4G); third uropod well developed (Fig. 4E) .... GAMMARIDAE, 4
    Accessory flagellum of first antenna absent or small, rudimentary, or consisting of one short and one long segment (Fig. 6D); third uropod usually reduced (Fig. 6G) **8**
4. Coxal gills with cylindrical appendages (Figs. 5B, C); up to 11 mm long; three uncommon species known from brackish waters and coastal streams in WA, OR, CA .... **Anisogammarus**
    Coxal gills without cylindrical appendages; up to 25 mm long; common and widely distributed; about six species are rare, are mostly hypogean, and have highly restricted distributions; four common species are epigean, widely distributed, and included in this key ... **Gammarus, 5**
5. Second antenna richly setose (Fig. 4J); up to 14 mm long; common in upper Mississippi drainage, Great Lakes drainage, and the Atlantic coastal plain to NC; lakes and rivers.
    Gammarus fasciatus Say
    Second antenna not richly setose (Fig. 4H) **6**
6. Posterior margin of basal segment of seventh pereiopod with long setae (Fig. 5A); up to 17 mm long; upper Mississippi River and Great Lakes drainages; chiefly in springs and streams.
    Gammarus pseudolimnaeus Bousfield
    Posterior margin of basal segment of seventh pereiopod with short setae (Fig. 5B) **7**

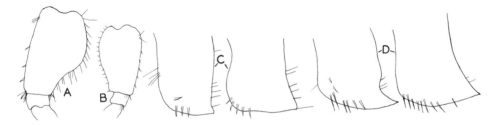

FIG. 5.–Structure of *Gammarus*. A, base of seventh pereiopod of *G. pseudolimnaeus*; B, base of seventh pereiopod of *G. minus*; C, second and third abdominal side plates of *G. minus*; D, second and third abdominal side plates of *G. lacustris*. (Modified from Holsinger, 1972.)

7. Posterior angle of second and third abdominal side plates weakly acuminate (Fig. 5C); up to 14 mm long; common in streams and springs from the central Appalachians to the Ozarks.
                                                                                                                    Gammarus minus Say
    Posterior angle of second and third abdominal side plates strongly acuminate (Fig. 5D); up to 22 mm long; a variable species found in a wide variety of habitats over the northern half of the United States, but also from the lower Colorado River .... **Gammarus lacustris** Sars
8. Last segment of sixth and sventh pereiopods with ventral spines (Fig. 6H); two species in hypogean habitats of south central OK and the Ozark plateau; up to 22 mm long.
                                                                                                        GAMMARIDAE, **Allocrangonyx**
    Last segment of sixth and seventh pereiopods without ventral spines **9**
9. Second antenna of mature male with paddlelike calceoli (Fig. 7A); chiefly epigean.
                                                                                                        CRANGONYCTIDAE, **10**
    Second antenna of mature male without calceoli; chiefly hypogean **21**
10. Third uropod uniramous, exopod short (Fig. 6J); 5 to 14 mm long; telson cleft (Fig. 6E); epigean.
                                                                                                        Synurella, 11
    Third uropod biramous, exopod long (Fig. 7G); 4 to 22 mm long; about 20 described U.S species plus many undescribed species; a difficult taxonomic group; this key includes only the U.S. epigean species that are found in more than one restricted locality .... **Crangonyx, 13**
11. Palmar margin of gnathopod concave (Fig. 7B); ponds, springs, and small streams of southeastern IN, southwestern OH, and central KS .... **Synurella dentata** Hubricht
    Palmar margin of gnathopod straight (Fig. 7C) **12**
12. Urosome segments fused; streams, springs, and ponds in AR, TN, south to the Gulf.
                                                                                                        Synurella bifurca (Hay)
    Urosome segments not fused; small streams and ponds in eastern VA, NC, SC.
                                                                                                        Synurella chamberlaini (Ellis)
13. Palmar margin of second gnathopod of female lined with small, weak spines (Fig. 8D) **14**
    Palmar margin of second gnathopod of female lined with strong, notched spines (Fig. 8G) **16**
14. Posterior corner of first abdominal side plate strongly mucronate (Fig. 7E); swamps and pools; FL to eastern LA .... **Crangonyx floridanus** Bousfield
    Posterior corner of first abdominal side plate weakly mucronate **15**
15. Mature males without a row of comb spines on outer ramus of second uropod; chiefly in the Great Lakes drainage basin; probably a species complex .... **Crangonyx gracilis** Smith
    Mature males with a row of comb spines on outer ramus of second uropod (Fig. 7G); uncommon; distribution spotty but probably widely dispersed in the northeastern states; possibly a species complex .... **Crangonyx pseudogracilis** Bousfield
16. Inner margin of dactyl of female second gnathopod with a row of bladelike spines (Fig. 7H) **17**
    Inner margin of dactyl of female second gnathopod with fine setae only (Fig. 7D) **19**
17. Palmar margin of second gnathopod straight; dactyls of third to seventh pereiopods with two or three short, stiff setae on inner margins (Fig. 7J); cold streams and ponds in lower MI, IN, OH, KY.
                                                                                                        Crangonyx setodactylus Bousfield
    Palmar margin of second gnathopod slightly concave (Fig. 7H); dactyls of third to seventh pereiopods with one stiff seta on inner margin **18**

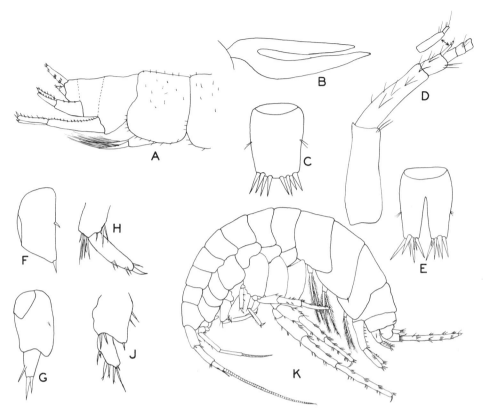

FIG. 6.–Structure of Gammaridae. A, posterior end of *Stygonectes pizzini* (Shoemaker); B, lateral sternal gill of sixth thoracic segment of *S. pizzini*; C, telson of *Stygonectes balconis* Hubricht; D, basal portion of first antenna of *S. flagellatus* (Benedict); E, telson of *Synurella dentata*; F, third uropod of *Apocrangonyx subtilis* Hubricht; G, third uropod of *Stygobromus mackini* Hubricht; H, end of seventh pereiopod of *Allocrangonyx pellucidus* (Mackin); J, third uropod of *Synurella*; K, *Allocrangonyx pellucidus,* a blind species, ×3. (A and B modified from Shoemaker, 1942; C and E–G modified from Hubricht, 1943; D redrawn from Weckel, 1907; H and J redrawn from Holsinger, 1972.)

18. Mature female with 12 to 15 spines on anterior margin of the basal segment of seventh pereiopod (Fig. 7K); small streams and ponds in IN, IL, IA, MO .... **Crangonyx minor** Bousfield
    Mature female with about eight small spines on anterior margin of the basal segment of seventh pereiopod (Fig. 7L); springs, small streams, and bogs in MD, DC, VA.
                                                              Crangonyx shoemakeri (Hubricht and Mackin)
19. Basal segments of pereiopods five, six, and seven broad and with numerous minute serrations (Fig. 7M); springs, and spring-fed streams in adjacent areas of IL, OH, KY.
                                                              Crangonyx anomalus Hubricht
    Basal segments of pereiopods five, six, and seven not expanded and with few serrations 20
20. Telson short, with shallow cleft and three or four apical spines per lobe (Fig. 7N); a variable species of streams and ponds in MO, IL, IN, OH, KY ... **Crangonyx forbesi** (Hubricht and Mackin)
    Telson about as long as broad, with two or three apical spines per lobe (Fig. 7O); probably a complex species group; widely distributed in eastern half of the United States in all types of waters.
                                                              Crangonyx obliquus-richmondensis
21. Third uropod with a single short ramus, or ramus absent (Figs. 6F, G; 7Q) 22
    Third uropod biramous 24

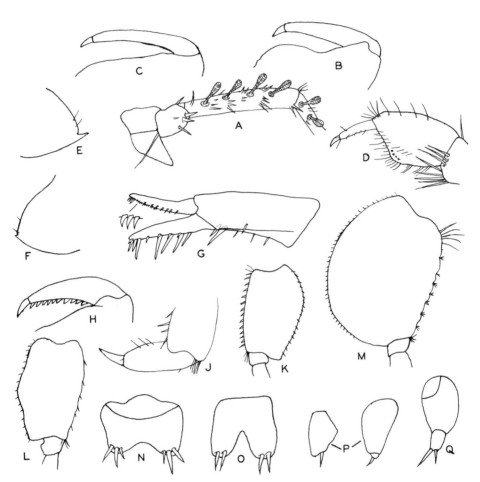

FIG. 7.–Structure of Amphipoda. A, basal segments of male second antenna of *Crangonyx*; B, tip of gnathopod of *Synurella dentata* (all setae and spines omitted); C, tip of gnathopod of *Synurella* (all setae and spines omitted); D, tip of second gnathopod of female *Crangonyx*; E, posterior corner of first abdominal side plate of *C. floridanus*; F, posterior corner of first abdominal side plate of *Crangonyx*; G, male uropod of *C. shoemakeri*: H, tip of second gnathopod of female *C. shoemakeri*; J, dactyl of seventh pereiopod of *C. setodactylus*; K, basal segment of seventh pereiopod of *C. minor*; L, basal segment of seventh pereiopod of *C. shoemakeri*; M, basal segment of seventh pereiopod of *C. anomalus*; N, telson of *C. forbesi*; O, telson of *Crangonyx obliquus-richmondensis* group; P, third uropods of *Allocrangonyx*; Q, third uropod of *Stygobromus*. (Mostly modified from Holsinger, 1972.)

22. Inner ramus of third uropod vestigial (Fig. 8A); several hypogean species in OH to KS and OK. .... CRANGONYCTIDAE, **Bactrurus**
    Inner ramus of third uropod not vestigial 23
23. Telson without apical spines; adult less than 2 mm long; a single rare species from artesian wells in San Marcos, TX .... SEBIDAE, **Seborgia relicta** Holsinger
    Telson with apical spines (Fig. 6C, E); adults more than 3 mm long; more than 100 species; caves, wells, seeps, and springs, and cavernicolous and interstitial situations; eyeless, unpigmented; generally distributed except in northcentral states .... CRANGONYCTIDAE, **Stygobromus**
24. First antennae longer than body .... HADZIIDAE, 25
    First antennae shorter than body 26
25. Segment six of fifth and sixth pereiopods with long setae on posterior margins (Fig. 8H); rare in TX artesian wells; one species .... **Allotexiweckelia**
    Segment six of fifth and sixth pereiopods without such long setae (Fig. 8J); several rare species in TX artesian wells .... **Texiweckelia**
26. Bases of pereiopods five and six greatly expanded (Fig. 9A); one rare species in TX artesian wells. .... ARTESIIDAE, **Artesia**
    Bases of pereiopods five and six very small (Fig. 9B); one rare species in TX artesian wells; with many close relatives in Central and South America .... BOGIDIELLIDAE, **Parabogidiella**

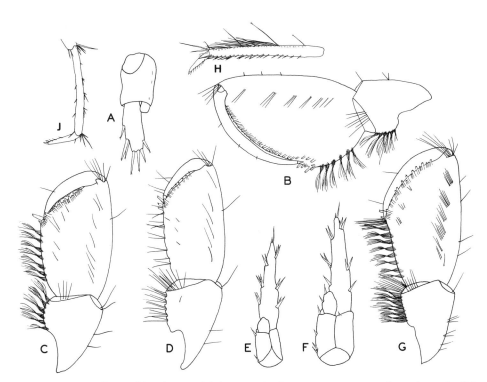

FIG. 8.–Structure of Amphipoda. A, third uropod of *Bactrurus*; B, second gnathopod of female *Crangonyx*; C, second gnathopod of female *C. shoemakeri*; D, second gnathopod of female *C. gracilis*; E, third uropod of male *C. gracilis*; F, third uropod of female *C. gracilis*; G, second gnathopod of male *C. forbesi* (fine setae of dactyl not visible); H, tip of pereiopod 6 of *Allotexiweckelia*; J, tip of pereiopod 6 of *Texiweckelia*. (A–G modified from Hubricht and Mackin, 1940; H and J modified from Holsinger and Longley, 1980.)

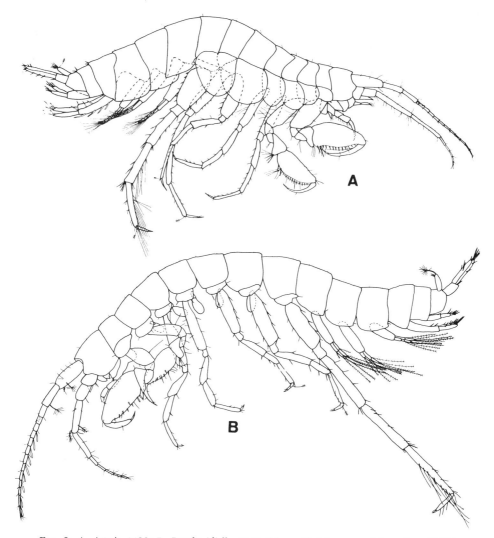

Fig. 9.–A, *Artesia,* ×22; B, *Parabogidiella,* ×47. (From Holsinger and Longley, 1980.)

REFERENCES

Adamstone, F. B. 1928. Relict amphipods of the genus Pontoporeia. *Trans. Am. Microsc. Soc.* **47**:366–371.

Barnard, J. L. 1958. Index to the families, genera, and species of the gammaridean Amphipoda (Crustacea). *Occas. Pap. Allan Hancock Found. Publ.* **19**:1–145.

Barnard, J. R., and C. M. Barnard. 1983. *Freshwater Amphipoda of the World. 2. Handbook and Bibliography.* Hayfield Associates, Mt. Vernon, VA, pp. 359–830.

Barnard, J. L., and W. S. Gray. 1968. Introduction of an amphipod crustacean into the Salton Sea, California. *Bull. S. Calif. Acad. Sci.* **67**:219–232.

Bousfield, E. L. 1958. Fresh-water amphipod crustaceans of glaciated North America. *Can. Field Nat.* **72**:55–113.

———. 1977. A new look at the systematics of gammaroidean amphipods of the world. *Crustaceana (Suppl.)* **4**:282–316.

Bowman, T. E., and F. Phillips. 1984. Bioluminescence in the freshwater amphipod, Hyalella azteca, caused by pathogenic bacteria. *Proc. Biol. Soc. Wash.* **97**:526–528.

Clampitt, P. T. 1965. Dispersal related to density in the amphipods Hyalella azteca and Gammarus pseudolimnaeus. *Proc. Iowa Acad. Sci.* **71**:474–484.

Cole, G. A. 1970. The epimera of North American freshwater species of Gammarus (Crustacea: Amphipoda). *Proc. Biol. Soc. Wash.* **83**:333–348.

———. 1985. Analysis of the Gammarus–Pecos complex (Crustacea: Amphipoda) in Texas and New Mexico, USA. *J. Ariz. Nev, Acad. Sci.* **20**:93–103.

Cole, G. A., and R. L. Watkins. 1977. Hyalella montezuma, a new species (Crustacea: Amphipoda) from Montezuma Well, Arizona. *Hydrobiologia* **52**:175–184.

Cooper, W. E. 1965. Dynamics and production of a natural population of a fresh-water amphipod, Hyalella azteca. *Ecol. Monogr.* **35**:377–394.

Creaser, E. P. 1934. A new genus and species of blind amphipod with notes on parallel evolution in certain amphipod genera. *Mus. Zool. Univ. Mich. Occas. Pap.* **282**:1–6.

Embody, G. C. 1912. A preliminary study of the distribution, food and reproductive capacity of some fresh-water amphipods. *Int. Rev. (Suppl)* **4**:1–33.

Gaylor, D. 1922. A study of the life history and productivity of Hyalella knickerbockeri Bate. *Proc. Ind. Acad. Sci.* (1921):239–250.

Geisler, Sister F. S. 1944. Studies on the post-embryonic development of Hyallela azteca (Saussure). *Biol. Bull.* **86**:6–22.

Gledhill, T. 1977. Numerical fluctuations of four species of subterranean amphipods during a five year period. *Crustaceana (Suppl.)* **4**:44–152.

Hargrave, B. T. 1970. Distribution, growth, and seasonal abundance of Hyalella azteca (Amphipoda) in relation to sediment microflora. *J. Fish. Res. Board Can.* **27**:685–699.

Holsinger, J. R. 1969. The systematics of the North American subterranean amphipod genus Apocrangonyx (Gammaridae), with remarks on ecology and zoogeography. *Am. Midl. Nat.* **81**:1–28.

———. 1969a. Biogeography of the freshwater amphipod crustaceans (Gammaridae) of the Central and Southern Appalachians. *Va. Polytech. Inst. Res. Div. Monogr.* **1**:19–50.

———. 1971. A new species of the subterranean amphipod genus Allocrangonyx (Gammaridae) with a redescription of the genus and remarks on its zoogeography. *Int. J. Speleol.* **3**:317–331.

———. 1972. The freshwater amphipod crustaceans (Gammaridae) of North America. *Biota of Freshwater Ecosystems, Identification Manual* **5**:1–89.

———. 1974. Systematics of the subterranean amphipod genus Stygobromus (Gammaridae), Part I; species of the western United States. *Smithson. Contrib. Zool.* **160**:1–63.

———. 1977. A review of the systematics of the Holarctic amphipod Family Crangonyctidae. *Crustaceana (Suppl.)* **4**:244–281.

———. 1978. Systematics of the subterranean amphipod genus Stygobromus (Crangonyctidae), Part II: species of the eastern United States. *Smithson. Contrib. Zool.* **266**:1–144.

———. 1981. Stygobromus canadensis, a troglobitic crustacean from Castleguard Cave, with remarks on the concept of cave glacial refugia. *Proc. 8th Int. Cong. Speleol.* **1**:93–95.

———. 1986. Zoogeographic patterns of North American subterranean amphipod crustaceans. *Crustacean Biogeography* 85–106. A. A. Balkema, Rotterdam.

Holsinger, J. R., and G. W. Dickson. 1977. Burrowing as a means of survival in the troglobitic amphipod crustacean Crangonyx antennatus Packard (Crangonyctidae). *Hydrobiologia* **54**:195–199.

Holsinger, J. R., and G. Longley. 1980. The subterranean amphipod crustacean fauna of an artesian well in Texas. *Smithson. Contrib. Zool.* **308**:1–62.

Holsinger, J. R., J. S. Mort, and A. D. Recklies. 1983. The subterranean crustacean fauna of Castleguard Cave, Columbia icefields, Alberta, Canada, and its zoogeographic significance. *Arctic Alpine Res.* **15**:543–549.

Hubricht, L. 1943. Studies in the Nearctic fresh-water Amphipoda, III. Notes on the fresh-water Amphipoda of eastern United States, with descriptions of 10 new species. *Am. Midl. Nat.* **29**:683–712.

Hubricht, L., and J. G. Mackin. 1940. Descriptions of nine new species of fresh-water amphipod Crustaceans with notes and new localities for other species. *Am. Midl. Nat.* **23**:187–218.

Hultin, L. 1971. Upstream movements of Gammarus pulex pulex (Amphipoda) in a South Swedish stream. *Oikos* **22**:329–347.

Juday, C., and E. A. Birge. 1927. Pontoporeia and Mysis in Wisconsin lakes. *Ecology* **7**:445–452.

Karaman, G. S. 1974. Contribution to the knowledge of the Amphipoda. Revision of the genus Stygobromus Cope 1872 (Fam. Gammaridae) from North America. *Glas. Repub. Zavoda Zast. Prir.—Prir. Muz. Titograd* **7**:97–125.

Kunkel, B. W. 1918. The Arthrostraca of Connecticut. *Bull. Conn. Geol. Nat. Hist. Surv.* **26**:1–261.

Marzolf, G. R. 1965. Substrate relations of the burrowing amphipod Pontoporeia affinis in Lake Michigan. *Ecology* **46**:579–591.

Mathias, J. A. 1971. Energy flow and secondary production of the amphipods Hyalella azteca and Crangonyx richmondensis occidentalis in Marion Lake, British Columbia. *J. Fish. Res. Board Can.* **28**:711–726.

Minckley, W. L. 1964. Upstream movements of Gammarus (Amphipoda) Doe Run, Meade County, Kentucky. *Ecology* **45**:195–197.

Minckley, W. L., and G. A. Cole. 1963. Ecological and morphological studies on gammarid amphipods (Gammarus spp.) in spring-fed streams of northern Kentucky. *Occ. Pap. C. C. Adams Center Ecol. Stud.* **10**:1–35.

Rees, C. P. 1972. The distribution of the amphipod Gammarus pseudolimnaeus Bousfield as influenced by oxygen concentration, substratum, and current velocity. *Trans. Am. Microsc. Soc.* **91**:514–528.

Sanders, H. O. 1969. Toxicity of pesticides to the crustacean Gammarus lacustris. *Tech. Pap. Bur. Sport Fish. Wildl.* **25**:1–18.

Schmitz, E. H. 1967. Visceral anatomy of Gammarus lacustris lacustris Sars (Crustacea: Amphipoda). *Am. Midl. Nat.* **78**:1–54.

Schmitz, E. H. and P. M. Scherrey. 1983. Digestive anatomy of Hyalella azteca (Crustacea, Amphipoda). *J. Morph.* **175**:91–100.

Segerstråle, S. G. 1937. Studien über die Bodentierwelt in südfinnländischen Küstengewässern III. Zur Morphologie und Biologie des Amphipoden Pontoporeia affinis, nebst einer Revision der Pontoporeia-Systematic. *Soc. Sci. Fenn. Comment. Biol.* **7**:1–183.

———. 1954. The freshwater amphipods. Gammarus pulex (L.) and Gammarus lacustris G. O. Sars, in Denmark and Fennoscandia-a contribution to the late- and post-glacial immigration history of the aquatic fauna of northern Europe. *Ibid.* **15**:1–91.

———. 1971. On summer breeding in populations of Pontoporeia affinis (Crustacea Amphipoda) living in lakes of North America. *Ibid.* **44**:1–18.

Shoemaker, C. R. 1938. A new species of fresh-water amphipod of the genus Synpleonia, with remarks on related genera. *Proc. Biol. Soc. Wash.* **51**:137–142.

———. 1942. Notes on some American fresh-water amphipod crustaceans and descriptions of a new genus and two new species. *Smithson. Misc. Coll.* **101**:1–31.

———. 1944. Description of a new species of Amphipoda of the genus Anisogammarus from Oregon. *J. Wash. Acad. Sci.* **34**:89–93.

Smith, W. E. 1972. Culture, reproduction, and temperature tolerance of Pontoporeia affinis in the laboratory. *Trans. Am. Fish. Soc.* **101**:253–256.

Wagner, V. T., and D. W. Blinn. 1987. A comparative study of the maxillary setae for two coexisting species of Hyalella (Amphipoda), a filter feeder and a detritus feeder. *Arch. Hydrobiol.* **109**:409–419.

Ward, J. V. 1977. First records of subterranean amphipods from Colorado (Crangonyctidae). *Trans. Am. Microsc. Soc.* **86**:452–466.

Weckel, A. L. 1907. The fresh-water Amphipoda of North America. *Proc. U.S. Nat. Mus.* **32**:25–58.

Wilder, J. 1940. The effects of population density upon growth, reproduction, and survival of Hyallela azteca. *Physiol. Zool.* **13**:439–462.

22

DECAPODA
(Crayfishes, Shrimps)

Of the great array of species constituting the Order Decapoda or Reptantia, the vast majority are marine. In the United States only the Astacidae and Cambaridae (crayfishes), about 12 species of Palaemonidae (fresh-water prawns and river shrimps), and four species of Atyidae are found in fresh waters. These representatives, totaling about 350 species, are characteristic and common inhabitants of a wide variety of environments, including most types of running waters, shallows of lakes, ponds, sloughs, swamps, underground waters, and even wet meadows where there is no open water. They are absent from the greater portion of the Rocky Mountain region, although one species occurring in Pacific slope drainages has migrated as far eastward as the Great Basin and the headwaters of the Snake, Yellowstone, and Missouri rivers. Compared with the United States, Europe has an impoverished crayfish fauna. Crayfish do not occur in Africa.

To this list should be added at least one grapsoid crab, *Platychirograpsus typicus* Rathbun, which has become well established in the Hillsboro River, FL. Possibly this river crab occurs also in Gulf coast streams of other states since it is common along the east coast of Mexico.

General characteristics. Crayfish, crawfish, crawdads, or crabs, as they are locally and variously known, are all more or less cylindrical, and the body and appendages are strongly sclerotized. The compound eyes are large, stalked, and movable. Trunk length, not including the antennae, ranges from 10 to 150 mm. The six abdominal segments are all distinct, but the head and thoracic segments are fused to form a large cephalothorax. A carapace covers the cephalothorax, and there is a large gill chamber on each side between the body and the downward extensions of the carapace. The cervical groove of the carapace roughly delimits the head and thoracic regions, and the areola is a narrow more or less hourglass-shaped median area of the carapace just posterior to the cervical groove. The anterior end of the carapace forms an elongated rostrum which bears a short spine (acumen) at its extremity.

The 19 pairs of serially homologous appendages are basically biramous but modified for a wide variety of functions. Detailed accounts of their morphology may be found in most zoology texts and they will be only briefly discussed here. Beginning at the anterior end, each of the first antennae (antennules) has two slender, segmented, sensory flagella. The dorsal surface of the basal segment of the first antenna bears the opening of the statocyst, an organ of balance. The second antenna has a flattened basal scale and a single long flagellum.

The next five pairs of appendages greatly overlap one another and are used chiefly for handling and mincing the food. The heavy, crushing mandibles lie on either side of the mouth. The first maxillae, which follow, are small, flattened, and delicate. Each of the second maxillae has a long, flat extension projecting laterally beyond the point of attachment of this appendage. This is the "bailer;" its rapid beating creates a posterior–anterior current of water through the gill chamber, thereby ensuring a continuous supply of oxygenated water for the gills in the gill chamber. The first, second, and third maxillipeds are progressively more robust and clearly biramous. The third maxillipeds are used for cleaning the antennae.

The last five pairs of appendages of the cephalothorax are large walking legs, or pereiopods (hence the name "Decapoda"). Each consists of seven segments. Beginning at the base, these are as follows: coxa, basis, ischium, merus, carpus, propodus, and dactylus. Sometimes the suffix "podite" is used for each segment; thus: coxopodite, basiopodite, and so on. The coxa and basis are always short and compact. The first three pereiopods are clawed, or chelate, the dactylus being the movable "finger" of the claw, and the propodus constituting the "hand" and immovable finger. The first leg, or chela, is greatly enlarged in the crayfish and is used for crushing the food and as an offensive and defensive weapon. It is of little use in locomotion, walking being chiefly a function of the last four pairs of legs. The second and third legs are also used in handling and mincing the food.

Near the base of the ischium of each walking leg and marked by a double crease is a "breaking joint." When a crayfish leg is seized or irritated, reflex action causes strong contraction of special muscles at the breaking joint, and the whole leg breaks off at this point. Consequently the collector must exercise care if he is to secure suitable specimens. Perfect specimens are uncommon since appendages are often lost as the result of crayfish fighting among themselves. Missing legs are regenerated in miniature at the subsequent molt but later become larger.

The first five abdominal segments bear pleopods, or swimmerets. In the female they are all more or less biramous and are used as places of attachment for the incubating eggs, but in the male the first two pairs are modified and used during copulation. The last abdominal segment bears a flat terminal telson and a pair of flat biramous uropods. The telson and

Fig. 1.–A, dorsal view of female *Orconectes,* ×0.7; B, dorsal view of male *Orconectes,* ×0.7.

two uropods collectively form the broad tail fan.

Mature crayfish exhibit such sexual dimorphism that there is no difficulty in distinguishing the two sexes. Males usually have the larger chelae and narrower abdomen. In the Cambarinae the third segment of some of the male pereiopods bears a small hook used during copulation (Fig. 13E). In the female the genital pores are on the basal segments of the third pereiopods, but in the male they are on the fifth pereiopods. The annulus ventralis, or seminal receptacle, is a grooved, elliptical, calcified area in the midline between the bases of the fourth and fifth pereiopods of the female (Fig. 5). The first pleopods of the female are usually similar to the others, but in the male they have lost their biramous character and are heavily sclerotized, grooved, and specialized at the tip for sperm transfer. The male second pleopods are also somewhat specialized, and both pairs are bent forward so that they lie against the median ventral surface of the body between the bases of the pereiopods.

General body coloration ranges from blackish through brown, red, orange, green, and occasionally blue, with innumerable intermediate shades. Mottling is common, and the pereiopods are often tipped or banded with brighter colors than the rest of the animal. Many species show a wide range of color. *Orconectes immunis*, for example, may be bluish, blackish, brown, greenish, or red. In general, newly molted specimens are more brightly colored than older specimens, especially since the latter usually have an accumulated coat of dirt, debris, and algae.

The coloration of a species often varies in accordance with the coloration of the substrate. In most crayfish such color adaptations are decidedly limited and occur only over a period of weeks or months. In *Palaemonetes paludosus*, however, variable coloration and background resemblance have become developed to a striking degree. This species has

FIG. 2.–A, ventral view of female *Orconectes*, ✕0.7; B, ventral view of male *Orconectes*, ✕0.7.

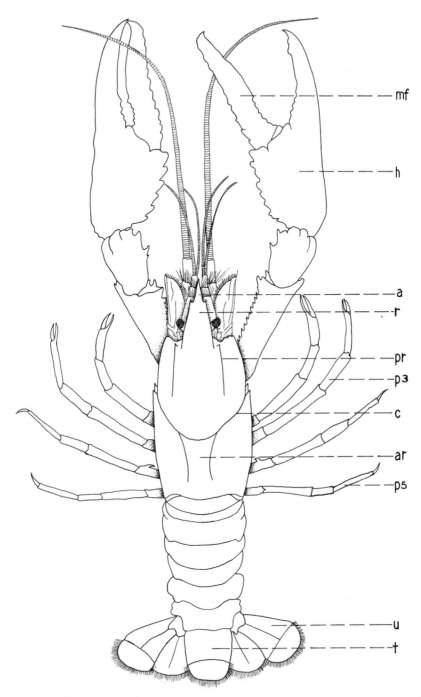

FIG. 3.–Structure of a typical crayfish. a, acumen; ar, areola; c, cervical groove; h, hand; mf, movable finger of chela; p3, third leg (pereiopod); p5, fifth leg (pereiopod); pr, postorbital ridge; r, rostrum; t, telson; u, uropod. (Modified from Hobbs, 1942a).

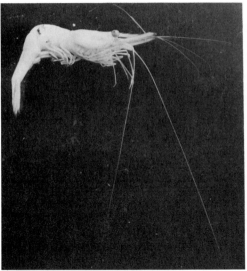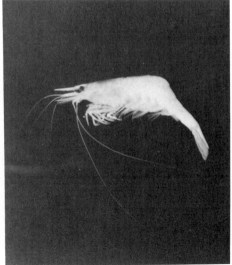

FIG. 4.–Fresh-water prawns. Left, *Palaemonetes paludosus,* ×2.5; right, *Syncaris pacifica,* ×1.2.

four types of chromatophores containing white, red, yellow, and blue pigment granules, respectively, and general body coloration is determined by the relative concentration or dispersion of these granules in various combinations. Color adaptation is completed in less than 24 hours after being placed on a new substrate.

Locomotion. Undisturbed crayfish walk or climb about slowly with the pereiopods; they may move forward, backward, or sideways. When alarmed, however, they quickly dart backward by ventrally doubling up the tail fan and the posterior end of the abdomen. They may also leave the substrate and move through the water for short distances by such abdominal contractions. In addition to these two modes of locomotion, the Palaemonidae are distinct swimmers and may move continuously forward through the water by movements of the pleopods.

Food, feeding. In general, crayfish are omnivorous but seldom predaceous, although they commonly feed on snails and small fish. There is also some evidence that crayfish will at times feed on living aquatic insects. They eat all kinds of succulent aquatic vegetation, and animal food is usually a minor part of the diet when there is abundant vegetation. They also prefer fresh to stale meat, and in the laboratory they have been fed raw and cooked meats of all kinds, prepared fish foods, hay, whole seeds, cottonseed meal, and soybean meal. Ecologically, they are usually considered scavengers. Occasionally they feed on sponge tissues.

The chelate appendages are used for crushing, picking up the food, and tearing it into pieces. The pieces are then passed forward where they are further cut and masticated by the maxillipeds. The maxillae strain out larger particles and mince the smaller ones further. The mandibles are of limited use in grinding and chewing; mostly they hold food while the maxillipeds tear off fragments.

Internal anatomy, physiology. These topics are treated rather completely in most zoology texts and will only be outlined here.

A short esophagus enters a large cardiac chamber of the stomach which, in turn, enters the smaller pyloric chamber. These chambers contain the gastric mill, which is composed of heavy ossicles for grinding the food and rows of stiff setae for straining out larger particles. The long intestine extends to the anus on the ventral side of the telson. Two large glands are associated with the digestive tract in the cephalothorax; they function in digestion, absorption, and food storage.

The pentagonal heart lies in the dorsal portion of the cephalothorax. It receives blood from the surrounding pericardial sinus through three pairs of ostia and sends it out anteriorly, posteriorly, and ventrally through arteries. The blood passes out of the fine terminal endings of the blood vessels and into the various organs and hemocoel. It then circulates back to the sternal sinus, thence to the gills, and back to the pericardial sinus. The blood is almost colorless and contains amoebocytes and a dissolved hemocyanin.

Seventeen or 18 featherlike gills lie within the branchial chamber on each side. They are attached on or near the bases of the maxillipeds and walking legs and extend dorsally. Blood circulates along definite channels through the central stalk and minute lateral processes of each gill.

Excretion and osmotic control are maintained by two complicated, compact green glands which are situated in the anterior end of the cephalothorax and open ventrally at the base of the second antennae. A crayfish does not drink, but quantities of water are continuously diffusing the blood through the gill surfaces. Blood (minus its proteins) is absorbed by the distal coelomic sac and labyrinth of the green-gland apparatus. As it moves on through the nephridial canal, the useful materials, such as salts, carbohydrates, and some water, are reabsorbed back into the surrounding blood. The material remaining in the canal is a very dilute urine; it passes into the urinary bladder, from which it is periodically voided to the outside through a small pore. There is also some evidence to show that the gills regulate the internal concentration of certain ions, such as sodium.

The nervous system is of the usual crustacean plan, with a ventral nerve cord and chain of ganglia, two circumesophageal connectives, and dorsal brain. The basal segment of each first antenna contains a statocyst, or organ of balance. Taste receptors are especially abundant on the mouth parts and antennae. The antennae are also sensitive tactile organs. The compound eyes are large and consist of hundreds of ommatidia.

In addition to the optic nerve and ganglia, each eyestalk contains a minute sinus gland, an endocrine gland whose secretions influence many important aspects of crayfish behavior and physiology. Among other things, the sinus gland, to a greater or lesser extent, governs: chromatophore contraction, frequency of molting, metabolic rate, growth, viability, and light adaptation of the eyes.

Reproduction. With reference to the structure of the first pleopods, there are two distinct types of adult male Cambarinae, first form and second form. The first pleopods of a first form male are corneous, hard, and distinctly sculptured at the tip. In a second form male they are soft and generally lacking in sculpturing. These differences are shown in Figs. 13A, B. Only a first form male is capable of copulating and transferring sperm to the body of a female. Though the instar that precedes the adult first form male is a juvenile instar, the first pleopods are almost indistinguishable from those of a true second form male and for all practical purposes this stage is considered second form. The true second form male, however, is the instar immediately following the first form. Depending on the species and ecological conditions, a true second form male may undergo a further molt to become a first form male again before death. *Orconectes immunis*, for example, is commonly first form during the spring, when copulation occurs, second form during the summer, and first form again beginning in the autumn; copulation may also occur in the autumn. *Pacifastacus* males do not have first and second forms.

Careful work has been done on less than a dozen species in regard to breeding habits, but the available data indicate that time of copulation is variable and occurs between early spring and autumn. In *Orconectes immunis* copulation begins as early as July. *Cambarus diogenes* copulates in the spring. *Palaemonetes paludosus* is thought to copulate in both spring and autumn. *Cambarus bartoni, Orconectes propinquus*, and a few other species probably copulate during spring, summer, and autumn, though spring copulation presumably does not occur in the more northern states.

Copulation seems to be more or less a matter of chance, although sex pheromones have been demonstrated in at least one species.

FIG. 5.–Annulus ventralis of female *Procambarus blandingi*. (Redrawn from Hobbs, 1942a.)

The male has no power of sex discrimination, and during the mating season he seizes and turns over every crayfish coming his way. Another male will always resist strongly, but a female will either resist or remain passive and receptive.

The actual process of copulation, sperm transfer, and egg deposition has been described most carefully for *Orconectes limosus* and *Pacifastacus trowbridgi*. The male seizes the female and turns her over on her back. He mounts on her ventral side and holds all her clawed appendages securely with his two chelae. The tips of the first two pleopods are then inserted into the chink of the annulus ventralis, and the tips of the two vasa deferentia are extruded into the bases of the grooves that extend along the first pleopods. Sperm move along these grooves in macaronilike cords and are deposited in the chink. The animals then separate, and either the male or female may copulate later with additional individuals. A single copulation may take from a few minutes to as much as 10 hours. The second pleopods are closely applied to the first during sperm transfer, but their exact function is unknown. After copulation a white waxy plug projects from the chink of the annulus ventralis. In *Pacifastacus* there is no annulus, and the spermatophores are deposited at random on the ventral surface of the female.

The female lays her eggs several weeks to several months after copulation, depending on the season of the year, but before laying them she cleans the ventral side of the abdomen thoroughly with the tips of the pereiopods. Immediately preceding egg extrusion she secretes a sticky, cementlike substance (glair) from ventral glands. The glair covers the ventral surface of the abdomen, tail fan, and pleopods. The sperm plug breaks and sperm are released into the glair. She then lies on her back, curls the abdomen, and releases the eggs from the genital pores. By curious turning movements of the entire body, the eggs become dispersed through the glair, fertilized, and securely attached to the pleopods by a short stalk and surrounding capsule. The female is then said to be "in berry." Most females of nearly all species are seen carrying eggs between March and June (Fig. 6). A female may carry as few as 10 eggs and as many as 700. In general, within a species, the larger the female, the greater the number of eggs. Rhythmic movements of the pleopods effectively aerate the developing eggs. Few eggs die.

Development, life cycle. The incubation period usually ranges from 2 to 20 weeks, depending largely on temperature, and during this interval the eggs change from dark and opaque to pale and translucent. Hatching involves the rupture and shedding of the embryonic skin and egg membrane, both of which, however, remain attached to the parent pleopod. A newly hatched first instar young clings to the egg stalk with the chelae, and the posterior end remains attached to the cast exoskeleton by a membrane (Fig. 6). Although clearly recognizable as a crayfish, the first instar differs markedly from advanced instars and adults. The cephalothorax is proportionately very large; the rostrum is large, pointed, and bent downward; the eyes are large; the chelae are slender; and there are no first and sixth abdominal appendages. The first instar usually persists for 2 to 7 days, and when it molts the posterior connection with the remains of the egg is lost. The second instar lasts for about 4 to 12 days and remains attached to the pleopods of the mother with its chelae and pereiopods. Third instar crayfish leave the parent pleopods intermittently and then permanently. The young of burrowing species leave the burrow of the mother when they have a body length of 8 to 20 mm. Individuals hatching in the spring usually have a total of 6 to 10 molts by autumn, when growth and molting cease. In the fresh-water

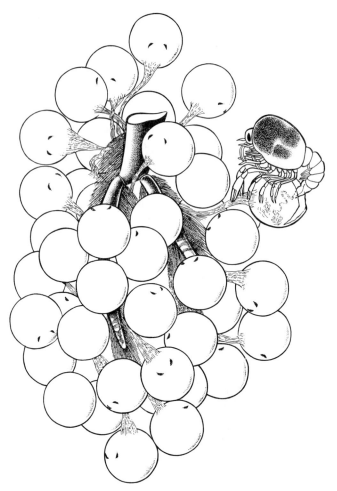

FIG. 6.–Diagram of one pleopod of a female crayfish with eggs attached. One egg has hatched and the stage I juvenile is grasping the egg stalk with its hooked chelipeds. The pigmented eye regions of juveniles just ready to hatch show through the shells of other eggs. (From Crocker and Barr, 1968, modified from Andrews, 1907.)

Palaemonidae there are five to eight larval molts. During these early instars the body and its appendages gradually assume adult proportions and anatomical details.

Great differences in growth rate may be found within a single pond or stream, the larger specimens being twice or more as long as the smaller ones by autumn. Such differential growth rates are probably due mainly to varying activity and amounts of food consumed.

By autumn many females are sexually mature and many males are first form, so that copulation commonly occurs in the first autumn of life. Some females, however, do not become sexually mature until after a molt early the following spring; similarly, some males do not become first form until after one or two spring molts. Such individuals copulate in the spring, and the female is in berry after a very short interval. Some females that copulate in the autumn are not in berry until the following spring. Other species, especially burrowers, copulate and are in berry in the autumn. The young leave the burrow some time during the following spring or summer.

Subsequent to the first mating season, most crayfish have only two to four molts before they die. Following copulation, a female does not molt until after her third-instar young have left her. If a male is first form in its first autumn, it usually remains first form during the winter, molts to second form early in the following spring, and after a very short interval molts back again to first form. Such males may therefore copulate again during their second summer or autumn. Among the few species on which life history studies have been made, the normal life span of both male and female crayfish is usually less than 24 months (see Payne, 1972), and a few survive their second winter. Females that pass through a second winter may be in berry for the second time the following spring. Lyle (1938) reports an exceptional longevity for *Procambarus hagenianus* (Faxon) in Mississippi and Alabama, the species attaining an age of 6 or 7 years.

The molting processes are a complex physiological cycle that is modified and controlled chiefly by temperature, light, and hormonal interrelationships.

Ecology. Adults remain hidden in their burrows, under stones or debris, or half buried in small depressions in the substrate during the daytime, but between dusk and dawn they feed and move about, even to the extent of coming out on banks of streams and ponds. In shaded streams, however, and especially on cloudy days, adults may venture into open water in daylight. Immature specimens of most species, on the other hand, regularly crawl about actively on the bottom and in vegetation during the daytime.

There is some evidence to show that the 24-hour cycle of activity is governed by an endogenous mechanism. Solar and lunar rhythms have also been suggested.

Many crayfish have specific habitat requirements. Some are burrowers in wet meadows and do not enter bodies of water; some occur only in muddy ponds and ditches; some are found only in lakes or slow rivers; others are restricted to swift, stony streams. The preferences of representative species are included in the key that follows.

Closely related species sometimes have similar ranges but are ecologically isolated within these ranges. For example, *Orconectes virilis* and *O. immunis* are often found in the same area but the former is restricted to streams and lake margins, and the latter to ponds and sloughs.

Crayfish are generally inhabitants of shallow waters, seldom being found deeper than a meter. An exceptional record is the occurrence of *Orconectes virilis* at a depth of 30 m in Lake Michigan. Most species tolerate normal but

FIG. 7.–Typical crayfish chimney.

wide ranges in temperature, hydrogen-ion concentration, and free and bound carbon dioxide, though stream species are usually less tolerant than lake and pond species. During winter months crayfish are found in deeper waters of lakes. Similar migrations during the warm months are thought to be associated with gonad maturation. In streams, however, some investigators have found that most individuals have a home range of less than 30 m.

Population densities vary greatly, depending on the species and habitat. Pond populations usually amount to less than 100 lb/acre, but in exceptional cases may attain 500, 1000, or even 1500 lb/acre. In general, stream populations are less dense, although there is one record of 1176 lb of crayfish per acre. Abrahamson and Goldman (1970) calculated an average of 0.9 adult *Pacifastacus leniusculus* per square meter in Lake Tahoe. Momot et al. (1978) found as high as 15 crayfish per square meter.

Several laboratory studies have shown that in groups of four to eight crayfish, dominance–subdominance relationships soon become established in a linear fashion, and that the dominance order is based on size and learned recognition of individuals. Such work has been done on several species of *Orconectes, Procambarus,* and *Cambarellus.*

A few studies have shown that crayfish have a remarkable daily cycle in their oxygen consumption. Fingerman (1955), for example, showed that *Cambarellus shufeldti* had its highest oxygen consumption around 6 A.M., with a secondary maximum at 3 to 6 P.M. Minima occurred from 9 A.M. to noon and from 9 P.M. to midnight.

Burrows. Some species, such as *Procambarus gracilis* and *Cambarus carolinus,* habitually burrow in wet pastures and marshy areas where there is no open water. Species in a second group live in bodies of water during

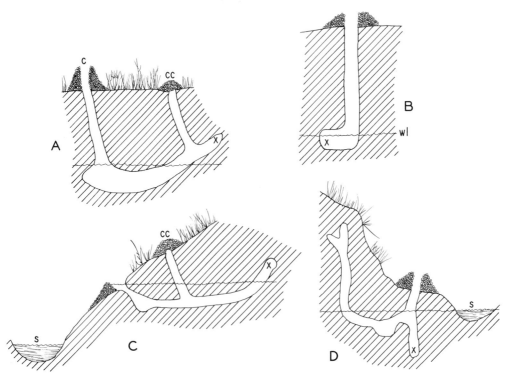

FIG. 8.–Typical crayfish burrows. c, chimney and opening of burrow; cc, closed chimney; s, stream; wl, ground water level; x, place where crayfish were taken. (A, C, and D modified from Ortmann, 1906.)

the greater part of the year, but when streams or ponds dry up in late summer or when the temperature drops in the autumn they construct burrows along the margins and live in them until the water level rises and the weather becomes warmer. *Procambarus simulans, P. blandingi,* and *Orconectes immunis* are included in this group. Species of a third group live continuously in permanent waters and do not make burrows.

Burrows differ widely in construction, depending on the species, the soil, and the depth of the water table (Fig. 8). Usually there is only one entrance, though there may be as many as three. The tunnel leading from the entrance may proceed vertically, at an angle, or almost laterally in a sloping bank. Sometimes the tunnels are branched or irregular, but there is always a chamber at the lower end where the crayfish remains during the hours of daylight. Occasionally a burrow has a lateral chamber. The depth of a burrow ranges from 5 cm to as much as 3 m and is

partially determined by the level of the water table, since the chamber must contain water to keep the gills wet. Burrows close to the edge of a pond or stream are shallow, those farther away are deeper.

Except during the breeding season, each burrow houses a single crayfish. Burrows are constructed only at night, and the crayfish brings up pellets of mud and deposits them at the entrance to form a chimney. Such chimneys are usually about 15 cm high, but a few as high as 45 cm have been reported. They do not serve any particular purpose, but are simply the result of the safest and most convenient method of disposing of the mud pellets.

Geographic distribution, dispersal. The Cambarinae are restricted to North America, and Ortmann (1905) postulated the origin of this subfamily between Cretaceous and Tertiary times in central Mexico. From this area

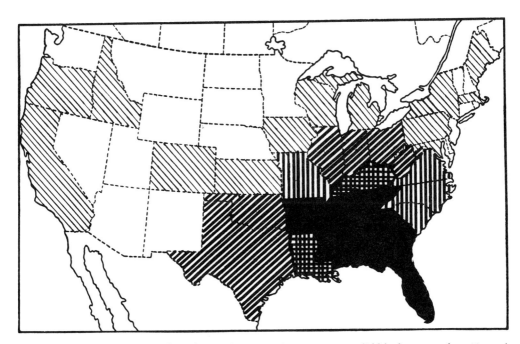

FIG. 9.–Relative distribution of crayfish in the 48 contiguous states. Solid black—more than 40 species and subspecies; cross hatched—31 to 40; vertical lines—21 to 30; coarse diagonal lines—11 to 20; fine diagonal lines—5 to 10; unshaded—less than 5. (Data from Hobbs, 1972.)

there was a general spread into the southern United States, where several centers of dispersal arose.

Ancestors of the present species of *Procambarus* are thought to have originated in Mexico and migrated to the general region of Kansas, Oklahoma, Texas, Arkansas, and Louisiana. From there, one group of species migrated northward as far as Wisconsin. A second group migrated eastward to the southern Appalachians where a new large center of dispersal later expanded out extensively up and down the Atlantic coast, as far north as Michigan and Ohio, and eventually as far west as Texas and Oklahoma.

Orconectes is thought to have arisen at the junctions of the Ohio, Mississippi, and Missouri rivers and radiated out in all directions as far as Colorado, Wyoming, Utah, southern Canada, Lake Superior, the St. Lawrence valley, and the Gulf states. A few species even crossed the Divide and reached Atlantic coast drainages.

Cambarus is thought to have arisen in Tennessee (Hobbs, 1969). Early migration occurred mainly in a northeasterly direction at high altitudes, but later some species invaded the lowlands, spread out in all directions, and attained a very wide distribution.

Cambarellus originated in Mexico or extreme southern United States and migrated eastward through the Gulf states.

The five American species of *Pacifastacus* are restricted to Pacific slope drainages, although two have migrated over to the Great Basin and headwaters of the Missouri River. The genus probably originated in the Old World, where a total of only about 11 species occur in northern Europe and Asia.

In general, speciation has proceeded most rapidly in the southeastern states, as shown in Fig. 9. Crocker (1979) lists only 10 species for New England, of which 3 are undoubtedly introduced by man and 3 others are possibly introduced naturally.

Dry land forms an effective barrier to the migration and geographical spread of lake and stream species, but burrowers such as *Cambarus diogenes* have migrated rapidly and are now found commonly in wet meadows east of the Rocky Mountains.

Male first form *Procambarus clarki* spawn in the autumn and then migrate overland by the thousands. Such exhausted males are weak and soon die; it is doubtful whether they contribute appreciably to the spread of the species.

Restriction to specific aquatic habitats forms a partial barrier for many species. *Procambarus blandingi*, for example, is restricted to sluggish streams and stagnant waters and cannot migrate upstream into rapid water drainage areas.

Orconectes virilis and *O. immunis* form an interesting example of ecological isolation. The two species are very closely related and are often in the same general geographic areas. Yet the former occurs only in clear, running waters with stony bottoms, whereas the latter is restricted to stagnant waters and mud bottoms. Very little is known about interspecific competition among crayfishes. Capelli (1982), however, has shown that the introduction of *Orconectes rusticus* into a lake results in the elimination of the original native species of crayfish. See also Capelli and Magnuson (1983).

Many species, especially in the southern states, are known only from a single state or single collecting locality.

Several American crayfishes have been intentionally introduced into Europe, dating back to the transportation of *Orconectes limosus* into France as early as 1890; it is now widely distributed in western Europe. *Pacifastacus leniusculus* was brought to Sweden and *Procambarus clarki* into both Hawaii and Japan.

In our estimation, several eastern stream-dwelling Cambaridae might be extremely valuable introductions into certain western trout streams, especially where the trout population is sparse and could benefit by this additional link in the food web.

Cave crayfish. Perhaps some of the most interesting examples of ecological isolation are to be found among those subterranean crayfish and shrimp that are restricted to the waters of single caves, cave systems, large springs, and sinks.

Cave species are strikingly modified for that environment (Fig. 10). They are mostly small, the body length ranging from 10 to 80 mm. The body is slender, the chelae are not well developed, and the antennae and other

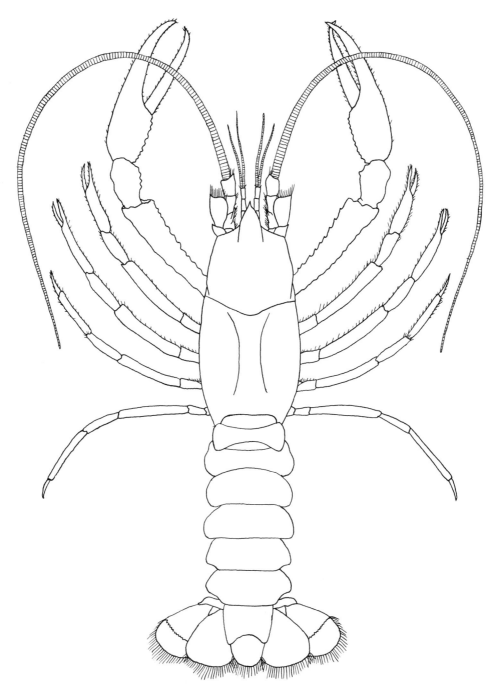

FIG. 10.–Male *Orconectes pellucidus* (Tellkampf), ×1.7, a cave species.

appendages are very long and highly specialized as tactile organs. Pigmentation is absent,
and the body is translucent or whitish. All are
blind, the eyes being atrophied and the
eyestalks reduced or rudimentary. About 25
North American crayfish species and subspecies are true troglobites.

In addition to true troglobitic species,
there are some which are normally found
above ground but have migrated into caves
where they have successfully persisted. With
the exception of one or two species that have
become differentiated into subspecies, these
forms are identical with individuals occurring
above ground.

Parasites, commensals, enemies. The
external surfaces of crayfish form a convenient
and favorable habitat for microorganisms,
and when the exoskeleton has not been shed
for a long time it is often covered with debris
and a dense growth of algae and sessile
Protozoa, especially at winter's end.

The Branchiobdellida (see page 307) are
small, leechlike, parasitic or commensal
oligochaetes that live on the gills, in the gill
chambers, on the general body surface, and
at the bases of the legs, antennae, and mouth
parts. A genus of ostracods, *Entocythere*, is
commensal on the gills (see page 456). *Macrobrachium* and *Palaemonetes* often carry parasitic isopods (*Probopyrus*) in their gill chambers
(see page 467).

Many species of crayfish east of the
Mississippi River may act as the second
intermediate host of *Paragonimus westermani*, a
lung fluke parasitic in man and many carnivores. The metacercariae of this parasite
become imbedded in the muscles and viscera
of the crayfish from which they may be
transmitted to the definitive host when the
raw or poorly cooked crayfish are used as
food.

Fish are the most important enemies of
crayfish, although wading birds, frogs, turtles,
raccoons, otter, and mink consume appreciable numbers.

Economic importance. In poorly drained
clay agricultural lands of West Virginia,
Mississippi, Alabama, and a few other southern states burrowing crayfish become so

abundant (as many as 30,000 burrows per
acre) that crops cannot be raised because of
the fact that the crayfish denude the land of
the young grain, sugarcane, and cotton
seedlings. Also, their chimneys clog farm
machinery. If the land cannot be drained, the
best control is the application of some poison
to the burrows. Pyrethrum, creosote, cyanide,
carbon disulfide, and unslaked lime are some
of the more useful poisons. They are most
effective when the water table is high. Bait
poisoned with DDT and spread over infested
areas is a more recent control measure. In
California, *Procambarus clarkii* is notorius
because of its depredations on rice seedlings.

Occasionally crayfish become a nuisance
in small reservoirs and irrigation ditches
when their lateral burrows through earthen
dams and dikes drain the reservoirs.

Europeans use crayfish for the table to a
much greater extent than do Americans.
Most of the edible crayfish in this country
come from the Great Lakes, certain parts of
the Mississippi River system, Louisiana, and
Oregon. The harvest has greatly declined
during the past 30 or 40 years, although in
1960 717,000 lb came from the Gulf coast,
609,000 lb from the Mississippi River drainage,
29,000 lb from Pacific coast drainages, and
only 2000 lb from the Great Lakes area. In
1978, however, the Atchafalaya Swamp crayfish harvest reached an astounding 36,000,000
lb.

In addition to the harvest of "wild crayfish,"
there is an increasing commercial production,
especially in Louisiana, where there is an
estimated total of 7000 hectares of crayfish
"farms." Only two species are cultured, *Procambarus clarkii* (the red crayfish) and *P. blandingi*
(the white crayfish). They are annual species,
which reach maturity in 6 months. Artificial
impoundments produce 80% of cultured
Louisiana crayfish, and rice fields produce
the remainder. Both species are euryokous
under natural conditions and variables. New
impoundments are originally stocked with
immature crayfish, but thereafter the population is usually self-sustaining. The food source
is mainly a wide variety of hydrophytes.
Harvesting occurs in November to June and
is done chiefly with chicken wire traps baited
with chunks of rough fish, canned dog food,
and other high-protein materials.

The giant fresh-water prawn, *Macrobrachium*, is raised commercially in southeastern Asia, but in the United States attempts to culture it commercially in the Gulf Coast states have thus far not been financially successful. Small species of crayfish have been raised experimentally at hatcheries for use as fish food, but the practice has not attained any importance.

On the other hand, crayfish often become a nuisance in fish-hatchery rearing ponds, where they may be excessively abundant and feed on small fish.

It has been shown that when it is introduced into weedy trout lakes, at least one species (*Orconectes causeyi*) can keep aquatic vegetation grazed down so that such lakes become much more favorable trout habitats, provided that the fish population does not decimate the crayfish population.

Collection, preparation. Stream and pond species may be collected with long-handled dipnets, minnow seines, and minnow traps. When crayfish are not abundant, however, it is often necessary to collect by hand and examine overhanging banks, masses of vegetation, and the undersides of submerged logs and stones.

A skillful collector with a headlamp may find it profitable to collect burrowers (as well as open-water species) at night, when they usually leave the burrow and roam about, but often there is no alternative but to dig up the burrow carefully with a spade or trowel until the specimen can be taken in its chamber. Surface traps baited with meat or peanut butter will sometimes take burrowing crayfish.

Crayfish should be killed in 5% formalin or 80% alcohol and stored in a change to 70% alcohol. This preservative keeps the joints pliable. Some workers remove the male first pleopods and the female annulus ventralis, dehydrate in absolute alcohol, clear in xylol, and mount on insect pins.

Taxonomy. In 1902 only 66 species of North American Cambarinae were known, but now there are about 350 from the United States alone, most of the additional species having been described from southern states. In addition, some of the common and widely distributed species are represented by several well-defined subspecies. Europe has only 15 species of crayfish.

Characters that are useful for identifying both males and females of certain genera and species include the structure of the rostrum, shape of the areola, structure of the chelae, spines on the cephalothorax, and general body proportions. Unfortunately, however, most species can be identified only from first form males, particularly by the copulatory hooks on some of the legs and the detailed morphology of the first pleopods. In making collections in the field it is therefore essential to get first form males whenever possible.

Beginning with the publication of Hagen's monograph in 1870, various classification schemes for the Cambarinae have been suggested. Most of these publications recognized only a single genus, *Cambarus*, with a series of subgenera or "groups." Beginning in 1942, however, Hobbs proposed a more logical and revised classification in which the Subfamily Cambarinae was subdivided and now contains seven genera. The following key is modified from Hobbs (1972) and includes all genera and the more common species. Total lengths extend from the tip of the acumen to the posterior edge of the telson.

KEY TO GENERA AND SPECIES OF DECAPODA

1. First three pairs of legs chelate; cephalothorax subcylindrical; abdomen more or less flattened dorsoventrally (Figs. 1, 2) 2
   First two pairs of legs chelate; cephalothorax and abdomen laterally compressed (Figs. 4, 17) 33
2. Ischia of pereiopods of male without hooks; male first pleopods simply rolled at tips; female without first pleopods; up to 110 mm long; except in the upper Missouri River, found in streams and lakes only west of the Continental Divide .... ASTACIDAE, **Pacifastacus, 3**

Ischia of one or more pairs of pereiopods with hooks (Fig. 13 E); male first pleopods bifid or toothed at tips (Fig. 12); female with first pleopods; chiefly east of the Continental Divide. (Beyond this point in this key most key characters apply only to first form males; females can be tentatively identified only by "association" with the males.) .... CAMBARIDAE, 7

3. Rostrum with a single pair of marginal tubercles or spines (Figs. 11 A–C); a complex species found in streams and lakes of WA, OR, CA, ID, NV .... **Pacifastacus leniusculus** Dana
   Rostrum with at least three pairs of marginal spines (Figs. 11 D, E, J) **4**
4. Palm of chela with two conspicuous dorsal patches of setae (Fig. 11 F) **5**
   Palm of chela without patches of setae **6**
5. Postorbital ridges with one or two posterior pairs of spines or tubercles (Fig. 11 J); ID, OR.

   Pacifastacus connectens (Faxon)

   Postorbital ridges without posterior spines or tubercles; streams and lakes of WA, OR, CA, ID, NV, UT, MT, WY .... **Pacifastacus gambeli** (Girard)
6. Palm of chela broad (Fig. 11 G); streams of Shasta County, CA; almost extinct.

   Pacifastacus fortis (Faxon)

   Palm of chela slim (Fig. 11 H); streams of San Francisco area; perhaps extinct.

   Pacifastacus nigrescens (Stimpson)
7. Ischia of second pereiopods without hooks .... CAMBARINAE, 8
   Ischia of second and third pereiopods with hooks; eight small species, usually 15 to 30 mm long; FL to TX and north to IL, TN, AR .... CAMBARELLINAE, **Cambarellus**
8. Body pigmented; eyes with facets and pigment **12**
   Albinistic species; eyes rarely faceted and pigment greatly reduced or absent; usually in underground waters **9**
9. Ischium of third maxilliped without teeth along inner margin; about 30 mm long; known only from a few subterranean waters in FL .... **Troglocambarus maclanei** Hobbs
   Ischium of third maxilliped with teeth along inner margin **10**
10. First pleopod terminating in two large elements that are strongly bent (Fig. 12A); eight species in White River, MO, TN, AL, GA, OK, FL .... **Cambarus**
    First pleopod with two or more small terminal elements, bent only slightly (Figs. 12B, C) ... **11**
11. First pleopod with two terminal elements (Fig. 12B); four species in KY, TN, AL, IN .... **Orconectes**
    First pleopod with two or more elements (Fig. 12C); if only two, then cephalic surface with strong shoulder (Fig. 12D); 4 species in AL, FL .... **Procambarus**
12. Second antennae conspicuously fringed (Fig. 13K); known only from the Barren and Green River systems in KY, TN; under large rocks; coloration bright green with cream or scarlet ridges and tubercles .... **Barbicambarus cornutus** (Faxon)
    Second antennae not fringed **13**
13. First pleopod terminating in two elements, of which one is very long and slender and the other is no more than half as long (Fig. 12E); three species in sluggish waters of SC to FL and west to OK, TX.

    Faxonella

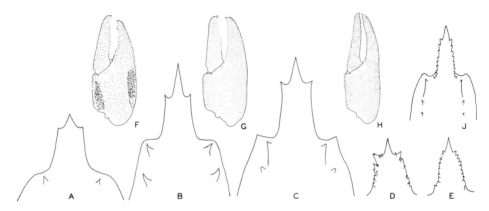

FIG. 11.–Structure of *Pacifastacus*. A to C, anterior end of *P. leniusculus*; D, rostrum of *P. nigrescens*; E, rostrum of *P. gambeli*; F, chela of *P. gambeli*; G, chela of *P. fortis*; H, chela of *P. nigrescens*; J, anterior end of *P. connectens*. (F–H from Hobbs, 1972.)

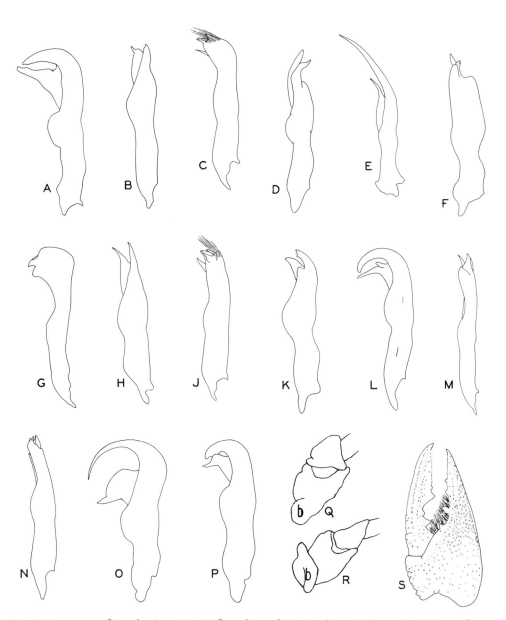

FIG. 12.–Structure of Cambarinae. A to P, first pleopods. A, *Cambarus*; B, *Orconectes*; C, *Procambarus*; D, *Procambarus*; E, *Faxonella*; F and G, *Procambarus*; H, *Hobbseus*; J, *Procambarus*; K, *Cambarus*; L, *Fallicambarus*; M and N, *Procambarus*; O, *Cambarus*; P, *Fallicambarus*; Q and R, base of fourth leg of *Procambarus* showing boss (*b*); S, right chela of *Fallicambarus*. (A–P, S from Hobbs, 1972.)

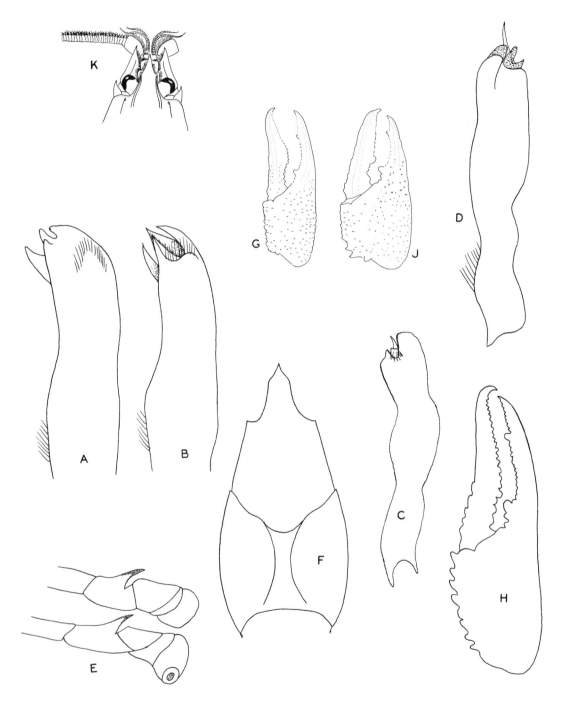

FIG. 13.–Structure of *Procambarus*. A, first pleopod of second form male *Procambarus blandingi*; B, first pleopod of first form male of *P. blandingi*; C, left first pleopod of *P. clarki*; D, first pleopod of male *P. simulans*; E, basal portions of third and fourth legs of male *Procambarus*; F, carapace of *P. simulans*; G, chela of *P. simulans*; H, chela of *P. clarki*; J, chela of *P. gracilis*; K, cephalic region of *Barbaricambarus*. (A and B modified from Hobbs, 1942a; D redrawn from Creaser and Ortenburger, 1933; E, modified from Turner, 1926; G, J, and K from Hobbs, 1972.)

First pleopod terminating in two or more elements; if there are two elements, then the shorter element is more than half as long as the other **14**

14. First pleopod terminating in two elements (Figs. 12A, O, P; 13C, D) **15**
    First pleopod terminating in more than two elements (Figs. 12C, J, L–N) **19**
15. First pleopod generally as in Figs. 13C, D; usually with short processes; about 80 taxa in this genus, mostly with restricted distributions in southeastern states; the four most widely distributed and common species are keyed here .... **Procambarus, 21**
    First pleopod with two prominent processes, more or less curved (Figs. 12A, O, P) **16**
16. Coxa of fourth pereiopod without a boss **17**
    Coxa of fourth pereiopod with a boss (Figs. 12Q, R) **18**
17. In resting position, first pleopods deep along the midline, hidden by a dense mat of setae at the base of the pereiopods; five species in MS, AL .... **Hobbseus**
    In resting position, first pleopods not concealed along the midline; about 70 species representing 10 subgenera; mostly with local distributions in southeastern states, but seven widely distributed species are keyed here .... **Orconectes, 24**

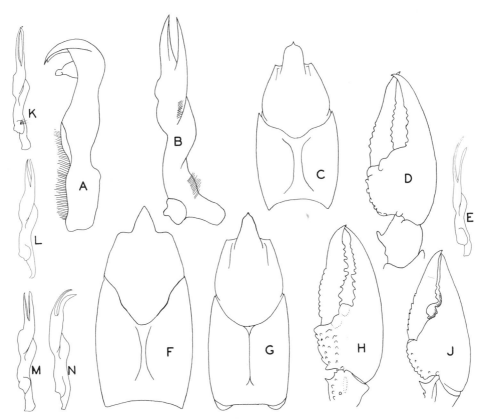

FIG. 14.–Structure of *Cambarus* and *Orconectes*. A, first pleopod of male *Cambarus bartoni*; B, first pleopod of male *Orconectes propinquus*; C, carapace of *Cambarus bartoni*; D, chela of *C. bartoni*; E, left first pleopod of *Orconectes virilis*; F, carapace of *Cambarus carolinus*; G, carapace of *C. diogenes*; H, chela of *C diogenes*; J, chela of *Fallicambarus fodiens*; K, left first pleopod of *Orconectes neglectus*; L, left first pleopod of *O. propinquus*; M, left first pleopod of *O. obscurus*; N, left first pleopod of *O. immunis*. (A–C redrawn from Turner, 1926; D modified from Ortmann, 1906; G redrawn from Hobbs, 1942a; H and J modified from Huntsman; E, K–N from Hobbs, 1972.)

18. Dactyl of chela with a prominent excision on the basal half (Fig. 14 J); about 12 species in LA, AR, TN, MS, FL, MD, SC, and one species [*F. fodiens* (Cottle)] more generally distributed east of the Mississippi River .... **Fallicambarus**
    Dactyl of chela without a prominent excision on the basal half (Fig. 14D); about 60 species, mostly with restricted distributions in southeastern states, but 4 widely distributed species are keyed here.
    Cambarus, 30
19. Central projection of first pleopod bladelike, always directed caudally, and bearing a notch (Fig. 12K) (see couplet 18 information) .... **Cambarus, 30**
    Central projection of first pleopod rarely bladelike, but if present then directed laterodistally or lacking a notch (Figs. 12C, F, G, J, L) **20**
20. First pleopod with large curved processes (Fig. 12L) (see couplet 18 distribution data).
    Fallicambarus
    First pleopod with small processes (Figs. 12C, F, G, J) (see couplet 15 information).
    Procambarus, 21
21. Cephalic surface of first pleopod with a prominent shoulder (Fig. 13C); hooks on ischia of both third and fourth pereiopods (Fig. 13E); lentic and lotic habitats from IL to TX, CA; introduced in CA, NV, VA, ID, UT, and probably elsewhere; the common "swamp crayfish."
    Procambarus clarki (Girard)
    Cephalic surface of first pleopod without such a prominent shoulder **22**
22. Hooks on ischia of third pereiopods only; common burrowers **23**
    Hooks on ischia of third and fourth pereiopods (Fig. 13E); streams and lentic habitats from ME to FL and westward from MN to TX .... **Procambarus acutus** (Girard)
23. Chela relatively slender (Fig. 13G); lentic and lotic habitats in NM, CO, KS, OK, AR, LA.
    Procambarus simulans (Faxon)
    Chela relatively stout (Fig. 13J); in muddy streams and ponds; WI south to AR, TX.
    Procambarus gracilis (Bundy)
24. Terminal processes of first pleopod less than one quarter of the total length (Figs. 14B, L, N; 15A, B).
    25
    Terminal processes of first pleopod more than one quarter of the total length (Figs. 14E, K) **28**
25. Terminal processes of first pleopod strongly curved (Fig. 14N); sluggish streams and ponds; New England west to WY and south to AL; a burrower; the "mud crayfish."
    Orconectes immunis (Hagen)
    Terminal processes of first pleopod not strongly curved **26**
26. Terminal processes of first pleopod divergent (Figs. 15A, B); general color olive green to olive yellow; seldom in burrows; common in larger rivers of Atlantic coastal plain from ME to VA.
    Orconectes limosus (Raf.)
    Terminal processes of first pleopod not divergent (Figs. 14B, L) **27**

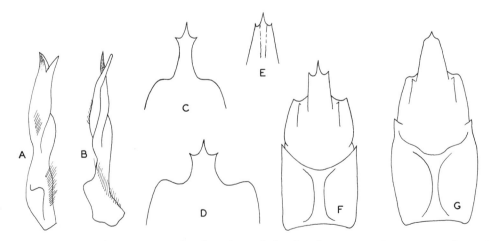

Fig. 15.–Structure of *Orconectes*. A and B, first pleopod of male *O. limosus*; C and D, rostrum of *O. rusticus*; E, rostrum of *O. propinquus*; F, carapace of *O, virilus*; G, carapace of *O. immunis*. (A and B redrawn from Ortmann, 1906; F and G redrawn from Turner, 1926.)

27. Rostrum with median keel (Fig. 15E); general color gray to olive green; streams, rivers, and lakes, especially on stony substrates; Great Lakes crayfish; gray rock crayfish; northeastern quarter of the United States, especially MI, OH, and Great Lakes drainages.

Orconectes propinquus (Girard)

Rostrum without median keel; general color olive green; Allegheny crayfish; large streams of upper Ohio drainage, NY to OH, WV .... **Orconectes obscurus** (Hagen)

28. Sides of rostrum slightly concave (Figs. 15C, D); streams and lakes in New England to IL, KY; introduced elsewhere .... **Orconectus rusticus** (Girard)

Sides of rostrum not concave (Figs. 15E, F) **29**

29. Rostrum with median keel (Fig. 15E); coloration yellowish; clear lotic waters in AR, MO, OK, KS, NE, CO .... **Orconectes neglectus** (Faxon)

Rostrum without a keel (Fig. 15F); northern crayfish; lakes, streams, and rivers in northeastern states but widely introduced elsewhere .... **Orconectes virilis** (Hagen)

30. Areola wide (Fig. 14C); general color green to brownish; burrows only rarely; in springs and small streams, occasional in large streams; common and generally distributed east of Mississippi River except for southern states .... **Cambarus bartoni** (Fabr.)

Areola narrow, linear, or obliterated in middle (Figs. 14F, G) **31**

31. Areola narrow but not linear (Fig. 14F); the red crayfish; burrows in swampy pastures and fields; Allegheny Mountains and foothills .... **Cambarus carolinus** (Erichson)

Areola linear or obliterated (Fig. 14G) **32**

32. Areola obliterated at midlength (Fig. 14G); general color greenish to brownish; burrows in wet fields and marshy areas; resorts to ponds and streams only during breeding season; the solitary crayfish; common and generally distributed east of the Continental Divide.

Cambarus diogenes (Girard)

Areola linear; lotic waters from NY to IL and south to TN, NC, VA ...... **Cambarus robustus** Girard

33. Second chelae larger than first, and without terminal hair tufts (Figs. 17H, J).

PALAEMONIDAE, **37**

First and second chelae subequal, with terminal hair tufts (Fig. 17K) .... ATYIDAE, **34**

34. Eyestalks rudimentary, no trace of facets or pigments; translucent; 10 to 23 mm long.

Palaemonias, 35

Eyes well developed; usually 20 to 50 mm long .... **Syncaris, 36**

35. Rostrum with about 30 dorsal and eight ventral spines; Mammoth Cave, KY.

Palaemonias ganteri Hay

Rostrum with 5 to 13 dorsal and zero to two ventral spines; Shelta Cave, Huntsville, AL.

Palaemonias alabamae Smalley

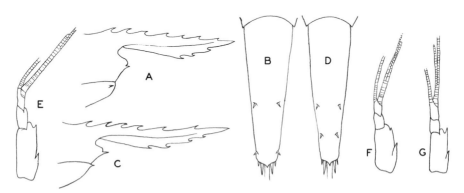

FIG. 16.–Structure of *Palaemonetes*. A, lateral view of anterior end of carapace of *P. kadiakensis*; B, dorsal view of telson of *P. kadiakensis*; C, lateral view of anterior end of carapace of *P. paludosus*; D, dorsal view of telson of *P. paludosus*; E, basal portion of first antenna of *P. paludosus*; F, basal portion of first antenna of *P. hiltoni*, a southern CA brackish species; G, basal portion of first antenna of *P. vulgaris*, a brackish species. (All modified from Holthuis, 1952.)

36. Upper margin of rostrum with one or two spines, lower margin with five to nine spines; supraorbital spine present (Fig. 17L); small coastal streams north of San Francisco Bay.

    Syncaris pacifica (Holmes)

    Upper margin of rostrum without spines, lower margin with few spines; supraorbital spine present or absent; last recorded from coastal streams near Los Angeles, California, but probably now extinct . . . . **Syncaris pasadenae** (Holmes)
37. Second legs only slightly longer than first (Fig. 17A); usually 25 to 45 mm long; fresh-water prawns, glass shrimps . . . . **Palaemonetes, 38**

    Second legs much longer than first (Fig. 17F); the giant river shrimps . . . . . **Macrobrachium, 43**
38. Eyes degenerated; coloration whitish; cave species **39**

    Eyes and coloration normal; epigean species **41**
39. Lower margin of rostrum with teeth; second legs longer than first legs; known only from a small solution sink in Alachua County, FL . . . . **Palaemonetes cummingi** Chace

    Lower margin of rostrum without teeth; second legs equal in length to first pair **40**
40. Telson tapered posteriorly, with two pairs of spines on posterior margin; known only from hypogean waters of Hays County, TX . . . . **Palaemonetes antrorum** Benedict

    Posterior margin of telson widened, with eight to 12 spines on posterior margin; known only from Ezell's Cave, San Marcos, TX . . . . **Palaemonetes holthuisi** Strenth
41. Rostrum slightly curved downward (Figs. 16A, C) **42**

    Rostrum straight; known only from rivers of Hays and Comal counties, TX.

    Palaemonetes texanus Strenth
42. Branchiostegal spine on anterior margin of carapace, just below branchiostegal groove (Fig. 16C); posterior pair of dorsal spines of telson situated midway between anterior pair and posterior margin of telson (Fig. 16D); up to 46 mm long; common in Atlantic coast drainages from NJ to FL; rare in central United States and southern CA . . . . . **Palaemonetes paludosus** (Gibbes)

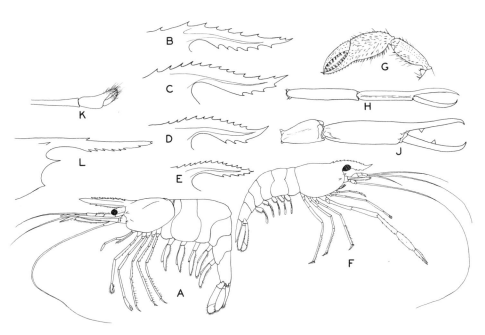

Fig. 17.–Structure of Palaemonidae and Atyidae. A, *Palaemonetes paludosus,* ×1.6; B, rostrum of *Macrobrachium acanthurus*; C, rostrum of *M. carcinus*; D, rostrum of *M. ohione*; E, rostrum of *M. olfersi*; F, *M. ohione,* ×0.6; G, large chela of male *M. olfersi*; H, chela of *M. acanthurus* (spines and setae omitted); J, chela of *M. carcinus* (spines and setae omitted); K, second chela of *Syncaris pacifica*; L, anterior end of carapace of *S. pacifica*. (A modified from Creaser, 1931; B–F redrawn from Hedgpeth, 1949; G–J modified from Hedgpeth, 1949.)

Branchiostegal spine removed from anterior margin of carapace, below branchiostegal groove (Fig. 16A); posterior pair of dorsal spines of telson placed close to posterior margin of telson (Fig. 16B); up to 50 mm long; sporadic in central third of the United States, west of Alleghenies, and from Canada to the Gulf .... **Palaemonetes kadiakensis** Rathbun

43. Carpopodite of second leg shorter than claw; rostrum with about three teeth posterior to margin of orbit and three to five ventral teeth (Figs. 17C–E) **44**

Carpopodite of second leg about as long as claw (Fig. 17H); rostrum usually with a single tooth posterior to margin of orbit and about six ventral teeth (Fig. 17B); body up to 170 mm long; rivers of FL to TX; sometimes in brackish waters ...... **Macrobrachium acanthurus** (Wiegmann)

44. Claw of second leg without prominent teeth or hairy covering (Fig. 17F); up to 95 mm long; NC to TX, and Mississippi and Ohio rivers and their main tributaries as far north as OK, MO, IL, IN, VA, OH .... **Macrobrachium ohione** (Smith)

Claw of second leg with hair or teeth **45**

45. Claw of second leg similar in both sexes, with a broad triangular tooth on each finger (Fig. 17J); body up to 240 mm long; rivers, spring brooks, and springs in FL, TX.

Macrobrachium carcinus (L.)

Claw of second leg on one side in male greatly enlarged, armed with spines and tufts or bristles (Fig. 17G); body up to 90 mm long; vicinity of St. Augustine, FL.

Macrobrachium olfersi (Wiegmann)

REFERENCES

Abrahamson, S. A. A., and C. R. Goldman. 1970. Distribution, density and production of the crayfish Pacifastacus leniusculus Dana in Lake Tahoe, California–Nevada. *Oikos* 21:83–91.

Ameyaw-Akumfi, C., and B. A. Hazlett. 1975. Sex recognition in the crayfish Procambarus clarkii. *Science* 190:1225–1226.

Andrews, E. A. 1904. Breeding habits of crayfish. *Am. Nat.* 38:165–206.

_____. 1907. The young of the crayfishes Astacus and Cambarus. *Smithson. Contrib. Knowl.* 35:5–79.

Black, J. B. 1963. Observations on the home range of stream-dwelling crawfishes. *Ecology* 44:592–595.

Bovbjerg, R. V. 1952. Comparative ecology and physiology of the crayfish Orconectes propinquus and Cambarus fodiens. *Physiol. Zool.* 25:34–56.

_____. 1953. Dominance order in the crayfish Orconectes virilis (Hagen). *Physiol. Zool.* 26:173–178.

_____. 1956. Some factors affecting aggressive behavior in crayfish. *Ibid.* 29:127–136.

_____. 1970. Ecological isolation and competitive exclusion in two crayfish (Orconectes virilis and Orconectes immunis). *Ecology* 51:225–236.

Broad, A. C., and J. H. Hubschman. 1963. The larval development of Palaemonetes kadiakensis M. J. Rathbun in the laboratory. *Trans. Am. Microsc. Soc.* 82:185–197.

Caine, E. A. 1974. Zoogeography of the Florida troglobitic crayfishes, genus Procambarus. *Am. Midl. Nat.* 92:487–492.

Capelli, G. M. 1982. Displacement of northern Wisconsin crayfish by Orconectes rusticus (Girard). *Limnol. Oceanogr.* 27:741–744.

Capelli, G. M., and J. J. Magnuson. 1983. Morphoedaphic and biogeographic analysis of crayfish distribution in northern Wisconsin. *J. Crust. Biol.* 3:548–564.

Creaser, E. P. 1931. The Michigan decapod crustaceans. *Pap. Mich. Acad. Sci. Arts and Lett.* 13:257–276.

Creaser, E. P., and A. I. Ortenburger. 1933. The decapod crustaceans of Oklahoma. *Publ. Univ. Okla. Biol. Surv.* 5:14–80.

Crocker, D. W. 1957. The crayfishes of New York State (Decapoda, Astacidae). *Bull. N.Y. State Mus.* 355:(1):1–97.

_____. 1979. The crayfishes of New England. *Proc. Biol. Soc. Wash.* 92:225–252.

Crocker, D. W., and D. W. Barr. 1968. Handbook of the crayfishes of Ontario. *Life. Sci. Misc. Publ. R. Ont. Mus.*, pp. 1–158.

Dean, J. L. 1969. Biology of the crayfish Orconectes causeyi and its use for control of aquatic weeds in trout lakes. *U.S. Dept. Inter., Bur. Sport Fish. Wildl., Tech. Pap.* 24:1–15.

Faxon, W. 1885. Revision of the Astacidae, Part I. The genera Cambarus and Astacus. *Mem. Mus. Comp. Zool.* 10:1–186.

_____. 1914. Notes on the crayfishes in the United States National Museum and the Museum of Comparative Zoology with descriptions of new species and subspecies to which is appended a catalogue of the known species and subspecies. *Mem. Mus. Comp. Zool. Harvard Coll.* 40:347–427.

Fingerman, M. 1955. Factors influencing the rate of oxygen consumption of the dwarf crawfish, Cambarellus shufeldtii (Decapoda, Astacidae). *Tulane Stud. Zool.* 3:103–116.

Fingerman, M., and A. D. Lago. 1957. Endogenous twenty-four hour rhythms of locomotor activity and oxygen consumption in the crawfish Orconectes clypeatus. *Am. Midl. Nat.* 58:383–393.

Fitzpatrick, J. F. 1967. The propinquus group of the crawfish genus Orconectes (Decapoda: Astacidae). *Ohio J. Sci.* **67**:129–177.

_____. 1987. The subgenera of the crawfish genus Orconectes (Decapoda; Cambaridae). *Proc. Biol. Soc. Wash.* **100**:44–74.

Fitzpatrick, J. F., and J. F. Payne. 1968. A new genus and species of crawfish from the southeastern United States (Decapoda, Astacidae). *Proc. Biol. Soc. Wash.* **81**:11–22.

Franz, R., and D. S. Lee. 1982. Distribution and evolution of Florida's troglobitic crayfishes. *Bull. Fla. State Mus. Biol. Sci.* **28**:53–78.

Gunter, G. 1937. Observations on the river shrimp, Macrobrachium ohionis (Smith). *Am. Midl. Nat.* **18**:1038–1042.

Guyselman, J. B. 1957. Solar and lunar rhythms of locomotor activity in the crayfish Cambarus virilis. *Physiol. Zool.* **30**:70–87.

Hagen, H. A. 1870. Monograph of the North American Astacidae. *Illus. Cat. Mus. Comp. Zool. Harvard Coll.* **3**:1–109.

Harris, J. A. 1903. An ecological catalogue of the crayfishes belonging to the genus Cambarus. *Kans. Univ. Sci. Bull.* **2**:51–187.

Hay, W. P. 1903. Observations on the crustacean fauna of the region about the Mammoth Cave, Kentucky, *Proc. U.S. Nat. Mus.* **25**:223–226.

Hazlett, B., D. Rittschof, and D. Rubenstein. 1974. Behavioral biology of the crayfish Orconectes virilis I. Home range. *Am. Midl. Nat.* **92**:301–319.

Hedgpeth, J. W. 1947. River shrimps. *Prog. Fish Cult.* **9**:181–184.

_____. 1949. The North American species of Macrobrachium (river shrimp). *Tex. J. Sci.* **1**:28–38.

_____. 1968. The atyid shrimp of the genus Syncaris in California. *Int. Rev. gesamten Hydrobiol.* **53**:511–524.

Hobbs, H. H., Jr. 1942. A generic revision of the crayfishes of the Subfamily Cambarinae (Decapoda, Astacidae) with the description of a new genus and species. *Am. Midl. Nat.* **28**:334–357.

_____. 1942a. The crayfishes of Florida. *Univ. Fla. Publ. Biol. Sci. Ser.* **3**:1–179.

_____. 1945. Two new species of crayfishes of the genus Cambarellus from the Gulf Coast states, with a key to the species of the genus (Decapoda, Astacidae). *Am. Midl. Nat.* **34**:466–474.

_____. 1967. A new crayfish from Alabama caves with notes on the origin of the genera Orconectes and Cambarus (Decapoda: Astacidae). *Proc. U.S. Nat. Mus.* **123**:1–17.

_____. 1969. On the distribution and phylogeny of the crayfish genus Cambarus. *Va. Polytech. Inst. Res. Div. Monogr.* **1**:93–178.

_____. 1972. Crayfishes (Astacidae) of North and Middle America. *Biota of Freshwater Ecosystems, Identification Manual* **9**:1–173.

_____. 1974. A checklist of the North and Middle American crayfishes (Decapoda: Astacidae and Cambaridae). *Smithson. Contrib. Zool.* **166**:1–161.

_____. 1981. The crayfishes of Georgia. *Smithson. Contrib. Zool.* **318**:1–549.

Hobbs, H. H., et al. 1977. A review of the troglobitic decapod crustaceans of the Americas. *Smithson. Contr. Zool.* **244**:1–183.

Holthuis, L. B. 1949. Note on the species of Palaemonetes (Crustacea Decapoda) found in the United States of America. *Proc. Konink. Ned. Akad. Wet.* **52**:87–95.

_____. 1952. A general revision of the Palaemonidae (Crustacea Decapoda Natantia) of the Americas. II. The subfamily Palaemoninae. *Occas. Pap. Allan Hancock Found.* **12**:1–396.

Huntsman, A. G. 1915. The fresh-water Malacostraca of Ontario. *Contrib. Canad. Biol. ii (Suppl.)* **47**:145–163.

Huxley, T. H. 1906. *The crayfish.* 2d ed. 371 pp. Kegan Paul, London.

Lagler, K. F., and M. J. Lagler. 1944. Natural enemies of crayfishes in Michigan. *Pap. Mich. Acad. Sci. Arts and Lett.* **29**:293–303.

Lowe, M. E. 1956. Dominance-subordinance relationships in the crawfish Cambarellus shufeldtii. *Tulane Stud. Zool.* **4**:139–170.

Lyle, C. 1938. The crawfishes of Mississippi, with special reference to the biology and control of destructive species. *Iowa State Coll. J. Sci.* **13**:75–77.

Maluf, N. S. R. 1939. On the anatomy of the kidney of the crayfish and on the absorption of chloride from freshwater by this animal. *Zool. Jahrb. Abt. allg. Zool. Physiol. Tiere* **59**:515–534.

Mason, J. C. 1970. Egg-laying in the western North American crayfish, Pacifastacus trowbridgii (Stimpson) (Decapoda, Astacidae). *Crustaceana* **19**:37–44.

_____. 1970a. Copulatory behavior of the crayfish Pacifastacus trowbridgii (Stimpson). *Can. J. Zool.* **48**:969–976.

_____. 1970b. Maternal-offspring behavior of the crayfish, Pacifastacus trowbridgii (Stimpson). *Am. Midl. Nat.* **84**:463–473.

Meehean, O. L. 1936. Notes on the freshwater shrimp Palaemonetes paludosa (Gibbes). *Trans. Am. Microsc. Soc.* **55**:433–441.

Merkel, E. L. 1969. Home range of crayfish Orconectes juvenalis. *Am. Midl. Nat.* **81**:228–235.

Mobberly, W. C. 1963. Hormonal and environmental regulation of the molting cycle in the crayfish Faxonella clypeata. *Tulane Stud. Zool.* **11**:79–96.

Momot, W. T. 1966. Upstream movements of crayfish in an intermittent Oklahoma stream. *Am. Midl. Nat.* **75**:150–159.

Momot, W. T., and H. Gowing. 1972. Differential seasonal migration of the crayfish, Orconectes virilis (Hagen) in marl lakes. *Ecology* **53**:479–483.

Momot, W. T. et al. 1978. The dynamics of crayfish and their role in ecosystems. *Am. Midl. Nat.* **99**:10–35.

Ortmann, A. E. 1902. The geographical distribution of the freshwater decapods and its bearing upon ancient geography. *Proc. Am. Philos. Soc.* **41**:267–400.

_____. 1905. The mutual affinities of the species of the genus Cambarus, and their dispersal over the United States. *Ibid.* **44**:91–136.

_____. 1906. The crawfishes of the state of Pennsylvania. *Mem. Carnegie Mus.* 2:343–521.

_____. 1931. Crawfishes of the southern Appalachians and the Cumberland Plateau. *Ann. Carnegie Mus.* 20:61–160.

Payne, J. F. 1972. The life history of Procambarus hayi. *Am. Midl. Nat.* **87**:25–35.

Pearse, A. S. 1909. Observations on copulation among crayfishes. *Am. Nat.* **43**:746–753.

Penn, G. H. 1942. Observations on the biology of the dwarf crawfish, Cambarellus shufeldtii (Faxon). *Am. Midl. Nat.* **28**:644–647.

_____. 1943. A study of the life history of the Louisiana red-crawfish, Cambarus clarkii Girard. *Ecology* **24**:1–18.

_____. 1957. Variation and subspecies of the crawfish Orconectes palmeri (Faxon). *Tulane Stud. Zool.* **5**:229–262.

_____. 1959. An illustrated key to the crawfishes of Louisiana with a summary of their distribution within the state (Decapoda, Astacidae). *Tulane Stud. Zool.* **7**:3–20.

Penn, G. H., and J. F. Fitzpatrick. 1963. Interspecific competition between two sympatric species of dwarf crawfishes. *Ecology* **44**:793–797.

Penn, G. H., and H. H. Hobbs. 1958. A contribution toward a knowledge of the crawfishes of Texas (Decapoda, Astacidae). *Tex. J. Sci.* **10**:452–483.

Rhoades, R. 1944. The crayfishes of Kentucky, with notes on variation, distribution and descriptions of new species and subspecies. *Am. Midl. Nat.* **31**:111–149.

Riegel, J. A. 1959. The systematics and distribution of crayfishes in California. *Calif. Fish and Game* **45**:29–50.

Roberts, T. W. 1944. Light, eyestalk chemical, and certain other factors as regulators of community activity for the crayfish, Cambarus virilis Hagen. *Ecol. Monogr.* **14**:359–392.

Schmitt, W. L. 1933. Notes on shrimps of the genus Macrobrachium found in the United States. *J. Wash. Acad. Sci.* **23**:312–317.

Scudamore, H. H. 1947. The influence of the sinus glands upon molting and associated changes in the crayfish. *Physiol. Zool.* **20**:187–208.

_____. 1948. Factors influencing molting and the sexual cycles in the crayfish. *Biol. Bull.* **95**:229–237.

Smith, E. W. 1953. The life history of the crawfish, Orconectes (Faxonella) clypeatus (Hay). *Tulane Stud. Zool.* **1**:79–96.

Steele, M. 1902. The crayfish of Missouri. *Univ. Cinci. Bull.* **10**:1–53.

Strenth, N. E. 1976. A review of the systematics and zoogeography of the freshwater species of Palaemonetes Heller of North America (Crustacea: Decapoda). *Smithson. Contrib. Zool.* **228**:1–27.

Tack, P. I. 1941. The life history and ecology of the crayfish Cambarus immunis Hagen. *Am. Midl. Nat.* **25**:420–446.

Turner, C. L. 1926. The crayfish of Ohio. *Bull. Ohio Biol. Surv.* **13**:145–195.

Van Deventer, W. C. 1937. Studies on the biology of the crayfish, Cambarus propinquus Girard. *Ill. Biol. Monogr.* **15**:1–67.

Williams, A. B. 1954. Speciation and distribution of the crayfishes of the Ozark plateaus and Ouachita provinces. *Univ. Kans. Sci. Bull.* **36**:803–918.

Williams, A. B., and A. B. Leonard. 1952. The crayfishes of Kansas. *Univ. Kans. Sci. Bull.* **34**:961–1012.

23

HYDRACARINA
(Water Mites)

The class Arachnoidea consists chiefly of terrestrial and parasitic members, including such familiar arthropods as spiders, scorpions, ticks, and mites. Only the water mites, forming the Hydracarina (or Hydrachnellae), have become generally adapted to fresh waters. Strictly speaking, "Hydracarina" is a term of convenience, and not a specific taxonomic term. It is an aggregation of families in the Suborder Trombidiformes that are restricted to fresh-water habitats and are further characterized by immature forms that differ markedly from adults in form and habits. Furthermore, considering the animal kingdom as a whole, the Hydracarina is one of the few characteristic fresh-water groups, being almost restricted to fresh waters; only a few species are found in brackish and salt waters, and none are terrestrial.

Although Hydracarina are found in almost all types of fresh-water habitats, they are most abundant and characteristic of streams, rivers, ponds, and the littoral region of lakes, especially where there are quantities of rooted aquatic vegetation. Their bright colors, globular to ovoid shape, and clambering and swimming habits identify them unmistakably.

General characteristics. Superficially, water mites appear to be minute spiders, but they differ most significantly from the true spiders in having the cephalothorax and abdomen fused into a single mass, all evident segmentation being lost. A decided minority of genera depart from the usual globular or ovoid shape and are dorsoventrally or laterally compressed. Most Hydracarina, and especially the more primitive genera, are soft bodied, but in many forms the cuticle is leathery or thickened to form a series of heavily sclerotized plates. Sometimes the body is almost completely enclosed by two plates, one dorsal and one ventral, whereas in other genera there are numerous smaller plates. The surface of the body is smooth, finely marked, granular, striated, or papillated. A variable number of setae are borne on the body. Total length usually ranges between 0.4 and 3.0 mm.

The six pairs of arachnid appendages are always present, the most conspicuous being the last four pairs, or legs. Each leg is long and consists of six segments that are variously supplied with spines, setae, and long hairs. The last segment usually bears two terminal movable claws. Each of the eight legs originates from a sclerotized ventral plate called a coxal plate, or epimera. At least some of the epimera are adjacent and more or less fused.

The first two pairs of appendages are borne on an anterior and ventral headlike structure called the capitulum, or maxillary organ. The mouth is located at the anterior end of the capitulum, often at the extremity of a straight or curved rostrum of variable length. The ventral surface of the capitulum is termed the maxillary shield. The chelicerae (usually called mandibles in the Hydracarina) are small and, except during feeding, are kept withdrawn into the capitulum. In a few genera they are stilettolike or saberlike (Fig. 7E), but usually there is a separate, terminal, curved, clawlike segment (Fig. 7G). The five-segmented pedipalps (or "palps") are inserted laterally or at the anterolateral angles of the maxillary shield and are customarily curved ventrally.

The only other conspicuous external feature is the genital field, a group of closely associated structures located on the ventral

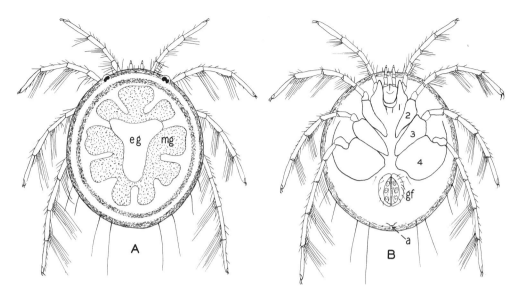

FIG. 1.–*Mideopsis orbicularis* (Müller) female, ×40. A, dorsal; B, ventral. 1, 2, 3, and 4, first, second, third, and fourth epimera; a, anus; eg, excretory gland; gf, genital field; mg, midgut. (Modified from Soar and Williamson, 1925–1929.)

median line between or behind the epimera and sometimes at the extreme posterior end of the body. The genital pore is a small longitudinal slit in the center of the field; it is flanked by two plates that bear few to many curious knoblike or cuplike acetabula of unknown function. Sometimes there is a pair of longitudinal movable or immovable genital flaps that may or may not cover the acetabula (Figs. 1B, 7J).

The two pigmented eyes are near the anterior margin of the body and are usually widely separated, but in a few genera they lie close together along the median line. Each eye is characteristically paired.

The so-called anus is an inconspicuous pore near the posterior end; sometimes it is borne on a special small plate.

Two minute spiracles are located on the dorsal surface of the capitulum near its base and are therefore hidden from view by the body and can be seen only when the capitulum is dissected off.

In contrast to fresh-water invertebrates in general, water mites are particularly striking because of their coloration. The majority of species are some shade of red or green, but there are many others that are blue, yellow, tan, or brown. The basic body coloration is caused by pigmentation of the body wall, but yellow, brown, or blackish markings and patterns, especially in dorsal view, are produced by the coloration of internal organs, especially the digestive tract and excretory gland (Fig. 1A). *Axonopsis complanata* is one of the most spectacular species; the appendages and general ventral surface are green; the epimera and much of the dorsal surface are blue; the capitulum, genital field, and central part of the dorsal surface are yellowish; and the large excretory gland produces a brown dorsal Y-shaped area. Some species of *Arrenurus* have equally striking combinations of green, blue, and orange. Stream species of water mites are characteristically brownish. Coloration of some species is somewhat variable, especially in *Limnesia,* and different specimens of the same species taken at the same place and time may exhibit various shades of red or green. These variations can be at least partially ascribed to differences in age and varying contents of the digestive tract and excretory gland.

Locomotion. Swimming is effected by relatively uncoordinated flailing movements of the legs, and the ability to swim is variously developed from one genus to another. The

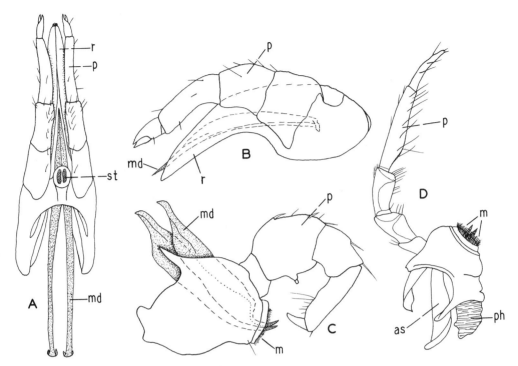

FIG. 2.–Hydracarina capitulum and associated structures. A, dorsal view of capitulum of *Hydrachna*; B, lateral view of capitulum of *Hydrachna* (only left palp shown); C, lateral view of capitulum of *Tyrrellia* (only right palp shown); D, lateral view of capitulum of *Eylais* (only right palp shown). as, air sac; m, mouth (with projecting mandibles); md, mandibles (shaded); p, palp; ph, pharynx; r, rostrum; st, stigma (cross hatched). (A and D modified from Viets; C modified from Marshall, 1940.)

best swimmers have rows of long swimming hairs on the legs, especially the more posterior pairs (Fig. 1). At the other extreme are many species that are strictly creepers and crawlers on the bottom or among vegetation; their legs are supplied with only short spines and setae. Swimming ability is lost in many stream forms.

Irrespective of the degree of swimming development, Hydracarina are clearly associated with the substrate and rarely stray far from it. Specific gravity of the body is usually high, and unless they cling to some object, they quickly fall to the bottom when not in motion. *Unionicola crassipes* is probably the only American species that strays far enough from vegetation and a substrate to be considered an accidental plankton organism.

In addition to locomotion, the legs are used periodically for cleaning debris from the body and from each other.

Food, feeding. The great majority of water mites are carnivorous or parasitic, their food consisting chiefly of entomostraca, small insects, various kinds of worms, or host tissues. Figure 3 shows a mite feeding on an ostracod. Sometimes the more active and voracious species are cannibalistic. Sluggish species usually feed on dead animals, and a few investigators have found that they may utilize vegetable material and detritus, thus making them omnivores.

The prey is seized and held with the palps, with or without the aid of the legs, and the body is quickly pierced with the mandibles. Only fluids are ingested, and it is probable that secretions of the several pairs of salivary glands are injected into the prey, where they predigest some of the more solid tissues so that they may be sucked up along with the body fluids. The exoskeleton and indigestible remnants of the prey are then discarded.

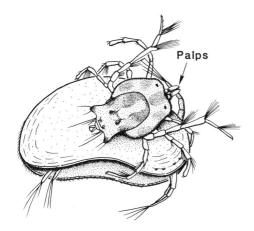

FIG. 3.–*Arrenurus* feeding on an ostracod. (Redrawn from Gledhill.)

Internal anatomy, physiology. Ingested fluids pass through the small cavity of the suctorial pharynx and into the midgut, the latter being a capacious, lobed, digestive and absorptive organ filling one third to one half of the hemocoel. The midgut usually has about 12 caeca, although there may be as many as 30 in some species. In all the Hydracarina whose anatomy has been carefully studied, it has been found that the midgut and its caeca end blindly, and there is no connection with the "anus." Presumably indigestible food residues are simply regurgitated.

A heart and blood vessels are lacking, but the blood is circulated about freely in the hemocoel by the movements of the muscles and digestive tract.

Some Hydracarina have a tracheole system (not visible in preserved material). Each of the two minute prostigmata opens into an elongated air sac (Fig. 2D). At the posterior end of this sac is a single, short tracheole trunk which branches into many fine, threadlike tracheoles ramifying and anastomosing to all parts of the body.

Although the tracheole system is always filled with a gas (air), no adult water mite has ever been observed taking air at the surface film in the manner of many aquatic insects. The prostigmata are thought to be covered with gas-permeable but waterproof membranes. When larvae or developing nymphs parasitize insects that leave the water and become aerial, however, it is probable that they take in air through the spiracles. In some species there seems to be no morphological link between the tracheole system and the prostigmata, which leaves the prostigmata with unknown functions. There is also the possibility that air could be taken into the tracheole system when the mandibles and a good portion of the capitulum are thrust into the air-containing tissues of aquatic plants. Nymphs have been observed in this attitude, but it is doubtful whether adults do it, with the possible exception of females of a few species that may assume the attitude as a preliminary to oviposition. Furthermore, mites are found in many habitats where there are no rooted aquatics.

Aside from the question of the physiological significance of the water mite tracheole system and the source of its contained gas, all investigators seem to agree that most of the oxygen utilized in metabolism is absorbed through the general body surface and that carbon dioxide diffuses outward through it. Those species with a thick body wall usually have scattered thin porous areas that presumably facilitate the gaseous exchange.

Although many species survive temporary conditions where the dissolved oxygen concentration may be below 1 ppm, no species

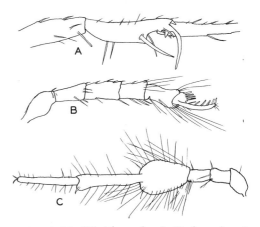

FIG. 4.–Modified legs of male Hydracarina. A, middle segments of third legs of *Hydrochoreutes ungulatus*; B, fourth leg of *Forelia*; C, fourth leg of *Acercus*. (A modified from Marshall, 1927; B modified from Marshall, 1935; C modified from Marshall, 1937.)

are known to remain active under anaerobic conditions.

The excretory gland is a large dorsal T- or Y-shaped organ (Fig. 1A) that connects with the ventral "anus," or "excretory pore," by a duct. Many typical mites have seven pairs of oval or pear-shaped epidermal glands opening over the dorsal and lateral surfaces of the body, and it is possible that these structures have an accessory excretory function.

The nervous system consists of a large central ganglionic mass with little indication of the method of fusion. It is perforated by the pharynx and has radiating nerves. Each of the two lateral eyes consists of two elements that may be clearly separated or more or less fused. An element consists of a pigmented bulb and cuticular lens. Many species have an additional fifth, or median, eye (Fig. 7D). Other senses are poorly developed; the palps are tactile and probably detect substances in solution, and most of the body setae are probably tactile.

In the male there are two testes, two vasa deferentia, and a single penis, which can be extruded through the genital pore. The female has two fused ovaries, two oviducts, and a single vagina, or uterus.

Reproduction. Sometimes the two sexes can be distinguished only by slight differences in the anatomy of the genital field, palps, third legs, and fourth legs, but more commonly sexual dimorphism is pronounced. Males frequently have much larger fourth epimera, and genital fields that differ considerably from those of females. The third and fourth legs of males are sometimes specialized for copulation and the transfer of spermatophores; often the fourth legs are elongated, and in a few genera the fourth segment is flattened and platelike (Fig. 4C) or the terminal segment is roughly sickle shaped (Fig. 4B). The third legs may be modified for clasping the female (Fig. 4A), and sometimes the distal segments form a club for sperm transfer. In *Arrenurus* sexual dimorphism is extreme, the males being characterized by a large posterior prolongation, or cauda (Figs. 9B, C). Unfortunately there is no single general anatomical character or several characters that may be used in distinguishing the sexes throughout the order.

Judging from the few species whose reproductive habits have been carefully described, it appears that Hydracarina exhibit a wide variety of clasping and spermatophore transfer habits. Occasionally the male is carried on the back of the female for a varying time. The males of a few species drag the female about in a haphazard manner. In some genera the genital pores of the two copulating individuals are almost in direct contact during spermatophore transfer, but more commonly the bodies are placed at various angles to each other (often at right angles), while the male securely holds the first two pairs of female legs with his fourth legs. Regardless of the specific copulatory position, the male picks up the packetlike spermatophores from the tip of his extruded penis with his third legs and transfers them to the region of the female genital pore. Spermatophore transfer takes from less than a minute to as long as an hour.

Statements concerning the breeding season vary widely. Some authorities state that breeding and egg deposition may occur at any time during the year; others maintain that eggs are usually deposited from May through July. It therefore seems logical to assume that the breeding season may vary widely from one species to another and also that it may vary within a single species according to latitude, temperature, and season.

As the female releases the eggs from her genital pore they are fertilized by the sperm contained in the spermatophores previously placed there by the male. Fertilized eggs are rarely deposited singly; usually they are extruded in groups of 20 to 400 onto stones, vegetation, and debris. Each egg may have its own individual gelatinous covering, or the whole group of eggs may be imbedded in a common gelatinous mass. Eggs are usually red. *Hydrachna* oviposits in the tissues of aquatic plants.

Development, life history. Depending on temperature and the particular species, the eggs develop into inactive prelarvae within the egg in 1 to 6 weeks. After emergence, the active larvae have functional mouthparts, small epimera, three pairs of legs, and no genital field (Figs. 5A, B). After a short free-swimming period, the larvae become attached

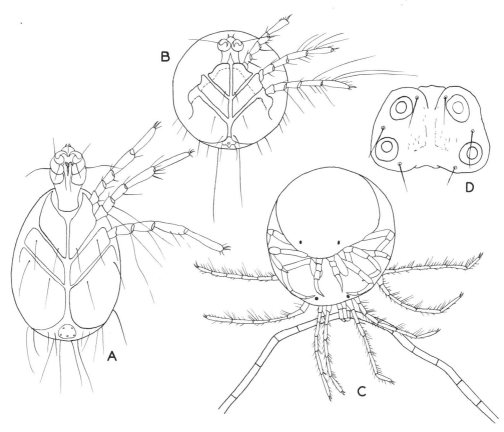

FIG. 5.–Immature Hydracarina. A, larva of *Arrenurus,* ×260; B, larva of *Mideopsis,* ×130; C, adult *Piona* developing within nymphal exoskeleton, attached to filamentous alga, ×40; D, genital field of *Limnesia* nymph. (A and B modified from Soar and Williamson, 1925–1929; C modified from Uchida; D redrawn from Viets.)

to aquatic insects with their capitulum and assume a parasitic existence. Wesenberg-Lund (1919) favors the theory of some broad preference by larvae for certain insect hosts; thus, he states that the Limnocharidae usually parasitize hydrometrids, that Eylaidae mostly parasitize aerial insects, that Hydryphantidae usually occur on culicids, and that several other families appear to be almost restricted to insects that do not leave the water. In view of more recent ecological observations, it is possible that his conclusions may be a reflection of restricted mite and insect faunas in the ponds he studied. It would be highly desirable to test this question experimentally in order to find out definitely whether or not a particular larval mite has any specific host preference. It is significant that Ephemeroptera and hy-

drophilid and dytiscid beetles are rarely parasitized, and that Trichoptera are uncommon hosts. The usual hosts are Plecoptera, Odonata, Diptera, and Hemiptera. Chironomid Diptera are the most frequent hosts. The larval mite may come in contact with its host on the substrate or at the surface film. Often a single insect may carry 10 or 20 mite parasites.

After a variable feeding period on the host, the larva begins to metamorphose to the inactive protonymphal stage. Feeding ceases, and the larva shrinks from its exoskeleton, which then becomes a loose, baglike puparium (still attached to the host). The contained larva loses its definitive features and becomes more or less reconstituted to form the deutonymph stage. Soon the deutonymph ruptures the puparium and escapes into the

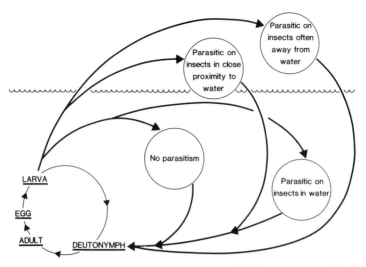

FIG. 6.–Types of parasitism shown by water mites. (From Gledhill, 1985 as modified from Böttger.)

water. The period of inactivity between cessation of feeding by the larva and the escape of the nymph is comparable to a pupal stage. The active deutonymph stage lasts for periods ranging from 5 days to 6 months.

The deutonymph is considerably larger than the larva and almost as large as the adult. It has the same general structures and proportions as the adult, but there is no genital opening in the genital field. Unlike the larva, the deutonymph is free living and has the same carnivorous food habits as the adult.

At the end of a variable period of activity, the deutonymph attaches itself to an algal filament, rooted aquatic, or bottom debris with its mouth parts and enters a second pupal stage, or tritonymph. Again, the nymphal exoskeleton becomes a baglike puparium with the developing adult within (Fig. 5C). This quiescent period usually lasts only 7 to 10 days, after which the tritonymphal exoskeleton ruptures and the sexually mature adult emerges.

By way of recapitulation, the stages in the life history of a typical water mite may be outlined as follows: egg, inactive prelarva, active larva, inactive protonymph, active deutonymph, inactive tritonymph, active adult. Some authors use other terminologies for the various stages in the life history stages. The active larval and parasitic larval stages are morphologically indistinguishable.

If the larva attaches to an insect that is aquatic until the termination of the nymphophan, the developed nymph simply breaks out of its puparium into the water. If the insect host is an adult beetle or hemipteran, however, it may leave the water intermittently with its attached larval parasite exposed to the air. Furthermore, the host may metamorphose to an adult that leaves the water and is aerial for the rest of its life (Odonata, Diptera, and Plecoptera, for example). In such instances the fully developed nymph breaks out of the puparium and may fall into the water fortuitously, or it may reenter the water when its host oviposits in the water. Nymphs that fall to the ground from an aerial host are presumed to die very shortly.

Figure 6 shows some of the common variations in mite life histories.

There are many exceptions to the generalized type of life history outlined in the preceding paragraphs. A few water mites deposit their eggs in the tissues of sponges or the gelatinous matrix of colonial Protozoa. Others are thought to leave the insect host while still in the larval stage and attach to the substrate for the transformation to the nymph. Certain stream species of *Piona* and *Limnesia* do not have a parasitic stage but remain in the gelatinous egg mass until they become nymphs.

The species of *Unionicola* have unusual life histories. The majority pass their entire lives in the tissues or cavities of mussels, but in a few species the adults either are found in mussels or are free swimming, and one or two are parasitic in sponges. It is thought that eggs are produced the year round; they are inserted into the tissues of the mussel gills by means of a special ovipositor. The larvae are parasitic on the gills, but the nymphs and adults clamber about actively in the gill or mantle cavities. The parasitic adults have short legs and no swimming hairs. Sometimes the great majority of mussels in a bed are infected, and although infestations may be heavy, the mites appear to do little damage to the hosts. It is possible that the nymphs and adults are commensals rather than true parasites. In addition to *Unionicola, Najadicola ingens,* an uncommon species, also occurs in mussels.

Complete life histories are known for relatively few water mites, and it is likely that many additional variations, exceptions, and complications will appear as further studies are made.

Almost nothing is known about the duration of the life cycle or number of generations per year, but it is thought that the great majority of species have a life span of less than 1 year. Overwintering probably occurs most frequently in the adult stage, although parasitic larvae, tritonymphs, and deutonymphs have also been reported.

Ecology. Unlike most groups of freshwater invertebrates, water mites may be collected in all seasons, even under ice, but the greatest numbers of adults are to be found in late spring and early autumn.

They are invariably most abundant in species and individuals among the heaviest growths of rooted aquatics where the water is only a meter or two deep. Populations in excess of 2000 mites per square meter of bottom have been reported from such habitats. For some unknown reason, however, water mites are rare in beds of *Chara* and *Elodea.* Acid lakes, though superficially an ideal collecting ground, contain very few Hydracarina. Wave-beaten sand and gravel shores are also poor places to collect. Many species are capable of clambering and swimming about with surprising agility, whereas others, devoid of swimming hairs, are sluggish bottom creepers. Between these two extremes are all gradations in crawling, swimming, and clambering ability. Mites are active only during the hours of daylight, and sometimes they remain fixed in the same position all night long.

As the water deepens offshore, the fauna becomes progressively depauperate, and even though Hydracarina may be collected from the bottom at depths of 10 to 100 m, such populations are usually scanty and composed of only one to several species. There is some evidence for migration into the weedy shallows in the late spring and back to deeper waters in the autumn, when vegetation disappears, but this problem needs further investigation.

Careful collecting in a favorable lake or weedy pond commonly yields as many as 20 to 40 species. Young (1969) collected 99 species in one county in Colorado, of which 72 were new state records and three were new for North America. Angelier et al. (1985) found 86 species in running waters of the Pyrenees, and Modlin and Gannon (1973) collected 32 species in the St. Lawrence Great Lakes.

Some genera, notably *Thyas, Hydryphantes, Acercus, Piona,* and *Limnesia,* are characteristic of temporary ponds. Probably they are capable of withstanding periods of drought by being buried in wet mud and debris of the substrate.

Brook and stream species are often conspicuously modified for withstanding the force of the current. Dorsoventral flattening, small size, strong claws, and the absence of swimming hairs are the usual adaptations. Stream species produce comparatively small numbers of eggs. Few species may be found in both running and standing waters.

Recent studies of the interstitial and phreatic waters of stream gravels have shown a new habitat for small species of Hydracarina. Sometimes these populations are remarkably dense and diverse. Especially in Europe and Japan, new genera and species are being described with great frequency.

Coloration is sometimes correlated with the background, though there is little evidence that the resemblance has any protective value. Elton (1923) found that hungry sticklebacks refused to eat bright scarlet mites, and he postulated possible Müllerian mimicry and warning coloration.

Geographical distribution, dispersal.

Although the geographical distribution and ecological affinities are well known for many species in Europe, collecting in the United States has been so highly regional and generally neglected that we have little reliable information on the distribution of most species. In view of this fact, it has been largely necessary to omit distributional data for the genera and species included in the key beginning on page 525. The majority of species reported from the United States appear to be restricted to this country, and many are known only from a single locality or two or three widely separated localities. However, only a single rare genus, *Najadicola,* appears to be restricted to North America. On the other hand, a surprising number of American species are cosmopolitan and widely distributed throughout the Northern and even the Southern hemispheres. Furthermore, it is striking that in each of about half of the genera occurring in the United States one species (and only one) is widely distributed and common in both North America and Europe. Some of these species are *Piona rotunda* (Koen.), *Unionicola crassipes* (Müller), *Hydryphantes ruber* (DeGeer), *Hydrodroma despiciens, Libertia porosa* Thor, *Hygrobates longipalpis* (Herm.), *Hydrochoreutes ungulatus* (Koch), and *Mideopsis orbicularis* (Müller).

Dispersal is both active and passive, and those species that are parasitic on insects are undoubtedly easily distributed overland to new drainage systems.

Enemies.

Hydracarina are preyed on by a wide variety of aquatic invertebrates, especially coelenterates and carnivorous insects. They are not ordinarily an important element in the diet of fishes, but occasionally a fish stomach is found in which mites predominate. Several new species have been described from the stomach contents of trout.

Collection, preparation.

A successful way to collect quantities of mites is to stir up bottom debris and catch the suspended material with a small dipnet. Hand picking, dipnetting, and washing out vegetation in an enameled pan are also useful techniques. Barr (1979) describes collecting mites at night in traps with chemoluminescent bait. During July and early August the large majority of specimens collected are immature. Living specimens may be kept in the laboratory and fed entomostraca or nonpredatory aquatic insects, but crowding is disastrous, since they are often cannibalistic.

Because they are so active, it is best to examine living specimens in a compression slide or by anesthetizing and relaxing them with chloretone or in water saturated with chloroform vapor. Binoculars, an opaque white background, and a bright spotlight are essential.

A good killing and preserving agent consists of 5 parts by volume of glycerin, 4 parts water, and 1 part glacial acetic acid. Wolcott recommends the following solution: 2 parts glycerin, 3 parts water, 2 parts glacial acetic acid, and 1 part absolute alcohol.

Suitably translucent whole mounts may sometimes be made by cutting a small incision in the dorsal body wall and pressing out the contents, but it is first necessary to clear dense and hard species with acetic corrosive in order to render them sufficiently translucent for work with the compound microscope. It consists of equal parts (by weight) of glacial acetic acid, chloral hydrate, and water. For critical study and accurate identification, the mouth parts, capitulum, and legs should be dissected off and mounted separately. Some workers prefer permanent euparal or similar mounting media (see Lavers, 1945); others use glycerin or glycerin jelly.

Polyvinyl alcohol and Hoyer's fluid are excellent mounting media that also have clearing properties. See Mitchell and Cook (1952) for a further discussion of making slide mounts.

Taxonomy.

More than 90% of all published information on our American species is based on the work of four careful investigators: Wolcott, whose publications extend from 1899 to 1918; Ruth Marshall, whose numerous papers appeared between 1903 and 1944; Cook, beginning in 1953; and Mitchell, also beginning in 1953. The potential number of species in the United States must be large, Marshall having reported 140 species and varieties from Wisconsin alone. Viets (1936) lists 444 species from Germany, and Soar and Williamson (1925–1929) list 247 for the British Isles. Mitchell (1954) lists 291 species from North America, and Crowell (1961) lists 457 taxa from North America, but the number of species reported to date for the United States is probably more than 600. Very little is known about the Hydracarina of rapid streams, high altitudes, states west of the Great Plains, and southeastern states. Viets (1987) lists 4,000 species and subspecies world wide.

Identification to genus and species is based primarily on the detailed adult anatomy of the epimera, genital field, capitulum, and palps. The taxonomic status and limits of several families and genera are still controversial, and the following classification of American fresh-water genera used in this manual is modified from the widely accepted one proposed by Viets (1936). It is likely that better limitations of Hydracarina families will be proposed in the future. We have the impression that specialists have been too anxious and zealous to establish new families.

Anisitsiellidae
  Bandakia
  Utaxatax
Arrenuridae
  Arrenurus
Athienemanniidae
  Athienemannia, Chappuisides
Aturidae
  Aturus, Kongsbergia
Axonopsidae
  Albia, Axonopsis, Brachypoda
Clathrosperchonidae
  Clathrosperchon
Eylaidae
  Eylais
Feltriidae
  Feltria
Hydrachnidae
  Hydrachna

Hydrodromidae
  Hydrodroma
Hydrovolziidae
  Hydrovolzia
Hydryphantidae
  Hydryphantes
Hygrobatidae
  Atractides, Hygrobates
Krendowskiidae
  Geayia, Krendowskia
Lebertiidae
  Lebertia
Limnesiidae
  Kawamuracarus, Limnesia, Neomamersa, Tyrrellia
Limnocharidae
  Limnochares
Mideidae
  Midea
Mideopsidae
  Horreolanus, Mideopsis, Xystonotus, Yachatsia
Momoniidae
  Stygomomonia
Neoacaridae
  Neoacarus, Volsellacarus
Omartacaridae
  Omartacarus
Oxidae
  Frontipoda, Gnaphiscus, Oxus
Piersigiidae
  Piersigia
Pionidae
  Forelia, Huitfeldtia, Hydrochoreutes, Nautarachna, Neotiphys, Piona, Pseudofeltria, Tiphys, Wettina
Protziidae
  Calonyx, Partnuniella, Wandesia
Sperchonidae
  Sperchon, Sperchonopsis
Thermacaridae
  Thermacarus
Thyasidae
  Euthyas, Panisopsis, Panisus, Thyas, Thyasides, Thyopsella, Thyopsis, Trichothyas
Torrenticolidae
  Testudacarus, Torrenticola
Uchidastygacaridae
  Uchidastygacarus
Unionicolidae
  Koenikea, Najadicola, Neumania, Unionicola

Because of the vagaries of variations within many of the ill-defined families, the following key is highly artificial. Except for a few rare forms, essentially all the U.S. genera are included, especially where the published literature includes specific collecting records. There are, however, additional potentially new genera in the collections of specialists, and there are also many unpublished U.S. records of species and genera that are well known elsewhere, especially in Europe. Anyone who collects in clean phreatic waters and gravel bars of streams will find many small species he cannot identify.

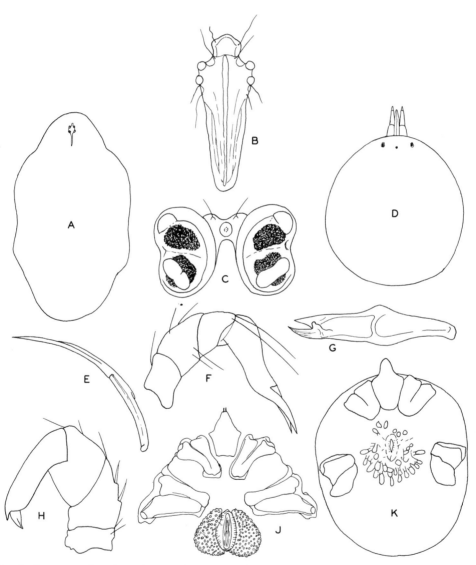

FIG. 7.–Structure of Hydracarina. A, dorsal view of *Limnochares* showing median ocular plate, ×15; B, ocular plate of *Limnochares*; C, ocular plate of *Eylais*; D, dorsal view of *Hydrachna,* ×15; E, mandible of *Hydrachna*; F, palp of *Hydrodroma despiciens*; G, mandible of *H. despiciens*; H, palp of *Hydryphantes*; J, ventral plates of *Hydrodroma despiciens*; K, ventral view of *Protzia,* ×45. (B, C, F, H, J, and K modified from Marshall, 1940.)

KEY TO GENERA OF HYDRACARINA

1. Lateral eyes of the two sides close together at the median lines and borne on a common plate (Figs. 7B, C); simple, red, soft-bodied species; up to 7 mm long **6**
   Lateral eyes widely separated (Figs. 9, 11, 12), or eyes absent **2**
2. Body wormlike (Fig. 8A); up to 2 mm long; rare; in hyporheic cold stream gravels; several species.
   Wandesia
   Body not wormlike **3**
3. Distal extremity of the fourth palp segment produced into a point well beyond the point of insertion of segment five, the two segments together usually resembling a pair of open or closed shears (Figs. 7F, H) **4**
   Distal extremity of fourth palp segment not pointed and produced, or only slightly produced and blunt (Figs. 9A; 10A); segment five, free, tapered, and with the tip pointed, clawed, or toothed.
   19
4. Mandible one segmented, the terminal portion straight and stilettolike (Fig. 7E); body globular, soft, and papillated; capitulum produced as a rostrum well beyond anterior end of body (Fig. 7D); with swimming hairs; usually red, orange, or brown; 1 to 8 mm long; common everywhere in standing waters; many species .... **Hydrachna**
   Mandible one segmented, the terminal segment curved, clawlike, and often very small (Fig. 7G); capitulum only slightly or not at all produced beyond anterior end of body **5**
5. Last two epimera widely separated from first two epimera (Fig. 7K); bottom forms; no swimming hairs; soft bodied; two species in cold mountain streams (possibly should be called *Calonyx*).
   Protzia
   Last two epimera not so widely separated from first two epimera (Figs. 9, 11) **7**
6. Eye capsules on either side of an elongated plate in the midline (Fig. 7B); mouth at the tip of a protruding rostrum; usually creepers; several common and widely distributed species in standing and slowly flowing waters .... **Limnochares**
   Eye capsules connected by a narrow bridge (Fig. 7C); mouth not at tip of a protruding rostrum; swimmers in quiet waters; several species .... **Eylais**
7. Distal extremity of fourth segment of palp short (Fig. 7H) **8**
   Distal extremity of fourth segment of palp long (Figs. 7F; 8B, C) **10**
8. Median eye present but sometimes unpigmented; 1 to 2.5 mm long; red; not in hot springs **12**
   Median eye absent **9**
9. Distal portion of fourth palp segment a large rounded cap; rare; known only from gravel bars in WY streams .... **Chappuisides**
   Distal portion of fourth palp segment otherwise; rare, in western hot springs and their effluents; (sometimes called *Wandesia*) .... **Partnuniella**
10. Soft red body; 2 mm long; with long swimming hairs; widely distributed and common.
    Hydrodroma despiciens (Müller)
    Body not soft and red; less than 0.8 mm long; several uncommon species in stream gravels **11**
11. Fifth segment of palp a long and slender claw (Fig. 8B) .... **Volsellacarus**
    Fifth segment of palp shorter and more robust (Fig. 8C) .... **Neoacarus**
12. Swimming hairs present; several widely distributed species in standing waters ... **Hydryphantes**
    Swimming hairs absent, only short spines on legs **13**
13. Dorsal surface covered by a large shield (Fig. 8D) **14**
    Dorsal surface not covered by a large shield **15**
14. Second pair of genital acetabula near posterior end of genital flaps (Fig. 8J); temporary ponds, seepages, and springs; one rare species .... **Thyopsis**
    Second pair of genital acetabula almost midway between the ends of the genital flaps (Fig. 8K); several uncommon species in springs and seepages .... **Thyopsella**
15. With a small, spindle-shaped median frontal plate on the dorsal surface (Fig. 8E); one rare species in temporary ponds .... **Euthyas**
    Without a small, spindle-shaped median frontal plate **16**
16. Genital flaps extending beyond the first genital acetabula (Figs. 8L, N) **17**
    Genital flaps not extending beyond the first genital acetabula (Fig. 8M) **18**
17. Genital flaps broad at posterior end (Fig. 8N); two species in streams, cold springs, and seepages.
    Panisus

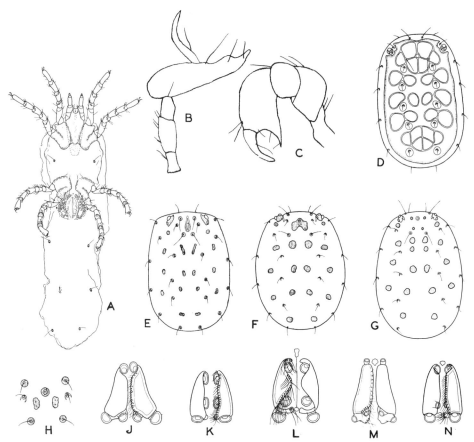

FIG. 8.–Structure of Hydracarina. A, ventral view of *Wandesia*; B, palp of *Volsellacarus*; C, palp of *Neoacarus*; D, dorsal view of male *Thyopsella*; E, dorsal view of male *Euthyas*; F, dorsal view of male *Thyasides*; G, dorsal view of female *Thyas*; H, median eye region of female *Thyasides*; J, genital field of male *Thyopsis*; K, genital field of male *Thyopsella*; L, genital field of female *Panisopsis*; M, genital field of female *Thyas*; N, genital field of female *Panisus*. (B and C modified from Cook, 1963; D–N from Cook, 1959.)

Genital flaps narrow at posterior end (Fig. 8L); several species in cold springs, streams, and seepages.

                                                      Panisus

18. Median eye surrounded by one large plate (Fig. 8F), or by a very small plate with two adjacent separate plates (Fig. 8H); rare in temporary bog waters .... **Thyasides**

    Median eye lying free in integument (Fig. 8G); about eight species in temporary ponds .... **Thyas**

19. Fifth segment of palp pointed, clawlike, and opposable to the projecting margin of segment four (Figs. 9A; 10A) 20

    Fifth segment of palp not opposable to fourth, and bearing distal claws, setae, or teeth 23

20. Genital field lying posterior to fourth epimera (Fig. 9C) 21

    Genital field lying between fourth epimera (Fig. 9E); running waters 22

21. Males usually with a large posterior prolongation, or cauda (Figs. 9B, C); body wall heavily sclerotized; common everywhere, especially in vegetation; usually red or greenish; the largest American genus; about 120 species .... **Arrenurus**

    Males without a cauda; not red or green; rare in sandy springs; (sometimes called *Chelomideopsis*).

                                                      Athienemannia

22. With six genital acetabula (Fig. 9E); dorsal shield large (Fig. 9F); reported from WI, IL, IN, MI.

                                                    Krendowskia similis Viets

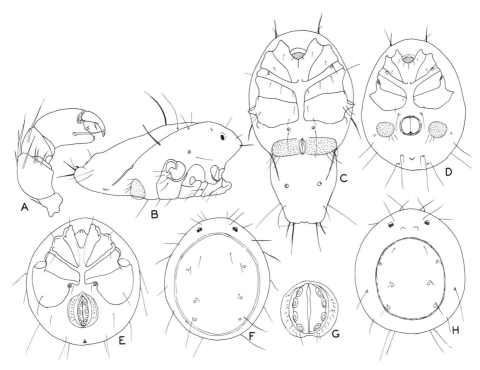

FIG. 9.–Structure of Hydracarina. A, palp of *Arrenurus*; B, lateral view of male *Arrenurus,* ×37; C, ventral view of male *Arrenurus*; D, ventral view of female *Arrenurus,* ×27; E, ventral view of male *Krendowskia similis,* ×33; F, dorsal view of *K. similis,* ×33; G, genital field of female *Geayia ovata*; H, dorsal view of *G. ovata,* ×37. (A–D modified from Lavers, 1945; E–H modified from Marshall, 1940.)

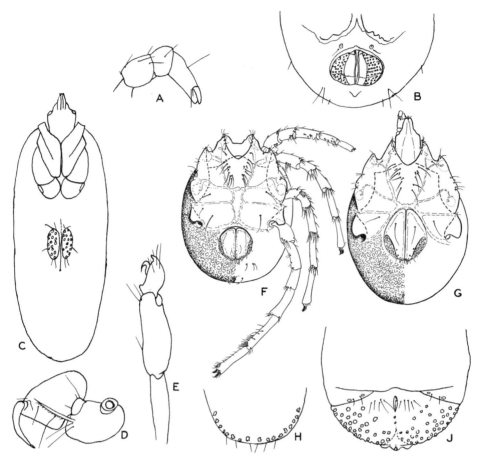

FIG. 10.–Structure of Hydracarina. A, palp of *Athienemannia*; B, posterior end of *Athienemannia*; C, ventral view of *Omartacarus* (appendages omitted); D, palp of *Uchidastygacarus*; E, terminal segments of first leg of *Stygomomonia*; F, ventral view of male *Thermacarus minuta* Mitchell, ×32; G, ventral view of female *T. minuta*, ×32; H, posterior end of *Aturus* showing distribution of acetabula; J, posterior end of *Feltria* showing distribution of acetabula. (All modified: A, Cook, 1961; C, Cook, 1963c; D, Cook, 1963d; Fand G, Mitchell, 1963; J, Cook, 1963.)

32. Genital field in the posterior half of the ventral surface, sometimes along the posterior margin 33
    Genital field in the anterior half of the ventral surface; about eight species in running waters.
                                                         Hygrobates
33. Dorsal surface with many transverse wrinkles; several species in streams .... **Kongsbergia**
    Dorsal surface relatively smooth; several species in running waters .... **Brachypoda**
34. Fifth and sixth segments of first leg swollen, with tarsal claw in a large dorsal incision (Fig. 10E); eight
      rare species in gravels of cold streams .... **Stygomomonia**
    Fifth and sixth segments of first leg not swollen 35
35. Anterior ends of first epimera projecting well beyond the anterior margin of body, leaving a large bay
      for the capitulum (Fig. 12A) 36
    Anterior ends of first epimera not projecting beyond anterior margin of body (Fig. 12D) 37
36. With four small anterior dorsal plates (Fig. 12C); without swimming hairs; usually in cold streams;
      widely distributed; about 20 species .... **Torrenticola**
    With three small anterior dorsal plates (Fig. 12B); two rare species reported from western streams.
                                                        Testudacarus

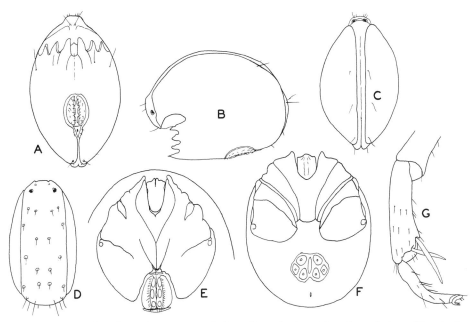

FIG. 11.–Structure of Hydracarina. A, ventral view of *Frontipoda,* ×80; B, lateral view of *Frontipoda,* ×80; C, dorsal view of *Frontipoda,* ×70; D, dorsal view of *Gnaphiscus,* ×47; E, ventral plates of *Libertia*; F, ventral view of *Atractides,* ×45; G, terminal portion of first leg of *Atractides.* (A and C modified from Marshall, 1914; E modified from Marshall, 1932; F and G modified from Marshall, 1943.)

37. Occurring only in western hot springs; 1 rare species and one common species (Figs. 10F, G).
     Thermacarus
     Not in hot springs 38
38. Genital field lying in a bay bounded by all the epimera (Fig. 12F); 1 rare species reported from WI.
     Midea
     Genital field not lying in a bay bounded by all the epimera 39
39. With numerous acetabula (Figs. 10H, J; 12D, E) **40**
     With three or four pairs of genital acetabula (Fig. 12G) **43**
40. Genital field near posterior margin of body (Figs. 10H, J) **41**
     Genital field not near posterior margin of body; fourth epimera not especially large (Fig. 12D); about 10 widely distributed species in rivers and streams .... **Koenikea**
41. Genital acetabula abundant (Figs. 10H, J) **42**
     Genital acetabula not abundant (Fig. 12E); several widely distributed species in ponds, lakes, and slow streams .... **Albia**
42. Genital acetabula arranged along the posterior margin (Fig. 10H); less than 5 mm long; several uncommon stream species in eastern states .... **Aturus**
     Genital acetabula not confined to body margin (Fig. 10J); about 10 species in stream gravels.
     Feltria
43. Four pairs of genital acetabula; genital field at extreme posterior end of body (Fig. 12G); 0.45 mm long; brightly colored in green, blue, and yellow; 1 common species, about 20 rare species; northern lakes and ponds .... **Axonopsis complanata** (Müller)
     Three pairs of genital acetabula; genital field not at extreme posterior end of body **44**
44. Swimming hairs present; several rare species in standing waters .... **Xystonotus**
     Swimming hairs absent **45**
45. Third pair of genital acetabula posterior to the edge of the gonopore area; rare; known only from gravel and sand bars of clear streams in OR .... **Yachatsia**

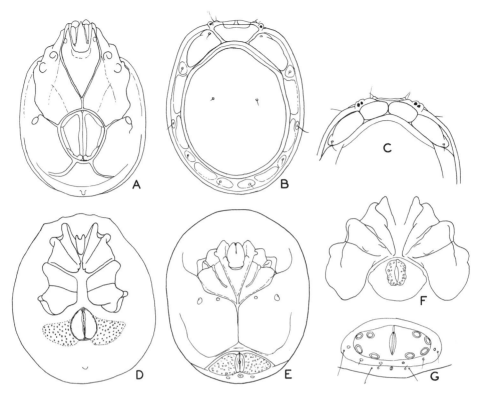

FIG. 12.–Structure of Hydracarina. A, ventral view of *Torrenticola,* ×80; B, dorsal view of *Testudacarus,* ×60; C, dorsal anterior region of *Torrenticola*; D, ventral view of *Koenikea,* ×65; E, ventral view of *Albia,* ×60; F, ventral plates of *Midea*; G, genital field of *Axonopsis complanata.* (A–C modified from Marshall, 1943; D and E modified from Marshall, 1935; G modified from Viets, 1936.)

Third pair of genital acetabula in the normal position; about eight species, of which only one is common and widely distributed .... **Mideopsis**

46. Third and fourth legs inserted at periphery of body (Fig. 13A); rare; reported only from IL; genital acetabula absent .... **Hydrovolzia**

Third and fourth legs not inserted at periphery of body **47**

47. Each genital flap composed of two complete or partial sclerites (Figs. 13B, C) **48**

Each genital flap, if present, composed of one sclerite **50**

48. Acetabula numerous (Fig. 13B) **49**

With only six acetabula (Fig. 13C); several hyporheic species in stream gravels ..... **Kawamuracarus**

49. Ventral surface with six and dorsal surface with 16 large oval latticelike or grid areas; very rare.

Clathrosperchon

No such sculpturing; several hyporheic species in stream gravels .... **Neomamersa**

50. Genital field anterior, at least between fourth epimera; no ancoral process on capitulum (Figs. 13F; 14C) **51**

Genital field posterior to fourth epimera; ancoral process present on capitulum (Fig. 15) **58**

51. With six or seven prominent sclerites (lateroglandularia) along each lateral margin, visible in both dorsal and ventral views (Figs. 13E, F); legs abundantly supplied with pectinate setae; several uncommon species in muddy swamps and marshes .... **Piersigia**

Without such prominent lateral sclerites, although dorsal and ventral sclerites may be prominent.

52

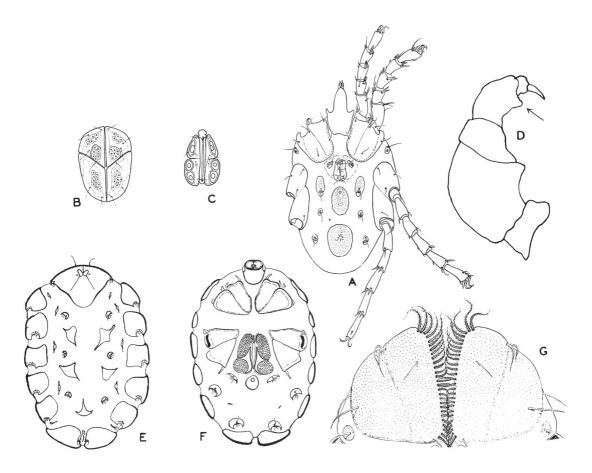

FIG. 13.–Structure of Hydracarina. A, ventral view of female *Hydrovolzia,* ×60; B, genital field of *Neomamersa*; C, genital field of *Kawamuracarus*; D, palp of female *Bandakia*; E, dorsal view of *Piersigia crusta* Mitchell; F, ventral view of *P. crusta*; G, anterior ventral surface of *Trichothyas* showing setae of first coxal plate. (A from Mitchell, 1954a; B and C from Cook, 1963; G modified from Mitchell, 1953.)

52. Mesial margin of first coxa with numerous heavy pinnate setae (Fig. 13G); legs heavy, adapted for crawling; in moss or algal masses on cold, wet rocks; rare, reported only from IL.

                                    Trichothyas

 Mesial margin of first coxa not setose 53

53. Genital acetabula borne on uncovered plates; genital flaps absent (Fig. 14C) 54

 Genital acetabula more or less covered by genital flaps (Fig. 14G); swimming hairs absent; sluggish bottom forms 55

54. Fourth leg with two terminal claws (Fig. 14A); swimming hairs absent; sluggish; several uncommon species . . . . **Tyrrellia**

 Fourth leg without terminal claws (Fig. 14B); swimming hairs usually present; in a variety of habitats but most common in standing waters; about 20 species . . . . **Limnesia**

55. Fourth segment of palp with two small processes on flexor surface, or processes absent (Fig. 14D); in northern states and mountainous areas; about 20 species in cold streams . . . . **Sperchon**

 Fourth segment of palp with one large process on flexor surface (Figs. 13D; 14F) 56

56. Process on fourth palp segment proximal (Fig. 14F); common and widely distributed in cold lakes.

                         Sperchonopsis verrucosa (Protz)

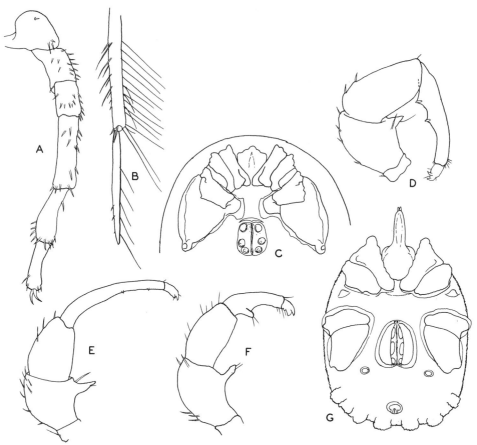

FIG. 14.–Structure of Hydracarina. A, fourth leg of *Tyrrellia*; B, fifth and sixth segments of fourth leg of *Limnesia*; C, ventral plates of *Limnesia*; D, palp of female *Sperchon*; E, palp of male *Sperchon*; F, palp of *Sperchonopsis verrucosa*; G, ventral view of male *S. verrucosa,* ×115. (A modified from Marshall, 1940a; B modified from Viets; C modified from Marshall, 1932; D–G modified from Marshall, 1943.)

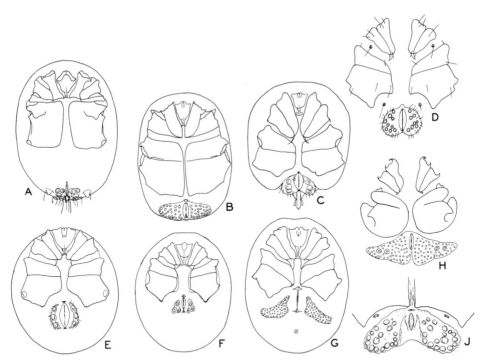

FIG. 15.–Structure of Hydracarina. A, ventral view of *Unionicola,* ×36; B, ventral view of *Neumania,* ×35; C, ventral view of male *Hydrochoreutes ungulatus,* ×65; D, ventral plates of *Huitfeldtia rectipes*; E, ventral view of female *Hydrochoreutes ungulatus,* ×40; F, ventral view of female *Tiphys torris* (Müller), ×25; G, ventral view of *Forelia,* ×44; H, ventral plates of *Najadicola ingens* Koen.; J, genital field of *Piona.* (All modified: A and B from Marshall, 1933; C and E from Marshall, 1927; D from Viets, 1936; F from Marshall, 1937; G and J from Marshall, 1935; H from Wolcott, 1905.)

61. Most distal medial seta on fourth segment of palp of normal shape 62
    Most distal medial seta on fourth segment of palp peglike, spinelike, and borne on a tubercle, or absent 63
62. No leg segments modified; very rare in sandy, cold springs and streams .... **Wettina**
    Fourth segment of third leg with a distal concavity (Fig. 4A); palps and legs very long; male with a petiole (Fig. 15C); two species in standing water .... **Hydrochoreutes**
63. Fourth segment of fourth leg with a concavity bearing numerous blunt, peglike setae 64
    Fourth segment of fourth leg never with a concavity bearing numerous peglike setae 65
64. First coxal plate bearing two to four heavy setae anteriorly; brightly colored; rare ...... **Nautarachna**
    First coxal plate lacking heavy setae anteriorly; brightly colored; common and widely distributed; many species .... **Piona**
65. Acetabular plates fused with the posterior edges of the fourth coxal plates along their entire anterior edges 66
    Acetabular plates never fused with the posterior edges of the fourth coxal plates 67
66. Leg setae short and heavy; rare .... **Pseudofeltria**
    Long, fine swimmng hairs present on some leg segments; several widely distributed species in lakes and ponds .... **Forelia**
67. Eight to 10 acetabula on each side (Fig. 15D); uncommon in northern lakes.
    Huitfeldtia rectipes Thor
    Three to six acetabula on each side 68
68. Fifth segment of fourth leg curved, with the dorsal surface broadly concave; about 5 species in lotic habitats .... **Tiphys**
    Fifth segment of fourth leg convex dorsally; rare in eastern states .... **Neotiphys**

HALACARIDAE

The Halacaridae are predominantly a marine family of predaceous mites in the Order Acari, and although several of the 30 genera are found in fresh waters, the Halacaridae are not usually considered as belonging in the Hydracarina.

Morphologically, halacarids are distin-

guished from Hydracarina by the fact that their palps are only three or four segmented. In addition, the rostrum and palps project well beyond the anterior end of the body. Body length is less thn 0.5 mm. The exoskeleton is leathery and reinforced with four dorsal plates and four ventral plates. There are no separate epimera, and the legs are attached laterally. These mites cannot swim and are usually found creeping about on aquatic plants; on the bottom of ponds, lakes, and streams; in the interstitial waters of sandy and gravel substrates of fresh waters; and in subterranean and hyporheic waters. Sometimes populations reach densities of 5000 individuals per square meter.

Halacarids are unusual in that some forms, especially *Limnohalacarus* and *Soldanellonyx*, frequently are found as members of the lake and pond pleuston association.

In general, halacarids are seldom collected, and their U.S. distribution and ecology are poorly known. The following genera are

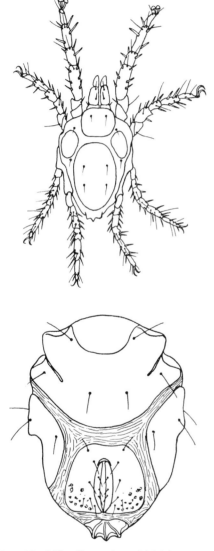

FIG. 16.–*Soldanellonyx chappuisi* Walter. upper, dorsal view, ×85; lower, ventral view (appendages not shown). (From Imamura, 1968.)

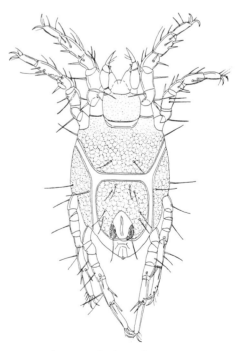

FIG. 17.–Ventral view of *Porohalacarus,* ×275. (From Ramazotti and Nocentini, 1960.)

either known to occur here or are likely to be taken: *Copidognathus, Halacarus, Hamohalacarus, Limnohalacarus, Lobohalacarus, Parasoldanellonyx, Porohalacarus* (Fig. 17), *Porolohmannella,* and *Soldanellonyx* (Fig. 16). *Halacarus* and *Copidognathus* are two genera in transition between fresh and salt waters. They are taken in wells and springs near marine shores.

Until much more information and field collecting records are available for the United States, it seems fruitless to draw up a key based on our fragmentary data.

ORIBATEI

The oribatids (soil mites) include many terrestrial families, only one very small family, comprising 3 genera, being restricted to the littoral benthos of fresh waters. Otherwise, there are a few scattered species throughout eight other families and about 15 genera that either are aquatic or live on moist soil or on *Sphagnum* and other hydrophytes that may be damp or intermittently submerged.

Morphologically, the Division Oribatei is separated most obviously from the Halacaridae and the true water mites by the modified "tear-drop" shape of the body owing to the broad triangular gnathosoma, or capitulum. Oribatids are also more coriaceous and dark colored.

Undoubtedly, *Hydrozetes* (Figs. 18, 19) (in the Family Eremaeidae) is the most common truly aquatic North American genus. It occurs chiefly in small permanent ponds where the mites are restricted to crevices on submerged logs, twigs, and debris (Fig. 20). Sometimes it

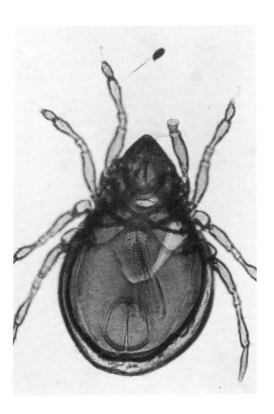

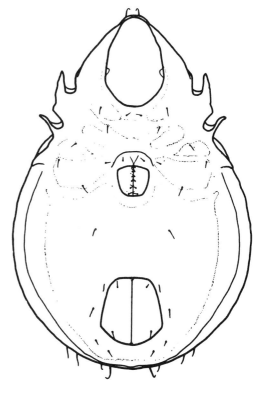

FIG. 18.–Adult female *Hydrozetes, 400* μm long. (Courtesy of J. Bushnell and D. Burford.)

FIG. 19.–Adult female *Hydrozetes,* ventral, legs omitted, 400 μm long. (Courtesy of J. Bushnell and D. Burford.)

FIG. 20.–Aggregation of *Hydrozetes* crowded into a crevice in a twig. (Courtesy of J. Bushnell and D. Burford.)

occurs in the pleuston, especially when it carries an air plastron. In addition, *Hydrozetes* can swallow a small bubble into the midgut from the plastron, which, by levitation, brings the mite to the surface, especially at night and when there are mechanical stimuli. In the spring it sometimes forms enormous dense aggregations, as large as 2 mm thick, 10 mm wide, and 2 m long! Populations are scattered during the winter months, and individuals appear to exist in a kind of cold narcosis state. Food consists of wood fibers, fungi, algae,

and detritus. *Hydrozetes* may be easily cultured in the laboratory, where the life cycle usually lasts 30 to 80 days. Instars include larva, protonymph, deutonymph, tritonymph, and adult (Fig. 21). Damselfly nymphs (especially *Enallagma* and *Tetragoneuria*) are important predators on *Hydrozetes*.

The American aquatic oribatids are too sketchily known to merit the construction of a key. About 20 species are known worldwide, and 6 species of *Hydrozetes* have been reported from the United States.

SPIDERS

Among the true spiders, or Araneae, there are no American species that are strictly aquatic, although a European and Asiatic species, *Argyroneta aquatica,* has become curiously adapted to the aquatic environment; it spins a dense, flat, underwater web in vegetation and brings small air bubbles down from the surface film and releases them beneath the web, thus forming a large bell-shaped bubble held in place by the web; the spider spends most of its time within the bubble. Many American species, however, are semiaquatic. They run about along the beach or among shore vegetation and often skitter out over the surface of ponds and marshes, where they may feed on Hemiptera and other in-

sects supported on the surface film. Some even dive beneath the surface when alarmed. Many build their webs on emergent vegetation. Some of the genera having common semiaquatic species are *Dolomedes, Lycosa* (wolf spiders), *Pirata, Pachygnatha,* and *Tetragnatha* (orb weavers). *Dolomedes* is a colorful and spectacular genus, often reaching a diameter of 5 cm. It dives with difficulty and remains submerged by hanging on to objects, but may remain below the surface as long as 45 minutes. Oxygen is probably obtained from air bubbles adhering to the body near the respiratory openings. *Dolomedes* feeds chiefly on aquatic insects, but it has also been reported catching small fish.

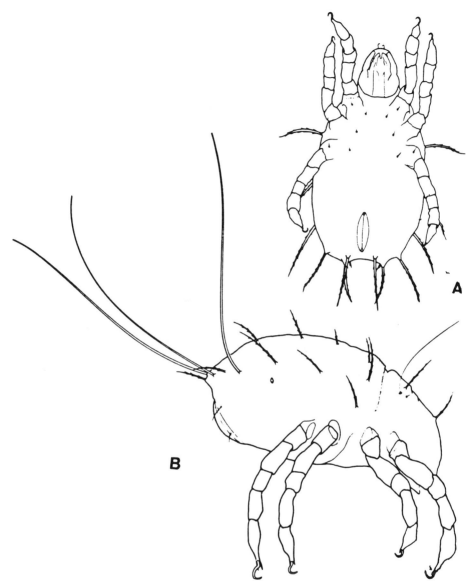

FIG. 21.–Immature *Hydrozetes*. A, larva, ventral, 250 μm; B, protonymph, lateral, 300 μm. (Courtesy of J. Bushnell and D. Burford.)

REFERENCES

Angelier, E., M.-L. Angelier, and J. Lauga. 1985. Recherches sur l'écologie des Hydracariens (Hydrachnellae, Acari) dans les eaux courantes. *Ann. Limnol.* 21:25–64.

Arndt, W., and K. Viets. 1938. Die biologischen (parasitologischen) Beziehungen zwischen Arachnoideen und Spongien. *Z. Parasitenkd.* 10:67–93.

Baker, E. W., and G. W. Wharton. 1952. *An introduction to acarology.* 465 pp. Macmillan, New York.

Balogh, J. 1972. *The oribatid genera of the world.* 188 pp. Akadémiai Kiadó, Budapest.

Barr, D. 1979. Water mites (Acari, Parasitengona) sampled with chemoluminescent bait in underwater traps. *Int. J. Acarol.* 5:187–194.

Bartsch, I. 1982. Halacaridae (Acari) im Süsswasser von Rhode Island, U.S.A., mit einer Diskussion über Verbreitung und Abstammung der Halacaridae. *Gewässer und Abwässer* 68/69:41–58.

———. 1987. Zur Biologie, Ökologie und Verbreitung der süsswasserbewohnenden Halacaride Porohalacarus alpinus (Acari). *Arch. Hydrobiol.* 111:83–93.

Böttger, K. 1970. Die Ernährung der Wassermilben (Hydrachnellae Acari). *Int. Rev. gesamten Hydrobiol.* 55:895–912.

———. 1972. Vergleichend biologisch-ökologische Studien zum Entwicklungszyklus der Süsswassermilben (Hydrachnellae, Acari). I. Der Entwicklungszyklus von Hydrachna globosa und Limnochares und Limnochares aquatica. *Int. Rev. gesamten Hydrobiol.* 57:109–152.

———. 1977. The general life cycle of fresh water mites. *Acarologia* 18:496–502.

Cook, D. R. 1954. Preliminary list of the Arrenuri of Michigan, Part I. The subgenus Arrenurus. *Trans. Am. Microsc. Soc.* 73:39–58.

———. 1954a. Preliminary list of the Arrenuri of Michigan, Part II. The Subgenus Megaluracarus. *Ibid.* 367–380.

———. 1955. Two new genera of Hydracarina from a spring in northern Michigan. *Am. Midl. Nat.* 53:412–418.

———. 1955a. Preliminary list of the Arrenuri of Michigan, Part III. The subgenera Micruracarus and Truncaturus. *Trans. Am. Microsc. Soc.* 74:60–67.

———. 1955b. Preliminary studies of the Hydracarina of Michigan: The Subfamily Foerliinae (Acarina: Pionidae). *Ann. Entomol. Soc. Am.* 48:299–307.

———. 1959. Studies on the Thyasinae of North America (Acarina: Hydryphantidae). *Am. Midl. Nat.* 62:402–428.

———. 1961. New species of Bandakia, Wettina, and Athienemannia from Michigan (Acarina, Hydracarina). *Proc. Entomol. Soc. Wash.* 63:262–268.

———. 1963. Studies on the phreaticolous water mites of North America: the genera Neomamersa Lundblad and Kawamuracarus Uchida. *Am. Midl. Nat.* 70:300–308.

———. 1963a. Studies on the phreaticolous water mites of North America: the Family Neoacaridae. *Ann. Entomol. Soc. Am.* 56:481–487.

———. 1963b. Studies on the phreaticolous water mites of North America: the genus Feltria (Acarina: Feltriidae). *Ibid.* 488–500.

———. 1963c. Omartacaridae, a new family of water mites from the ground waters of North America. *Entomol. News* 74:37–43.

———. 1963d. Studies on the phreaticolous water mites of North America: new or unreported genera of Mideopsoidea and Acalyptonotoidea. *Ibid.* 63–70.

———. 1968. Water mites of the genus Stygomomonia in North America (Acarina: Momoniidae). *Proc. Entomol. Soc. Wash.* 70:210–224.

———. 1968a. New species of Neoacarus Halbert and Volsellacarus Cook from North America (Acarina: Neoacaridae). *Ibid.* 67–74.

———. 1969. New studies on the water mite genera Neomamersa Lundblad and Kawamuracarus Uchida (Acarina, Limnesiidae) from North America. *Am. Midl. Nat.* 81:29–38.

———. 1974. North American species of the genus Axonopsis (Acarina: Aturidae: Axonopsinae). *Great Lakes Entomol.* 7:55–79.

Crowell, R. M. 1960. The taxonomy, distribution and developmental stages of Ohio water mites. *Bull. Ohio Biol. Surv.* 1:1–77.

———. 1961. Catalogue of the distribution and ecological relationships of North American Hydracarina. *Can. Entomol.* 93:321–359.

Elton, C. S. 1923. On the colours of water-mites. *Proc. Zool. Soc. London* (1922):1231–1239.

Fernandez, N., and J. Trave. 1984. La variabilité chaetotaxique et la neotrichie gastronotique des Hydrozetidae (Oribates). *Acarologia* 25:407–417.

Gledhill, T. 1985. Water mites—predators and parasites. *Rep. Freshwater Biol. Assoc.* 1985:45–59.

Husmann, S., and D. Teschner. 1970. Okologie, Morphologie und Verbreitungsgeschichte subterraner Wassermilben (Limnohalacaridae) aus Schweden. *Arch. Hydrobiol.* 67:242–267.

Imamura, T. 1957. Erste Mitteilung über Porohalacaridae aus unterirdischen Gewässern in Japan. *Abh. Naturw. Ver. Bremen* 35:53–62.

———. 1957a. Subterranean water-mites of the Middle and Southern Japan. *Arch. Hydrobiol.* 53:350–391.

———. 1959. Check list of the troglobiontic Trombidiidae, Porohalacaridae and Hydrachnellae of Japan. *Bull. Biogeogr. Soc. Jpn.* 21:63–66.

———. 1968. Results of the speleological survey in South Korea 1966. IX. Halacaridae (Acari) found in a limestone cave of South Korea. *Bull. Nat. Sci. Mus.* 11:281–284.

———. 1970. Some psammobiotic water mites of Lake Biwa. *Annot. Zool. Jpn.* 43:200–206.

Imamura, T., and R. Mitchell. 1967. The ecology and life cycle of the water mite, Piersigia limophila Protz. *Annot. Zool. Jpn.* 40:37–44.

Jrantz, G. W., and G. T. Baker. 1982. Observations on the plastron mechanism of Hydrozetes sp. (Acari: Oribatida: Hydrozetidae). *Acarologia* 23:273–278.

Lavers, C. H., Jr. 1945. The species of Arrenurus of the state of Washington. *Trans. Am. Microsc. Soc.* 44:228–264.

Lundblad, O. 1935. Die nordamerikanischen Arten der Gattung Hydrachna. *Ark. Zool.* 28:1–44.

Marshall, R. 1914. Some new American water mites. *Trans. Wis. Acad. Sci. Arts and Lett.* 17:1300–1304.

———. 1927. Hydracarina of the Douglas Lake region. *Trans. Am. Microsc. Soc.* 46:268–285.

———. 1932. Preliminary list of the Hydracarina of Wisconsin. Part II. *Trans. Wis. Acad. Sci. Arts and Lett.* 27:339–358.

———. 1933. Preliminary list of the Hydracarina of Wisconsin. Part III. *Ibid.* 28:37–61.

———. 1935. Preliminary list of the Hydracarina of Wisconsin. IV. *Ibid.* 29:273–298.

———. 1937. Preliminary list of the Hydracarina of Wisconsin. V. *Ibid.* 30:225–252.

———. 1940. Preliminary list of the Hydracarina of Wisconsin. VI. *Ibid.* 32:135–165.

_____. 1940a. The water mite genus Tyrellia. *Ibid.* 383–389.

_____. 1943. Hydracarina from California. Part I. *Trans. Am. Microsc. Soc.* 62:306–324.

_____. 1943a. Hydracarina from California. Part II. *Ibid.* 404–415.

Mitchell, R. D. 1953. A new species of Lundbladia and remarks on the Family Hydryphantidae (water mites). *Am. Midl. Nat.* 49:159–170.

_____. 1954. Water-mites of the genus Aturus (Family Axonopsidae). *Trans. Am. Microsc. Soc.* 73:350–367.

_____. 1954a. A description of the water-mite, Hydrovolzia gerhardi new species, with observations on the life history and ecology. *Nat. Hist. Misc.* 134:1–9.

_____. 1954b. Check list of North American water-mites. *Fieldiana: Zool.* 35:27–70.

_____. 1955. Anatomy, life history, and evolution of the mites parasitizing fresh-water mussels. *Misc. Publ. Mus. Zool. Univ. Mich.* 89:1–28.

_____. 1957. Evolutionary lines in water mites. *Syst. Zool.* 6:137–148.

_____. 1957a. The mating behavior of pionid water-mites. *Am. Midl. Nat.* 58:360–366.

_____. 1960. The evolution of thermophilous water mites. *Evolution* 14:361–377.

_____. 1963. A new water mite of the Family Thermacaridae from hot springs. *Trans. Am. Microsc. Soc.* 82:230–233.

_____. 1964. A study of sympatry in the water mite genus Arrenurus (Family Arrenuridae). *Ecology* 45:546–558.

_____. 1965. Population regulation of a water mite parasitic on unionid mussels. *J. Parasitol.* 51:990–996.

_____. 1967. Host exploitation of two closely related water mites. *Evolution* 21:59–75.

Mitchell, R. D., and D. R. Cook. 1952. The preservation and mounting of water-mites. *Turtox News* 30:169–172.

Modlin, R. F., and J. E. Gannon. 1973. A contribution to the ecology and distribution of aquatic Acari in the St. Lawrence Great Lakes. *Trans. Am. Microsc. Soc.* 92:217–224.

Newell, I. M. 1945. Hydrozetes Berlese (Acari, Oribatoidea): the occurrence of the genus in North America and the phenomenon of levitation. *Trans. Conn. Acad. Sci. Arts and Lett.* 36:253–268.

_____. 1947. A systematic and ecological study of the Halacaridae of eastern North America. *Bull. Bingham Oceanogr. Coll.* 10:1–232.

Nocentini, A. M. 1961. Primiritrovamenti di Porohalacaridae (Acari) nel Lago Maggiore. *Mem. Ist. Ital. Idrobiol.* 13:127–138.

Petrova, A. 1974. Sur la migration des Halacariens dans les eaux douces et la position systematique des Halacariens et Limnohalacariens. *Vie et Milieu* 24:87–96.

Pieczynski, E. 1960. (Formation of groupings of water mites (Hydracarina) in different environments of Lake Wilkus.) *Ekol. Pol. A,* 8:169–198.

_____. 1961. Numbers, sex ratio, and fecundity of

several species of water mites (Hydracarina) of Mikolajskie Lake. *Ibid.* 9:219–228.

_____. 1963. Some regularities in the occurrence of water mites (Hydracarina) in the littoral of 41 lakes in the River Krutynia basin and the Mikolajki District. *Ibid.* 11:141–157.

Ramazzotti, G., and A. M. Nocentini. 1960. Porohalacaridae (Hydracarina) del Lago di Mergozzo. *Mem. Ist. Ital. Idrobiol.* 12:185–200.

Riessen, H. P. 1982. Predatory behavior and prey selectivity of the pelagic water mite Piona constricta. *Can. J. Fish. Aquat. Sci.* 39:1569–1579.

Schmidt, U. 1935. Beiträge zur Anatomie und Histologie der Hydracarinen, besonders von Diplodontus despiciens O. F. Müller, *Z. Morphol. Oekol. Tiere* 30:99–176.

Schwoerbel, J. 1955. Über einige Porohalacariden (Acari) aus dem südlichen Schwarzwald. *Zool. Anz.* 155:146–150.

_____. 1959. Ökologische und tiergeographische Untersuchungen über die Milben (Acari, Hydrachnellae) der Quellen und Bäche des südlichen Schwarzwaldes und seiner Randgebiete. *Arch. Hydrobiol. Suppl.* 24:385–546.

Smith, I. M. 1976. A study of the systematics of the water mite family Pionidae (Prostigmata: Parasitengona). *Mem. Ent. Soc. Canada* 98:1–299.

Smith, I. M., and D. R. Oliver. 1976. The parasitic associations of larval water mites with imaginal aquatic insects, especially Chironomidae. *Can. Ent.* 108:1427–1442.

Smith, J. 1979. A review of the water mites of the family Anisitsiellidae (Prostigmata, Lebertioidea) from North America. *Can. Ent.* 111:529–550.

Soar, C. D., and W. Williamson. 1925–1929. *British Hydracarina.* 3 vols. 612 pp. Ray Society, London.

Sokolow, I. 1924. Untersuchungen über die Eiablage und den Laich der Hydracarina. I. *Arch. Hydrobiol.* 15:383–405.

_____. 1925. Untersuchungen über die Eiablage und den Laich der Hydracarina. II. *Z. Morphol. Oekol. Tiere* 4:301–332.

Szalay, L. 1949. Ueber die Hydracarinen der unterirdischen Gewässer. *Hydrobiologia* 2:141–179.

Uchida, T. 1932. Some ecological observations on water mites. *J. Fac. Sci. Hokkaido Imp. Univ. Ser. VI. Zool.* 1:143–165.

Viets, K. 1936. Wassermilben oder Hydracarina. *Die Tierwelt Deutschl.* 31 and 32:1–574.

_____. 1938. Über die verscheiden Biotope der Wassermilben, besonders über solche mit anormalen Lebensbedingungen und über einige neue Wassermilben aus Thermalgewässern. *Verh. Int. Ver. Limnol.* 8:209–224.

_____. 1950. Porohalacaridae (Acari) aus der Grundwasserfauna des Maingebietes. *Arch. Hydrobiol.* 43:247–257.

_____. 1955. In subterranen Gewässern Deutschlands lebende Wassermilben (Hydrachnellae, Porohalacaridae und Stygothrombiidae). *Ibid.* 50:33–63.

———. 1955a. Wassermilben aus Nordbayern (Hydrachnellae und Porohalacaridae Acari). *Abh. Bayer. Akad. Wiss. N. S.* **73**:1–106.

———. 1955b. *Die Milben des Süsswassers und des Meeres. Teil I, Bibliographie.* 476 pp. Gustav Fischer, Jena, Germany.

———. 1956. *Die Milben des Süsswassers und des Meeres. Hydrachnellae et Halacaridae (Acari). Teil II und III, Katalog und Nomenklatur.* 870 pp. Gustav Fischer, Jena, Germany.

———. 1982. *Die Milben des Süsswassers. Hydrachnellae und Halacaridae (Part), Teil 1: Bibliographie.* 116 pp. Paul Parey, Hamburg, Germany.

———. 1987. *Die Milben des Süsswassers. Hydrachnellae und Halacaridae (Part), Teil 2: Katalog.* 1,012 pp. Paul Parey, Hamburg, Germany.

Wesenberg-Lund, C. 1919. Contributions to the knowledge of the postembryonal development of the Hydracarina. *Vidensk. Medd. Dansk. Naturhist. For.* **70**:5–57.

Wiles, P. R. 1984. Watermite respiratory systems. *Acarologia* **25**:27–32.

Wolcott, R. H. 1900. New genera and species of North American Hydrachnidae. *Trans. Am. Microsc. Soc.* **21**:177–200.

———. 1905. A review of the genera of water mites. *Ibid.* **26**:161–243.

Wooley, T. A. 1969. Two new species of Hydrozetes, extant and fossil (Acari: Cryptostigmata, Hydrozetidae). *J. N.Y. Entomol. Soc.* **77**:250–256.

Young, W. C. 1968. New species of Midea, Piona, and Sperchon (Hydracarina) from Colorado. *Trans. Am. Microsc. Soc.* **87**:165–177.

———. 1969. Ecological distribution of Hydracarina in north central Colorado. *Am. Midl. Nat.* **82**:367–401.

Young, W. C., and A. C. Rhodes. 1974. The influence of dissolved oxygen concentrations on three species of water mites (Hydracarina). *Am. Midl. Nat.* **92**:115–129.

24

GASTROPODA
(Snails, Limpets)

On a global basis, only 15 out of a total of about 300 families of gastropods are partially or wholly found in fresh waters of the United States. Nevertheless, almost every type of fresh-water habitat, from the smallest ponds and streams to the largest lakes and rivers, has its characteristic population of snails, or univalve mollusks. Only the coldest alpine lakes, grossly polluted waters, acid lakes, and saline inland waters are predictably without snails. Some localities may yield only one or two species, whereas others may contain dozens. Sometimes the collector needs to hunt assiduously in order to find a few specimens; under other circumstances they may be present in enormous numbers. Snails are animals of the substrate and are found creeping about on all types of submerged surfaces, chiefly in water 10 cm to 2 m deep.

Harman (1972) demonstrates definite species selection of four general substrate types by gastropods: clean cobble, silt and detritus, hydrophytes and hydrophyte detritus, and allochthonous organic matter.

General characteristics. The vast majority of fresh-water gastropods have a spiral or discoidal coiled shell. However, in seven American genera (the limpets), the shell is in the form of a very low cone. The range in the maximum dimensions of the shell (length or width) of the mature animal is about 2 to 70 mm.

The muscular portion of the animal which projects from the shell is called the foot. For the most part it is inconspicuously colored, usually being grayish, brownish, or blackish, and often flecked or mottled with yellowish or whitish. The ventral surface of the foot is flat and there is a more or less prominent head at the anterior end. The head bears two tentacles which range in shape from short, blunt, or conical to long and filiform. The eyes are on or near the base of the tentacles. The mouth is on the ventral surface of the head at the extreme anterior end and in contact with the substrate in most species, but in the Viviparidae, Pleuroceridae, Hydrobiidae, and Valvatidae it is at the end of a muscular proboscis, or rostrum. In the Ampullariidae and Neritidae this rostrum is divided into two long, tentaclelike lobes. In the Lymnaeidae the head itself is widened into two flat lateral lobes, constituting the velum, a structure that persists from the larval stage.

The foot consists mainly of muscle tissue, but it contains the anterior portions of the digestive tract and reproductive system, as well as most of the nervous system. The rest of the internal organs form the visceral hump that lies within the shell, where its shape is a counterpart of the coils. Forming a lining for the shell and enclosing the viscera is a thin layer of specialized tissue, the mantle. The shell grows by the addition of the shell material secreted by the edge of the mantle (collar) at the orifice of the shell.

Although a few snails, such as *Valvata,* have a delicate external gill, most fresh-water gastropods have either internal gills that are specialized folds of the mantle for aquatic respiration or an internal air-filled "lung" consisting of a special portion of the mantle cavity surrounded by highly vascularized mantle. The respiratory opening (pneumostome) to either the internal gills or lung is usually situated on the right side at the edge of the shell. In those aquatic snails with a lung the edge of the mantle is often elongated into a respiratory siphon through which air is obtained at the surface film.

The exposed portion of the body of a snail is securely fastened to the shell internally by a large columellar muscle, and by the contraction of this muscle the foot may be doubled up and withdrawn from view. Those snails with gills have a discoidal sclerotized (rarely calcareous) operculum attached dorsally on the posterior part of the foot, and when the anmal is withdrawn into the shell the operculum fits into the opening and effectively closes it. Figure 2 shows the most common types of opercula.

In some families external reproductive structures may be seen. In the Hydrobiidae and Valvatidae, for example, more or less laterally in the region of the neck, is the verge, a projecting male copulatory organ of variable structure and size. In the Viviparidae the right tentacle of the male is much larger than the left and serves as a penis sheath.

Shell. Compared with marine and tropical land snails, the fresh-water forms are a drab-colored lot. A few species are rich green or yellow with red tinges or bright color bands, but the great majority are nondescript gray, tan, brown, blackish, or "horn colored," sometimes with indistinct bands or a mottled appearance. The shell proper is covered with a thin organic epidermis in which the pigments are deposited. The epidermis has the important function of protecting the underlying chalky white calcium carbonate from erosion.

In most of the common and widespread species the surface of the shell is superficially smooth, but if it is examined closely, fine longitudinal growth lines and spiral sculpture may be distinguished. Sometimes these lines can be detected only with the aid of a lens. At the other extreme are a few species whose shells are pronouncedly costate, ridged, carinate, or tuberculate.

When a shell is held with the opening toward the observer and with the tip upward, the shell is said to be dextral if the opening is on the observer's right and sinistral if on the left. The great majority of species are dextral. Some genera contain both dextral and sinistral species, but only in a very few species

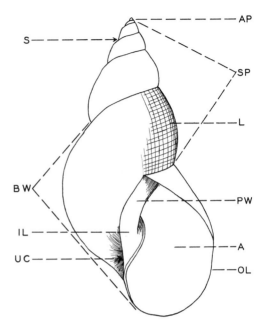

FIG. 1.–Anatomical features of a typical *Lymnaea* shell, ×2. A, aperture; AP, apex; BW, body whorl; IL, inner lip reflected over columella; L, spiral and growth lines; OL, outer lip; PW, parietal wall; S, suture; SP, spire; UC, umbilical chink. (Modified from Baker, 1928.)

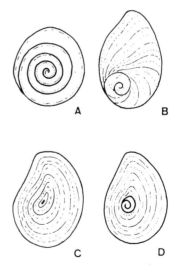

FIG. 2.–Types of opercula. A, multispiral; B, paucispiral; C, concentric; D, concentric with spiral nucleus. (From Burch, 1982.)

are both dextral and sinistral specimens known.

The important features of a typical gastropod shell are shown in Fig. 1. The opening of the shell is called the aperture, and the main portion of the shell above the aperture is the spire. The first large coil, or whorl, of the shell is the body whorl, and the nucleus, or protoconch, is the small round knob of from 1¼ to 1½ whorls without distinct sculpturing at the apex of the shell. The whorls are coiled around a central axis, or columella. An inner lip at the columellar margin of the aperture is usually present and reflected over the columellar region. The inner surface of the shell immediately adjacent to the inner lip is termed the parietal wall. The outer lip, or peristome, lies on the opposite side of the aperture; it is often reflected back on itself. The umbilicus may be a small chink or narrow slit between the reflected inner lip and the body whorl; it may be a definite hole one whorl deep; or it may be complete and well developed so that all of the larger whorls revolve around a thin-walled hollow tube occupying the position of the typical solid columella. In many instances the umbilicus is sealed, and the shell is said to be imperforate.

Considerable variations in size, shape, shell thickness, and surface markings occur within many genera, and sometimes even within a single species. These variations are particularly striking in the Pleuroceridae. In some cases variations within a species are clearly correlated with certain ecological features of the environment. Thickness of the shell, for example, may be directly related to the amount of calcium in the water. In highly alkaline waters, especially in the western states, many species show increased tendencies toward being ridged, plicate, or rugose. Specimens in such waters are also generally smaller than those in other habitats. On wave-beaten shores and in rapid streams there is usually an increase in the relative size of the aperture and a decrease in the length of the spire.

Locomotion. In spite of many close observations, the detailed mechanism of the usual gliding movements is not well understood. As a snail moves along on the under-side of the surface film or on the substrate, it leaves a familiar "slime track," a thin, flat ribbon of mucus secreted by the ventral surface of the foot, especially near the anterior end. The secretion of mucus is apparently necessary for locomotion, perhaps only as a lubricant, but it is not definitely known whether the actual movement of the snail is entirely due to indistinct waves of muscular contraction on the ventral surface of the foot, to ciliary action in the same area, or to both.

Another method of locomotion has been called "hunching." It involves obvious muscular contractions of the foot and a jerky pulling forward of the shell. Such movements are not normal, however, and usually occur when the snail is entangled in vegetation or is out of water.

A third method of locomotion, called "spinning," is utilized chiefly by some species of *Physella, Lymnaea,* and *Helisoma*; it occurs when the snail is moving through the water rather than over the substrate. For moving upward in this manner the snail decreases its specific gravity. A thread of mucus is fastened to the substrate at the point of leaving, and the snail moves upward with the lateral margins of the foot brought together and leaving a vertical mucous thread behind. Similarly, a snail at the surface film or on some object above the bottom may fasten a thread and move downward, trailing a mucous track. The same thread may be used a number of times by the same or other individuals, but it soon becomes brittle and breaks. For the most part, those snails that spin have slender, tapered, or pointed feet.

Feeding, digestive system. The great majority of fresh-water gastropods are normally vegetarians. The coating of living algae, which covers most submerged surfaces form the chief food, but dead plant material is frequently ingested, and occasionally dead animal material is eaten by some species. *Physella* and *Lymnaea,* for example, are good scavengers and essentially omnivorous. There are even a few records of lymnaeids feeding on living animals. When first deposited, mucous trails are sticky and accumulate debris and microscopic organisms, and as snails move about they incidentally utilize it and its

FIG. 3.–Representative American fresh-water gastropod shells. *Top, Pomacea* and its operculum, ×1; *second row,* left to right, ×1.5: *Hydrobia,* (a European genus), *Fluminicola, Amnicola, Pomatiopsis, Valvata,* and *Gyraulus* (2); *third row,* left to right, ×1: *Helisoma* (2), *Lymnaea, Campeloma, Viviparus,* and *Lioplax; bottom row,* left to right, ×1; *Lanx* (2), *Carinifex, Physella, Goniobasis, Pleurocera,* and *Neritina.*

contents as food. The entire foot and head areas are chemoreceptive for taste but are decreasingly sensitive laterally and posteriorly. Food in the form of animal tissues elicits responses much more quickly than plant tissues. It has been clearly shown that the higher the temperature, the greater the activity and food intake.

Just inside the dorsal part of the mouth is a

set of one, two, or three small sclerotized jaws (Fig. 6), used in cutting off bits of food. One jaw is dorsal and (or) two are lateral. Immediately behind the jaws the digestive tract is swollen to form a large buccal mass. The ventral portion of this organ is greatly thickened and contains stout, movable buccal cartilages to which muscles are attached. Dorsally, the buccal cartilages and their

FIG. 4.–External features of typical Gastropoda. A, female *Campeloma,* ×1.2; B, ventral view of *Physa,* ×1.7; C, *Pomatiopsis,* ×.8; D, *Valvata,* ×5, showing plumose left external gill, rudimentary right gill, verge, tentacles, rostrum, and divided foot; E, *Helisoma,* ×1.7; F, ventral view of *Ferrissia,* ×8; G, *Lymnaea,* ×1.5. F, foot; M, mouth; MA, mantle; O, operculum; P, pseudobranch; PN, pneumostome; R, rostrum; S, shell; T, tentacle; V, velum; VG, vertical groove of foot. (A–F modified from Baker, 1928; G modified from Baker, 1911.)

muscles are covered by the radula, one of the most characteristic features of the Gastropoda. It is essentially a longitudinal, toothed, straplike, chitinoid structure which occupies a position analogous with that of the human tongue. The radula is moved back and forth very rapidly by means of the underlying cartilages and muscles, and in this way pieces of food are thoroughly ground between it and the roof of the buccal cavity. The radula wears away as a result, especially at the anterior end, but it is continuously replaced since it is formed in a radular sac at the posterior end of the buccal mass and grows outward much like the human fingernail.

The teeth of the radula are fastened to the flat radular membrane in transverse rows. The number of teeth in a transverse row

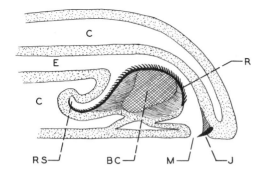

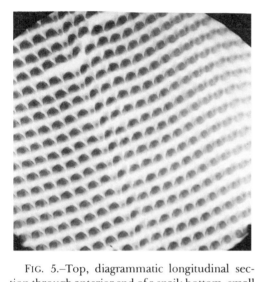

FIG. 5.–Top, diagrammatic longitudinal section through anterior end of a snail; bottom, small central part of the radula ribbon of *Lymnaea*. BC, buccal cartilages; C, body cavity; E, esophagus; J, sclerotized jaw; M, mouth; R, radula; RS, radula sac.

ranges from 7 to 175 (Fig. 5). The number of transverse rows is usually large, some species having more than 150. Thus the total number of teeth in a radula may range into the thousands. The morphology of the teeth varies greatly from one family and genus to another (Fig. 7), but all teeth are fundamentally similar; each has a base of attachment and a reflected portion bearing the cutting points, or cusps.

In each transverse row three general types of teeth are usually recognized: (1) a single, distinctive, median central tooth, (2) a series of lateral teeth on each side of it, and (3) a series of marginal teeth located outside of the laterals. Thus, in those families having seven

teeth in a transverse series, they are named in order: second marginal, first marginal, lateral, central, lateral, first marginal, and second marginal. In the other families there are correspondingly more lateral and marginal teeth, and the morphological gradations between adjacent teeth are very gradual. In the Lymnaeidae it is customary to designate a series of transitional teeth that lie between typical laterals and marginals.

The esophagus, which leaves the buccal mass, is quite long; it passes from the foot into the visceral mass contained within the shell, where it may or may not be dilated to form a crop. A pair of salivary glands may be found along the esophagus or on the crop. Shortly behind the crop is the dilated stomach, which may have a gizzard associated with it. In some genera the gizzard contains sand, presumably to macerate the food further. The stomach is followed by the long intestine, whose posterior end is dilated to form the rectum. The anus opens into the mantle cavity near the edge of the mantle and shell. A very large digestive gland, the so-called liver, empties into the stomach.

Circulatory system. The heart is situated in the lower part of the visceral hump, usually on the left side. In mature individuals with thin shells and in immature individuals the beating may be seen through the shell. The heart consists of one auricle (rarely two) and a ventricle. The ventricle pumps the blood through a short aortic trunk to all parts of the body via a series of arteries and capillaries. From the capillaries the blood passes into spaces, or sinuses, in the tissues collectively called the hemocoel. From the hemocoel it passes into veins and thence back to the auricle.

The blood of gastropods contains a dissolved, complex, copper-containing compound, hemocyanin, which, like hemoglobin, is capable of transporting oxygen. Oxidized hemocyanin imparts a definite bluish color to the blood; in the reduced state it is colorless. In the Planorbidae, however, hemoglobin is the respiratory pigment.

Respiration. In the Ampulariidae, Bithyniidae, Viviparidae, Valvatidae, Pleuroceridae, Thiaridae, Hydrobiidae, and Neritidae

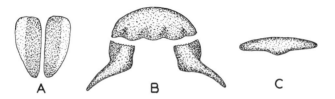

FIG. 6.–Sclerotized jaws of typical Gastropoda. A, *Pleurocera*; B, *Lymnaea*; C, single dorsal jaw of *Aplexa*.

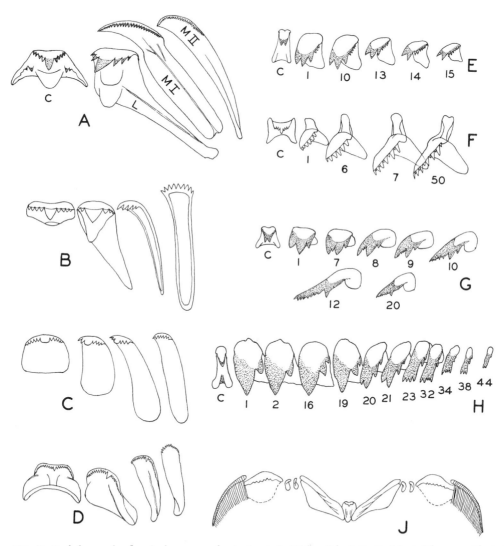

FIG. 7.–Radular teeth of typical gastropods. A, *Amnicola* (Bithyniidae); B, *Goniobasis* (Pleuroceridae); C, *Viviparus* (Viviparidae); D, *Valvata* (Valvatidae); E, *Ferrissia* (Ancylidae); F, *Physella* (Physidae); G, *Helisoma* (Planorbidae); H, *Lymnaea* (Lymnaeidae); J, *Neritina* (Neritinidae). C, central tooth; L, lateral tooth; MI and MII, first and second marginal teeth; teeth designated by number are a few laterals and marginals selected from the complete series. (A modified from Baker, 1928; B–G redrawn from Baker, 1928; H modified from Baker, 1911.)

respiration is strictly aquatic and occurs through an internal gill, or ctenidium, to which the surrounding water has easy access. This structure consists of a series of narrow, flat leaflets well supplied with blood and arranged like the teeth of a comb. It is located on the surface of the mantle within the mantle cavity of the body whorl. In some species there are over 100 leaflets in the ctenidium. In *Valvata* the ctenidium is external and plumose (Fig. 4D). The gilled snails may obtain a small amount of oxygen through the general body surface.

Most members of the Physidae, Lymnaeidae, Planorbidae, Ancylidae, Lancidae, and Acroloxidae do not have gills, but obtain oxygen through a "lung," or pulmonary cavity. This is an air filled, saclike, highly vascularized portion of the mantle cavity that occupies as much as one half of the body

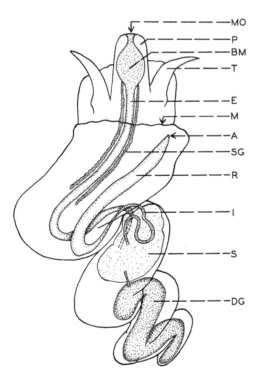

FIG. 8.–Diagrammatic dorsal view of head and visceral mass of *Pleurocera,* showing location of digestive organs. A, anus; BM, buccal mass; DG, digestive gland; E, esophagus; I, intestine; M, edge of mantle; MO, mouth; P, proboscis; R, rectum; S, stomach; SG, salivary gland; T, tentacle. (Modified from Magruder, 1935.)

whorl. The pneumostome, a small opening to the pulmonary cavity, is situated where the edge of the mantle and shell meet the foot. In the Lymnaeidae and Physidae the edge of the mantle is frequently drawn out into a long, muscular respiratory siphon around the pneumostome. It is probable that the mantle border is especially significant in all snail respiration.

At variable intervals most pulmonates come to the surface of the water where the pneumostome is brought into contact with the atmosphere and a fresh supply of air is taken into the pulmonary cavity. In a very quiet laboratory the opening of the pneumostome at the surface may be heard as a faint clicking sound. The migrations to the surface by spinning or on emergent vegetation appear to be largely governed by temperature and the amount of dissolved oxygen in the water, and under average conditions may occur every few minutes to several hours. Under both natural and experimental conditions may occur every few minutes to several hours. Under both natural and experimental conditions, however, it has been found that many pulmonates rarely or never come to the surface for air. This is particularly true of certain species of *Lymnaea, Physella,* and *Helisoma,* which may pass their entire life cycle without access to the surface. Some pulmonates have been collected in water more than 15 m deep, and Forel collected *Lymnaea* at a depth of 250 m in Lake Geneva, Switzerland. Certainly under such circumstances there are no migrations to the surface for air. Although there seems to be little specific information available (Cheatum, 1934), it appears that pulmonates that occur at great depths and some that remain submerged for protracted periods in the shallows fill the pulmonary cavity with water and use it as a gill. In many species, however, it is well established that all oxygen absorption may occur through the general body surface in both immature and mature individuals. Indeed, in the limpets the respiratory cavity is vestigial. Individuals in this family, as well as in the Planorbidae, have a pseudobranch. This is a short, blunt to long, conical projection from the dorsal or lateral portion of the foot near the edge of the shell. It is highly vascularized and undoubtedly functions as an accessory gill (Fig. 4F). Under experimental conditions it has been found that few snails can tolerate anaerobic conditions more than 48 hours. Students in the writer's laboratory have found that *Physella gyrina* Say from standing water habitats has a minimum tolerance of 1.8 ppm dissolved oxygen, but specimens from running waters have a minimum tolerance of 1.5 ppm.

Excretion. The renal organ is spongy and of varying size and shape. It is situated near the heart and respiratory cavity or gill, and its small duct opens near the anus.

Muscles. Aside from the foot, which is composed mostly of muscle tissue, there are

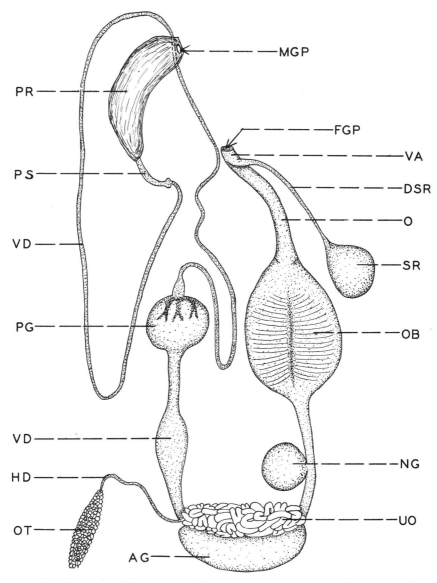

FIG. 9.–Reproductive system of *Lymnaea stagnalis appressa* Say; protractors and retractors of praeputium and penis sheath not shown. AG, albumen gland; DSR, duct of seminal receptacle; FGP, female genital pore; HD, hermaphroditic duct; MGP, male genital pore; NG, nidamental gland; O, oviduct; OB, oviducal bulb; OT, ovotestis; PG, prostate gland; PR, praeputium; PS, penis sheath; SR, seminal receptacle; UO, uterine portion of oviduct; VA, vagina; VD, vas deferens. (Modified from Baker, 1911.)

several important free muscles. The columellar muscle is attached to the shell internally and serves to withdraw the animal. Sets of muscles protract, retract, or depress the buccal mass. Those parts of the reproductive system directly concerned with copulation are usually supplied with protractors and retractors.

Nervous system. The greater portion of the gastropod nervous system (the "brain") usually consists of nine large ganglia, eight of which are paired. They are connected with each other by commissures and are arranged around the esophagus just behind the buccal mass. Large branching nerves originating in these ganglia innervate all parts of the body. Several small accessory ganglia are associated with some of the sense organs.

Sensory areas and organs. The eyes, situated at the base of the tentacles, are well developed, although little is known concerning the powers of vision. The ability to detect certain substances in solution is probably centered in the osphradium, a small specialized area of the mantle cavity. The sense of hearing, or perhaps more appropriately the sense of equilibrium and the ability to detect vibrations, is centered in two statocysts. Each of these is a minute sac closely associated with the central nervous system; it contains a fluid in which are suspended a variable number of calcareous bodies called statoliths. Although the general body surface reacts to touch, the tentacles are especially sensitive. It seems logical to assume that taste is centered in the mouth region where this sense has been attributed to Semper's organ.

Reproduction. The anatomy of the reproductive system varies widely in the fresh-water gastropods and is coming to be of increasing taxonomic significance, especially for the separation and identification of difficult species having nearly identical shells.

The Physidae, Lymnaeidae, Planorbidae, Ancylidae, Valvatidae, Acroloxidae, and Lancidae are hermaphroditic, both the male and female reproductive organs being in the same individual, but in all other fresh-water

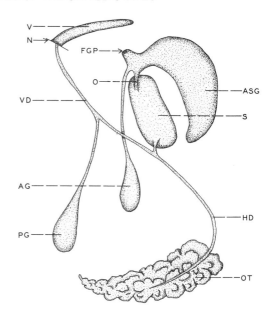

FIG. 10.–Reproductive system of *Valvata tricarinata. ASG,* accessory shell gland; N, wall of neck; S, sphermatheca; V, verge; see legend of foregoing figure for other structures.

families the sexes are separate. Figures 9–12 illustrate the reproductive systems of some typical species. The basal portions of the ducts lie in the foot and body whorl; the distal organs, however, occur in the smaller whorls, with the ovotestis, ovary, or testis being more or less imbedded in the digestive gland at the tip of the spire.

The male genital pore is usually located near the base of the right tentacle. It is commonly situated at the end of a muscular, protrusible, intromittent copulatory organ, or penis, which is withdrawn into the body except during copulation. In the Viviparidae the right tentacle is modified as a penis sheath. In the Hydrobiidae and Valvatidae the copulatory organ, usually called a verge, remains protruded and cannot be retracted. Sometimes the verge is long and thin; in other genera it is blunt, lobed, or divided. There is no male copulatory organ in the Pleuroceridae and the male genital pore lies at the edge of the mantle cavity.

The female genital pore is unspecialized and usually lies at the base of the neck near

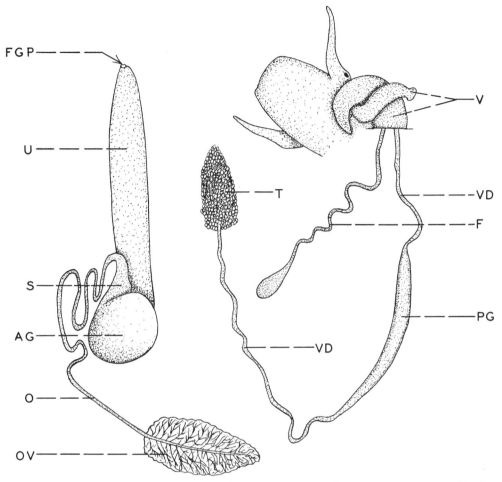

FGP

U

S

AG

O

OV

V

VD

F

PG

T

VD

FIG. 11.–Reproductive systems of male and female individuals of *Bithynia tentaculata* L. F, flagellum; OV, ovary; T, testis; U, uterus; see legends of foregoing figures for other structures. (Modified from Baker, 1928.)

the pulmonary aperture or at the edge of the mantle cavity.

Although surprisingly little accurate work has been done, the available information seems to indicate that a variety of conditions exist regarding fertilization in the hermaphroditic fresh-water Gastropoda. In some species isolated individuals have been observed to produce young; under such circumstances both mature sperm and mature eggs are produced simultaneously, and self-fertilization, rather than self-copulation, occurs in the reproductive tract. In other cases two such individuals may copulate and ex-

change sperm, each animal acting as male and female during the process; this is thought to be the common condition. In still other instances of protandry, the ovotestis may produce eggs at one time and sperm at another, and therefore an individual may act as either male or female during copulation, but not both.

Parthenogenesis is uncommon, but it is known to occur in *Hydrobia* and in *Campeloma*, especially in the northern states.

Oviposition usually occurs in the spring, although it may continue into the summer and early fall. Some species produce few eggs;

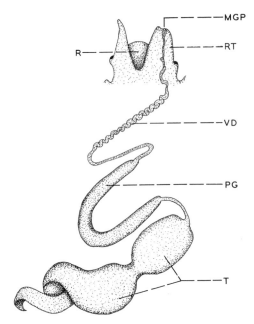

FIG. 12.–Reproductive system of male *Campeloma integrum*. R, rostrum; RT, right tentacle; see legends of foregoing figures for other structures. (Modified from Baker, 1928.)

others hundreds at a time. They are almost invariably deposited in a gelatinous mass on some substrate.

The early developmental stages occur within the egg mass, and by the time the young snail leaves, it has taken on the basic morphological features of the adult, and the shell has one to two whorls. The Viviparidae are ovoviviparous, the individuals being well developed at birth.

It is thought that in the great majority of species the usual length of life is 9 to 15 months. Some species, however, have two or three generations per year, especially in southern states, and notably among limpets and *Physella*. Length of life is often determined by period(s) of enforced aestivation. Some of the Lymnaeidae, Planorbidae, *Viviparus,* and *Goniobasis* are known to live as long as 2 to 4 years.

Predators. Perhaps the greatest natural enemies of snails are the fishes. It has been estimated that about 20% of our fresh-water species feed to a greater or lesser extent on mollusks. For only a few forms, however, do snails form a significant portion of the diet; some of these are the suckers, perch, sheepshead, pumpkinseed, and whitefish. Some ducks, shore birds, and occasional amphibians may eat snails. Among the invertebrates their most important predators are leeches, beetle larvae, and Hemiptera and Odonata nymphs. At least in the laboratory, it has been found that ostracods will attack small snails and destroy their eggs.

Many species of snails are known to serve as the intermediate hosts of trematodes, but it is thought that such infections are seldom fatal, although the reproductive capacity may be greatly decreased and the digestive gland severely damaged.

Ecology. It has been said that it is difficult to "disentangle the importance of quality of water from geographical distribution, the physical character of the habitat, and available food materials." In a study of field data for 91 species and subspecies of boreal and Arctic North American fresh-water gastropods, Clarke (1979) found that only 6 were indicators of general trophic lake stages; *Valvata sincera,* for example, was characteristic of oligotrophy. Nevertheless, it is possible to point out certain ecological factors of the environment that have a pronounced influence in the determination of the habitat and activities.

One of the most important of these factors is the amount of dissolved salts in the water, especially calcium carbonate, which is the essential material for shell construction. It is generally true that soft waters contain few species and individuals, whereas hard waters contain many species and individuals. The majority of Lymnaeidae, for example, occur in water high in carbonates (more than 15 ppm of bound carbon dioxide). A few species, however, are striking in their ability to thrive in soft waters that are low in carbonates. The Valvatidae are most common in waters containing less than 8 ppm. *Campeloma* has been found in abundance in water containing only 1 ppm. It is difficult to understand how calcium carbonate can be absorbed and utilized at such low concentrations. Many species are, of course, adapted to a wide range of

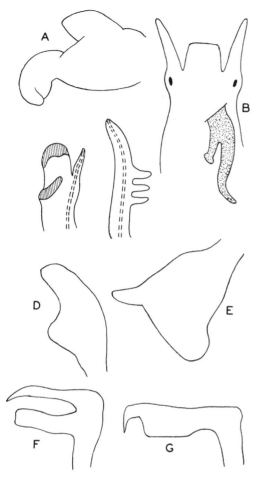

FIG. 13.–Verges of typical gastropods. A, *Somatogyrus*; B, *Amnicola*; C, hydrobiid; D, *Amnicola*; E, *Amnicola*; F, *Amnicola*; G, *Pomatiopsis,* H, hydrobiid. (A and C to G modified from Berry, 1943; B modified from Baker, 1928; C and H from Burch, 1979.)

carbonate content; *Amnicola limosa porata* has been collected from lakes having a bound carbon dioxide content ranging from 1 to 30 ppm; and other species undoubtedly have still wider limits. Some investigators have shown that on substrates composed of mixed high- and low-carbonate rubble that the majority of individuals will select the high-carbonate pieces.

There is some evidence to show that calcium is stored in connective tissues of the mantle edge, pulmonary region, and the foot. In general, the ionic uptake mechanisms have a high affinity for calcium, and, at least in several species investigated, the calcium content of the blood remains remarkably constant, in spite of varying ecological factors. Furthermore, it has been shown that in some species there is no relation between the calcium: tissue ratio and the level of environmental calcium. As a limiting factor, however, it appears that calcium levels may be instrumental in preventing development of the gastropod egg.

Hydrogen-ion concentration is closely associated with, and partly determined by, the carbon dioxide content, and lakes low in carbonates are usually toward the acid side of the scale (below pH 7.0), whereas those high in carbonates are almost always alkaline. It follows, then, that the great majority of species and the largest numbers of individuals occur under alkaline conditions. All the Valvatidae and nearly all the Lymnaeidae, for example, are confined to waters having pH readings of 7.0 or above. Nevertheless, there are some striking exceptions in other families. *Ferrissia parallelus* has been found in waters ranging from pH 6.0 to 8.4, and *Amnicola limosa porata* from pH 5.7 to 8.3. In general, however, snails are uncommon in lakes and streams whose surface waters are more acid than pH 6.2. They are never found in true acid sphagnum bogs.

Dissolved oxygen is another important limiting factor, most pulmonate and gilled species requiring rather high concentrations. For this reason severely polluted rivers and the deeper parts of lakes that become oxygen deficient during the summer and winter are usually devoid of gastropods. Limpets seem to be found only where the water remains almost saturated. For common species of *Physella*, 2 ppm is about the limiting level for dissolved oxygen.

The great majority of species and individuals occur in the shallows, especially in water less than 3 m deep. At greater depths river and lake faunas tend to become more and more depauperate. The fact that the shallows are a favorite habitat is probably correlated with the abundance of food in this zone. Although a few genera, such as *Lymnaea,* occur in a wide variety of habitats, many forms are restricted to particular types of substrates. *Pleurocera* is usually found on rocky or sandy shoals, the Viviparidae are most common on sandy bottoms, and the Ampullariidae are mud lovers. *Physella* occurs in greatest abundance where there is a moderate amount of aquatic vegetation and organic debris, and it is rare among dense mats of vegetation. The Hydrobiidae are usually found among aquatic plants, sometimes in enormous numbers. *Fontigens nickliniana,* the "water cress snail," is restricted to the dense mats of water cress and other vegetation of cool springs. Swift streams with sand or gravel bottoms and wave-swept beaches are generally poor places for collecting.

Large bodies of water usually have many species and small bodies have few, the reason presumably being the fact that the former have a greater variety of specific subhabitats suitable for the individual species. A few snails have become isolated in caves and underground streams, especially *Physella* and some Hydrobiidae. Such species are blind and poorly pigmented.

Although some gastropods may be active down to the freezing point of water, such conditions are not favorable, and perennially cold lakes and streams and bodies of water in high, mountainous areas contain few species and individuals. At the other extreme, only a few U.S. species can tolerate a continuous temperature of 30°C. Some lake pulmonates exhibit seasonal migrations correlated with temperature; in the fall they move into deeper waters and in spring into the shallows. *Physella* is the dominant genus in thermal waters.

A single bioluminescent snail is known from streams in northern New Zealand. Light is generated in secreted mucus, and is probably the result of bacterial metabolism.

Hibernation, aestivation. Populations in shallow ponds that freeze solid are able to overwinter by burrowing into the mud and debris on the bottom and hibernating in the frozen substrate. In intermittent streams and ponds that are dry for a short period in the summer many pulmonates burrow into the mud up to a depth of several inches and aestivate during the unfavorable period. The most effective seal and protection is afforded by a mud bottom that has a high percentage of clay. In *Lymnaea* the formation of an epiphragm has been studied closely. This structure is usually produced when the snail is imbedded in the mud. It is a sheet of mucus which is formed just within the aperture by the foot. On drying, it hardens and forms an effective seal that aids in protecting the animal during the drought period. This device may account for the presence of *Lymnaea* in some waters to the exclusion of other genera. Viable aestivating *Planorbis* and *Lymnaea* have been kept in the laboratory for more than 3 years.

Juvenile *Lymnaea* usually move out onto the adjacent shore and aestivate in detritus for 3 to 4 months of cold or dry conditions. Mature *Lymnaea,* however, usually move into deeper water and rarely leave the water.

Overwintering nonhibernating gastropods seldom ingest food, and consequently they usually suffer loss of weight.

Geographic distribution, dispersal. The fresh-water gastropod fauna of the United States is rich in species (about 500), rich in individuals, and rich in diversity. Not counting additional varieties, Baker (1928) lists 95 species from Wisconsin; Goodrich and van der Schalie (1944) list 99 from Indiana; Harman and Berg (1970) report 38 species from the Finger Lakes region of New York; Taylor (1981) records 67 species from California, and Winslow (1926) lists 142 from Michigan. Even Texas, which is by no means a favorable area for the development of an extensive fauna, has 96 species (Murray and Roy, 1968). It is estimated that more than 90% of the species and half of the genera found in North American waters are endemic.

Some genera, such as *Lymnaea, Helisoma, Gyraulus,* and *Ferrissia,* occur almost everywhere from coast to coast. The Pleuroceridae are confined largely to the states east of the Mississippi except for a few species in the Pacific coast states. Native species of Viviparidae are not found west of the Mississippi drainage, but two imported species are found in California. The Neritinidae occur only in Alabama, Florida, and along the Gulf coast, and the fresh-water species of the Ampullariidae are found only in Florida and Georgia.

Some genera are similarly restricted. *Lanx* occurs only in the west coast states, *Aplexa* in the northern states, *Lyogyrus* in the Atlantic coast states, and *Tulotoma* only in the Alabama River and its tributaries. Perhaps the most interesting case of all, *Gyrotoma, Amphigyra,* and *Neoplanorbis* are (were) confined to the Coosa River, Alabama.

Bithynia tentaculata, a small introduced euryokous European species, is spreading very rapidly over the United States. It is now the dominant species in Oneida Lake, New York, where the snail fauna dropped from 36 species in 1918 to 31 species in 1970.

Many species in the Mississippi drainage and in other eastern drainages undoubtedly would find suitable ecological conditions in many areas west of the Continental Divide, but their rate of geographic spread across the central plains and over the Divide since the last glaciation has been much too slow, except where man has made intentional transplants.

Because of increasing levels of pollution during the past 60 years, populations of many of our American stenokous and unusual species have been disappearing, so that many forms that were formerly abundant in restricted geographical ranges are now uncommon or rare and difficult to find. Many species now have ranges that are more and more restricted to headwater streams. Clark (1970) reports that out of 103 species in the Ohio River drainage, 41 are rare and endangered, and probably 8 of these are extinct. This problem is especially acute in some of the Texas and southeastern drainages, such as the Coosa River, Alabama, where *Amphigyra, Neoplanorbis,* and other exotic forms have undoubtedly become extinct. The information on geographic ranges of unusual species in this chapter should not therefore be accepted at face value.

The general problem of an explanation of the present geographic distribution of American fresh-water snails is extremely complicated, poorly understood, and beyond the scope of this volume. It involves, among other things, a careful consideration of past geological, geographic, and climatic changes.

Presumably it is relatively easy for a species to migrate slowly and extensively throughout an entire, connected drainage system provided

that unfavorable environmental conditions are not encountered during such migrations. Movements between normally isolated bodies of water, or between inaccessible parts of the same drainage system, however, are necessarily passive and dependent on outside agencies. Occasional flood waters may leave their normal channels and carry eggs or adults to distant places. There are numerous instances recorded in the literature where small snails have been found on the feathers or in mud on the feet and legs of ducks and shore birds, and most authorities are agreed that this is a most important means of increasing the geographical range. Goodrich and van der Schalie (1944) have aptly summarized the essentials of distribution:

> Whatever the mode of distribution, it has to be remembered that continued existence of a mollusk in any spot to which it may penetrate depends upon whether that spot is environmentally favorable, particularly for reproduction. It is to be suspected that in times without number migration has proved a failure.

Several species native to Europe, Japan, and the general Pacific area have been introduced into the United States by aquarium fanciers and have "escaped" to natural habitats. *Bithynia tentaculata*, the "faucet snail," was imported to the Great Lakes region from northern Europe in the late 1870s. *Tarebia* is native to the Far East and western Pacific islands, but was introduced into Florida, where it is now locally abundant. *Viviparus stelmaphorus* is a Japanese species in springs, lakes, and streams near San Francisco, Boston, Philadelphia, Niagara Falls, St. Petersburg, Florida, and other areas. Other introduced fresh-water species have been collected in various parts of the country, and the list will undoubtedly grow with time. Many unreported tropical species have "escaped" into the wild but are not yet common or abundant.

Collection, preparation. The nature of collecting apparatus depends on the type of environment being visited. In the shallows of lakes and streams simple hand picking may be used. In deeper waters a long-handled net or tin strainer is effective. In water more than a meter or two deep it is usually necessary to work from a boat with a dredge. Bottom debris and vegetation should be sifted and examined on shore. Limpets should be carefully removed from the substrate with a small knife blade. Wide-mouth bottles may be used to transport specimens to the laboratory.

For anatomical work the animal should preferably be killed and fixed in boiling water, although 75 to 95% alcohol is also useful. In the Hydrobiidae and Valvatidae, where species identification may be based on the morphology of the undistorted verge, it is best first to narcotize the animal by sprinkling menthol or magnesium sulfate crystals on the surface of the water and wait until it becomes very sluggish (which may be a whole day later), then kill and fix in Bouin's solution. Sixty to 75% alcohol is a suitable preservative. Formaldehyde should not be used, as it makes the specimens brittle and damages the shell.

Either freshly killed or alcoholic specimens may be used for dissection. The foot and the attached visceral hump should be kept intact by carefully pulling the latter from the shell with a pin or hairpin. If the specimen is small, the shell may be broken away from the underlying parts bit by bit, or the shell may be dissolved in dilute hydrochloric or sulfuric acid. The only instruments necessary for the dissection itself are a fine scalpel, forceps, scissors, and needles. First, the foot should be pinned down in a small dish containing a thick layer of black paraffin on the bottom. Beginning at the head, a median, dorsal, longitudinal cut should be made through the body wall; it should be carried well into the mantle area, using great care not to injure the underlying parts. The dorsal flaps of the body wall and mantle may then be pinned back, thus exposing most of the viscera.

The radula may be isolated and cleaned by simply placing the whole buccal mass in cold 10% potassium hydroxide for several hours to a day or by heating the solution for a few minutes. A quicker method of preparing the radula for mounting involves dropping a few drops of household Clorox on the isolated buccal mass. Within a few minutes the radula will be free of tissues and is ready for further processing. After a thorough rinsing in dilute acetic acid and twice in distilled water, the radula should be placed between two slides with a strip of paper on each side of the radula to prevent crushing. The two slides should be bound together with string. While in this flattened position, the radula should be dehydrated in alcohol and cleared. One slide is then removed and the radula is covered with mounting fluid and a cover slip. It is sometimes advantageous to stain it in strong chromic acid or carmine. It is wise to mount the radula with the teeth up and to make several transverse cuts so the surface of the teeth can be more closely examined at the ends of the cuts. The teeth at the anterior end of the radula are too worn to be diagnostic. For small radulae, Mikkelsen (1985) recommends staining and mounting in CMCP in shallow depression slides.

If the dry shell is to be preserved, the animal should first be pulled out and then the shell cleaned. It should be rubbed internally with a bit of sponge on a wire and then rinsed with a syringe. The external surface is often encrusted with algae, debris, and calcium carbonate or iron oxide deposits, and the appearance may be much improved by scrubbing with a stiff toothbrush and water, or by placing the shell in a weak solution of oxalic acid for a half hour or longer. Finally the shell should be well rinsed and dried before storage. It is not necessary to remove the animal from the shell in the minute species, but the whole specimen should be kept in 70% alcohol for a day or two and then simply allowed to dry without any resulting offensive odor. Some workers place the live snail in a stoppered shell vial in the freezing compartment of the refrigerator. After it is frozen solid, thaw it; the whole visceral mass then can be easily pulled out of the shell.

Opercula should be cleaned in oxalic acid, dried, and then placed in the shell whose aperture is plugged with cotton.

Suggestions for storing, arranging, and cataloging a shell collection are given in a book by Abbott (1955).

Taxonomy. Three orders of Gastropoda are represented in American fresh waters. The Mesogastropoda includes the families Ampullariidae, Thiaridae, Bithyniidae, Hydrobiidae, Viviparidae, Valvatidae, Micromelaniidae, Pomatiopsidae, and Pleuroceridae. All these gastropods except the Valvatidae are dioecious and have an internal gill; all have an operculum, a heart with a single auricle, and a radula with few teeth in a transverse series; the two visceral nerves are crossed, forming an 8-shaped loop. The Order Neretinacea, including the single Family Neritinidae, is similar to the preceding except that the heart has two auricles and the radula has many teeth in a transverse series. The Limnophila (pulmonates), on the other hand, consisting of the Physidae, Lymnaeidae, Planorbidae, Acroloxidae, and Ancylidae, are hermaphroditic and have an internal pulmonary cavity, a heart with a single auricle, and a radula consisting of many teeth in a transverse row. The operculum is absent, and the visceral nerves do not cross, but form a simple loop.

Much of the generic and specific nomenclature of the past is confused because of descriptions based on one or a few shells taken from one or a few localities, but even with our present knowledge there are many species that are still troublesome to the specialists. This situation is due mainly to individual variations, age differences, and ecological variations within single species. The shells of some species of Pleuroceridae, for example, are said to be "infinitely variable."

Basch (1963) has shown a wide range of shell distortions in *Ferrissia* kept in aquariums for long periods.

It has become increasingly apparent that more accurate, critical identifications can be made by using the anatomical details of the radula and the internal and external reproductive organs as criteria. Since none of these structures show any considerable variations within a species, they are proving to be particularly valuable in such difficult families as the Hydrobiidae, where the shells are minute and not particularly distinctive.

Some investigators divide the Subfamily Lymnaeinae into seven genera, including *Acella, Bulimnaea, Fossaria, Lymnaea, Pseudosuccinea, Radix,* and *Stagnicola,* but others consider them only as subgenera of the single genus *Lymnaea.* The former practice has been followed in this manual.

The key that follows includes essentially all the known genera occurring in the United States; only a few rare and poorly known forms are omitted. For example, this key does not include three genera of minute snails reported from phreatic waters in the Edwards aquifer in Texas. (See Herschler and Longley, 1986.) Only fresh-water forms are indicated, those restricted to brackish waters being omitted. In general, the taxonomy follows that of Burch (1982).

KEY TO FAMILIES AND GENERA OF GASTROPODA

1. Without an operculum, and mantle cavity a lung, or with a small pseudobranch outside the mantle cavity; mouth opening lateral; pulmonates 51
   With an operculum; gills present; mantle opening facing anteriorly; prosobranchs 2
2. Shell globose, with a very low spire, up to 20 mm tall (Fig. 15A); one species in FL, GA, MS.
   NERITINIDAE, **Neritina**
   Shell variously shaped, but if shaped like *Neritina* then less than 5 mm tall 3
3. Shell less than 8 mm in diameter, spire depressed (Figs. 16A, B); operculum circular and multispiral (Fig. 16C); gill external (Fig. 4D); with a pallial tentacle (Fig. 4D); about 10 species generally distributed in a variety of habitats; round-mouthed snails . . . . VALVATIDAE, **Valvata**
   Shell small to large, spire depressed to elongate; operculum various; pallial tentacle absent . . . 4
4. Operculum multispiral or paucispiral (Fig. 2) 5
   Operculum concentric (Fig. 2) **44**
5. Shell less than 7 mm long; males with a verge (Fig. 13) 6
   Shell more than 10 mm long; males without a verge 36
6. Shell sculptured with numerous ridges; one species in caves and streams in KY, IN.
   MICROMELANIIDAE, **Antroselatus spiralis** Hubricht
   Shell usually smooth; if sculpturing is present it does not consist of spiral ridges 7

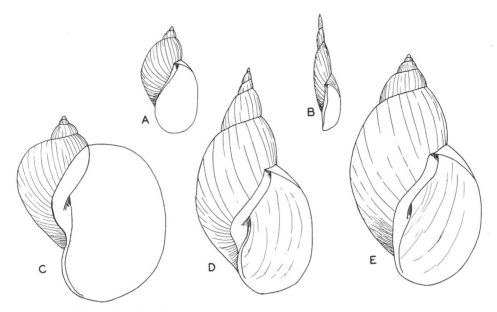

FIG. 14.–Typical Lymnaeidae. A, *Pseudosuccinea columella* (Say), ×3; B, *Acella haldemani* (Binn.), ×1.3; C, *Radix auricularia* (L.), ×2; D, *Lymnaea stagnalis,* ×1.5; E, *Bulimnaea megasoma* (Say), ×1.5.

7. Shell high-spired (Fig. 21B) foot divided by a vertical groove; 5 to 10 mm long; usually amphibious but sometimes found on submerged substrates; 3 species in coastal CA and 3 species in eastern and southern states .... POMATIOPSIDAE, **Pomatiopsis**
   Shell small, high spired to depressed; foot not divided by a groove; strictly aquatic.
                                                                           HYDROBIIDAE, **8**
8. Verge of male simple, cylindrical **9**
   Verge of male with two or three main branches **33**
9. Verge simple, with no accessory lobes **10**
   Verge having accessory lobes or glandular crests (Fig. 13) **16**
10. Shell neritiform and thin (Figs. 15B, C); 1 rare species in Coosa and Cahaba rivers, AL.
                                                                  Lepyrium showalteri (Lea)
    Shell not neritiform **11**
11. Shell depressed, heliciform, with spiral brown bands.
                                                  Cochliopina riograndensis (Pilsbry and Ferriss)
    Shell conical to subglobose; no spiral color bands **12**
12. Shell imperforate or narrowly perforate (Fig. 21J) **13**
    Shell umbilicate **15**
13. Rapid streams of Pacific drainages; about 12 species .... **Fluminicola**
    Mississippi, Gulf, and Atlantic drainages **14**
14. Shell thin; columella not thickened; NJ to SC .... **Gillia altilis** (Lea)
    Shell thick; columella thickened (Fig. 21D); MS and Gulf drainages; about 40 species.
                                                                           Somatogyrus
15. Shell less than 2 mm long; MO .... **Antrobia culveri** Hubricht
    Shell 2.5 to 3.0 mm long; Coosa and Cahaba rivers, AL; two rare species .... **Clappia**
16. Verge of male with accessory lobes (Fig. 13) **17**
    Verge of male with glandular tips (Fig. 13) **23**
17. Top of shell spire flattened; widely distributed in eastern states ...... **Probythinella lacustris** (Baker)
    Top of shell spire not flattened **18**
18. Deep waters of Lake Michigan; shell 3.0 to 3.5 mm long .... **Hoyia sheldoni** (Pilsbry)
    Southern and western distribution **19**

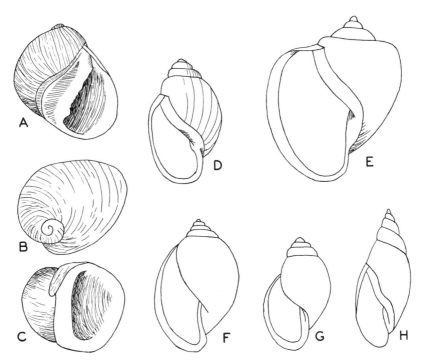

FIG. 15.–Typical Neritinidae, Hydrobiidae, and Physidae. A, *Neritina,* ×1.5; B and C, *Lepyrium showalteri,* ×6.5; D, *Physella gyrina,* ×1.7; E, *P. parkeri,* ×2; F, *Physella,* ×2; G. *P. integra,* ×2.5; H, *Aplexa elongata,* ×2.5.

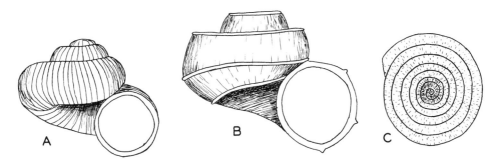

FIG. 16.–Valvatidae. A, *Valvata sincera,* ×8; B, *V. tricarinata,* ×11; C, operculum of *V. tricarinata.* (C modified from Baker, 1928.)

19. Five species in TX, AZ, CA . . . . **Tryonia**
    Found in GA, FL **20**
20. Verge with 7 to 50 papillae on right margin and 4 to 11 papillae on distal third of left margin; 2 species
    in FL, GA . . . . **Littoridinops**
    Verge with 1 to 7 papillae on right margin and 1 or 2 on left margin; FL **21**
21. Shell sculptured with fine spiral lines **22**
    Shell without fine spiral sculpturing; about 10 species . . . . **Aphaostracon**
22. Spiral sculpturing composed of raised threads . . . . **Pyrgophorus platyrachis** Thompson
    Spiral sculpturing composed of fine incised striations; 2 species (sometimes included in *Tryonia*).
    Hyalopyrgus

23. Shell almost entirely uncoiled (Fig. 21T); springs on Edwards Plateau, TX .... **Orygoceras**
    Shell coiled **24**
24. Shell up to 10 mm long; common in central states .... **Birgella subglobosa** (Say)
    Shell up to 5 mm long **25**
25. Shell turban shaped, up to 1.5 mm long; Alabama River drainage .... **Stiobia nana** Thompson
    Shell with another shape **26**
26. Verge with a simple glandular pattern (Fig. 21W) **27**
    Verge with elaborate patterns of many glands (Fig. 21V) **28**
27. Shell elongated (Fig. 21N); common and widely distributed; 8 species, sometimes included in
    Pyrgulopsis .... **Marstonia**
    Shell subglobose; Chipola River system, FL .... **Rhapinema dacryon** Thompson
28. Shell subglobose or broadly ovate; 4 to 8 mm long; rivers in AL, GA, FL; 2 species .... **Notogillia**
    Shell conic or ovate **29**
29. Occurring only east of the Continental Divide **30**
    Occurring only west of the Continental Divide **32**
30. Penis large, spatulate; rivers of FL, GA; 3 species .... **Spilochlamys**
    Penis small, slender, conical; widely distributed **31**
31. Shell elongate, conical .... **Pyrgulopsis**
    Shell broadly conical to ovate; about 15 species .... **Cincinnatia**
32. Shell elongately conical; whorls flat sided, angulate, or carinate; Upper Klamath Lake, OR and
    Pyramid and Walker lakes, NV; 2 species .... **Pyrgulopsis**
    Shell conical; whorls rounded, not angulate or carinate; ten species in western states; sometimes
    included in *Pyrgulopsis* .... **Fontelicella**
33. Males with bipartite verge **34**
    Males with tripartite verge; small species; widely distributed in caves and springs; 10 species.
                                                                                        Fontigens
34. Shell ovate to globose; widely distributed and very common (Fig. 21E); about 20 species.
                                                                                        Amnicola

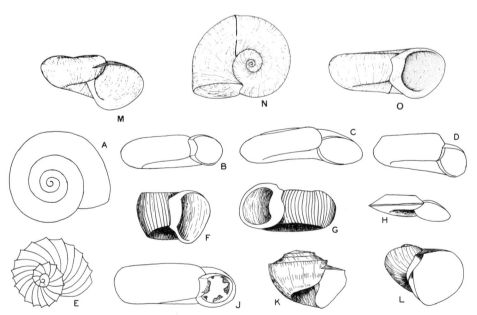

FIG. 17.–Typical Planorbidae. A and B, *Gyraulus circumstriatus,* ×6; C, *G. deflectus,* ×6; D, *G. altissimus,* ×5; E, *Armiger crista,* ×6; F, *Helisoma antrosa,* ×1.7; G, *H. trivolvis,* ×1.2; H, *Promenetus exacuous,* ×4; J, *Planorbula,* ×4; K, *Helisoma,* ×.9; L, *Vorticifex,* ×4; M–O, *Menetus.* (A and B redrawn from Baker, 1928; M–O from Burch, 1979.)

Shell discoidal or subdiscoidal; TX **35**
35. Shell discoidal (Fig. 21M) .... **Hauffenia micra** (Pilsbry and Ferriss)
    Shell subdiscoidal (Fig. 21U) .... **Horatia nugax** (Pilsbry and Ferriss)
36. Mantle edge smooth; males present; shells thick and solid; operculum corneous and paucispiral.
    PLEUROCERIDAE, **38**
    Mantle edge papillate; males absent; introduced tropical species .... THIARIDAE, **37**
37. Shell sculptured with spiral threads and grooves; FL, TX, AZ ..... **Melanoides tuberculata** (Müller)
    Shell sculptured with spiral rows of beads and nodules; FL, TX .... **Thiara granifera** (Lamarck)
38. Shell very large, with periphery of whorls angulated or inflated and with bosses or blunt spines (Fig.
    20D); 1 highly variable species in Tennessee River system .... **Io fluvialis** (Say)
    Shell small to large, conical to subglobose, without prominent peripheral bosses or spines **39**
39. Terminal whorl with a broad slit along the sutural juncture (Fig. 20A); about six highly variable species
    in the Coosa River, AL .... **Gyrotoma**
    Terminal whorl without such a broad slit **40**
40. Lateral radular teeth with broad, bluntly rounded median cusps; shell globose, conical, or ovate;
    southeastern states but most abundant in AL drainages; about 25 species ...... **Leptotoxis**

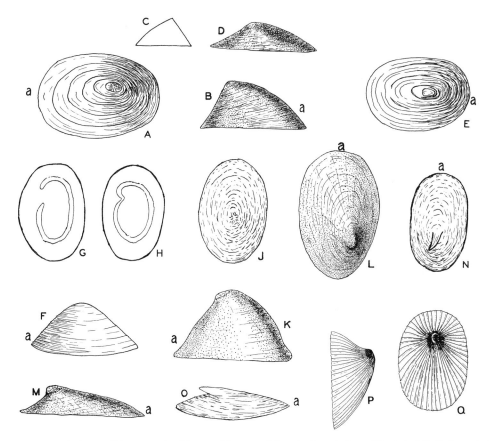

FIG. 18.–Typical Ancylidae, Acroloxidae, and Lymnaeidae. A, dorsal view of *Ferrissia*, ×9; B, lateral view of *F. rivularis,* ×6; C, cross section of *F. rivularis*; D, lateral view of *Laevapex*; E, dorsal view of *Lanx newberryi*, ×3; F, lateral view of *Lanx*; G, ventral surface of shell of *Fisherola* showing muscle scar; H, same for *Lanx*; J, dorsal view of *Rhodacmea*; K, lateral view of *Rhodacmea*; L, dorsal view of *Hebetancylus*; M, lateral view of *Hebetancylus*; N, dorsal view of *Acroloxus coloradensis*; O, lateral view of *A. coloradensis*; P and Q, *Rhodacmea*; a, anterior. (B, D, K, L, and M from Basch, 1963; P and Q from Burch, 1979.)

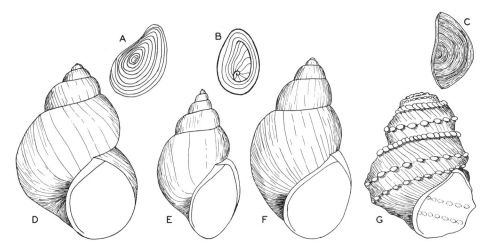

FIG. 19.–Typical Viviparidae. A, operculum of *Viviparus*; B, operculum of *Lioplax subcarinata*; C, operculum of *Tulotoma*; D, *Viviparus subpurpureus,* ×1.5; E, *Lioplax subcarinata*; F, *Campeloma decisum,* ×1.7; G, *Tulotoma magnifica,* ×1.

Lateral radular teeth with narrow, pointed, spade shaped, or triangular median cusps; shell elongated or conic 41

41. Shell of medium size, subglobose to cylindrical, sculptured with low spines or nodules; parietal wall thickened above and below (Fig. 20E); OH, TN, Black, Spring, and Big Black river systems; 10 species .... **Lithasia**
Shell small to large, elongated, rarely nodular; no thickening of parietal wall 42

42. Basal end of aperture auger shaped (Figs. 20F, G); Mississippi River, Great Lakes drainages, and Hudson River drainage; about 20 species .... **Pleurocera**
Basal end of aperture not auger shaped (Fig. 20H) 43

43. Found only east of the Continental Divide; about 80 species; (same as *Goniobasis*) .... **Elimia**
Found only west of the Continental Divide in Pacific Slope and Great Basin drainages; 8 species. **Juga**

44. Shell more than 20 mm long; operculum corneous 45
Shell less than 15 mm long (Fig. 21C); operculum calcareous; Great Lakes drainages; usually in aquatic vegetation; introduced from Europe .... BITHYNIIDAE, **Bithynia tentaculata** (L.)

45. Shell globose or discoidal, width more than 40 mm (Fig. 3); apple snails; AL, FL, GA.
AMPULARIIDAE, **46**
Shell subglobose to turreted; ovoviviparous; medium to large; widely distributed and common.
VIVIPARIDAE, **47**

46. Shell discoidal or planispiral; yellowish brown to greenish brown, with a flared aperture; an introduced tropical species in southern FL .... **Marisa cornuarietis** (L.)
Shell subglobose; the largest American fresh-water snail (Fig. 3); usually in muddy substrates; 2 species in FL, GA, AL .... **Pomacea**

47. Shell thin, usually 35 to 50 mm long; introduced; scattered distribution; 2 species.
Cipangopaludina
Shell thick, less than 35 mm long 48

48. Shell with spiral rows of nodules (Fig. 19G); Coosa–Alabama drainage.
Tulotoma magnifica (Conrad)
Shell without rows of nodules 49

49. Operculum concentric, but with a spiral nucleus (Fig. 19B); six species in eastern and southern states.
Lioplax
Operculum entirely concentric 50

50. Shell subconic, thin; width and length of aperture nearly equal (Fig. 19D); with or without spiral color bands; 3 species in eastern and southeastern states .... **Viviparus**

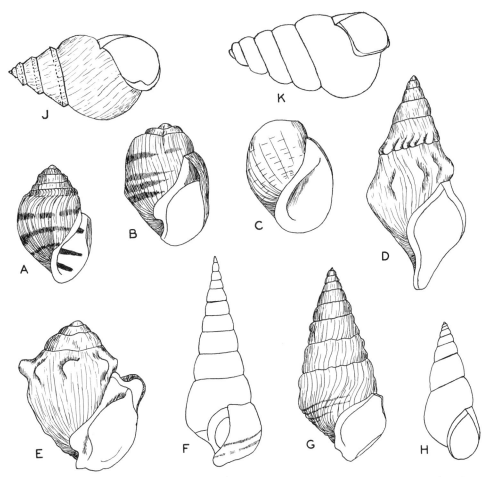

FIG. 20.–Typical Pleuroceridae. A, *Gyrotoma amplum,* ×2; B, *Lithasia anthonyi,* ×1.5; C, *Anculosa picta,* ×2; D, *Io fluvialis,* ×1.2; E, *Lithasia,* ×2.2; F, *Pleurocera acuta,* ×2; G, *Pleurocera,* ×1.5; H, *Elimia livescens,* ×2; J and K, *Anculosa,* ×2. (A redrawn from Goodrich, 1924; C modified from Goodrich, 1922; F modified from Goodrich and van der Schalie, 1944.)

57. Shell thin and fragile; spire small (Fig. 14A); eastern half of the United States.
        Pseudosuccinea columella (Say)
    Shell not thin and fragile; spire normal **58**
58. Shell more than 35 mm long **59**
    Shell less than 35 mm long **62**
59. Body whorl relatively narrow; about 25 species .... **Stagnicola**
    Body whorl wider **60**
60. Shell with a narrow, pointed spire (Fig. 14D); from CO east and northeast.
        Lymnaea stagnalis L.
    Shell with a wider spire **61**
61. Shell spire depressed; one rare species in ME lakes .... **Stagnicola**
    Shell more elongated (Fig. 14E); Great Lakes and St. Lawrence drainages.
        Bulimnea megasoma (Say)
62. Shell usually more than 13 mm long, sculptured with fine spiral striations; widely distributed and
        common; about 25 species .... **Stagnicola**
    Shell usually less than 13 mm long, sculpturing absent; about 12 species; common and widely
        distributed .... **Fossaria**
63. Animal and shell dextral; apex of shell sharply acute (Fig. 18N); 1 rare relict species in Rocky
        Mountain lakes. .... ACROLOXIDAE, **Acroloxus coloradensis** (Henderson)
    Animal and shell sinistral (Fig. 18); generally distributed .... ANCYLIDAE, **64**
64. Shell tall, apex in midline (Figs. 18J, K, P, Q); apex with pinkish coloration; southeastern states; 3
        species, of which 2 are probably extinct .... **Rhodacmea**
    Shell elevated or depressed; apex in midline or to the right, and same coloration as rest of shell; widely
        distributed **65**
65. Shell usually elevated (Fig. 18B); apex with fine radial striations (Fig. 18A); 5 species .... **Ferrissia**
    Shell usually depressed; apex not striated **66**
66. Apex distinctly to the right of the midline (Fig. 18L); FL to TX.
        Hebetancylus excentricus (Morelet)
    Apex almost in the midline; 2 species east of the Mississippi River .... **Laevapex**
67. Shell with a raised spire; mantle margin digitate or lobed (Fig. 4B) .... PHYSIDAE, **68**
    Shell discoidal, with a sunken spire (Fig. 17); mantle margin simple ...... PLANORBIDAE, **72**
68. Mantle edge digitate **69**
    Mantle without digitations, but there may be serrations **70**
69. With digitations on both sides of mantle (Fig. 4B); 2 species in northern states .... **Physa**
    With digitations on only one side of mantle; the most common and abundant genus everywhere (Figs.
        15D–G); about 30 species .... **Physella**
70. Mantle edge smooth and not extending beyond edge of shell apertural lip **71**
    Mantle edge serrated and partly overlapping the shell; 2 species introduced into TX .... **Stenophysa**
71. Shell spindle shaped (Fig. 15H); northern states in swales, intermittent streams, and stagnant pools.
        Aplexa elongata (Say)
    Shell globular; spire short; UT .... **Physella zionis** (Pilsbry)
72. Shell less than 8 mm in diameter **73**
    Shell more than 8 mm in diameter **85**
73. Shell heavily ridged, less than 3 mm in diameter (Fig. 17E); northern states ...... **Armiger crista** (L.)
    Shell not heavily ridged **74**
74. Shell less than 2 mm in diameter; known only from the Coosa River, AL **75**
    Shell more than 2 mm in diameter **76**
75. Shell limpetlike, with a small apical coil .... **Amphigyra alabamensis** Pilsbry
    Shell not limpetlike; 4 species .... **Neoplanorbis**
76. Shell greatly compressed; aperture without teeth **77**
    Shell not greatly compressed; aperture with teeth (Fig. 17J); widely distributed in eastern states.
        Planorbula
77. Shell either extremely flattened and multiwhorled or with numerous low ridges; 3 species in FL, TX,
        AZ; (sometimes divided into one *Antillorbis* and two *Fossulorbis*) .... **Drepanotrema**
    Shell not extremely flattened; without ridges **78**
78. Spire-pit shallow and wide **79**
    Spire-pit deep and narrow **80**
79. Height of body whorl rapidly increasing toward aperture (Figs. 17M–O); IL, MO, AR ...... **Menetus**

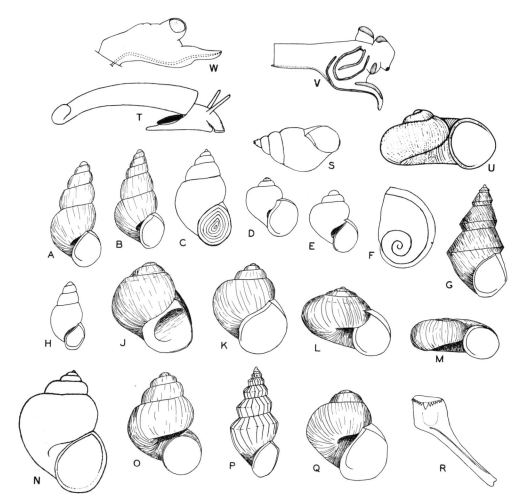

FIG. 21.–Typical Hydrobiidae, Pomatiopsidae, and Bithyniidae. A, *Fontigens,* ×7; B, *Pomatiopsis lapidaria,* ×4; C, *Bithynia tentaculata* with operculum, ×2.5; D, *Somatogyrus,* ×2; E, *Amnicola limosa,* ×4; F, operculum of *A. limosa* with fine striations and growth lines omitted; G, *Pyrgulopsis nevadensis,* ×5; H, *P. letsoni,* ×5; J, *Fluminicola columbiana,* ×2.5; K, *Gillia altilis,* ×2.5; L, *Cochliopina riograndensis,* ×3; M, *Hauffenia micra,* ×8; N, *Marstonia;* O, *Amnicola browni,* ×10; P, *Tryonia clathrata,* ×5; Q, *Clappia,* ×8; R, lateral tooth of *Clappia;* S, *Littoridinops,* ×10; T, *Orygoceras;* U, *Horatio nugax;* V, verge of *Cincinnatia;* W, verge of *Marstonia.* (R redrawn from Baker, 1928; N, T–W from Burch, 1979.)

Height of body whorl nearly equal from side to side (Figs. 17A–D); widely distributed and common.
Gyraulus
80. Shell with a ridged periphery 81
Shell periphery not ridged 83
81. Western species .... **Menetus**
East of the Continental Divide 82
82. Relative height of body whorl rapidly broadening toward the aperture; OH, AL ...... **Menetus**
Relative height of body whorl nearly equal from side to side; widely distributed ... **Promenetus**
83. Relative height of body whorl rapidly increasing toward aperture (Fig. 17M); widely distributed.
Promenetus
Relative height of body whorl nearly equal from side to side 84

84. Periphery of body whorl more or less angular; West Coast states .... **Menetus**
    Periphery of body whorl rounded; western, but reported also from OK, OH, NY .... **Menetus**
85. Shell thin, fragile; body whorl depressed **86**
    Shell thick; body whorl depressed or not **87**
86. Found in FL to TX and AZ; 2 species .... **Biomphalaria**
    Northern and western distribution .... **Planorbula**
87. Body whorl containing teeth (Fig. 17J) .... **Planorbula**
    Body whorl without teeth **88**
88. Shell with few whorls, but body whorl very large (Figs. 17K, L); two western species (=*Parapholyx*).
                                                                          Vorticifex
    Few to many whorls, but body whorl not especially large **89**
89. Shell spire strongly inverted (Figs. 17F, G); widely distributed; four species, especially in quiet waters.
                                                                            Helisoma
    Shell spire not strongly inverted; widely distributed; about 12 species .... **Planorbella**

REFERENCES

Abbott, R. T. (ed.). 1955. *How to Collect Shells. A Symposium.* 75 pp. American Malacological Union, Buffalo, New York.

Adamstone, F. B. 1923. The distribution and economic importance of Mollusca in Lake Nipigon. *Univ. Toronto Stud. Biol. Ser.* 22:67–119.

Aho, J. 1966. Ecological basis of the distribution of the littoral freshwater molluscs in the vicinity of Tampere, South Finland. *Ann. Zool. Fenn.* 3:287–322.

Aldridge, D. W. 1983. Physiological ecology of freshwater prosobranchs. In W. D. Russell-Hunter (ed.), *The Mollusca,* Vol. 6, Ecology, pp. 329–358. Academic Press, Orlando, Florida.

Baker, F. C. 1911. The Lymnaeidae of North and Middle America. *Spec. Publ. Chic. Acad. Sci.* 3:1–539.

_____. 1918. The productivity of invertebrate fish food on the bottom of Oneida Lake, with special reference to mollusks. *Ibid.* 9:1–233.

_____. 1928. *The fresh water Mollusca of Wisconsin. Part I. Gastropoda.* 507 pp. Trans. Wis. Acad. Sci., Arts and Lett. Madison, Wis.

_____. 1945. *The molluscan family Planorbidae.* Urbana, Illinois, 530 pp.

Basch, P. F. 1959. The anatomy of Laevapex fuscus, a freshwater limpet (Gastropoda: Pulmonata). *Misc. Publ. Mus. Zool. Univ. Mich.* 108:1–56.

_____. 1963. Environmentally influenced shell distortion in a fresh-water limpet. *Ecology* 44:193–194.

_____. 1963a. A review of the recent freshwater limpet snails of North America (Mollusca: Pulmonata). *Bull. Mus. Comp. Zool.* 129:401–461.

Berg, K., and K. W. Ockelmann. 1959. The respiration of freshwater snails. *J. Exp. Biol.* 36:690–708.

Berry, E. G. 1943. The Amnicolidae of Michigan: distribution, ecology, and taxonomy. *Mus. Zool. Univ. Mich. Misc. Publ.* 57:1–68.

Boss, K. J. 1974. Oblomovism in the Mollusca. *Trans. Am. Microsc. Soc.* 93:460–481.

Bovbjerg, R. V. 1968. Responses to food in lymnaeid snails. *Physiol. Zool.* 4:412–423.

_____. 1975. Dispersal and dispersion of pond snails in an experimental environment varying to three factors, singly and in combination. *Ibid.* 48:203–215.

Branson, B. A. 1970. Checklist and distribution of Kentucky aquatic gastropods. *Ken. Fish. Bull.* 54:1–20 (mimeographed).

Bryce, G. W. 1970. Rediscovery of the limpet, Acroloxus coloradensis (Basommatophora: Acroloxidae), in Colorado. *Nautilus* 83:105–108.

Burch, J. B. 1979. Genera and subgenera of recent freshwater gastropods of North America (north of Mexico). *Malacol. Rev.* 12:96–99.

_____. 1982. *Freshwater snails (Mollusca: Gastropoda) of North America.* EPA Report 600/3-82-026, 294 pp.

Burky, A. J. 1971. Biomass turnover, respiration, and interpopulation variation in the stream limpet Ferrissia rivularis (Say). *Ecol. Monogr.* 41:235–251.

Carriker, M. R. 1946. Morphology of the alimentary system of the snail Lymnaea stagnalis appressa Say. *Trans. Wis. Acad. Sci. Arts and Lett.* 38:1–88.

Chamberlin, R. V., and D. T. Jones. 1929. A descriptive catalog of the Mollusca of Utah. *Bull. Univ. Utah* 19:1–203.

Cheatum, E. P. 1934. Limnological investigations on respiration, annual migratory cycle, and other related phenomena in fresh-water pulmonate snails. *Trans. Am. Microsc. Soc.* 53:348–407.

Clampitt, P. T. 1970. Comparative ecology of the snails Physa gyrina and Physa integra (Basommatophora: Physidae). *Malacologia* 10:113–151.

Clarke, A. H. (ed.). 1970. Papers on the rare and endangered mollusks of North America. *Malacologia* 10:1–56.

_____. 1973. The freshwater mollusks of the Canadian interior basin. *Ibid.* 13:1–509.

_____. 1979. Gastropods as indicators of trophic lake stages. *Nautilus* 94:138–142.

Clench, W. J., and R. D. Turner. 1956. Fresh-water mollusks of Alabama, Georgia, and Florida from the Escambia to the Suwanee River. *Bull. Fla. State Mus. Biol. Sci.* 1:97–239.

Dawson, J. 1911. The biology of Physa. *Behavior Monogr.* 1:1–120.

Dazo, B. C. 1965. The morphology and natural history of Pleurocera acuta and Goniobasis livescens (Gastropoda: Ceritheacea: Pleuroceridae). *Malacologia* 3:1–80.

DeWitt, R. M. 1954. Reproduction, embryonic development, and growth in the pond snail, Physa gyrina Say. *Trans. Am. Microsc. Soc.* 73:124–137.

Gillespie, D. M. 1968. Population studies of four species of Molluscs in the Madison River, Yellowstone National Park. *Limnol. Oceanogr.* 14:101–114.

Goodrich, C. 1922. The Anculosae of the Alabama River drainage. *Mus. Zool. Univ. Mich. Misc. Publ.* 7:1–57.

———. 1924. The genus Gyrotoma. *Ibid.* 12:1–32..

———. 1936. Goniobasis of the Coosa River, Alabama. *Ibid.* 31:1–60.

———. 1941. Distribution of the gastropods of the Cahaba River, Alabama. *Mus. Zool. Univ. Mich. Occas. Pap.* 428:1–30.

Goodrich, C., and H. van der Schalie. 1939. Aquatic mollusks of the Upper Peninsula of Michigan. *Misc. Publ. Mus. Zool. Univ. Mich.* 43:1–45.

———. 1944. A revision of the Mollusca of Indiana. *Am. Midl. Nat.* 32:257–326.

Greenaway, P. 1971. Calcium regulation in the freshwater mollusc, Limnaea stagnalis (L.) (Gastropoda: Pulmonata). I. The effect of internal and external calcium concentration. *J. Exp. Biol.* 54:199–214.

Gunter, G. 1936. Radular movements in gastropods. *J. Wash. Acad. Sci.* 26:361–365.

Hanna, G. D. 1966. Introduced mollusks of western North America. *Occas. Pap. Calif. Acad. Sci.* 48:1–108.

Harman, W. N. 1972. Benthic substrates: their effect on fresh-water Mollusca. *Ecology* 53:271–277.

Harman, W. N., and C. O. Berg. 1970. Fresh-water Mollusca of the Finger Lakes region of New York. *Ohio J. Sci.* 70:146–170.

———. 1971. The freshwater snails of central New York with illustrated keys to the genera and species. *Search, Agriculture* 1:1–68.

Harman, W. N., and J. L. Forney. 1970. Fifty years of change in the molluscan fauna of Oneida Lake, New York. *Limnol. Oceanogr.* 15:454–460.

Henderson, J. 1924. Mollusca of Colorado, Utah, Montana, Idaho, and Wyoming. *Univ. Colo. Studies* 13:65–223.

———. 1929. Non-marine Mollusca of Oregon and Washington. *Ibid.* 17:47–190.

———. 1936. Mollusca of Colorado, Utah, Montana, Idaho, and Wyoming—Supplement. *Ibid.* 23:81–145.

———. 1936a. The non-marine Mollusca of Oregon and Washington. Supplement. *Ibid.* 251–280.

Herschler, R., and G. Longley. 1986. Phreatic hydrobiids (Gastropoda: Prosobranchia) from the Edwards (Balcones Fault Zone) aquifer, south-central Texas. *Malacologia* 27:127–172.

Herschler, R., and F. G. Thompson. 1987. North American Hydrobiidae (Gastropoda: Rissoacea): re-

description and systematic relationships of Tryonia Stimpson, 1965 and Pyrgulopsis Call and Pilsbry, 1886. *Nautilus* 101:25–32.

Hoff, C. C. 1940. Anatomy of the ancylid snail, Ferrissia tarda (Say). *Trans. Am. Microsc. Soc.* 59:224–242.

Hubendick, B. 1951. Recent Lymnaeidae. Their variation, morphology, taxonomy, nomenclature, and distribution. *K. Svenska Vetenska. Handl.* 3:1–225.

Jokinen, E. H. 1978. The aestivation pattern of a population of Lymnaea elodes (Say) (Gastropoda: Lymnaeidae). *Am. Midl. Nat.* 100:43–53.

Jones, W. C., and B. A. Branson. 1964. The radula, genital system, and external morphology in Mudalia potosiensis (Lea) 1841 (Gastropoda: Prosobranchiata: Pleuroceridae) with life history notes. *Trans. Am. Microsc. Soc.* 83:41–62.

Macan, T. T. 1950. Ecology of fresh-water Mollusca in the English Lake District. *J. Anim. Ecol.* 19:124–146.

Magruder, S. R. 1935. The anatomy of the fresh-water prosobranchiate gastropod, Pleurocera canaliculatum undulatum (Say). *Am. Midl. Nat.* 16:883–912.

Malone, C. R. 1965. Dispersal of aquatic gastropods via the intestinal tract of water birds. *Nautilus* 78:135–139.

McMahon, R. F. 1975. Effects of artificially elevated water temperatures on the growth, reproduction and life cycles of a natural population of Physa virgata Gould. *Ecology* 56:1167–1175.

———. 1983. Physiological ecology of freshwater pulmonates. In W. D. Russell-Hunter (ed.), *The Mollusca*, Vol. 6, Ecology; Academic Press, Orlando, Florida, pp. 359–430.

McNeil, C. W. 1963. Winter survival of Stagnicola palustris nuttalliana and Physa propinqua. *Ecology* 44:187–191.

Mikkelsen, P. S. 1985. A rapid method for slide mounting of minute radulae, with a bibliography of radula mounting techniques. *Nautilus* 99:62–65.

Murray, H. D., and E. C. Roy. 1968. Checklist of freshwater and land mollusks of Texas. *Sterkiana* 30:25–42.

Nekrassow, A. D. 1928. Vergleichende Morphologie der Laiche von Süsswasser-Gastropoden. *Z. Morphol. Oekol. Tiere* 13:1–35.

Noland, L. E., and E. Reichel. 1943. Life cycle of Lymnaea stagnalis completed at room temperature without access to air. *Nautilus* 57:8–13.

Pilsbry, H. E. 1934. Review of the Planorbidae of Florida, with notes on other members of the family. *Proc. Acad. Nat. Sci.* 86:29–66.

Ross, L. F., and A. D. Harrison. 1977. Effects of environmental calcium deprivation on the egg masses of Physa marmorata Guilding (Gastropoda: Physidae) and Biomphalaria glabrata Say (Gastropoda: Planorbidae). *Hydrobiologia* 55:45–48.

Russell-Hunter, R. D., and W. W. Lull. 1977. Physiologic and environmental factors influencing the calcium-to-tissue ratio in populations of three species of freshwater pulmonate snails. *Oecologia* 29:205–218.

Russell-Hunter, R. D. et al. 1984. Overwinter tissue degrowth in natural populations of freshwater pulmonate snails (Helisoma trivolvis and Lymnaea palustris). *Ecology* **65**:223–229.

Shoup, C. S. 1943. Distribution of fresh-water gastropods in relation to total alkalinity of streams. *Nautilus* **56**:130–134.

Taylor, D. W. 1981. Freshwater mollusks of California: a distributional checklist. *Calif. Fish and Game* **67**:140–163.

Walker, B. 1918. A synoposis of the classification of the freshwater Mollusca of North America, north of Mexico, and a catalogue of the more recently described species, with notes. *Mus. Zool. Univ. Mich. Misc. Publ.* **6**:1–213.

Wenz, W. 1938–1944. *Gastropoda. Band 1: Prosobranchia und Allgemeiner Teil.* 1639 pp. Koeltz, Koenigstein-Taunus, Germany.

———. 1960. *Gastropoda. Band 2: Euthyneura.* 834 pp. Koeltz, Koenigstein-Taunus, Germany.

Winslow, M. L. 1926. A revised check list of Michigan Mollusca. *Mus. Zool. Univ. Mich. Occas. Pap.* **181**:1–28.

Wood, D. H. 1982. The aquatic snails (Gastropoda) of the Savannah River plant, Aiken, South Carolina. *Publ. DOEs Savannah R. Plant, Nat. Environ. Res. Park,* 46 pp.

Wurtz, C. B. 1956. Fresh-water mollusks and stream pollution. *Nautilus* **69**:96–100.

25

PELECYPODA
(Clams, Mussels)

Unlike the Gastropoda, the bivalve mollusks are all aquatic, and although they occur in nearly all types of fresh-water habitats, they are most abundant and varied in our larger rivers.

General characteristics. The mussels, or clams, of our inland waters range from about 2 to 250 mm in length. The two valves of the shell are securely attached to each other dorsally by an elastic hinge ligament, and under natural conditions the valves gape slightly to permit the protrusion of the muscular hatchet- or axe-shaped foot at the anteroventral margin and the inhalent and exhalent siphons at the posterior margin. There are no tentacles, head, or eyes. The animal lies obliquely with the ventral half (or more) hidden in the substrate; in some species the shell may be entirely hidden from view with only the siphons appearing at the surface.

Locomotion. Mussels move over and through the substrate by means of a series of contractions of the intrinsic muscle fibers of the foot. The foot is first elongated and thrust forward in the substrate; then the distal portion of the foot becomes more or less swollen transversely so that it obtains a purchase, and at the same time the entire foot is shortened; as a result the main portion of the body and its enclosing shell are pulled forward slightly. These repeated muscular contractions therefore result in locomotion that is essentially a series of short "hunching" movements. Some species are known to move several feet per hour and the resulting long, troughlike tracks of larger mussels may often be seen on sandy lake bottoms where there is

little wave action. The stimuli responsible for extensive movements of clams are not generally known, but it is thought that stagnation and fall in water level are of primary importance. Thin-shelled species are generally more active than heavy-shelled species.

Shell, muscles. Mussel shells exhibit a variety of shapes. Some are decidedly elongated or oval; others are subcircular, rhomboidal, quadrate, trapezoidal, or subtriangular. At the dorsal margin of each valve, just anterior to the hinge ligament, is a raised area, the umbo or beak, which denotes the point at which growth began in the juvenile mussel. Surrounding each umbo and extending out to the edges of the valves is a series of concentric lines which are formed at the edge of the shell as the animal grows. At intervals these rings are more closely grouped, forming a more or less distinct ridge. In many species it can be shown that these markings clearly denote limits of annual growth and winter rest periods, especially if an adequate sample from a particular locality is studied. The annual growth-ring picture is often confused, however, by the presence of accessory rings formed in response to a variety of temporary unfavorable environmental conditions, such as lack of food, low oxygen supply, and fall in water level. In addition to these concentric markings, many genera bear pustules, rays, knobs, or wrinkles.

External coloration is whitish to brown in the Sphaeriidae, and light yellow, light green, dark green, brown, or blackish in other families. In older shells, especially near the umbones, the outer protective colored layer (periostracum) is often eroded and abraided away, exposing the chalky white calcium

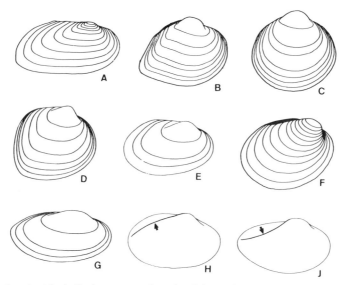

FIG. 1.–Typical unionid shell shapes. A, rhomboidal or elongated; B, triangular; C, circular; D, quadrate; E and F, oval; G, elliptical; H, posterior ridge convex; J, posterior ridge concave. (Modified from Burch, 1973.)

carbonate layer underneath. The internal surface of the shell is thickly coated with nacre, or mother of pearl, which ranges in color from silvery white through pink to dark purple; it is composed of extremely thin alternating laminae of calcium carbonate and an organic substance. Between the nacre and the periostracum is the prismatic layer; it consists of minute, closely packed, prismlike blocks of calcium carbonate (Fig. 4).

The important features of the inner surface of a typical shell are shown in Fig. 5. Near the dorsal margin there is usually a series of projecting and interlocking hinge teeth which aid in keeping the two valves in juxtaposition. In the Unionidae the pseudocardinal teeth are in the anterior part of the shell below the umbones, and the long, narrow, ridgelike lateral teeth are more posterior and below the region of the ligament. In the Sphaeriidae there are lateral teeth both anterior and posterior to the true cardinals.

The hinge teeth serve as a fulcrum, and the springlike hinge ligament normally keeps the shell slightly open. When the foot and siphons are withdrawn, however, the shell may be tightly closed by the contraction of the two large but short transverse muscles fastened to the inner surface of the valves, the anterior and posterior adductors.

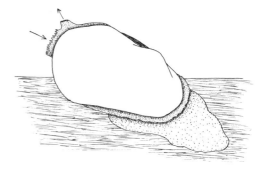

FIG. 2.–Female *Lampsilis siliquoidea* (Barnes) in natural position in river bottom, ×3. Arrows indicate currents of water. (Modified from Baker, 1928.)

In addition to the scars produced by the attachment of the adductor muscles, each valve typically contains the scars of three other muscles. Above, and more or less closely associated with the posterior adductor scar, is a small scar produced by the attachment of the posterior retractor of the foot. The anterior adductor has a comparable associated scar for the anterior retractor of the foot, as well as an additional scar for the protractor muscle of the foot. There are also

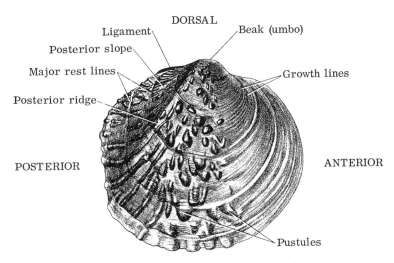

FIG. 3.–Shell structure of *Cyclonais*. (From Burch, 1973.)

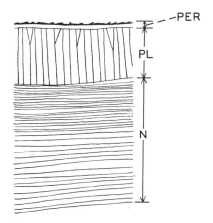

FIG. 4.–Small section of mussel shell. PER, periostracum; PL, prismatic layer; N, nacre.

some small and indistinct scars in the cavity of the umbo from muscles that aid in holding the visceral mass in place.

Except where muscles are attached, the inner surface of the shell is completely lined with a sheet of glandular tissue, the mantle. This structure is securely fastened near the edges of the valves along the muscular pallial line. Such an attachment prevents foreign particles from getting between mantle and shell. An indentation in the pallial line may

be present near the posterior end of the shell; it is called the pallial sinus and is produced by intrinsic muscles that retract the siphons. The entire lateral surface of the mantle secretes the nacre, but the periostracum and prismatic layer are formed only by the border of the mantle at the periphery of the shell.

Gross visceral anatomy. When one valve of a mussel is removed, the underlying mantle can be seen covering the viscera. At the posterior end of a typical animal the edges of the two lobes of the mantle are modified to form three slitlike openings. Uppermost is the smooth supra-anal opening; below this is the anal (exhalent) opening, which may be smooth or crenulated; lowermost is the distinctly papillose branchial (inhalent) opening. In the Unionidae and Margaritiferidae the edges of the mantle around the branchial opening project to form a more or less distinct siphon. Some Sphaeriidae have long, cylindrical, protrusible anal and branchial siphons (Fig. 7); in other species only the anal siphon is prominent. Neither the Margaritiferidae nor the Sphaeriidae have a supra-anal opening.

When the mantle is removed from one side, the two long flat gills of that side are

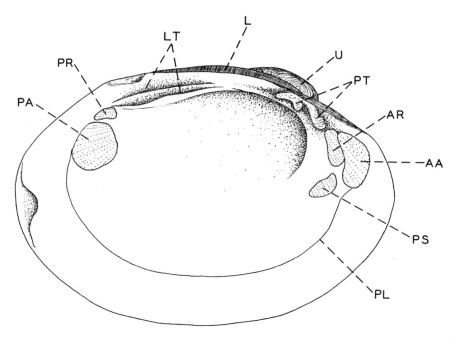

FIG. 5.–Diagrammatic view of inner surface of left valve of *Elliptio,* ×1. AA, anterior adductor muscle scar; AR, anterior retractor muscle scar; L, ligament; LT, lateral teeth; PA, posterior adductor muscle scar; PL, pallial line; PR, posterior retractor muscle scar; PS, scar of protractor muscle of foot; PT, pseudocardinal teeth; U, umbo.

exposed to view. Each gill is composed of two lamellae that are united at the lower but not the upper margin, the two gills together being roughly W shaped in cross section. The outer gill is connected at the outside and top to the mantle, and the inner gill is attached at the inside to the outer surface of the main median visceral mass; the base of the inner lamella of the outer gill and the base of the outer lamella of the inner gill are attached to each other. The cavity between the two lamellae of each gill is subdivided by vertical partitions into many narrow chambers or water tubes, closed below but opening above into a longitudinal space along the dorsal portion of the gill called the suprabranchial chamber. At the posterior end of the animal the suprabranchial chambers are fused to form a single cavity, the cloacal chamber. Near the anterior end of the gills and surrounding the small slitlike mouth are two flat labial palps on each side. The upper lip is continuous with the left and right outer palps, and the lower lip is

continuous with the left and right inner palps. The greater portion of the foot lies between the gills in the anterior part of the mantle cavity. It tapers posteriorly, and the median visceral mass of the posterior part of the animal is confined to the dorsal part of the mantle cavity.

Feeding, digestive system. The food of a mussel consists chiefly of fine organic detritus dislodged from the substrate and momentarily in suspension. Plankton forms only a minor element of the diet. The feeding processes are specialized for the removal of suspended microscopic particles from the water. It has been found that some Sphaeriidae can remove particles as small as 1 μm from the water. The inner surface of the mantle, the gills, and the visceral mass are covered with cilia that beat in such a coordinated way as to draw a stream of water in through the branchial siphon. This water passes into the mantle

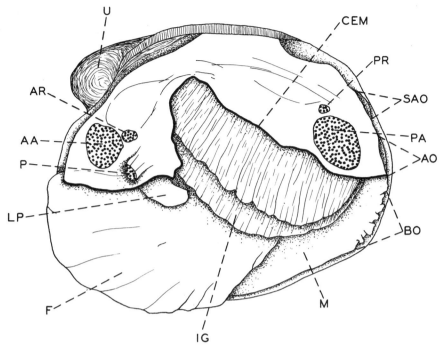

FIG. 6.–Diagrammatic partial dissection of *Fusconaia,* ×1.4. AA, anterior adductor muscle; AO, anal opening; AR, anterior retractor muscle; BO, branchial opening; CEM, cut edge of mantle; F, foot; IG, inner gill; LP, labial palps; M, mantle; P, protractor muscle of foot; PA, posterior adductor muscle; PR, posterior retractor muscle; SAO, supra-anal opening; U, umbo. (Modified from Lefevre and Curtis, 1912.)

cavity and enters the many minute openings, or ostia, of the gill lamellae into the gill chambers, then upward to the suprabranchial chambers, and finally posteriorly through the cloacal chamber and out the exhalent opening. As this continuous stream of water passes over the outer surface of the gills, however, the food particles become entangled in their mucous covering. The mucus with its contained food is driven by ciliary action along the edge of the gills toward the mouth where it is picked up by cilia on the labial palps and directed into the mouth in a constant ribbon-like stream.

Sphaeriidae burrowing well below the mud surface must rely completely on organic detritus for their food. Indeed, it has recently been established that subsurface mud sphaeriids have an interstitial suspension-feeding mechanism. As they burrow through the mud, organic particles and bacteria are placed in suspension momentarily, to be filtered out promptly and taken into the digestive tract.

FIG. 7.–*Musculium,* showing branchial and anal siphons, ×2. Arrows indicate currents of water. (Modified from Baker, 1928.)

Although some inorganic silt may be mixed with the organic food taken into the mouth, a large portion of the inedible material is separated out beforehand. The mechanisms involved in this selection and separation have been the subject of a long controversy, and it appears that the labial palps are most important, and that by some means much of the indigestible material is separated out on their surface and carried to the ventral edges where it drops off and is carried backward by ciliary action and expelled between the valves just below the inhalent siphon. It is likely that some additional selection and separation of food and inorganic debris is effected by the edge of the inhalent siphon and the mouth. The entire digestive tract is imbedded in the main visceral mass. The mouth opens into a short esophagus which leads to the more or less bulbous stomach situated in the anterior dorsal portion of the foot. The stomach is surrounded by a large green gland, or "liver." The long narrow intestine has several coils behind and below the stomach before it proceeds to the dorsal region and runs posteriorly above the posterior adductor muscle and opens into the mantle cavity through the anus just above the exhalent opening of the mantle cavity.

The anterior part of the intestine of the Unionidae has a lateral diverticulum, or groove, containing the curious crystalline style. This is a cylindrical structure of a hyaline, milky, or brownish color and of a dense gelatinous consistency. During normal feeding activities it is revolved on its long axis by the ciliary epithelium of the style sac. The anterior end of the style usually projects into the stomach, where it rubs against a cartilage-like shield and is constantly being eroded away. This erosion liberates a polysaccharide-digesting enzyme and is thought to function also in separating food from foreign particles. The style disappears during periods of starvation but is regenerated as a response to ingestion. It is formed only slowly at low temperatures.

Circulation. The circulatory system consists of a single ventricle, two auricles, blood vessels, and spaces or sinuses in the tissues. Dorsally and posterior to the stomach is the thin-walled pericardial cavity, containing the heart and through which the posterior part of the intestine passes longitudinally. The ventricle is an oval organ surrounding the intestine, and the two flat auricles lie at the sides of the ventricle and open into it just below the middle of each side.

Arising from the ventricle are an anterior and a posterior artery whose branches supply all of the soft parts of the animal, the mantle and gills being supplied with a particularly large quantity of blood. From the smaller arteries the blood passes into numerous sinuses in the tissues, and thence into veins and back to the auricles.

The blood is colorless or slightly bluish owing to dissolved hemocyanin, a respiratory pigment. The heart rate is usually less than 20 beats per minute, but a rate as high as 100 beats per minute has been recorded.

Respiration. Although the entire body surface is in contact with water and probably functions in respiration, the greater portion of the oxygen–carbon dioxide exchange undoubtedly occurs in the mantle and gills. Hibernating bivalves rely heavily on the edges of the mantle for respiration.

Excretion. The two kidneys, or organs of Bojanus, lie in the visceral mass immediately below or behind the pericardial cavity. Each kidney is a dark-colored more or less convoluted tubule with the opening at one end in the pericardial cavity and the opening at the other (external) in a suprabranchial chamber.

Nervous system. Unlike snails, mussels have a simple nervous system, with only three pairs of ganglia. The cerebropleural ganglia lie immediately posterior to the anterior adductor muscle, one on each side of the esophagus, and are connected by a short commissure passing over the esophagus; the pedal ganglia are deeply imbedded in the foot and are more or less fused; the visceral ganglia lie just anterior to the posterior adductor and are also partially fused. Connecting the cerebropleural ganglia with the pedal and visceral ganglia are two pairs of long commissures. Each ganglion is the source of nerve fibers to the adjacent organs.

Sense organs. Maintenance of equilibrium is effected by a pair of minute statocysts situated near the pedal ganglia. Each statocyst is an ovoid or spherical fluid-filled cavity lined with sensory cells and containing a spherical concretion, the statolith.

The osphradia are two small areas of specialized epithelium on the roof of the cloacal chamber. Their function has never been successfully established, but most writers

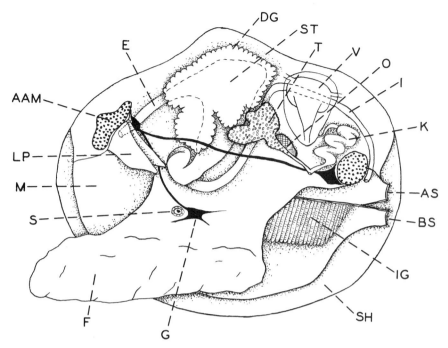

FIG. 8.–Diagrammatic longitudinal section of *Musculium,* ×12. AAM, anterior adductor muscle; AS, anal siphon; BS, branchial siphon; DG, digestive gland; E, esophagus; F, foot; G, pedal ganglion; I, intestine; IG, inner gill; K, kidney; LP, labial palp; M, mantle; O, ovary; S, statocyst; SH, shell; ST, stomach; T, testis; V, ventricle of heart.

assume that they are useful in detecting dissolved foreign materials in the water.

In addition to being sensitive to touch, the projecting edges of the mantle are capable of detecting pronounced changes in light intensity, and the valves can be caused to shut by casting a shadow or bright light on the mantle.

Reproduction. The Sphaeriidae are uniformly hermaphroditic and ovoviviparous, but in the Unionidae the sexes are separate except for a few species that are sporadically or consistently hermaphroditic. The ovaries and testes are imbedded in the upper part of the foot and are connected with the suprabranchial chambers by short ducts.

The details of reproduction differ markedly in these two families. In the Sphaeriidae it is thought that self-fertilization occurs in the reproductive ducts. The zygotes pass into the suprabranchial chambers and then downward into the water tubes of the inner pair of

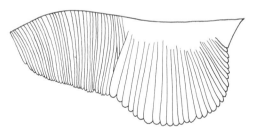

FIG. 9.–Outer gill of female *Lampsilis ventricosa* (Barnes), showing posterior marsupium.

gills. Here they develop and grow, and the gills, becoming greatly distended, are then called marsupia. An adult may contain from 1 to an astounding 60 young in various stages of development. When released from the marsupia, the immature individuals are fully formed, having the shell and all morphological features of adults; sometimes they are surprisingly large, often one quarter to one third as long as the parent. Reproduction is thought to continue throughout the year, although very few young are released during the winter months. Many species exhibit a wide range of phenotypic plasticity in the features of their life history.

Some Unionidae are dioecious, others are monoecious, and still others are herma-phroditic. The unfertilized eggs pass into the suprabranchial chambers and then downward into the water tubes of the gills. Depending on the species, all four gills, the two outer gills, or only special parts of the outer gills (Fig. 9) are utilized. Sperm of the male are swept out of the animal through the exhalent opening into the surrounding water. Some of the sperm are then drawn fortuitously into the incurrent siphon of a mature female in the vicinity; they pass through the ostia of the gill lamellae and fertilize the eggs in the water tubes. The embryos are retained only for the early stages of development, but growth is pronounced, and the gills (marsupia) soon become distended. The number of embryos present at one time ranges from several thousand in the smaller species to more than 3 million in some of the largest mussels.

Depending on the species, breeding may begin in the first to eighth year of life. The Ambleminae are short-term breeders and are gravid between April and August; the Unioninae are long-term breeders, the eggs being fertilized in midsummer and carried until the following spring or summer. Gametogenesis may continue throughout the year.

Marine bivalves and *Corbicula* have characteristic ciliated free-swimming larvae, but all other American fresh-water mussels have an entirely different type of larva, the glochidium (Fig. 10), which becomes a temporary and obligatory parasite on a fish. The mature glochidium, which is released from the marsupium of the female mussel ranges from 0.05 to 0.50 mm in diameter. Superficially, it appears to be a small edition of an adult in its general morphology. There are two chitinoid valves joined at their bases along a hinge, lined by a layer of cells comprising the larval mantle, and held together by a single adductor muscle. The mantle bears several tufts of sensory hairs. In species of *Unio, Anodonta,* and *Quadrula* the glochidium has a peculiar long thread that is thought to be useful in becoming attached to the host fish (Fig. 11A). The general shape of the valves may be simple and oval or rounded, roughly triangular, or ax-head shaped (in *Proptera* only). Sometimes the periphery of the shell opposite the hinge bears teeth.

In the Ambleminae and some Unioninae

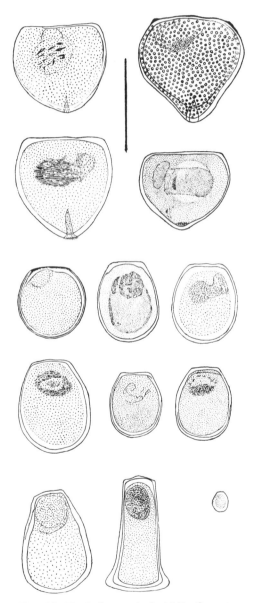

FIG. 10.–Typical mussel glochidia, face views. Scale line is 0.3 mm. (From Baker, 1928.)

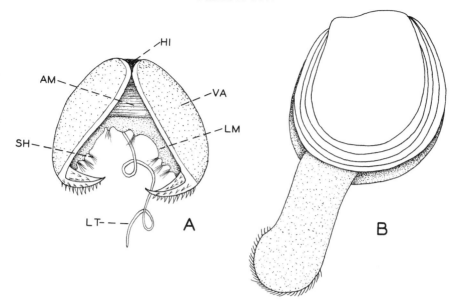

FIG. 11.–Immature Pelecypoda. A, end view of glochidium of *Anodonta,* ×130. B, lateral view of juvenile stage of *Actinonaias,* ×150. AM, adductor muscle; HI, hinge; LM, larval mantle; LT, larval thread; SH, tuft of sensory hairs; VA, valve. (A modified from Tucker; B modified from Lefevre and Curtis, 1912.)

the glochidia leave the female via the supra-branchial chambers and exhalent siphon; they may be free or more or less imbedded in mucus. In other Unioninae irregular masses break through temporary openings in the water tubes. In either case the glochidia become scattered and sink to the bottom where they remain with the valves gaping upward. If they do not come in contact with the body of a fish they usually die within several days. Fish become infected by brushing against the bottom, by stirring up the bottom with fin movements, and by taking in water that contains the glochidia in temporary suspension. The larvae are exceedingly sensitive and clamp tightly to the superficial tissues immediately on contacting the fish. The hookless species usually become attached to the gills, hooked species to the fins or general body surface. Some forms are host specific and infect only one species of fish; others are capable of infecting several to many species of fish. Sometimes the glochidia become attached to an unsuitable host but soon leave voluntarily.

Within a short time the tissues of the fish begin to grow over the glochidium, and in a day or two the latter is completely encased and forms a small projecting nodule. This parasitic stage usually lasts from 10 to 30 days, although some records indicate as long as 190 days; undoubtedly water temperatures influence the duration of the parasitic period.

During encystment the glochidial structures change markedly, and most of the adult organ systems become more or less definitely formed. The single adductor muscle is replaced by two; the mouth and intestine are present; the gills, kidneys, heart, and foot are represented by functional rudiments; and the mantle has taken on a more definitive structure. Usually, however, there is little increase in size.

At the close of the parasitic period the young mussel breaks out of its cyst, falls to the bottom, and enters a period appropriately called the juvenile stage (Fig. 11B). The organ systems begin to develop rapidly, and the beginning of the adult shell appears beneath the glochidial shell. The foot is long and ciliated and provides a means of locomotion. A long, sticky, hyaline thread, or byssus, is secreted by the posterior part of the foot in many species; it apparently facilitates attachment to the substrate and is usually shed in a few weeks. The smaller juvenile mussels burrow into the bottom and for this reason are seldom collected. As growth proceeds, however, the adult habits are assumed, and the organs take on their adult characteristics. Strictly speaking, the juvenile stage lasts from 1 to as much as 8 years, that is, until the animal is sexually mature.

Several species are known to be able either to complete development to the juvenile stage in the brood chamber or to leave the parent as a glochidium and parasitize a fish in the usual manner. One species, *Simpsoniconcha ambigua,* parasitizes *Necturus,* the mud puppy.

Even though more than 3000 glochidia have been found parasitizing a single adult fish, there is little evidence that they do any direct harm in natural habitats. Nevertheless, heavy infections (100 or more) have been shown to kill 30-mm fingerling trout by causing secondary bacterial infections.

Growth, longevity. The period of growth extends from about April through September, and the actual rate of growth is dependent on numerous factors, including temperature, food supply, current, and the specific chemical nature of the water. Mussels usually grow to a length of 30 to 80 mm in two growing seasons. Some of the Anodontidae live for 10 to 15 years, but most Sphaeriidae are thought to live no longer than 12 to 18 months. At least one species of *Sphaerium* has a complete life cycle in 1 month or less. At the other extreme, some sphaeriids live as long as 3 years.

Enemies. The Sphaeriidae are eaten regularly by numerous fishes, including the gizzard shad, whitefish, buffalo, sucker, redhorse, and perch, and soft-shelled species of the Anodontidae are eaten to a limited extent by catfish and sheepshead. Muskrats are undoubtedly the most important mammalian enemy of mussels. They usually select thin-shelled forms and either let them die on shore when the shell gapes, or force the shell open with their teeth. Piles of discarded shells near muskrat colonies are said to be an excellent source for some of the uncommon species. Mussels are also occasionally eaten by mink, otter, raccoon, turtles, and hell benders (*Cryptobranchus*).

The branchial and mantle cavities are often infested with parasitic or commensal Hydracarina. Larval flukes are sometimes found in the branchial and pericardial cavities. Both of these groups may be responsible for the production of small irregular pearls and blisters on the inner surface of the host shell, and sometimes the nacre may be discolored.

Economic importance. As a result of the utilization of marine shells, satisfactory synthetic substitutes, and the gradual depletion of mussel beds, the fresh-water pearl button industry has been declining in importance for many years, although the value of the products has held up well. The industry was founded in 1891, and by 1912 there were 196 factories in 20 states along the larger rivers of the Mississippi system and a few rivers entering the Gulf east of the Mississippi. In that year these plants produced $6,173,486 worth of merchandise. In 1941, with far fewer factories, the corresponding figure was about $2,500,000, but in 1944 it was $4,306,353, and in 1949 it was $3,696,452. About 20,300 tons of shell were harvested in 1944 with a value, before processing, of $892,915; Tennessee, Kentucky, and Arkansas rivers yielded 78% of the total.

There are about 40 useful species and 17 species of first importance in the manufacture of buttons; these are included in the following genera: *Fusconaia, Megalonaias, Amblema, Quadrula, Actinonaias, Plagiola, Ligumia,* and *Lampsilis.* The mussels are collected in several ways. One widely used device is the crowfoot dredge. It consists of a long iron bar to which are attached 100 to 200 four-pronged hooks on short lengths of cord. When the dredge is lowered from a flat-bottomed boat and dragged along the bottom, the hooks become lodged between the open valves of mussels, and the animals clamp tightly on them. At intervals the dredge is raised and the mussels are removed. Other devices include rakelike tongs, stout rakes with large attached bags or nets, and forks for shallow water collecting. Some states outlaw forking in the interests of conservation. Indeed, the harvesting of mussels is very carefully controlled by Federal and state governmental agencies.

The mussels are killed by steaming in a cooker, and the meat is removed. The shells are then sold by the ton to the buyers. In the factories or along the river's edge they are sorted and soaked in water before the blanks are cut with a tubular saw. The blanks are ground to uniform thickness, drilled, finished, sorted, and carded or packed for sale.

The beginning of World War II marked the demise of the pearl button industry, and subsequently plastics began to be used almost exclusively for buttons. In the early 1950s, however, the cultured pearl industry of Japan found that small pellets from American thick-shelled mussels formed excellent nuclei for cultured pearls, and the U.S. mussel industry had a revival, especially in the Ohio, Tennessee, and Green rivers. In the 1950s Tennessee, Alabama, Arkansas, Kentucky, and Indiana were the chief shell-producing states. During the 1950s and 1960s musselmen overharvested the natural production, so that by the 1970s the industry again had a demise. In the 1960s mussel shells sold for as much as $100 to $240 per ton. In 1966 the commercial mussel shell harvest in Illinois rivers amounted to more than 3500 tons.

Mussels are seldom collected primarily for their pearls, the fresh-water pearl industry being carried on in conjunction with, and in addition to, the preparation of shells for the button factories. The pearls are more or less imbedded in the soft tissues, and after removal from the shell these parts are examined visually and slipped between the fingers to detect hidden pearls. Small pearls are sometimes recovered by allowing the meats to rot. When reduced to a pulp, the mass is rubbed through a fine sieve that retains the pearls.

Pearls are formed by the secretion of nacre around a foreign body that becomes lodged against the mantle or other tissues. This secretion takes place in a sac lined with epithelium physiologically and histologically similar to the shell-secreting epidermis of the mantle. The foreign body may be a bit of debris, sand grains, or even some kind of a parasite. Though the layman usually thinks of pearls as being "balls," "pears," or "buttons," there are many other irregular shapes habitually produced by certain species. Some of these are called "petal," "lily," "leaf," and "arrowhead," according to their shape. Such curious but natural shapes are determined by pressures and movements of surrounding organs during their formation. Spherical pearls, on the other hand, occur only in a part of the body that is free from unequal pressures and extensive movements; button pearls are formed near the outer surface of the mantle, where they are pressed against the shell while nacre deposition is occurring.

The formation of real freak pearls, or "baroques," usually instigated by the presence of a parasite or by a button or ball pearl becoming lodged in a new location. Baroques may be either attached to the shell or free. They are of irregular shape, but rare specimens may simulate a bird, insect, other object, or even a face. Single unusual and large baroques and perfectly formed regular pearls command high prices, hundreds and even thousands of dollars having been paid for many specimens. One Arkansas pearl, weighting 3.4 g, is reputed to have been sold for $6750. Unfortunately, the finding of really fine pearls is an event of increasing rarity because of the great depletion of our mussel resources.

The reported total value of mussel pearls collected in the entire Mississippi River drainage system in 1931 was only $11,436.

In view of our rapidly declining mussel populations, U.S. government research workers have intermittently been trying to develop effective and economical methods of artificially propagating mussels ever since 1894.

Corbicula fluminea, introduced from Asia, sometimes is so abundant as to be a nuisance by plugging waterways in pumping plants, irrigation systems, and the production of sand and gravel. There has been some success in eradicating this species by the addition of 1 ppm of chlorine to the water. In some areas the juveniles are consumed by crayfish. *Rangia* is occasionally a nuisance in industrial water systems.

Ecology. Fresh-water bivalves occur in all types of unpolluted habitats but they are most abundant and varied in the larger rivers. Small creeks and spring brooks are devoid of mussels, although they contain Sphaeriidae. Similarly, small lakes, especially seepage lakes, contain few species, and large drainage lakes contain many. Lake Pepin, for example, which is really a wide place on the Mississippi, contains 32 species of mussels.

Like snails, bivalves are most abundant in the shallows, especially in water less than 2 m deep. In the largest rivers and some lakes, however, mussels occur as deep as 7 m. The record depth collection is *Anodonta,* which was taken at 30 m in Lake Michigan. The specimens were 7 to 14 years old but less than 53 mm long. Sphaeriidae are regularly found in great abundance on some lake bottoms more than 30 m below the surface. These small bivalves, especially the immature ones, are often active burrowers and have been taken up to 25 cm deep in soft substrates. Under such conditions the sphaeriids are said to be in a "resting stage."

The occurrence of the largest species and largest numbers of mussels in rivers where there is a good current has not yet been satisfactorily explained from an ecological standpoint. It is possible that abundant food supply, the specific type of substrate, an abundance of oxygen, water chemistry, or a combination of factors may be significant. River genera have thicker shells than lake genera, and individuals of a single species inhabiting both types of environments usually have heavier shells in running waters. Also, in a single species, flatter and less inflated varieties may occur in upper tributaries, more swollen varieties in the lower parts of a river system. Mussels inhabiting exposed lake shoals may be somewhat stunted and more heavy-shelled than specimens from protected waters.

Bare rock bottoms and shifting sands and muds, with their accompanying high turbidities, are unsuitable for bivalves, whereas stable gravel, sand, and substrates composed of sand or gravel mixed with other materials support the largest populations, some such beds having more than 50 mussels per square meter. Though soft mud bottoms are generally uninhabited, there are some common "mud-loving" species of *Anodonta.* Customarily, mussels inhabit substrates free of rooted vegetation, but there are numerous exceptions. The Sphaeriidae are less specific in their occurrence and are found on all types of bottoms except clay and rock; in favorable situations populations of more

than 10,000 individuals per square meter may be counted.

To a large extent the geographical distribution of the fresh-water bivalves appears to be determined by the chemical features of waters. The Anodontidae are rarely found in acid waters (pH below 7.0) or in waters having a bound carbon dioxide content of less than 15.0 mg/l. Decidedly alkaline waters and an abundance of calcium carbonate for shell construction favor large numbers of individuals and species. The Sphaeriidae, on the other hand, have become adapted to a wider range of conditions, and some species are common in lakes having a pH as low as 6.0 and a bound carbon dioxide content of only 2.0 mg/l. It is difficult to understand how a calcium carbonate shell can be secreted and maintained under such unfavorable conditions. Salts appear to be actively absorbed directly from the water. Biomass calculations sometimes show that the calcium in unionid shells may be 50% as much as the total calcium dissolved in the water of a lake in which the clams occur. A few Sphaeriidae species are indicators of lake chemistry trophic conditions, but, at least in the Sphaeriidae, there is little correlation between $CaCO_3$ in the shell and water hardness of the habitat. Potassium seems to be relatively toxic, and waters having more than 7 mg/l of potassium are generally devoid of mussels.

Because of the apparent disappearance of mussels from the shallows during the winter months, it was formerly inferred that they migrated to deeper waters in the fall. It is now believed, however, that migrations are unimportant and that most individuals simply burrow more deeply into the bottom of their summer habitat so that only the inconspicuous edges of the siphons project. The siphons are opened infrequently, and the animals remain in a dormant state throughout the winter. Ten degrees is thought to be an average critical hibernation temperature.

Many Sphaeriidae are known to burrow into the drying substrate in times of drought, and among the Unionidae, species of *Uniomerus, Carunculina, Ligumia,* and *Amblema* have all been found buried in substrates which were moist or dry for several months.

Fresh-water mussels are thought to have first developed in the New World, and, more specifically, in the general area of the Mississippi drainage basin, for it is here that the greatest diversity of species occurs. *Anodonta* is found also in Europe and Asia, but all other U.S. genera of Anodontidae are endemic. The Margaritanidae are widely distributed throughout the Northern Hemisphere, and the Sphaeriidae are worldwide. The other families mentioned in this chapter are predominantly marine, only a few species having invaded brackish waters and the coastal areas of rivers.

The Mississippi drainage includes more than half of the area of the United States, from the Continental Divide on the west to the Atlantic region on the east, and from Canada to the Gulf. It has an extremely rich naiad fauna. The Upper Mississippi River system alone has about 50 naiad species. The Pacific region includes those drainages entering the Pacific Ocean and is characterized mainly by the presence of *Anodonta, Gonidea,* and *Margaritifera.*

Not counting many additional varieties, there are about 230 species of Unionidae and perhaps 45 species of Sphaeriidae known from the 48 contiguous states. The Cumberland River has (had) more than 80 species of Unionidae; Michigan has 66 Unionidae; Dane County, Wisconsin, has 34; Ohio has 70; Kansas about 70; Texas has 78; Oklahoma 67; Big Black River, Tennessee, 31 now; California only 24. The known naiad fauna of the Great Lakes comprises 6 species from Lake Superior, 10 from Michigan, 17 from Huron, 40 from Erie, and 6 from Ontario. The Atlantic coast states, Great Plains, Rocky Mountain states, and the Pacific coast area have few species of Anodontidae. The Olympic Peninsula has only 2 unionids and 4 sphaeriids. The literature contains many additional papers listing species of bivalves by state of by drainage system. By way of contrast, all of Europe has only 15 unionids.

By migrating naturally and extensively into suitable habitats, a species may, over a long period of time, come to occupy much of a large drainage area. More rapid migration is undoubtedly effected by fishes carrying the glochidial stage of mussels, and by birds and large insects transporting Sphaeriidae and immature Unionidae overland in mud or debris on their legs and feet and in the feathers.

Greater numbers of species are usually found in the lower waters of river systems rather than in the small tributaries. The greater abundance of species in large, shallow lakes is probably due to the fact that such environments have a greater variety of subhabitats or niches than small lakes.

Although sphaeriids are common, there are no mussels in high Rocky Mountain lakes, probably chiefly because these lakes do not contain the proper host fishes for the glochidia. Fundamentally, however, such lakes are usually poor in calcium and could not support mussels anyway.

Pollution, silting of rivers, shell harvesting, and the construction of dams are decidedly inimical to our American mussel fauna. Many of the less common species are fast disappearing, and reduced populations appear to be unsuccessful in increasing their present restricted geographical distribution.

Recent surveys, compared with early surveys, give some idea of the rate at which our fauna is becoming depauperate. Isom (1969), for example, found that annual harvests in the Tennessee River decreased from 10,000 tons of shells in the 1940s and 1950s to 2000 tons from 1964 through 1967. The present mussel fauna in this river now consists of only 44 species, as contrasted with 100 species before the 1936 Tennessee Valley Authority impoundment. Matteson and Dexter (1966) report a "drastic decline" in the mussel fauna of the Big Vermilion River between 1918 and 1962. Isom and Yokley (1968) report only 48 species in the Duck River, Tennessee, as contrasted with 63 reported originally by Ortmann. In the Illinois River, Starrett (1971) found only 24 species in 1966–1969 (5 species represented by single specimens) as contrasted with 45 in 1906–1912. Suloway (1981) found only 15 unionids in the Kankakee River, Illinois, in contrast to 29 species reported in 1906. Only three mussel beds in the Kentucky River are now said to be commercially valuable. Since it usually takes a long time for field data to make their way into the literature, it is probable that the commercial mussel situation in the United States is much more serious than we realize.

Collection, preparation. Little special apparatus is necessary for field collecting. The larger species may easily be handpicked in the shallows; small forms may be washed from vegetation or sifted from bottom debris. A long-handled rake may be used for collecting mussels in water of moderate depths, and dredges are necessary for sampling in deep lakes.

If mussels are placed in warm water and heated to boiling, the valves will separate, the viscera can be easily removed, and the epidermis will not crack so easily. The judicious use of dilute oxalic or hydrochloric acid and a toothbrush are valuable for removing surface incrustations and restoring the natural brilliance of the nacre. After treatment with acid, shells should be rinsed thoroughly in clear water. The surface of study specimens should be varnished or greased lightly with petroleum jelly before the two valves are tied together. The Sphaeriidae should be placed in 40% alcohol for a day or two, the viscera should then be removed, and the shell wrapped in tissue paper to prevent gaping while they are allowed to dry thoroughly.

Preservation for the study of soft parts should be in 70 or 80% alcohol, not formaldehyde.

Taxonomy. Except for *Margaritifera* and a few representatives of marine families that have invaded brackish waters and coastal regions of rivers, the fresh-water bivalves of the United States are all included in the Unionidae and Sphaeriidae. In the former family, the subfamilies and the great majority of genera and species are well established, although some forms have a lengthy synonymy dating back more than a century. Many species have several subspecific designations, some of which are useful ecologically. According to Baker (1928), for example, *Strophitus rugosus* (Swainson) occurs in large and medium rivers, *S. rugosus pavonius* (Lea) is a creek form, *S. rugosus pepinensis* Baker a river–lake form, *S. rugosus winnebagoenis* Baker a small river–lake form, and *S. rugosus lacustris* Baker a small lake form. *Quadrula quadrula* (Raf.) is even more variable.

To the pearl fisherman and button industry, mussels are known by highly descriptive vernacular names. Some of these are niggerhead, pimple back, pig toe, washboard, pocketbook, elephant's ear, slop bucket, spectacle case, paper shells, three ridge, maple leaf, warty back, monkey face, lady finger, floater, and snuff box.

During the past 40 years many of the longstanding taxonomic and nomenclatural problems of our fresh-water bivalves appear to have been resolved. In general, this key follows the suggestions of Burch and other workers, but is highly modified and simplified from these sources.

KEY TO GENERA OF PELECYPODA (=BIVALVIA)

1. Ligament external 2
   Ligament internal; in mild to strongly brackish waters 8
2. Shell with true cardinal teeth (Figs. 12B, 34) 5
   Shell with well-developed pseudocardinal teeth, occasionally vestigial or absent (Figs. 12A, 17A, 18A). 3
3. Shell elongated, laterally compressed (Fig. 12A); dorsoposterior mantle margins not united to form a separate anal opening; epidermis blackish; 80 to 175 mm long; lotic.
   MARGARITIFERIDAE, **4**
   Shell subcircular, oval, subtriangular, or elongated; dorsoposterior mantle margins forming one or more separate openings; edge of mantle thickened between incurrent and excurrent openings; lotic and lentic; pearly mussels or naiads .... UNIONIDAE, **9**
4. Shell thin and elongated (Fig. 14B); Cumberland and Tennessee river systems, and OH, IL, IA, WI, NE .... **Cumberlandia monodonta** (Say)
   Shell thicker, not elongated (Fig. 12A); *M. margaritifera* (L.) in northern third of states from PA to upper Missouri River system; *M. falcata* (Gould) in Pacific drainages; *M. hembeli* (Conrad) in AL, LA; *M. marrianae* Johnson in AL .... **Margaritifera**
5. Hinge with anterior and posterior lateral teeth (Figs. 34F, G; 12B) 6
   Hinge without lateral teeth; shell circular and thin; 10 to 15 mm long; a single species in brackish waters of FL .... CYRENELLIDAE, **Cyrenella floridana** Dall

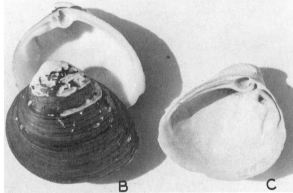

FIG. 12.–Representative Pelecypoda. A, *Margaritifera margaritifera,* ×0.7; B, *Polymesoda caroliniana,* ×0.8; C, *Rangia cuneata,* ×0.8; D, *Mytilopsis leucophaeta,* ×1.4.

6. Shell thin and fragile, length ranging from 2 to 20 mm but rarely more than 10 mm; concentric growth rings not prominent; pallial line simple; widely distributed and common; seed shells, pea, pill, and fingernail clams .... SPHAERIIDAE, **66**

    Shell heavy, more than 10 mm long, with prominent concentric growth rings (Figs. 12B, 14A); pallial line sinuate **7**

7. More than 25 mm long (Fig. 12B); a single species in streams and brackish waters near the coast from SC to TX .... CYRENIDAE, **Polymesoda caroliniana** (Bosc)

    Ten to 50 mm long; periostracum blackish (Fig. 14A); introduced from Asia and spreading in a phenomenal fashion; sloughs and streams; from CA north to WA and UT and east to FL, and also in the greater Mississippi River drainage; sometimes called *C. manilensis* or *C. laena;* able to tolerate 13⁰/₀₀ salinity, and prefers fine sand substrates.

                                            CORBICULIDAE, **Corbicula fluminea** (Müller)

8. Hinge with cardinal and lateral teeth; 25 to 70 mm long (Fig. 12C); shell thick; DE, MD to FL to TX.

                                                MACTRIDAE, **Rangia cuneata** Gray

    Hinge without distinct teeth; shell mytiliform; with a byssus; 10 to 15 mm long (Fig. 12D); MD to FL.

                                     DREISSENIIDAE, **Mytilopsis leucopheata** Conrad

9. Marsupium formed of all four gills, swollen in gravid females .... AMBLEMINAE, **10**

    Marsupium formed of only the outer gill on each side, swollen in gravid females.

                                                   UNIONINAE, **22**

FIG. 13.–Representative Unionidae. A, *Anodonta,* ×0.7; B, *Arcidens confragosus,* ×0.7.

10. Hinge teeth well developed 11
    Hinge teeth rudimentary or lacking; shell smooth, elongated, and usually with a high, sharp, posterior ridge (Fig. 17A); Pacific coast drainages .... **Gonidea angulata** (Lea)
11. Shell corrugated posteriorly (Fig. 14C) 12
    Shell not corrugated posteriorly (Figs. 14D, 17B) 20
12. Posterior edge steep and short (Fig. 14C); AL .... **Quadrula stapes** Lea
    Posterior edge not especially steep and short (Fig. 15) 13
13. Shell pustulated (Figs. 14E, 16A) 14
    Shell not pustulated (Fig. 14D) but may be ridged 16
14. Shell essentially circular (Fig. 16A); Tennessee River .... **Quadrula intermedia** (Conrad)
    Shell elongated 15

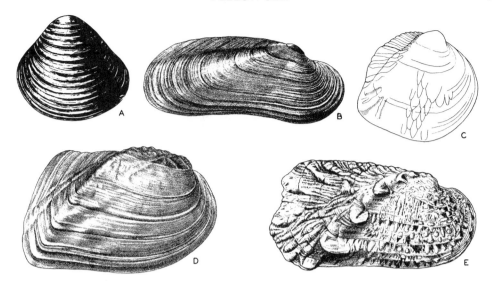

FIG. 14.–Representative Pelecypoda. A, *Corbicula manilensis,* ×1.3; B, *Cumberlandia monodonta,* ×0.5; C, *Quadrula stapes;* D, *Plectomerus dombeyanus,* ×0.5; E, *Quadrula cylindrica,* ×0.5. (C from Burch, 1975, by permission; A, B, D, and E from Burch, 1973.)

FIG. 15.–*Tritogonia verrucosa.* A, female, ×0.6; B, male, ×0.6.

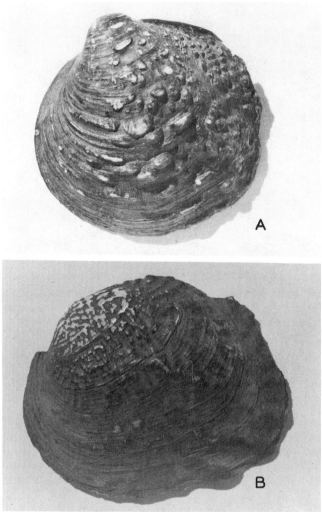

FIG. 16.–Representative Unionidae. A, *Quadrula,* ×0.9; B, *Megalonaias gigantea,* ×0.9.

15. Shell with a well-developed posterior ridge; pustules small (Fig. 15); MS and Gulf drainages.
Tritogonia verrucosa (Raf.)
Posterior ridge low; with a row of large pustules (Fig. 14E); Ohio, Cumberland, and Tennessee river systems, but west to NE, OK . . . . **Quadrula cylindrica** (Say)

16. Shell elongated, rhomboidal (Fig. 14D); nacre violet to bronze 17
Shell elongated, oval to circular (Figs. 16B, 18A); nacre white 18

17. Anterior adductor muscle scar smooth; GA, FL, KY . . . . **Elliptoideus sloatianus** (Lea)
Anterior adductor muscle scar rough; AL to TN, TX . . . . **Plectomerus dombeyanus** (Val.)

18. Shell less than 7 cm long, smooth; FL to TX; three species . . . . **Quincuncina**
Shell large and corrugated (Fig. 16B) 19

19. Beak sculptured with strong zigzag lines; shell elongate quadrate, very heavy; 80 to 250 mm long (Fig. 16B); in large, deep rivers; Mississippi system generally and from AL to TX.
Megalonais gigantea (Barnes)
Beak not sculptured (Fig. 18A); eastern half of the United States; small to large rivers; 2 species.
Amblema

FIG. 17.–Representative Unionidae. A, *Gonidea angulata,* ×0.9; B, *Fusconaia,* ×0.8.

20. Shell rhomboidal, with or without a posterior ridge (Fig. 14D) 17
    Shell circular, triangular, or oval (Figs. 14E, 17B) 21
21. Shell pustulose (Fig. 14E); Mississippi drainage as far west as MN, TX; about 8 species .... **Quadrula**
    Shell smooth (Fig. 17B); common and generally distributed in eastern half of the United States; about
    12 species .... **Fusconaia**
22. Glochidia hooked (Fig. 11A); each vertical water tube of gill of gravid female subdivided into three
    compartments (Fig. 20A) 31
    Glochidia not hooked; vertical water tubes of gills of gravid female not subdivided 23
23. Each outer gill (marsupium) swollen throughout its length; shells of males and females similar 24
    Each outer gill locally swollen (usually the posterior half or in alternate pouches) (Fig. 10); shell
    sexually dimorphic 44
24. Glochidia present only in fall, winter, and spring; 2 species; southern FL and southern TX.
                                                                Popenaias
    Glochidia present only in summer 25

Fɪɢ. 18.–Representative Unionidae. A, *Ambelema,* ×0.8; B, *Cyclonaias tuberculata,* ×1.

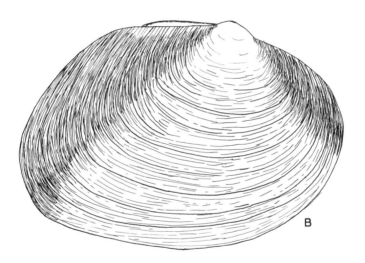

FIG. 19.–Representative Unionidae. A, *Plethobasus,* ×1; B, *Lexingtonia,* ×1.8.

29. Shell elongate, rhomboidal; beak low, not arched forward (Figs. 26B, C); many species in a wide variety of lotic and lentic habitats throughout the eastern half of the United States; about 20 species .... **Elliptio**
    Shell another shape; beak high, arched forward (Figs. 19B, 20D) 30

30. Outer gills (marsupia) swollen, orange or red; Tennessee River drainage and North River, VA; 2 species .... **Lexingtonia**
    Outer gills not swollen; white, gray, yellow, or pale orange; about 30 species, mostly in the southeastern states, but a few in drainages in the northeastern quarter of the United States.
                                                                   Pleurobema

31. Hinge without teeth, or only with pseudocardinals present (Figs. 13A, 20E, F); shell thin, fragile 32
    Hinge teeth present, occasionally poorly developed (Fig. 21B); shell not fragile 41

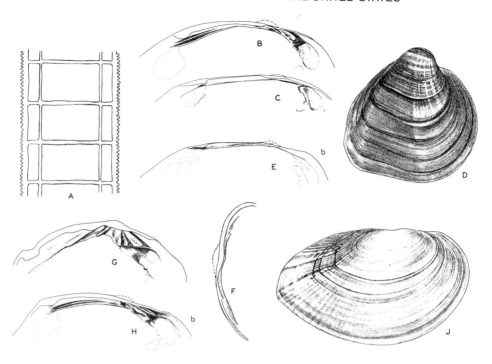

FIG. 20.–Structure of Unionidae. A, horizontal section of a bit of marsupium of *Lasmigona* showing each main water tube divided into three chambers; B, teeth of *Uniomerus tetralasmus*; C, teeth of *Hemistema lata*; D, *Pleurobema*, ×0.6; E, teeth of *Anodonta*; F, teeth of *Alasmidonta*; G and H, teeth of *Lasmigona*; J, *Anodonta*, ×0.5. (A modified from Burch, 1973; B–E, G–J from Burch, 1973; F redrawn from Burch, 1975.)

32. Pseudocardinal teeth absent (Figs. 13A, 20E) 33
    Pseudocardinal teeth rudimentary (Fig. 20F) 37
33. Restricted to drainages east of the Continental Divide 34
    Restricted to Pacific drainages .... **Anodonta**
34. Umbos extending only slightly or not at all above the dorsal margin (Fig. 20J); widely distributed and common .... **Anodonta**
    Umbos extending well above the dorsal margin (Fig. 13A) 35
35. Beak sculpture double looped (Fig. 22A); widely distributed and common .... **Anodonta**
    Beak sculpture concentric (Fig. 22B) 36
36. Beak sculpture coarse (Fig. 22C); widely distributed .... **Strophitus undulatus** (Say)
    Beak sculpture fine (Fig. 22D); widely distributed .... **Anodontoides ferussacianus** (Lea)
37. Pseudocardinal teeth thin, bladelike (Fig. 20F) 38
    Pseudocardinal teeth tubercular (Fig. 21D) 39
38. Shell rhomboidal, with a posterior ridge (Fig. 22F); St. Lawrence River to SC.
                                Alasmidonta varicosa (Lamarck)
    Shell elongated (Fig. 21A); AL, GA, FL .... **Anodontoides radiatus** (Conrad)
39. Posterior slope slightly corrugated (Fig. 22F); eastern states; common .... **Alasmidonta**
    Posterior slope not corrugated 40
40. Shell about twice as long as high (Fig. 23B); Ohio River system, extending to AR, IA, MI, TN.
                                  Simpsoniconcha ambigua (Say)
    Shell length/height ratio 1.8 or less (Fig. 21D); FL to MS .... **Strophitus subvexus** (Conrad)
41. Shell with large corrugations on either or both the central disc or posterior slope (Figs. 13B, 22H) ... 42
    Shell without large corrugations in these areas (Figs. 21B, C) 43
42. Central disc not sculptured (Fig. 22H); AR, OK ..... **Arcidens wheeleri** (Ortmann and Walker)
    Central disc sculptured (Fig. 13B); states adjoining Mississippi River, TX, AL.
                                  Arcidens confragosus (Say)

FIG. 21.–Representative Unionidae. A, *Anodontoides,* ×0.9; B, *Alasmidonta,* ×0.9; C, *Alasmidonta,* dorsal, ×0.7; D, *Strophitus,* ×0.9.

FIG. 22.–Structure of Unionidae. A, double-looped beak sculpture of *Anodonta*; B, concentric beak sculpture; C, coarse beak sculpture of *Strophitus*; D, fine beak sculpture of *Anodontoides ferrusacianus*; E, pseudocardinal teeth of *Alasmidonta*; F, *Alasmidonta varicosa*, ×0.8; G, *Alasmidonta*, ×0.5; H, *Arcidens wheeleri*, ×0.3. (From Burch, 1973.)

FIG. 23.–Representative Unionidae. A, *Lasmigona*, ×0.7; B, *Simpsoniconcha ambigua*, ×1.7.

FIG. 24.–Representative Unionidae. A, *Uniomerus,* ×1.7; B, *Ptychobranchus,* ×0.8; C, *Obliquaria reflexa,* ×0.8; D, *Cyprogenia,* ×1; E, female *Medionidus,* ×1.3.

FIG. 25.–*Lampsilis*. A, male, ×1; B, female, ×1.2.

43. Beak sculpture concentric (Fig. 22 B); widely distributed east of the Continental Divide, but especially abundant in southeastern states; about 12 species (Figs. 21 B, C) .... **Alasmidonta**
    Beak sculpture double-looped (Fig. 22 A); six widely distributed species in eastern half of the United States (Fig. 23 A) .... **Lasmigona**
44. Entire outer gill serving as a marsupium 45
    Marsupium confined to the central or posterior part of the outer gill, as shown by swollen portion.
                                                                                                      46
45. Ventral edge of marsupium not folded; southeastern TX .... **Cyrtonaias tampicoensis** (Lea)
    Ventral edge of marsupium folded; 5 species; Ohio River system and south to FL and west to OK (Fig. 24 B) .... **Ptychobranchus**
46. Marsupium confined to middle part of outer gill 47
    Marsupium confined to posterior part of outer gill (Fig. 9) 48
47. Shell with several large lateral tubercles (Fig. 24 C); Mississippi River system.
                                                                      Obliquaria reflexa Raf.
    Shell without such large tubercles (Fig. 24 D); 2 species; Ohio, Cumberland, and Tennessee river systems, and west to OK, KS .... **Cyprogenia**
48. Marsupium confined to lower part of posterior part of outer gill; Tennessee and Cumberland river systems .... **Dromus dromas** Lea
    Marsupium in both upper and lower halves of posterior part of outer gill 49

FIG. 26.–Representative Unionidae. A, *Pleurobema,* ×0.7; B, *Elliptio,* ×0.9; C, dorsal view of *Elliptio,* ×0.8.

49. Posterior slope corrugated or radially grooved (Fig. 27A) 50
Posterior slope smooth (Figs. 28A, D) 51
50. Shell with radiating grooves attaining the posteroventral margin; Tennessee River drainage.
Lemiox rimosus Raf.
Shell with another kind of sculpturing (Figs. 24E, 27A); AL, FL, MS; 5 species .... **Medionidus**
51. Posterior half of pseudocardinal teeth forming a series of deep, parallel, rough lamellae (Fig. 27C);
slow streams; FL to TX .... **Glebula rotundata** (Lamarck)
Without such deep lamellae 52
52. Shell high, compressed, and with a smooth arc (Fig. 28A); hinge teeth very large and heavy; PA to MN
and AL to OK .... **Ellipsaria lineolata** Raf.
With another combination of characters 53
53. Females with a wartlike caruncle on the inner edge of each side of the mantle in front of the branchial
opening; total length of adults less than 45 mm (Fig. 28D); probably 2 species; NY to MN and
south to FL, AR .... **Carunculina**
Females without such a caruncle; total length of adults less than 45 mm 54
54. Shell twice or more as long as high 55
Shell less than twice as long as high 56

FIG. 27.–Representative Unionidae. A, *Medionidus,* ×1.2; B, *Ellipsaria,* ×0.6; C, left pseudocardinal teeth of *Glebula rotundata*; D, female *Plagiola,* ×1; E, male *Plagiola,* ×1; F, *Lampsilis,* ×0.7. (From Burch, 1973.)

55. Inner edge of mantle with a ribbonlike flap (Fig. 32A); Mississippi drainage system and FL to TX; 2
    species (Fig. 32F) .... **Lampsilis**
    Inner edge of mantle papillate (Fig. 32C); generally distributed in eastern half of the United States;
    probably 2 species (Fig. 31B) .... **Ligumia**
56. Shell circular or high oval (Fig. 31C); teeth heavy; generally distributed in eastern half of the United
    States; 6 species .... **Obovaria**
    Shell with another shape; teeth heavy or not 57
57. Shell of mature mussel less than 6 cm long; sexually dimorphic (Figs. 27D, E); posterior end of some
    species with raised radiating ridges; about 18 species; generally distributed in eastern half of the
    United States but most species restricted to the Tennessee and Cumberland river drainages.
                                                                            Plagiola
    Shell with another combination of characters 58
58. Posterior ridge angular (Figs. 28B, 25) 59
    Posterior ridge rounded or absent (Fig. 29B) 60
59. Shell compressed; beak cavities shallow; color rays with or without V-shaped markings (Fig. 28B); 5
    cm long; 3 species; Mississippi and St. Lawrence river drainages, and TX and OK ..... **Truncilla**

FIG. 28.–Representative Unionidae. A, *Ellipsaria lineolata,* ×0.9; B, *Truncilla,* ×1; C, *Dysnomia,* ×1; D, *Carunculina,* ×1.2.

Shell inflated; beak cavities deep; color rays present or absent (Figs. 25, 32F); about 7 species; Mississippi and St. Lawrence drainages, and southeastern states .... **Lampsilis**
60. Pseudocardinal teeth poorly developed (Fig. 29B); 4 species in eastern half of the United States.
                                                                                              Leptodea
Pseudocardinal teeth well developed (Fig. 29A) 61
61. Shell with a well-developed wing (Fig. 29A); 3 species in a genus distributed widely in eastern half of the United States .... **Proptera alata** (Say)
Wing lacking or poorly developed 62
62. Shell extremely inflated .... **Proptera**
Shell not extremely inflated 63
63. Shell up to 115 mm long; purple nacre .... **Potamilus purpuratus** (Lamarck)
Nacre white; if nacre is pinkish purple, then shell is less than 60 mm long 64
64. Posterior inner mantle margin smooth or crenulated, not papillate or ribbonlike; 3 species in eastern half of the United States (Fig. 30) .... **Actinonaias**
Posterior inner mantle margin papillate or ribbonlike (Figs. 32A–C) 65
65. Posterior inner mantle margin papillate (Figs. 32B, C); a poorly defined genus consisting of about 15 species; mostly southeastern, but at least 1 species extending as far north as VA, MI (Fig. 32E).
                                                                                              Villosa
Posterior inner mantle margin ribbonlike (Fig. 32A); about 20 species in the eastern half of the United States (Figs. 27F, 32F) .... **Lampsilis**

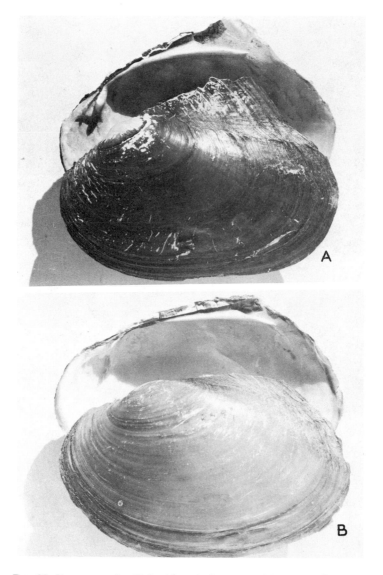

FIG. 29.–Representative Unionidae. A, *Proptera,* ×0.9; B, *Leptodea,* ×0.7.

66. Shell with two cardinal teeth in one valve, and one cardinal tooth in the other valve (Figs. 34F, G);
      generally distributed **72**
      Shell with only one feeble cardinal tooth in each valve, shell subrhomboidal and mottled (Fig. 33D); 7
      mm long; NC to TX .... **Eupera cubensis** (Prime)
67. Shell sculptured with coarse striae or widely spaced striae (Fig. 34A); nepionic valves not inflated and
      separated from subsequent portion of shell; 3 species .... **Sphaerium**
      Shell sculptured with fine striae; nepionic valves usually inflated and separated from subsequent
      portion of shell by a sulcus (Figs. 34B, E) **68**

FIG. 30.–*Actinonaias,* ×0.8.

FIG. 31.–Representative Unionidae. A, *Villosa,* ×0.9; B, *Ligumia,* ×0.8; C, *Obovaria,* ×0.9.

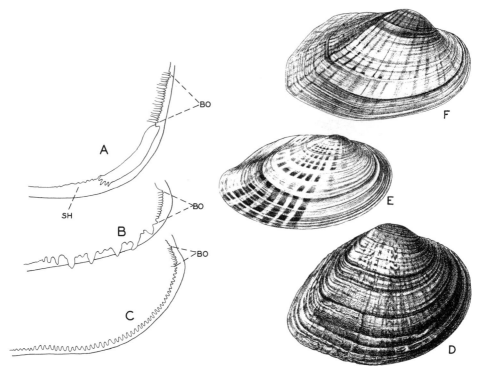

FIG. 32.–Structure of Unionidae. A, ventro-posterior edge of mantle and shell in a female *Lampsilis ventricosa*; B, same for *Micromya iris*; C, same for *Ligumia recta*; D, *Truncilla,* ×1; E, *Villosa,* ×0.6; F, *Lampsilis,* ×0.6. (D–F from Burch, 1973.)

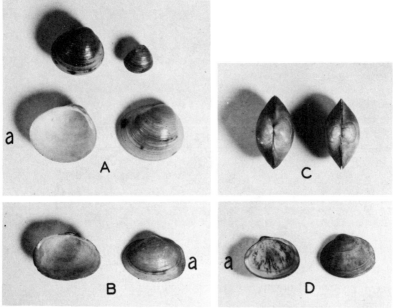

FIG. 33.–Representative Sphaeriidae. A, *Pisidium,* ×1.5; B, *Musculium,* ×1.5; C, *Musculium,* dorsal, ×1.5; D, *Eupera,* ×1.7.

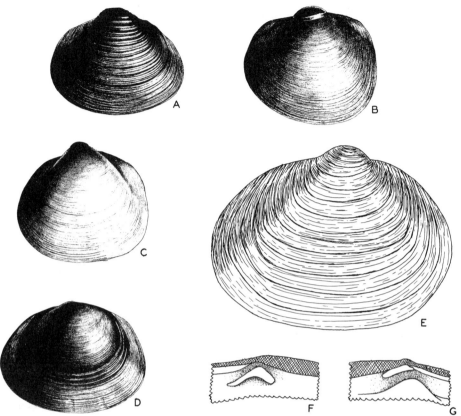

FIG. 34.–Structure of Sphaeriidae. A, left valve of *Sphaerium,* ×4; B, left valve of *Sphaerium,* ×4; C, left valve of *Musculium,* ×3; D, left valve of *Sphaerium,* ×3; E, right valve of *Sphaerium simile,* ×5; F, cardinal tooth of right valve of *S. simile*; G, cardinal teeth of left valve of *S. simile.* (A–D from Burch, 1973; F and G modified from Baker, 1928.)

68. Mature shell more than 8 mm long 69
    Mature shell less than 8 mm long 70
69. Beaks prominent (Fig. 33C); 3 species .... **Musculium**
    Beaks not prominent (Fig. 34D); 3 species .... **Sphaerium**
70. Posterior end nearly at right angles to the dorsal margin; 2 species .... **Musculium**
    Posterior end and dorsal margin rounded or forming an obtuse angle 71
71. Beaks swollen; 1 species .... **Musculium**
    Beaks not swollen; 3 species .... **Sphaerium**
72. Shell nearly equipartite, beaks subcentral, anterior of center (Fig. 34), with both anal and branchial
    siphons (Fig. 7) 67
    Shell inequipartite, beaks posterior of center (Fig. 33A); with anal siphon only; usually 2 to 6 mm long,
    but a few species as long as 11 mm; generally distributed in many types of habitats; about 25
    species .... **Pisidium**

REFERENCES

Allen, W. R. 1921. Studies of the biology of freshwater mussels. Experimental studies of the food relations of certain Unionidae. *Biol. Bull.* **40**:210–242.

Arey, L. B. 1932. The formation and structure of the glochidial cyst. *Biol. Bull.* **62**:212–221.

Baker, F. C. 1898. The Mollusca of the Chicago area. The Pelecypoda. *Bull. Nat. Hist. Surv. Chica. Acad. Sci.* **3**:1–130.

_____. 1916. The relation of mollusks to fish in Oneida Lake. *N.Y. State Coll. For., Tech. Publ.* **4**:1–366.

_____. 1918. The productivity of invertebrate fish food on the bottom of Oneida Lake, with special reference to mollusks. *Ibid.* 9:1–233.

_____. 1918a. The relation of shellfish to fish in Oneida Lake, New York. *N.Y. State Coll. For., Circ.* 21:1–34.

_____. 1928. *The fresh water Mollusca of Wisconsin. Part II. Pelecypoda.* 495 pp. Wis. Acad. Sci, Arts and Lett. Madison, Wisconsin.

Bedford, J. W., et al. 1968. The freshwater mussel as a biological monitor of pesticide concentrations in a lotic environment. *Limnol. Oceanogr.* 13:118–126.

Boycott, A.E. 1936. The habitats of fresh-water Mollusca in Britain. *J. Animal Ecol.* 5:116–186.

Brown, C. J. D., C. Clark, and B. Gleissner. 1938. The size of certain naiades from western Lake Erie in relation to shoal exposure. *Am. Midl. Nat.* 19:682–701.

Burch, J. B. 1972. Freshwater sphaeriacean clams (Mollusca: Pelecypoda) of North America. *Biota of Freshwater Ecosystems, Identification Manual* 3:1–31.

_____. 1973. Freshwater Unionacean clams (Mollusca: Pelecypoda) of North America. *Ibid.* 11:1–176.

_____. 1975. *Freshwater sphaeriacean clams (Mollusca: Pelecypoda) of North America.* 99 pp. Malacological Publications, Hamburg, Mich.

_____. 1975a. *Freshwater unionacean clams (Mollusca: Pelecypoda) of North America.* 204 pp. Malacological Publications, Hamburg, Michigan.

Burky, A. J. 1983. Physiological ecology of freshwater bivalves. In W. D. Russell-Hunter (ed.), *The Mollusca,* vol. 6, Ecology, pp. 281–327. Academic Press, Orlando, Florida.

Burky, A. J., et al. 1979. The ratio of calcareous and organic shell components of freshwater sphaeriid clams in relation to water hardness and trophic conditions. *J. Moll. Stud.* 45:312–321.

Chamberlin, R. V., and D. T. Jones. 1929. A descriptive catalog of the Mollusca of Utah. *Bull. Univ. Utah* 19:1–203.

Churchill, E. P., Jr., and S. I. Lewis. 1924. Food and feeding in fresh-water mussels. *Bull. U.S. bur. Fish.* 39:437–471.

Clarke, A. H. (ed.). 1970. Papers on the rare and endangered mollusks of North America. *Malacologia* 10:1–56.

Clarke, A. H. 1979. Sphaeriidae as indicators of trophic lake stages. *Nautilus* 94:178–184.

_____. 1981. The Tribe Alasmidontini (Unionidae: Anodontinae). Part I: Pegias, Alasmidonta, and Arcidens. *Smithson. Contrib. Zool.* 326:1–101.

Clarke, A. H., and C. O. Berg. 1959. The fresh-water mussels of central New York. *Mem. Cornell Univ. Agr. Exp. Sta.* 367:1–79.

Coker, R. E. 1919. Freshwater mussels and mussel industries of the United States. *Bull. U.S. Bur. Fish.* 36:13–89.

Coker, R. E., A. F. Shira, H. W. Clark, and A. D. Howard. 1921. Natural history and propagation of fresh-water mussels. *Bull. U.S. Bur. Fish.* 37:76–181.

Dundee, D. S. 1977. Catalog of introduced molluscs of eastern North America (north of Mexico). *Sterkiana* 55:1–37.

Dussart, G. B. J. 1979. Sphaerium corneum (L.) and Pisidium spp. Pfeiffer—the ecology of freshwater bivalve molluscs in relation to water chemistry. *J. Moll. Stud.* 45:19–34.

Foster, T. D. 1932. Observations on the life history of a fingernail shell of the genus Sphaerium. *J. Morphol.* 53:473–497.

Gale, W. F., and R. L. Lowe. 1971. Phytoplankton ingestion by the fingernail clam, Sphaerium transversum (Say) in Pool 19, Mississippi River. *Ecology* 52:507–513.

Gilmore, R. J. 1917. Notes on reproduction and growth in certain viviparous mussels of the family Sphaeriidae. *Nautilus* 31:16–30.

Goodrich, C., and H. van der Schalie. 1939. Aquatic mollusks of the Upper Peninsula of Michigan. *Mus. Zool. Univ. Mich. Misc. Publ.* 43:1–45.

_____. 1944. A revision of the Mollusca of Indiana. *Am. Midl. Nat.* 32:257–326.

Green, R. H. 1980. Role of a unionid clam population in the calcium budget of a small Arctic lake. *Can. J. Fish. Aquat. Sci.* 37:219–224.

Hanna, G. D. 1966. Introduced mollusks of western North America. *Occas. Pap. Calif. Acad. Sci.* 48:1–108.

Hartfield, P. D., and R. G. Rummel. 1985. Freshwater mussels (Unionidae) of the Big Black River, Mississippi. *Nautilus* 99:116–119.

Heard, W. H. 1962. Distribution of Sphaeriidae (Pelecypoda) in Michigan, USA. *Malacologia* 1:139–160.

_____. 1965. Comparative life histories of North American pill clams (Sphaeriidae: Pisidium). *Ibid.* 2:381–411.

_____. 1970. Reproduction of fingernail clams (Sphaeriidae): Sphaerium and Musculium). *Malacologia* 10:421–455.

_____. 1975. Sexuality and other aspects of reproduction in Anodonta (Pelecypoda: Unionidae). *Ibid.* 15:81–103.

_____. Reproduction of fingernail clams (Sphaeriidae): Sphaerium and Musculium). *Malacologia* 10:421–455.

Henderson, J. 1924. Mollusca of Colorado, Utah, Montana, Idaho, and Wyoming. *Univ. Colo. Studies* 13:65–223.

_____. 1929. Non-marine Mollusca of Oregon and Washington. *Ibid.* 17:47–190.

_____. 1936. Mollusca of Colorado, Utah, Montana, Idaho, and Wyoming—supplement. *Ibid.* 23:81–145.

Hornbach, D. J., and D. L. Childers. 1987. Life-history variation in a stream population of Musculium partumeium (Bivalvia: Pisidiidae). *J. N. Am. Benth. Soc.* 5:263–271.

Hudson, R. G., and B. G. Isom. 1984. Rearing juveniles of the freshwater mussels (Unionidae) in a laboratory setting. *Nautilus* 98:129–137.

Imlay, M. J. 1973. Effects of potassium on survival and distribution of freshwater mussels. *Malacologia* 12:97–113.

Isom, B. G. 1969. The mussel resources of the Tennessee River. *Malacologia.* 7:397–425.

Isom, B. G., and R. G. Hudson. 1982. In vitro cultures of parasitic freshwater mussel glochidia. *Nautilus* 96:147–151.

Isom, B. G., and P. Yokley. 1968. The mussel fauna of Duck River in Tennessee, 1965. *Am. Midl. Nat.* 80:34–42.

Johnson, R. I. 1978. Systematics and zoogeography of Plagiola (+Dysnomia = Epioblasma), an almost extinct genus of freshwater mussels (Bivalvia: Unionidae) from Middle North America. *Bull. Mus. Comp. Zool.* **148**:239–320.

_____. 1980. Zoogeography of North American Unionacea (Mollusca: Bivalvia) north of the maximum Pleistocene glaciation. *Ibid.* **149**:77–189.

Jones, R. O. 1949. Propagation of fresh-water mussels. *Prog. Fish-Cult.* **12**:13–25.

Kunz, G. F. 1898. The freshwater pearls and pearl fisheries of the United States. *Bull. U.S. Fish Comm.* **17**:373–426.

Lauritsen, D. D. 1986. Filter-feeding in Corbicula fluminea and its effect on seston removal. *J. N. Am. Benthol. Soc.* **5**:165–172.

Lefevre, G., and W. C. Curtis. 1912. Studies of the reproduction and artificial propagation of freshwater mussels. *Bull. U.S. Bur. Fish.* **30**:105–201.

Lopez, G. R., and I. J. Holopainen. 1987. Interstitial suspension-feeding by Pisidium spp. (Pisidiidae: Bivalvia): a new guild in the lentic benthos. *Bull. Am. Malacol. Union* **5**:21–29.

Mackie, G. L. 1979. Dispersal mechanisms in Sphaeriidae (Mollusca: Bivalvia). *Bull. Am. Malacol. Union* (**1979**):17–21.

_____. 1981. Nearctic freshwater Sphaeriacea (Bivalvia). *Ibid.* **50**:49–52.

Mathiak, H. A. 1979. *A River Survey of the Unionid Mussels of Wisconsin.* 76 pp. Sand Shell Press, Horicon, Wisconsin.

Matteson, M. R., and R. W. Dexter. 1966. Changes in pelecypod populations in Salt Fork of Big Vermilion River, Illinois, 1918–1962. *Nautilus* **79**:96–101.

McKee, P. M., and G. L. Mackie. 1983. Respiratory adaptations of the fingernail clams Sphaerium occidentale and Musculium securis to ephemeral habitats. *Can. J. Fish. Aquatic Sci.* **40**:783–791.

McMahon, R. F. 1982. The occurrence and spread of the introduced Asiatic freshwater clam, Corbicula fluminea (Müller), in North America: 1924–1982. *Nautilus* **96**:134–141.

Meier-Brook, C. 1969. Substrate relations in some Pisidium species (Eulamellibranchiata: Sphaeriidae). *Malacologia* **9**:121–125.

Monk, C. R. 1925. The anatomy and life-history of a fresh-water mollusk of the genus Sphaerium. *J. Morphol. Physiol.* **45**:473–501.

Morrison, J. P. E. 1932. A report on the Mollusca of the northeastern Wisconsin lake district. *Trans. Wis. Acad. Sci. Arts and Lett.* **27**:359–396.

Murphy, G. 1942. Relationship of the fresh-water mussel to trout in the Truckee River. *Calif. Fish and Game* **28**:89–102.

Murray, H. D., and E. C. Roy. 1968. Checklist of freshwater and land mollusks of Texas. *Sterkiana* **30**:25–42.

Nelson, T. C. 1918. On the origin, nature, and function of the crystalline style of lamellibranchs. *J. Morphol.* **31**:53–111.

Ortmann, A. E. 1912. Notes upon the families and genera of the Najades. *Ann. Carnegie Mus.* **8**:222–365.

_____. 1919. A monograph of the Naiades of Pennsylvania. Part III. Systematic account of the genera and species. *Mem. Carnegie Mus.* **8**:1–384.

Reigle, N. J. 1967. An occurrence of Anodonta (Mollusca, Pelecypoda) in deep water. *Am. Midl. Nat.* **78**:530–531.

Roscoe, E. J., and S. Redelings. 1964. The ecology of the fresh-water pearl mussel Margaritifera margaritifera (L.). *Sterkiana* **16**:19–32.

Sepkoski, J. J., and M. A. Rex. 1974. Distribution of fresh-water mussels: coastal rivers as biogeographic islands. *Syst. Zool.* **23**:165–188.

Simpson, C. T. 1914. *A descriptive catalogue of the naiades, or pearly freshwater mussels.* 1540 pp. Bryant Walker. Detroit, Michigan.

Sinclair, R. M., and B. G. Isom. 1963. *Further studies on the introduced Asiatic clam (Corbicula) in Tennessee.* Tenn. Dept. Public Health, Tenn. Stream Pollution Control Board, 76 pp. (mimeographed).

Starrett, W. C. 1971. A survey of the mussels (Unionacea) of the Illinois River, a polluted stream. *Bull. Ill. Nat. Hist. Surv.* **30**:266–403.

Strecker, J. K. 1931. The distribution of the naiades or pearly fresh-water mussels of Texas. *Baylor Univ. Spec. Bull.* **2**:3–71.

Suloway, L. 1981. The unionid (Mollusca: Bivalvia) fauna of the Kankakee River in Illinois. *Am. Midl. Nat.* **105**:233–239.

Surber, T. 1912. Identification of the glochidia of freshwater mussels. *U.S. Bur. Fish. Doc.* **771**:1–10.

Thiele, J. 1935. *Handbuch der systematischen Weichtierkunde.* Zweiter Band, pp. 779–1154. Gustav Fischer, Jena.

Tucker, M. E. 1928. Studies on the life cycles of two species of fresh water mussels belonging to the genus Anodonta. *Biol. Bull.* **54**:117–127.

Utterback, W. I. 1915–1916. The naiades of Missouri. *Am. Midl. Nat.* **4**:41–53, 97–152, 181–204, 244–273, 311–327, 339–354, 387–400, 432–464.

Van Cleave, H. J. 1940. Ten years of observation on a fresh-water mussel population. *Ecology* **21**:363–370.

Van der Schalie, H. 1938. The naiad fauna of the Huron River, in southeastern Michigan. *Mus. Zool. Univ. Mich. Misc. Publ.* **40**:1–83.

_____. 1940. Aestivation of fresh-water mussels. *Nautilus* **53**:137–138.

_____. 1945. The value of mussel distribution in tracing stream confluence. *Pap. Mich. Acad. Sci. Arts and Lett.* **30**:355–373.

_____. 1970. Hermaphroditism among North American freshwater mussels. *Malacologia* **10**:93–112.

Van der Schalie, H., and A. Van der Schalie. 1950. The mussels of the Mississippi River. *Am. Midl. Nat.* **44**:448–466.

Walker, B. 1918. A synopsis of the classification of the freshwater Mollusca of North America, north of Mexico, and a catalogue of the more recently described species, with notes. *Mus. Zool. Univ. Mich. Misc. Publ.* **6**:1–213.

Williams, J. C. 1969. *Mussel fishery investigation. Tennessee, Ohio, and Green rivers.* Final Report. Ken. Dept. Fish and Wildl. Res. 107 pp. (mimeographed).

Wilson, C. B., and H. W. Clark. 1914. The mussels of the Cumberland River and its tributaries. *Bur. Fish. Doc.* **781**:1–63.

Wood, E. M. 1974. Development and morphology of the glochidium larva of Anodonta cygnea (Mollusca: Bivalvia). *J. Zool. London* **173**:1–13.

Appendix

REAGENTS, SOLUTIONS, AND LABORATORY ITEMS

For the most part, this brief appendix lists only those reagents and solutions that are mentioned more than once in the text portion of this volume. Information concerning additional reagents and procedures, including fixation, imbedding, sectioning, staining, and so on, can be obtained by consulting any of the standard microtechnique handbooks.

In the writer's laboratory all transfers of small specimens from one solution to another are performed in the same vial. When solutions are added they are squirted into the vial with a pipette so that there is thorough mixing; when solutions are removed they are sucked up with a pipette after the specimens have all settled to the bottom of the vial. Although two or three such portions are usually necessary to effect a complete change of solutions in a vial, the small additional time is well compensated for by the convenience and compactness of the method. Furthermore, there is little danger of drying, spilling, or loss of specimens. We now use only one-piece discardable plastic pipettes.

The following list includes a variety of anesthetics. In general, such solutions should be slowly and gradually added to the water containing the specimens. Experience will determine proper amounts.

Acetic acid. Usually supplied as glacial acetic acid, which is 99.5% CH_3COOH. Two to 10 parts of glacial acetic acid are often added to 100 parts saturated mercuric chloride as a fixative.

Alcohol. Wherever "alcohol" is indicated in this manual, ethyl alcohol, or C_2H_5OH, is meant. It is used in varying concentrations as a fixative and preservative. Alcohol is usually supplied commercially in about 95% concentration (190 proof). All alcohol dilutions are made by measuring out the number of milliliters of 95% alcohol equivalent to the percentage of alcohol desired and diluting to 95 ml with water. For example, to make a 60% solution add 35 ml of water to 60 ml of 95% alcohol. Absolute (essentially 100%) alcohol is expensive to buy but may be easily made in the laboratory by dehydrating 95% alcohol with anhydrous copper sulfate. Crystals of copper (cupric) sulfate should first be heated in an evaporating dish until the water of crystallization is driven off and a white powder remains. Place some of this powder in a bottle and add 95% alcohol. The water in the alcohol immediately unites with the copper sulfate, turning it blue. Add more anhydrous copper sulfate until it no longer turns blue. Then quickly filter the alcohol into a dry bottle and stopper with a tight cork or ground glass stopper.

Alum cochineal. An excellent whole-mount stain for many small invertebrates. Add 6 g of potassium alum and 6 g of powdered cochineal to 90 ml of distilled water and boil for 30 minutes. Allow the fluid to settle, decant the supernatant, add a little water, and boil until about 90 ml of solution remains. Filter when cool and add 0.5 g thymol to inhibit molds. Depending on size and density of specimens, stain for 4 to 30 hours. After staining, wash specimens in several changes of water for 15 to 30 minutes to extract the alum. The material is then dehydrated and mounted in balsam, permount, or some similar nonaqueous material.

Amman's fluid. (See p. 296.)

Balsam. (Canada balsam). A general permanent mounting medium for dehydrated whole specimens or microtome sections. It is easiest to buy a thick solution and dilute with xylol (xylene) to the proper consistency. Add a marble chip to keep the solution neutral.

Benzamine hydrochloride, benzamine lactate. Anesthetics for certain small metazoans; usually used in 2% solutions.

Bismarck brown. An intravitam stain useful for protozoans. Boil 1 g of the stain in 100 ml of water and filter.

Borax carmine. This is a most useful stain for whole mounts of small metazoans. Boil 2 g of carmine in 100 ml of 4% borax until dissolved. Then add 100 ml of 70% alcohol; filter after 24 hours. Stain objects for 10 hours to several days, depending on size and density. Transfer to 70% acid alcohol and keep there until the stain no longer forms clouds. Wash the specimens in neutral alcohol.

Bottom fauna samplers. Many different devices are described in the literature. Boulton (1985) lists some of these references.

Bouin's fluid. A general fixative consisting of 75 ml of saturated aqueous picric acid, 25 ml of formalin, and 5 ml of glacial acetic acid. The normal fixing time is 2 to 16 hours, but tissues are not harmed if left in the fluid up to several weeks.

Butyn, butyn sulfate. Anesthetics for certain small metazoans; usually used in a 0.1 to 2% solution.

Carmine. (See Gower's carmine.)

Carnoy's fluid. Especially useful for killing and fixing fragile immature insects. Consists of 10% acetic acid, 60% alcohol, and 30% chloroform. Fix for 12 to 24 hours; wash in 70% alcohol; store in 70% alcohol.

Chloral hydrate. A general anesthetic for small metazoans. Make up as a 10% solution.

Chloretone. A common anesthetic (chlorbutanol) usually used in 0.1 to 0.6% solution.

Chlorine removal. Tap water is generally not suitable for fresh-water invertebrates because of its chlorine content. If allowed to stand for a day or two, with frequent stirring, it will usually be usable. Otherwise, add 0.5 mg/l of sodium bisulfite or 2 to 3 mg/l of sodium thiosulfate.

Chloroform. Occasionally used as a saturated aqueous solution (1.0%) for anesthetizing small metazoans. Chloroform vapors are sometimes useful if the organisms are concentrated in a small amount of water under a bell jar beside a small beaker of chloroform.

Chloroplatinic acid. This compound, $H_2PtCl_6 \cdot 6H_2O$, is especially valuable to preserve and keep osmic acid solutions stable. A gram should be added to every 100 ml of osmic acid solution.

Cleaning solution. Dissolve 60 to 65 g of sodium or potassium bichromate by heating in 30 to 65 ml of water. Cool to room temperature and then *slowly* add concentrated sulfuric acid to make 1 l of solution. Handle this solution with great care. After using it for cleaning glassware, rinse the glassware thoroughly.

Clorox. A commercial caustic useful for isolating small sclerotized structures from musculature and connective tissues, and especially for isolating and cleaning trophi of rotifers. Usually diluted 1 to 10 with water.

CMCP (nonresinous mounting medium). This is a permanent mounting material useful for whole mounts of aquatic insects, crustaceans, and other small metazoans. It is available in a thick, syrupy solution in which specimens may be mounted directly and satisfactorily from water, formalin preservative solutions, alcohols, and glycerin. It dries hard (but slowly) and has the additional advantage of being a clearing agent. For real permanency, however, cover slips should be ringed with Murrayite or some similar cement after the mounts have dried for several days. CMCP may be thinned with water. It is just as permanent as balsam, and is indeed an important contribution to biological technique and slide making. Some researchers tint CMCP with acid fuchsin or anilin blue. The product is marketed by Polysciences.*

Cocaine, novocain. Excellent anesthetics for small metazoans but difficult to obtain legally(!) because of narcotics regulations. Best used as a 1% solution in 95% alcohol or water.

Corn syrup mounting medium. Microzoans, and especially plankton material, may be satisfactorily mounted on glass slides in a solution consisting of 70% aqueous 4% formalin, and 30% commercial "light" corn syrup obtainable in grocery stores everywhere. Slides should be stored horizontally for several days as the medium forms a seal at the edge of the coverslip. If the mountant draws away from the edge, simply add another drop or two. After the seal is completely hardened, slides may be stored on edge. The writer has had such plankton slides in satisfactory condition for 20 years.

Culturing. See the appropriate sections in the chapters in this volume. Also see Boulton (1985), Galtsoff et al. (1937), and Sudia (1951).

Eosin. General purpose cytoplasmic stain for small metazoans. Usually used as a 0.5% solution in 95% alcohol or water.

Euparal. A permanent slide mounting medium. Specimens may be transferred directly from 95% alcohol into euparal.

FAA (formalin-aceto-alcohol). A general purpose fixative in which material may

be kept for several days without harm. Numerous modifications of this fixative are used; a typical one consists of 40 ml of distilled water, 2 ml of glacial acetic acid, 50 ml of 95% alcohol, and 10 ml of formalin.

Formalin. A common and cheap preservative consisting of a 37 to 40% aqueous solution of formaldehyde, H_2CO. Depending on the nature of the preserved material, 2 to 6% solutions of formalin (*not* 2–6% formaldehyde) are usually used. Commercial formalin is slightly acid and may be neutralized by maintaining a slight deposit of magnesium carbonate or household borax in the bottom of the reagent bottle. When formalin dries it forms a whitish irreversible polymer.

Fuchsin (basic). A nuclear stain sometimes useful in temporary whole mounts of micrometazoans. Make up a 0.5% aqueous solution.

Gilson's fluid. (See p. 138.)

Glycerin. An excellent medium for dissecting and for temporary and permanent mounts for many kinds of micrometazoans. Glycerin is miscible with water and alcohol at all concentrations and has the advantage of being a slight clearing agent. Because glycerin is nondrying, the writer regularly adds about 5% of it to stored vials of alcoholic and formalin specimens. If such vials should break, or the water and alcohol should dry, the specimens are still available in good condition. Specimens to be used in concentrated glycerin should first be placed in a 5 to 20% solution of glycerin in 50 to 95% alcohol. If the container is uncovered but protected from dust), the alcohol and water evaporate in a few days, leaving the specimens in pure glycerin.

Glycerin jelly, glycerin jelly slide mounts. A valuable permanent mounting medium that is solid at ordinary temperatures. It is sold by most biological supply firms, but may be easily made in the laboratory. Soak 1.5 g of clean gelatin overnight in 100 ml of water,

*Polysciences, 400 Valley Road, Warrington, PA 18976.

then dissolve with gentle heat. Add half its volume of clean glycerin and 0.5 g of carbolic acid. Warm *gently* and stir for 10 minutes until homogeneous. Store in a small wide-mouth bottle with a screw cap. Specimens to be mounted in glycerin jelly should first be in at least 50% glycerin. Melt the glycerin jelly gently in warm water or on a hot plate. (If it is heated too much it polymerizes and is useless.) Place the specimen on a slide in a droplet of glycerin, drop on one to several drops of liquid glycerin jelly with a thin glass rod, and quickly orient the specimen with a warm needle. Then pick up a clean circular cover slip with a forceps, breathe on the undersurface, and gently put it in place on top of the fluid. Set the mount aside to congeal for at least 30 minutes. Later the excess glycerin jelly may be removed from the edge of the cover slip with a razor blade and clean cloth. For real permanence the cover slip should be ringed on a turntable with several coats of Murrayite.

Gower's carmine stain. This stain is especially useful for staining small fresh-water invertebrates to be mounted in glycerin or glycerin jelly. It will not "bleed" into the mounting medium. Add 10 g of carmine to 100 ml of 45% acetic acid. Heat to a boil, cool, filter, dry, and save the residue. Then to 1 g of dry residue add 10 g of alum and 200 ml of distilled water. Heat to dissolve, cool, filter, and add one crystal of thymol. (See Staining with Gower's carmine.)

Harper's fluid. (See p. 139.)

Heat absorbing fluid. To be used in a glass container between a light source and heat sensitive living animals being studied on the stage of a binocular dissecting microscope. Use 2% aqueous calcium chloride; or dissolve 200 g of ferrous ammonium sulfate in water, filter, and add 1.7 ml of concentrated sulfuric acid.

Helly's fluid. (See p. 139.)

Hematoxylin. Delafield's hematoxylin is a general purpose nuclear stain. Dissolve 1 g

of hematoxylin crystals in 10 ml of absolute alcohol and add it drop by drop to 100 ml of a saturated aqueous solution of aluminum ammonium sulfate. Expose this mixture to air and light for several weeks to "ripen." Then filter, add 25 ml of glycerin, and 25 ml of methyl alcohol.

Hertwig-Schneider's fluid. (See p. 119.)

Hoyer's mounting medium. Dissolve 30 g of pulverized gum arabic in 50 ml of water. Add 20 ml of glycerin. Add and dissolve 125 g of chloral hydrate. Add 2 g of iodine crystals. Add 1 g of potassium iodide. Filter through cotton wool.

Hydroxylamine hydrochloride. Anesthetic for certain small metazoans; usually used in a 2% aqueous solution.

Irwin loop. This item is indispensable for sorting, picking up, and transferring microscopic species. It replaces needles, toothpicks, bristles, and fine pipettes.*

Isopropyl alcohol. Occasionally useful as an anesthetic, especially for certain protozoans. Sometimes used in histological work as a substitute for ethyl alcohol.

Jeweler's forceps. These are extremely fine and well-made forceps that are indispensable for dissections. Obtainable from your local jeweler or jewelers supply house.

Kahle's fluid. Useful for killing and fixing fragile immature insects. Consists of 11% formalin, 28% alcohol, 2% percent glacial acetic acid, and 59% water. Drain off after a week, rinse in 70% alcohol, and store the specimens in 70% alcohol.

Lactophenol. A special mounting medium with index of refraction especially favorable for cuticular details. Made up of 20 g

---

*Order from Mark D. Schram, 208 Baxter Lane, Fayetteville, AR 72701.

of phenol crystals, 40 g of glycerin, 20 g of lactic acid, and 20 ml of distilled water.

Lugol's solution. An excellent fixative and preservative for microzoans, especially for cellular details. Consists of 5 g of iodine, 10 g of potassium iodide, 10 g of glacial acetic acid, and 100 ml of water. Add 1 ml of this solution to each 100 ml of water containing the specimens.

Magnesium sulfate. Occasionally used as an anesthetic for fresh-water invertebrates but more useful for marine forms.

Menthol. An anesthetic occasionally used for fresh-water invertebrates. Sprinkle fine crystals on the surface of the water containing the specimens.

Menthol–chloral hydrate narcotic. General-purpose narcotic for small invertebrates. Grind 48 g menthol and 52 g of chloral hydrate with a mortar and pestle until an oily fluid results. Place a few drops of this liquid on the surface of the water containing the animals.

Methyl cellulose. A fluffy white substance, easily soluble in water, and especially useful for slowing the movements of protozoans. Usually sold as 1 or 10% syrupy water solutions. Add a drop or two to water containing protozoans on a slide, mix with a needle, and add a cover slip. Keep it away from the end of the microscope objective!

Methylene blue. An excellent intravitam stain for translucent forms. Make up a stock of 0.01 g of methylene blue in 100 ml of absolute alcohol. Add enough of this solution to the water containing the organisms to tinge it a light blue.

Monk's fluid. This is an aqueous mounting medium that is useful for making permanent slides of copepods and other small arthropods. It consists of 5 ml of white corn syrup, 5 ml of Certo (fruit pectin), and 3 ml of water. Add a crystal of thymol as a preservative. The mixture sets in a few minutes. Dry to hardness over gentle heat, preferably a hot plate.

Murrayite. A synthetic cement that is dissolved in benzol to a thin, syrupy consistency. Since it is permanently waterproof, alcohol proof, and glycerin proof, it is an excellent sealing cement for museum jars and cover slips. If circular cover slips are ringed on a turntable, several coats should be applied with a camel's hair brush, allowing at least 10 hours between coats.

Neosynephrin hydrochloride. Anesthetic for small metazoans; usually used in a 1% water solution.

Neutral red. An intravitam stain useful for protozoans. Dissolve 1 g of the powder in 100 ml of water.

Newcomer's fixative. A good general fixative for small invertebrates. Six parts isopropyl alcohol, 3 parts propionic acid, 1 part petroleum ether, 1 part acetone, and 1 part dioxane. Specimens may be stored in this fixative.

Nicotine sulfate. This salt is occasionally used as an anesthetic for small fresh-water invertebrates in about 2% solution.

Nitric acid. Concentrated nitric acid is used in cleaning and isolating sponge spicules and gemmules. Two percent nitric acid is useful for fixing turbellarians, and a 1% solution is used for dissociating tissues of hydras.

Novolac. One of several related alcohol-soluble resins suitable for high quality permanent mounts of a wide variety of small aquatic invertebrates. See Crumpton and Wetzel (1980).

Osmic acid. Really osmium tetroxide, OsO_4, this compound is poisonous and extremely volatile. Used in a 1% solution in distilled water, it is an excellent and rapid fixative for a variety of small metazoans. Prepare the solution by crushing the cleaned hermetically sealed glass container in an appropriate amount of water and allowing the released osmic acid crystals to dissolve. This compound is easily oxidized by light and dust and should therefore be kept in a blackened dropping bottle. Solutions have much greater permanence if they contain 1% chloroplatinic acid. If specimens are in a drop of water on a slide, fixation is accomplished in a few seconds if the slide is inverted over the mouth of the osmic acid bottle. A few drops of osmic acid are sufficient for fixing specimens in 5 ml to 10 ml of water. All fixed material should be washed in several changes of water. If the material is colored dark brown or black by the action of the osmic acid, place in commercial hydrogen peroxide until sufficiently bleached. Osmic acid is unfortunately very expensive.

Perforated pipe technique. Several types of perforated pipe samplers are in use for taking interstitial hyporheic and phreatic water samples. If the pipe is sturdy and has a pointed end it may be driven into the substrate with a sledge hammer. Otherwise it is placed in an excavation dug with a shovel, and the hole is backfilled (Fig. 1). If the substrate is saturated, water may be pumped out immediately and passed through a fine net to recover the contained micrometazoans. The pipe may also be left in place for a variable time before any pumping is done. Sampling pipes up to 1.5 m or more long are effective in suitable substrates. The Bou-Rouche sampler is an elaborate version.

Phenol. Melt 400 g of solid phenol by immersing container in hot water; then add 10 ml of water.

Potassium hydroxide. A 2 to 10% solution of potassium hydroxide is often used for clearing opaque chitinous insects and other arthropods. Specimens may be heated gently in the solution for a few minutes or left in it cold for 8 to 48 hours. Specimens should then

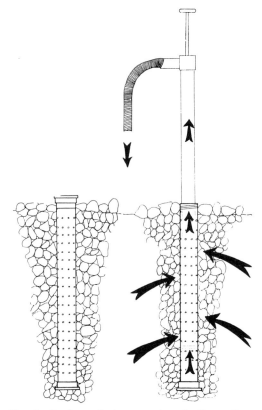

FIG. 1.–Perforated pipe sampler, similar to Bou-Rouch sampler. (From Gledhill.)

be neutralized with 10% acetic acid and washed with several changes of water.

Propylene phenoxetol. Same as hydroxyethyl phenyl ether. A useful narcotic for highly contractile aquatic invertebrates. Use only 0.2 ml/L of water. Requires 15 minutes to 2 hours to attain complete relaxation.*

Schaudinn's fixative. A general fixative consisting of 2 parts saturated mercuric chloride and 1 part 95% alcohol. Fix for 5 minutes to 3 hours, depending on size and density of the objects. Remove all the salt by washing in several changes of 70% alcohol for a total of 3 to 8 days. Salt removal may be

*Obtainable from Goldschmidt Chemical Corp., 153 Waverly Place, New York.

greatly speeded by adding a small amount of tincture of iodine to the alcohol. The action of this fixative is enhanced by the addition of sufficient glacial acetic acid to make a 1% solution immediately before using.

Sea water salt. Generally available in quantities at aquarium shops and biological supply houses. Not worth the effort involved in making synthetically from component salts.

Spoon microcompressor. A rotary compressor that permits delicately controlled compression of living microzoans between coverslip and slide. When mounted on a microscope stage, it is designed to gently slow and stop swimming organisms. See construction details in Spoon (1978).

Staining with Gower's Carmine. Transfer specimens to water. Add Gower's carmine solution to make 20 to 90% of the total volume. Let stand a half hour to 24 hours. Wash twice in water. If necessary, destain in 70% alcohol. Transfer to dilute glycerin in readiness for further processing.

Terpineol (synthetic oil of lilac). A good clearing agent, readily miscible with 90% alcohol. Does not make specimens brittle.

Urethane. A good general narcotic; used as a 2% aqueous solution.

Watchmaker's forceps. (See jeweler's forceps.)

Zut. Trade name of a cement that is especially useful for tacking or ringing temporary or permanent slides, especially glycerin jelly mounts. Thin with butyl acetate. Apply with a turntable and a small camel's hair brush.

REFERENCES

Boulton, A. 1985. A sampling device that quantitatively collects benthos in flowing or standing waters. *Hydrobiologia* 127:31–40.

Crumpton, W. G., and R. G. Wetzel. 1980. Novalacs: synthetic resins suitable for mounting biological materials. *Trans. Am. Microsc. Soc.* 99:347–348.

Galtsoff, P. W. et al. 1937. *Culture methods for invertebrate animals.* 590 pp. Dover, New York.

Gledhill, T. 1977. Numerical fluctuations of four species of subterranean amphipods during a five-year period. *Crustaceana, Suppl.* 4:144–152.

Higgins, R. P., and H. Thiel (eds.). 1988. *Introduction to the study of meiofauna.* 487 pp. Smithsonian Institution Press, Washington, D. C.

Lawrence, S. G. 1981. Manual for the culture of selected freshwater invertebrates. *Can. Spec. Publ. Fish. Aquatic Sci.* 54:1–169.

Lind, O. T. 1979. *Handbook of common methods in limnology.* 199 pp. Mosby, St. Louis.

Saur, J. F. T., et al. 1964. Sources of limnological and oceanographic apparatus and supplies. (3rd rev.) *Limnol. Oceanogr. Spec. Publ.* 1:i–xxxi.

Schwoerbel, J. 1966. *Methoden der Hydrobiologie.* 207 pp. Franckh'sche Verlagshandlung, Keller, Stuttgart.

Spoon, D. M. 1978. A new rotary microcompressor. *Trans. Am. Microsc. Soc.* 97:412–416.

Sudia, W. D. 1951. A device for rearing animals requiring a flowing water environment. *Ohio J. Sci.* 51:197–202.

Welch, P. S. 1948. *Limnological methods.* 381 pp. Blakiston, Philadelphia.

INDEX

The main entries of this index include all names of genera and higher taxa used in this volume, as well as significant anatomical and ecological terms. Where two or more species names in a genus are used, then such species are sublisted alphabetically under the appropriate generic names.

Entries are based chiefly on the earliest and/or most significant page references. There is no attempt to list *all* pages on which a particular item is mentioned.

Roman type page numbers indicate key or text locations. **Boldface** page numbers indicate the location of pertinent figures. An *italicized* page number means that both a figure and a text and/or key reference appear on that page.

Acanthocyclops:
    crassicaudis, 425
    robustus, 424
    vernalis, 424, **426**, 427
Acanthocystis, 70
Acanthodiaptomus denticornis, 421, **433**
Acantholeberis curvirostris, **373**, 396
Acari, 534
Acella haldemani, **558**, 563
Acetabulum, *515*
Acetic acid, 605
Achromadora, 240, 241
Achtheres, **416**, 423
Acineta, **86**, 87
Acoela, 124
Acrobeloides, 243, 245
Acroloxidae, 564
Acroloxus coloradensis, **561**, 564
Acroperus harpae, 399
Actinobdella, 326, **328**
    inequiannulata, 326
    triannulata, 326
Actinobolina, **72**, 73
Actinocoma, 69, 70
Actinolaimus, 242, 327
Actinomonas, **63**, 64
Actinonaias, 597
Actinophrys sol, **34**
Acyclus inquietus, *219*
Adductor muscle, 570
Adhesive organ, *127*
Adineta, 221, 222
Adinetidae, 196
Aeolosoma, 291, 294, 298, **299**
    headleyi, *299*
    hemprichi, *299*
    leidyi, *299*
    tenebrarum, 299
    variegatum, 299
Aeolosomatidae, 291, 298
Aesthetask, 412, 446, 477

Aestivation, 15, 554
Aglaodiaptomus:
    clavipes, 431
    clavipoides, **433**, 436
    conipedatus, **435**, 436
    dilobatus, **433**, 436
    forbesi, *434*
    kingsburyae, 431
    leptopus, *430*
    lintoni, 436
    marshianus, **433**, 436
    saskatchewanensis, 436
    spatulocrenatus, **435**, 436
    stagnalis, **435**, 436
Alae, 227
Alaimus, 242
Alasmidonta varicosa, 590, **592**
Albertia, 215, 217
Albia, 529, **530**
Alboglossiphonia heteroclita, **328**, 329
Alcohol, 605
Alkali faunas, 22
Allen's B-15 fixative, 139
Allocrangonyx, 482, **483**
Alloeocoela, 124, 140
Allonais pectinata, 301
Allotexiweckelia, *485*
Alona, 400
    barbulata, 401
    bicolor, *401*
    circumfimbriata, 401
    costata, **400**, 401
    guttata, **400**, 401
    monacantha, **400**, 401
    quadrangularis, **400**, 401
    rustica, *401*
    setulosa, 400
Alonella, 401
    dadayi, **405**, 406
    dentifera, **403**, 406
    diaphana, **400**, 401, **402**

    excisa, 404, **405**
    exigua, 404, **405**
    fitzpatricki, 399
    globulosa, 402
    karua, **400**, 406
    nana, **403**, 404, 406
Alonopsis:
    americana, 399
    aureola, *399*
Alum cochineal, 605
Amblema, 586, **588**
Ambleminae, 583
Amictic egg, 183, 185
Amictic female, 182
Amitosis, 42
Amman's fluid, 296
Amnicola, 560
Amoeba, 34, 65
Amoebida, 65
Amoebocyte, 94, 95
Amphibolella, 147, 149
Amphichaeta, 300
Amphid, *232*
Amphidelus, 242, 243
Amphidinium, 50, 51
Amphidisc, 98
Amphigyra alabamensis, 564
Amphileptus, 75
Amphimonas, 60, **61**
Amphionidacea, 337
Amphipoda, 474
Amphitrema, 67, **69**
Ampulariidae, **544**, 562
Anabiosis, 13, 262
Anaerobiosis, 16
Anaplectus, 240, **243**
Anaspidacea, 342
Anatonchus, 237, **238**
Ancestrulae, 281
Anchistropus minor, **403**, 404
Ancylidae, 564